Lecture Notes in Artificial Intelligence 9978

Subseries of Lecture Notes in Computer Science

LNAI Series Editors

Randy Goebel
  University of Alberta, Edmonton, Canada
Yuzuru Tanaka
  Hokkaido University, Sapporo, Japan
Wolfgang Wahlster
  DFKI and Saarland University, Saarbrücken, Germany

LNAI Founding Series Editor

Joerg Siekmann
  DFKI and Saarland University, Saarbrücken, Germany

More information about this series at http://www.springer.com/series/1244

Van-Nam Huynh · Masahiro Inuiguchi
Bac Le · Bao Nguyen Le
Thierry Denoeux (Eds.)

Integrated Uncertainty in Knowledge Modelling and Decision Making

5th International Symposium, IUKM 2016
Da Nang, Vietnam, November 30 – December 2, 2016
Proceedings

Springer

Editors
Van-Nam Huynh
Japan Advanced Institute of Science
    and Technology
Nomi, Ishikawa
Japan

Masahiro Inuiguchi
Graduate School of Engineering Science
Osaka University
Toyonaka, Osaka
Japan

Bac Le
University of Science
Ho Chi Minh City
Vietnam

Bao Nguyen Le
Duy Tan University
Da Nang
Vietnam

Thierry Denoeux
Université de Technologie de Compiègne
Compiègne
France

ISSN 0302-9743   ISSN 1611-3349 (electronic)
Lecture Notes in Artificial Intelligence
ISBN 978-3-319-49045-8   ISBN 978-3-319-49046-5 (eBook)
DOI 10.1007/978-3-319-49046-5

Library of Congress Control Number: 2016955999

LNCS Sublibrary: SL7 – Artificial Intelligence

Printed on acid-free paper

This Springer imprint is published by Springer Nature
The registered company is Springer International Publishing AG
The registered company address is: Gewerbestrasse 11, 6330 Cham, Switzerland

Preface

This volume contains the papers that were presented at the 5th International Symposium on Integrated Uncertainty in Knowledge Modelling and Decision Making (IUKM 2016) held in Da Nang, Vietnam, from November 30 to December 2, 2016.

The IUKM symposia aim to provide a forum for the exchange of research results and ideas, and experiences in application among researchers and practitioners involved with all aspects of uncertainty modelling and management. Previous editions of the conference were held in Ishikawa, Japan (originally under the name of International Symposium on Integrated Uncertainty Management and Applications – IUM 2010), Hangzhou, China (IUKM 2011), Beijing, China (IUKM 2013), and Nha Trang, Vietnam (IUKM 2015).

IUKM 2016 was jointly organized by Duy Tan University (Da Nang, Vietnam), Japan Advanced Institute of Science and Technology (JAIST), and Belief Functions and Applications Society (BFAS).

The organizers received 78 submissions. Each of which was peer reviewed by at least two members of the Program Committee. While 35 papers were accepted after the first round of reviews, 22 others were conditionally accepted and underwent a rebuttal stage in which authors were asked to revise their paper in accordance to the reviews, and prepare an extensive response addressing the reviewers' concerns. The final decision was made by the program chairs. Finally, 57 papers were accepted for presentation at IUKM 2016 and publication in the proceedings. Invited talks presented at the symposium are also included in this volume.

As a follow-up of the symposium, a special volume of the *International Journal of Approximate Reasoning* is anticipated to include a small number of extended papers selected from the symposium as well as other relevant contributions received in response to subsequent open calls. These journal submissions will go through a fresh round of reviews in accordance with the journal's guidelines.

The IUKM 2016 symposium was partially supported by the National Foundation for Science and Technology Development of Vietnam (NAFOSTED) and Duy Tan University. The IUKM 2016 best student paper award was sponsored by Elsevier. We are very thankful to Dr. Gia-Nhu Nguyen and his local organizing team from Duy Tan University for their hard work and efficient services and for the wonderful local arrangements.

We would like to express our appreciation to the members of the Program Committee for their support and cooperation in this publication. We are also thankful to Alfred Hofmann, Anna Kramer, and their colleagues at Springer for providing a meticulous service for the timely production of this volume. Last, but certainly not the least, our special thanks go to all the authors who submitted papers and all the attendees for their contributions and fruitful discussions that made this conference a great success.

December 2016

Van-Nam Huynh
Masahiro Inuiguchi
Bac Le
Bao N. Le
Thierry Denoeux

Organization

General Co-chairs

Bao N. Le — Duy Tan University, Da Nang, Vietnam
Thierry Denoeux — University of Technology of Compiègne, France

Honorary Co-chairs

Michio Sugeno — European Center for Soft Computing, Spain
Hung T. Nguyen — New Mexico State University, USA;
Chiang Mai University, Thailand
Co C. Le — Duy Tan University, Da Nang, Vietnam
Sadaaki Miyamoto — University of Tsukuba, Japan

Program Co-chairs

Van-Nam Huynh — JAIST, Japan
Masahiro Inuiguchi — University of Osaka, Japan
Bac Le — University of Science, VNU-Ho Chi Minh, Vietnam

Local Arrangements Chair

Gia-Nhu Nguyen — Duy Tan University, Da Nang, Vietnam

Publication and Financial Chair

Van-Hai Pham — Pacific Ocean University, Nha Trang, Vietnam

Program Committee

Byeong-Seok Ahn — Chung-Ang University, Korea
Yaxin Bi — University of Ulster, UK
Bernadette Bouchon-Meunier — Université Pierre et Marie Curie, France
Lam Thu Bui — Le Quy Don Technical University, Vietnam
Humberto Bustince — Universidad Publica de Navarra, Spain
Tru Cao — Ho Chi Minh City University of Technology, Vietnam
Fabio Cuzzolin — Oxford Brookes University, UK
Tien-Tuan Dao — University of Technology of Compiègne, France
Bernard De Baets — Ghent University, Belgium
Yong Deng — Xian Jiaotong University, China

Thierry Denoeux	University of Technology of Compiègne, France
Sebastien Destercke	University of Technology of Compiègne, France
Karim El Kirat	University of Technology of Compiègne, France
Zied Elouedi	LARODEC, ISG de Tunis, Tunisie
Tomoe Entani	University of Hyogo, Japan
Lluis Godo	IIIA - CSIC, Spain
Fernando Gomide	University of Campinas, Brazil
Peijun Guo	Yokohama National University, Japan
Enrique Herrera-Viedma	University of Granada, Spain
Marie-Christine Ho Ba Tho	University of Technology of Compiègne, France
Katsuhiro Honda	Osaka Prefecture University, Japan
Tzung-Pei Hong	National University of Kaohsiung, Taiwan
Van Nam Huynh	JAIST, Japan
Masahiro Inuiguchi	Osaka University, Japan
Radim Jirousek	University of Economics, Czech Republic
Janusz Kacprzyk	Polish Academy of Sciences, Poland
Gabriele Kern-Isberner	Technische Universität Dortmund, Germany
Etienne Kerre	Ghent University, Belgium
Laszlo T. Koczy	Budapest University of Technology and Economics, Hungary
Vladik Kreinovich	University of Texas at El Paso, USA
Rudolf Kruse	University of Magdeburg, Germany
Yasuo Kudo	Muroran Institute of Technology, Japan
Yoshifumi Kusunoki	Osaka University, Japan
Jonathan Lawry	University of Bristol, UK
Anh-Cuong Le	Ton Duc Thang University, Vietnam
Bac Le	University of Science, VNU-Ho Chi Minh, Vietnam
Churn-Jung Liau	Academia Sinica, Taipei, Taiwan
Chin-Teng Lin	National Chiao-Tung University, Taiwan
Jun Liu	University of Ulster, UK
Weiru Liu	Queen's University Belfast, UK
Anitawati Mohd Lokman	Universiti Teknologi MARA (UiTM) Malaysia
Tieju Ma	East China University of Science and Technology, China
Catherine K. Marque	University of Technology of Compiègne, France
Luis Martinez	University of Jaen, Spain
Radko Mesiar	Slovak University of Technology in Bratislava, Slovakia
Tetsuya Murai	Hokkaido University, Japan
Canh Hao Nguyen	Kyoto University, Japan
Thanh Binh Nguyen	Duy Tan University, Vietnam; IIASA, Austria
Le Minh Nguyen	JAIST, Japan
Hung Son Nguyen	University of Warsaw, Poland
Xuan Hoai Nguyen	Hanoi University, Vietnam
Thanh Hien Nguyen	Ton Duc Thang University, Vietnam
Akira Notsu	Osaka Prefecture University, Japan

Vilem Novak	Ostrava University, Czech Republic
Nikhil Pal	Indian Statistical Institute, India
Irina Perfilieva	Ostrava University, Czech Republic
Tuan Phung-Duc	Tokyo Institute of Technology, Japan
Zengchang Qin	Beihang University, China
Yasuo Sasaki	JAIST, Japan
Hirosato Seki	Osaka University, Japan
Dominik Slezak	University of Warsaw and Infobright Inc., Poland
Noboru Takagi	Toyama Prefectural University, Japan
Yongchuan Tang	Zhejiang University, China
Phantipa Thipwiwatpotjana	Chulalongkorn University, Thailand
Vicenc Torra	University of Skovde, Sweden
Seiki Ubukata	Osaka University, Japan
Bay Vo	HUTECH, Vietnam
Guoyin Wang	Chongqing University of Posts and Telecom., China
Thanuka Wickramarathne	University of Massachusetts, USA
Zeshui Xu	Sichuan University, China
Hong-Bin Yan	East China University of Science and Technology, China
Chunlai Zhou	Renmin University of China

Local Organizing Committee

Viet Hung Dang	Duy Tan University, Da Nang, Vietnam
Van Son Phan	Duy Tan University, Da Nang, Vietnam
Phung Hoi Phan	Duy Tan University, Da Nang, Vietnam
Thanh Duong Nguyen	Duy Tan University, Da Nang, Vietnam
Duc Man Nguyen	Duy Tan University, Da Nang, Vietnam
Thoai My Ho	Duy Tan University, Da Nang, Vietnam
Ngoc Trung Dang	Duy Tan University, Da Nang, Vietnam
Vu Tien Truong	Duy Tan University, Da Nang, Vietnam
Dac Nhuong Le	Hai Phong University, Hai Phong, Vietnam

Sponsoring Institutions

 The National Foundation for Science and Technology Development of Vietnam (NAFOSTED)

 Duy Tan University, Da Nang, Vietnam

 Japan Advanced Institute of Science and Technology

Invited Speakers

Machine Learning Applications: Past, Present and Future

Hiroshi Mamitsuka

Kyoto University, Kyoto, Japan

Short biography: Dr. Hiroshi Mamitsuka is a Professor of Bioinformatics Center, Institute for Chemical Research, Kyoto University, being jointly appointed as a Professor of School of Pharmaceutical Sciences of the same university. Also currently he is a FiDiPro (Finland Distinguished Professor Program) Professor of Department of Computer Science, Aalto University, Finland. His present research interest includes a variety of aspects of machine learning and diverse applications, primarily cellular- or molecular-level biology, chemistry and medical sciences. He has published more than 100 scientific papers, including those appearing in top-tier conferences or journals in machine learning and bioinformatics, such as ICML, KDD, ISMB, Machine Learning, Bioinformatics, etc. Also he has served program committee member of numerous conferences and associate editor of several well-known journals of the related fields. Prior to joining Kyoto University, he worked in industry for more than ten years, mainly data analytics in business sectors, for example, customer/revenue churn, web-access pattern, campaign management, collaborative filtering, recommendation engine, etc. So after moving to Kyoto University, he has worked as research advisor on data mining of several enterprises.

Summary: Machine learning is data-driven and so application-driven technology always seeking real problems. Interestingly in the beginning, techniques currently considered as part of machine learning, were developed in other domains mainly. A typical example is hidden Markov model (HMM), which was extensively studied for speech recognition while not necessarily in machine learning community. HMM is now technically well matured and commonly used not only for speech but in many other applications including biological sequence alignment. This type of "machine learning" can be found in classical applications, such as natural language processing, computer vision and pattern recognition. Currently, due to the era of Internet, big data and big science, machine learning applications are much broader, covering numerous fields, such as science, engineering, economics and other many aspects of our society. Particularly commercial or business-oriented applications are weighted more, being part of or data mining itself. The question is on future. In this talk, I'd like to review the past and present applications of machine learning, and also shed light on possible and promising future applications, which are already gradually coming out.

On Evidential Measures of Support
for Reasoning with Integrated Uncertainty

Hung T. Nguyen

New Mexico State University, Las Cruces, USA
Chiang Mai University, Chiang Mai, Thailand

Short biography: Prof. Hung T. Nguyen received the BS degree (1967) from University of Paris XI, the Master degree (1968) from University of Paris VI, and the PhD degree (1975) from University of Lille (France), all in Mathematics. After spending several years at the University of California, Berkeley and the University of Massachusetts, Amherst, he joined the faculty of Mathematical Sciences, New Mexico State University (USA), where he is currently a Professor Emeritus. He is also an Adjunct Professor of Economics, Chiang Mai University, Thailand, and was on the LIFE Chair on Fuzzy Theory at Tokyo Institute of Technology (Japan) during 1992-1993, Distinguished Visiting Lukacs Professor of Statistics at Bowling Green State University, Ohio, in 2002, Distinguished Fellow of the American Society of Engineering Education (ASEE), Fellow of the International Fuzzy Systems Association (IFSA). He has published 16 books, 6 edited books and more than 100 papers. Dr. Nguyen's current research interests include Fuzzy Logics and their Applications, Random Set Theory, Risk Analysis, and Casual Inference in Econometrics.

Summary: In view of the recent ban of the use of P-values in statistical inference, since they are not qualified as information measures of support from empirical evidence, we will not only take a closer look at them, but also embark on a panorama of more promising ingredients which could replace P-values for statistical science as well as for any fields involving reasoning with integrated uncertainty. These ingredients include the recently developed theory of Inferential Models, the emergent Information Theoretic Statistics, and of course Bayesian statistics. One main focus of our analysis of information measures is its logical aspect where emphasis will be placed upon conditional (event) logic, probability logic, possibility distributions, and some fuzzy sets.

Autonomous Systems: Many Possibilities and Challenges

Akira Namatame

National Defense Academy of Japan, Yokosuka, Japan
Asian Office of Aerospace Research & Development (AOARD),
US Air Force Research Laboratory (AFRL)

Short biography: Dr. Akira Namatame is Professor emeritus of National Defense Academy, Japan. He is now Scientific Advisor, Asian Office of Aerospace Research & Development of US Air Force Research Laboratory. His research interests include multi-agent systems, complex networks, artificial intelligence, computational social science, and game theory. In the past ten years, he has given over 35 invited talks, and over 15 tutorial lectures in international conferences and workshops, and academic institutions. He has organized more that 30 international conferences and workshops, and special sessions. He is the editor-in-chief of Springer's Journal of Economic Interaction and Coordination (JEIC), editor in Modeling and Simulation Society Letter. He has published more than 300 refereed scientific papers, together with eight books on multi-agent systems, agent modeling and network dynamics, collective systems and game theory. More detail information can be obtained through http://www.nda.ac.jp/ ~nama.

Summary: Our lives have been immensely improved by decades of automation technologies. Most manufacturing equipment, home appliance, cars and other physical systems are somehow automated. We are more comfortable, more productive and safer than ever before. Without automation, they are more troublesome, more time consuming, less convenient, and far less safe. Systems that can change their behavior in response to unanticipated events during operation are called autonomous. Autonomous systems generally are those that take actions automatically under certain conditions. They can be thought of as self-governing systems capable of acting on their own within programmed boundaries. Depending on a system's purposes and required actions, autonomy may occur at different scales and degrees of sophistication. The capability of such autonomous systems and their domains of application have expanded significantly in recent years. These successes have also been accompanied by failures that compellingly illustrate the real technical difficulties associated with seemingly natural behavior specification for truly autonomous systems. The autonomous technology also holds the potential for enabling entirely new capabilities in environments where direct human control is not physically possible. For all of these reasons, autonomous systems technology is as an important element of its science and technology vision and a critical area for future development.

Autonomy is a growing field of research and application. Specialized robots in hazardous environments and medical application under human supervisory control for space and repetitive industrial tasks have proven successful. However, research in areas of self-driving cars, intimate collaboration with humans in manipulation tasks, human control of humanoid robots for hazardous environments, and social interaction with robots is at initial stages. Autonomous systems are in their infancy and are capable only of performing well-defined tasks in predictable environments. Advances in technologies enabling autonomy are needed for these systems to respond to new situations in complex, dynamic environments. Research on autonomy includes many challenging problems and has the potential to produce solutions with positive social impact. Its interdisciplinary nature also requires that researchers in the field understand their research within a broader context. In this talk, I will discuss autonomous technologies that promise to make humans more proficient in addressing such needs. The current status of autonomy research is reviewed, and key current research challenges for the human factors community are described. I will also present a unified treatment of autonomous systems, identify key themes, and discuss challenge problems that are likely to shape the science of autonomy.

Fuzzy Co-clustering and Application to Collaborative Filtering

Katsuhiro Honda

Osaka Prefecture University, Sakai, Japan

Short biography: Katsuhiro Honda is a professor of the Department of Computer Science and Intelligent Systems, Graduate School of Engineering, Osaka Prefecture University, Japan. His research interests include hybrid techniques of fuzzy clustering and multivariate analysis, data mining with fuzzy data analysis, and neural networks. He has published more than 100 scientific papers, including those appearing in such journals as IEEE Transactions on Fuzzy Systems, International Journal of Approximate Reasoning, etc. He received the Outstanding Book Award (2010), the Best Paper Award (2002, 2011, 2012) and so on from the Japan Society for Fuzzy Theory and Intelligent Informatics, and delivered a tutorial lecture at the 2004 IEEE International Conference on Fuzzy Systems.

Summary: Cooccurrence information analysis became more popular in many web-based system analysis such as document analysis or purchase history analysis. Rather than the conventional multivariate observations, each object is characterized by its cooccurrence degrees with various items, and the goal is often to extract co-cluster structures among objects and items, such that mutually familiar object-item pairs form a co-cluster. A typical application of co-cluster structure analysis can be seen in collaborative filtering (CF). CF is a basic technique for achieving personalized recommendation in various web services by considering the similarity of preferences among users. In this talk, I'd like to introduce a fuzzy co-clustering model, which is motivated from a statistical co-clustering model, and demonstrate its applicability to CF tasks following a brief review of the CF framework.

Contents

Biomedical and Image Applications

Statistical Methods

Econometric Applications

Invited Papers

On Evidential Measures of Support
for Reasoning with Integrated Uncertainty:
A Lesson from the Ban of P-values
in Statistical Inference

Hung T. Nguyen[1,2(✉)]

[1] Department of Mathematical Sciences, New Mexico State University,
Las Cruces, USA
hunguyen@nmsu.edu
[2] Faculty of Economics, Chiang Mai University, Chiang Mai, Thailand

Abstract. In view of the recent ban of the use of P-values in statistical inference, since they are not qualified as information measures of support from empirical evidence, we will not only take a closer look at them, but also embark on a panorama of more promising ingredients which could replace P-values for statistical science as well as for any fields involving reasoning with integrated uncertainty. These ingredients include the recently developed theory of Inferential Models, the emergent Information Theoretic Statistics, and of course Bayesian statistics. The lesson learned from the ban of P-values is emphasized for other types of uncertainty measures, where information measures, their logical aspects (conditional events, probability logic) are examined.

Keywords: Bayesian statistics · Conditional events · Entropy inference procedures · Information measures · Information theoretic statistics · Integrated uncertainty probability logic P-values · Testing hypotheses

1 Introduction

The recent ban on the use of the notion of P-values in hypothesis testing (Trafimow and Marks [32]) triggered a serious reexamination of the way we used to conduct inference in the face of uncertainty. Since statistical uncertainty is an important part of an integrated uncertainty system, a closer look at what went wrong with statistical inference is necessary to "repair" the whole inference machinery in complex systems.

Thus, this paper is organized as follows. We start out, in Sect. 2, by elaborating on the notion of P-values as a testing procedure in null hypothesis signficance testing (NHST). In Sect. 3, within the context of reasoning with uncertainty where logical aspects and information measures are emphasized, we elaborate on why p-values should not be used as an inference procedure anymore. Section 4 addresses the question "What are the items in statistical theory

© Springer International Publishing AG 2016
V.-N. Huynh et al. (Eds.): IUKM 2016, LNAI 9978, pp. 3–15, 2016.
DOI: 10.1007/978-3-319-49046-5_1

which are affected by the removal of P-values?". Section 5 points out alternative inference procedures in a world without P-values. We reserve the last Sect. 6 for a possible "in defense of P-values".

2   The Notion of P-values in Statistical Inference

It seems useful to trace back a bit of Fisher's great achievements in statistical science. The story goes like this. A lady claimed that she can tell whether a cup of tea with milk was mixed with tea or milk first, R. Fisher designed an experiment in which eight cups of mixed tea/milk (four of each kind) was presented to her (letting her know that four cups are mixed with milk first, and the other four are mixed with tea first) in a random sequence, and asked her to taste and tell the order of mixture of all cups. She got all eight correct identifications. How did Fisher arrive at the conclusion that the lady is indeed skillful? See Fisher [10], also Salbursg [29]. This kind of testing problem is termed Null Hypothesis Significance Testing (NHST), due to Fisher [9].

The important question is "Could we use P-values to carry out NHST?". You may ask "what is the rationale for using P-value to make inference?". Well, don't you know the answer? It could be the *Cournot's principle* (see, e.g., Shafer and Vork,[31], pp. 44+), according to which, it is practically certain that predicted events of small probabilities will not occur. But it is just a "principle", not a theorem! It does have some flavor of logic (for reasoning), but which logic? See also Gurevich and Vovk [14] , where two "interesting" things to be noted: First, to carry out a test, one just "adopts" a "convention", namely "for a given test, smaller values provide stronger impugning evidence"! And secondly, it is a fact that "every test statistic is equivalent to a unique exact P-value function".

3   Why P-values Are Banned?

Starting with NHST, the unique way to infer conclusions from data is the traditional notion of P-values. However, there is something fishy about the use of P-values as a "valid" inference procedure, since quite sometimes serious problems with them arised, exemplified by Cohen [6], Schervish [30], Goodman [13], Hurlbert and Lombardi [15], Lavine [16], and Nuzzo [28].

Having relied upon P-value as the inference procedure to carry out NHST (their bread and butter research tool) for so long, the Psychology community finally has enough of its "wrong doings", and without any reactions from the international statistical community (which is responsable for inventing and developing statistical tools for all other sciences to use), decided, on their own, to ban NHST-Procedure (meaning P-values), Trafimow and Marks [32]. While this is a ban only for their Basic and Applied Social Psychology Journal, the impact is worldwide. It is not about the "ban", it is about "what wrong with P-values?" that we should all be concerned. For a flavor of doing wrong statistics, see e.g., Wheelan [34].

Even, the ban gets everybody's attention now, what happend since last year? Nothing! Why? Even after the American Statistical Association issued a "statement" about P-values (ASA News [2]), and Wassertein and Lazar [33], not banning P-values (why not?), but "stating" six "principles".

What do you read and expect from the above "statement"? Some literature search reveals stuff like this. "Together we agreed that the *current culture* of statistical significance testing, interpretation, and reporting *has to go*, and that adherence to a minimum of six principles can help to pave the way forward for science and society". And in the *Sciences News*, for laymen, "P-value ban: small step for a journal, giant leap for sicence". See also, Lavine [16].

There are three theoretical facts which make P-values undesirable for statistical inference:

(i) P-values are not model probabilities.

First, observe that a hypothesis is a statistical model. The P-value $P(T_n \geq t|H_o)$ is the probability of observing of an extreme value t if the null hypothesis is true. It is not $P(H_o|T_n \geq t)$ even when this "model probability given the data" makes sense (e.g., as in the Bayesian framework where H_o is viewed as a random event). Note that when $P(H_o|T_n \geq t)$ makes sense and is available, it is legitimate to use it for model selection (a valid inference procedure from at least a common sense standpoint). In a frequentist framework, there is no way to convert $P(T_n \geq t|H_o)$ to $P(H_o|T_n \geq t)$. As such, the P-value $P(T_n \geq t|H_o)$, alone, is useless for inference, precisely as "stated" in the sixth principle of the ASA.

(ii) The reasoning with P-values is based on an invalid logic.

As mentioned by Cohen [6] and in the previous section, the use of P-values to reject H_o seems to be based on a form of Modus Tollens in logic, since after all, reasoning under uncertainty is inference ! and, each mode of reasoning is based upon a logic. Now, thanks to Artificial Intelligence (AI), we are exposed to a variety of logics, such as probability logic, conditional probability logic, fuzzy logics...which are logics for reasoning under various types of uncertainty. See a text like Goodman, Nguyen and Walker [12]. In particular, we could face rules that have exceptions (see e.g., Bamber, Goodman and Nguyen, [3]). The famous "penguin triangle" in AI can be used to illustrate well the invalidity of Modus Tollens in uncertain logics.

While we focus in this address on reasoning with P-values in probabilistic systems, perhaps few words about reasoning with more complex systems in which several different types of uncertainty are involved (integrated uncertain systems) should be mentioned. To create machines capable of ever more sophisticated tasks, and of exhibiting ever more human-like behavior, we need knowledge representation and associated reasoning (logic). In probabilistic systems, no additional mathematical tools are needed, since we are simply dealing with probability distributions, and the logic used is classical two-valued logic. For general integrated uncertain systems, new mathematical tools such as conditional events, possibility theory, fuzzy logics are needed. See, e,g., Nguyen and Walker [25], Nguyen and Walker [26], Nguyen [27].

(iii) As set-functions, P-values are not information measures of model support.

Schervish [30], while discussing the "usual" use of P-values to test hypotheses (in both NHST and Neyman-Pearson tests), "discovered" that "a common informal use of P-values as measures of support or evidence for hypotheses has serious logical flaws". We will elaborate on his "discovery" in the context of information theory.

Essentially, the reason to use P-values, in the first place, although not stated explicitly as such, to "infer" conclusions from data, is that they seems to be "information measures of location" derived from data (evidence) in support of hypotheses. Is that true? Specifically, Given a null hypothesis H_o and a statistic T_n and the observed value $T_n = t$, the P-value $p(H_o) = P(T_n \geq t|H_o)$, as a function of H_o, for fixed T_n and the observed value $T_n = t$, is "viewed" as a measure of support that the observed value t lends to H_o (or amount of evidence in favor of H_o) since large values of $p(H_o) = P(T_n \geq t|H_o)$ make it harder to reject H_o (whereas, small values reflect non-support for H_o, i.e., rejection). But this "practice" is always informal, and "no theory is ever put forward for what properties a measure of support or evidence should have".

What is an information measure? Information decreases uncertainty. Qualitative information is high if surprise is high. When an event A is realized, it provides an information. Clearly, in the context of "statistical information theory", information is a decreasing function of probability: the smaller the probability for A to occur, the higher the information obtained when A is realized. If A stands for "snowing", then when A occured, say, in Bangkok, it provides a "huge" amount of information $I(A)$. Put it mathematically (as in Information Theory, see e.g., Cover and Thomas, [7], $I(A) = -\log P(A)$. For a general theory of information without probability, but keeping the intuitive behavior that information should be a decreasing function of events, see, e.g., Nguyen [24]. This intuitive behavior is about a specific aspect of the notion of information that we are considering in uncertainty analysis, namely, *information of localization.*

In the context of testing about a parametric model, say, $f(x|\theta), \theta \in \Theta$, each hypothesis H_o can be identified with a subset of Θ, still denoted as $H_o \subseteq \Theta$. An information measure of location on Θ is a set-function $I : 2^\Theta \to \mathbb{R}^+$ such that $A \subseteq B \implies I(A) \geq I(B)$. The typical probabilistic information measure is $I(A) = -\log P(A)$. This is the appropriate concept of information measure in support of a subset of Θ (a hypothesis). Now, consider the set function $I(H_o) = P(T_n \geq t|H_o)$ on 2^Θ. Let $H_o' \supseteq H_o$. If we use P-values to reject null hypotheses or not, e.g., rejecting H_o' (i.e., the true $\theta_o \notin H_o'$) when, say, $I(H_o') \leq \alpha = 0.05$, then since $H_o' \supseteq H_o$, we also reject H_o, so that $I(H_o) \leq \alpha$, implying that $H_o' \supseteq H_o \implies I(H_o') \geq I(H_o)$ which indicates that $I(.)$ is not an information measure (derived from empirical evidence/ data) in support of hypotheses, since it is an increasing rather than a decreasing set function. *P-values are not measures of strength of evidence.*

4 Are Neyman-Pearson Testing Theory Affected?

So far we have just talked about NHST. How Neyman-Pearson (NP) testing framework differs from NHST? Of course, they are "different", but now, in view of the ban of P-values in NHST, you "love" to know if that ban "affects" your routine testing problems where in teaching and research, in fact, you are using NP tests instead? Clearly the findings are extremely important: either you can continue to proceed with all your familiar (asymptotic) tests such as Z-test, t- test, $\mathcal{X}^2$-test, KS- test, DF- test, ...or... you are facing "the final collapse of the Neyman-Pearson decision theoretic framework "(as announced by Hurlbert and Lombardi [15]! And in the latter (!), are you panic?

In accusing Fisher's work on NHST as "worse than useless", Neyman and Pearson embarked on shaping Fisher's testing setting into a decision framework as we all know and use so far, although science is about discovery of knowledge, and not about decision-making. The "improved" framework is this. Besides a hypothesis, denoted as H_o (although it is not for nullifying, but in fact for acceptance), there is a specified alternative hypothesis H_a (to choose if H_o happens to be rejected). It is a model selection problem, where each hypothesis corresponds to a statistical model. The NP testing is a decision-making problem: using data to reject or accept H_o. By doing so, two types of errors might be committed: false positive: $\alpha = P(\text{reject } H_o | H_o \text{ is true})$, false negative $\beta = P(\text{accept } H_o | H_o$ is false). It "improves" upon Fisher's arbitrary choice of a statistic to compute the P-value to reach a decision, namely a most powerful "test" at a fixed α-level. Note that while the value of α could be the same in both approaches, say 0.05 (a "small" number in $[0, 1]$ for Fisher), its meaning is different, as $\alpha = 5\%$ in NP approach (the probability of making the wrong decision of the first kind).

Let's see how NP *carry out* their tests? As a test is usually based on some appropriate statistic $T_n(X_1, X_2, ..., X_n)$ (though technically not required) where, say, $X_1, X_2, ..., X_n$ is a random sample, of size n, drawn from the population, so that we select a set B in the sample space of T_n as a rejection (critical) region.

The most important question is: What is the **rationale** for selecting a set B as a rejection region? Since data (values of the statistics T_n) in B lead to rejection of H_o (i.e., on the basis of elements of B we reject H_o), this *inference procedure* has to have a "plausible" explanation for people to trust! Note that an inference procedure is not a mathematical theorem! In other words, why a data in B provides evidence to reject H_o? Clearly, this has something to do with the statistic $T_n(X_1, X_2, ..., X_n)$!

Given α and a (test) statistic T_n, the rejection region R_α is determined by

$$P(T_n \in R_\alpha | H_o) \leq \alpha$$

Are P-values left out in this determination of rejection regions? (i.e., the rationale of inference in NP tests does not depend on P-values of T_n?). Put it differently: How to "pick" a region R_α to be a rejection region?

Note that the P-value statistic $p(T_n) = 1 - F_{T_n | H_o}(T_n)$ corresponds to a N-P test. Indeed, let α be given as the type-I error. Then, the test, say, S_n, which rejects H_o if and only if $p(T_n) \leq \alpha$ has $P(\text{rejecting } H_o | H_o) = P(p(T_n) \leq \alpha)$ which will be $\leq \alpha$, if the random variable $p(T_n)$ dominates (first) stochastically the uniform distribution on $[0, 1]$ (so-called a "valid" P-value statistic).

Thus, in summary, the statistics used for significance tests (Fisher) and hypothesis tests (N-P) are the same, and hypothesis tests can be carried out via p-values (as the rationale for rejection), where significance levels (but not p-values) are taken as error probabilities. Thus, in improving Fisher's NHST setting, NP did not "improve" Fisher's intended inference procedure (i.e., P-values), so that both NHST and NP tests share the same (wrong) inference procedure. See Lehmann [17], Lehmann and Romano [18].

In this respect, I cannot resist to put down the following from Freedman, Pisani, and Purves [11], (pp. 562–563):

"Nowadays, tests of significance are extremely popular. One reason is that the tests are part of an impressive and well-developed mathematical theory... This sounds so impressive, and there is so much mathematical machniery clanking around in the background, that tests seem truly scientific-even when they are complete nonsense. St. Exupery understood this kind of problem very well:

"When a mistery is too overpowering, one dare not disobey" (*The Little Prince*).

The basic questions for inference in testing are these. What is the rationale for basing our conclusions (decisions for reject or accept) on a statistic T_n? Are all rejection regions should be determined this way? Based upon which logic (or rationale) we construct rejection regions, from which we reach decisions (i.e., from data to conclusions)?

With all the fancy mathematics to come up with, say, an asymptotic distribution of a statistic T_n (e.g., the Dickey-Fuller test for stationarity in AR(1) model), the goal is to say this. If the observed value of $T_n = t$ is "large", then reject H_o, where "large" is "defined" as the critical value c determined by $P(T_n \geq c | H_o) \leq \alpha$. Don't you see that is clearly equivalent to the P-value of T_n being less than α? Specifically:

$$t \geq c \Leftrightarrow P(T_n \geq t | H_o) \leq \alpha$$

Thus, this kind of inference (or logic!) is exactly the same as what Fisher had suggested for NHST. In other words, all fancy mathematical works aim at providing the necessary stuff for computing p-values! from which to jump to rejection conclusion, just like in NHST.

Freedman, Pisani, and Purves [11] seemed to feel something wrong about the "logic of p-values" /(In fact the "*logic of the z-test*"), pp. 480–481, but did not dare to go all the way to say that it's silly to use it to make inference. They said things like

"It is an argument by contradiction, designed to show that the null hypothesis will lead to an absurd conclusion and must be rejected". Wrong, we are not in binary logic! And

"P is not the chance of the null hypothesis being right", "Even worse, according to the frequentist theory, there is no way to define the probability of the null hypothesis being right".

Remember: if you must choose (or select) two things under uncertainty, you would choose the one with higher "probability" to be right, given your evidence (data). But you do not have it if it does not make sense to consider such a probability (since a hypothesis is not a random event), let alone with p-values. Perhaps, because of this "difficulty" that statisticians "play around" with a seemingly OK logic of p-values? and no one has catched it. Should we continue to use this wrong logic or try to find a better (correct) one? There is no possible choice anymore: the P-value logic is now officially banned!

It seems we cannot "repair" the p-value logic. We must abandon it completely. Clearly, the Bayesian approach to statistical inference does not have this problem. Note that, there are no *names* of tests in Bayesian statistics, since their is only one way to conduct tests (similar to estimation method), which is in fact a selection problem, based on Bayes factors (no test statistics, no sampling distributions!). The crucial thing is that, hypotheses are random events and hence it makes sense to consider "probabilities of hypotheses/ given data" which form a common sense reasoning for reaching conclusions.

However, the following observation seems interesting? Most of NP rejection regions are nested, i.e. of the form $(T_n > c)$, and as such their associated testing procedure is equivalent to using P-values whose threshold is taken as the size of the test. Thus, from a logical point of view, there is no difference between NHST and NP-testing as far as "inference" is concerned. However, there is something interesting here which could "explain" the meaning of NP-rejection regions. While the null hypothesis is rejected by using the "logic of P-values", this inference is controlled by the type-I error α (Noting that there is no such guarantee in NHST framework). In other words, while the inference based on P-value might not be "logical", it could be "plausible": decisions using P-values in NP-testing, say, by NP Theory are enforced by error probabilities. Specifically the use of P-values in NP-testing is in fact carried out *together* with error probabilities, and not *alone*. The implication is that decisions for rejecting hypotheses are controlled with specified error probabilities in advance.

5 Some Alternatives to P-values

Clearly, Bayesian selection (testing) is valid in this sense. The basic "ingredients" are priors and Bayes' theorem which are used precisely to obtain model probabilities for selection purposes. See a Text like Koch (2007).

With respect to "Statistics of the 21st century", David Draper (2009) already claimed that it is Bayesian statistical reasoning, in

Bayesian Statistical Reasoning:
An Inferential, Predictive and Decision-Making Paradigm for the 21st Century
(www.ams.ucsc.edu/~draper)

Well, not so fast! Even *now* with the possible "final collapse of the Neyman-Pearson decision theoretic framework and the rise of the neoFisherian" (Hurlbert and Lombardi [15]) the Bayesian statistics is not the unique "super power", thanks to the existence (since 1959) of the *non-Bayesian Information Theoretic (IT) Statistics!*

Another approach to statistics which posseses also two similar ingredients, allowing us to reach model probabilities for selection, is the *Information Theoretic (IT)Statistics*. These are "a chosen class of possible models" (playing the role of prior distributions in Bayesian approach, but not subjectively in nature), and the notion of "cross entropy" (playing the role of Bayes' formula). See Kullback [21], Anderson [1], Burnham and Anderson [5], Konishi and Kitagawa [20], Cumming [8], Cover Thomas [7].

Testing of hypotheses is a special case of model selection. The IT approach to model selection is a sort of posterior analysis without subjective priors. What could be considered as priors (a priori) is our choice of a class of competing models for selecting the "best" model for prediction purpose. The Kullback-Liebler divergence (together with its estimate/ the AIC statistic) plays the role of Bayes' formula to arrive at model probabilities which serve as a valid inference for decisions on selections (as opposed to P-values).

Given data $X_1, X_2, ..., X_n$ from a population X, with, of course, unkown probability density $f(.)$, we consider a class of models $\mathcal{M} = \{M_j : j = 1, 2, ..., k\}$ where each $M_j = \{g_j(.|\theta_j) : \theta_j \in \Theta_j\}$, to be possible candidates for approximating $f(.)$.

Somewhat similar to Fisher's idea of finding a statistic to measure the incompatibility between the data and a hypothesis (here, a model), from which we can figure out a way to "reject" it, the IT approach proceeds as follows.

First, in general terms: Suppose we use a density $g(.)$ to approximate an unknown $f(.)$. How to measure the "lost of information?". Well, remember how you answer such a question in your simple linear regression? You use the coefficient of determination as a measure of how much the linear model captures the real variation of the true model, or equivalently, one minus that coefficient as how much information is lost when approximating the true model by a linear model. The IT approach is more general as it addresses directly to the models themselves, by considering a sort of distance between distributions. Note another similarity with P-values! A distance between the true model and an approximate one, measuring the loss of "information", could be used to judge whether the approximating model is reasonable.

As we will see shortly, this idea is much better than P-values, and serves as the fundamentals for valid statistical inference (in the sense that it can provide model probabilities for ranking alternatives, whereas P-values cannot).

Now, given densities f, g, there are many possible ways to define (real) distances between them (just as in functional analysis). We seek a kind of distance which measures a loss of information.

But what is *information*? it is right here that we need a theory of information!

Intuitively, information is a decrease of uncertainty. Uncertainty involves probabilities. Thus, any measure of information should be a function of probability (?). Clearly, an event A gives us less information when its probability $P(A)$ is high: how much information you learn for $A=$"it will be real hot in April in

Chiang Mai next year"? Since $P(A) \simeq 1$, your information is zero (no surprise). How about $B =$ "it could be snowing"? Well, big surprise, lot of information: $P(B) \simeq 0$, information is infinity! Thus, the information provided by the realization of an event A is of the form $I(A) = -\log P(A)$, i.e., a non increasing function of probability.

Based upon information theory, via Shannon's entropy, Kullback and Liebler (see Kullback, 1968) considered a "pseudo" distance (relative entropy, or divergence) between two probability density f and g as

$$I(f|g) = \int f(x) \log \frac{f(x)}{g(x)} dx = E[\log \frac{f(X)}{g(X)}]$$

measuring the loss of information when using g to approximate f. For a good explanation of this "loss of information", see Benish [4].

The expectation is of course with respect to the distribution f of the random variable X, so that sometimes we write $I(f|g) = E_f[\log \frac{f(X)}{g(X)}]$ to be explicit. It is a pseudo distance in the sense that $I(f|g) \geq 0$ and $I(f|g) = 0$ if and only if $g(.) = f(.)$.

The purpose to consider such a distance is to compare different $g(.)$ as possible candidates to be used as approximations of $f(.)$. Given, say, an i.i.d. sample $X_1, X_2, ..., X_n$ drawn from X (with true, unknown distribution f), we cannot compute, or even estimate $I(f|g)$, even if $g(.)$ is (completely) known (specified). Indeed,

$$I(f|g) = \int f(x) \log f(x) dx - \int f(x) \log g(x) dx$$

If $g(.)$ is known, then since

$$- \int f(x) \log g(x) dx = -E[\log g(X)]$$

this term can be estimated consistently, for large n, via the strong law of large numbers, by

$$-\frac{1}{n} \sum_{i=1}^{n} \log g(X_i)$$

But, the first term $\int f(x) \log f(x) dx$ is unknown. Fortunately, while it is unknown, it does not involve the candidate g, so that it is the same for any g. Thus, let $\int f(x) \log f(x) dx = C$, the comparison of $I(f|g)$ among various different g, only involves the term $-E[\log g(X)]$, namely

$$I(f|g) - C = -E[\log g(X)]$$

With this meaning of $I(f|g)$, clearly we seek

$$\arg \min_{g \in \mathcal{G}} [-E[\log g(X)]]$$

A problem arises when each $g(.)$ is a statistical model, i.e., $g(.)$ is only specified up to some unknown parameter (possibly vector) $\theta \in \Theta$: $g(.|\theta)$, so that we are facing

$$\min_{g \in \mathcal{G}}[-E[\log g(X|\theta)]]$$

which cannot be estimated anymore since θ is unklnown. We handle this problem by replacing $\theta \in \Theta$ by its MLE

$$\theta_n(X_1, X_2, ..., X_n) = \arg\max_{\theta \in \Theta} L_g(X_1, X_2, ..., X_n|\theta)$$

For i.i.d. sample $L_g(X_1, X_2, ..., X_n|\theta) = \prod_{i=1}^n g(X_i|\theta)$.

Thus, replacing $-E[\log g(X|\theta)]$ for $\theta \in \Theta$ by $-E[\log g(X|\theta_n(X_1, X_2, ..., X_n))]$ which is a random variable (since it depends on the values of the sample $(X_1, X_2, ..., X_n) = Y$). An estimate of it could be simply its mean. So, finally, we are led to estimating

$$E_Y E_X[\log g(X|\theta_n)]$$

from which, a reasonable selection criterion is

$$\max_{g \in \mathcal{G}} E_Y E_X[\log g(X|\theta_n)]$$

It turns out that, for large sample size n (see an appendix), $E_Y E_X[\log g(X|\theta_n)]$ can be estimated by $\log L_g(\theta_n|Y) - k$ where k is the number of estimated parameters (dimension of Θ), resulting in the so-called Akaike Information Criterion (AIC), for model g:

$$AIC(g) = -2\log L_g(\theta_n|Y) + 2k$$

Note that $L_g(\theta_n|Y)$ is the value of the likelihood function of g evaluated at the MLE estimator θ_n.

Thus, for n large, the selection problem (containg hypothesis testing as a special case) is carried out simply by computing $AIC(g)$ for all $g \in \mathcal{G}$, and pick the one with smallest AIC value.

Let $AIC_{\min}$ be the smallest AIC value, corresponding to some model. Then, let

$$\Delta_g = AIC(g) - AIC_{\min}$$

and make a transformation, e.g., $\Delta_g \rightarrow e^{-\frac{\Delta_g}{2}}$ (as suggested by Akaike) to obtain "model likelihood), we arrive at Akaike's weights (of evidence supporting models)

$$w_g = \frac{e^{-\frac{\Delta_g}{2}}}{\sum_{g \in \mathcal{G}} e^{-\frac{\Delta_g}{2}}}$$

which are interpreted as evidence in favor of model g being the best approximating model in the chosen set of models $\mathcal{G}$, viewing as "model probability" given the data, needed for explaining the rationale in the selection process (where P-values are lacking), noting that

$$w_h = \arg_{g \in \mathcal{G}} w_g \Leftrightarrow AIC(h) = \arg\min_{g \in \mathcal{G}} AIC(g)$$

How about "small sample size?". Remember how you consider small sample sizes in your introductory statistical courses? Here, we are talking about small size for AIC approximation to K-L statistic. As a "rule of thumb", the sample size n is considered as small, relative to the number of parameters k in the model g, when $n < 40\,k$. In that case, the AIC is "corrected" to be

$$AIC_c(g) = AIC(g) + \frac{2k(k+1)}{n-k-1}$$

6 Plausibilities in Inferential Models

In discussing this "crisis" in (frequentist) statistical inference with many colleagues, I received "mixed signals". Almost all agreed on the second "principle" of ASA's statement, namely " P-values do not measure the probability that the studied hypothesis is true", but stopped short of saying anything more! Some mentioned that ASA did not ban P-values. What does that means? Does that mean that you still can publish research papers using P-values in statistical journals (but not in psychology journals!)? I don't think so, since the statistical journals' editors have to take into account of public reactions (everybody was aware of the wrong doings/ not just the misuses of P-values from Sciences News), unless they can clarify their actions, scientifically. Almost all did not "feel" that NP testing is affected (and hence will "survive" this crisis) without explaning why, although Gurevich and Vovk [14] proved that "Every test statistic is equivalent to a unique (exact) p-function". Remember also: An inference procedure is valid only if it is based on a firm logical basis. The "future" will settle the matter soon, I guess.

Meanwhile, you could ask "Rather than throwing P-values away, can we find a way to save them so that they can contribute to statistical inference?" I happened to read Martin and Liu [22] in which they proposed a way to save P-values, in the framework of their theory of Inferential Models, Martin and Liu [23]. Below are the essentials.

Let the null hypothesis H_o be identified with a subset Θ_o of the parameter space Θ (in a statistical model $X \sim F_\theta(.), \theta \in \Theta$). For a test statistic T_n, the P-value of $T_n = t$ under Θ_o is extended to $Pv(\Theta_o|t) = \sup_{\theta \in \Theta_o} P(T_n \geq t|\theta)$. While $Pv(\Theta_o|t)$ is not the probability that Θ_o is true when we observe t, it could be equated to another concept of uncertainty, namely "plausibility", to be used as a new inference engine. But what is a *plausibility measure* (rather than a probability measure)? It is the capacity functional of a random set (see Nguyen, [27]). Specifically, if S is a random set with values in 2^Θ, then its capacity functional (or plausibility measure) $Pl_S(.) : 2^\Theta \to [0,1]$ is defined as $Pl_S(A) = P(S \cap A \neq \varnothing)$.

The result of Martin and Liu [22] is this. Any $Pv(.)$ on 2^Θ can be written as $Pl_S(.)$ for some random set S on 2^Θ. The construction of a suitable random set S is carried out within the Inferential Model framework, Martin and Liu [23]. Note that the computation of plausibilities does not require one to assume that the null hypothesis is true, as opposed to P-values, so that it does make sense to take $Pl_S(\Theta_o|t)$ as the plausibility that Θ_o is true. The intention is to transfer P-values to plausibilities (not to probabilities) and use plausibilities to make inferences. It remains to take a closer look at this proposal, especially with respect to the objections similar to those of P-values.

References

1. Anderson, D.R.: Model Based Inference in the Life Sciences. Springer, Heidelberg (2008)
2. ASA News, American Statistical Association releases statement on statistical significance and p-value, ASA News, March 2016
3. Bamber, D., Goodman, I.R., Nguyen, H.T.: Robust reasoning with rules that have exceptions: from second-order probability to argumentation via upper envelopes of probability and possibility plus directed graphs. Ann. Math. Artif. Intell. **45**, 83–171 (2005)
4. Benish, W.A.: Relative entropy as a measure of diagnostic information. Med. Decis. Making **19**, 202–206 (1999)
5. Burnham, K.P., Anderson, D.R., Selection, M., Inference, M.: A Practical Information Theoretic Approach. Springer, New York (2002)
6. Cohen, J.: The earth is round. Am. Psychlog. **49**(12), 997–1003 (1994)
7. Cover, T.M., Thomas, J.A.: Elements of Information Theory. Wiley, New York (2006)
8. Cumming, G.: Understanding the New Statistics. Routledge, New York (2012)
9. Fisher, R.A.: Statistical Methods for Research Workers. Oliver and Boyd, Edinburgh (1925)
10. Fisher, R.A.: Mathematics of the lady tasting tea, the world of mathematics. In: Newman, J.R. (ed.) (Part VIII): Statistics and the Design of Experiments, vo. III, pp. 1514–1521. Simon and Schuster (1956)
11. Freedman, D., Pisani, R., Purves, R.: Statistics. W.W. Norton, New York (2007)
12. Goodman, I.R., Nguyen, H.T., Walker, E.A., Inference, C.: Logic for Intelligent Systems: A Theory of Measure-Free Conditioning. Hardcover, North-Holland (1991)
13. Goodman, S.: A dirty dozen: twelve p-value misconceptions. Semin. Hematol. **45**, 135–140 (2008)
14. Gurevich, Y., Vovk, V., Fundamentals of P-values: introduction. Bull. Euro. Assoc. Theor. Comput. Sci. (2016, to appear)
15. Hurlbert, S.H., Lombardi, C.M.: The final collapse of the Neyman-Pearson decision theoretic framework and the rise of the neoFisherian. Ann. Zool. Fenn. **46**, 311–349 (2009)
16. Lavine, M.: Comment on Murtaugh. Ecology **93**(5), 642–645 (2014)
17. Lehmann, E.L.: The fisher, Neyman-pearson theories of testing hypotheses: one theory or two? J. Am. Stat. Assoc. **88**(424), 1242–1249 (1993)
18. Lehmann, E.L., Romano, J.P.: Testing Statistical Hypotheses. Springer, New York (2005)

19. Kock, K.R.: Introduction to Bayesian Statistics. Springer, Heidelberg (2007)
20. Konishi, S., Kitagawa, G.: Information Criteria and Statistical Modeling. Springer, New York (2008)
21. Kullback, S.: Information Theory and Statistics. Dover, New York (1968)
22. Martin, R., Liu, C.: A note on P-values interpreted as plausibilities. Stat. Sinica **24**, 1703–1716 (2014)
23. Martin, R., Liu, C.: Inferential Models. Chapman and Hall/CRC Press, Boca Raton (2016)
24. Nguyen, H.T.: Sur les measures d'information de type Inf. In: Nguyen, H.T. (ed.) Theories de l'Information, vol. 398, pp. 62–75. Springer, Heidelberg (1974)
25. Nguyen, H.T., Walker, E.A.: A history and introduction to the algebra of conditional events and probability logic. IEEE Trans. Man Syst. Cybern. **24**(2), 1671–1675 (1996)
26. Nguyen, H.T., Walker, E.A.: A First Course in Fuzzy Logic. Chapman and Hall/CRC Press, Boca Raton (2005)
27. Nguyen, H.T.: An Introduction to Random Sets. Chapman and Hall/CRC Press, Boca Raton (2006)
28. Nuzzo, R.: Statistical errors. Nature **506**, 150–152 (2014)
29. Salbursg, D.: The Lady Teasting Tea. A.W.H Freeman, New York (2001)
30. Schervish, M.J.: P values: what they are and what they are not. Am. Stat. **50**(3), 203–206 (1996)
31. Shafer, G., Vovk, V.: Probability and Finance: It's only a Game. Wiley, New York (2001)
32. Trafimow, D., Marks, E.: Basic and applieds. Soc. Psychol. **37**, 1–2 (2015)
33. Wassertein, R.L., Lazar, N.A.: The ASA's statement on P-value: context, process and purpose. Am. Stat. **70**, 129–133 (2016)
34. Wheelan, C.: Naked Statistics. W.W Norton, New York (2013)

Fuzzy Co-Clustering and Application to Collaborative Filtering

Katsuhiro Honda$^{(\boxtimes)}$

Osaka Prefecture University, Osaka, Sakai 599-8531, Japan
honda@cs.osakafu-u.ac.jp

Abstract. Cooccurrence information analysis became more popular in many web-based system analyses such as document analysis or purchase history analysis. Rather than the conventional multivariate observations, each object is characterized by its cooccurrence degrees with various items, and the goal is often to extract co-cluster structures among objects and items, such that mutually familiar object-item pairs form a co-cluster. A typical application of co-cluster structure analysis can be seen in collaborative filtering (CF). CF is a basic technique for achieving personalized recommendation in various web services by considering the similarity of preferences among users. This paper introduces a fuzzy co-clustering model, which is motivated from a statistical co-clustering model, and demonstrates its applicability to CF tasks following a brief review of the CF framework.

Keywords: Co-clustering · Fuzzy clustering · Collaborative filtering

1 Introduction

Cooccurrence information analysis becomes much more popular especially in recent web-based systems such as document-keyword frequencies in web document analysis and customer-product purchase transactions in web market analysis. Cooccurrence information among objects and items are often summarized into some co-cluster structures, where mutually familiar object-item pairs are grouped into co-clusters. Then, the goal of co-clustering is to extract pairwise clusters of dual partitions among both objects and items, rather than the conventional multivariate observations on objects, where cluster structures are extracted by focusing on mutual similarities among objects only.

In simultaneously extracting dual partitions of objects and items, we have roughly two different partition concepts. First, if we assume that each co-cluster should be formed by familiar objects and items without considering the priority between objects and items in analysis, both objects and items are handled under common constrains. Non-negative Matrix Factorization (NMF) [1] and its extension to co-clustering [2] is typical examples of this approach, where each of both objects and items is often forced to be assigned to a solo co-clusters such that both of them form non-overlapping clusters. This category also includes such

© Springer International Publishing AG 2016
V.-N. Huynh et al. (Eds.): IUKM 2016, LNAI 9978, pp. 16–23, 2016.
DOI: 10.1007/978-3-319-49046-5_2

other model as sequential co-cluster extraction [3] and coVAT [4], where objects and items are mixed and an object × item cooccurrence matrix is enlarged into an (object + item) × (object + item) matrix.

Second, if our purpose is mainly to find object clusters in conjunction with associating typical items to each cluster, we can utilize the conventional clustering concepts under the cluster-wise Bag-of-Words concept. Typical examples of this category are multinomial mixture models (MMMs) [5] or its extensions [6], which utilize multinomial distribution or Dirichlet distribution of items as component densities and try to find object clusters under the mixture density concept. In these mixture models, each component density is estimated in each cluster under the constraint of sum-to-one probabilities of items, and objects are exclusively assigned to nearest clusters under the constrains of sum-to-one probability of objects. The traditional k-means clustering model [7] and fuzzy c-means (FCM) [8] have also been extended to co-clustering models. Fuzzy clustering for categorical multivariate data (FCCM) [9] and fuzzy clustering for document and keywords (Fuzzy CoDoK) [10] modified the within-cluster-error measure of FCM by introducing two types of fuzzy memberships for objects and items, and adopted an aggregation measure to be maximized in each cluster under the similar partition constrains to MMMs. Considering the conceptual similarity among MMMs and FCCM, Honda et al. proposed fuzzy co-clustering induced by MMMs concept (FCCMM) [11], which introduced a mechanism for tuning intrinsic fuzziness of the MMMs pseudo-log-likelihood function and demonstrated the advantage of tuning intrinsic fuzziness in MMMs.

Some hybrid models of the above two concepts have also been proposed by extending FCM-based frameworks. By introducing the exclusive nature of item memberships to FCCMM, we can produce dual exclusive partition [12,13]. It was also tried to apply the FCM-like exclusive condition to both objects and items [14,15].

In this paper, the characteristics of fuzzy co-clustering models are discussed considering to the applicability to collaborative filtering (CF) [16,17], which is a promising application of co-clustering models. CF is a recommendation system for reducing information overloads in recent network societies, where promising items are recommended to users by analyzing user × item evaluation matrices. A traditional approach of neighborhood-based model is often implemented through a two-stage procedure: user neighborhood searching and evaluation averaging in neighborhood. Then, the process is expected to be imitated through user clustering in conjunction with estimating item typicalities in each cluster. In this sense, the second co-clustering concept of 'object clustering with item typicality estimation' seems to be plausible in this task.

The remaining parts of this paper are organized as follows: Sect. 2 gives a brief survey of co-clustering based on soft partition, which includes comparison of two partition concepts. Section 3 introduces the basic concept of CF and discusses the applicability of fuzzy co-clustering to the task. A summary conclusion is presented in Sect. 4.

2   Co-clustering Based on Soft Partition

Assume that we have an $n \times m$ rectangular relational information matrix $R = \{r_{ij}\}$ among n objects and m items, where r_{ij}, $i = 1, \ldots, n$, $j = 1, \ldots, m$ is the degree of mutual connectivity among object i and item j such that a large value means a high connectivity. A typical example is the cooccurrence frequency among documents and keywords in document analysis, e.g., a large r_{ij} means a high frequency of appearance of keyword j in document i, where documents and keywords correspond to objects and items, respectively. This kind of datasets is also popular in network societies such as market analysis based on purchase transaction history.

A promising approach for summarizing the intrinsic information of such rectangular relational data is to extract pairwise clusters composed of mutually familiar object-item pairs. Such clustering tasks as extracting C disjoint pairwise clusters is called co-clustering (or bi-clustering). In this section, a brief survey on some co-clustering based on soft partition concepts is presented.

2.1   Co-clustering Models Supported by Dual Exclusive Partition Nature

When we have multivariate numerical observations on n objects such as an $n \times m$ numerical matrix X, a popular approach for low-dimensional summarization is principal component analysis (PCA), where the mathematical goal is to estimate an singular value decomposition $X \approx FA^{\top}$ such that the reconstruction error is minimized. The low-dimensional factors F can be often utilized for visually summarizing mutual relation among objects on the feature space spanned by basis A.

NMF [1] extended the matrix decomposition concept to non-negative cooccurrence information such that R is decomposed into $R \approx UW^{\top}$, where U and W are non-negative feature matrices of objects and items, respectively. In order to represent the cluster belongingness of objects and items, all elements of U and W are forced not to be negative, and the columns of U and W are mutually almost orthogonal such that each object and item has high feature values in at most one cluster. Oja $et\ al.$ [2] has further extended NMF for emphasizing the co-cluster information of component matrices U and W. Because these models equally force the columns of U and W to have zero inner products, both objects and items are exclusively assigned to clusters without considering the priority under partition purposes.

Besides the singular value decomposition approach, objects and items can be combined into one set in such a way that an enlarged square relational information of $(n + m) \times (n + m)$ matrix.

$$S = \begin{pmatrix} G & R \\ R^{\top} & H \end{pmatrix} \tag{1}$$

Honda $et\ al.$ [3] adopted the sequential fuzzy cluster extraction (SFCE) [18], which was designed for square relational data, to a modified relational data matrix. In this model, missing parts of the full-square matrix were given by $G = H = O$. Bezdek *et al.* [4] adopted visual assessment technique (VAT) to rectangular relational data, which was originally designed for square relational data, estimating G and H by calculating mutual similarity among rows or columns of R.

These co-clustering models are generally useful for summarizing the intrinsic connectivity features. However, if we have a primal target in clustering tasks, these models may not suit the goal of the task.

2.2 Co-clustering Models Considering Primal Target in Objects vs. Items

In such tasks as document clustering, the goal is often mainly to find document clusters in conjunction with selecting their cluster-wise typical keywords. That is, keywords are expected to contribute to characterizing each document cluster and are not intended to be exclusively assigned to clusters. Indeed, some common keywords may frequently occurred in various documents and should have larger typicalities in multiple clusters. So, the typicality of each keyword should be independently evaluated in each cluster.

MMMs [5] is a basic statistical model for co-cluster analysis, in which each component density is defined by multinomial distribution. Multinomial distribution is a multi-category extension of binomial distribution, where the probability of occurrence of each item is estimated following the relative frequencies of item appearances. Mixtures of C component multinomial distributions are estimated through the expectation-maximization (EM) algorithm [19], where the expectation step estimates the probability u_{ci} of each object i drawn from component density c and the maximization step estimates multinomial component densities of probability w_{cj} of item j in component c reflecting object assignments. Although the both two steps are based on a maximum likelihood estimation scheme with a common likelihood function, the partition concepts for objects and items are not common supported by different constraints. The item probability is estimated under the intra-cluster constraint of $\sum_{j=1}^{m} w_{cj} = 1$ so that it represents the relative typicality of item j in component c. On the other hand, the object probability is estimated under the inter-cluster constraint of $\sum_{c=1}^{C} u_{ci} = 1$ so that it represents the exclusive assignment of object i to C clusters. In this sense, Objects are regarded as the primal target in the clustering task and item probability plays a role just for characterizing each object cluster.

Similar co-clustering models have been also proposed supported by k-means-like clustering concepts. In order to estimate the compactness of clusters by considering the intra-cluster typicalities of items instead of centroids of k-means families, FCCM [9] and Fuzzy CoDoK [10] defined the objective function to be maximized with the following aggregation measure:

$$cluster\ aggregation = \sum_{c=1}^{C} \sum_{i=1}^{n} \sum_{j=1}^{m} u_{ci} w_{cj} r_{ij}, \qquad (2)$$

under the same constraints with MMMs. Co-clusters are extracted in such a way that familiar object i and item j having large cooccurrence r_{ij} tend to have large memberships u_{ci} and w_{cj} in a same cluster c. Besides the crisp k-means-like linear objective function of Eq. (2), FCCM and Fuzzy CoDoK achieved fuzzy partition by introducing such fuzzification schemes as the entropy-based regularization [20,21] and the quadratic term-based regularization [21,22]. Although the regularization-based fuzzy partition models have some advantages against the traditional statistical models such as MMMs in arbitrarily tuning the fuzziness degrees of object and item partitions, it is often difficult to find the plausible fuzziness degrees for dual partitions without guidelines or comparative statistical models.

Considering the conceptual similarity among MMMs and FCCM, Honda *et al.* proposed FCCMM [11], which introduced a mechanism for tuning intrinsic fuzziness of the MMMs pseudo-log-likelihood function. The objective function to be maximized was defined as follows:

$$L_{fccmm_1} = \sum_{c=1}^{C} \sum_{i=1}^{n} \sum_{j=1}^{m} u_{ci} r_{ij} \log w_{cj} + \lambda_u \sum_{c=1}^{C} \sum_{i=1}^{n} u_{ci} \log \frac{\alpha_c}{u_{ci}}, \qquad (3)$$

where α_c represents the volume of cluster c such that $\sum_{c=1}^{C} \alpha_c = 1$. Eq. (3) is equivalent to the pseudo-log-likelihood function for MMMs when $\lambda_u = 1$, and the intrinsic fuzziness degree of object partition can be arbitrarily tuned by adjusting the penalty weight λ_u. A larger λ_u ($\lambda_u > 1$) brings a fuzzier partition rather than MMMs while a smaller λ_u ($\lambda_u < 1$) causes a crisper one. Indeed, $\lambda_u \to 0$ implies a crisp partition.

Additionally, the fuzziness degree of item partition can be also tuned by adjusting the non-linearity with respect to w_{cj} such as:

$$L_{fccmm2} = \sum_{c=1}^{C} \sum_{i=1}^{n} \sum_{j=1}^{m} \frac{1}{\lambda_w} u_{ci} r_{ij} \left((w_{cj})^{\lambda_w} - 1 \right) + \lambda_u \sum_{c=1}^{C} \sum_{i=1}^{n} u_{ci} \log \frac{\alpha_c}{u_{ci}}, \quad (4)$$

where the additional weight λ_w is responsible for tuning the item partition fuzziness. $\lambda_w \to 0$ implies the same fuzziness degree with MMMs and a smaller λ_w brings a fuzzier partition.

Because FCCMM was designed supported by the MMMs concept, the plausible fuzziness degrees of dual partitions are expected to be achieved under the guideline of MMMs. That is, the plausible partition can be often searched for by slightly tuning MMMs partitions.

2.3   Hybrid Approaches of Object Targeting Partition and Exclusive Item Partition

Even if the primal target is object partitioning like FCCM and FCCMM, exclusive partition of items is also expected to be useful in order to improve the interpretability of dual partitions. Then, some researches tried to introduce exclusive natures into the object targeting co-clustering models.

In order for each item can have large membership w_{cj} at most in one cluster, FCCM and FCCMM were modified by introducing an additional sharing penalty on items such as $\sum_{t \neq c} w_{cj} w_{tj}$ into the conventional objective function [12,13]. This model was demonstrated to be useful for improving the interpretability by selecting cluster-wise unique items in each object cluster. It is also possible to force the exclusive nature only to some selected items not to be shared by multiple clusters. By carefully selecting the specific items, partition quality can be improved [23].

Another challenge is to adopt the FCM-like exclusive condition to both object and item partitions such that $\sum_{c=1}^{C} u_{ci} = 1$ and $\sum_{c=1}^{C} w_{cj} = 1$. Generally, the objective functions of the object targeting co-clustering models have a trivial solution, where all objects and items are gathered into one cluster. Then, Tjhi and Chen [14,15] introduced a heuristic-based algorithm and demonstrated some advantages in revealing plausible item features.

3   Applicability of Co-clustering to Collaborative Filtering Tasks

This section considers the applicability of co-clustering to CF tasks. CF is a promising technique for reducing information overloads through personalized recommendation based on collaborative data analysis among users [16].

Assume that $R = \{r_{ij}\}$ is an $n \times m$ evaluation matrix, where r_{ij} is an implicit or explicit rating for item j given by user i such that a larger r_{ij} implies a better availability of item j for user i. Each element can be not only an explicit evaluation such as an active rating given by a user but also an implicit evaluation such as purchase transaction of an item by a user. The goal of personalized recommendation in CF can be identified with the task of predicting the availability of missing items [17], e.g., items the active user have not purchased or evaluated. Then, the recommendation system recommends the items, whose availability are expected to be maximum. Here, it should be noted that a missing element in an implicit rating matrix may be not only an unknown situation but also a negative feeling. For example, in purchase history, missing elements can be either of 'negative evaluation' and 'positive but not yet bought'.

The representative memory-based algorithm is GroupLens [16], which virtually implements *word-of-mouth* in network societies and is composed of main two phases: neighborhood users search and preferences averaging. Using the mutual similarity among users, the applicability of item j for an active user a is predicted as follows:

$$y_{aj} = \frac{\sum_{i=1}^{n} (r_{ij} - \bar{r}_i) \times s_{ai}}{\sum_{i=1}^{n} s_{ai}} + \bar{r}_a, \tag{5}$$

where $\bar{r}_i$ is the mean of evaluations by user i and s_{ai} is the mutual similarity measure among user i and the active user a such as Pearson correlation coefficient of two users' evaluations. In GroupLens, the deviations from mean evaluation of each user are considered for eliminating the influences of users' evaluation tendencies.

Because the process of co-clustering can be regarded as summarization of preference tendencies on items in each homogeneous user (object) clusters, a potential application of co-clustering is found in CF tasks [3, 12, 24]. Here, from the viewpoint of realization of *word-of-mouth*, co-clustering-based CF should consider users (objects) as the clustering target, where item preferences are independently summarized in each cluster. In this sense, MMMs-induced co-clustering models are expected to be available in various CF applications.

4   Conclusions

This paper presented a brief survey on co-clustering models, in which many algorithms were summarized into three categories considering their constraints: Dual Exclusive Partition Models, Object Targeting Partition Models and their Hybrid Models. Besides their algorithmic frameworks, they have different applicability to many real world tasks under different constraints on dual partitions.

In this paper, document-keyword analysis and CF are introduced as typical applications of Object Targeting Partition Models, where the goal is mainly to extract object clusters in conjunction with characterization emphasizing their representative items. However, we have also other application areas, where objects and items should be equally handled. For example, in a factory management problem, where many products are manufactured in several machining centers utilizing a wide range of machine tools, we should assign products to machining centers such that each machining center manufactures their products using only similar (common) tools. In this case, product-tool relational data should be summarized into product-tool co-clusters without targeting neither of products and tools.

Rectangular relational data becomes much more popular in various network societies but they should be processed under plausible constraints reflecting the characteristics of tasks. It is expected that many co-clustering researches can contribute to tackling various real applications.

Acknowledgment. This work was supported by JSPS KAKENHI Grant Number JP26330281.

References

1. Lee, D.-D., Seung, H.-S.: Learning the parts of objects by non-negative matrix factorization. Nature **401**(6755), 788–791 (1999)
2. Oja, E., Ilin, A., Luttinen, J., Yang, Z.: Linear expansions with nonlinear cost functions: modeling, representation, and partitioning. In: 2010 IEEE World Congress on Computational Intelligence, Plenary and Invited Lectures, pp. 105–123 (2010)
3. Honda, K., Notsu, A., Ichihashi, H.: Collaborative filtering by sequential user-item co-cluster extraction from rectangular relational data. Int. J. Knowl. Eng. Soft Data Paradigms **2**(4), 312–327 (2010)
4. Bezdek, J.C., Hathaway, R.J., Huband, J.M.: Visual assessment of clustering tendency for rectangular dissimilarity matrices. IEEE Trans. Fuzzy Syst. **15**(5), 890–903 (2007)

5. Rigouste, L., Cappé, O., Yvon, F.: Inference and evaluation of the multinomial mixture model for text clustering. Inf. Process. Manag. **43**(5), 1260–1280 (2007)
6. Holmes, I., Harris, K., Quince, C.: Dirichlet multinomial mixtures: generative models for microbial metagenomics. PLoS ONE **7**(2), e30126 (2012)
7. MacQueen, J.B.: Some methods of classification and analysis of multivariate observations. In: Proceedings of the 5th Berkeley Symposium on Mathematical Statistics and Probability, pp. 281–297 (1967)
8. Bezdek, J.C.: Pattern Recognition with Fuzzy Objective Function Algorithms. Plenum Press, New York (1981)
9. Oh, C.-H., Honda, K., Ichihashi, H.: Fuzzy clustering for categorical multivariate data. In: Proceedings of Joint 9th IFSA World Congress and 20th NAFIPS International Conference, pp. 2154–2159 (2001)
10. Kummamuru, K., Dhawale, A., Krishnapuram, R.: Fuzzy co-clustering of documents and keywords. In: Proceedings of 2003 IEEE International Conference Fuzzy Systems, vol. 2, pp. 772–777 (2003)
11. Honda, K., Oshio, S., Notsu, A.: Fuzzy co-clustering induced by multinomial mixture models. J. Adv. Comput. Intell. Intell. Inf. **19**(6), 717–726 (2015)
12. Honda, K., Oh, C.-H., Matsumoto, Y., Notsu, A., Ichihashi, H.: Exclusive partition in FCM-type co-clustering and its application to collaborative filtering. Int. J. Comput. Sci. Netw. Secur. **12**(12), 52–58 (2012)
13. Honda, K., Nakano, T., Oh, C.-H., Ubukata, S., Notsu, A.: Partially exclusive item partition in MMMs-induced fuzzy co-clustering and its effects in collaborative filtering. J. Adv. Comput. Intell. Intell. Inf. **19**(6), 810–817 (2015)
14. Tjhi, W.-C., Chen, L.: A partitioning based algorithm to fuzzy co-cluster documents and words. Pattern Recogn. Lett. **27**, 151–159 (2006)
15. Tjhi, W.-C., Chen, L.: A heuristic-based fuzzy co-clustering algorithm for categorization of high-dimensional data. Fuzzy Sets Syst. **159**, 371–389 (2008)
16. Konstan, J.A., Miller, B.N., Maltz, D., Herlocker, J.L., Gardon, L.R., Riedl, J.: Grouplens: applying collaborative filtering to usenet news. Commun. ACM **40**(3), 77–87 (1997)
17. Herlocker, J.L., Konstan, J.A., Borchers, A., Riedl, J.: An algorithmic framework for performing collaborative filtering. In: Proceedings of 22nd Conference on Research and Development in Information Retrieval, pp. 230–237 (1999)
18. Tsuda, K., Minoh, M., Ikeda, K.: Extracting straight lines by sequential fuzzy clustering. Pattern Recogn. Lett. **17**, 643–649 (1996)
19. Dempster, A.P., Laird, N.M., Rubin, D.B.: Maximum likelihood from incomplete data via the EM algorithm. J. R. Stat. Soc. B **39**, 1–38 (1977)
20. Miyamoto, S., Mukaidono, M.: Fuzzy c-means as a regularization and maximum entropy approach. In: Proceedings of the 7th International Fuzzy Systems Association World Congress, vol. 2, pp. 86–92 (1997)
21. Miyamoto, S., Ichihashi, H., Honda, K.: Algorithms for Fuzzy Clustering. Springer, Heidelberg (2008)
22. Miyamoto, S., Umayahara, K.: Fuzzy clustering by quadratic regularization. In: Proceedings 1998 IEEE International Conference Fuzzy Systems and IEEE World Congress Computational Intelligence, vol. 2, pp. 394–1399 (1998)
23. Nakano, T., Honda, K., Ubukata, S., Notsu, A.: MMMs-induced fuzzy co-clustering with exclusive partition penalty on selected items. Integrated Uncertainty in Knowledge Modelling and Decision Making **9376**, 226–235 (2015)
24. Honda, K., Muranishi, M., Notsu, A., Ichihashi, H.: FCM-type cluster validation in fuzzy co-clustering and collaborative filtering applicability. Int. J. Comput. Sci. Netw. Secur. **13**(1), 24–29 (2013)

Evidential Clustering: A Review

Thierry Denœux[1]([⊠]) and Orakanya Kanjanatarakul[2]

[1] Sorbonne Universités, Université de Technologie de Compiègne,
CNRS, UMR 7253 Heudiasyc, Compiègne, France
Thierry.Denoeux@utc.fr
[2] Faculty of Management Sciences,
Chiang Mai Rajabhat University, Chiang Mai, Thailand
orakanyaa@gmail.com

Abstract. In evidential clustering, uncertainty about the assignment of objects to clusters is represented by Dempster-Shafer mass functions. The resulting clustering structure, called a credal partition, is shown to be more general than hard, fuzzy, possibility and rough partitions, which are recovered as special cases. Three algorithms to generate a credal partition are reviewed. Each of these algorithms is shown to implement a decision-directed clustering strategy. Their relative merits are discussed.

1 Introduction

Clustering is one of the most important tasks in data analysis and machine learning. It aims at revealing some structure in a dataset, so as to highlight groups (clusters) of objects that are similar among themselves, and dissimilar to objects of other groups. Traditionally, we distinguish between *partitional* clustering, which aims at finding a partition of the objects, and *hierarchical* clustering, which finds a sequence of nested partitions.

Over the years, the notion of partitional clustering has been extended to several important variants, including fuzzy [3], possibilistic [12], rough [17] and evidential clustering [8,9,15]. Contrary to classical (hard) partitional clustering, in which each object is assigned unambiguously and with full certainty to one and only one cluster, these variants allow ambiguity, uncertainty or doubt in the assignment of objects to clusters. For this reason, they are referred to as *soft* clustering methods, in contrast with classical, *hard* clustering [18].

Among soft clustering paradigms, evidential clustering describes the uncertainty in the membership of objects to clusters using a *Dempster-Shafer mass functions* [20]. Roughly speaking, a mass function can be seen as a collection of sets with corresponding masses. A collection of such mass functions for n objects is called a credal partition. Evidential clustering consists in constructing such a credal partition automatically from the data, by minimizing a cost function.

This research was supported by the Labex MS2T, which was funded by the French Government, through the program "Investments for the future" by the National Agency for Research (reference ANR-11-IDEX-0004-02).

© Springer International Publishing AG 2016
V.-N. Huynh et al. (Eds.): IUKM 2016, LNAI 9978, pp. 24–35, 2016.
DOI: 10.1007/978-3-319-49046-5_3

In this paper, we provide a comprehensive review of evidential clustering algorithms, implemented in the R package evclust[1] [7]. Each of the main algorithms to date can be seen as implementing a decision-directed clustering strategy: starting from an initial credal partition and an evidential classifier, the classifier and the partition are updated in turn, until the algorithm has converged to a stable state.

The rest of this paper is structured as follows. In Sect. 2, the notion of credal partition is first recalled, and some relationships with other clustering paradigms are described. The main evidential clustering algorithms are then reviewed in Sect. 3. Finally, Sect. 4 concludes the paper.

2 Credal Partition

We first recall the notion of credal partition in Sect. 2.1. The relation with other clustering paradigms is analyzed in Sect. 2.2, and the problem of summarizing a credal partition is addressed in Sect. 2.3.

2.1 Credal Partition

Assume that we have a set $\mathcal{O} = \{o_1, \ldots, o_n\}$ of n objects, each one belonging to one and only one of c groups or clusters. Let $\Omega = \{\omega_1, \ldots, \omega_c\}$ denote the set of clusters. If we know for sure which cluster each object belongs to, we have a (hard) partition of the n objects. Such a partition may be represented by binary variables u_{ik} such that $u_{ik} = 1$ if object o_i belongs to cluster ω_k, and $u_{ik} = 0$ otherwise.

If objects cannot be assigned to clusters with certainty, then we can quantify cluster-membership uncertainty by mass functions $m_1, \ldots, m_n$, where each mass function m_i is a mapping from 2^Ω to $[0, 1]$, such that $\sum_{A \subseteq \Omega} m_i(A) = 1$. Each mass $m_i(A)$ is interpreted as a degree of support attached to the proposition "the true cluster of object o_i is in A", and to no more specific proposition. A subset A of Ω such that $m_i(A) > 0$ is called a *focal set* of m_i. The n-tuple $\mathcal{M} = (m_1, \ldots, m_n)$ is called a *credal partition* [9].

Example 1. *Consider, for instance, the "Butterfly" dataset shown in Fig. 1(a). Figure 1(b) shows the credal partition with $c = 2$ clusters produced by the Evidential c-means (ECM) algorithm [15]. In this figure, the masses $m_i(\emptyset)$, $m_i(\{\omega_1\})$, $m_i(\{\omega_2\})$ and $m_i(\Omega)$ are plotted as a function of i, for $i = 1, \ldots, 12$. We can see that $m_3(\{\omega_1\}) \approx 1$, which means that object o_3 almost certainly belongs to cluster ω_1. Similarly, $m_9(\{\omega_2\}) \approx 1$, indicating almost certain assignment of object o_9 to cluster ω_2. In contrast, objects o_6 and o_{12} correspond to two different situations of maximum uncertainty. Object o_6 has a large mass assigned to Ω: this reflects ambiguity in the class membership of this object, which means that it might belong to ω_1 as well as to ω_2. The situation is completely different*

---

[1] This package can be downloaded from the CRAN web site at https://cran.r-project.org/web/packages.

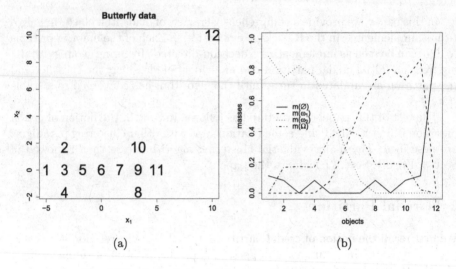

Fig. 1. Butterfly dataset (a) and a credal partition (b).

for object o_{12}, for which the largest mass is assigned to the empty set, indicating that this object does not seem to belong to any of the two clusters.

2.2   Relationships with Other Clustering Paradigms

The notion of credal partition boils down to several alternative clustering structures when the mass functions composing the credal partition have some special forms (see Fig. 2).

Hard partition: If all mass functions m_i are *certain* (i.e., have a single focal set, which is a singleton), then we have a hard partition, with $u_{ik} = 1$ if $m_i(\{\omega_k\}) = 1$, and $u_{ik} = 0$ otherwise.

Fuzzy partition: If the m_i are *Bayesian* (i.e., they assign masses only to singletons, in which case the corresponding belief function becomes additive), then the credal partition is equivalent to a fuzzy partition; the degree of membership of object i to cluster k is $u_{ik} = m_i(\{\omega_k\})$.

Fuzzy partition with a noise cluster: A mass function m such that each focal set is either a singleton, or the empty set may be called an *unnormalized Bayesian mass function*. If each mass function m_i is unnormalized Bayesian, then we can define, as before, the membership degree of object i to cluster k a $u_{ik} = m_i(\{\omega_k\})$, but we now have $\sum_{k=1}^{c} u_{ik} \leq 1$, for $i = 1, \dots, n$. We then have $m_i(\emptyset) = u_{i*} = 1 - \sum_{k=1}^{c} u_{ik}$, which can be interpreted as the degree of membership to a "noise cluster" [5].

Possibilistic partition: If the mass functions m_i are *consonant* (i.e., if their focal sets are nested), then they are uniquely described by their contour functions

$$pl_i(\omega_k) = \sum_{A \subseteq \Omega, \omega_k \in A} m_i(A), \tag{1}$$

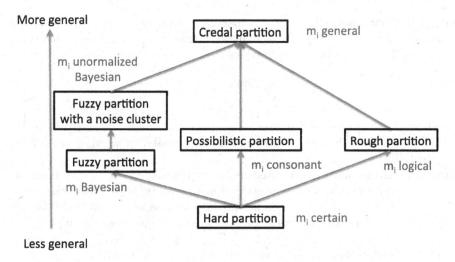

Fig. 2. Relationship between credal partitions and other clustering structures.

which are possibility distributions. We then have a possibilistic partition, with $u_{ik} = pl_i(\omega_k)$ for all i and k. We note that $\max_k pl_i(\omega_k) = 1 - m_i(\emptyset)$.

Rough partition: Assume that each m_i is *logical*, i.e., we have $m_i(A_i) = 1$ for some $A_i \subseteq \Omega$, $A_i \neq \emptyset$. We can then define the *lower approximation* of cluster ω_k as the set of objects that *surely* belong to ω_k,

$$\omega_k^L = \{o_i \in \mathcal{O} | A_i = \{\omega_k\}\}, \tag{2}$$

and the *upper approximation* of cluster ω_k as the set of objects that *possibly* belong to ω_k,

$$\omega_k^U = \{o_i \in \mathcal{O} | \omega_k \in A_i\}. \tag{3}$$

The membership values to the lower and upper approximations of cluster ω_k are then, respectively, $\underline{u}_{ik} = Bel_i(\{\omega_k\})$ and $\overline{u}_{ik} = Pl_i(\{\omega_k\})$. If we allow $A_i = \emptyset$ for some i, then we have $\overline{u}_{ik} = 0$ for all k, which means that object o_i does not belong to the upper approximation of any cluster.

2.3   Summarization of a Credal Partition

A credal partition is a quite complex clustering structure, which often needs to be summarized in some way to become interpretable by the user. This can be achieved by transforming each of the mass functions in the credal partition into a simpler representation. Depending on the representation used, each of clustering structures mentioned in Sect. 2.2 can be recovered as different partial views of a credal partition. Some of the relevant transformations are discussed below.

Fuzzy and hard partitions: A fuzzy partition can be obtained by transforming each mass function m_i into a probability distribution p_i using the plausibility-probability transformation defined as

$$p_i(\omega_k) = \frac{pl_i(\omega_k)}{\sum_{\ell=1}^{c} pl_i(\omega_\ell)}, \quad k = 1, \ldots, c, \tag{4}$$

where pl_i is the contour function associated to m_i, given by (1). By selecting, for each object, the cluster with maximum probability, we then get a hard partition.

Fuzzy partition with noise cluster: In the plausibility-probability transformation (4), the information contained in the masses $m_i(\emptyset)$ assigned to the empty set is lost. However, this information may be important if the dataset contains outliers. To keep track of it, we can define an unnormalized plausibility transformation as $\pi_i(\omega_k) = (1 - m_i(\emptyset))p_i(\omega_k)$, for $k = 1, \ldots, c$. The degree of membership of each object i to cluster k can then be defined as $u_{ik} = \pi_i(\omega_k)$ and the degree of membership to the noise cluster as $u_{i*} = m_i(\emptyset)$.

Possibilistic partition: A possibilistic partition can be obtained from a credal partition by computing a consonant approximation of each of the mass functions m_i [11]. The simplest approach is to approximate m_i by the consonant mass function with the same contour function, in which case the degree of possibility of object o_i belonging to cluster ω_k is $u_{ik} = pl_i(\omega_k)$.

Rough partition: As explained in Sect. 2.2, a credal partition becomes equivalent to a rough partition when all mass functions m_i are logical. A general credal partition can thus be transformed into a rough partition by deriving a set A_i of clusters from each mass function m_i. This can be done by selecting a focal set A_i such that $m_i(A_i) \geq m_i(A)$ for any subset A of Ω, as suggested in [15]. Alternatively, we can use the following *interval dominance decision rule*, and select the set A_i^* of clusters whose plausibility exceeds the degree of belief of any other cluster,

$$A_i^* = \{\omega \in \Omega | \forall \omega' \in \Omega, pl_i^*(\omega) \geq m_i^*(\{\omega'\})\}, \tag{5}$$

where pl_i^* and m_i^* are the normalized contour and mass functions defined, respectively, by $pl_i^* = pl_i/(1 - m_i(\emptyset))$ and $m_i^* = m_i/(1 - m_i(\emptyset))$. If the interval dominance rule is used, we may account for the mass assigned to the empty set by defining A_i as follows,

$$A_i = \begin{cases} \emptyset & \text{if } m_i(\emptyset) = \max_{A \subseteq \Omega} m_i(A) \\ A_i^* & \text{otherwise.} \end{cases} \tag{6}$$

3   Review of Evidential Clustering Algorithms

Three main algorithms have been proposed to generate credal partitions: the Evidential c-means (ECM) [15,16], EK-NNclus [8], and EVCLUS [9,10]. These algorithms are described in the next sections.

3.1   Evidential c-means

In contrast to EVCLUS, the Evidential c-means algorithm (ECM) [15] is a prototype-based clustering algorithm, which generalizes the hard and fuzzy c-means (FCM) algorithm. The method is suitable to cluster attribute data. As in FCM, each cluster ω_k is represented by a prototype v_k in the attribute space. However, in contrast with FCM, each non-empty set of clusters $A_j \subseteq \Omega$ is also represented by a prototype $\overline{v}_j$, which is defined as the center of mass of all prototypes v_k, for $\omega_k \in A_k$ (Fig. 3). Formally,

$$\overline{v}_j = \frac{1}{c_j} \sum_{k=1}^{c} s_{kj} v_k, \tag{7}$$

where $c_j = |A_j|$ denotes the cardinality of A_j, and $s_{kj} = 1$ if $\omega_k \in A_j$, $s_{kj} = 0$ otherwise.

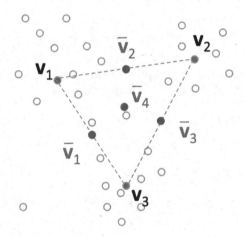

Fig. 3. Representation of sets of clusters by prototypes in the ECM algorithm.

Let Δ_{ij} denote the distance between a vector x_i and prototype $\overline{v}_j$, and let the distance between any vector x_i and the empty set by defined as a fixed value δ. The ECM algorithm is based on the idea that $m_{ij} = m_i(A_j)$ should be high if x_i is close to $\overline{v}_j$, i.e. if Δ_{ij} is small. Furthermore, if x_i is far from all prototypes $\overline{v}_j$, then $m_{i\emptyset} = m_i(\emptyset)$ should be large. Such a configuration of mass functions and prototypes can be achieved by minimizing the following cost function,

$$J_{\mathrm{ECM}}(\mathcal{M}, V) = \sum_{i=1}^{n} \sum_{\{j|A_j \neq \emptyset, A_j \subseteq \Omega\}} c_j^{\alpha} m_{ij}^{\beta} \Delta_{ij}^2 + \sum_{i=1}^{n} \delta^2 m_{i\emptyset}^{\beta}, \tag{8}$$

subject to

$$\sum_{\{j|A_j \subseteq \Omega, A_j \neq \emptyset\}} m_{ij} + m_{i\emptyset} = 1 \quad \forall i = 1, n, \tag{9}$$

where $\mathcal{M}$ is the credal partition and $V = (v_1, \ldots, v_c)$ is the matrix of prototypes. This cost function depends on three coefficients: β controls the hardness of the evidential partition as in the FCM algorithm; δ controls the amount of data considered as outliers, as in the Davé's Noise Clustering algorithm [5]; finally parameter α controls the specificity of the evidential partition, larger values of α penalizing subsets of clusters with large cardinality.

As in FCM, the minimization of the cost function J_{ECM} can be achieved by alternating two steps: (1) minimize $J_{\mathrm{ECM}}(\mathcal{M}, V)$ with respect to $\mathcal{M}$ for fixed V, and (2) minimize $J_{\mathrm{ECM}}(\mathcal{M}, V)$ with respect to V for fixed $\mathcal{M}$. The first step is achieved by the following update equations,

$$m_{ij} = \frac{c_j^{-\alpha/(\beta-1)} \Delta_{ij}^{-2/(\beta-1)}}{\sum_{A_k \neq \emptyset} c_k^{-\alpha/(\beta-1)} \Delta_{ik}^{-2/(\beta-1)} + \delta^{-2/(\beta-1)}}, \tag{10}$$

for $i = 1, \ldots, n$ and for all j such that $A_j \neq \emptyset$, and

$$m_{i\emptyset} = 1 - \sum_{A_j \neq \emptyset} m_{ij}, \tag{11}$$

for $i = 1, \ldots, n$. The second step implies solving a system of the form $HV = B$, where B is the matrix of size $c \times p$ with general term

$$B_{lq} = \sum_{i=1}^{n} x_{iq} \sum_{A_j \ni \omega_l} c_j^{\alpha-1} m_{ij}^{\beta} \tag{12}$$

and H the matrix of size $c \times c$ given by:

$$H_{lk} = \sum_{i} \sum_{A_j \supseteq \{\omega_k, \omega_l\}} c_j^{\alpha-2} m_{ij}^{\beta}. \tag{13}$$

We can observe that Eqs. (10) and (11) define an evidential classifier: given the matrix V of prototypes, they make it possible to compute a mass function for any new instance. The prototype-updating step can then be seen as a training phase, where the classifier is fitted to the data. ECM can thus be seen as a decision-directed clustering algorithm.

The Relational Evidential c-Means (RECM), a version of ECM for dissimilarity data, was introduced in [16]. In this version, we assume that the data consist in a square matrix $D = (d_{ij})$ of dissimilarities between n objects, so that ECM cannot be used directly. However, if we assume the dissimilarities d_{ij} to be metric, i.e., to be squared Euclidean distances in some attribute space, we can still compute the distances Δ_{ij} in (8) without explicitly computing the vectors x_i and v_k, which allows us to find a credal partition $\mathcal{M}$ minimizing (8). Although the convergence of RECM is not guaranteed when the dissimilarities are not metric, the algorithm has been shown to be quite robust to violations of this assumption.

3.2   E*K*-NNclus

The E*K*-NNclus algorithm [8] is another decision-directed clustering procedure based on the evidential *k*-nearest neighbor (E*K*-NN) rule [6]. The E*K*-NN rule works as follows. Consider a classification problem in which an object o has to be classified in one of c groups, based on its distances to n objects in a dataset. Let $\Omega = \{\omega_1, \ldots, \omega_c\}$ be the set of groups, and d_j the distance between the object to be classified and object o_j in the dataset. The knowledge that object o is at a distance d_j from o_j is a piece of evidence that can be represented by the following mass function on Ω,

$$m_j(\{\omega_k\}) = u_{jk}\varphi(d_j), \quad k = 1, \ldots, c \tag{14a}$$
$$m_j(\Omega) = 1 - \varphi(d_j), \tag{14b}$$

where φ is a non-increasing mapping from $[0, +\infty)$ to $[0, 1]$, and $u_{jk} = 1$ if o_j belongs to class ω_k, $u_{jk} = 0$ otherwise. In [6], it was proposed to choose φ as $\varphi(d_j) = \alpha_0 \exp(-\gamma d_j)$ for some constants α_0 and γ. Denoting by N_K the set of indices of the K nearest neighbors of object o is the learning set, the K mass function m_j, $j \in N_K$ are then combined by Dempster's rule [20] to yield the combined mass function

$$m = \bigoplus_{j \in N_K} m_j. \tag{15}$$

A decision can finally be made by assigning object o to the class ω_k with the highest plausibility. We can remark that, to make a decision, we need not compute the combined mass function m explicitly. The contour function pl_j corresponding to m_j in (14) is

$$pl_j(\omega_\ell) = (1 - \varphi(d_j))^{1-u_{j\ell}}, \tag{16}$$

for $\ell = 1, \ldots, c$. The combined contour function is thus

$$pl(\omega_\ell) \propto \prod_{j \in N_K} (1 - \varphi(d_j))^{1-u_{j\ell}}, \tag{17}$$

for $\ell = 1, \ldots, c$. Its logarithm can be written as

$$\ln pl(\omega_\ell) = \sum_{j=1}^{n} w_j u_{j\ell} + C, \tag{18}$$

where C is a constant, and $w_j = -\ln(1 - \varphi(d_j))$ if $j \in N_K$, and $w_j = 0$ otherwise.

The E*K*-NNclus algorithm implements a decision-directed approach, using the above E*K*-NN rule as the base classifier. We start with a matrix $D = (d_{ij})$ of dissimilarities between n objects. To initialize the algorithm, the objects are labeled randomly (or using some prior knowledge if available). As the number of clusters is usually unknown, it can be set to $c = n$, i.e., we initially assume that there are as many clusters as objects and each cluster contains exactly one object. If n is very large, we can give c a large value, but smaller than n, and initialize the object labels randomly. As before, we define cluster-membership binary variables u_{ik} as $u_{ik} = 1$ is object o_i belongs to cluster k, and $u_{ik} = 0$ otherwise. An iteration of the algorithm then consists in updating the object labels in some random order, using the EKNN rule. For each object o_i, we compute the logarithms of the plausibilities of belonging to each cluster (up to an additive constant) using (18) as

$$s_{ik} = \sum_{j \in N_K(i)} w_{ij} u_{jk}, \quad k = 1, \ldots, c, \tag{19}$$

where $w_{ij} = -\ln(1 - \varphi(d_{ij}))$ and $N_K(i)$ is the set of indices of the K nearest neighbors of object o_i in the dataset. We then assign object o_i to the cluster with the highest plausibility, i.e., we update the variables u_{ik} as

$$u_{ik} = \begin{cases} 1 & \text{if } s_{ik} = \max_{k'} s_{ik'}, \\ 0 & \text{otherwise.} \end{cases} \tag{20}$$

If the label of at least one object has been changed during the last iteration, then the objects are randomly re-ordered and a new iteration is started. Otherwise, we move to the last step described below, and the algorithm is stopped. We can remark that, after each iteration, some clusters may have disappeared. To save computation time and storage space, we can update the number c of clusters, renumber the clusters from 1 to c, and change the membership variables s_{ik} accordingly, after each iteration. After the algorithm has converged, we can compute the final mass functions

$$m_i = \bigoplus_{j \in N_K(i)} m_{ij}, \tag{21}$$

for $i = 1, \ldots, n$, where each m_{ij} is the following mass function,

$$m_{ij}(\{\omega_k\}) = u_{jk}\varphi(d_{ij}), \quad k = 1, \ldots, c \tag{22a}$$
$$m_{ij}(\Omega) = 1 - \varphi(d_{ij}). \tag{22b}$$

As compared to EVCLUS, EK-NNclus yields a credal partition with simpler mass functions, whose focal sets are the singletons and Ω. A major advantage of EK-NNclus is that it does not require the number of clusters to be fixed in advance. Heuristics for tuning the two parameters of the algorithm, K and γ, are described in [8]. Also, EK-NNclus is applicable to non-metric dissimilarity data.

3.3   EVCLUS

The EVCLUS algorithm [9,10] applies some ideas from Multidimensional Scaling (MDS) [4] to clustering. Let $D = (d_{ij})$ be an $n \times n$ dissimilarity matrix, where d_{ij} denotes the dissimilarity between objects o_i and o_j. To derive a credal partition $\mathcal{M} = (m_1, \ldots, m_n)$ from D, we assume that the plausibility pl_{ij} that two objects

o_i and o_j belong to the same class is a decreasing function of the dissimilarity d_{ij}: the more similar are two objects, the more plausible it is that they belong to the same cluster. Now, it can be shown [10] that the plausibility pl_{ij} is equal to $1 - \kappa_{ij}$, where κ_{ij} is the degree of conflict between m_i and m_j. The credal partition $\mathcal{M}$ should thus be determined in such a way that similar objects o_i and o_j have mass functions m_i and m_j with low degree of conflict, whereas highly dissimilar objects are assigned highly conflicting mass functions. This can be achieved by minimizing the discrepancy between the pairwise degrees of conflict and the dissimilarities, up to some increasing transformation. In [10], we proposed to minimize the following stress function,

$$J(\mathcal{M}) = \eta \sum_{i<j} (\kappa_{ij} - \delta_{ij})^2, \tag{23}$$

where $\eta = \left(\sum_{i<j} \delta_{ij}^2 \right)^{-1}$ is a normalizing constant, and the $\delta_{ij} = \varphi(d_{ij})$ are transformed dissimilarities, for some fixed increasing function φ from $[0, +\infty)$ to $[0, 1]$. A suitable choice for φ is a soft threshold function, such as $\varphi(d) = 1 - \exp(-\gamma d^2)$, where γ is a user-defined parameter. A heuristic for fixing γ is described in [10]. The stress function (23) by the Iterative Row-wise Quadratic Programming (IRQP) [10,21]. The IRQP algorithm consists in minimizing (23) with respect to each mass function m_i at a time, leaving the other mass functions m_j fixed. At each iteration, we thus solve

$$\min_{m_i} \sum_{j \neq i} (\kappa_{ij} - \delta_{ij})^2, \tag{24}$$

such that $m_i(A) \geq 0$ for any $A \subseteq \Omega$ and $\sum_{A \subseteq \Omega} m(A) = 1$, which is a linearly constrained positive least-square problem that can be solved efficiently. We can remark that this algorithm can be seen as a decision-directed procedure, where each object o_i is classified at each step, using its distances to all the other objects. The IRQP algorithm has been shown to be much faster than the gradient procedure, and to reach lower values of the stress function.

A major drawback of the EVCLUS algorithm as originally proposed in [9] is that it requires to store the whole dissimilarity matrix D, which precludes its application to very large datasets. However, there is usually some redundancy in a dissimilarity matrix. In particular, if two objects o_1 and o_2 are very similar, then any object o_3 that is dissimilar from o_1 is usually also dissimilar from o_2. Because of such redundancies, it might be possible to compute the differences between degrees of conflict and dissimilarities, for *only a subset of randomly sampled dissimilarities*. More precisely, let $j_1(i), \ldots, j_k(i)$ be k integers sampled at random from the set $\{1, \ldots, i-1, i+1, \ldots, n\}$, for $i = 1, \ldots, n$. Let J_k the following stress criterion,

$$J_k(\mathcal{M}) = \eta \sum_{i=1}^{n} \sum_{r=1}^{k} (\kappa_{i,j_r(i)} - \delta_{i,j_r(i)})^2, \tag{25}$$

where, as before, η is a normalizing constant. Obviously, $J(\mathcal{M})$ is recovered as a special case when $k = n - 1$. However, in the general case, the calculation of $J_k(\mathcal{M})$ requires only $O(nk)$ operations. If k can be kept constant as n increases, or, at least, if k increases slower than linearly with n, then significant gains in computing time and storage requirement could be achieved [10].

4   Conclusions

The notion of credal partition, as well as its relationships with alternative clustering paradigms have been reviewed. Basically, each of the alternative partitional clustering structures (i.e., hard, fuzzy, possibilistic and rough partitions) correspond to a special form of the mass functions within a credal partition. A credal partition can be transformed into a simpler clustering structure for easier interpretation. Recently, evidential clustering has been successfully applied in various domains such as machine prognosis [19], medical image processing [13,14] and analysis of social networks [22]. Three main algorithms for generating credal partitionsand implemented in the R package evclust have been reviewed. Each of these three algorithms have their strengths and limitations, and the choice of an algorithm depends on the problem at hand. Both ECM and EK-NN are very efficient for handling attribute data. EK-NN has the additional advantage that it can determine the number of clusters automatically, while EVCLUS and ECM produce more informative outputs (with masses assigned to any subsets of clusters). EVCLUS was shown to be very effective for dealing with non metric dissimilarity data, and the recent improvements reported in [10] make it suitable to handle very large datasets. Methods for exploiting additional knowledge in the form of pairwise constraints have been studied in [1,2], and the problem of handling a large number of clusters has been addressed in [10].

In future work, it will be interesting to perform detailed comparative experiments with these algorithms using a wide range of attribute and dissimilarity datasets. Such a study will require the definition of performance indices to measure the fit between a credal partition and a hard partition, or between two credal partitions. This approach should provide guidelines for choosing a suitable algorithm, depending on the characteristics of the clustering problem.

References

1. Antoine, V., Quost, B., Masson, M.-H., Denoeux, T.: CEVCLUS: evidential clustering with instance-level constraints for relational data. Soft Comput. 18(7), 1321–1335 (2014)
2. Antoine, V., Quost, B., Masson, M.-H., Denoeux, T.: CECM: constrained evidential c-means algorithm. Comput. Stat. Data Anal. 56(4), 894–914 (2012)
3. Bezdek, J.: Pattern Recognition with Fuzzy Objective Function Algorithm. Plenum Press, New York (1981)
4. Borg, I., Groenen, P.: Modern Multidimensional Scaling. Springer, New York (1997)

5. Davé, R.: Characterization and detection of noise in clustering. Pattern Recognit. Lett. **12**, 657–664 (1991)
6. Denœux, T.: A *k*-nearest neighbor classification rule based on Dempster-Shafer theory. IEEE Trans. Syst. Man Cybern. **25**(05), 804–813 (1995)
7. Denœux, T.: evclust: Evidential Clustering, R package version 1.0.2 (2016)
8. Denœux, T., Kanjanatarakul, O., Sriboonchitta, S.: E*K*-NNclus: a clustering procedure based on the evidential *k*-nearest neighbor rule. Knowl.-Based Syst. **88**, 57–69 (2015)
9. Denœux, T., Masson, M.-H.: EVCLUS: evidential clustering of proximity data. IEEE Trans. Syst. Man Cybern. B **34**(1), 95–109 (2004)
10. Denœux, T., Sriboonchitta, S., Kanjanatarakul, O.: Evidential clustering of large dissimilarity data. Knowl.-Based Syst. **106**, 179–195 (2016)
11. Dubois, D., Prade, H.: Consonant approximations of belief measures. Int. J. Approximate Reasoning **4**, 419–449 (1990)
12. Krishnapuram, R., Keller, J.: A possibilistic approach to clustering. IEEE Trans. Fuzzy Syst. **1**, 98–111 (1993)
13. Lelandais, B., Ruan, S., Denœux, T., Vera, P., Gardin, I.: Fusion of multi-tracer PET images for dose painting. Med. Image Anal. **18**(7), 1247–1259 (2014)
14. Makni, N., Betrouni, N., Colot, O.: Introducing spatial neighbourhood in evidential c-means for segmentation of multi-source images: application to prostate multiparametric MRI. Inf. Fusion **19**, 61–72 (2014)
15. Masson, M.-H., Denoeux, T.: ECM: an evidential version of the fuzzy c-means algorithm. Pattern Recogn. **41**(4), 1384–1397 (2008)
16. Masson, M.-H., Denœux, T.: RECM: relational evidential c-means algorithm. Pattern Recogn. Lett. **30**, 1015–1026 (2009)
17. Peters, G.: Is there any need for rough clustering? Pattern Recogn. Lett. **53**, 31–37 (2015)
18. Peters, G., Crespo, F., Lingras, P., Weber, R.: Soft clustering: fuzzy and rough approaches and their extensions and derivatives. Int. J. Approximate Reasoning **54**(2), 307–322 (2013)
19. Serir, L., Ramasso, E., Zerhouni, N.: Evidential evolving Gustafson-Kessel algorithm for online data streams partitioning using belief function theory. Int. J. Approximate Reasoning **53**(5), 747–768 (2012)
20. Shafer, G.: A Mathematical Theory of Evidence. Princeton University Press, Princeton (1976)
21. ter Braak, C.J., Kourmpetis, Y., Kiers, H.A., Bink, M.C.: Approximating a similarity matrix by a latent class model: a reappraisal of additive fuzzy clustering. Comput. Stat. Data Anal. **53**(8), 3183–3193 (2009)
22. Zhou, K., Martin, A., Pan, Q., Liu, Z.-G.: Median evidential c-means algorithm and its application to community detection. Knowl.-Based Syst. **74**, 69–88 (2015)

Uncertainty Management and Decision Support

Non-uniqueness of Interval Weight Vector to Consistent Interval Pairwise Comparison Matrix and Logarithmic Estimation Methods

Masahiro Inuiguchi$^{(\boxtimes)}$

Graduate School of Engineering Science, Osaka University,
Toyonaka, Osaka 560-8531, Japan
inuiguti@sys.es.osaka-u.ac.jp

Abstract. In this paper we investigate the interval priority weight estimation from a given interval pairwise comparison matrix. The lower and upper models as well as goal programming model were proposed. It has been expected that the sum of widths of interval weights estimated by lower model is not greater than that estimated by upper model. We show that this expectation is not always hold. Especially when the given interval comparison matrix is totally consistent, it is possible that the solution is not unique and the reverse inequality relation holds. We investigate uniqueness conditions of interval comparison matrices and the influence of non-uniqueness in the comparison of alternatives. Based on the results of investigation above, we propose interval weight estimation methods from interval pairwise comparison matrix.

Keywords: Interval AHP · Interval priority weight · Interval matrix · Linear programming · Logarithmic transformation

1 Introduction

AHP (Analytic Hierarchy Process) [1] is one of well applied decision aiding methods. It evaluates the priority weights of alternatives and criteria from a pairwise comparison matrix (PCM) in multiple criteria decision problems. The PCM is given by a decision maker and the (i, j) component shows the relative importance of the i-th alternative/criterion to the j-th one. From a given PCM, the priority weights are estimated by maximum eigenvalue method [1] or by geometric mean method [2]. However, the given PCM is seldom consistent each other because of the vagueness of human judgements. The degree of inconsistency is measured by the consistency index and the estimated priority weights are adopted for decision analysis when the consistency index is in the allowable range.

To cope with the inaccuracy, imprecision and vagueness of human judgements, fuzzy and interval techniques are applied to AHP. In many of fuzzy approaches, a PCM with fuzzy components is used to represent the vagueness of

This work was partially supported by JSPS KAKENHI Grant Number 26350423.

© Springer International Publishing AG 2016
V.-N. Huynh et al. (Eds.): IUKM 2016, LNAI 9978, pp. 39–50, 2016.
DOI: 10.1007/978-3-319-49046-5_4

human judgment on the relative importance between criteria. Then a fuzzy priority weight vector has been estimated so as to approximate the fuzzy PCM [3–5]. Similarly, to represent the vagueness of human judgment, a PCM with interval components is also considered. However, a crisp priority weight vector consistent or nearly consistent to a given interval PCM is estimated (see [6]).

On the other hand, in Interval AHP [7], an interval priority weight vector is estimated from a given crisp PCM. This approach is based on the idea that the decision maker's evaluation is vague so that the weights of criteria are expressed by intervals and that the PCM is obtained by judgments with arbitrarily selected values from the intervals. An interval priority weight vector covering the given PCM is estimated so as to minimize its total spreads. Later Interval AHP is extended to treat a PCM with interval components (see [8]). We focus on the interval priority weight estimation from a given interval PCM.

To estimate suitable interval priority weight vector from a given interval PCM $\mathcal{A}$, a lower model and an upper model have been proposed. In the lower model [8], the interval weights $[v_i^{\mathrm{L}}, v_i^{\mathrm{R}}]$, $i = 1, 2, \ldots, n$ are estimated so as to maximize the total spreads under constraints where interval ratios $[v_i^{\mathrm{L}}/v_j^{\mathrm{R}}, v_i^{\mathrm{R}}/v_j^{\mathrm{L}}]$ are included in the (i, j) component $[a_{ij}^{\mathrm{L}}, a_{ij}^{\mathrm{R}}]$ of $\mathcal{A}$ for $i \neq j$. In the upper model [8], the interval weights $[w_i^{\mathrm{L}}, w_i^{\mathrm{R}}]$, $i = 1, 2, \ldots, n$ are estimated so as to minimize the total spreads under constraints where interval ratios $[w_i^{\mathrm{L}}/w_j^{\mathrm{R}}, w_i^{\mathrm{R}}/w_j^{\mathrm{L}}]$ include $[a_{ij}^{\mathrm{L}}, a_{ij}^{\mathrm{R}}]$ for $i \neq j$. It is expected that the sum of widths of $[w_i^{\mathrm{L}}, w_i^{\mathrm{R}}]$, $i = 1, 2, \ldots, n$ is bigger than the sum of widths of $[v_i^{\mathrm{L}}, v_i^{\mathrm{R}}]$, $i = 1, 2, \ldots, n$. On the other hand, goal programming approach [9] was proposed to estimate the interval weights so as to minimize the differences between $(A^{\mathrm{L}} - I)\boldsymbol{u}^{\mathrm{R}}$ and $(n-1)\boldsymbol{u}^{\mathrm{L}}$ and differences between $(A^{\mathrm{R}} - I)\boldsymbol{u}^{\mathrm{L}}$ and $(n-1)\boldsymbol{u}^{\mathrm{L}}$, where I is the $n \times n$ identity matrix, $A^{\mathrm{L}} = (a_{ij}^{\mathrm{L}})$ is the matrix composed of lower bounds of interval components of $\mathcal{A}$ and $A^{\mathrm{R}} = (a_{ij}^{\mathrm{R}})$ is the matrix composed of upper bounds of interval components of $\mathcal{A}$. When $\mathcal{A}$ is consistent, those differences vanish.

In this paper, we first demonstrate that the expected relation between the lower and upper models does not always hold. We show the reverse relation, i.e., the sum of widths of $[w_i^{\mathrm{L}}, w_i^{\mathrm{R}}]$, $i = 1, 2, \ldots, n$ is smaller than the sum of widths of $[v_i^{\mathrm{L}}, v_i^{\mathrm{R}}]$, $i = 1, 2, \ldots, n$ when $\mathcal{A}$ is consistent. When $\mathcal{A}$ is consistent, we tend to expect that an interval priority weights vector is uniquely determined. The given example disclaims this expectation, too. Then we investigate conditions for the interval weight vector to be uniquely estimated from a consistent interval PCM. Considering the unexpected properties found in this paper, we propose estimation methods for interval priority weight vector from a given interval PCM.

This paper is organized as follows. In next section, we briefly review the previous estimation methods for interval priority weights. A counter example to the expected relation between lower and upper models is given in Sect. 3. Moreover, conditions for a given interval PCA to be consistent are investigated. In Sect. 4, the effect by the non-uniqueness of interval priority weight vector is examined and the necessity of normalization is discussed. Based on the obtained results, we propose estimation methods for interval priority weights in Sect. 5.

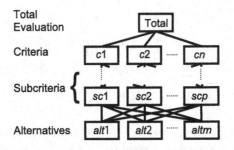

Fig. 1. Hierarhcy of criteria, subcriteria and alternatives

2 Interval Weights Estimated from Interval Matrix

We introduce Interval AHP with a PCM with interval components for multiple criteria decision problem with n criteria. For simplicity, we define $N = \{1, 2, \ldots, n\}$, $M = \{1, 2, \ldots .m\}$ and $N \backslash j = N \backslash \{j\} = \{1, 2, \ldots, j-1, j+1, \ldots, n\}$ for $j \in N$.

In AHP, the decision problem is structured hierarchically as criteria, subcriteria and alternatives as shown in Fig. 1. In the simplest hierarchical tree, we only have criteria and alternatives. At each node except leaf nodes of the hierarchical tree, a priority weight vector for criteria or for alternatives is obtained from a PCM A. The (i, j) component a_{ij} of A shows the relative importance of the i-th criterion/alternative to the j-th one. From this meaning, we assume the reciprocity of A, i.e., $a_{ij} = 1/a_{ij}$, $i \neq j$, $i, j \in N$. In the conventional AHP. each component a_{ij} of A is given by the decision maker as a real number. However, the human evaluation is not very accurate so that giving correct a_{ij} is a difficult task and usually includes some errors. Then fuzzy and interval evaluations for a_{ij}, $i \neq j$, $i, j \in N$ are proposed. We consider the decision maker is allowed to give an interval value $[a_{ij}^{L}, a_{ij}^{R}]$ for the evaluation of a_{ij}. Therefore, we obtain an interval PCA $\mathcal{A} = ([a_{ij}^{L}, a_{ij}^{R}])$, where we assume $a_{ii}^{L} = a_{ii}^{R} = 1$, $i \in N$ and $a_{ij}^{L} = 1/a_{ji}^{R}$, $i, j \in N$. From this interval PCA, we estimate interval priority weights $[w_{i}^{L}, w_{i}^{R}]$, $i \in N$. Theoretically, the (i, j) component of $\mathcal{A}$ $(i \neq j)$ represents $[w_{i}^{L}/w_{j}^{R}, w_{i}^{R}/w_{j}^{L}]$. However, because of the vagueness of human evaluation, we obtain $\mathcal{A}$ only as an approximation of the matrix whose (i, j) component is $[w_{i}^{L}/w_{j}^{R}, w_{i}^{R}/w_{j}^{L}]$ if $i \neq j$ and 1 otherwise.

To estimate interval priority weights $[w_{i}^{L}, w_{i}^{R}]$, $i \in N$ from $\mathcal{A}$, Sugihara et al. [8] proposed the following two models:

$$\text{maximize} \sum_{i \in N} (v_{i}^{R} - v_{i}^{L}),$$
$$\text{subject to } v_{i}^{L} - a_{ij}^{L} v_{j}^{R} \geq 0, \ v_{i}^{R} - a_{ij}^{R} v_{j}^{L} \leq 0, \ i < j, \ i, j \in N,$$
$$v_{i}^{L} + \sum_{j \in N \backslash i} v_{i}^{R} \geq 1, \ v_{i}^{R} + \sum_{j \in N \backslash i} v_{i}^{L} \leq 1, \ i \in N, \tag{1}$$
$$v_{i}^{R} \geq v_{i}^{L} \geq \epsilon, \ i \in N,$$

$$\text{minimize} \sum_{i \in N} (w_i^{\mathrm{R}} - w_i^{\mathrm{L}}),$$

$$\text{subject to } w_i^{\mathrm{L}} - a_{ij}^{\mathrm{L}} w_j^{\mathrm{R}} \leq 0, \ w_i^{\mathrm{R}} - a_{ij}^{\mathrm{R}} w_j^{\mathrm{L}} \geq 0, \ i < j, \ i, j \in N,$$

$$w_i^{\mathrm{L}} + \sum_{j \in N \setminus i} w_i^{\mathrm{R}} \geq 1, \ w_i^{\mathrm{R}} + \sum_{j \in N \setminus i} w_i^{\mathrm{L}} \leq 1, \ i \in N, \tag{2}$$

$$w_i^{\mathrm{R}} \geq w_i^{\mathrm{L}} \geq \epsilon, \ i \in N,$$

where ϵ is a very small positive number. In those problems, the second group of constraints expresses the normality condition for interval weight vectors (see [8]), i.e., the sum of weights equals to one. The third group of constraints expresses the non-negativity and normal conditions for intervals $[v_i^{\mathrm{L}}, v_i^{\mathrm{R}}]$ and $[w_i^{\mathrm{L}}, w_i^{\mathrm{R}}]$, $i \in N$. The first group of constraints of Problem (1) shows $[v_i^{\mathrm{L}}/v_j^{\mathrm{R}}, v_i^{\mathrm{R}}/v_j^{\mathrm{L}}] \subseteq [a_{ij}^{\mathrm{L}}, a_{ij}^{\mathrm{R}}]$, $i \neq j$, $i, j \in N$ while the first group of constraints of Problem (2) shows $[w_i^{\mathrm{L}}/w_j^{\mathrm{R}}, w_i^{\mathrm{R}}/w_j^{\mathrm{L}}] \supseteq [a_{ij}^{\mathrm{L}}, a_{ij}^{\mathrm{R}}]$, $i \neq j$, $i, j \in N$. Because of those inclusion relations, Problems (1) and (2) are called the lower model and the upper model, respectively. We note that to avoid the confusion, we use interval priority weights $[v_i^{\mathrm{L}}, v_i^{\mathrm{R}}]$, $i \in N$ for the lower model, interval priority weights $[w_i^{\mathrm{L}}, w_i^{\mathrm{R}}]$, $i \in N$ for the upper model and interval priority weights $[u_i^{\mathrm{L}}, u_i^{\mathrm{R}}]$, $i \in N$ for the goal programming approach described later.

From the inclusion relations in the constraints of Problems (1) and (2), we expect that the sum of the widths of the interval priority weights $[v_i^{\mathrm{L}}, v_i^{\mathrm{R}}]$, $i \in N$ at an optimal solution to Problem (1) is not larger than the sum of the widths of the interval priority weights $[w_i^{\mathrm{L}}, w_i^{\mathrm{R}}]$, $i \in N$ at an optimal solution to Problem (2) (see [8]). However, as we will show later, this is not always true.

As another approach to estimate the interval weight vector, a goal programming approach is proposed (see [9]). Let $A^{\mathrm{L}} = (a_{ij}^{\mathrm{L}})$ and $A^{\mathrm{R}} = (a_{ij}^{\mathrm{R}})$ be the matrices composed of lower and upper bounds of interval components of $\mathcal{A}$, respectively. Let $\boldsymbol{u}^{\mathrm{L}} = (u_1^{\mathrm{L}}, \dots, u_n^{\mathrm{L}})^{\mathrm{T}}$ be the vector of lower bounds of interval priority weights and $\boldsymbol{u}^{\mathrm{R}} = (u_1^{\mathrm{R}}, \dots, u_n^{\mathrm{R}})^{\mathrm{T}}$ be the vector of upper bounds of interval priority weights. Then theoretically, we have $(A^{\mathrm{L}} - I)\boldsymbol{u}^{\mathrm{R}} = (n-1)\boldsymbol{u}^{\mathrm{L}}$ and $(A^{\mathrm{R}} - I)\boldsymbol{u}^{\mathrm{L}} = (n-1)\boldsymbol{u}^{\mathrm{R}}$. From this point of view, interval priority weights $[u_i^{\mathrm{L}}, u_i^{\mathrm{R}}]$, $i \in N$ are estimated so as to minimize the differences between $A^{\mathrm{L}}\boldsymbol{u}^{\mathrm{R}}$ and $(n-1)\boldsymbol{u}^{\mathrm{L}}$ as well as differences between $A^{\mathrm{R}}\boldsymbol{u}^{\mathrm{L}}$ and $(n-1)\boldsymbol{u}^{\mathrm{R}}$. Namely, we obtain the following problem:

$$\text{minimize} \sum_{i \in N} (p_i^{\mathrm{L}} + n_i^{\mathrm{L}} + p_i^{\mathrm{U}} + n_i^{\mathrm{U}}),$$

$$\text{subject to } \sum_{j \in N \setminus i} a_{ij}^{\mathrm{L}} u_j^{\mathrm{R}} - (n-1) u_i^{\mathrm{L}} - p_i^{\mathrm{L}} + n_i^{\mathrm{L}} = 0, \ i \in N,$$

$$\sum_{j \in N \setminus i} a_{ij}^{\mathrm{R}} u_j^{\mathrm{L}} - (n-1) u_i^{\mathrm{R}} - p_i^{\mathrm{U}} + n_i^{\mathrm{U}} = 0, \ i \in N, \tag{3}$$

$$u_i^{\mathrm{L}} + \sum_{j \in N \setminus i} u_i^{\mathrm{R}} \geq 1, \ u_i^{\mathrm{R}} + \sum_{j \in N \setminus i} u_i^{\mathrm{L}} \leq 1, \ i \in N,$$

$$u_i^{\mathrm{R}} \geq u_i^{\mathrm{L}} \geq \epsilon, \ p_i^{\mathrm{L}} \geq 0, \ n_i^{\mathrm{L}} \geq 0, \ p_i^{\mathrm{R}} \geq 0, \ n_i^{\mathrm{R}} \geq 0, \ i \in N.$$

Table 1. Interval matrix $\mathcal{A}$ and obtained interval priority weights

Interval matrix $\mathcal{A}$			Lower model	Upper model	Goal program
i	1	2	3	$[v_i^L, v_i^R]$	$[w_i^L, w_i^R]$	$[u_i^L, u_i^R]$
1	1	$[1,2]$	$[2/3,4/3]$	$[0.3158, 0.4211]$	$[0.3, 0.4]$	$[0.3, 0.4]$
2	$[1/2,1]$	1	$[4/9,1]$	$[0.2105, 0.3158]$	$[0.2, 0.3]$	$[0.2.0.3]$
3	$[3/4,3/2]$	$[1,9/4]$	1	$[0.3158, 0.4737]$	$[0.3, 0.45]$	$[0.3, 0.45]$

3 A Counter Example and Properties in Consistent Cases

We give the following example showing that the sum of the widths of the interval priority weights $[v_i^L, v_i^R]$, $i \in N$ at an optimal solution to Problem (1) can be larger than the sum of the widths of the interval priority weights $[w_i^L, w_i^R]$, $i \in N$ at an optimal solution to Problem (2).

Example 1. Consider a 3×3 interval PCM $\mathcal{A}$ given in Table 1. Applying lower and upper models as well as goal programming approach, we obtain the interval priority weights shown in the last three columns. We obtain $v_i^R - v_i^L > w_i^R - w_i^L$, $i = 1, 2, 3$. Namely, the sum of spreads of $[v_i^L, v_i^R]$, $i = 1, 2, 3$ is larger than the sum of spreads of $[w_i^L, w_i^R]$, $i = 1, 2, 3$. In this example, we recognized that $\mathcal{A}$ is consistent because we obtain $[a_{ij}^L, a_{ij}^R] = [w_i^L/w_j^R, w_i^R/w_j^L]$, $i \neq j$, $i, j \in N$. This holds also for $[v_i^L, v_i^R]$ and $[u_i^L, u_i^R]$, $i = 1, 2, 3$ although we have some round-off errors in $[v_i^L, v_i^R]$, $i = 1, 2, 3$. Theoretically, the optimal value of goal programming approach is zero and many optimal solutions exist. However, by the round-off errors, the obtained optimal value becomes very small positive number and the obtained solution is as shown in Table 1.

As shown in Example 1, the expected relations between lower and upper models does not always hold. If we analyze the obtained solution more deeply, we found that $[v_i^L, v_i^R]$, $i = 1, 2, 3$ are the widest interval priority weights $[x_i^L, x_i^R]$, $i = 1, 2, 3$ satisfying

$$
\begin{cases}
a_{ij}^L x_j^R = x_i^L, \ a_{ij}^R x_j^L = x_i^R, \ i \in N, \\
x_i^L + \sum_{j \in N \setminus i} x_j^R \geq 1, \ x_i^R + \sum_{j \in N \setminus i} x_j^L \leq 1, \ i \in N, \\
x_i^R \geq x_i^L \geq \epsilon, \ i \in N.
\end{cases}
\tag{4}
$$

On the other hand, $[w_i^L, w_i^R]$, $i = 1, 2, 3$ are the narrowest interval priority weights satisfying (4). Generally, let $[x_i^L, x_i^R]$, $i \in N$ satisfy the first group of conditions in (4). Then, $[\lambda x_i^L, \lambda x_i^R]$, $i \in N$ $(\lambda > 0)$ satisfying

$$
\lambda \left(x_i^L + \sum_{j \in N \setminus i} x_j^R \right) \geq 1, \ \lambda \left(x_i^R + \sum_{j \in N \setminus i} x_j^L \right) \leq 1, \ i \in N.
\tag{5}
$$

are feasible solutions satisfying constraints of Problems (1), (2) and (3). When given $\mathcal{A}$ is consistent, we easily show that an optimal solution is expressed as

$[\lambda x_i^{\mathrm{L}}, \lambda x_i^{\mathrm{R}}]$, $i \in N$. Therefore, the differences of the obtained interval priority weights are in the selection of parameter λ. Indeed, we have $v_i^{\mathrm{L}}/w_i^{\mathrm{L}} = v_i^{\mathrm{R}}/w_i^{\mathrm{R}} = v_j^{\mathrm{L}}/w_j^{\mathrm{L}} = v_j^{\mathrm{R}}/w_j^{\mathrm{R}}$ and $u_i^{\mathrm{L}}/w_i^{\mathrm{L}} = u_i^{\mathrm{R}}/w_i^{\mathrm{R}} = u_j^{\mathrm{L}}/w_j^{\mathrm{L}} = u_j^{\mathrm{R}}/w_j^{\mathrm{R}}$, $i \neq j$, $i, j \in N$. In the lower model, we maximize the sum of widths $\lambda(x_i^{\mathrm{R}} - x_i^{\mathrm{L}})$ and the largest λ satisfying (5) is selected. In the upper model, we minimize the sum of widths $\lambda(x_i^{\mathrm{R}} - x_i^{\mathrm{L}})$ and the smallest λ satisfying (5) is selected. In the goal programming approach. for any λ, the optimal value 0 is achieved theoretically. However, because of the round-off error, some of $|a_{ij}^{\mathrm{L}} u_j^{\mathrm{R}} - (n-1)u_i^{\mathrm{L}}|$ and $|a_{ij}^{\mathrm{R}} u_j^{\mathrm{L}} - (n-1)u_i^{\mathrm{R}}|$, $i \in N$ may take a very small positive value, so that the smallest λ is selected.

As described above, even when the given information is totally consistent, the solutions of Problems (1), (2) and (3) are not always unique. In other words, the solution of (4) is not always unique. The following proposition gives a necessary and sufficient condition for the uniqueness of the solution of (4). We assume $\epsilon > 0$ is sufficiently small so that it does not influence the optimal solutions to Problems (1), (2) and (3).

Proposition 1. *Assume that $[x_i^{\mathrm{L}}, x_i^{\mathrm{R}}]$, $i \in N$ satisfy (4). Then if the following equations are satisfied, $\{[x_i^{\mathrm{L}}, x_i^{\mathrm{R}}], \ i \in N\}$ is a unique solution satisfying (4):*

$$\min_{i \in N} \left(x_i^{\mathrm{L}} + \sum_{j \in N \setminus i} x_j^{\mathrm{R}} \right) = 1, \ \max_{i \in N} \left(x_i^{\mathrm{R}} + \sum_{j \in N \setminus i} x_j^{\mathrm{L}} \right) = 1. \tag{6}$$

Proposition 1 is not directly applicable to $\mathcal{A}$ for checking the uniqueness of the solution $[x_i^{\mathrm{L}}, x_i^{\mathrm{R}}]$, $i \in N$ satisfies (4). Then we further consider the conditions for the uniqueness. We have the following proposition.

Proposition 2. *There exists a solution to $a_{jk}^{\mathrm{L}} x_k^{\mathrm{R}} = x_j^{\mathrm{L}}$, $a_{jk}^{\mathrm{R}} x_k^{\mathrm{L}} = x_j^{\mathrm{R}}$, $j, k \in N$, $j \neq k$ if and only if $\mathcal{A}$ satisfies the following equalities; for an arbitrarily chosen $i \in N$,*

$$a_{ij}^{\mathrm{L}} a_{jk}^{\mathrm{R}} a_{ki}^{\mathrm{L}} = a_{is}^{\mathrm{L}} a_{sr}^{\mathrm{R}} a_{ri}^{\mathrm{L}}, \ j \neq k, \ s \neq r, \ j, k, s, r \in N \setminus i. \tag{7}$$

Even when $\mathcal{A}$ satisfies (7) and $a_{jk}^{\mathrm{L}} \leq a_{jk}^{\mathrm{R}}$, $j, k \in N$, it is possible that there is no solution satisfying $a_{jk}^{\mathrm{L}} x_k^{\mathrm{R}} = x_j^{\mathrm{L}}$, $a_{jk}^{\mathrm{R}} x_k^{\mathrm{L}} = x_j^{\mathrm{R}}$, $j, k \in N$, $j \neq k$ and $x_j^{\mathrm{L}} \leq x_j^{\mathrm{R}}$, $j \in N$ as shown in the following example.

Example 2. Consider $\mathcal{A}$ defined by $a_{12}^{\mathrm{L}} = 1/2$, $a_{12}^{\mathrm{R}} = 2/3$, $a_{13}^{\mathrm{L}} = 1/3$, $a_{13}^{\mathrm{R}} = 2$, $a_{14}^{\mathrm{L}} = 5/4$, $a_{14}^{\mathrm{R}} = 4$, $a_{23}^{\mathrm{L}} = 1$, $a_{23}^{\mathrm{R}} = 2$, $a_{24}^{\mathrm{L}} = 15/4$, $a_{24}^{\mathrm{R}} = 4$, $a_{34}^{\mathrm{L}} = 5/4$, $a_{34}^{\mathrm{R}} = 6$ and $a_{ij}^{\mathrm{L}} = 1/a_{ji}^{\mathrm{R}}$, $a_{ij}^{\mathrm{R}} = 1/a_{ji}^{\mathrm{L}}$, $i > j$, $i, j \in N = \{1, 2, 3, 4\}$. For any $i \in N$, (7) is satisfied. We obtain a solution to $a_{ij}^{\mathrm{L}} x_j^{\mathrm{R}} = x_i^{\mathrm{L}}$, $a_{ij}^{\mathrm{R}} x_j^{\mathrm{L}} = x_i^{\mathrm{R}}$, $i \in N$ as $x_1^{\mathrm{L}} = 1$, $x_1^{\mathrm{R}} = 2$, $x_2^{\mathrm{L}} = 3$, $x_2^{\mathrm{R}} = 2$, $x_3^{\mathrm{L}} = 1$, $x_3^{\mathrm{R}} = 3$, $x_4^{\mathrm{L}} = 1/2$, $x_4^{\mathrm{R}} = 4/5$. We have $x_2^{\mathrm{L}} > x_2^{\mathrm{R}}$.

Toward a necessary and sufficient condition of the uniqueness of the solution to (4), Proposition 2 is modified in the following theorems.

Proposition 3. *There exists a solution to $a_{jk}^{\mathrm{L}} x_k^{\mathrm{R}} = x_j^{\mathrm{L}}$, $a_{jk}^{\mathrm{R}} x_k^{\mathrm{L}} = x_j^{\mathrm{R}}$, $j, k \in N$, $j \neq k$ and $x_j^{\mathrm{L}} \leq x_j^{\mathrm{R}}$, $j \in N$ if and only if, for an arbitrarily chosen $i \in N$, $\mathcal{A}$*

satisfies (7) and the following inequalities; for an arbitrarily chosen pair (j,k) such that $j \neq k$, $j,k \in N\backslash i$,

$$a_{ij}^{\mathrm{L}} a_{jk}^{\mathrm{R}} a_{ki}^{\mathrm{L}} \leq 1, \ a_{si}^{\mathrm{L}} \leq a_{sr}^{\mathrm{R}} a_{ri}^{\mathrm{L}}, \ s \neq r, \ s,r \in N\backslash i. \tag{8}$$

Proposition 4. *There exists a solution to (4) if and only if, for an arbitrarily chosen $i \in N$, $\mathcal{A}$ satisfies (7), (8) and the following inequalities; for an arbitrarily chosen pair (j,k) such that $j \neq k$, $j,k \in N\backslash i$,*

$$\min \left(a_{ij}^{\mathrm{L}} a_{jk}^{\mathrm{R}} a_{ki}^{\mathrm{L}} + a_{jk}^{\mathrm{R}} a_{ki}^{\mathrm{L}} + \sum_{r \in N\backslash\{i,j\}} a_{rj}^{\mathrm{R}} a_{ji}^{\mathrm{L}}, \ \min_{r \in N\backslash i} \left(a_{ri}^{\mathrm{L}} + 1 + \sum_{s \in N\backslash\{i,r\}} a_{sr}^{\mathrm{R}} a_{ri}^{\mathrm{L}} \right) \right)$$

$$\geq \max \left(1 + \sum_{r \in N\backslash i} a_{ri}^{\mathrm{L}}, \ a_{jk}^{\mathrm{R}} a_{ki}^{\mathrm{L}} + a_{ij}^{\mathrm{L}} a_{jk}^{\mathrm{R}} a_{ki}^{\mathrm{L}} + \sum_{r \in N\backslash\{i,j\}} a_{ri}^{\mathrm{L}}, \right.$$

$$\left. \max_{r \in N\backslash\{i,j\}} \left(a_{rj}^{\mathrm{R}} a_{ji}^{\mathrm{L}} + a_{ij}^{\mathrm{L}} a_{jk}^{\mathrm{R}} a_{ki}^{\mathrm{L}} + \sum_{s \in N\backslash\{i,r\}} a_{ri}^{\mathrm{L}} \right) \right).$$

$$\tag{9}$$

The solution to (4) is unique if and only if $\mathcal{A}$ satisfies (7), (8), and (9) with equality.

Proposition 3 shows a necessary and sufficient condition for the existence of a solution to $a_{ij}^{\mathrm{L}} x_j^{\mathrm{R}} = x_i^{\mathrm{L}}$, $a_{ij}^{\mathrm{R}} x_j^{\mathrm{L}} = x_i^{\mathrm{R}}$, $i \in N$ and $x_j^{\mathrm{L}} \leq x_j^{\mathrm{R}}$, $j \in N$. Proposition 4 shows a necessary and sufficient condition for the existence of a solution to (4), and moreover, a necessary and sufficient condition for the uniqueness of a solution to (4). In Example 2, we have $a_{21}^{\mathrm{R}} a_{14}^{\mathrm{L}} < a_{24}^{\mathrm{L}}$, for example. Thus we did not have $x_2^{\mathrm{L}} \leq x_2^{\mathrm{R}}$.

4   Consideration of Non-uniqueness and Normalization

As shown in the previous section, even when a given interval PCM $\mathcal{A}$ is totally consistent, the estimated interval priority weight vector is not unique. Moreover, the expected relation of interval priority weight vectors estimated from the lower and upper models does not always hold.

However, we observed that

$$\exists \lambda > 1, \ v_i^{\mathrm{L}} = \lambda w_i^{\mathrm{L}}, \ v_i^{\mathrm{R}} = \lambda w_i^{\mathrm{R}}, \ i \in N, \tag{10}$$

where $[v_i^{\mathrm{L}}, v_i^{\mathrm{R}}]$, $i \in N$ are the interval priority weights estimated by the lower model and $[w_i^{\mathrm{L}}, w_i^{\mathrm{R}}]$, $i \in N$ are the interval priority weights estimated by the upper model. From (10), we have the following property.

Proposition 5. *Assume that evaluations of alternatives in each criterion are given. Namely, we need an interval priority weight estimation only once from a comparison matrix among criteria (top level of Fig. 1). We rank alternatives by the estimated interval priority weights of criteria. If (10) holds, ranking alternatives by $[v_i^{\mathrm{L}}, v_i^{\mathrm{R}}]$, $i \in N$ is equivalent to ranking alternatives by $[w_i^{\mathrm{L}}, w_i^{\mathrm{R}}]$, $i \in N$.*

Proposition 5 implies the selection of λ of interval priority weights of criteria, $[\lambda x_i^L, \lambda x_i^R]$, $i \in N$ for $[x_i^L, x_i^R]$, $i \in N$ satisfying (4) is not important in ranking alternatives when evaluations of alternatives in each criterion are given, as far as they satisfy (5). Therefore, any previous estimation methods are good when $\mathcal{A}$ is totally consistent. This fact implies that the non-uniqueness of interval priority weights of criteria in the form $[\lambda x_i^L, \lambda x_i^R]$, $i \in N$ does not cause any difficulty. Namely, we can chose any $\lambda > 0$ which makes $[\lambda x_i^L, \lambda x_i^R]$, $i \in N$ satisfy (4) and (5).

On the other hand, the non-uniqueness of interval priority weights become problematic when some interval priority weight estimations are required at lower levels of the hierarchical tree of criteria, subcriteria and alternatives. For example, let us consider the simplest tree case with 2 criteria and 3 alternatives. Let $[x_i^{Lc}, x_i^{Rc}]$, be the interval priority weight of criterion c_i ($i = 1, 2$) and $[y_{ij}^{La}, y_{ij}^{Ra}]$ the interval priority weight of alternative alt_j in view of criterion c_i ($i = 1, 2$, $j = 1, 2, 3$). For simplicity, we assume $x_i^{Lc} = x_i^{Rc} = x_i^c$, $i = 1, 2$. Then to confirm whether the decision maker prefers alt_j to alt_k, we solve

$$\text{minimize } x_1^c(y_{1j} - y_{1k}) + x_2^c(y_{2j} - y_{2k}),$$
$$\text{subject to } y_{ij}^{La} \leq y_{ij} \leq y_{ij}^{Ra}, \ i = 1, 2, \ j = 1, 2, 3,$$
$$y_{i1} + y_{i2} + y_{i3} = 1, \ i = 1, 2.$$

Then we estimate that alt_j is preferred to alt_k if the optimal value is positive. If $[y_{ij}^{La}, y_{ij}^{Ra}]$ is not unique, i.e., $[y_{ij}^{La}, y_{ij}^{Ra}] = [\lambda_i z_{ij}^{La}, \lambda_i z_{ij}^{Ra}]$ ($i = 1, 2$, $j = 1, 2, 3$), with λ_i in some range, the estimation of the preference can be different. For example, if $z_{1j} < z_{1k}$, by taking a large λ_1, it is easier for the optimal value of Problem (4) to be negative.

From this fact, suitable scaling among criteria and that among subcriteria are necessary. In other words, λ's should be selected properly. When λ's are properly determined, the scores (obtained as priority weights or by their linear calculations) in different criteria are commensurable and thus the normalization is not necessary.

5   The Proposed Estimation Methods

5.1   Estimation of Interval Priority Weights at Leaf Nodes

The method proposed in this subsection is applied to the interval priority weight estimation at all leaf nodes of the hierarchical tree like Fig. 1. Thus, we compare m alternatives in pairs. From the fact that ratios x_i^L / x_j^R, $i \in M$ are more meaningful than interval priority weights $[x_i^L, x_i^R]$, the minimization of differences between x_i^L / x_j^R and a_{ij}^L and differences between x_i^R / x_j^L and a_{ij}^R is employed as the objective. For example, corresponding to the lower model, we consider

$$\frac{v_i^L}{v_j^R} - \bar{p}_{ij}^L = a_{ij}^L, \ \frac{v_i^R}{v_j^L} + \bar{n}_{ij}^R = a_{ij}^R, \ \bar{p}_{ij}^L \geq 0, \ \bar{n}_{ij}^R \geq 0,$$

for a pair $(i,j) \in M \times M$ such that $i \neq j$. $v_i^L/v_j^R \geq a_{ij}^L$ is equivalent to $v_j^R/v_i^L \leq a_{ji}^R$ owing to the reciprocity $a_{ij}^L = 1/a_{ij}^R$. Therefore, only one of them is used in the lower model (1). However, $\bar{p}_{ij}^L = v_i^L/v_j^R - a_{ij}^L$ is different from $\bar{n}_{ji}^R = a_{ji}^R - v_j^R/v_i^L$. When $a_{ij}^L > 1$, we have $a_{ji}^R < 1$ and difference p_{ij}^L bigger than difference n_{ji}^R. Because both of these differences caused by the evaluations of a_{ij}^L, v_i^L and v_j^R, the difference between v_i^L/v_j^R and a_{ij}^L equals to the difference between a_{ji}^R and v_j^R/v_i^L. From this point of view, we consider the differences in the logarithmic space, i.e.,

$$\log v_i^L - \log v_j^R - p_{ij}^L = \log a_{ij}^L, \ \log v_i^R - \log v_j^L + n_{ij}^R = \log a_{ij}^R, \ p_{ij}^L \geq 0, \ n_{ij}^R \geq 0.$$

In this way, we have $p_{ij}^L = \log v_i^L - \log v_j^R - \log a_{ij}^L = \log a_{ji}^R - \log v_j^R + \log v_i^L = n_{ji}^R$. Therefore, corresponding to the lower, upper and goal programming models, we formulate the following inner, outer and average models, respectively:

$$\text{minimize} \sum_{i=1}^{m-1} \sum_{j=i+1}^{m} (p_{ij}^L + n_{ij}^R),$$
$$\text{subject to } \check{v}_i^L - \check{v}_j^R - p_{ij}^L = \check{a}_{ij}^L, \ \check{v}_i^R - \check{v}_j^L + n_{ij}^R = \check{a}_{ij}^R, \ i < j, \ i,j \in M, \tag{11}$$
$$\check{v}_i^R \geq \check{v}_i^L, \ i \in M, \ p_{ij}^L \geq 0, \ n_{ij}^R \geq 0, \ i < j, \ i,j \in M,$$

$$\text{minimize} \sum_{i=1}^{m-1} \sum_{j=i+1}^{m} (n_{ij}^L + p_{ij}^R),$$
$$\text{subject to } \check{w}_i^L - \check{w}_j^R + n_{ij}^L = \check{a}_{ij}^L, \ \check{w}_i^R - \check{w}_j^L - p_{ij}^R = \check{a}_{ij}^R, \ i < j, \ i,j \in M, \tag{12}$$
$$\check{w}_i^R \geq \check{w}_i^L, \ i \in M, \ n_{ij}^L \geq 0, \ p_{ij}^R \geq 0, \ i < j, \ i,j \in M,$$

$$\text{minimize} \sum_{i=1}^{m-1} \sum_{j=i+1}^{m} (p_{ij}^L + n_{ij}^L + p_{ij}^R + n_{ij}^R),$$
$$\text{subject to } \check{u}_i^L - \check{u}_j^R - p_{ij}^L + n_{ij}^L = \check{a}_{ij}^L,$$
$$\check{u}_i^R - \check{u}_j^L - p_{ij}^R + n_{ij}^R = \check{a}_{ij}^R, \ i < j, \ i,j \in M, \tag{13}$$
$$\check{u}_i^R \geq \check{u}_i^L, \ i \in M, \ p_{ij}^L, \ p_{ij}^R, \ n_{ij}^L, \ n_{ij}^R \geq 0, \ i < j, \ i,j \in M,$$

where $\check{a}_{ij}^L$ and $\check{a}_{ij}^R$ are $\log a_{ij}^L$ and $\log a_{ij}^R$, respectively. Similarly, variables $\check{v}_i^L$, $\check{v}_i^R$, $\check{w}_i^L$, $\check{w}_i^R$, $\check{u}_i^L$ and $\check{u}_i^R$ represent $\log v_i^L$, $\log v_i^R$, $\log w_i^L$, $\log w_i^R$, $\log u_i^L$ and $\log u_i^R$, respectively. We note that variables $\check{v}_i^L$, $\check{v}_i^R$, $\check{w}_i^L$, $\check{w}_i^R$, $\check{u}_i^L$ and $\check{u}_i^R$ can be negative. Considering priority weights are not greater than one, we can assume they are non-positive. Moreover, Problem (11) may have no solution. In this case, we can obtain a solution close to the feasible region by reformulating the problem.

Problems (11), (12) and (13) are linear programming problems. Solving those problems, we obtain interval priority weights by $v_i^L = \lambda^{\text{in}} \exp \hat{v}_i^L$, $v_i^R = \lambda^{\text{in}} \exp \hat{v}_i^R$, $w_i^L = \lambda^{\text{out}} \exp \hat{w}_i^L$, $w_i^R = \lambda^{\text{out}} \exp \hat{w}_i^R$, $u_i^L = \lambda^{\text{av}} \exp \hat{u}_i^L$ and $u_i^R = \lambda^{\text{av}} \exp \hat{u}_i^R$, $i \in I$ with some multipliers λ^{in}, λ^{out} and λ^{av}. In Problems (11), (12) and (13), the normality condition of interval priority weights is not introduced.

If we introduce the normality condition, the problems become nonlinear and the constraints become stronger so that the optimal value may become worse.

Thus, the solutions to Problems (11), (12) and (13) are not always optimal solutions to the problems with the normality condition. However, when the decision maker evaluates $[a_{ij}^{\mathrm{L}}, a_{ij}^{\mathrm{R}}]$, $i, j \in M$, s/he does not consider possible weights of all alternatives satisfying the normality but only weights of i-th and j-th alternatives without consciousness of the normality. From this point of view, the normality condition cannot be added to Problems (11), (12) and (13) for reflecting the decision maker's preference. As the result, multipliers λ^{in}, λ^{out} and λ^{av} can be decided appropriately. For example, they can be determined so that the center values of interval priority weights are normalized.

5.2   Estimation of Interval Priority Weights at Non-leaf Nodes

When evaluations (scores) of alternatives in each criterion are given, the nonuniqueness of interval priority weights of criteria does not influence on the total evaluations. On the contrary, as in the hierarchical tree like Fig. 1, when Interval AHP is applied for evaluations of alternatives in each criterion, the nonuniqueness of interval priority weights as evaluations of alternatives is problematic. To overcome this difficulty, we can determine multipliers λ_i, properly. In this subsection, we propose a method for estimating scaled interval priority weights. In this method, proper multipliers λ_i are estimated indirectly as products of higher priority weights and multipliers. Therefore, this method is applied to the interval priority weight estimation at all non-leaf nodes of the hierarchical tree like Fig. 1.

We assume that one of Problems (11), (12) and (13) is solved from a given pairwise comparison matrix in view of each criterion/subcriterion at a leaf node of the hierarchical tree like Fig. 1. For simply, we assume that we obtain interval priority weights as $[\lambda_k x_i^{k\mathrm{L}}, \lambda_k x_i^{k\mathrm{R}}] = [\lambda_k \exp \hat{x}_i^{k\mathrm{L}}, \lambda_k \exp \hat{x}_i^{k\mathrm{R}}]$, $i \in M$, $k \in N$, where we introduced index k to indicate k-th criterion/subcriterion and $x_i^{k\mathrm{L}}$ and $x_i^{k\mathrm{R}}$ show $\exp \hat{v}_i^{\mathrm{L}}$ and $\exp \hat{v}_i^{\mathrm{R}}$, $\exp \hat{w}_i^{\mathrm{L}}$ and $\exp \hat{w}_i^{\mathrm{R}}$ or $\exp \hat{u}_i^{\mathrm{L}}$ and $\exp \hat{u}_i^{\mathrm{R}}$. Then we determine multipliers λ_k, $k \in N$, appropriately.

To this end, we define a reference alternative for each criterion/subcriterion. In this paper, we select an alternative with the largest $x_i^{k\mathrm{L}}$ as the reference alternative for k-th criterion/subcriterion. Let $ind(k) \in N$ be the index of the reference alternative for k-th criterion/subcriterion. Then, the decision maker is requested to make pairwise comparisons answering the relative importance of the goodness of $alt_{ind(k)}$ in view of k-th criterion/subcriterion to the goodness of $alt_{ind(l)}$ in view of l-th criterion/subcriterion. By this way, we obtain an interval PCA $\mathcal{B} = ([b_{kl}^{\mathrm{L}}, b_{kl}^{\mathrm{R}}])$ from the decision maker. The (k, l)-component $[b_{kl}^{\mathrm{L}}, b_{kl}^{\mathrm{R}}]$ shows the decision maker's evaluation of $\left[(\lambda_k^{\mathrm{L}} x_{ind(k)}^{k\mathrm{L}})/(\lambda_l^{\mathrm{R}} x_{ind(l)}^{k\mathrm{R}}), (\lambda_k^{\mathrm{R}} x_{ind(k)}^{k\mathrm{R}})/(\lambda_l^{\mathrm{L}} x_{ind(l)}^{k\mathrm{L}}) \right]$. The evaluation of $[b_{kl}^{\mathrm{L}}, b_{kl}^{\mathrm{R}}]$ includes not only multipliers λ_k and λ_l but also priority weights of the k-th and l-th criteria/subcriteria.

From the description above, the scaled interval priority weights $[y_k^{\mathrm{L}}, y_k^{\mathrm{R}}]$, $k \in N$ are estimated by solving the following linear programming problems corresponding to inner, outer and average models, respectively:

$$\text{minimize} \quad \sum_{k=1}^{n-1} \sum_{l=k+1}^{n} (p_{kl}^{\mathrm{L}} + n_{kl}^{\mathrm{R}}),$$

subject to

$$\check{y}_k^{\mathrm{inL}} - \check{y}_l^{\mathrm{inR}} - p_{ij}^{\mathrm{L}} = \check{b}_{kl}^{\mathrm{L}} - \hat{v}_{ind(k)}^{kL} + \hat{v}_{ind(l)}^{kR}, \ k < l, \ k, l \in N,$$

$$\check{y}_k^{\mathrm{inR}} - \check{y}_l^{\mathrm{inL}} + n_{ij}^{\mathrm{R}} = \check{b}_{kl}^{\mathrm{R}} - \hat{v}_{ind(k)}^{kR} + \hat{v}_{ind(l)}^{kL}, \ k < l, \ k, l \in N,$$

$$\check{y}_k^{\mathrm{inR}} \geq \check{y}_k^{\mathrm{inL}}, \ k \in N, \ p_{kl}^{\mathrm{L}} \geq 0, \ n_{kl}^{\mathrm{R}} \geq 0, \ k < l, \ k, l \in N,$$

(14)

$$\text{minimize} \quad \sum_{k=1}^{n-1} \sum_{l=k+1}^{n} (n_{kl}^{\mathrm{L}} + p_{kl}^{\mathrm{R}}),$$

subject to

$$\check{y}_k^{\mathrm{outL}} - \check{y}_l^{\mathrm{outR}} + n_{kl}^{\mathrm{L}} = \check{b}_{kl}^{\mathrm{L}} - \hat{w}_{ind(k)}^{kL} + \hat{w}_{ind(l)}^{kR}, \ k < l, \ k, l \in N,$$

$$\check{y}_k^{\mathrm{outR}} - \check{y}_l^{\mathrm{outL}} - p_{kl}^{\mathrm{R}} = \check{b}_{kl}^{\mathrm{R}} - \hat{w}_{ind(k)}^{kR} + \hat{w}_{ind(l)}^{kL}, \ k < l, \ k, l \in N,$$

$$\check{y}_k^{\mathrm{outR}} \geq \check{y}_k^{\mathrm{outL}}, \ k \in N, \ n_{kl}^{\mathrm{L}} \geq 0, \ p_{kl}^{\mathrm{R}} \geq 0, \ k < l, \ k, l \in N,$$

(15)

$$\text{minimize} \quad \sum_{k=1}^{n-1} \sum_{l=k+1}^{n} (p_{kl}^{\mathrm{L}} + n_{kl}^{\mathrm{L}} + p_{kl}^{\mathrm{R}} + n_{kl}^{\mathrm{R}}),$$

subject to

$$\check{y}_k^{\mathrm{avL}} - \check{y}_l^{\mathrm{avR}} - p_{kl}^{\mathrm{L}} + n_{kl}^{\mathrm{L}} = \check{b}_{kl}^{\mathrm{L}} - \hat{u}_{ind(k)}^{kL} + \hat{u}_{ind(l)}^{kR}, \ k < l, \ k, l \in N,$$

$$\check{y}_k^{\mathrm{avR}} - \check{y}_l^{\mathrm{avL}} - p_{kl}^{\mathrm{R}} + n_{kl}^{\mathrm{R}} = \check{b}_{kl}^{\mathrm{R}} - \hat{u}_{ind(k)}^{kR} + \hat{u}_{ind(l)}^{kL}, \ k < l, \ k, l \in N,$$

$$\check{y}_k^{\mathrm{avR}} \geq \check{y}_k^{\mathrm{avL}}, \ k \in N, \ p_{kl}^{\mathrm{L}}, \ p_{kl}^{\mathrm{R}}, \ n_{kl}^{\mathrm{L}}, \ n_{kl}^{\mathrm{R}} \geq 0, \ k < l, \ k, l \in N,$$

(16)

where $\hat{v}_{ind(k)}^{kL}$, $\hat{v}_{ind(k)}^{kR}$, $\hat{w}_{ind(k)}^{kL}$, $\hat{w}_{ind(k)}^{kR}$, $\hat{u}_{ind(k)}^{kL}$ and $\hat{u}_{ind(k)}^{kR}$, $k \in N$ are fixed as the solutions to Problems (11), (12) and (13) with respect to the k-th criterion/subcriterion.

In the case of the simplest hierarchical tree, i.e., no subcriteria, under the scaled interval priority weights $[y_k^{\mathrm{L}}, y_k^{\mathrm{R}}]$, $k \in N$ and the interval priority weights $[x_i^{kL}, x_i^{kR}]$, $i \in N$, $k \in N$, the dominance relation between i-th and j-th alternatives are checked in the following way: (i) Calculate $df_{ij}^k = x_i^{kL} - x_j^{kR}$, $k \in N$. (ii) Calculate the minimum total difference $td_{ij} = \sum_{k:df_{ij}^k \leq 0} y_k^{\mathrm{R}} df_{ij}^k + \sum_{k:df_{ij}^k > 0} y_k^{\mathrm{L}} df_{ij}^k$. (iii) If $td_{ij} > 0$, the i-th alternative is surely better than the j-th alternative. We note that the dominance relation is not complete but transitive and reflexive. Moreover, because the scaled interval priority weights do not satisfy the normality conditions, the calculations for checking the dominance become simpler than those with normality conditions (see [10]).

Example 3. Consider an imaginary care selection problem. There are three different cars: car A, car B and car C. Among those we choose a car and buy it. As criteria, we take into account three criteria: fuel consumption, performance and design. To score three cares in view of each criterion, we make pairwise comparisons so that we obtain three interval PCMs as shown in Tables 2, 3 and 4. We apply the inner model. The obtained interval priority weights are shown in the rightmost column in each table. To obtain suitable multipliers and priority weights on criteria, we give another PCM as shown in Table 5. The scaled interval priority weights, as the range of the product of multiplier and priority weights,

Table 2. Fuel consumption

fuel	car A	car B	car C	score
car A	1	$[1/3, 2/3]$	$[1/3, 1/2]$	$[1/3, 1/2]$
car B	$[3/2, 3]$	1	$[2/3, 1]$	$[3/4, 1]$
car C	$[2, 3]$	$[1, 3/2]$	1	$[1, 1]$

Table 3. Performance

perfor	car A	car B	car C	score
car A	1	$[2/3, 1]$	$[3/2, 3]$	$[3/2, 9/4]$
car B	$[1, 3/2]$	1	$[2, 3]$	$[9/4, 9/4]$
car C	$[1/3, 2/3]$	$[1/3, 1/2]$	1	$[3/4, 1]$

Table 4. Design

design	car A	car B	car C	score
car A	1	$[1, 3/2]$	$[1, 3/2]$	$[3/2, 3/2]$
car B	$[2/3, 1]$	1	$[1, 3/2]$	$[1, 3/2]$
car C	$[2/3, 1]$	$[2/3, 1]$	1	$[1, 1]$

Table 5. Criteria

cri.	car A	car B	car C	multiplier
car A	1	$[3/2, 3]$	$[6/5, 3/2]$	$[9/5, 9/4]$
car B	$[1/3, 5/6]$	1	$[2/3, 4/5]$	$[4/9, 8/15]$
car C	$[2/3, 5/6]$	$[5/4, 3/2]$	1	$[1, 1]$

Table 6. Evaluation of dominance relation

(i, j)	(A,B)	(A,C)	(B,A)	(B,C)	(C,A)	(C,B)
td_{ij}	-1.9000	-0.7778	-0.0500	-0.5069	-0.4000	-1.3000

are shown in the rightmost column of Table 5. We evaluate the dominance relation between cars. The minimum total difference td_{ij} are shown in Table 6. From Table 6, no strong dominance relation holds. However, if we enforced to select one, car B would be a good solution considering td_{ij} values.

References

1. Saaty, T.L.: The Analytic Hierarchy Process. McGraw-Hill, New York (1980)
2. Saaty, T.L., Vargas, C.G.: Comparison of eigenvalue, logarithmic least squares and least squares methods in estimating ratios. Math. Model. **5**, 309–324 (1984)
3. van Laarhoven, P.J.M., Pedrycz, W.: A fuzzy extension of Saaty's priority theory. Fuzzy Sets Syst. **11**, 199–227 (1983)
4. Buckley, J.J.: Fuzzy hierarchical analysis. Fuzzy Sets Syst. **17**, 233–247 (1985)
5. Wang, Y.-M., Elhag, T.M.S., Hua, Z.: A modified fuzzy logarithmic least squares method for fuzzy analytic hierarchy process. Fuzzy Sets Syst. **157**, 3055–3071 (2006)
6. Arbel, A.: Approximate articulation of preference and priority derivation. Eur. J. Oper. Res. **43**, 317–326 (1989)
7. Sugihara, K., Tanaka, H.: Interval evaluations in the analytic hierarchy process by possibilistic analysis. Comput. Intell. **17**, 567–579 (2001)
8. Sugihara, K., Ishii, H., Tanaka, H.: Interval priorities in AHP by interval regression analysis. Eur. J. Oper. Res. **158**, 745–754 (2004)
9. Wang, Y.-M., Elhag, T.M.S.: A goal programming method for obtaining interval weights from an interval comparison matrix. Eur. J. Oper. Res. **177**, 458–471 (2007)
10. Entani, T., Inuiguchi, M.: Pairwise comparison based interval analysis for group decision aiding with multiple criteria. Fuzzy Sets Syst. **271**, 79–96 (2015)

Sequential Decision Process Supported by a Compositional Model

Radim Jiroušek[1,2]($\boxtimes$) and Lucie Váchová[1]

[1] Faculty of Management, University of Economics,
Jindřichův Hradec, Czech Republic
radim@utia.cas.cz, vachova@fm.vse.cz
[2] Institute of Information Theory and Automation,
Czech Academy of Sciences, Prague, Czech Republic

Abstract. The goal of the paper is to describe a classical sequential decision process that is often used for both medical and technical diagnosis making in a relatively new theoretical setting. For this, we represent the background knowledge, which is assumed to be expressed in the form of a multidimensional probability distribution, as a compositional model. Though we do not perform a detailed analysis of its computational complexity, we show that the whole process is easily tractable for probability distributions of very high dimensions in the case that the distribution is represented as a compositional model of special properties.

Keywords: Sequential decision-making · Multidimensional models · Probability distributions · Composition · Information theory

1 Introduction

The basic idea, upon which a sequential decision process is based, is simple. The goal is to find out evidence supporting a decision with a required certainty. To express this idea more precisely let us use the language of probability theory. Let Z denote a decision variable whose values correspond to individual decisions (e.g., diagnoses). If κ is a probability distribution describing a relationship among the feature variables and the decision variable, we want to find out a subset E of feature variables whose values $\mathbf{e}$ (evidence) observed in the situation in question yields

$$\kappa(Z = \mathbf{d}|E = \mathbf{e}) \geq \varepsilon$$

for some decision $\mathbf{d}$, and a required reliability threshold ε. In a sequential process, the set E is gradually constructed by adding (usually) one variable at each step. Naturally, we want to keep the set E as small as possible, or generally, if some weights (or costs) connected with individual variables are given, we want to get the cheapest possible decision process.

Quite often, sequential decision processes are represented in a form of a decision tree. However, in many practical problems a fixed tree does not suit the

© Springer International Publishing AG 2016
V.-N. Huynh et al. (Eds.): IUKM 2016, LNAI 9978, pp. 51–63, 2016.
DOI: 10.1007/978-3-319-49046-5_5

situations. For example, in case of a medical diagnosis making, a patient visits a physician with some complains, i.e., with symptoms (values of variables) that need not be at the beginning of the decision tree, and still they should be included in the set E. Similarly, faults of a technical device manifest at the very beginning in different ways (or, it may be observed by different users in different ways). Not speaking about a newly appointed manager whose task is to strengthen a company. To diagnose weaknesses of the enterprise they always start with different prior knowledge. Moreover, costs assigned to variables may differ from case to case. A patient may come to a specialist with the results of some expensive laboratory tests that were already performed by another physician, so the specialist has it free. This is why we prefer not to compute a single decision tree in advance but to construct a sequential decision process based on a multidimensional probability distribution at time of its application.

When speaking about dimensionality of probability distributions, it is clear that in practical situations we should have in mind rather thousands than tens of variables. Therefore we will use a tool enabling us to represent and compute with such large distributions.

For this purpose, Bayesian networks [5] that were developed in 1980s, or other graphical models [10] are often used. In contrast to this, in this paper we want to enhance application of another, non-graphical approach, so called compositional models that were proposed for multidimensional probability distributions; see [6].

The paper is organized as follows: Sect. 2 introduces a necessary notation and serves as a brief introduction to compositional models (more on this topic can be found in [6]). In Sect. 3 we show that it is possible to compute conditionals even for probability distributions of very high dimensions when the distributions are represented in the form of compositional models of special properties. Section 4 is devoted to the description of the sequential decision process.

2 Compositional Model

In this text we use the notation from the paper [7] presented at IUKM 2015. We deal with a finite system of finite-valued random variables $N = \{X_1, \ldots, X_n, Z\}$. A set of values of feature variable X_i, which is denoted $\mathbb{X}_i$, is assumed to have at least two elements. The same is assumed also about set $\mathbb{Z}$ of values of decision variable Z. The set of all combinations of the considered values is denoted $\mathbb{X}_N = \mathbb{X}_1 \times \mathbb{X}_2 \times \ldots \times \mathbb{X}_n \times \mathbb{Z}$. Analogously, for $K \subset N \setminus \{Z\}$, $\mathbb{X}_K = \times_{X_i \in K} \mathbb{X}_i$, and for $Z \in K \subset N$, $\mathbb{X}_K = \times_{X_i \in K} \mathbb{X}_i \times \mathbb{Z}$.

Distributions of the considered variables are denoted by Greek letters (π, κ, μ) with possible indices; thus for $K \subseteq N$, we can consider a distribution $\pi(K)$, which is a $|K|$-dimensional distribution and $\pi(x)$ denotes a value of probability distribution π for state $x \in \mathbb{X}_K$.

For a probability distribution $\pi(K)$ and $J \subset K$, its *marginal distribution* $\pi^{\downarrow J}$ is computed for all $x \in \mathbb{X}_J$ by

$$\pi^{\downarrow J}(x) = \sum_{y \in \mathbb{X}_K : y^{\downarrow J} = x} \pi(y),$$

where $y^{\downarrow J}$ denotes the *projection* of $y \in \mathbb{X}_K$ into $\mathbb{X}_J$. For computation of marginal distributions we do not exclude situations when $J = \emptyset$, in this case $\pi^{\downarrow \emptyset} = 1$.

Having two distributions $\pi(K)$ and $\kappa(K)$, we say that κ dominates π (in symbol $\pi \ll \kappa$) if for all $x \in \mathbb{X}_K$

$$\kappa(x) = 0 \implies \pi(x) = 0.$$

One of the most important notions supporting an efficient representation of multidimensional probability distributions is a famous concept of conditional independence (se e.g. [10] or [13]).

For a probability distribution $\pi(K)$ and three disjoint subsets $L, M, R \subseteq K$ such that both $L, M \neq \emptyset$, we say that groups of variables L and M are *conditionally independent* given R (in symbol $L \perp\!\!\!\perp M | R \; [\pi]$) if

$$\pi^{\downarrow L \cup M \cup R} \pi^{\downarrow R} = \pi^{\downarrow L \cup R} \pi^{\downarrow M \cup R}.$$

2.1 Operator of Composition

For the compositional models, the key notion is that of a composition.

Definition 1. *For arbitrary distributions* $\pi(K)$ *and* $\kappa(L)$, *for which* $\pi^{\downarrow K \cap L} \ll \kappa^{\downarrow K \cap L}$, *their* composition *is, for each* $x \in \mathbb{X}_{(L \cup K)}$, *given by the following formula*

$$(\pi \triangleright \kappa)(x) = \frac{\pi(x^{\downarrow K}) \kappa(x^{\downarrow L})}{\kappa^{\downarrow K \cap L}(x^{\downarrow K \cap L})}.$$

In case $\pi^{\downarrow K \cap L} \not\ll \kappa^{\downarrow K \cap L}$, *the composition remains undefined.*

Let us briefly repeat the basic properties of this operator that were discussed in more details in [6] and also in [7].

Lemma 1. *Suppose* $\pi(K)$, $\kappa(L)$ *and* $\mu(M)$ *are such probability distributions that all the following expressions (compositions) are defined. Then the following statements hold.*

1. *$\pi \triangleright \kappa$ is a probability distribution of variables $(L \cup K)$ and its marginal distribution for variables K equals π:*

$$(\pi \triangleright \kappa)^{\downarrow K} = \pi.$$

2. *The composition is not commutative, therefore in general: $\pi \triangleright \kappa \neq \kappa \triangleright \pi$, however, π and κ are consistent, i.e., $\pi^{\downarrow K \cap L} = \kappa^{\downarrow K \cap L}$, if and only if*

$$\pi \triangleright \kappa = \kappa \triangleright \pi.$$

3. *The composition is not associative, therefore in general $(\pi \triangleright \kappa) \triangleright \mu \neq \pi \triangleright (\kappa \triangleright \mu)$, however if $K \supset (L \cap M)$, or, $L \supset (K \cap M)$, then*

$$(\pi \triangleright \kappa) \triangleright \mu = \pi \triangleright (\kappa \triangleright \mu).$$

4. If $(K \cap L) \subseteq R \subseteq K \cup L$, *then*

$$(\pi \triangleright \kappa)^{\downarrow R} = \pi^{\downarrow K \cap R} \triangleright \kappa^{\downarrow L \cap R}.$$

5. If $K \supset (L \cap M)$, *then* $(\pi \triangleright \kappa) \triangleright \mu = (\pi \triangleright \mu) \triangleright \kappa$.

The above presented assertion expresses all the properties of the operator of composition we will need in this paper. For example, from Property 1 one can easily see that it enables us to construct a more-dimensional distribution from two low-dimensional ones. The result is an *extension* of the first argument. On the other hand side, having a more-dimensional distribution one can, in some situations, factorize this distribution into two (or more) low-dimensional ones. This property is precisely expressed in the following assertion, the proof of which can also be found in [6].

Lemma 2. *Consider a probability distribution* $\pi(K)$, *and three disjoint subsets* $L, M, R \subseteq K$ *such that both* $L, M \neq \emptyset$. *Then* $L \perp\!\!\!\perp M | R \ [\pi]$ *if and only if*

$$\pi^{\downarrow L \cup M \cup R} = \pi^{\downarrow L \cup R} \triangleright \pi^{\downarrow M \cup R}.$$

2.2 Generating Sequences

In the rest of the paper we will deal with sequences of low-dimensional probability distributions. To avoid some technical problems and the necessity of repeating some assumptions to excess, let us make the following three conventions.

First, whenever we speak about a distribution π_k, it will be a distribution $\pi_k(K_k)$. Thus, formula $\pi_1 \triangleright \ldots \triangleright \pi_m$, if it is defined, determines the distributions of variables $K_1 \cup \ldots \cup K_m$.

Since the operator of composition is not associative (see Property 3 of Lemma 1), the formulas like $\pi_1 \triangleright \ldots \triangleright \pi_m$ are, strictly speaking, ambiguous. Therefore, the second convention avoids this ambiguity saying that we always apply the operators from left to right. Thus

$$\pi_1 \triangleright \pi_2 \triangleright \pi_3 \triangleright \ldots \triangleright \pi_m = (((\pi_1 \triangleright \pi_2) \triangleright \pi_3) \triangleright \ldots \triangleright \pi_m),$$

and the parentheses will be used only when we want to change this default ordering. Therefore, to construct a multidimensional distribution it is sufficient to determine a sequence – we call it a *generating sequence* – of low-dimensional distributions.

The third aforementioned convention is of a rather technical nature. Considering a generating sequence $\pi_1, \pi_2, \ldots, \pi_m$ we will always assume that all the included operators of composition are well defined, which means that we assume

$$\pi_1^{\downarrow K_2 \cap K_1} \ll \pi_2^{\downarrow K_2 \cap K_1},$$
$$(\pi_1 \triangleright \pi_2)^{\downarrow K_3 \cap (K_1 \cup K_2)} \ll \pi_3^{\downarrow K_3 \cap (K_1 \cup K_2)},$$

and so on. Therefore, we assume that $\pi_1 \triangleright \pi_2 \triangleright \pi_3 \triangleright \ldots \triangleright \pi_m$ is defined.

Another important notion we will need in this paper is that of a perfect sequences that is defined in the following way (see also [6]).

Definition 2. *A generating sequence of probability distributions* $\pi_1, \pi_2, \ldots, \pi_m$ *is called* perfect *if*

$$\pi_1 \triangleright \pi_2 = \pi_2 \triangleright \pi_1,$$
$$\pi_1 \triangleright \pi_2 \triangleright \pi_3 = \pi_3 \triangleright (\pi_1 \triangleright \pi_2),$$
$$\vdots$$
$$\pi_1 \triangleright \pi_2 \triangleright \ldots \triangleright \pi_m = \pi_m \triangleright (\pi_1 \triangleright \ldots \triangleright \pi_{m-1}).$$

The importance of perfect sequences becomes clear from the following characterization theorem (for the proofs of all the assertions presented in the rest of this section see [6]).

Theorem 1. *A sequence of distributions* $\pi_1, \pi_2, \ldots, \pi_m$ *is perfect iff all the distributions from this sequence are marginals of the distribution* $\pi_1 \triangleright \pi_2 \triangleright \ldots \triangleright \pi_m$.

Another advantageous property of the perfect sequences says that perfect sequences represent a unique distribution in the following sense.

Theorem 2. *If a sequence* $\pi_1, \pi_2, \ldots, \pi_m$ *and its permutation* $\pi_{i_1}, \pi_{i_2}, \ldots, \pi_{i_m}$ *are both perfect, then*

$$\pi_1 \triangleright \pi_2 \triangleright \ldots \triangleright \pi_m = \pi_{i_1} \triangleright \pi_{i_2} \triangleright \ldots \triangleright \pi_{i_m}.$$

In the rest of the paper we will need two more facts expressed in the following assertions. The first says that any generating sequence can be transformed into a perfect sequence without influencing the resulting multidimensional distribution.

Theorem 3. *For any generating sequence* $\pi_1, \pi_2, \ldots, \pi_m$, *the sequence* $\kappa_1, \kappa_2, \ldots, \kappa_m$ *computed by the following process*

$$\kappa_1 = \pi_1,$$
$$\kappa_2 = \kappa_1^{\downarrow K_2 \cap K_1} \triangleright \pi_2,$$
$$\kappa_3 = (\kappa_1 \triangleright \kappa_2)^{\downarrow K_3 \cap (K_1 \cup K_2)} \triangleright \pi_3,$$
$$\vdots$$
$$\kappa_m = (\kappa_1 \triangleright \ldots \triangleright \kappa_{m-1})^{\downarrow K_m \cap (K_1 \cup \ldots \cup K_{m-1})} \triangleright \pi_m$$

is perfect, and

$$\pi_1 \triangleright \ldots \triangleright \pi_m = \kappa_1 \triangleright \ldots \triangleright \kappa_m.$$

The process of perfectization described in Theorem 3 is simple. Unfortunately, not from the point of view of computational complexity. This is because the computation of $(\kappa_1 \triangleright \ldots \triangleright \kappa_{r-1})^{\downarrow K_r \cap (K_1 \cup \ldots \cup K_{r-1})}$ may be computationally very expensive [2]. Therefore, in the next sections we will take advantage of special generating sequences, namely those whose sequences of variables $K_1, K_2, \ldots, K_m$ meet the so called *running intersection property* (RIP):

$$\forall r = 2, \ldots, m \ \exists s (1 \le s < r) \ \left(K_r \cap \left(\bigcup_{k=1}^{r-1} K_k \right) \subseteq K_s \right).$$

Consider a perfectization process applied to such a RIP generating sequence $\pi_1, \pi_2, \ldots, \pi_m$. In this case, either $K_1 \supseteq K_3 \cap (K_1 \cup K_2)$, or $K_2 \supseteq K_3 \cap (K_1 \cup K_2)$, which means that $K_3 \cap (K_1 \cup K_2)$ equals either $K_1 \cap K_3$ or $K_2 \cap K_3$. Since the sequence $\kappa_1, \ldots, \kappa_m$ is constructed in the way that κ_1 and κ_2 are consistent, i.e., $\kappa_1^{\downarrow K_2 \cap K_1} = \kappa_2^{\downarrow K_2 \cap K_1}$, due to Property 2 of Lemma 1 $\kappa_1 \triangleright \kappa_2 = \kappa_2 \triangleright \kappa_1$ and therefore, using Property 1 of the same Lemma, $(\kappa_1 \triangleright \kappa_2)^{\downarrow K_1} = \kappa_1$ and $(\kappa_1 \triangleright \kappa_2)^{\downarrow K_2} = \kappa_2$. Thus we got that $(\kappa_1 \triangleright \kappa_2)^{\downarrow K_3 \cap (K_1 \cup K_2)}$ equals either $\kappa_1^{\downarrow K_3 \cap K_1}$, or $\kappa_2^{\downarrow K_3 \cap K_2}$. Analogously, we can see that for any $r = 3, \ldots, m$ there exists s such that

$$(\kappa_1 \triangleright \ldots \triangleright \kappa_{r-1})^{\downarrow K_r \cap (K_1 \cup \ldots \cup K_{r-1})} = \kappa_s^{\downarrow K_r \cap K_s},$$

which makes the perfectization process computationally simple.

Another advantageous property concerning the RIP generating sequences is expressed in the following theorem [6].

Theorem 4. *If $\pi_1, \ldots, \pi_m$ is a sequence of pairwise consistent probability distributions such that $K_1, \ldots, K_m$ meets RIP, then this sequence is perfect.*

3   Conditioning

In [3] we showed that even the conditional distribution can be computed using the operator of composition. Consider any variable $Y \in N$ and its value $\mathbf{a} \in \mathbb{X}_{\{Y\}}$. Define a *degenerated* one-dimensional probability distribution $\delta_{\mathbf{a}}(Y)$ as a distribution of variable Y achieving probability 1 for value $Y = \mathbf{a}$, i.e.,

$$\delta_{\mathbf{a}}(Y) = \begin{cases} 1 & \text{if } Y = \mathbf{a}, \\ 0 & \text{otherwise.} \end{cases}$$

Now, compute $(\delta_{\mathbf{a}}(Y) \triangleright \kappa)^{\downarrow \{X\}}$ for a probability distribution $\kappa(L)$ with $X, Y \in L$. For $\mathbf{b} \in \mathbb{X}_{\{X\}}$

$$(\delta_{\mathbf{a}}(Y) \triangleright \kappa)^{\downarrow \{X\}}(\mathbf{b}) = ((\delta_{\mathbf{a}}(Y) \triangleright \kappa)^{\downarrow \{X,Y\}})^{\downarrow \{X\}}(\mathbf{b}) = (\delta_{\mathbf{a}}(Y) \triangleright \kappa^{\downarrow \{X,Y\}})^{\downarrow \{X\}}(\mathbf{b})$$

$$= \sum_{x \in \mathbb{X}_{\{Y\}}} (\delta_{\mathbf{a}}(Y) \triangleright \kappa^{\downarrow \{X,Y\}})(\mathbf{b}, x)$$

$$= \sum_{x \in \mathbb{X}_{\{Y\}}} \frac{\delta_{\mathbf{a}}(Y) \cdot \kappa^{\downarrow \{X,Y\}}(\mathbf{b}, x)}{\kappa^{\downarrow \{Y\}}(x)}$$

$$= \frac{\kappa^{\downarrow \{X,Y\}}(\mathbf{b}, \mathbf{a})}{\kappa^{\downarrow \{Y\}}(\mathbf{a})} = \kappa^{\downarrow \{X,Y\}}(\mathbf{b}|\mathbf{a}).$$

Analogously, we can easily show that for $\kappa(L)$

$$\kappa(L \setminus \{Y\} | Y = \mathbf{a}) = (\delta_{\mathbf{a}}(Y) \triangleright \kappa)^{\downarrow L \setminus \{Y\}},$$

and for $E = \{Y_1, Y_2, \ldots, Y_k\}$, and $\mathbf{e} \in \mathbb{X}_E$,

$$\kappa(L \setminus E | E = \mathbf{e}) = (\delta_{\mathbf{e}^{\downarrow \{Y_1\}}}(Y_1) \triangleright (\ldots \triangleright (\delta_{\mathbf{e}^{\downarrow \{Y_k\}}}(Y_k) \triangleright \kappa)))^{\downarrow L \setminus E}.$$

Now, we want to show that the last expression can be effectively computed for a multidimensional distribution κ represented in a form of a RIP perfect generating sequence: $\kappa = \pi_1(K_1) \triangleright \pi_2(K_2) \triangleright \ldots \triangleright \pi_m(K_m)$.

Let us, first, consider the computation

$$\delta_{\mathbf{e}\downarrow\{Y_k\}}(Y_k) \triangleright \kappa = \delta_{\mathbf{e}\downarrow\{Y_k\}}(Y_k) \triangleright (\pi_1 \triangleright \pi_2 \triangleright \ldots \triangleright \pi_m).$$

It is well known (and the reader can see it from the properties of joint trees described in Sect. 4.1) that for each K_i the RIP sequence $K_1, K_2, \ldots, K_m$ can be reordered so that the resulting sequence is also RIP and K_i is at the beginning of this new sequence. So, without loss of generality we can assume that $\pi_1, \pi_2, \ldots, \pi_m$ is such that $Y_k \in K_1$. The fact that the respective reordering does not influence the represented multidimensional distribution $\pi_1 \triangleright \pi_2 \triangleright \ldots \triangleright \pi_m$ follows from Theorems 4 and 2.

Now, consider

$$\delta_{\mathbf{e}\downarrow\{Y_k\}}(Y_k) \triangleright ((\pi_1 \triangleright \pi_2 \triangleright \ldots \triangleright \pi_{m-1}) \triangleright \pi_m).$$

Since Y_k is contained in K_1, it is contained also in $(K_1 \cup K_2 \cup \ldots \cup K_{m-1})$, and therefore, applying Property 3 of Lemma 1, we get

$$\delta_{\mathbf{e}\downarrow\{Y_k\}}(Y_k) \triangleright ((\pi_1 \triangleright \pi_2 \triangleright \ldots \triangleright \pi_{m-1}) \triangleright \pi_m)$$
$$= (\delta_{\mathbf{e}\downarrow\{Y_k\}}(Y_k) \triangleright ((\pi_1 \triangleright \pi_2 \triangleright \ldots \triangleright \pi_{m-1})) \triangleright \pi_m.$$

Repeating this reasoning $m-1$ times we eventually get

$$\delta_{\mathbf{e}\downarrow\{Y_k\}}(Y_k) \triangleright (\pi_1 \triangleright \pi_2 \triangleright \ldots \triangleright \pi_m) = (\delta_{\mathbf{e}\downarrow\{Y_k\}}(Y_k) \triangleright \pi_1) \triangleright \pi_2 \triangleright \ldots \triangleright \pi_m,$$

which means that computing a conditional from a distribution represented by a RIP perfect generating sequence results, again, in a distribution represented as a RIP sequence. The latter can be, as showed in the preceding section, efficiently transformed into a RIP perfect generating sequence (Theorem 3). Therefore, to compute $\kappa(L \setminus E | E = \mathbf{e})$ we have to successively apply the above idea for all $Y_j \in E$. Let us stress that for each Y_j we have to find a RIP ordering of $(K_1, K_2, \ldots, K_m)$ such that Y_j is in the first set of this RIP ordering. Generally, for different Y_j we have to use a different RIP ordering of $(K_1, K_2, \ldots, K_m)$.

4   Sequential Decision Process

A sequential decision process consists in a successive repetition of the following step:

Knowing state $\mathbf{e} \in \mathbb{X}_E$ of variables E, find a variable $X \in N \setminus (E \cup \{Z\})$ such that the detection of its value $\mathbf{a}$ increases (as much as possible) the chances of getting

$$\kappa(Z = \mathbf{d} | E = \mathbf{e}, X = \mathbf{a}) \geq \varepsilon$$

for some value $\mathbf{d} \in \mathbb{Z}$.

It is important to realize that in this step we search for a variable X whose value is to be ascertained next. It means that at the moment of looking for X we do not know the value $\mathbf{a} \in \mathbb{X}_{\{X\}}$. This value will be ascertain before the next sequential step is realized with new $E := E \cup \{X\}$.

For the selection of $X \in N \setminus (E \cup \{Z\})$ we can hardly find a better criterion than that used in the process of construction of efficient decision (or search) trees for many years [9, 12]. Using this criterion, in the process of data based construction of a decision tree, we look for the variable that splits the considered training data file into subfiles yielding the minimum expected Shannon entropy for the decision variable. In fact, it is nothing else than looking for a variable $X \in N \setminus (E \cup \{Z\})$ maximizing the expression

$$
\begin{aligned}
&MI_{\kappa(N \setminus E|E=\mathbf{e})}(X; Z) \\
&= \sum_{(x,z) \in \mathbb{X}_{\{X\}} \times \mathbb{Z}} \kappa^{\downarrow\{X,Z\}}(x, z|E = \mathbf{e}) \cdot \log \frac{\kappa^{\downarrow\{X,Z\}}(x, z|E = \mathbf{e})}{\kappa^{\downarrow\{X\}}(x|E = \mathbf{e})\kappa^{\downarrow\{Z\}}(z|E = \mathbf{e})}.
\end{aligned}
$$

Now, as the reader certainly expects, we take advantage of the results from the preceding section and assume that the conditional distribution $\kappa(N \setminus E|E = \mathbf{e})$ is represented in the form of a RIP perfect generating sequence. This assumption enables us not only to compute values $MI_{\kappa(N \setminus E|E=\mathbf{e})}(X; Z)$ for all $X \in N \setminus (E \cup \{Z\})$ in an efficient way, but also to speed up the computational process by indicating those variables, for which we need not to compute the value of mutual information because we can learn in advance that the respective conditional mutual information cannot achieve a maximal value.

4.1   Computations in Joint Trees

It is known from both data-base [1] and Bayesian network [5, 10] theories that a system of sets $K_1, K_2, \ldots, K_m$ can be ordered to meet RIP if and only if one can construct a structure called a *joint tree*. It is a tree having $K_1, K_2, \ldots, K_m$ for its nodes and possessing the following special property: If K_k lies on the path from K_r to K_s then $K_k \supseteq K_r \cap K_s$.

Recall that it is an easy task to construct a joint tree for a RIP sequence $K_1, K_2, \ldots, K_m$: For, each K_r $(r = 2, \ldots, m)$ the joint tree contains an edge connecting K_r with that K_s, which meets the RIP condition

$$
1 \leq s < r \ \& \ K_r \cap (K_1 \cup \ldots \cup K_{r-1}) \subseteq K_s.
$$

If there are more such nodes K_s, then K_r is connected to only one of them. The tree contains no other edges than those specified above.

To compute $MI_{\kappa(N \setminus E|E=\mathbf{e})}(X; Z)$ for all $X \in N \setminus (E \cup \{Z\})$, start enumerating this mutual information for all variables from K_r, for which $Z \in K_r$ (if there are more sets meeting this condition, consider all of them). Since we assume that

$$
\kappa(N \setminus E|E = \mathbf{e}) = \pi_1 \triangleright \pi_2 \triangleright \ldots \triangleright \pi_m,
$$

and that $\pi_1, \pi_2, \ldots, \pi_m$ is a perfect sequence, due to Theorem 1 we know that π_r is marginal of $\kappa(N \setminus E | E = \mathbf{e})$. This means that we can compute the required conditional mutual information just from π_r, which is simple.

After having evaluated the required mutual information for all the variables from $K_r \setminus (E \cup \{Z\})$, we start processing variables from the neighboring nodes, i.e., from nodes K_s that are adjacent to K_r in the joint tree. Now, it is important to realize two facts that makes the computation very efficient.

First, since K_s is adjacent to K_r in the joint tree, it is possible to find a RIP ordering of $K_1, K_2, \ldots, K_m$ such that it starts $K_r, K_s, \ldots$, and therefore (thanks to Theorem 2 and Property 1 of Lemma 1) we know that

$$\kappa((K_r \cup K_s) \setminus E | E = \mathbf{e}) = \pi_r \triangleright \pi_s.$$

The other fact that can speed up the computational process follows from a famous property of mutual information. It is known from any textbook on information theory (e.g. [4]) that if $X \perp\!\!\!\perp Z | M [\mu]$, for some set of variables M then

$$MI_\mu(X; Z) \leq MI_\mu(M; Z).$$

Therefore, applying this property and Lemma 2 to this situation we get that for $X \in K_s \setminus K_r$ (recall that $Z \in K_r \setminus K_s$ because all variables from K_r were treated in the previous step)

$$MI_{\pi_r \triangleright \pi_s}(X; Z) \leq MI_{\pi_r \triangleright \pi_s}(K_r \cap K_s; Z).$$

This means that if there is variable $X \in K_r$ for which

$$MI_{\pi_r \triangleright \pi_s}(K_r \cap K_s; Z) \leq MI_{\pi_r \triangleright \pi_s}(X; Z)$$

then we do not need to compute mutual information $MI_{\pi_r \triangleright \pi_s}(X'; Z)$ for variables $X' \in K_s \setminus K_r$ because we know that it cannot achieve the looked for maximum.

In a similar way we can compute $MI_{\kappa(N \setminus E | E = \mathbf{e})}(X; Z)$ for all the remaining variables from $N \setminus (E \cup \{Z\})$. First we have to find the shortest path in the considered joint tree connecting nodes containing X and Z (realize that in a tree, two nodes are always connected by a unique path, but both the considered variables X and Z may be in several nodes). Denote this path $K_{j_1}, K_{j_2}, \ldots, K_{j_k}$ in the way that $X \in K_{j_1}$ and $Z \in K_{j_k}$. Then from the perfectness of $\pi_1, \pi_2, \ldots, \pi_m$ and the RIP property we get that

$$\kappa((K_{j_1} \cup K_{j_2} \cup \ldots \cup K_{j_k}) \setminus E | E = e) = \pi_{j_1} \triangleright \pi_{j_2} \triangleright \ldots \triangleright \pi_{j_k},$$

and thus we compute

$$MI_{\kappa(N \setminus E | E = \mathbf{e})}(X; Z) = MI_{\pi_{j_1} \triangleright \pi_{j_2} \triangleright \ldots \triangleright \pi_{j_k}}(X; Z).$$

However, the longer path $K_{j_1}, K_{j_2}, \ldots, K_{j_k}$ the greater chances that among the variables, for which the mutual information has been evaluated, we can find variable X', for which

$$MI_{\pi_{j_1} \triangleright \pi_{j_2} \triangleright \ldots \triangleright \pi_{j_k}}(K_{j_i} \cap K_{j_{i+1}}; Z) \leq MI_{\pi_{j_1} \triangleright \pi_{j_2} \triangleright \ldots \triangleright \pi_{j_k}}(X'; Z),$$

for some $i \in \{1, 2, k - 1\}$, and therefore the computation of $MI_{\pi_{j_1} \triangleright \pi_{j_2} \triangleright \ldots \triangleright \pi_{j_k}}(X; Z)$ is wasteful.

4.2   An Algorithm

Up to now, we have described and theoretically supported all the individual parts of the proposed sequential decision process. Let us (rather informally) summarize it in several steps.

Initialization. We assume that the probability distribution represented by a RIP generating sequence is given. If it is not perfect then apply the perfectization procedure described in Theorem 3 so that the resulting generating sequence $\pi_1, \pi_2, \ldots, \pi_m$ meets RIP condition and is perfect. Thus, in the sequel we compute with the distribution

$$\kappa(N) = \pi_1 \triangleright \pi_2 \triangleright \ldots \triangleright \pi_m.$$

Set $E := \emptyset$.

Preliminary Evidence Processing. E_0 denotes the set of variables whose values e_0 are given before the sequential process starts. If there is no preliminary evidence, i.e. $E_0 = \emptyset$, skip the rest of this step. Otherwise for all variables X from E_0 realize the following conditioning procedure.

Conditioning. Assign a new value to $E := E \cup \{X\}$, and extend the point $e := e_0^{\downarrow E}$. Reorder (renumber) the given generating sequence in the way that the new ordering $K_1, K_2, \ldots, K_m$ meets RIP and $X \in K_1$. Apply the perfectization procedure (Theorem 3) to the sequence: $(\delta_{e_0^{\downarrow X}}(X) \triangleright \pi_1), \pi_2, \ldots, \pi_m$, and assign the result, after marginalizing variable X out, as a new value to $\pi_1, \pi_2, \ldots, \pi_m$. So that now,

$$\kappa(N \setminus E | E = e) = \pi_1 \triangleright \pi_2 \triangleright \ldots \triangleright \pi_m.$$

Sequential Procedure. Perform the following process consisting of three steps (**Variable selection**, **Application** and **Conditioning**) repeatedly until $\pi_r^{\downarrow \{Z\}}(\mathbf{d}) \geq \varepsilon$ for some $\mathbf{d} \in \mathbb{Z}$. (Take any r such that $Z \in K_r$.)

Variable selection. Set $L := N \setminus (E \cup \{Z\})$ (L is a set of variables, for which we should compute the value of mutual information in this step).
For all $X \in L$, for which there exists $r \in \{1, 2, \ldots, m\}$ such that both $X, Z \in K_r$, compute $MI_{\pi_r}(X; Z)$, and reset $L := L \setminus \{X\}$.
  Now, denote X the variable achieving the maximal value of the mutual information, and repeat the following step until $L = \emptyset$.

Computation in a joint tree. Find the shortest path $K_{j_1}, K_{j_2}, \ldots, K_{j_k}$ from the respective joint tree meeting the following three properties: (1) $Z \in K_{j_k}$, (2) $K_{j_1} \cap L \neq \emptyset$, and (3) $(K_{j_2} \cup \ldots \cup K_{j_k}) \cap L = \emptyset$.
If

$$MI_{\pi_{j_2} \triangleright \ldots \triangleright \pi_{j_k}}(K_{j_1} \cap K_{j_2}; Z)$$
$$> MI_{\pi_{j_2} \triangleright \ldots \triangleright \pi_{j_k}}(X; Z)$$

Then for all $Y \in K_{j_1} \cap L$ compute

$$MI_{\pi_{j_1} \triangleright \dots \triangleright \pi_{j_k}}(Y; Z),$$

and reset $L := L \setminus K_{j_1}$.

If the maximum from the computed values of mutual information is higher than

$$MI_{\pi_{j_2} \triangleright \dots \triangleright \pi_{j_k}}(X; Z)$$

then reset X to the variable with the highest mutual information.

Else reset L by removing from L all such K_s for which the path from K_s to K_{j_k} goes through K_{j_1}.

Application. Ask the user to ascertain the value of variable X. Denote the result **a**.

Conditioning. Reset $E := E \cup \{X_i\}$, and extend the point **e** so that the new value is from $\mathbb{X}_E$, and $e^{\downarrow\{X\}} = \mathbf{a}$. Reorder (renumber) generating sequence $\pi_1, \pi_2, \dots, \pi_m$ in the way that the respective new ordering $K_1, K_2, \dots, K_m$ meets RIP and $X \in K_1$. Apply the perfectization procedure (Theorem 3) to the sequence: $(\delta_{\mathbf{a}}(X) \triangleright \pi_1), \pi_2, \dots, \pi_m$, and the result, after marginalizing variable X out, assign as a new value to $\pi_1, \pi_2, \dots, \pi_m$. So that now,

$$\kappa(N \setminus E | E = \mathbf{e}) = \pi_1 \triangleright \pi_2 \triangleright \dots \triangleright \pi_m.$$

5 Summary and Conclusions

We have described a sequential diagnosis making process based on a knowledge represented by a multidimensional probability distribution. Unfortunately, because of the restrictions on number of pages we are not able to include an illustrative example that will be presented in the conference lecture. Nevertheless, the idea of the process is simple and can be easily understood from the algorithm description. The reader certainly realized that the controlling rule aims for the least number of variables whose values are to be ascertain. As said in Introduction, it is really not difficult to modify the variable selection rule so that some weights of the variables are taken into account. In this case, however, one can hardly rely upon the fact that the longer path $K_{j_1}, K_{j_2}, \dots, K_{j_k}$ is constructed in the **Computation in a joint tree** step the greater chances to cut off the nodes of the joint tree that cannot contain a variable optimizing the selection criterion.

Applying just basic properties of the operator of composition and perfect generating sequences recalled in Sect. 2 we showed that all the necessary computations, including the computation of the required conditionals, can be performed *locally*, and therefore very efficiently. Application of local computational procedures, based on original Lauritzen and Spiegelhalter ideas [11], was made possible due to specific properties of perfect RIP generating sequences.

As it can be seen from the previous text, it is the very application of perfect compositional models that have one great advantage visible in comparison with application of Bayesian networks. Each low-dimensional distribution, which the model is constructed from, is marginal to the multidimensional model. This makes verification of some conditions, like for example the stoping rule from the Algorithm, very simple. More generally, there is a whole class of sets of variables, for which the marginals can easily be computed. These are the sets that can be got as a union of nodes of a subtree (connected subgraph) of the respective joint tree. This property was exploited in the **Computation in a joint tree** step of the Algorithm, where we considered marginals $\pi_{j_1} \triangleright \ldots \triangleright \pi_{j_k}$.

Though it is beyond the scope of this paper, let us mention yet another advantage of the considered compositional models. The operator of composition was introduced also in possibility theory [14] and recently even in Shenoy's Valuation Based Systems [8], which, as a generic uncertainty calculus covers many other calculi, such as Spohn's epistemic belief theory, and D-S belief function theory. This makes it possible to apply compositional models, and all the methods based on the compositional models like the described sequential decision process, in all these alternative uncertainty theories (naturally under the assumption that we have a function with the properties of mutual information at our disposal).

Acknowledgement. This research was partially supported by GAČR under Grant No. 15-00215S.

References

1. Beeri, C., Fagin, R., Maier, D., Yannakakis, M.: On the desirability of acyclic database schemes. J. ACM **30**(3), 479–513 (1983)
2. Bína, V., Jiroušek, R.: Marginalization in multidimensional compositional models. Kybernetika **42**(4), 405–422 (2006)
3. Bína, V., Jiroušek, R.: On computations with causal compositional models. Kybernetika **51**(3), 525–539 (2015) .
4. Csiszár, I., Korner, J.: Information Theory: Coding Theorems for Discrete Memoryless Systems. Akademiai Kiado, New York (1997)
5. Jensen, F.V.: Bayesian Networks and Decision Graphs. IEEE Computer Society Press, New York (2001)
6. Jiroušek, R.: Foundations of compositional model theory. Int. J. Gen. Syst. **40**(6), 623–678 (2011)
7. Jiroušek, R., Krejčová, I.: Minimum description length principle for compositional model learning. In: Huynh, V.-N., Inuiguchi, M., Denoeux, T. (eds.) IUKM 2015. LNCS (LNAI), vol. 9376, pp. 254–266. Springer, Heidelberg (2015). doi:10.1007/978-3-319-25135-6_25
8. Jiroušek, R., Shenoy, P.P.: Compositional models in valuation-based systems. Int. J. Approx. Reasoning **53**(8), 1155–1167 (2012)
9. Landa, L.N.: Opit primenenija matematitcheskoj logiki i teorii informacii k nekotorym problemam obutchenija. Voprosy psichologii **8**, 19–40 (1962)
10. Lauritzen, S.L.: Graphical models. Oxford University Press, Oxford (1996)
11. Lauritzen, S.L., Spiegelhalter, D.J.: Local computations with probabilities on graphical structures and their application to expert systems. J. Royal Stat. Soc. Series B (Methodological) **50**, 157–224 (1988)

12. Quinlan, J.R.: Induction of Decision Trees. Mach. Learn. **1**(1), 81–106 (1986)
13. Studený, M.: Probabilistic Conditional Independence Structures. Springer, Heidelberg (2005)
14. Vejnarová, J.: Possibilistic independence and operators of composition of possibility measures. In: Prague Conference on Information Theory, Statistical Decision Functions and Random Processes, pp. 575–580 (1998)

A Theory of Modeling Semantic Uncertainty in Label Representation

Zengchang Qin[1]($\boxtimes$), Tao Wan[2]($\boxtimes$), and Hanqing Zhao[1,3]

[1] Intelligent Computing and Machine Learning Lab, School of ASEE,
Beihang University, Beijing, China
zcqin@buaa.edu.cn
[2] School of Biological Science and Medical Engineering,
Beihang University, Beijing, China
taowan@buaa.edu.cn
[3] École Centrale de Pékin, Beihang University, Beijing, China

Abstract. A new theory of modeling the uncertainty associated with vague concepts is introduced. We consider the problem of quantifying an agents uncertainty concerning which labels are appropriate to describe a given observation. This can be regarded as a simplified model of natural language communication. Semantic meaning conveyed by high-level knowledge representation is often inherently uncertain. Such uncertainty is referred to *semantic uncertainty* and dominated by fuzzy modeling. In this framework, from an epistemic point of view, labels are precise and uncertainty comes from the undecidable boundary between labels in agents conceptual space. In this framework the boundary is regarded as a random variable and it can be modeled by a probability distribution. We also propose a functional calculus to measure how appropriate of using a certain label to describe an observation. In this way, a vague concept can be represented by a distribution on the labels. The new theory is verified by applying it to the vague category game.

Keywords: Label differentiation · Boundary distribution · Linguistic label · Label image · Category game

1 Introduction

Natural language is imprecise but effective for communicating ideas. Due to the limitation of human cognitive capabilities, we cannot communicate through lines of precise codes like machines do, but use high-level knowledge representation to summarize obtained information and exchange ideas. A language is learned empirically through interactions with others and the environment. Such interactions build associations between words and the real-world facts, that is how come the words have meaning. This process can be simply interpreted as a communication model where speaker encodes his/her semantic meaning into a piece of language and pass to the hearer to decode the meaning. Philosophically, it is still controversial that whether language can be regarded as only a communication

© Springer International Publishing AG 2016
V.-N. Huynh et al. (Eds.): IUKM 2016, LNAI 9978, pp. 64–75, 2016.
DOI: 10.1007/978-3-319-49046-5_6

model that is independent to thinking. There are espousers of both extremes in *communicative conception* (supporting the existence of language is only for communication) and *cognitive conception* (supporting language is indispensible to human consciousness) and supporters lie in between [1]. In this research, we are not trying to unbury the deep entangled roots of language and thinking. We tend to use behavior approach to propose mathematical treatment to study communicative function of language. Hope such theories or models can be used to develop high-level language for machines. Or in a large extent, try to understand how human evolve the capability of communicate through compact but imprecise high-level knowledge representation (natural language) while machines cannot, yet. High-level knowledge representation like natural language is powerful, flexible and robust in communication, is also inherently uncertain and hard to be explicitly quantified. This is referred to *semantic uncertainty* by Lawry and Tang [2] contrast to stochastic uncertainty which studies the uncertainty about the state of the world.

Roughly speaking, there are two main streams of philosophy to understand uncertainty [3]. Epistemic uncertainties are due to things we could in principle know but don't in practice. This may be either because we cannot measure a quantity sufficiently accurate or because we neglect certain effects. Stochastic or statistical uncertainties come from the limitation of human cognitive abilities. Though there exist an ideological and undeniable fact which is the reason for a phenomenon, however, the observable evidence of this "fact" is incomplete and imprecise. Philosophically, uncertainty is ubiquitous and hard to be quantified. However, in the practice of science and engineering, what we are concerned is how to model uncertainty with a proper measure [3]. Probability is no doubt the most accepted and used uncertainty measure for modeling objective (frequentist approach) and subjective (Bayesian approach) uncertainties. Jaynes [4] considers probability is an extension of logic with incomplete information. de Finetti [5,6] even claim that probability is the optimal calculus for measures of uncertainty.

The inception of fuzzy logic [7] as a generalization of binary logic has greatly changed people's view towards uncertainty modeling and reasoning. Fuzzy logic and fuzzy set theory has revolutionized many engineering areas, especially in control engineering, inductive logical reasoning [8], data mining [3] and intelligent systems [9]. For semantic computing of natural language, Zadeh [10] proposed Computing with Words (CW) and the General Theory of Uncertainty [11]. The CW framework describes theoretical feasibility of semantic calculation but no applicable calculus were discussed in his monograph on CW [12]. Though there are some subsequent researches of applying CW to question-answering [13], most CW related research is still in infancy. Lawry [14] proposed a framework of *modeling with words* which is referred to *label semantics* by introducing linguistic labels defined by fuzzy membership functions. Different from the CW theory, label semantics puts the emphasis on uncertainty modeling and reasoning in practical applications. It has been applied to data mining and several new transparent data mining algorithms were proposed in [3]. Lawry and Tang [2] later on proposed prototype interpreted label semantics, and discussed its relation to fuzziness in [6]. Similar applications in data mining based on prototype inter-preted label semantics are discussed in [3,15]. In this research, we propose a simpler model without using fuzzy modeling.

It is undeniable that humans possess some special mechanism for decid-ing whether or not to make certain bivalent assertions [16]. Given a particular instance x and description θ, you have to assert that 'x is θ' or not, θ could lin-guistic labels from a finite set of basic labels. Given a population of agents and each agent has partial knowledge about what labels are appropriate to assert. In label semantics theory proposed in [17], agent assertion on appropriate labels are modeled with a random set, which means that an agent may choose a set of labels but not one single label as appropriate description of the vague concept. In our framework, we follow Williamson's assumption [18] that for the extensions of a vague concept there is a precise but unknown dividing boundary between it, although the exact boundary is virtually impossible for anyone to identify precisely. The basic labels can be exclusive and each agent has her own dividing boundaries of the label set. Therefore, given an observation, each agent can only have one appropriate label.

2   Label Differentiation Theory

In nature, it is common that there exist implicate or explicate orders between objective realities [19]. A set of observations can be described by ordered lin-guistic labels, such as object's length, human age and color in each dimension of the RGB space. Thus, we hope to develop a framework to provide an epistemic method for measuring the appropriateness of a certain label as a description of an observation in a continuous universe Ω, which should have the following two properties:

- **Totally Ordered**: Ω possesses an order relation $\preccurlyeq$, where $\forall (x, y) \in \Omega^2$, we have either $x \preccurlyeq y$ or $y \preccurlyeq x$.
- **$\mathbb{R}$-Vector Space**: There exist an internal operation $+$ defined on Ω and an external operation $\bullet$ defined between Ω and $\mathbb{R}$. Making $(\Omega, +, \bullet)$ an $\mathbb{R}$-vector space.

2.1   Order Relation of Labels

Given a totally ordered continuous universe Ω, each linguistic label can be defined on a subset of Ω, furthermore, as different from existing frameworks for modeling linguistic uncertainties [2,3], we accept a priori assumption proposed by Fine [20] that the labels are mutually exclusive and exhaustive, for example, when we describe a person's height, we don't say "Aha!, this person is *tall* and *short*", an observation can and only can be described by one label among a set of basic labels. Based on the above assumption, in our proposed framework, each linguistic label $L_i \in LA$ corresponds to a subset of the observation space $S_i \subseteq \Omega$ called the *image* of label with the following properties.

Definition 1 *(Label Representation). The label representation Δ is an injective function defined on a set of labels LA to the power set of Ω, $\Delta : LA \to P(\Omega)$, thus, each label $L_i \in LA$ has a unique image $S_i \in P(\Omega)$, satisfying:*

- ***Exclusiveness.*** $S_i \cap S_j = \emptyset, \quad \forall(i,j) \in [\![1, card(LA)]\!]^2$
- ***Exhaustiveness.*** $\bigcup S_i = \Omega, \quad i \in [\![1, card(LA)]\!]$
- ***Ordering Preserving.*** $\forall(i,j) \in [\![1, card(LA)]\!]^2$, *we have either* $S_i \preccurlyeq_1 S_j$ *or* $S_j \preccurlyeq_1 S_i$.

Where $card(\cdot)$ represents the cardinality of a set and $\preccurlyeq_1$ is the order relation between label images, it can be defined as follows.

Definition 2 *(Order Relation of Label Images). Given two label images $(S_i, S_j) \in P(\Omega)^2$: we have $S_i \preccurlyeq_1 S_j$ if and only if*

$$x \preccurlyeq y, \quad \forall(x,y) \in S_i \times S_j$$

In which $\preccurlyeq$ is the order relation in the totally ordered universe Ω (e.g. the real number space $\mathbb{R}$). Consequently, given a set of labels LA being totally ordered, the order relation between labels $\preccurlyeq_2$ is defined by:

Definition 3 *(Order Relation of Labels). $\forall(L_i, L_j) \in LA^2$, given $\Delta(L_i) = S_i$ and $\Delta(L_j) = S_j$:*

- *$L_i \preccurlyeq_2 L_j$, if and only if $S_i \preccurlyeq_1 S_j$*

Furthermore, two adjacent label images (S_i, S_{i+1}) constitute an adjacent label image pair defined by:

Definition 4 *(Adjacent Label Image Pair). A pair of label images $(S_i, S_{i+1}), i \in [\![1, card(LA) - 1]\!]$ is a pair of adjacent label images, if and only if:*

- *$S_i \preccurlyeq_1 S_{i+1}$.*
- *$\forall j \in [\![1, card(LA)]\!] \backslash [\![i, i+1]\!]$, we have either $S_{i+1} \preccurlyeq_1 S_j$ or $S_j \preccurlyeq_1 S_i$.*

Intuitively, given a pair of adjacent images, there is no image exist between S_i and S_{i+1}. On step further, the adjacent label pair can be defined based on the above definition and label representation Δ in Definition 1:

Definition 5 *(Adjacent Label Pair). A pair of labels $(L_i, L_{i+1}), i \in [\![1, card(LA) - 1]\!]$ is a pair of adjacent labels, if and only if:*

- *$\Delta(L_i) = S_i, \Delta(L_{i+1}) = S_{i+1}$.*
- *(S_i, S_{i+1}) constitutes an adjacent label image pair.*

Each adjacent label image pair (S_i, S_{i+1}), $i \in [\![1, card(LA) - 1]\!]$ constitutes a generalized Dedekind cut (an exclusive, exhaustive and ordering preserving partition of a totally ordered continuous set into two non-empty parts) of the totally ordered continuous set $S_i \cup S_{i+1} \subseteq \Omega$. Hence, the boundary between labels can be defined according to the Dedekind cut theorem:

Definition 6 *(Boundary Between Labels). For each label image pair (S_i, S_{i+1}), according to the Dedekind cut theorem, there exist an unique element $b_{i,j} \in S_i \cup S_{i+1}$ named boundary between labels L_i and L_{i+1} satisfying:*

- $\forall x \in \{x \mid x \preccurlyeq b_{i,j}, \quad x \neq b_{i,j}\}, x \in S_i$.
- $\forall x \in \{x \mid b_{i,j} \preccurlyeq x, \quad x \neq b_{i,j}\}, x \in S_{i+1}$.

The boundary between labels has the following properties:

Theorem 1. $\forall (S_i, S_{i+1}) \subseteq \Omega, \quad i \in [\![1, card(LA) - 1]\!]$, *separated by a boundary between labels $b_{i,i+1} \in (S_i, S_{i+1})$ we have:*

$$b_{i,i+1} = sup(S_i) = inf(S_{i+1}) \tag{1}$$

In which $sup(S_i)$ indicates the supremum of set S_i, and $inf(S_{i+1})$ indicates the infimum of set S_{i+1}.

Proof.

- *Since $S_i \neq \emptyset$, from Definition 6, $\forall x \in S_i$, we have $x \preccurlyeq b_{i,j}$;*
- *Under the continuity of $S_i \cup S_{i+1}$, $\forall \varepsilon \mid 0_\Omega \preccurlyeq \varepsilon, \exists x' \mid b_{i,i+1} - \varepsilon \preccurlyeq x' \preccurlyeq b_{i,i+1}$, satisfies $x' \in S_i$.*

where 0_Ω is the zero element in the $\mathbb{R}$-vector space Ω, in conclusion, we have $b_{i,j} = sup(S_i)$, besides, $b_{i,j} = inf(S_{i+1})$ can be proved reciprocally.

2.2 Boundary Distribution for Uncertainty Modeling

In the theory of label differentiation, uncertainty lies in undecidable boundaries between labels across a population of agents. Intuitively, we assume that for each boundary $b_{i,i+1}$ (boundary between a pair of adjacent labels L_i and L_{i+1}), uncertainty can be represented by a probabilistic density function (PDF) $\delta_{i,i+1}$. Or formally:

Definition 7 *(Boundary Distribution). The boundary distribution for $b_{i,i+1}$ is a probability density function defined on $\delta_{i,i+1} : \Omega \to \mathbb{R}$, satisfying:*

$$\int_{I^+} \delta_{i,i+1}(\varepsilon) d\varepsilon = \int_{I^-} \delta_{i,i+1}(\varepsilon) d\varepsilon = 0.5,$$
$$I^+ = \{y \mid E(b_{i,i+1}) \preccurlyeq y\}, I^- = \{y \mid y \preccurlyeq E(b_{i,i+1})\} \tag{2}$$

where $E(b_{i,i+1})$ is the expectation of boundary $b_{i,i+1}$, defined by:

$$E(b_{i,i+1}) = \int_\Omega \varepsilon \bullet \delta_{i,i+1}(\varepsilon) d\varepsilon \tag{3}$$

Hence, given an observation $x \in \Omega$, for each adjacent label image pair (S_i, S_{i+1}), the boundary weights can be used to represent the appropriateness of each label as a description of x according to an agent's conceptual understanding, it can be calculated as the following.

Definition 8 *(Boundary Weight). Given an observation $x \in \Omega$ and an adjacent label image pair (L_i, L_{i+1}) for $(L_i, L_{i+1}) \in LA^2$. The boundary weights on x, $m_i(x) : \Omega \mapsto [0,1]$ represents the appropriateness of using label L_i to describe x:*

$$m_{i,i+1}(x,i) = \int_I \delta_{i,i+1}(\varepsilon)d\varepsilon, \quad I = \{y \mid x \preccurlyeq y\}$$

$$m_{i,i+1}(x,i+1) = 1 - \int_I \delta_{i,i+1}(\varepsilon)d\varepsilon, \quad I = \{y \mid x \preccurlyeq y\}$$

(4)

where $\preccurlyeq$ is the order relation in the totally ordered set Ω. Boundary weight $m_{i,i+1}(x,i)$ is the measure of how appropriate to use L_i to describe x given the boundary distribution $\delta_{i,i+1}$.

For example, given an universe $\Omega = \mathbb{R}$. $\forall x \in \mathbb{R}$ can be described by one of the three labels $\{L_1, L_2, L_3\}$ which are associated with three label image sets $S_1 = (-\infty, b_{1,2}), S_2 = [b_{1,2}, b_{2,3}), S_3 = [b_{2,3}, +\infty)$, suppose the boundary distribution defined on this universe is Gaussian:

$$\forall b_{i,i+1} \sim \delta_{i,i+1} = \mathcal{N}(\mu_i, \sigma_i) = \frac{1}{\sigma_i \sqrt{2\pi}} e^{-\frac{(x-\mu_i)^2}{2\sigma_i^2}}$$

(5)

Given $x = 155$, $\mu_1 = 140$, $\mu_2 = 170$, an illustration of boundary distributions between two label pairs and boundary weights on x is shown in Fig. 1. We first consider the differentiation between L_1 and L_2 (left-hand figure). There is a much bigger chance that the boundary $b_{1,2}$ (boundary between L_1 and L_2) falls into the light-shaded area marked by $m_{1,2}(x,2)$. In that case, $x = 155$ will be labeled as L_2. However, there is still a fair probability that $b_{1,2}$ falls into the dark-shaded area $m_{1,2}(x,1)$ where the label L_1 is appropriate for describing $x = 155$. The ratio of probability mass of two sides decides the appropriateness of using L_1 and L_2. Or formally, given an $x \in \Omega$, for each adjacent label pairs $(L_i, L_{i+1}) \in LA^2$, the ratios of boundary weights $\frac{m_{i,i+1}(x,i)}{m_{i,i+1}(x,i+1)}$ can be used as a measure of appropriateness. Similarly, we can obtain the appropriateness of using L_2 and L_3 (right-hand figure). By normalization, we can obtain the final appropriateness degrees as a probability distribution on the labels.

Definition 9 *(Appropriateness Measure). The appropriateness measure is a function defined on: $\mu : LA \times \Omega \mapsto [0,1]$. For $\forall(L_i, x) \in LA \times \Omega$, the appropriateness measure $\mu(L_i, x)$ represents the probability of using label L_i to describe x, which is the solution of the following equations:*

$$\begin{cases} \dfrac{\mu(L_i, x)}{\mu(L_{i+1}, x)} = \dfrac{m_{i,i+1}(x,i)}{m_{i,i+1}(x,i+1)}, & \forall i \in [\![1, card(LA) - 1]\!] \\ \displaystyle\sum_{i=1}^{card(LA)} \mu(L_i, x) = 1 \end{cases}$$

(6)

Based on the new proposed theory, given predefined boundary distribution between labels, a vague concept x can be represented by a probability distribution on labels. This is also referred to as the *label distribution* of x.

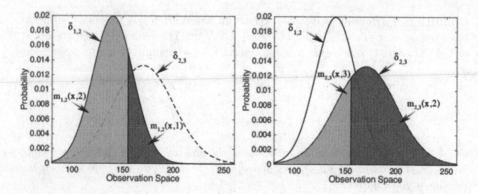

Fig. 1. Illustration of boundary distributions and boundary weights on a universe of 3 labels $\{L_1, L_2, L_3\}$. $\delta_{1,2}$ and $\delta_{2,3}$ are the PDFs of boundary distribution between L_1, L_2 (solid line) and L_2, L_3 (dotted line). The left-hand side figure shows the boundary weight ratio of L_1 and L_2. The right-hand figure shows the boundary weight ratio of L_2 and L_3.

3   Linguistic Reconstruction

Consider the case of two-agents communication where the agents share the same conceptual space (with the same boundary distributions). Given an observation $x \in \Omega$ and a set of labels LA, an agent can generate a description of x by the following way. At each round, the speaking agent samples one label based on his/her label distribution and pass it to the hearer. Such process is repeated for N times. Based on the frequency of receiving different labels, the hearing agent may approximate the true label distribution and try to reconstruct the observation x.

An observation x can be interpreted as an appropriate label by the speaker at each round in communication process, the appropriateness degrees can be approximated by the frequency of occurrences of labels, thus, the hearing agent should be able to "understand" based on his/her own conceptual space in terms of boundary distributions, and reconstruct this observation approximately. However, the boundary distributions for two agents may not be the same, it makes the reconstruction even harder. How to converge conceptual spaces of different agents will be discussed in Sect. 4. Given the same conceptual space for both speaking and hearing agent, we can represent the observation $x \in \Omega$ by a linear combination of boundary expectations defined by the following:

Definition 10 *(Boundary Expectation). For any adjacent label image pairs* (S_i, S_{i+1}), *the boundary expectation* $E_{i,i+1} : \Omega \times [\![i, i+1]\!] \mapsto \Omega$ *represents the most possible value of observation* $x \in \Omega$ *on each side of the boundary* $b_{i,i+1}$:

$$E_{i,i+1}(x, i) = 2 \bullet \int_{I^-} \varepsilon \bullet \delta(\varepsilon)d\varepsilon, \quad I^- = \{y \mid y \preccurlyeq E(b_{i,i+1})\}$$

$$E_{i,i+1}(x, i+1) = 2 \bullet \int_{I^+} \varepsilon \bullet \delta(\varepsilon)d\varepsilon, \quad I^+ = \{y \mid E(b_{i,i+1}) \preccurlyeq y\}$$

$$(7)$$

where $E(b_{i,i+1})$ is the expectation of boundary $b_{i,i+1}$ defined in Eq. (3).

Intuitively, from an agent's subjective perspective, the boundary expectation represents the meaning of each label on the basis of the given adjacent label pair. Furthermore, $\forall x \in \Omega$ can be reconstructed as a combination of boundary expectations, denoted x':

$$x' = \sum_{i=1}^{card(LA)-1} \alpha_i \bullet E_{i,i+1}(x,i) + \alpha_{i+1} \bullet E_{i,i+1}(x,i+1) \qquad (8)$$

where x' is the approximately reconstruction of $x \in \Omega$, $\bullet$ is the external operation in the $\mathbb{R}$-vector space $(\Omega, +, \bullet)$ as introduced in Sect. 2, and coefficients $\alpha_i \in \mathbb{R}, \forall i \in [\![1, card(LA) - 1]\!]$ are solutions of the following equations:

$$\begin{cases} \dfrac{\alpha_i}{\alpha_{i+1}} = \dfrac{\mu(L_i, x)}{\mu(L_{i+1}, x)}, & \forall i \in [\![1, card(LA) - 1]\!] \\ \alpha_1 + \alpha_{card(LA)} + 2 \times \displaystyle\sum_{i=2}^{card(LA)-1} \alpha_i = 1 \end{cases} \qquad (9)$$

By estimating α_i, we can reconstruct the observation from the hearer's conceptual space. Or the hearer can "understand" the speaker what does he/she mean. By this approach, by simulating human natural language, we can create a high-level *agent language* in terms of labels.

4   Experimental Study on Category Games

Firstly proposed by Steels and Belpaeme [21], the category game is a language game which models how a perceptual category (label) becomes a consensus across a population in order to achieve successful communication. For example, how do human height categories like *tall* or *short* become sufficiently shared so that after encountering a person in a scene, one agent from the population can use the label *tall* to let another agent generally understand the real height of the target person.

In a multi-agent system where category game is taken place, each agent has its own conceptual space contains the syntactic and semantic structure which determines the agent's use of labels, moreover, an agent can evolve its conceptual space by communication with others, based on this, agents can corporate to form similar conceptual spaces that also means to generate a common communication method (e.g. high level language).

Specifically, in this paper, we deal with a category game which focuses on observations in Ω as defined in Sect. 2, given set of labels LA. We consider the evolution of a multi-agent system in which agents' conceptual spaces $\mathbb{D}$ are sets of boundary distributions between adjacent labels in LA, that can be updated through interactions between agents.

4.1   Conceptual Space Evolution of Hearing Agent

In each interaction, a speaking agent describes an observation $x \in \Omega$ according to the information in its conceptual space and transfers the descriptions (labels) to the hearing agent. As introduced in Sect. 3, the transferred description enabled the hearing agent to reconstruct the observation x by x'. In order to achieve successful communication, the hearing agent has to evolve its conceptual space (defined by boundary distributions) with the objective of minimizing the communication loss function $J(x, x')$, as defined below:

Definition 11 *(Communication Loss). Communication loss is a function defined on $J : \Omega^2 \mapsto \Omega$, note 0_Ω is the zero element in the $\mathbb{R}$-vector space Ω, the target function satisfies:*

$$J(x, x) = 0_\Omega$$
$$0_\Omega \preccurlyeq J(x, y), \quad \forall (x, y) \in \Omega^2, x \neq y \tag{10}$$

where $\preccurlyeq$ is the order relation on the totally ordered universe Ω.

In practice, the gradient descent method can be applied into the evaluation of hearing agent's conceptual space. Suppose that $card(LA) = a$, and each boundary distribution function in agents' conceptual spaces $\delta_i \in \mathbb{D}$ is determined by a set of coefficients. For Gaussian distribution we can have: $\alpha = \{\alpha_{i,1}, \alpha_{i,2}, ..., \alpha_{i,n}\}$ for means and $\beta = \{\beta_{i,1}, \beta_{i,2}, ..., \beta_{i,n}\}$ for standard derivations, the hearing agent's conceptual space can be evolved iteratively as shown in Table 1.

For example, considering a two-agents system. Given an observation space $\Omega = \mathbb{R}$ which can be described by three labels $LA = \{L_1, L_2, L_3\}$. Suppose that for the speaking agent, three labels are defined by two Gaussian boundary distributions:

$$\delta_{1,2} = \mathcal{N}(\mu_1 = 150, \sigma_1 = 8)$$

Table 1. Pseudo-code of gradient descent in conceptual space evolution.

Given observation $x \in \Omega$, set of appropriateness measures $\mathbb{A} = \{\mu_1(x), ..., \mu_a(x)\}$. A set of randomly initialized boundary distributions $\mathbb{D} = \{\delta_1, ..., \delta_{a-1}\}$ with parameters α and β. A threshold $0_\Omega \preccurlyeq \varepsilon$, and a learning rate $\lambda \in \mathbb{R}$

Step 1. Reconstruct observation x as $x' \in \Omega$ according to $\mathbb{A}$ and $\mathbb{D}$

Step 2. While $\varepsilon \preccurlyeq J(x, x')$:
For each distribution $\delta_i \in \mathbb{D}$, evolve its coefficients α and β :
For each coefficient $\alpha_{i,j}$ and $\beta_{i,j}$:
$\alpha_{i,j} = \alpha_{i,j} - \lambda \bullet \frac{\partial}{\partial \alpha_{i,j}} J(x, x')$
$\beta_{i,j} = \beta_{i,j} - \lambda \bullet \frac{\partial}{\partial \alpha_{i,j}} J(x, x')$
Reconstruct observation x as $x' \in \Omega$ according to $\mathbb{A}$ and $\mathbb{D}$

$$\delta_{2,3} = \mathcal{N}(\mu_2 = 170, \sigma_2 = 9)$$

in contrast, the two boundary distributions of hearing agent are randomly generated, the communication loss function J is defined in Eq. (11).

$$J = \frac{1}{2}(x - x')^2, \quad (x, x') \in [120, 200]^2 \tag{11}$$

4.2 Experimental Results

Considering the experimental results of 200 interactions between in the two-agents system, during each interaction, an observation x is randomly selected from interval $[120, 200]$ and the hearing agent evolves its conceptual space based on the result of the last interaction, the pseudo-code is shown in Table 1. The results of the hearing agent's conceptual space evolution during 200 interactions is shown in Figs. 2 and 3.

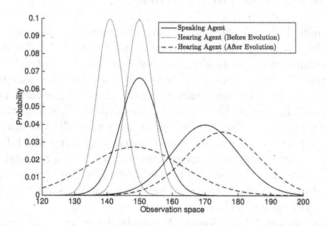

Fig. 2. Evolution of boundary distributions of the hearing agent in 200 interactions.

As we can see from Fig. 3, with the increase of the times of interactions, boundary distributions in hearing agent's conceptual space becomes similar to the speaking agent's, specifically, the former's coefficients (e.g. mean, standard deviation) are converged more closely to the latter's. While, due to the impact of linguistic uncertainties, the conceptual spaces of two agents can not become exactly the same, but to a certain extent, they can evolve to be sufficiently similar if we have more labels, in order to achieve effective communications.

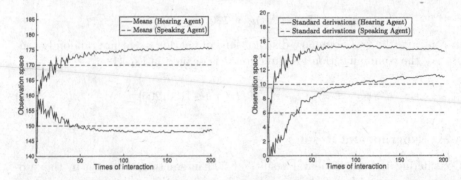

Fig. 3. Convergence of means and standard deviations of boundary distributions in 200 interactions. The means and stds are unchanged for speaking agents in the evolution (or learning process) of hearing agents.

5   Conclusions and Future Works

Uncertainty modeling and reasoning is an important topic in general artificial intelligence. In this research, we proposed a theory of label differentiation for modeling semantic uncertainties where linguistic labels are used to describe vague concepts. We argue that the label itself does not have semantic uncertainty but the agents who have uncertainty to decide the boundary between labels. The boundaries can be regarded as random variables in the observation space. A probability distribution can be used to model the distribution of the boundaries between labels. The appropriateness of using which label as the description of a vague concept is uniquely determined by the boundary distribution. We developed the calculus to obtain label distribution to represent a vague concept based on probability theory. The new model illustrates its capability and effectiveness in the category game.

Unlike other uncertainty reasoning theories based on multivariate logic, belief function or fuzzy membership function, the label differentiation theory is totally based on probability theory but gives a high-level knowledge representation. The future work can be done in the following aspects: (1) we have demonstrated how the theory can be used in the vague category game, by evolving the conceptual spaces of speaking-hearing agent pairs, to make them communicate through labels. How to develop high-level *Agent Language* still needs more work to realize more effective communication in a complex world. (2) Based on this framework, to a large extent, we can try to develop a cognitive model of agent's conceptual space, in this space, agent can learn using probabilistic or neural approaches.

Acknowledgements. This work is partially supported by the Natural Science Foundation of China under grant Nos. 61305047 and 61401012.

References

1. Carruthers, P.: Language, Thought and Consciousness: An Essay in Philosophical Psychology. Cambridge University Press, Cambridge (1996)
2. Lawry, J., Tang, Y.: Uncertainty modelling for vague concepts: a prototype theory approach. Artif. Intell. **173**(18), 1539–1558 (2009)
3. Qin, Z., Tang, Y.: Uncertainty Modeling for Data Mining: A Label Semantics Approach. Springer, Heidelberg (2014)
4. Jaynes, E.T.: Probability Theory: The Logic of Science. Cambridge University Press, Cambridge (2003)
5. de Finetti, B.: Sul significanto soggettivo della probabilita. Fundam. Math. **17**, 298–329 (1931)
6. Lawry, J., Tang, Y.: Probability, fuzziness and borderline cases. Int. J. Approximate Reasoning **55**, 1164–1184 (2014)
7. Zadeh, L.A.: Fuzzy sets. Inf. Control **8**, 338–353 (1965)
8. Baldwin, J.F., Martin, T.P., Pilsworth, B.W.: Fril-Fuzzy and Evidential Reasoning in Artificial Intelligence. Wiley, New York (1995)
9. Klir, G., Yuan, B.: Fuzzy Sets and Fuzzy Logic. Prentice Hall, Upper Saddle River (1995)
10. Zadeh, L.A.: Fuzzy logic = computing with words. IEEE Trans. Fuzzy Syst. **4**(2), 103–111 (1996)
11. Zadeh, L.A.: Toward a generalized theory of uncertainty. Inf. Sci. **172**(1–2), 1–40 (2005)
12. Zadeh, L.A.: Computing with Words: Principal Concepts and Ideas. Springer, Heidelberg (2012)
13. Thint, M., Beg, S., Qin, Z.: PNL-enhanced restricted domain question answering system. In: Proceedings of IEEE-FUZZ (2007)
14. Lawry, J.: A framework for linguistic modelling. Artif. Intell. **155**, 1–39 (2004)
15. Zhao, H., Qin, Z.: Clustering data and vague concepts using prototype theory interpreted label semantics. In: Huynh, V.-N., Inuiguchi, M., Denoeux, T. (eds.) IUKM 2015. LNCS (LNAI), vol. 9376, pp. 236–246. Springer, Heidelberg (2015). doi:10.1007/978-3-319-25135-6_23
16. Lawry, J.: Appropriateness measures: an uncertainty model for vague concepts. Synthese **161**, 255–269 (2008)
17. Lawry, J.: Modeling and Reasoning with Vague Concepts. Springer, New York (2006)
18. Williamson, T.: Vagueness. Routledge, London (1994)
19. Bohm, D., Park, D.: Wholeness and the implicate order. Am. J. Phys. **49**, 796–797 (1981)
20. Fine, K.: Vagueness, truth and logic. Synthese **30**(3), 265–300 (1975)
21. Steels, L., Belpaeme, T.: Coordinating perceptually grounded categories through language: a case study for colour. Behav. Brain Sci. **28**(4), 469–488 (2005)

A Probability Based Approach to Evaluation of New Energy Alternatives

Hong-Bin Yan[(✉)]

School of Business, East China University of Science and Technology,
Meilong Road 130, Shanghai 200237, People's Republic of China
hbyan@ecust.edu.cn

Abstract. Exploitation of new and innovative energy alternatives is a key means toward a sustainable energy system. This paper proposes a novel model for linguistic energy planning with an application to public transportation. In particular, a probability based approach is firstly introduced to perform group aggregation based on a random interpretation of an evaluation-team's judgments. Secondly, a weighted approach is proposed to perform multi-criteria aggregation based on the concept of stochastic dominance degree, which results with a matrix of overall stochastic dominance degrees of different alternatives. Thirdly, a choice function based on the idea of the PROMETHEE-II method is developed to rank and select the best/desired alternative(s), which can divide the alternatives into two classes: positive (acceptable) and negative (unacceptable) ones. Our model is used to re-evaluate the technological development of public transportation with 12 alternative fuel modes.

Keywords: Linguistic energy planning · Group aggregation · Multi-criteria aggregation · PROMETHEE-II · Fuel modes

1 Introduction

The exploitation of sustainable or new energy alternatives, referred to as *energy planning* [14], has become a key means of satisfying these objectives and has gained a great interest over the past two decades. The energy planning endeavor involves finding a set of energy alternatives so as to meet the energy requirements/demands of all the tasks in an optimal manner [6]. An energy planning decision involves a process of balancing diverse *environmental*, *social*, *technical*, and *economic* aspects over space and time, and such a balance is critical to the survival of nature and to the prosperity of energy dependent nations [8]. The diverse aspects are usually represented as multi-criteria that are often expressed by conflicting objectives, which makes multi-criteria decision making (MCDM) a valuable tool to provide the flexibility and capacity to assess the technologies' implications to economic, environmental and social framework and to the design of energy and environmental policies. Some of the most common MCDM methods used for assisting energy policy and planning towards sustainability are the

© Springer International Publishing AG 2016
V.-N. Huynh et al. (Eds.): IUKM 2016, LNAI 9978, pp. 76–88, 2016.
DOI: 10.1007/978-3-319-49046-5_7

Multi-Attribute Utility, the ELECTRE [13], the PROMETHEE, the Analytical Hierarchy Process (AHP) [12], and the TOPSIS [7].

The energy planning is a quite vague, uncertain, and complex process [3,11, 15]. Therefore, the crisp multi-criteria approaches to energy planning encapsulate or merely discard the uncertainty, vagueness and multiplicity of possible concerns, making the energy planning decision sound rationally but dysfunctional. Fuzzy set theory [20] offers a formal mathematical framework for energy planning in order to solve the vagueness, ambiguity and subjectivity of human judgment. A realistic approach to uncertain energy planning is the use of fuzzy linguistic approach [21], which deals with linguistic information in forms of linguistic labels represented by fuzzy sets. Consequently, a lot of fuzzy linguistic models have been proposed/applied to energy planning (referred to as *fuzzy linguistic energy planning*), such as fuzzy linguistic AHP based model [8] and fuzzy TOPSIS based model [2,17].

Despite their great success in dealing with both fuzzy uncertainty and complexity in energy planning, the fuzzy set based computational models simultaneously have an unavoidable limitation of the information loss, which consequently implies a lack of precision in the final result. In addition, the fuzzy set based semantics of each linguistic label is often defined subjectively and context-dependently, which can sensitively influence the final planning results. Finally, existing research only focuses on the ranking of alternatives, which is not enough for the decision support to decision-makers. A ranking with a classification may provide a more sufficient and transparent decision support to decision-makers.

Toward this end, the main focus of this paper is to propose an alternative linguistic model for energy planning with computations solely based on linguistic labels as well as providing a better decision support to the decision-makers. In particular, a probabilistic approach is firstly introduced to perform group aggregation based on a random interpretation of an evaluation-team' judgments. Secondly, a probabilistic approach is proposed to perform multi-criteria aggregation based on the concept of stochastic dominance degree, which results with the a matrix of overall stochastic dominance degrees of alternatives. Thirdly, a choice function based on the idea of the PROMETHEE-II method is developed to rank and select the desired alternative(s), which divides the alternatives into two classes: positive (acceptable) and negative (unacceptable) ones. As a case study, our model is used to evaluate public transportation with 12 alternative fuel modes.

The outline of this paper is as follows. Section 2 presents some basic knowledge of energy planning and formulates the research problem. Section 3 then proposes a probabilistic model for energy planning. As a case study, Sect. 4 evaluates the technological development of public transportation with 12 alternative fuel modes. Finally, some concluding remarks are given in Sect. 5.

2   Problem Formulation

The energy alternative has a broader meaning, it can be energy resources. Most popular energy supply resources are the alternatives based on solar energy (photovoltaic and thermal), wind energy, hydraulic energy, biomass, animal manure, combined heat and power, and ocean energy [8]. Despite their environmental drawbacks, nuclear and conventional energy alternatives like coal, oil and natural gas may be still included in the list of energy alternatives to be promoted. On the other hand, the energy alternative can also be fuel modes, e.g., compressed natural gas, liquid propane gas [16]. Even for the same energy planning problem, different energy alternatives may be used in different contexts [3,16,17]. Formally, let $\mathcal{A} = \{A_1, A_2, \ldots, A_M\}$ be a set of energy alternatives.

Due to the complexity of evaluating various energy alternatives, it is infeasible to compare the energy alternatives based on only a single aspect or a few criteria. The decision making process to determine the best energy policy is usually multi-dimensional, made up of a number of aspects such as economic, technical, environmental, political, and social ones. Note that the decision criteria in the process of energy planning are also dependent on the formulation of specific problem and particularly on the country's specific energy characteristics, its development needs and perspectives, and the energy actors engaged in the decision making process. Formally, let $\mathcal{C} = \{C_1, C_2, \ldots, C_N\}$ denote the planning decision criteria.

The inherent complexity and uncertainty necessitate the participation of many stakeholders (also called experts) in the process of energy planning. A group of experts is denoted by $\mathcal{E} = \{E_1, E_2, \ldots, E_K\}$, referred to as evaluation-team. Information will be gathered from the evaluation-team, processed and maintained, and presented to the decision-maker. It is a critical challenge for decision-makers to construct a high-quality evaluation-team for energy planning. Due to the uncertainty and vagueness in energy planning, the following two linguistic variables are used by the evaluation-team to provide their linguistic judgments.

– A linguistic variable $\mathcal{S}^{\mathrm{I}} = \{S_0^{\mathrm{I}}, S_1^{\mathrm{I}}, \ldots, S_G^{\mathrm{I}}\}$ is used to rate the energy alternatives with respect to different decision criteria $\mathcal{C}$.
– A linguistic variable for assessing the importance of different criteria is defined by $\mathcal{S}^{\mathrm{II}} = \{S_0^{\mathrm{II}}, S_1^{\mathrm{II}}, \ldots, S_G^{\mathrm{II}}\}$.

The evaluation-team is then asked to rate each energy alternative A_m regarding the selected criteria $\mathcal{C}$ making use of the linguistic variable $\mathcal{S}^{\mathrm{I}}$. The committee of experts would be also asked to provide their opinions on the absolute importance of the different criteria $\mathcal{C}$ using linguistic variable $\mathcal{S}^{\mathrm{II}}$. The linguistic data obtained by this way can be described as shown in Table 1, where

– each x_{mn}^k is used to denote the rating of energy alternative A_m on criterion C_n provided by expert E_k, where $m = 1, 2, \ldots, M, n = 1, 2, \ldots, N, k = 1, 2, \ldots, K$, and $x_{mn}^k \in \mathcal{S}^{\mathrm{I}}$;
– each y_n^k is used to denote the absolute importance of criterion C_n provided by expert E_k, where $n = 1, 2, \ldots, N, k = 1, 2, \ldots, K$, and $y_n^k \in \mathcal{S}^{\mathrm{II}}$.

Table 1. Linguistic assessments of energy alternative A_m on a set of criteria $\mathcal{C}$

Evaluation-team	Decision criteria			
	C_1	C_2	$\cdots$	C_N
E_1	$\left(x_{m1}^1, y_1^1\right)$	$\left(x_{m2}^1, y_2^1\right)$	$\cdots$	$\left(x_{mN}^1, y_N^1\right)$
E_2	$\left(x_{m1}^2, y_1^2\right)$	$\left(x_{m2}^2, y_2^2\right)$	$\cdots$	$\left(x_{mN}^2, y_N^2\right)$
$\cdots$	$\cdots$	$\cdots$	$\ddots$	$\cdots$
E_K	$\left(x_{m1}^K, y_1^K\right)$	$\left(x_{m2}^K, y_2^K\right)$	$\cdots$	$\left(x_{mN}^K, y_N^K\right)$

3 A Probabilistic Approach to Linguistic Energy Planning

3.1 A Probability Interpretation of the Evaluation-Team's Judgments

Note that a group of experts $\mathcal{E}$ (evaluation-team) is involved in our context of energy planning. We assume a subjective probability distribution $p_{\mathcal{E}}$ defined over the set of experts $\mathcal{E}$, which essentially underlies the calculating basis for the following proposed group aggregation function.

From a practical point of view, given a criterion, if there is an ideal expert, say E_I, whose judgment on the criterion (either the performance or importance) the decision-maker completely believes in, then it is enough for the decision-maker to use E_I's assessment as the performance or importance of the criterion. However, this is not generally the case in practice. A pool of multi-disciplinary and multi-functional experts is therefore called to express their judgments regarding the criterion, on one hand, to collect enough information for the energy planning problem from various points of view; and on the other hand, to reduce the subjectivity of the energy planning problem. In this regards, $p_{\mathcal{E}}(E_k)$, for each $k = 1, 2, \ldots, K$, may be interpreted as the probability that the decision-maker randomly selects expert E_k from the expert set $\mathcal{E}$ as a sufficient source of information for the purpose of energy planning. Such a probability distribution $p_{\mathcal{E}}$ may come from the decision-maker's knowledge about the experts. In addition, in the tradition of group decision analysis, a weighting vector $\omega = (\omega_1, \omega_2, \ldots, \omega_K)$ is also often associated with the set of experts $\mathcal{E}$ such that $\omega_k \in [0, 1]$ and $\sum_k \omega_k = 1$. In this sense, the set of experts plays the role of states of the world and the weighting vector ω serves as the subjective probabilities assigned to the states such that $p_{\mathcal{E}}(E_k) = \omega_k, k = 1, 2, \ldots, K$.

Returning back to Table 1, the set of experts $\mathcal{E}$ is chosen to rate energy alternative A_m using the linguistic sets $\mathcal{S}^I$ with respect to a set of criteria $\mathcal{C}$. When an expert E_k provides his/her judgment for alternative A_m on criterion C_n, a probability distribution of his/her opinion over the linguistic variable $\mathcal{S}^I$ can be elicited as

$$p_{\mathcal{S}^I}\left(S_g^I | A_m, C_n, E_k\right) = \begin{cases} 1, & \text{if } x_{mn}^k = S_g^I; \\ 0, & \text{otherwise.} \end{cases} \tag{1}$$

where $m = 1, \ldots, M, n = 1, \ldots, N, k = 1, \ldots, K, g = 0, \ldots, G$. The value $p_{\mathcal{S}^{\mathrm{I}}}\left(S_g^{\mathrm{I}}|A_m, C_n, E_k\right)$ can be seen as a prior probability that the decision-maker believes in that the linguistic label S_g^{I} is appropriate enough to describe the performance of energy alternative A_m on the criterion C_n, given the judgment of expert E_k.

In addition, the set of experts $\mathcal{E}$ is also chosen to provide their absolute linguistic judgments regarding the importance of the criteria set $\mathcal{C}$. Similar to the formulation of linguistic performance, a probability distribution of expert E_k's judgment with respect to the importance of criterion C_n over linguistic variable $\mathcal{S}^{\mathrm{II}}$ can be elicited as

$$p_{\mathcal{S}^{\mathrm{II}}}\left(S_g^{\mathrm{II}}|C_n, E_k\right) = \begin{cases} 1, \text{ if } y_n^k = L_g^{\mathrm{II}}; \\ 0, \text{ otherwise.} \end{cases} \tag{2}$$

where $n = 1, \ldots, N, k = 1, \ldots, K, g = 0, \ldots, G$.

3.2   Group Aggregation

Taking the subjective probability distribution $p_{\mathcal{E}}(E_k)$ into consideration, a posterior probability (collective probability) distribution of alternative A_m on criterion C_n over linguistic variable $\mathcal{S}^{\mathrm{I}}$ can be obtained as

$$p_{\mathcal{S}^{\mathrm{I}}}\left(S_g^{\mathrm{I}}|A_m, C_n\right) = \sum_{k=1}^{K} \omega_k \cdot p_{\mathcal{S}^{\mathrm{I}}}\left(S_g^{\mathrm{I}}|A_m, C_n, E_k\right) \tag{3}$$

where $m = 1, \ldots, M, n = 1, \ldots, N, g = 0, \ldots, G$. The derived probability distribution $X_{mn} = \left[p_{\mathcal{S}^{\mathrm{I}}}\left(S_0^{\mathrm{I}}|A_m, C_n\right), p_{\mathcal{S}^{\mathrm{I}}}\left(S_1^{\mathrm{I}}|A_m, C_n\right), \ldots, p_{\mathcal{S}^{\mathrm{I}}}\left(S_G^{\mathrm{I}}|A_m, C_n\right)\right]$ will be refereed to as *linguistic performance profile* of the energy alternative A_m on criterion C_n.

Similarly, a posterior probability distribution of criterion C_n over linguistic variable $\mathcal{S}^{\mathrm{II}}$ can be obtained by

$$p_{\mathcal{S}^{\mathrm{II}}}\left(S_g^{\mathrm{II}}|C_n\right) = \sum_{k=1}^{K} \omega_k \cdot p_{\mathcal{S}^{\mathrm{II}}}\left(S_g^{\mathrm{II}}|C_n, E_k\right) \tag{4}$$

where $g = 0, \ldots, G, n = 1, \ldots, N$. The derived probability distribution $Y_n = \left[p_{\mathcal{S}^{\mathrm{II}}}\left(S_0^{\mathrm{II}}|C_n\right), p_{\mathcal{S}^{\mathrm{II}}}\left(S_1^{\mathrm{II}}|C_n\right), \ldots, p_{\mathcal{S}^{\mathrm{II}}}\left(S_G^{\mathrm{II}}|C_n\right)\right]$ will be refereed to as *linguistic importance profile* of criterion C_n.

Remark 1. Essentially, Eqs. (1)–(4) are based on the assumption that there is mutual independence among the experts and among the criteria. As pointed out by [1], in any linguistic decision analysis, the procedure of asking each expert to provide his/her absolute linguistic evaluations for a set of alternatives is based on the mutual independence among the set of alternatives. As mentioned in [5], in fuzzy linguistic multi-criteria energy planning, the pool of experts is called to provide their opinions independently, therefore mutual independence among the experts is assumed naturally. In addition, Cruz [4] has pointed out that the set of experts provides their opinions on each criterion separately, therefore the criteria set $\mathcal{C}$ is considered to be mutually independent.

3.3   Multi-criteria Aggregation

Now we shall propose our procedure of multi-criteria aggregation based on the stochastic dominance introduced in our previous work [19].

Definition 1. *Given two linguistic importance profiles Y_n and Y_l expressed as probability distributions over a linguistic term set S^{II}, the stochastic dominance degree of Y_n over Y_l is defined as $D_{nl} = \Pr(Y_n \geq Y_l) - 0.5\Pr(Y_n = Y_l)$. where*

- *if $D_{nl} > 0.5$, then it indicates that Y_n is preferred to Y_l such that $Y_n > Y_l$;*
- *if $D_{nl} = 0.5$, then there is indifference between Y_n and Y_l such that $Y_n = Y_l$;*
- *if $D_{nl} < 0.5$, then it indicates that Y_n is preferred to Y_l such that $Y_n < Y_l$.*

Consequently, a matrix of stochastic dominance degrees can be derived from the linguistic importance profiles with respect to the set of criteria $\mathcal{C}$ as

$$
\mathbf{D} = \begin{array}{c} \\ Y_1 \\ Y_2 \\ \vdots \\ Y_N \end{array}
\begin{array}{cccc}
Y_1 & Y_2 & \dots & Y_N \\
0.5 & D_{12} & \dots & D_{1N} \\
D_{21} & 0.5 & \dots & D_{2N} \\
\vdots & \vdots & \ddots & \vdots \\
D_{N1} & D_{N2} & \dots & 0.5
\end{array},
\tag{5}
$$

where D_{ml} denotes the stochastic dominance degree of C_n over C_l.

A weighting vector $\mathbf{W} = (W_1, W_2, \dots, W_N)$ of the criteria set $\mathcal{C}$ can then be induced from the matrix of stochastic dominance degrees $\mathbf{D}$ in Eq. (5). In fact, a consistent weighting vector $\mathbf{W}$ of the criteria set $\mathcal{C}$ verifies the following condition: $W_n - W_l = D_{nl} - 0.5$, $\forall n, l = 1, \dots, N$. Even if the matrix $\mathbf{D}$ in Eq. (5) is not additive transitive, i.e., not perfectly consistent, we can also derive a weighting vector $\mathbf{W}$ of the criteria set as follows [18]:

$$
W_n = \frac{1}{N} \sum_{l=1}^{N} \frac{D_{nl}}{\sum_{m=1}^{N} D_{ml}}
\tag{6}
$$

Remark 2. The multiplicative preference relations play a basic role in the AHP [12]. If $\mathbf{D}$ is a matrix of multiplicative preference relations of a criteria set $\mathcal{C}$, the multiplicative preference relation D_{nl} is a preference ratio such that $D_{nl} = 1/D_{ln}$, $\forall n, l = 1, \dots, N$. A consistent weighting vector $\mathbf{W} = (W_1, W_2, \dots, W_N)$ of the criteria set $\mathcal{C}$ associated with the matrix of multiplicative preference relations verifies the following condition: $D_{nl} = W_n/W_l$. One commonly used method to derive the weighting vector from a multiplicative reciprocal preference matrix is the normalized principal eigenvector [12], refer to Eq. (6). When the preference matrix $\mathbf{D}$ is a matrix of fuzzy reciprocal preference relations of the criteria set $\mathcal{C}$, the normalized principal eigenvector can also be used to derive a weighting vector [18].

Similar to the process of deriving weighting vector of the criteria set $\mathcal{C}$, a matrix of stochastic dominance degrees $\mathbf{R}^n$ of alternative set $\mathcal{A}$ regarding criterion C_n can be obtained by

$$
\mathbf{R}^n = \begin{array}{c} \\ X_{1n} \\ X_{2n} \\ \vdots \\ X_{Mn} \end{array} \begin{array}{cccc} X_{1n} & X_{2n} & \dots & X_{Mn} \\ 0.5 & R_{12}^n & \dots & R_{1M}^n \\ R_{21}^n & 0.5 & \dots & R_{2M}^n \\ \vdots & \vdots & \ddots & \vdots \\ R_{M1}^n & R_{M2}^n & \dots & 0.5 \end{array} , \tag{7}
$$

where R_{ml}^n denotes the stochastic dominance degree (fuzzy preference relation) of alternative A_m over energy alternative A_l regarding criterion C_n.

Taking the weighting vector $\mathbf{W}$ into consideration, a matrix $\mathbf{R}$ of overall stochastic dominance degrees of the alternative set $\mathcal{A}$ is constructed as

$$
\mathbf{R} = \begin{array}{c} \\ A_1 \\ A_2 \\ \vdots \\ A_M \end{array} \begin{array}{cccc} A_1 & A_2 & \dots & A_M \\ 0.5 & R_{12} & \dots & R_{1M} \\ R_{21} & 0.5 & \dots & R_{2M} \\ \vdots & \vdots & \ddots & \vdots \\ R_{M1} & R_{M2} & \dots & 0.5 \end{array} \tag{8}
$$

where R_{ml} denotes the overall dominance degree of alternative A_m over alternative A_l and is calculated by

$$
R_{ml} = \sum_{n=1}^{N} W_n \cdot R_{ml}^n, \forall m, l = 1, \dots, M. \tag{9}
$$

It is easy to derive that $R_{ml} + R_{lm} = 1, R_{mm} = 0.5$, and $0 \leq R_{ml} \leq 1$, for all $m, l = 1, \cdots, M$.

3.4 Choice Function

With the matrix of overall stochastic dominance degrees of different alternatives obtained, we will develop a choice approach based on the idea of the PROMETHEE-II method as follows.

- Let Φ_m^+ be the dominant degree which is a measure that alternative A_m is dominating the other alternatives, referred to as *leaving flow* in the terminology of decision making.
- Let Φ_m^- be the non-dominant degree which is a measure that alternative A_m is dominated by the other alternatives, referred to as *entering flow* in the terminology of decision making.

Here, Φ_m^+ and Φ_m^- can be defined by the following formulas, respectively:

$$
\Phi_m^+ = \frac{1}{M-1} \sum_{l=1, l \neq m}^{M} R_{ml}, \quad \Phi_m^- = \frac{1}{M-1} \sum_{l=1, l \neq m}^{M} R_{lm} \tag{10}
$$

Moreover, let Φ_m be the relative dominant degree which measures the difference between dominant degree and non-dominant degree of alternative A_m such that

$$\Phi_m = \Phi_m^+ - \Phi_m^-, m = 1, \ldots, M. \tag{11}$$

Φ_m is referred to as *net flow* in the terminology of decision making. The greater Φ_m is, the better alternative A_m will be. When Φ_m is a positive value, A_m is acceptable; when Φ_m is a negative value, A_m is unacceptable. Therefore, according to net flows $\Phi_1, \Phi_2, \ldots, \Phi_M$, we can not only determine a ranking order of all the alternatives or select the most desirable alternative(s), but also divide the alternatives into acceptable and unacceptable classes.

4 Evaluations of Fuel Modes for Public Transportation

4.1 Problem Descriptions: Alternative Fuel Buses Selection

In order to reduce the negative impacts of regulated pollutants and considerably enhance the air quality in major cities, governments have enacted strict legislation on internal combustion engine emissions and fuel quality. During the development period, alternative fuel vehicles are also considered. The advantages of hybrid electric vehicles (HEVs) are regeneration of braking energy, engine shutdown instead of idling, and engine driving under high-load conditions; these advantages are more noticeable in city driving. On the other hand, the key weakness of electric vehicles (EVs) is that time is needed to recharge the batteries. Since the bus system possesses such features as stable depot, route, group of commuter, time of operation, and frequency, it is of high interest to find alternative fuel modes for public transportation. Hence, the purpose of this part is to evaluate the best alternative fuel buses suitable for the urban area, which will lead us to the direction of development in the future.

Morita [10] has thought that the leading types of automobiles in the 21st century are likely to be of the following four types: internal combustion engine vehicles (ICEVs), HEVs, EVs, and fuel cell vehicles (FCVs). McNicol et al. [9] have pointed out that the principal competitors of FCVs are EVs, HEVs, and advanced conventional ICE-powered vehicles (ICEVs). Based on the literature mentioned above, Vahdani et al. [17] have studied the problem of alternative fuel buses selection in Iran, which will be re-considered in this study. In their research, the following 12 energy technologies (fuel modes) are considered. A_1: Conventional diesel engine, A_2: Compressed natural gas (CNG), A_3: Liquid propane gas (LPG), A_4: Fuel cell (hydrogen), A_5: Methanol, A_6: Electric vehicle-opportunity charging, A_7: Direct electric charging, A_8: Electric bus with exchangeable batteries, A_9: Hybrid electric bus with gasoline engine, A_{10}: Hybrid electric bus with diesel engine, A_{11}: Hybrid electric bus with CNG engine, and A_{12}: Hybrid electric bus with LPG engine.

The evaluation of alternative fuel modes are performed according to four aspects: social, economic, technological, and transportation aspects; and divided into 11 sub-criteria as follows. C_1: Energy supply, C_2: Energy efficiency, C_3: Air

Fuel	Decision criteria										
modes	C_1	C_2	C_3	C_4	C_5	C_6	C_7	C_8	C_9	C_{10}	C_{11}
A_1	VG,VG,VG	G,MG,MG	MP,F,F	MG,MG,MG	G,G,G	F,F,F	MG,F,F	VG,VG,VG	VG,VG,VG	VG,VG,VG	MG,MG,G
A_2	VG,VG,G	VG,VG,G	VG,G,G	G,G,MG	G,G,G	F,F,MG	MG,MG,F	G,G,VG	VG,VG,G	VG,G,G	VG,G,G
A_3	VG,VG,VG	VG,G,VG	G,VG,G	G,G,G	G,G,G	MG,F,F	MG,F,MG	VG,G,G	VG,VG,G	G,VG,G	G,G,G
A_4	MG,MG,G	VG,G,G	VG,VG,VG	G,MG,MG	MG,G,G	MG,MG,MG	G,VG,VG	MG,MG,G	G,G,G	MG,MG,G	G,VG,G
A_5	G,G,MG	MG,G,MG	VG,G,G	G,G,G	G,MG,G	MG,MG,F	G,G,G	MG,MG,MG	G,G,G	G,G,MG	G,G,VG
A_6	G,G,VG	VG,VG,G	VG,VG,VG	G,G,G	VG,G,G	VG,VG,VG	VG,VG,G	MG,MG,F	F,F,MG	G,G,G	G,G,VG
A_7	VG,G,G	VG,VG,G	VG,VG,VG	G,G,G	G,G,G	VG,VG,VG	G,VG,VG	F,F,MG	MG,MG,F	VG,VG,VG	VG,G,G
A_8	VG,VG,G	G,VG,VG	VG,VG,VG	MG,G,G	G,G,G	VG,VG,VG	G,G,VG	MG,MG,F	MG,MG,MG	VG,VG,G	G,VG,G
A_9	G,G,VG	G,G,G	MG,MG,G	MG,MG,G	MG,G,G	G,G,G	G,G,G	G,G,VG	G,G,G	G,G,VG	G,G,VG
A_{10}	VG,VG,G	G,G,MG	MG,F,MG	MG,MG,MG	G,G,MG	G,VG,G	VG,G,G	G,G,G	G,VG,G	VG,G,G	VG,G,G
A_{11}	VG,G,VG	VG,G,G	VG,VG,VG	MG,G,G	MG,MG,G	G,G,VG	G,G,VG	VG,G,G	VG,G,G	G,G,G	VG,VG,G
A_{12}	G,G,VG	VG,G,VG	VG,VG,VG	MG,MG,G	G,G,MG	G,G,G	G,G,G	G,VG,G	G,G,G	G,G,G	G,G,VG

Fig. 1. Linguistic assessments of 12 fuel modes with respect to 11 decision criteria

Table 2. Linguistic assessments of importance with respect to 11 decision criteria

Experts $\mathcal{E}$	Decision criteria $\mathcal{C}$										
	C_1	C_2	C_3	C_4	C_5	C_6	C_7	C_8	C_9	C_{10}	C_{11}
E_1	H	VH	L	L	H	L	L	VH	H	VH	VH
E_2	H	VH	L	L	H	L	L	VH	H	VH	H
E_3	H	H	L	L	VH	L	L	VH	VH	VH	H

pollution, C_4: Noise pollution, C_5: Industrial relationship, C_6: Employment cost, C_7: Maintenance cost, C_8: Vehicle capability, C_9: Road facility, C_{10}: Speed of traffic flow, C_{11}: Sense of comfort. For more detailed explanation of the choice of decision criteria, refer to [16,17]. A group of three experts $\mathcal{E} = \{E_1, E_2, E_3\}$ from the electric bus manufacturing, academic institutes, and bus operations sectors with a weighting vector $\omega = (1/3, 1/3, 1/3)$ is involved in this problem. Two linguistic variables are used to represent the uncertain judgments provided by the three experts Eq. (12).

$$\mathcal{S}^I = \{\text{Very poor (VP), Poor (P), Medium poor (MP), Fair (F),}$$
$$\text{Medium good (MG), Good (G), Very good (VG)}\}$$
$$\mathcal{S}^{II} = \{\text{Very low (VL), Low (L), Medium low (ML), Medium (M),}$$
$$\text{Medium high (MH), High (H), Very high (VH)}\}$$

(12)

where $\mathcal{S}^I$ is used to represent the uncertain performance values with respect to different fuel modes on different decision criteria and $\mathcal{S}^{II}$ is used to represent the uncertain importance with respect to different decision criteria. The linguistic assessments are given in Fig. 1.

4.2 Evaluation of Fuel Mode Applications

Step 1: Probabilistic group aggregation. Due to the random interpretation of experts' judgments, we can elicit the probability distributions p_{S^I} of experts' linguistic judgments for the performance with respect to different criteria. Similarly, the probability distributions $p_{S^{II}}$ of experts' linguistic judgments for the criteria importance can also be elicited. With the weighting vector $\omega = (1/3, 1/3, 1/3)$ associated with the pool of three experts, we can derive the linguistic performance and importance profiles (collective probability distributions).

Step 2: Multi-criteria aggregation. We first calculate the matrix of stochastic dominance degrees for the linguistic importance profiles with respect to the 11 decision criteria as follows.

$$D = \begin{array}{c|ccccccccccc}
 & C_1 & C_2 & C_3 & C_4 & C_5 & C_6 & C_7 & C_8 & C_9 & C_{10} & C_{11} \\
\hline
C_1 & 0.50 & 0.167 & 1.0 & 1.0 & 0.333 & 1.0 & 1.0 & 0.0 & 0.333 & 0.0 & 0.333 \\
C_2 & 0.833 & 0.5 & 1.0 & 1.0 & 0.667 & 1.0 & 1.0 & 0.333 & 0.667 & 0.333 & 0.667 \\
C_3 & 0.0 & 0.0 & 0.5 & 0.5 & 0.0 & 0.5 & 0.5 & 0.0 & 0.0 & 0.0 & 0.0 \\
C_4 & 0.0 & 0.0 & 0.5 & 0.5 & 0.0 & 0.5 & 0.5 & 0.0 & 0.0 & 0.0 & 0.0 \\
C_5 & 0.667 & 0.333 & 1.0 & 1.0 & 0.5 & 1.0 & 1.0 & 0.167 & 0.5 & 0.167 & 0.5 \\
C_6 & 0.0 & 0.0 & 0.5 & 0.5 & 0.0 & 0.5 & 0.5 & 0.0 & 0.0 & 0.0 & 0.0 \\
C_7 & 0.0 & 0.0 & 0.5 & 0.5 & 0.0 & 0.5 & 0.5 & 0.0 & 0.0 & 0.0 & 0.0 \\
C_8 & 1.0 & 0.667 & 1.0 & 1.0 & 0.833 & 1.0 & 1.0 & 0.5 & 0.833 & 0.5 & 0.833 \\
C_9 & 0.667 & 0.333 & 1.0 & 1.0 & 0.5 & 1.0 & 1.0 & 0.167 & 0.5 & 0.167 & 0.5 \\
C_{10} & 1.0 & 0.667 & 1.0 & 1.0 & 0.833 & 1.0 & 1.0 & 0.5 & 0.833 & 0.5 & 0.833 \\
C_{11} & 0.667 & 0.333 & 1.0 & 1.0 & 0.5 & 1.0 & 1.0 & 0.167 & 0.5 & 0.167 & 0.5
\end{array} \tag{13}$$

Consequently, a weighting vector

$$(0.094, 0.132, 0.033, 0.033, 0.113, 0.033, 0.033, \mathbf{0.152}, 0.113, \mathbf{0.152}, 0.113)$$

is obtained, which indicates that criteria C_8: Vehicle capability and C_{10}: Speed of traffic flow are the most important ones.

Using the linguistic performance profiles of the 12 fuel modes on the 11 decision criteria, we are able to derive the individual matrix of stochastic dominance degrees of the 12 fuel modes regarding each decision criterion C_n, denoted by $\mathbf{R}^n$. With the weighting vector $\mathbf{W}$, a matrix $\mathbf{R}$ of collective stochastic dominance degrees of the 12 fuel modes is obtained as

$$\mathbf{R} = \begin{array}{c|cccccccccccc}
 & A_1 & A_2 & A_3 & A_4 & A_5 & A_6 & A_7 & A_8 & A_9 & A_{10} & A_{11} & A_{12} \\
\hline
A_1 & 0.5 & 0.494 & 0.479 & 0.623 & 0.669 & 0.552 & 0.495 & 0.51 & 0.564 & 0.586 & 0.555 & 0.575 \\
A_2 & 0.506 & 0.5 & 0.498 & 0.705 & 0.742 & 0.605 & 0.548 & 0.563 & 0.601 & 0.609 & 0.541 & 0.559 \\
A_3 & 0.521 & 0.502 & 0.5 & 0.697 & 0.739 & 0.607 & 0.55 & 0.565 & 0.604 & 0.611 & 0.543 & 0.561 \\
A_4 & 0.377 & 0.295 & 0.303 & 0.5 & 0.552 & 0.431 & 0.435 & 0.432 & 0.379 & 0.378 & 0.314 & 0.327 \\
A_5 & 0.331 & 0.258 & 0.261 & 0.448 & 0.5 & 0.402 & 0.389 & 0.381 & 0.319 & 0.316 & 0.265 & 0.298 \\
A_6 & 0.448 & 0.395 & 0.393 & 0.569 & 0.598 & 0.5 & 0.449 & 0.426 & 0.473 & 0.459 & 0.421 & 0.438 \\
A_7 & 0.505 & 0.452 & 0.45 & 0.565 & 0.611 & 0.551 & 0.5 & 0.477 & 0.536 & 0.522 & 0.491 & 0.501 \\
A_8 & 0.49 & 0.437 & 0.435 & 0.568 & 0.619 & 0.574 & 0.523 & 0.5 & 0.516 & 0.502 & 0.47 & 0.48 \\
A_9 & 0.436 & 0.399 & 0.396 & 0.621 & 0.681 & 0.527 & 0.464 & 0.484 & 0.5 & 0.517 & 0.436 & 0.465 \\
A_{10} & 0.414 & 0.391 & 0.389 & 0.622 & 0.684 & 0.541 & 0.478 & 0.498 & 0.483 & 0.5 & 0.436 & 0.472 \\
A_{11} & 0.445 & 0.459 & 0.457 & 0.686 & 0.735 & 0.579 & 0.509 & 0.53 & 0.564 & 0.564 & 0.5 & 0.529 \\
A_{12} & 0.425 & 0.441 & 0.439 & 0.673 & 0.702 & 0.562 & 0.499 & 0.52 & 0.535 & 0.528 & 0.471 & 0.5
\end{array}$$

Table 3. Choice values

Flows	Fuel modes $\mathcal{A}$											
	A_1	A_2	A_3	A_4	A_5	A_6	A_7	A_8	A_9	A_{10}	A_{11}	A_{12}
Φ_m^+	0.555	0.589	0.591	0.384	0.333	0.461	0.515	0.51	0.493	0.492	0.551	0.527
Φ_m^-	0.445	0.411	0.409	0.616	0.667	0.539	0.485	0.49	0.507	0.508	0.449	0.473
Φ_m	0.11	0.177	**0.182**	−0.232	−0.333	−0.078	0.029	0.021	−0.014	−0.017	0.101	0.054

Step 3: Choice function Using the matrix **R** of collective stochastic dominance degrees of the 12 fuel modes, stochastic dominant degree of A_m dominating the other fuel modes and stochastic non-dominant degree of A_m dominated by the other alternatives can be calculated, as shown in Table 3, which indicates that A_3: Liquid propane gas is the best fuel mode.

4.3   Analysis of the Evaluation Results Obtained

Considering the results of the weights **W** with respect to various assessment criteria, the importance ranking of the 11 decision criteria is obtained as

$$\{C_8, C_{10}\} \succ \{C_2\} \succ \{C_5, C_9, C_{11}\} \succ \{C_1\} \succ \{C_3, C_4, C_6, C_7\},$$

which indicates the following interesting meanings. Although energy savings and carbon reduction are usually the objective of an energy policy, energy efficiency, air pollution, and noise pollution are not the most important criteria in this case study. It is clearly seen that "C_8: vehicle capability" and "C_{10}: speed of traffic flow" are the first most important factors influencing the use of different fuel modes in our current research. The main reason arises from fact that the three experts provide their linguistic judgements with respect to C_8 and C_{10} as 'VH' (the maximal value of $\mathcal{S}^{II}$), separatively. The criterion 'vehicle capability' represents the cruising distance, slope climbing, and average speed. The criterion 'speed of traffic flow' refers to the comparison of the average speed of alternative vehicles for certain traffic. The second most important factor is "C_2: Energy efficiency" representing the efficiency of fuel energy. The third most important factors reflecting the use of alternative fuel modes are "C_5: Industrial relationship, C_9: Road facility, C_{11}: Sense of comfort". The four most important factor is "C_1: Energy supply". The least important criteria are "C_3: Air pollution, C_4: Noise pollution, C_6: Costs of implementation, C_7: Costs of maintenance". The main reason of the importance ranking may come from the following facts. An energy policy toward a sustainable energy system is a long-term objective. In this example, the fuel modes are studied in Iran, which is a developing country. Iran is very rich in the traditional fossil energy.

In addition to the importance ranking, according to the results in Table 3, the ranking order of the fuel modes with positive net flows is $A_3 \succ A_2 \succ A_1 \succ A_{11} \succ A_{12} \succ A_7 \succ A_8$ which should be given serious considerations. Firstly, it is seen that only the first three fuel modes A_3, A_2, A_1 have the net flows greater than 0.1. The liquid propane gas and compressed natural gas are ranked first and second, reflecting the need for alternative fuel modes. The conventional diesel engine is ranked third, indicating that conventional diesel engine is preferred to other alternative fuel modes, except the liquid propane gas and compressed natural gas. Secondly, in comparison with conventional vehicles, vehicles with alternative fuel modes would contribute significantly to the improvement of air quality in Iran. However, the alternative fuel modes except the liquid propane gas and compressed natural gas, have the following characteristics. They demand recharging, and it remains uncompetitive for them with the fuel-engine vehicle due to their frequent recharging needs. Since the bus system owns such features as permanent terminal, route, group of user, time of operation, and frequency, it is likely to expect that the implementation of an alternative fuel bus might become a very important option for the development of public transportation. Compared with traditional products that use fossil energy, the high cost is an important factor hindering the proliferation of alternative fuel modes applications. Finally, the modes $A_4, A_5, A_6, A_9, A_{10}$ should be absolutely abandoned since they have negative net flows.

5 Concluding Remarks

This paper proposed a fuzzy linguistic multi-criteria group energy planning model with computations solely on linguistic labels as well as providing the decision-makers better decision support. In particular, a probabilistic approach was firstly introduced to perform group aggregation of linguistic judgments given by a set of experts. Secondly, a probabilistic approach was proposed to derive the weighting vector of different criteria and to induce the matrix of overall stochastic dominance degrees of different alternatives. Thirdly, a choice approach based on the idea of the PROMETHEE-II method was developed to rank and choose the most desired alternative(s). As a case study, our model was used to re-evaluate the technological development of public transportation with 12 alternative fuel-modes. The results showed that liquid propane gas and compressed natural gas are ranked first and second among the 12 fuel modes.

Acknowledgement. We would like to appreciate constructive comments and valuable suggestions from the anonymous referees. This study was partly supported by the National Natural Sciences Foundation of China (NSFC) under grant no. 71471063; sponsored by the Innovation Program of Shanghai Municipal Education Commission under grant no. 14ZS060; and supported by the Fundamental Research Funds for the Central Universities in China under grant no. WN1516009.

References

1. Bordogna, G., Fedrizzi, M., Passi, G.: A linguistic modeling of consensus in group decision making based on OWA operator. IEEE Trans. Syst. Man Cybern. Part A Syst. Hum. **27**(1), 126–132 (1997)
2. Cavallaro, F.: Fuzzy TOPSIS approach for assessing thermal-energy storage in concentrated solar power (CSP) systems. Appl. Energy **87**(2), 496–503 (2010)

3. Chang, P.L., Hsu, C.W., Lin, C.Y.: Assessment of hydrogen fuel cell applications using fuzzy multiple-criteria decision making method. Appl. Energy **100**, 93–99 (2012)
4. Cruz, J.M.: A sustainable policy making-energy system for Colombia. Interim Report IR-04-009, International Institute for Applied Systems Analysis (IIASA), Laxenburg, Austria (2004)
5. Haralambopoulos, D.A., Polatidis, H.: Renewable energy projects: structuring a multi-criteria group decision-making framework. Renew. Energy **28**(6), 961–973 (2003)
6. Hiremath, R.B., Shikha, S., Ravindranath, N.H.: Decentralized energy planning; modeling and application-a review. Renew. Sustain. Energy Rev. **11**(5), 729–752 (2007)
7. Hwang, C.L., Yoon, K.: Multiple Attribute Decision Making: Methods and Applications. Springer, New York (1981)
8. Kaya, T., Kahraman, C.: Multicriteria renewable energy planning using an integrated fuzzy VIKOR & AHP methodology: the case of Istanbul. Energy **35**(6), 2517–2527 (2010)
9. McNicol, B.D., Rand, D., Williams, K.R.: Fuel cells for road transportation purposeyes or no? J. Power Sources **100**(1), 47–59 (2001)
10. Morita, K.: Automotive power source in 21st century. JSAE Rev. **24**(1), 3–7 (2003)
11. Onar, S.C., Oztaysi, B., Otay, I., Kahraman, C.: Multi-expert wind energy technology selection using interval-valued intuitionistic fuzzy sets. Energy **90**(1), 274–285 (2015)
12. Saaty, T.L.: Axiomatic foundation of the analytic hierarchy process. Manage. Sci. **32**(7), 841–855 (1986)
13. Siskos, J., Hubert, P.: Multi-criteria analysis of the impacts of energy alternatives: a survey and a new comparative approach. Euro. J. Oper. Res. **13**(3), 278–299 (1983)
14. Taylor, R.: Strategic energy planning, project developmentand the importance of champions. Technical report, National Renewable Energy Laboratory of the U.S. Department of Energy Office of Energy Efficiency and Renewable Energy (2009)
15. Torrini, F.C., Souza, R.C., Oliveira, F.L.C., Pessanha, J.F.M.: Long term electricity consumption forecast in Brazil: a fuzzy logic approach. Soc. Econ. Plan. Sci. **54**, 18–27 (2016)
16. Tzeng, G.H., Lin, C.W., Opricovic, S.: Multi-criteria analysis of alternative-fuel buses for public transportation. Energy Policy **33**, 1373–1383 (2005)
17. Vahdani, B., Zandieh, M., Tavakkoli-Moghaddam, R.: Two novel FMCDM methods for alternative-fuel buses selection. Appl. Math. Model. **35**(3), 1396–1412 (2011)
18. Xu, Y., Da, Q., Liu, L.: Normalizing rank aggregation method for priority of a fuzzy preference relation and its effectiveness. Int. J. Approx. Reason. **50**(8), 1287–1297 (2009)
19. Yan, H.B., Ma, T., Li, Y.: A novel fuzzy linguistic model for prioritising engineering design requirements in quality function deployment under uncertainties. Int. J. Prod. Res. **51**(21), 6336–6355 (2013)
20. Zadeh, L.A.: Fuzzy sets. Inf. Control **8**(3), 338–353 (1965)
21. Zadeh, L.A.: The concept of a linguistic variable and its application to approximate reasoning-part I. Inf. Sci. **8**(3), 199–249 (1975)

Minimax Regret Relaxation Procedure of Expected Recourse Problem with Vectors of Uncertainty

Thibhadha Saraprang[✉] and Phantipa Thipwiwatpotjana

Faculty of Science, Department of Mathematics and Computer Science,
Chulalongkorn University, Bangkok 10330, Thailand
Thibhadha.n@gmail.com, phantipa.t@chula.ac.th

Abstract. In this work, uncertain entities in a linear program are probability intervals that can be represented as random sets. The uncertainty in our model has a special pattern; i.e., the coefficients and the right hand side of each constraint form vector realizations. Moreover, each constraint may have a few vectors realizations to represent its situation. We transform the linear program with special uncertainty into an interval expected recourse problem, then find the minimax regret of this issue by a relaxation procedure. The relaxation procedure finally has been improved by using the idea of ordering and the fact that we can reduce the size of the lower probability set of all possible ordering cases at each iteration.

Keywords: Probability interval · Minimax regret · Linear programming with uncertainty

1   Introduction

There are many types of information in the real world, for example, errors in forecasting future data and uncontrollable external occurrences, such as weather or the actions of independent data. We hope all information would be the same type or easy to categorize into groups because it may be simpler to use. So, this work focuses on the information of uncertainty that can be represented by a set of probability intervals with a random set properties. Readers could refer to [2,8,9] for the relationship between a probability interval and a random set.

We cannot find an exact optimal solution of a linear program with uncertainty since it is undefined, but we can represent the problem by using pessimistic, optimistic, or minimax regret approaches. The pessimistic approach is the opposite of the optimistic one. It provides the maximum of the expected recourse values when the original linear program with uncertainty is a minimizing problem. Pessimistic and optimistic solutions provide the boundary of the actual objective value when we do not know the exact realizations. The minimax regret approach provides the minimum of the maximum regret due to not knowing the actual probability to establish an expected recourse model. Minimax regret solution is

© Springer International Publishing AG 2016
V.-N. Huynh et al. (Eds.): IUKM 2016, LNAI 9978, pp. 89–98, 2016.
DOI: 10.1007/978-3-319-49046-5_8

to minimize the maximum difference between the actual and the best possible outcome under a particular scenario when we cannot forecast the future result. There are some literature reviews on the minimax regret approach of linear problems with uncertainty as follows.

- In 1995, Inuiguchi [4] studied a linear programming problem with interval objective function coefficients by using minimax regret criterion.
- In 1999, Mausser and Laguna [6] proposed a heuristic method to find the minimax regret for linear programs with interval objective function coefficients. This heuristic approach concentrates on the minimax regret problem which can be guaranteed solving the candidate maximum regret problem to optimality. It reduces the time per iteration, but the total time increases.
- In 2000, Averbakh [1] suggested a method for finding solutions of minimax regret for a group of combinatorial optimization problems with an objective function of minimax type and uncertain objective function coefficients. The approach base on reducing a problem with uncertainty to some problems without uncertainty. He described the method on bottleneck combinatorial optimization problems, minimax multi-facility location problems and maximum weighted delay scheduling problems with uncertainty.
- In 2013, Thipwiwatpotjana and Lodwick [10] presented comprehensive methods for handling linear programming under mixed uncertainty by using pessimistic and optimistic, and minimax regret approaches.

In this paper, we consider only the minimax regret approach and its relaxation procedure for a linear program problem with vector of uncertainty. We want to improve the original relaxation procedure of the minimax regret method by using the idea of ordering. We then compare the time of these approaches.

We limit number of uncertain constraints and each of uncertain vectors to be only five constraints and three realizations. For the case of one or two realizations, there is no differences between the original and the improved methods. For the case of more than three realizations, it is not difficult to see that if we apply our method to only partial of any three realizations and leave the rest normally, we still would be able to reduce the calculation time. Therefore, it is not necessary to expand our investigation to more than three realizations. We consider only five uncertain constraints because of the similar reasons.

2   Stochastic Expected Recourse Model

Let x and c be vectors of size of n. Consider the following a linear program with uncertainty:

$$
\begin{aligned}
\min \ & c^T x \\
\text{s.t} \ \ & \hat{A}_1^T x \geq \hat{b}_1 \\
& \vdots \\
& \hat{A}_m^T x \geq \hat{b}_m \\
& x \geq 0,
\end{aligned}
\tag{1}
$$

where $(\hat{A}_i, \hat{b}_i)$ is a vector of uncertainty with k_i realizations, for each $i = 1, \cdots,$ m. We can write all k_i realizations of $(\hat{A}_i, \hat{b}_i)$ as $(A_{i_1}, b_{i_1}), (A_{i_2}, b_{i_2}), \cdots,$ $(A_{i_{k_i}}, b_{i_{k_i}})$. Each jth realization of $(\hat{A}_i, \hat{b}_i)$ has probability $f_{i_j}, j = 1, \cdots, k_i$. For example, suppose x_1 and x_2 are the number of bags of brown and white sugar (at the some fixed price per bag), respectively. The ith constraint could refer to as a demand constraint, where $\hat{A}_i$ is an uncertain vector of weight (kg) for each bag of brown and white sugar and b_i is an uncertain demand. The productivities of both types of sugar may have three realizations (Below average, Average and Above average) depending on the weather. The demand b_i also has realizations in the same pattern under economic behavior. Therefore, three vector realizations are Below average (realization 1) : $(A_{i_1}, b_{i_1}) = (7, 8, 45)$ with probability f_{i_1},

Average (realization 2): $(A_{i_2}, b_{i_2}) = (8, 9, 50)$ with probability f_{i_2} and

Above average (realization 3): $(A_{i_3}, b_{i_3}) = (9, 10, 53)$ with probability f_{i_3}.

When linear programs are deterministic, they can be solved using a simplex method by adding slack/surplus variables. If a constraint is uncertain, we remodel the problem as a two-stage stochastic expected recourse model. The first stage is making the decision before knowing the realization of the uncertainty. The second stage considers an understanding of the stochastic elements of the problem when the first stage does not satisfy the requirements. We can rewrite (1) to be the stochastic expected recourse model as follows.

$$
\begin{aligned}
\min \ & c^T x + \sum_{i=1}^{m} s_i (f_i^T w_i) \\
\text{s.t} \ & A_{1_1}^T x + w_{1_1} \geq b_{1_1} \\
& A_{1_2}^T x + w_{1_2} \geq b_{1_2} \\
& \quad \vdots \\
& A_{1_k}^T x + w_{1_k} \geq b_{1_k} \\
& \quad \vdots \\
& A_{m_1}^T x + w_{m_1} \geq b_{m_1} \\
& A_{m_2}^T x + w_{m_2} \geq b_{m_2} \\
& \quad \vdots \\
& A_{m_k}^T x + w_{m_k} \geq b_{m_k} \\
& x \geq 0,
\end{aligned}
\tag{2}
$$

where x is the first stage decision vector and $w_i = (w_{i_1}, \cdots, w_{i_{k_i}})$ is the second stage (recourse) decision vector according to the original i^{th} constraint, with corresponding probability $f_i = (f_{i_1}, \cdots, f_{i_k})$. Each w_{i_j} defines as $w_{i_j} = max(b_{i_j} - A_{i_j}^T x, 0)$. The term $\sum_{i=1}^{m} s_i(f_i^T w_i)$ is the expected recourse value, when $s = (s_i, \cdots, s_m)$ is a penalty vector respected to the original constraints. We define the set of feasible solutions of (2) as Ω. Example 2.1 shows the corresponding stochastic expected recourse problem of the demand constraint of brown and white sugar with minimized transportation cost.

Example 2.1. A linear program P_1 with uncertain vector $(\hat{A}_1, \hat{b}_1)$

$$P_1: \min 3x_1 + 2x_2$$
$$\text{s.t} \quad \hat{A}_1^T x \geq \hat{b}_1$$
$$x \geq 0$$

Suppose we know realizations of $(\hat{A}_1, \hat{b}_1)$ and their probabilities as :
Below average (realization 1) : $(A_{1_1}, b_{1_1}) = (7, 8, 45)$, with probability f_{1_1},
Average (realization 2) : $(A_{1_2}, b_{1_2}) = (8, 9, 50)$, with probability f_{1_2} and
Above average (realization 3) : $(A_{1_3}, b_{1_3}) = (9, 10, 53)$, with probability f_{1_3}.
The corresponding stochastic expected recourse problem with recourse variables (w_1, w_2, w_3) of the demand constraint is

$$P_2 : \min \quad 3x_1 + 2x_2 + s_1 f_{1_1} w_{1_1} + s_1 f_{1_2} w_{1_2} + s_1 f_{1_3} w_{1_3}$$

$$\text{s.t } 7x_1 + 8x_2 + w_{1_1} \geq 45$$
$$8x_1 + 9x_2 + w_{1_2} \geq 50$$
$$9x_1 + 10x_2 + w_{1_3} \geq 53 \tag{3}$$
$$x, w_1 \qquad\qquad \geq 0.$$

⊠

How do we interpret a result of (3) if we do not know the exact probability, for example, if $f_{1_1} \in [\frac{1}{3}, \frac{1}{2}], f_{1_2} \in [\frac{1}{6}, \frac{2}{3}]$ and $f_{1_3} \in [\frac{1}{6}, \frac{1}{2}]$? One of the approaches is to use an idea of interval expected value, which will be presented in the next section.

3 Interval Expected Recourse Model

We know that a linear program with uncertainty cannot have an exact optimal solution. Some common approaches for such a problem are to find the maximum/minimum expected objective value or a minimum of maximum regret solution. In this paper, we consider only uncertainties that are in the form of interval probabilities which can be written as random sets. Thus, our general uncertain expected recourse problem with unknown probability in probability interval is stated as follows, using the same feasible set Ω as (2).

$$\min_{(x, w_i) \in \Omega} c^T x + \sum_{i=1}^{m} s_i (f_i^T w_i)$$
$$\text{where} \quad f_{i_j} \in [l_{i_j}, u_{i_j}]; \quad \forall i = 1, 2, \ldots, m, \quad j = 1, 2, \ldots, k_i. \tag{4}$$

Let $\hat{t}$ be an uncertain information of our concern, and $T = \{t_1, t_2, \cdots, t_k\}$ be the set of all realizations of $\hat{t}$ where $t_1 \leq t_2 \leq \cdots \leq t_k$. Boodgumarn [2] proved the conditions when a probability interval is a random set, which help us to conclude that our uncertainty in this paper has the following properties, for any $A \subseteq T$, as follows.

$$Bel(A) = l(A) = \max(\sum_{t_i \in A} l_i, 1 - \sum_{t_i \in A^c} u_i) \tag{5}$$

$$Pl(A) = u(A) = \min(\sum_{t_i \in A} u_i, 1 - \sum_{t_i \in A^c} l_i), \tag{6}$$

where probability of $\{t_i\}$ is bounded in $[l_i, u_i]$ and the function Bel and Pl are belief and plausibility measures defined in Dempster [3] and Klir and Yuan [5]. Moreover, Nguyen [7] showed a method to find the largest and the smallest expected values of $\hat{t}$ when we have a probability interval as a random set, i.e.,

$$\underline{f}(t_1) = Bel(\{t_1, t_2, \cdots, t_k\}) - Bel(\{t_2, t_3, \cdots, t_k\})$$

$$\vdots$$

$$\underline{f}(t_i) = Bel(\{t_i, t_{i+1}, \cdots, t_k\}) - Bel(\{t_{i+1}, t_{i+2}, \cdots, t_k\})$$

$$\vdots$$

$$\underline{f}(t_k) = Bel(\{t_k\})$$

and $\hspace{10cm}$ (7)

$$\overline{f}(t_1) = Bel(\{t_1\})$$

$$\vdots$$

$$\overline{f}(t_i) = Bel(\{t_1, t_2, \cdots, t_i\}) - Bel(\{t_1, t_2, \cdots, t_{i-1}\})$$

$$\vdots$$

$$\overline{f}(t_k) = Bel(\{t_1, t_2, \cdots, t_k\}) - Bel(\{t_1, t_2, \cdots, t_{k-1}\}),$$

where $\overline{f}$ and $\underline{f}$ are probabilities that provide the largest and the smallest expected values, respectively. Hence we can find $\overline{f}$ and $\underline{f}$ in the form of l_i, u_i and the boundary of expected value of $\hat{t}$ is $\left[\sum_{j=1}^{k} t_j \underline{f}(t_j), \sum_{j=1}^{k} t_j \overline{f}(t_j)\right]$. This boundary is called an interval expected value.

For the rest of the paper, we will work on only three realizations of uncertain vector for each uncertain constraint, as we mentioned at the end of our introduction. Thus, in the case of the three realizations $t_1 \leq t_2 \leq t_3$ of $\hat{t}$, we know that

$$\underline{f}(t_3) = Bel(\{t_3\}) = \max(l_3, 1 - u_1 - u_2)$$

$$\underline{f}(t_2) = Bel(\{t_2, t_3\}) - Bel(\{t_3\}) = \max(l_2 + l_3, 1 - u_1) - \max(l_3, 1 - u_1 - u_2)$$

$$\underline{f}(t_1) = Bel(\{t_1, t_2, t_3\}) - Bel(\{t_2, t_3\}) = 1 - \max(l_2 + l_3, 1 - u_1)$$

$\qquad$ and

$$\overline{f}(t_1) = Bel(\{t_1\}) = \max(l_1, 1 - u_2 - u_3)$$
$$\overline{f}(t_2) = Bel(\{t_1, t_2\}) - Bel(\{t_1\}) = \max(l_1 + l_2, 1 - u_3) - \max(l_1, 1 - u_2 - u_3)$$
$$\overline{f}(t_3) = Bel(\{t_1, t_2, t_3\}) - Bel(\{t_1, t_2\}) = 1 - \max(l_1 + l_2, 1 - u_3).$$

We can apply this idea to our vector of uncertainty when its realization can be ordering. Let $(\hat{A}_i, \hat{b}_i)$ be a vector of uncertainty with the set of realizations $\{(A_{i_1}, b_{i_1}), (A_{i_2}, b_{i_2}), \cdots, (A_{i_{k_i}}, b_{i_{k_i}})\}$ where $(A_{i_j}, b_{i_j}) \leq (A_{i_l}, b_{i_l})$ when $j \leq l$. please note that $(A_{i_j}, b_{i_j}) \leq (A_{i_l}, b_{i_l})$ if and only if $A_{i_j} \leq A_{i_l}$ and $b_{i_j} \leq b_{i_l}$.

Let us continue using Example 2.1, to explain an interval expected recourse model with interval probability (with random set properties) information using the uncertain demand constraint.

Example 3.1. Consider the expected recourse problem P_2 in Example 2.1, when we do not know the probability $\boldsymbol{f}_1 = (f_{1_1}, f_{1_2}, f_{1_3})$ for certain, but we know that $f_{1_1} \in [\frac{1}{3}, \frac{1}{2}], f_{1_2} \in [\frac{1}{6}, \frac{2}{3}]$, and $f_{1_3} \in [\frac{1}{6}, \frac{1}{2}]$.

Therefore, to find an interval expected value to the objective function of problem P_2 with unknown probability $(f_{1_1}, f_{1_2}, f_{1_3})$ we have to know the ordering of $w_{1_1}, w_{1_2}, w_{1_3}$. All possible ordering cases of $w_{1_1}, w_{1_2}, w_{1_3}$ with their corresponding lower/upper probabilities to get the boundary of interval expected value are in the following table.

Ordering cases	Lower probability $\underline{\boldsymbol{f}}_i$			Upper probability $\overline{\boldsymbol{f}}_i$		
	$\underline{f}_{1_1}$	$\underline{f}_{1_2}$	$\underline{f}_{1_3}$	$\overline{f}_{1_1}$	$\overline{f}_{1_2}$	$\overline{f}_{1_3}$
$w_{1_1} \leq w_{1_2} \leq w_{1_3}$	$\frac{1}{2}$	$\frac{1}{3}$	$\frac{1}{6}$	$\frac{1}{3}$	$\frac{1}{6}$	$\frac{1}{2}$
$w_{1_1} \leq w_{1_3} \leq w_{1_2}$	$\frac{1}{2}$	$\frac{1}{6}$	$\frac{1}{3}$	$\frac{1}{3}$	$\frac{1}{6}$	$\frac{1}{2}$
$w_{1_2} \leq w_{1_1} \leq w_{1_3}$	$\frac{1}{3}$	$\frac{1}{2}$	$\frac{1}{6}$	$\frac{1}{3}$	$\frac{1}{2}$	$\frac{1}{6}$
$w_{1_2} \leq w_{1_3} \leq w_{1_1}$	$\frac{1}{3}$	$\frac{1}{2}$	$\frac{1}{6}$	$\frac{1}{3}$	$\frac{1}{2}$	$\frac{1}{6}$
$w_{1_3} \leq w_{1_1} \leq w_{1_2}$	$\frac{1}{3}$	$\frac{1}{6}$	$\frac{1}{2}$	$\frac{1}{2}$	$\frac{1}{6}$	$\frac{1}{3}$
$w_{1_3} \leq w_{1_2} \leq w_{1_1}$	$\frac{1}{3}$	$\frac{1}{6}$	$\frac{1}{2}$	$\frac{1}{2}$	$\frac{1}{3}$	$\frac{1}{2}$

$\boxtimes$

However, a decision marker may prefer to have a solution that can present a reasonable regret of not knowing an exact probability, which leads us to a minimax regret approach.

4 Minimax Regret of Uncertain Expected Recourse Problem

Consider the stochastic expected recourse model (2) when $\boldsymbol{f} = (\boldsymbol{f}_1, \boldsymbol{f}_2, \cdots, \boldsymbol{f}_m)$ is an unknown probability vector. Let the set of all probabilities in probability intervals be $M = \{(\boldsymbol{f}_1, \boldsymbol{f}_2, \cdots, \boldsymbol{f}_m) \mid \boldsymbol{f}_i = (f_{i_1}, f_{i_2}, \cdots, f_{i_{k_i}})$ where

$f_{i_j} \in [l_{i_j}, u_{i_j}]\}$. A minimax regret of an uncertain expected recourse problem is a problem to minimize of maximum regret over all unknown probability vectors in M.

The objective function of (4) is to minimize

$$z(\boldsymbol{f},\boldsymbol{x},\boldsymbol{w}) := \boldsymbol{c}^T \boldsymbol{x} + \sum_{i=1}^{m} s_i \boldsymbol{f}_i^T \boldsymbol{w}_i, \text{ where } \boldsymbol{w} = (\boldsymbol{w}_1, \cdots, \boldsymbol{w}_m).$$

The best of the worst regret over all $\boldsymbol{f} \in M$ is

$$\min_{(\boldsymbol{x},\boldsymbol{w})\in\Omega} \max_{\boldsymbol{f}\in M} \left(z(\boldsymbol{f},\boldsymbol{x},\boldsymbol{w}) - \min_{(\bar{\boldsymbol{x}},\bar{\boldsymbol{w}})\in\Omega} z(\boldsymbol{f},\bar{\boldsymbol{x}},\bar{\boldsymbol{w}}) \right) \tag{8}$$

A method to find an optimal solution of minimax regret for an uncertain expected recourse problem has four general steps as follows. Reader may find the relaxation procedure (similar to algorithm 1) that deals with interval objective function in [4]

Algorithm 1. General relaxation procedure (see [10]).

1. Initialization. Choose $\boldsymbol{f}^{(1)} \in M$. Solve $\min_{(\boldsymbol{x},\boldsymbol{w})\in\Omega} z(\boldsymbol{f}^{(1)}, \boldsymbol{x}, \boldsymbol{w})$ and obtain its optimal solution $(\bar{\boldsymbol{x}}^{(1)}, \bar{\boldsymbol{w}}^{(1)})$ Set $p = 1$.
2. Solve the following current relaxed problem to obtain an optimal solution $(R^{(p)}; (\boldsymbol{x}^{(p)}, \boldsymbol{w}^{(p)}))$.

$$\min_{R;((\boldsymbol{x},\boldsymbol{w})\in\Omega)} R$$

$$\text{s.t. } R \geq 0$$

$$R \geq \left(z(\boldsymbol{f}^{(i)}, \boldsymbol{x}, \boldsymbol{w}) - z(\boldsymbol{f}^{(i)}, \bar{\boldsymbol{x}}^{(i)}, \bar{\boldsymbol{w}}^{(i)}) \right), i = 1, 2, \cdots, p.$$

3. Obtain an optimal solution $(\boldsymbol{f}^{(i+1)}, \bar{\boldsymbol{x}}^{(i+1)}, \bar{\boldsymbol{w}}^{(i+1)})$ where its optimal value $Z^{(p)}$ is

$$Z^{(p)} = \max_{\boldsymbol{f}\in M;((\bar{\boldsymbol{x}},\bar{\boldsymbol{w}})\in\Omega)} \left(z(\boldsymbol{f}, \boldsymbol{x}^{(p)}, \boldsymbol{w}^{(p)}) - z(\boldsymbol{f}, \bar{\boldsymbol{x}}, \bar{\boldsymbol{w}}) \right).$$

4. If $Z^{(p)} \leq R^{(p)}$, terminate the procedure. An optimal solution to the minimax expected regret model is $(R^{(p)}; \boldsymbol{x}^{(p)}, \boldsymbol{w}^{(p)})$. Otherwise, set $p = p + 1$ then return to step 2 . □

5   Relaxation Procedure by Ordering

Thipwiwatpojana and Lodwick [10] used the ordering idea to improve the general relaxation procedure by removing the set M to be $\overline{M}$, the set of lower probabilities of all possible ordering cases of realizations. In step 3, of the ith iteration of Algorithm 1, given probability $\boldsymbol{f} \in \overline{M}$, we have the sequence of probabilities generated upon the order of $(\boldsymbol{w} - \bar{\boldsymbol{w}})$ through the following relationship.

$$Z_f^{(i)} = \max_{(\bar{x},\bar{w})\in\Omega} \left(z(f, x^{(i)}, w^{(i)}) - z(f, \bar{x}, \bar{w}) \right)$$

$$= \max_{(\bar{x},\bar{w})\in\Omega} \left(c^T(x^{(i)} - \bar{x}) + s_1 \sum_{j=1}^{k_1} f_{1_j}(w_{1_j}^{(i)} - \bar{w}_{1_j}) + \cdots \right.$$

$$\left. + s_m \sum_{j=1}^{k_m} f_{m_j}(w_{m_j}^{(i)} - \bar{w}_{m_j}) \right)$$

$$= c^T(x^{(i)} - \bar{x}_f) + s_1 \sum_{j=1}^{k_1} f_{1_j}(w_{1_j}^{(i)} - \bar{w}_{1_j,f}) + \cdots$$

$$+ s_m \sum_{j=1}^{k_m} f_{m_j}(w_{m_j}^{(i)} - \bar{w}_{m_j,f})$$

$$\leq c^T(x^{(i)} - \bar{x}_f) + s_1 \sum_{j=1}^{k_1} h_{1_j}(w_{1_j}^{(i)} - \bar{w}_{1_j,f}) + \cdots$$

$$+ s_m \sum_{j=1}^{k_m} h_{m_j}(w_{m_j}^{(i)} - \bar{w}_{m_j,f}) \tag{9}$$

where $h = (h_1, h_2, \cdots, h_m) \in \overline{M}, h_i = (h_{i_1}, \cdots, h_{i_{k_i}})$ and
h_{i_j} is corresponding to the order of $(w_{i_j} - \bar{w}_{i_j,f})$

$$\leq c^T(x^{(i)} - \bar{x}_h) + s_1 \sum_{j=1}^{k_1} h_{1_j}(w_{1_j}^{(i)} - \bar{w}_{1_j,h})' + \cdots$$

$$+ s_m \sum_{j=1}^{k_m} h_{m_j}(w_{m_j}^{(i)} - \bar{w}_{m_j,h})$$

$$= \max_{(\bar{x},\bar{w})\in\Omega} (z(h, x^{(i)}, w^{(i)}) - z(h, \bar{x}, \bar{w}))$$

$$= Z_h^{(i)}.$$

The above statement helps to obtain a modify version of the relaxation procedure in Algorithm 2, see [10].

Algorithm 2. Relaxation procedure by ordering of realizations.

1. Initialization. Choose $f^{(1)} \in \overline{M}$. Solve $\min\limits_{(x,w)\in\Omega} z(f^{(1)}, x, w)$ and obtain its optimal solution $(\bar{x}^{(1)}, \bar{w}^{(1)})$ Set $p = 1$.
2. Solve the following current relaxed problem to obtain an optimal solution $(R^{(p)}; (x^{(p)}, w^{(p)}))$.

$$\min_{R;((x,w)\in\Omega)} R$$

$$\text{s.t. } R \geq 0 \tag{10}$$

$$R \geq \left(z(f^{(i)}, x, w) - z(f^{(i)}, \bar{x}^{(i)}, \bar{w}^{(i)}) \right), i = 1, 2, \cdots, p.$$

3. Start with $\boldsymbol{f}^{(p)}$ and work on the system (9) to find $\boldsymbol{h}$. Calculate $Z_h^{(p)}$ and its optimal solution $(\bar{\boldsymbol{x}}^{(p)}, \bar{\boldsymbol{w}}^{(p)})$.

4. If $Z_h^{(p)} > R^{(p)}$, set $\boldsymbol{h} = \boldsymbol{f}^{(p+1)}$. Set $p = p + 1$, then return to step 2.

5. If $Z_h^{(p)} \leq R^{(p)}$, select $\boldsymbol{f}$ that has not been used in this iteration of step 3, and reprocess the system (9) until we find $\boldsymbol{h}$ such that $Z_h^{(i)} > R^{(p)}$, then continue step 4. Otherwise, $\overline{M} = \emptyset$ and we terminate the procedure. An optimal solution to the minimax expected regret model (8) is $(R^{(p)}; \boldsymbol{x}^{(p)}, \boldsymbol{w}^{(p)})$. □

We continue modifying Algorithm 2, by removing $\boldsymbol{f}^{(i)}; i = 1, \cdots, p-1$, in step 2 from the set $\overline{M}$ before using it in step 3, since we had already knew the value of these $(z(\boldsymbol{f}^{(i)}, \boldsymbol{x}, \boldsymbol{w}) - z(\boldsymbol{f}^{(i)}, \bar{\boldsymbol{x}}^{(i)}, \bar{\boldsymbol{w}}^{(i)}))$ from (10), we only need to check Z_h for the rest of $\boldsymbol{h} \in \overline{M}$.

Algorithm 3. Modified relaxation procedure by ordering of realization and reducing the size of $\overline{M}$.

1. Initialization. Choose $\boldsymbol{f}^{(1)} \in \overline{M}$. Solve $\min\limits_{(\boldsymbol{x},\boldsymbol{w})\in\Omega} z(\boldsymbol{f}^{(1)}, \boldsymbol{x}, \boldsymbol{w})$ and obtain its optimal solution $(\bar{\boldsymbol{x}}^{(1)}, \bar{\boldsymbol{w}}^{(1)})$ Set $p = 1$.

2. Solve the following current relaxed problem to obtain an optimal solution $(R^{(p)}; (\boldsymbol{x}^{(p)}, \boldsymbol{w}^{(p)}))$.

$$\min_{R;((\boldsymbol{x},\boldsymbol{w})\in\Omega)} R$$
$$\text{s.t. } R \geq 0 \tag{11}$$
$$R \geq \left(z(\boldsymbol{f}^{(i)}, \boldsymbol{x}, \boldsymbol{w}) - z(\boldsymbol{f}^{(i)}, \bar{\boldsymbol{x}}^{(i)}, \bar{\boldsymbol{w}}^{(i)}) \right), i = 1, 2, \cdots, p.$$

3. If $\overline{M} = \emptyset$, $(R^{(p)}, \boldsymbol{x}^{(p)}, \boldsymbol{w}^{(p)})$ is an optimal solution. If $\overline{M} \neq \emptyset$, start with $\boldsymbol{f}^{(p)}$ and set $\overline{M} = \overline{M} \backslash \{\boldsymbol{f}^{(i)} \mid i = 1, \cdots, p\}$, then work on the system (9) to find $\boldsymbol{h}$. If $\boldsymbol{h} \notin \overline{M}$, $(R^{(p)}, \boldsymbol{x}^{(p)}, \boldsymbol{w}^{(p)})$ is an optimal solution for minimax regret problem. Else, calculate $Z_h^{(i)}$ and its optimal solution $(\bar{\boldsymbol{x}}^{(p)}, \bar{\boldsymbol{w}}^{(p)})$.

4. If $Z_h^{(p)} > R^{(p)}$, set $\boldsymbol{h}$ as $\boldsymbol{f}^{(p+1)}$. Set $p = p + 1$, then return to step 2.

5. If $Z_h^{(p)} \leq R^{(p)}$, set $M_1 = \overline{M}$. Select $\boldsymbol{f} \in M_1$ that has not been used in this iteration of step 3, and reprocess the system (9) until we find $\boldsymbol{h}$ such that $Z_h^{(i)} > R^{(p)}$, then continue step 4. Otherwise, $M_1 = \emptyset$ and we terminate the procedure with an optimal solution to the minimax expected regret model (8) as $(R^{(p)}; \boldsymbol{x}^{(p)}, \boldsymbol{w}^{(p)})$. □

By comparing the average time spending in these three algorithms for linear programs with five uncertain constraints where each constraint has three realization vectors, we obtain the results in the following table.

	Algorithm 1	Algorithm 2	Algorithm 3
Average Time (s.)	11.207	7.2055	6.3484

6   Conclusion

We applied the analysis written in [10] to our vectors of uncertainty. Therefore, the uncertain entities in this paper is said to be a special case of general uncertainty in [10]. In minimax regret approach, we clearly improved Algorithm 2. by reducing the set of $\overline{M}$ in each iteration.

References

1. Averbakh, I.: Minmax regret solutions for minimax optimization problems with uncertainty. Oper. Res. Lett. **27**(2), 57–65 (2000)
2. Boodgumarn, P., Thipwiwatpotjana, P., Lodwick, W.A.: When a probability interval is a random set. ScienceAsia **39**(3), 319–326 (2013)
3. Dempster, A.P.: Upper and lower probabilities induced by a multivalued mapping. Ann. Math. Stat. **38**, 325–339 (1967)
4. Inuiguchi, M., Sakawa, M.: Minimax regret solution to linear programming problems with an interval objective function. Eur. J. Oper. Res. **86**(3), 526–536 (1995)
5. Klir, G., Yuan, B.: Fuzzy Sets and Fuzzy Logic, vol. 4. Prentice hall, Upper Saddle River (1995)
6. Mausser, H.E., Laguna, M.: A heuristic to minimax absolute regret for linear programs with interval objective function coefficients. Eur. J. Oper. Res. **117**(1), 157–174 (1999)
7. Nguyen, H.T.: An Introduction to Random Sets. CRC Press, New York (2006)
8. Thipwiwatpotjana, P., Lodwick, W.A.: The relationship between a probability interval and a random set. In: 2011 Annual Meeting of the North American Fuzzy Information Processing Society (2011)
9. Thipwiwatpotjana, P., Lodwick, W.A.: A relationship between probability interval and random sets and its application to linear optimization with uncertainties. Fuzzy Sets Syst. **231**, 45–57 (2013)
10. Thipwiwatpotjana, P., Lodwick, W.A.: Pessimistic, optimistic, and minimax regret approaches for linear programs under uncertainty. Fuzzy Optim. Decis. Mak. **13**(2), 151–171 (2014)

Bottom Up Review of Criteria in Hierarchically Structured Decision Problem

Tomoe Entani[✉]

University of Hyogo, 7-1-28 Minatojima-minamimachi, Chuo, Kobe 6500047, Japan
entani@ai.u-hyogo.ac.jp

Abstract. A decision problem is structured hierarchically by criteria and alternatives in AHP. The goal of this paper is to review criterion importance from the alternative evaluations and assign the crisp weights to the criteria. Usually, a decision maker gives the pairwise comparison matrix of the criteria and then the criterion importance is obtained. In addition, this paper uses some example alternatives to measure how much uncertain the decision maker's thinking on the criterion is. First, the uncertainty is measured by the shared weights obtained from the comparison matrix of the alternatives under the criterion, since it is not sure which alternative has the weight. Then, the criterion importance weight is obtained from the optimistic viewpoint as the sum of the sure and possible weights from the comparison matrix of the criteria. It is based on the idea that the more uncertain a criterion is, the less the possible weight becomes. The model to derive the crisp weights of the criteria reflecting their uncertainties measured from the alternative evaluations is proposed.

Keywords: Pairwise comparison · Uncertainty · Interval AHP · Review

1 Introduction

A typical decision problem requires us to choose the best among several alternatives with respect to a set of criteria. It often includes internal and external complexities. To make a good decision, people cannot simply rely on their feelings. One of the ways is to assign a number to a thing, even when there are no scale to measure. The purpose of thinking about decision making is to help people make decisions systematically. The decision making support is useful when a decision maker feels it according to his/her own thinking, such as values, beliefs and convictions. One of the approaches to support decision making is to break a decision problem down into smaller judgments about the criteria and alternative of that decision. Based on this idea, a decision problem is structured hierarchically in AHP [11]. In AHP, we derive the priorities from these given judgements and synthesize the priorities to get an overall outcome for the alternatives. A decision maker gives the judgments of criteria and those of alternatives under each criterion. In addition to the judgements of alternatives, the judgements of

© Springer International Publishing AG 2016
V.-N. Huynh et al. (Eds.): IUKM 2016, LNAI 9978, pp. 99–109, 2016.
DOI: 10.1007/978-3-319-49046-5_9

criteria are needed, since the importance weights of all criteria are seldom equal. The overall synthesized alternative priorities and/or ranking of alternatives may be changed when the importance weights of criteria are a little different. It is worth deriving the criterion importance carefully. Therefore, this paper reviews criterion importance based on the hierarchical structure of a decision problem. Analytic Network Process (ANP) has been proposed to capture the dependence of criteria from alternatives [13]. The criterion importance depends on the alternative by using a supermatrix consisting of criteria and alternatives. In this setting, there are various importance weights of a criterion. This paper aims to assign a number to a criterion, instead of assigning a number to an alternative in usual or various numbers to a criterion in ANP.

The other reason to focus on the criterion importance is from the viewpoint of a group decision making. Everyday, we face various kinds of decision making situations. Although some of them can be done individually, most of them need to be done by a group. In case of an individual, no one may blame for his/her individual decision, since it may not impact on the surroundings. A decision can be impulsive since there is no need for a decision maker to explain his/her thinking to the others. However, in case of a group, the communication is unavoidable. It encourages a group decision making to support interpersonal information exchange [6]. The development of various communication technologies makes it easy for us to get together not only physically but also virtually. It increases the opportunities of making decisions by a group. A group decision making is a way to address some of the higher-level needs in Maslow's hierarchy of needs theory [12]. The sense of love and belonging that comes from being a member of a social and well performing group. Each member is expected to apply his/her individual expertise to contribute to the group outcome. The solution of helping a group is not only to reach a consensus but also to enhance the members' communication. It often happens that a conflict among individuals can be productive process, since it shows them a diversity of ideas [18]. This step is necessary for improving a group decision making performance. It is important to express members' views clearly to the others to reach the satisfied group evaluations of the alternatives. The alternative and criterion preferences seem to be a result and a reason, respectively. To enhance a discussion and support a decision process, it is more useful for the group members to share the criterion preference on its way than the final alternative preference. It is worth focusing on the criterion importance of a decision maker. In this sense, this paper makes the criterion importance of a decision maker clear.

This paper is organized as follows. The next section introduces a structure of a decision problem and a pairwise comparison matrix. In Sect. 3, the uncertainty of each criterion is measured through the alternative evaluations. Then, in Sect. 4, the crisp importance weight of a criterion reflecting its uncertainty defined in the previous section is obtained from the optimistic viewpoint. Section 5 is a numerical example to show the role of criterion uncertainty in inducing its optimistic weight. The last section is a conclusion.

2   Preliminary

2.1   Hierarchical Structure of a Decision Problem

For instance, a man is plan to change his job. The structure of his decision problem is illustrated as in Fig. 1, where jobs and viewpoints for thinking about jobs are alternatives and criteria, respectively. It is noted that two kinds of final decisions can be considered. One is to choose a job from a list in left box and the other is to think about viewpoints for choosing jobs in the right box. In both cases, what a decision maker does is almost the same. He compares the criteria at the middle level and the alternatives at the bottom level under each criterion. By synthesizing these evaluations, he gets to know the total evaluation scores of alternatives for the left final decision. These alternative priorities show a decision maker the ranking of the jobs in the list. On the other hand, this study uses the hierarchical structure for the right final decision. It is to review the importance of criteria through the alternatives, instead of deriving alternative priorities for the left one. For this purpose, the alternatives need not to be the real ones but to be familiar to him. The criterion importance weights show a decision maker his attitude to the decision problem. When changing jobs is his personal issue, it is not necessary to make his attitude to the jobs clear. However, when his team is plan to change its jobs, it is helpful for the team members to understand their own attitudes to the jobs each other. From the viewpoint of a group decision making, it is more reasonable to work for the consensus of the viewpoints than for the listed jobs.

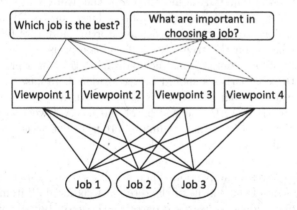

Fig. 1. Decision problem structure

2.2   Pairwise Comparison Matrix

In case of n compared items, a pairwise comparison matrix is denoted as follows.

$$A = \begin{bmatrix} 1 & \cdots & a_{1n} \\ \vdots & a_{ij} & \vdots \\ a_{n1} & \cdots & 1 \end{bmatrix}, \tag{1}$$

where a_{ij} represents the importance ratio of item i to item j and often chosen from the list: $\{1/9, 1/7, 1/5, 1/3, 1, 3, 5, 7, 9\}$, whose element corresponds to linguistic evaluation such as much more important than, equal to, or less important than. The advantage of pairwise comparison technique is that a decision maker focuses on the compared pair of items and needs not care each item separately. A decision maker compares $n(n-1)/2$ pairs of items because of $a_{ii} = 1$ (identical) and $a_{ij} = 1/a_{ji}$ (reciprocal). They are consistent if and only if $a_{ij} = a_{il}a_{lj}, \forall i, j, l$. The comparisons are often inconsistent, since the importance ratio may be out of the list. Furthermore, a part of the inconsistency stems from the uncertainty in a decision maker's thinking, since item i is compared $(n-1)$ times and has $(n-1)$ kinds of comparisons in A. However, it is difficult to discern which item and how much causes the uncertainty. The comparison is represented as interval or fuzzy to treat the imprecision and vagueness of human judgement in some studies, which derive interval or fuzzy weights [2–4, 10] and crisp weights [5]. This study handles the crisp comparisons and the proposed method can be easily extended into the interval or fuzzy comparisons.

The inconsistency among the comparisons is a well-known topic and various approaches have been proposed. One of the ways is to measure it and several measurements are proposed. In AHP, consistency index, $C.I. = (\lambda - n)/(n - 1)$, is used [11]. λ is the principal eigenvalue by eigenvector method $A\boldsymbol{w} = \lambda\boldsymbol{w}$, where $\boldsymbol{w} = (w_1, \ldots, w_n)^T$. The weights of items are the normalized elements of the eigenvector; $\sum_i w_i = 1$. The weight is assigned to one of the items so that the plausible preferences are obtained. The more inconsistent $a_{ij}, \forall i, j$ are, the greater $C.I.$ becomes. The $C.I.$ is modified into the average of $C.I.$s which are obtained from the small matrices by dividing a large matrix [9]. Several other measurements which are not based on eigenvector method have been proposed. One of them is based on the transitivity of the comparisons [14] and another is based on the column comparisons' normalization [15]. In a practice, in case of $C.I. > 0.1$, the comparisons should be reconsidered. In case of many items, the threshold of a principal eigenvalue is determined by regression [1]. On the other hand, in Interval AHP, the widths of the interval weights measure the uncertainty of the decision maker's thinking [16, 17]. The weight of an item is denoted as interval to reflect uncertain decision maker's thinking on the item. It is based on the idea that a decision maker may use a real value in the interval weight when s/he gives a comparison. Then, the interval weights are obtained by minimizing the sum of the widths of the interval weights of all items. Instead of the width, the uncertainty is measured based on the lower bounds of the interval weights [7, 8]. The sum of the lower bounds of all items represents the weight surely assigned to one of the items. Accordingly, the left weight is not sure which item has it, so that it can be shared by all items. Such a shared weight measures the uncertainty of a decision maker's thinking on the decision problem. The less uncertain a decision maker's thinking is, the more the sum of the lower bounds becomes. The width represents an item's uncertainty and the shared weight represents a problem's uncertainty. This study follows the latter uncertainty measurement, since a decision maker can give a comparison without assuming the weights of items in his/her mind. It puts more emphasis on uncertainty of a comparison matrix than an item and directly measures uncertainty of a comparison matrix.

3   Uncertainty of Criteria

This paper aims at assigning a number to a criterion. This section measures the uncertainty of a criterion through the example alternatives. The criterion is reviewed by the bottom up estimations from the alternatives, which are at the lower level than the criteria as in Fig. 1. The role of the alternatives in this study's setting is to reveal how uncertain a decision maker's thinking on the criterion is. Therefore, they need not to be in a real list but to be familiar to the decision maker. It is often easier for a decision maker to compare the substantial alternatives than the abstract criteria. The criterion importance is obtained not only from the criterion comparisons but also from alternative comparisons.

There assume to be n alternatives. They are compared from criterion k's viewpoint and the pairwise comparison matrix as $A_k = [a_{kij}]$ is given. Then, the weights of alternatives, $w_{ki}, \forall i$, and the uncertainty of matrix A_k, e_k, are obtained by the following problem.

$$
\begin{aligned}
&\min e_k, \\
s.t.\ &\frac{w_{ki}}{w_{kj} + e_k} \leq a_{kij} \leq \frac{w_{ki} + e_k}{w_{kj}}, \forall i, j, i > j, \\
&\sum_i w_{ki} + e_k = 1,
\end{aligned}
\tag{2}
$$

where the variables are $w_{ki}, \forall i$ and e_k. The uncertainty measurement e_k is the weight that is not sure which alternative has it. In the sense that all alternatives can share such a possible weight, it reflects the inconsistency in A_k caused by the uncertain decision maker's thinking. As for the weight of alternative i under criterion k, it is surely more than w_{ki}, at least, and is possibly $(w_{ik} + e_k)$, at most. By the first kind of constraints, the given comparison is included in the interval ration of the weights of the corresponding pair of alternatives. By the second constraint, the sum of all the weights and uncertainty is normalized to be 1. There are two kinds of weights, such as each alternative's sure weight and all alternatives' possible weight. The inconsistency among the given comparisons are reflected to the latter possible weight, which is shared by all alternatives. Although it does not discern uncertainty of each item, it focuses on uncertainty of whole problem.

In case of 5 alternatives under a criterion, $n = 5$, the weights by (2) are illustrated as a pie graph in Fig. 2. The black area is not sure which alternatives have the weight and measures uncertainty of a decision maker's thinking on the criterion under which 5 alternatives are compared.

4   Crisp Importance Weights of Criteria

In the previous section, the uncertainty of a decision maker's thinking on a criterion is measured by the shared weight in the given comparison matrix of

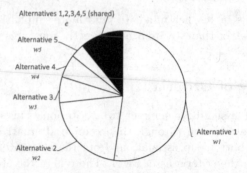

Fig. 2. Sure weights of alternatives and shared weight by all alternatives

alternatives under the criterion. The criteria are placed in order of their uncertainty measurements from the most uncertain to the least uncertain. Subscription k to distinguish criteria is renumbered so as to be $e_1 \geq e_2 \geq \ldots \geq e_m$. There are m criteria and the decision maker's thinking on criterion 1 is the most and that on criterion m is the least uncertain among all criteria, respectively. A decision maker gives a comparison matrix on m criteria, $A = [a_{ij}]$. Its elements are $a_{ij}, \forall i, j, i > j$, which represent how much more important criterion i comparing to criterion j in the decision problem. Similarly to (2), the importance weights of criteria $w_i, \forall i$ are obtained from A by the following problem.

$$
\begin{aligned}
& \min \sum_i e_i, \\
s.t. \quad & \frac{w_i}{w_j + e_j} \leq a_{ij} \leq \frac{w_i + e_i}{w_j}, \forall i, j, i > j, \\
& \sum_i (w_i + e_i) = 1, \\
& e_1 \leq e_2 \leq \ldots \leq e_m,
\end{aligned}
\tag{3}
$$

where variables w_i and $e_i, \forall i$ are the sure and possible importance weights of all criteria, respectively. In replace of the shared weight of all alternatives e_k in (2), the possibly assigned weights to the criteria, $e_i, \forall i$, are used in (3). Then, the optimistic importance weight of criterion i is denoted as $(w_i + e_i)$ and their sum is normalized by the second constraint. The uncertainty in the given comparison matrix A of the criteria is $\sum_i e_i$, which is possible to be shared by all criteria. When it is forced to be distributed to any of the criteria from the optimistic viewpoint, it is reasonable for criterion i to have e_i. By the last constraint with respect to uncertainty of criteria by (2), $\forall k$, the more uncertain a criterion is, the less the possible weight of the criterion becomes. In the same way as in (2), by the first kind of constraints, the given comparison is included in the interval ration of the weights of the corresponding pair of criteria. The proposed model (3) gives a decision maker two kinds of information. One is how much important a criterion for him/her in a decision problem is, w_i, and the other is how much uncertain his/her thinking on the criterion is, e_i. Then, as a final decision, the importance of the criteria reflecting their uncertainty are obtained as their optimistic crisp weights, $(w_i + e_i), \forall i$.

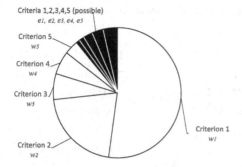

Fig. 3. Sure weights of criteria and possible weights of criteria

In case of 5 criteria, $m = 5$, their importance weights by (3) are illustrated as a pie graph in Fig. 3. Differently from Fig. 2, the black area is divided into 5 criteria. The assigned possible weights to criteria increases from criterion 1 to criterion 5. Each reflects the uncertainty of the decision maker's thinking on the criterion and represents the weight the criterion may have.

5   Numerical Example

5.1   Pairwise Comparison Matrix of Viewpoints

The example situation is that a client got an unexpected job offer and then started thinking about working for him/her. The criteria are the viewpoints for working, such as C1:salary, C2:work environment, C3:achievement, and C4:spontaneity. The alternatives are the positions, such as A1:programmer, A2:engineer, A3:researcher, and A4:manager. This study shows how important each viewpoint for him/her is to make him/her be aware of his/her attitude to jobs and working. A client is asked such a question; how much more important for your job is the salary than the spontaneity? It may be difficult for him/her to give these judgments on the viewpoints, since the criteria are often abstract and vague. Therefore, the information on the criteria through the example alternatives in the next section is used supplementary.

The pairwise comparison matrix of four viewpoints is given as in Table 1. Next to the matrix, the importance weights of the viewpoints are shown. They are obtained by (3) without the third constraint so that the uncertainty of the criteria is not considered. These weights reveal that a client evaluates the viewpoints in order, such as C1 > C3 > C2 = C4 from the most important to the least important, from the optimistic viewpoint. A crisp weight includes the possible weight in the parenthesis. From the pessimistic viewpoint, by excluding the possible weights form the optimistic weights, the ranking changes into C1 > C2 > C3 > C4, since most of the shared weight is assigned to C3. It is not sure whether this order reverse is reasonable or not, since we cannot discern which viewpoint causes the inconsistency of the comparisons. In the following

Table 1. Comparison matrix of viewpoints and importance weights without uncertainty information

	C1	C2	C3	C4	Importance (possible)
C1	1	5	3	5	0.577 (0.000)
C2	1/5	1	5	7	0.115 (0.000)
C3	1/3	1/5	1	5	0.192 (0.169)
C4	1/5	1/7	1/5	1	0.115 (0.099)
					uncertainty (0.268)

section, at first the criterion uncertainty is measured through the alternatives. Then, a criterion importance is obtained as a crisp weight reflecting its uncertainty in the client's thinking.

5.2   Uncertainty of Viewpoints

At first, the uncertainty of viewpoint for working is measured through four example positions at the lower level. They need not to be in his/her considering positions in the usual AHP, although they should be well-known and familiar to him/her. Then, s/he gives a pairwise comparison matrix of the jobs under each viewpoint by answering such a question; how much better from the achievement viewpoint is the researcher position than the engineer position? The pairwise comparison matrices on the positions from four viewpoints are given in Table 2. The judgments on the positions supplement the direct judgments on the viewpoints in Table 1, since it may be easy for him/her to compare the substantial positions than the conceptual viewpoints. It is reasonable to use the example alternatives to measure the criterion uncertainty.

The shared weights by all positions under each viewpoint are obtained by (2) and shown at the bottom rows of Table 2. They represent uncertainty of the comparison matrices since it remains unclear which positions have them. Then, the client's thinking on the viewpoints is in order, C1, C2, C3, and C4 from certain to uncertain. In addition, the sure weights of the alternatives are shown next to each matrix, although, they are not important information from the viewpoint of how much uncertain the client thinks on the criteria.

Let compare the comparison matrices under C1 and C4, whose shared weights are 0.057 and 0.323, respectively, so that they are the least uncertain and the most uncertain among four viewpoints. It looks apparent that position A4 is the best from viewpoint C1, since the comparisons at the 4th row of its matrix A_1 are all more than 1. Moreover, the ranking of the comparisons at four rows are the same as $a_{1*3} \geq a_{1*1} \geq a_{1*2} \geq a_{1*4}$. The comparisons in the same row represent the importance ratio of the position to the other positions. While, in the comparison matrix from viewpoint C4, it is different in the rows. Because of these difference, C4 is more uncertain than C1 so that C4 has more shared weight than C1.

Table 2. Comparison matrices of positions from viewpoints and shared weights

A_1 under C1	A1	A2	A3	A4	weight	A_2 under C2	A1	A2	A3	A4	weight
A1	1	1/3	3	1/7	0.068	A1	1	1	5	5	0.439
A2	3	1	5	1/5	0.148	A2	1	1	5	3	0.355
A3	1/3	1/5	1	1/7	0.041	A3	1/5	1/5	1	5	0.088
A4	7	5	7	1	0.685	A4	1/5	1/3	1/5	1	0.034
	uncertainty				0.057		uncertainty				0.084
A_3 under C3	A1	A2	A3	A4	weight	A_4 under C4	A1	A2	A3	A4	weight
A1	1	7	7	5	0.676	A1	1	1/3	1/9	1/3	0.079
A2	1/7	1	1/3	9	0.076	A2	3	1	1/5	7	0.142
A3	1/7	3	1	5	0.113	A3	9	5	1	1	0.389
A4	1/5	1/9	1/5	1	0.021	A4	3	1/7	1	1	0.066
	uncertainty				0.114		uncertainty				0.323

5.3   Importance of Viewpoints

Next, the criterion uncertainty obtained by the previous section is introduced into inducing its importance weight. The given comparisons of the viewpoints in Table 1 are inconsistent each other. In the same way as the positions in the previous section, there are the sure weight of each viewpoint and the shared weight by all viewpoints. As for criteria, the shared weight is forced to be assigned one of the criteria from the optimistic viewpoint. The possible weight of the least uncertain viewpoint, C1, is more than that to the most uncertain one, C4. The third constraint of (3) is $e_1 \geq e_2 \geq e_3 \geq e_4$. The crisp importance weights of the viewpoints are obtained by (3) and shown at the left column of Table 3. They include the possible weights in the parentheses. The viewpoints for working are in order from the most to least important as $C1 > C2 > C3 > C4$, which equals to uncertainty measurements' order.

In order to compare the influence of the criterion uncertainty on its weight through the alternative evaluations, the crisp weights of the criteria under the different assumption are shown at the right column of Table 3. It assumes the reverse order of the viewpoints, i.e., C1 and C4 are the most and least uncertain, so that the third constraint of (3) is $e_1 \leq e_2 \leq e_3 \leq e_4$, in replace. In this case, the viewpoints are in order from the most to the least important as $C1 > C3 > C4 > C2$, where C3 and C4 are almost equal. The importance weight order is not equal to the uncertainty order.

The sum of the weights in the parentheses is increased from 0.268 in Table 1 to 0.374 or 0.316 in Table 3. By taking the viewpoint uncertainty into consideration in (3), the inconsistency among the comparisons cannot be estimated as small as that only from the relative relations of the comparisons. However, it is acceptable, since it is not sure the given comparisons of the viewpoints represent the client's thinking as precisely as possible or they may be given by chance or mistake. At

Table 3. Crisp importance weights of viewpoints

	From uncertian criteria to certain criteria	
	C4, C3, C2, C1	C1, C2, C3, C4
	$e_4 \leq e_3 \leq e_2 \leq e_1$	$e1 \leq e_2 \leq e_3 \leq e_4$
C1	0.549 (0.105)	0.539 (0)
C2	0.215 (0.105)	0.108 (0)
C3	0.148 (0.105)	0.180 (0.158)
C4	0.089 (0.059)	0.174 (0.158)
	uncertainty (0.374)	uncertainty (0.316)

the left column of Table 3, viewpoint C2 is less uncertain than C3 and C4 in the client's thinking, so that its importance weight becomes larger than them. It is because that the less uncertain the criterion is, the more the possible weight it is assigned. On the other hand, when viewpoint C2 is assumed to be the second uncertain at the right column, it becomes the least important viewpoint among all. In this way, from the same given comparison matrices, the different optimistic crisp importance weights are obtained depending on the uncertainty of the criteria in the decision maker's thinking.

6   Conclusion

This paper shows a decision maker his/her thinking on the criteria in a hierarchically structured decision problem. It focuses on the importance of the criteria, instead of the evaluations of alternatives in the conventional AHP. A decision maker can barely give the comparisons of the criteria. However, it is not easy to discern how uncertain his/her thinking on each criterion is only from the given comparisons of the criteria. It is reasonable that the more uncertain the criterion for him/her is, the less importance it becomes. In order to reflect such a criterion uncertainty into its importance weight, this paper uses the evaluations of the alternatives at the lower level under the criterion. When the alternatives are familiar to a decision maker, it is easy for him/her to give their comparisons under a criterion. The criterion uncertainty is measured by the shared weights from the comparison matrix of the alternatives under it. Then, the criterion importance is obtained so as to add the more weight to the less uncertain criterion to its sure weight. This paper introduces the criterion uncertainty and considers it into deriving its crisp optimistic importance weight.

References

1. Alonso, J.A., Lamata, M.T.: Consistency in the analytic hierarchy process: a new approach. Uncertainty, Fuzziness Knowl.-Based Syst. **14**, 445–459 (2006)
2. Arbel, A., Vargas, L.: Interval judgments and euclidean centers. Eur. J. Oper. Res. **46**(7–8), 976–984 (2007)
3. Buckley, J.J., Feuring, T., Hayashi, Y.: Fuzzy hierarchical analysis revisited. Eur. J. Oper. Res. **129**(1), 48–64 (2001)
4. Dopazo, E., Lui, K., Chouinard, S., Guisse, J.: A parametric model for determining consensus priority vectors from fuzzy comparison matrices. Fuzzy Sets Syst. **246**, 49–61 (2014)
5. Dutta, B., Guha, D.: Preference programming approach for solving intuitionistic fuzzy AHP. Int. J. Comput. Intell. Syst. **8**(5), 977–991 (2015)
6. Dyer, R.F., Forman, E.H.: Group decision support with the analytic hierarchy process. Decis. Support Syst. **8**, 94–124 (1992)
7. Entani, T., Inuiguchi, M.: Maximum lower bound estimation of fuzzy priority weights from a crisp comparison matrix. In: Huynh, V.-N., Inuiguchi, M., Denoeux, T. (eds.) IUKM 2015. LNCS (LNAI), vol. 9376, pp. 65–76. Springer, Heidelberg (2015). doi:10.1007/978-3-319-25135-6_8
8. Entani, T., Sugihara, K.: Uncertainty index based interval assignment by interval AHP. Eur. J. Oper. Res. **219**(2), 379–385 (2012)
9. Pelaez, J., Lamata, M.: A new measure of consistency for positive reciprocal matrices. Comput. Math. Appl. **46**, 1839–1845 (2003)
10. Ramík, J., Perzina, R.: A method for solving fuzzy multicriteria decision problems with dependent criteria. Fuzzy Optim. Decis. Making **9**(2), 123–141 (2010)
11. Saaty, T.L.: The Analytic Hierarchy Process. McGraw-Hill, New York (1980)
12. Saaty, T.L.: Group Decision Making and the AHP. McGraw-Hill, New York (1989)
13. Saaty, T.L., Vargas, L.G.: Decision Making with the Analytic Network Process. Springer, US (2006)
14. Salo, A.A., Hämäläinen, R.P.: On the measurement of preferences in the analytic hierarchy process. J. Multi-Criteria Decis. Anal. **6**, 309–319 (1997)
15. Stein, W.E., Mizzi, P.J.: The harmonic consistency index for the analytic hierarchy process. Eur. J. Oper. Res. **177**, 488–497 (2007)
16. Sugihara, K., Ishii, H., Tanaka, H.: Interval priorities in AHP by interval regression analysis. Eur. J. Oper. Res. **158**(3), 745–754 (2004)
17. Sugihara, K., Tanaka, H.: Interval evaluations in the analytic hierarchy process by possibilistic analysis. Comput. Intell. **17**(3), 567–579 (2001)
18. Wenger, E.: Communities of Practice: Learning, Meaning, and Identity. Cambridge University Press, Cambridge (1999)

A Two-Stage Fuzzy Quality Function Deployment Model for Service Design

Hong-Bin Yan, Shaojing Cai, and Ming Li[✉]

School of Business, East China University of Science and Technology,
Meilong Road 130, Shanghai 200237, People's Republic of China
hbyan@ecust.edu.cn, shaojingcai@126.com,
limingl10228@163.com

Abstract. Over the past decades, a great interest of quality function deploy-ment (QFD) application has shift from industrial product to service sector. However, most previous studies focus on the first-phase of QFD or multi-phase model in service design, which is incomplete and not systematic. Moreover, the inherent vagueness or impreciseness in QFD presents a special challenge to the effective results. Toward this end, this paper tries to propose a two-phase fuzzy QFD model for service industry based on the theory of strategic, tactical and operational (STO). By this, on one hand, our proposed model can deal with the ill-defined nature of the linguistic judgments. On the other hand, it divides the decision and plan process into three correlated level, strategic, tactical and operational. In particular, this paper addresses detailed implementation proce-dures of the two-phase model, which can translate the customer needs (CNs) to service characteristics (SCs) to service process (SP) elements.

Keywords: Service design · QFD · STO model · Two-phase fuzzy QFD model

1 Introduction

Service firms have been exposed to the significant levels of competition, which has put great pressure on service firms to provide innovative service and to deliver services with a higher quality [8]. However, service planning in service improvement is still a challenging task for service organizations. Decision makers need to allocate limited resources to improve proper service measure and gain maximal benefit, which is in fact a multi-criteria decision-making process. Moreover, the service planning is derived by satisfying CNs which is usually conceptually vague. Consequently, it thus requires a systematic method to strategically deploy service planning into more specific level, especially to operational level.

QFD is a systematic planning process used by cross-functional teams to identify and resolve the issues involved in providing products, processes, services, and strategies that enhance customer satisfaction [20]. Over the past several decades, we have witnessed many studies on QFD and its applications, as reviewed in Sect. 2.1. Since the service may be viewed as a special kind of product, there are several studies on applications of QFD to service design. For example, Jeong et al. [11] have proposed

© Springer International Publishing AG 2016
V.-N. Huynh et al. (Eds.): IUKM 2016, LNAI 9978, pp. 110–123, 2016.
DOI: 10.1007/978-3-319-49046-5_10

a six-step House of Quality (HOQ) model and elaborated a hotel case study. Chan et al. [5] have built a systematic approach to prioritize service quality techniques for a fried Chinese vegetable. Essentially, the above-mentioned studies have made good use of the HOQ model, but HOQ model alone is incomplete and unsystematic for the service process. Therefore, Paryani et al. [20] have modified the conventional four-phase, manufacturing-based QFD process and applied it to the hotel service improvement. It is concluded that most studies focus on either the first phase (HOQ) or four-phase QFD model. Unfortunately, none of them is applicable to the service product development, since the service delivers process is not same as manufacture product process.

Traditional organization management theory divides the process of decision-making into three levels: strategic, tactical and operational level, known as STO framework. Commonly, the strategic decisions can be transformed into tactical decisions which will be transformed to some specific operational decisions later based on STO framework. In service innovation or improvement process, decisions can be divided into three levels: CNs, SCs, and SP elements, which are examples of strategic, tactical and operational decision, respectively. In this sense, the STO framework can be used to map the QFD process to conduct a two-phase model—from CNs to SCs and further to SP elements.

In traditional QFD, most of the input variables are assumed to be precise and are treated as crisp numerical data [10]. However, the inherent vagueness or impreciseness in QFD presents a special challenge to the effective results.

- The environment is marked by uncertainty and rapid changes in technologies and markets, and therefore multiple functional groups, each of which with a different perspective, are involved in a QFD system.
- The QFD process involves various linguistic information, e.g. human perception, judgment, evaluation on importance of CNs.
- Formal mechanisms for translating CNs (which are generally qualitative) into specific operations (which are usually quantitative) are lacking.

Due to the inherent vagueness or impreciseness in QFD, fuzzy linguistic variables semantically represented by fuzzy sets may seem more appropriate to describe those inputs in QFD [25], see the review in Sect. 2.1.

Toward this end, the main focus of this paper is to conduct a consistent two-phase fuzzy QFD model, which decouples the general strategic goals (CNs) to several tactical goals (SCs) and then further to detail operational plans (SP elements). By this, a systematic and complete QFD model is conducted to provide guidelines for decision makers in enterprises. By this, the service providers are able to know their goals and tasks clearly and what actions they should do to achieve the strategic goals. The rest of this paper is organized as follows. Section 2 briefly introduces the theories framework and proposes a conceptual two-phase QFD model of this paper. The detailed methodology based on the conceptual model is presented in Sect. 3. Finally, this paper is concluded with some future research in Sect. 4.

2 Preliminaries and Literature Review

2.1 Review of QFD

QFD frameworks are useful tools for constructing a new product development plan that enables the clear itemization of the voice customer requirements and for translating them into the various stages of product planning, design, engineering, and manufacturing by systematically evaluating each solution to maximize customer satisfaction [24]. Generally, a most commonly seen QFD consists of four inter-linked phases based on the manufacturing process in product sectors: Phase 1 to translate CNs into design requirements (DRs); Phase 2 to translate important DRs into product characteristics; Phase 3 to translate important product characteristics into manufacturing operations; and Phase 4 to translate key manufacturing operations into operations and control. Each phase is described by a matrix of WHATs and HOWs and that each phase's important outputs (HOWs), generated from the phase's inputs (WHATs), are converted into the next phase as its inputs (new WHATs) [5]. Specially, the first phase of QFD, HOQ, is of fundamental and strategic importance in the QFD framework, which usually contains some basic elements, such as voice of customer (VOC), market environment, goal, and financial or technological restriction.

Over the past several decades, we have witnessed many studies on QFD and its wide applications. For example, Chan and Wu [6] in their review of QFD theory and applications have noted wide range successful applications of QFD from product development, customer needs analyses to decision making. Because of its wide applicability, QFD has been used in various fields, such as, determining customer needs, developing priorities, manufacturing strategies [7], technology foresight [12] and supply chain management. QFD has also been applied successfully for service design, such as, logistics service [3], hotel service [20] and healthcare service [15].

To model and manage the inherent uncertainty and vagueness of linguistic descriptors, the fuzzy linguistic approach based on fuzzy set theory has been extensively used [18]. The use of linguistic information enhances the reliability and flexibility of classical decision models [17]. For example, Bevilacqua et al. [2] illustrated their fuzzy QFD based methodology for characterizing customer rating of food products. Chan et al. [5] applied fuzzy number and entropy methods to derive the importance of CNs, respectively, and combined the results to obtain the final importance of CNs. Bottani et al. [3] proposed an original fuzzy QFD methodology for strategic management of logistics service. Büyüközkan and Çifçi [4] proposed a fuzzy logic-based group decision making approach, which can be used for quality function deployment in the development of product improvement strategies. Su and Lin [21] used fuzzy QFD to identify the critical determinants relating to customer satisfaction. However, their model concentrates on apply fuzzy logic to the first phase of QFD, the subsequent phases (part deployment, process planning, and production planning) are rarely addressed. Thus, Liu [16] proposed an extended fuzzy QFD approach, which expends the research scope, from product planning to part deployment.

2.2   The STO Framework

Enterprises are highly complex systems in which one or more sectors share a definite mission, goals and objectives to offer a product or service. As a result, decision-making in the enterprise becomes a highly challenging task [17]. In order to deal with the system complexity, it is necessary to decouple the system across a hierarchy of appropriately chosen levels without disregarding the interrelationship that exists among them. Traditionally, enterprise management has been divided in three decision levels: strategic, tactical and operational, which can decouple the complex systems efficiently and consistently [9]. Therefore, we believe that one important step to support managers' decision-making is to conduct the STO framework, as shown in Fig. 1. The strategic level is a high-level plan and forms the foundational basis of a policy and will dictate decisions in the long-term. The tactical plan describes the procedures planned to achieve the ambitions goals of the strategic plan. It is a short-range plan. If the strategic plan is an answer to "What?" the tactical plan responds to "How?" The operational plan defines the day-to-day activities. The operational plan provides an approach to achieve the tactical goals within a realistic period of time. This plan is highly specific with focus on short-term goals [1]. Therefore, the general goals and plans can be divided into specific operations step by step based on the STO framework, everyone in each level knows their goals and tasks and what actions they should do to achieve the strategic goals clearly.

The STO framework is integrated in various models to solve practice problems. Thomason [23] discussed various demands based on the strategic, tactical and operational level and gave guidelines for demand management. Arababadi et al. [1] assessed the European Union (EU) 2020 energy through mapping energy policies to the STO framework. Laínez et al. [14] combined the strategic, tactical and operational supply chain decision levels with a model predictive control framework to tackle unforeseen events in the supply chain planning problem in chemical process industries. As a conclusion, the STO framework is helpful to deal with systematic problems and make them clear and continuous.

Fig. 1. The STO framework

2.3   Two-Phase Service Design Model Based on QFD

In service organization, many and varied decisions involved in designing and delivering a service are made at several levels – from the strategic level to the tactical level and to operational levels. It is very challenging for service organizations to ensure that decisions at each of these levels are made consistently, and focused on delivering the correct service to targeted customers. Fortunately, QFD is a strategic planning approach for ensuring consistency between the target customer needs and the service bundle elements, to more detailed issues that ensure each service component is delivered in a consistent and effective manner.

With the STO framework to the service enterprise in mind, the mission of strategic level for service designer can be regard as figuring out their CNs. In tactical level, designers have to decide through what kind of service to satisfy their CNs, in a professional word, SCs. After that, in the operational level, tactics are express or translated into more detailed SP elements. By this, designers are able to understand that what exactly they need to do so as to fulfill the CNs.

In our proposed model, two related matrices are used to correlate the three planning level and analyze the information, decomposed into a more specific and contrastable attribute: PhaseImatrix is used to translate general strategy into more several explicit tactics, and then Phase II is used to translate tactic into detailed operation, as shown in Fig. 2.

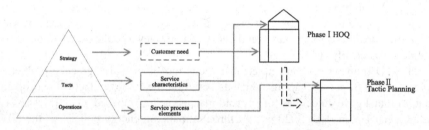

Fig. 2. The framework of proposed method

3   Proposed Methodology

3.1   Framework of the Proposed Methodology

In our proposed model, we constructed two consecutive matrices. The first matrix, the HOQ, translates emotional CNs into SCs, which are transferred to the second matrix named "service planning matrix" and subsequently translated into operational service. Figure 3 illustrates the overall structure and different components of the two matrices. It is worth noting that most information involved in the HOQ process is generated from human beings' perceptions and linguistic assessments that are quite subjective and vague. The symmetrical triangular fuzzy numbers (STFNs) are used to model the uncertainty and vagueness of the following four kinds of elements:

- the relative importance ratings of CNs;
- the relationships between CNs and SCs, and the ones between SCs and SP elements;
- the correlations among SCs; and
- the subjective evaluation of probability factors.

3.2   Detailed Procedures of the Proposed Methodology

3.2.1   Phase 1: From CNs to SCs

In this phase, the general CNs expressed in terms of emotional word are identified. In order to obtain the final importance ratings of CNs, the following three related attributes are required: customer's perceptions of relative important degrees of CNs, the competitive priority of CNs, and the expected improve rate of CNs. Then, the CNs are translated to the SCs through the correlation between CNs and SCs, and correlation among SCs.

- **Step 1-1: Identify the CNs**

    At the beginning of the process, understanding what customers need for a service is important for the company, otherwise you cannot know how to satisfy your customers and thus how to keep your business successful. Available methods to collect CNs include focus group, individual interviews, listening and watching, and using existing information. Suppose M CNs are identified, denoted as $CN_m, m = 1, \ldots, M$,

- **Step 1-2 : Determine the relative importance ratings of CNs**

    In this step, a survey is conducted to achieve the importance degrees of CNs. K customers are required to use the fuzzy linguistic terms in Table 1 to rate the importance degrees of CNs. The kth customer's rating of the importance degrees on CN_m is denoted as $\tilde{W}_m^k$. The relative importance rating of CN_m can be obtained as follows.

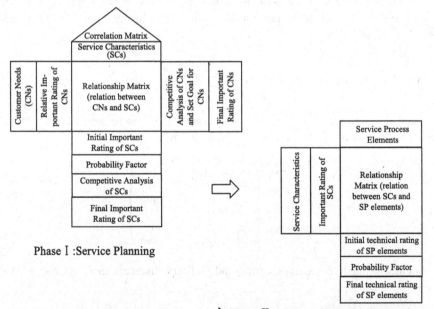

Phase I :Service Planning

Phase II : Process characteristics matrix

Fig. 3. The two-phase QFD service design model

Table 1. Rating scale for important rating of CNs

Linguistics term	Rating scale	TFNs
Very low	1	(0, 0, 0.3)
Low	2	(0, 0.3, 0.5)
Moderate	3	(0.2, 0.5, 0.8)
High	4	(0.5, 0.7, 1)
Very high	5	(0.7, 1, 1)

$$\tilde{W} = \frac{1}{K}\sum_{k=1}^{k} \tilde{W}_m^k \tag{1}$$

- **Step 1-3: Identify the competitive priority of CNs**

Similar services existing in the market should be identified by the company under study. Knowing the strengths and constraints of the company in question and its main competitors in all aspects is essential for a company if it wishes to improve its competitiveness. This kind of information can be obtained through questionnaire by asking the customers to rate the relative performance of the company and its competitors. Denote the company in question as C_1 and its main competitors as $C_2, \ldots, C_p$. Then K customers are requested to provide their perceptions on the performance of these P companies' services in terms of the M CNs. Suppose that customer k supplies a rating x_{mp}^k on company C_P's performance in terms of CN_m using scale (3), where x_{mp}^k is a crisp number in scale (3). Then the performance rating of company C_p on customer need CN_m is given as

$$x_{mp} = \sum_{k=1}^{k} x_{mp}^k \tag{2}$$

Thus, the companies' performance ratings on the CNs can be denoted by an $M \times L$ matrix, called customer comparison matrix, as follows.

$$x = \begin{array}{c} CN_1 \\ CN_2 \\ \vdots \\ CN_M \end{array} \begin{pmatrix} C_1 & C_2 & \cdots & C_p \\ x_{11} & x_{12} & \cdots & x_{1P} \\ x_{21} & x_{22} & \cdots & x_{2P} \\ \vdots & \vdots & \ddots & \vdots \\ x_{M1} & x_{M2} & \cdots & x_{MP} \end{pmatrix}$$

For measuring the companies' current and goal performance in terms of CNs and SCs:

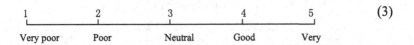

$$(3)$$

Then we can conduct the competitive priority ratings of each CN through entropy method (see Appendix). Suppose e_m is company C_1's priority rating on CN_m. According to the company C_1's current performance and its main competitor's performance, performance goals on the CNs can be set for the company. For customer need CN_m, a proper performance goal g_m has been set according to scale (3). In most cases, each goal performance level should not be lower than current performance level, implying the need or desire for further improvement. From this we can also set the company's improvement ratio for CN_m as

$$g'_m = g_m / x_{m1} \qquad (4)$$

- **Step 1-4: Determine the final importance ratings of CNs**

   The CNs with higher relative importance perceived by customers, higher competitive priorities, and improvement ratios should receive greater attention. Thus, the final importance rating of customer need CN_m is determined jointly by its relative importance $\tilde{W}_m$, competitive priority e_m, and improvement ratio g'_m as

$$\tilde{W}'_m = e_m \times g'_m \times \tilde{W}_m \qquad (5)$$

- **Step 1-5: Identify the SCs**

   The company's service design team should develop a set of tactical measures to satisfy the CNs. In service industry, SCs could be generated from current service standards or selected by ensuring through cause-effect analysis that the SCs are the first-order causes for the CNs. Assume N SCs have been developed, denoted as $SC_1, SC_2, \ldots, SC_N$.

- **Step 1-6: Determine the relationships between CNs and SCs**

   This is an essential step in HOQ which is determined jointly by several experts, who are required to evaluate the relationships between CNs and SCs. The relationships are expressed with graphic symbols which are able to indicate to what extent each SC could technically relate to and influence the CNs. Usually, symbols express three degrees of strength (weak, medium, strong), which can be translated to TFNs, as shown in Table 2. Let the relationship value between SC_n and CN_m be determined as $\tilde{r}_{mn}$. Then we can obtain the relationship matrix $[\tilde{r}_{mn}]$ between the SCs and the CNs.

Table 2. The relationship between CNs and SCs

Linguistics term	Symbol	TFNs
Weak	[△]	(0, 0, 0.3)
Medium	[○]	(0.3, 0.5, 0.7)
Strong	[◎]	(0.7, 1, 1)

Table 3. The correlation between SCs

Linguistics term	Symbol	TFNs
Strong negative	[■]	(0, 0, 0.3)
Negative	[□]	(0, 0.3, 0.5)
Positive	[○]	(0.5, 0.7, 1)
Strong positive	[●]	(0.7, 1, 1)

- **Step 1-7: Determine the correlations among SCs**

In this step, the relationships between each pairs of SCs are captured. Similar to the Step 1-6, several experts are asked to evaluate the correlations among SCs using graphic symbols in Table 3, denoted as $\tilde{C}_{nl}$ where $n, l = 1,...,N$. A positive relationship indicates that two SCs can complement or improve each other, while a negative one suggests that tradeoffs are required.

- **Step1-8: Determine the initial technical ratings of SCs**

In this step, the initial technical ratings of SCs are decided by three factors, final importance ratings of CNs, the relationships between CNs and SCs and the correlations among SCs. The importance ratings of SCs can be calculated as the following equation:

$$\tilde{i}_n = \frac{1}{M} \sum_{m=1}^{M} \tilde{W}_m \times \tilde{r}_{mn} \tag{6}$$

In order to quantitatively measure the correlations among SCs, we adopt the approach in Tang et al. [22] In particular, the correlation $\tilde{C}_{nl}$ can be interpreted as the incremental change of the degree attainment of the SC_n when attainment of the SC_l is unitary increased. Under this definition, the initial technical ratings of SCs can be computed as follows:

$$\tilde{i}_n^* = \tilde{i} + {}_n \sum_{l=n}^{N} \tilde{c}_{ln} \times \tilde{i}_n \tag{7}$$

- **Step 1-9: Evaluate the competitive priority of SCs**

Similar to the CNs competitive priority rating process (Step 1-3), several experts are required to evaluate the performance of SCs of each company by using scale (3). Suppose the evaluation score on company C_P's performance in terms of SC_n is denoted as y_{np}. Then we have the SCs comparison matrix Y as follows.

$$Y = \begin{array}{c} SC_1 \\ SC_2 \\ \vdots \\ SC_N \end{array} \begin{pmatrix} C_1 & C_2 & \cdots & C_p \\ y_{11} & y_{12} & \cdots & y_{1P} \\ y_{21} & y_{22} & \cdots & y_{2P} \\ \vdots & \vdots & \ddots & \vdots \\ y_{N1} & y_{N2} & \cdots & y_{NP} \end{pmatrix}$$

From this matrix, competitive priority of SCs can be obtained for C_1 company using the entropy method as described in the Appendix. We denote the competitive priority of SC_n as e'_n.

- **Step 1-10: Assess the probability factors for SCs**

In this step, the probability factors that affect the implementation of SCs are considered to complete analysis process. Two probability factors, financial cost and technological constrain are included in our study. Several experts are asked to assess the financial cost and technical constraint using the TFNs in Table 4. We use parameters $\tilde{\delta}_n^1$ and $\tilde{\delta}_n^2$ to denote the financial cost and technical constraint in implementing SC_n.

Table 4. Rating scale for probability factor and TFN

Financial cost (Technical constraint)	TFN
Very low (Very easy)	(0, 0, 2)
Low (Easy)	(0, 0.2, 0.4)
Medium	(0.3, 0.5, 0.7)
High (Difficult)	(0.6, 0.8, 1)
Very high (Very difficult)	(0.8, 1, 1)

- **Step 1-11: Obtain the final importance ratings of SCs**

Those SCs with higher initial technical rating $\tilde{t}_n^*$, higher technical competitive priorities e'_n, and high probability factor $\tilde{\delta}_n^1, \tilde{\delta}_n^2$ indicate working focuses and market opportunities for the company. The final importance ratings of SCs can be computed using follow equation:

$$\tilde{t}_n = \tilde{t}_n^* \times \frac{e'_n}{\tilde{\delta}_n^1 \tilde{\delta}_n^1} \tag{8}$$

3.2.2   Phase 2: From SCs to SP Elements

After the first phase analysis, service designers are able to recognize what kind of SCs should be fulfilled to satisfy their customers, but it is still difficult to apply the analysis result to operational level, that is the reason why we extend the analysis into second phase. In the second phase, SCs are able to be translated to SP elements, which capture the SCs in measurable and operable technical actions.

- **Step 2-1: Generate SP elements**

  In order to provide guidelines for service designers, SP elements (denoted as SP_i) that contribute to SCs have to be identified. Moreover, SP elements are measureable attributes usually generate from service standard or service specification. And the improvement direction of SP elements should be decided by service designers.

- **Step 2-2: Determine the relationships between SCs and SP elements**

  Similar to the evaluation of the relationships between CNs and SCs in step 1-6, the relationships between SCs and SP elements are assessed by several experts with the previous defined measure scale (Table 2). And we denote the relationships between SP_i and SC_n as r'_{ni}.

- **Step 2-3: Evaluate the initial technical ratings of SP elements**

  After getting the relationships between SCs and SP elements, the initial technical ratings of SP elements can be computed by the following equation:

$$\tilde{u}_i = \frac{1}{N} \sum_{n=1}^{N} \tilde{t}_n \times \tilde{t}'_{ni} \tag{9}$$

- **Step 2-4: Assess the probability factors for SP elements**

  Similar to Phase 1, the probability factors for SP elements are also considered in Phase 2, we defined γ_i^1 as financial cost and γ_i^2 as the technical constrain in implementing SP_i. The evaluation method is same as the method in Step 1-10.

- **Step 2-5: Determine the final importance ratings of SP elements**

  Taking the initial technical ratings and probability factors for SP elements into account, final technical ratings of SP elements can be computed by following equation:

$$\tilde{u}'_i = \frac{\tilde{u}'_i}{\tilde{\gamma}_i^1 \tilde{\gamma}_i^2} \tag{10}$$

In order to make $\tilde{u}'_i$ comparable and rank the results, defuzzified values should be computed. There are numerical methods related to the defuzzification of fuzzy numbers and is no one method univocally accepted. In this work, we suggest a commonly used method of choosing the convex combination between pessimistic and optimistic method that were applied to a TFN FN = $(\mathrm{FN}_\alpha, \mathrm{FN}_\beta, \mathrm{FN}_\gamma,)$ [13]. This produces a score identified by the value:

$$\text{score} = \frac{\mathrm{FN}_\alpha + 4 \cdot \mathrm{FN}_\beta + \mathrm{FN}_\gamma}{4} \tag{11}$$

4 Concluding Remarks

In this paper, we have developed a two-phase fuzzy QFD model for service design based on the STO framework and fuzzy theory. By implementing the proposed methodology, the final importance ratings of SCs and SP elements can be easily obtained, which can provide service designers useful guidelines to pay more attention to the SPs that really affect the CNs. Although we propose a new systematic model and a detail methodology, a real case study and a comparative study are still needed to evaluate the efficiency and effectiveness of our proposed methodology. However, this is left for the future work.

Acknowledgement. We would like to appreciate constructive comments and valuable suggestions from the anonymous referees. This study was partly supported by the National Natural Sciences Foundation of China (NSFC) under grant no. 71471063; sponsored by the Innovation Program of Shanghai Municipal Education Commission under grant no. 14ZS060; and supported by the Fundamental Research Funds for the Central Universities in China under grant no. WN1516009.

Appendix: Entropy Method

The application of entropy begins with the construction of comparison matrix, step 1-3 is to obtain and analyze the follow comparison matrix X, similar to the step 1-3, step 1-9 is to obtain SCs comparison matrix Y.

$$
X = \begin{array}{c} CN_1 \\ CN_2 \\ \vdots \\ CN_M \end{array}
\begin{pmatrix}
C_1 & C_2 & \cdots & C_p \\
x_{11} & x_{12} & \cdots & x_{1P} \\
x_{21} & x_{22} & \cdots & x_{2P} \\
\vdots & \vdots & \ddots & \vdots \\
x_{M1} & x_{M2} & \cdots & x_{MP}
\end{pmatrix}
\quad
Y = \begin{array}{c} SC_1 \\ SC_2 \\ \vdots \\ SC_N \end{array}
\begin{pmatrix}
C_1 & C_2 & \cdots & C_p \\
y_{11} & y_{12} & \cdots & y_{1P} \\
y_{21} & y_{22} & \cdots & y_{2P} \\
\vdots & \vdots & \ddots & \vdots \\
y_{N1} & y_{N2} & \cdots & y_{NP}
\end{pmatrix}
$$

Take example of comparison X, where x_{mp} is the performance of company p's service on customer need CN_m. The managerial mean of applied entropy method: If company C_1 performs much better than many companies in terms of a customer need CN_m, then further improvement may not be urgently needed and thus a lower priority could be assigned to CN_m. At the other extreme, if C_1 performs much worse than many other companies on CN_m, then it may be difficult for C1 to build a competitive advantage within a short period of time. In both cases, CNs could be assigned a lower priority rating. However, if most companies perform quite similarly on CNs, not too much improvement effort from C_1 may result in a better performance of its service and give C_1 a unique competitive advantage.

Entropy is a measure for the amount of information (or uncertainty, variations) represented by a discrete probability distribution, $q_1, q_2, \ldots, q_L$:

$$E(q_1, q_2, \ldots, q_p) = -\emptyset_p \sum\nolimits_{p=1}^{p} q_p \ln(q_p) \tag{A.1}$$

where $\phi_p = 1/\ln(P)$ is normalization constant to guarantee $0 \leq E(q_1, q_2, \ldots, q_p) \leq 1$. Larger entropy or $E(q_1, q_2, \ldots, q_p)$ value implies smaller variations among the Q_p's and hence less information contained in the distribution.

Take the computation of CNs competitive priority for an example, for the mth row of customer comparison matrix X corresponding to the customer need CN_m, x_{m1}, x_{m2}, ..., x_{mp}, let $x_m = \sum_{p=1}^{p} x_{mp}$ be the total score with respect to CN_m, then according to (B.1), the normalized ratings $q_{mp} = x_{mp}/x_m$ for p = 1,2,...,P can be viewed as the "probability distribution" of CN_m on the P companied with entropy as

$$E(q_1, q_2, \ldots, q_p) = -\phi_p \sum\nolimits_{p=1}^{p} q_p \ln(q_p) = -\phi_p \sum\nolimits_{p=1}^{p} \left(\frac{x_{mp}}{x_m}\right) \ln\left(\frac{x_{mp}}{x_m}\right) \tag{A.2}$$

It is clear that larger the $E(CN_m)$ value, the less information contained in CNs or smaller variations among the qmp's (or x_{ml}'s). So $E(CN_m)$ can be used to reflect the relative competitive advantage in terms of customer need CN_m. All these E(CNs) value, after normalizations:

$$E(q_1, q_2, \ldots, q_p) = -\emptyset_p \sum\nolimits_{p=1}^{p} q_p \ln(q_p) \tag{A.3}$$

where e_m can be considered as the competitive priority rating for company C_1 on the M customer needs. It is similar to compute the competitive priority of SCs.

References

1. Arababadia, R., Parrisha, K., Asmara, M.E.: Waging war on climate change: mapping energy policies to their strategic, tactical, and operational levels. Procedia Eng. **145**, 11–17 (2016)
2. Bevilacqua, M., Ciarapica, F.E., Marchetti, B.: Development and test of a new fuzzy-QFD approach for characterizing customers rating of extra virgin olive oil. Food Qual. Prefer. **24**, 75–84 (2012)
3. Bottani, E., Rizzi, A.: Strategic management of logistics service: a fuzzy QFD approach. Int. J. Prod. Econ. **103**, 585–599 (2006)
4. Büyüközkan, C., Çifçi, G.: An extended quality function deployment incorporating fuzzy logic and GDM under different preference structures. Int. J. Comput. Intell. Syst. **8**, 433–454 (2015)

5. Chan, L.K., Wu, M.L.: A systematic approach to quality function deployment with a full illustrative example. Omega **33**, 119–139 (2005)
6. Chan, L.K., Wu, M.L.: Quality function deployment: a literature review. Eur. J. Oper. Res. **143**(3), 463–497 (2002)
7. Crowe, T.J., Cheng, C.C.: Using quality function deployment in manufacturing strategic planning. Int. J. Oper. Prod. Manage. **16**(4), 35–48 (1996)
8. Geuma, Y.J., Kwak, R., Park, Y.: Modularizing service A modified HOQ approach. Comput. Ind. Eng. **62**, 579–590 (2012)
9. Hitt, M.A.: Spotlight on strategic management. Bus. Horiz. **49**, 349–352 (2006)
10. Ho, E.S.S.A., Lai, Y.J., Chang, S.I.: An integrated group decision-making approach to quality function deployment. IIE Trans. **31**(6), 553–567 (1999)
11. Jeong, M., Oh, H.: Quality function deployment: An extended framework for service quality and customer satisfaction in the hospitality industry. Hospitality Manage. **17**, 375–390 (1998)
12. Ju, Y.H., Sohn, Y.S.: Patent-based QFD framework development for identification of emerging technologies and related business models: A case of robot technology in Korea. Technol. Forecast. Soci. Chang. **94**, 44–64 (2015)
13. Chan, L.-K., Kao, H.P., Ng, A., M-L, Wu: Rating the importance of customer needs in quality function deployment by fuzzy and entropy methods. Int. J. Prod. Res. **37**, 2499–2518 (1999)
14. Laínez, J.M., Kopanos, K.M., Badell, M. et al.: Integrating strategic, tactical and operational supply chain decision levels in a model predictive control framework. In: 18th European Symposium on Computer Aided Process Engineering (2008)
15. Lee, C.K.M., Ru, C.T.Y., Yeung, C.L., Choy, K.L., Ip, W.H.: Analyze the healthcare service requirement using fuzzy QFD. Comput. Ind. **74**, 1–15 (2015)
16. Liu, H.: The extension of fuzzy QFD: From product planning to part deployment. Expert Syst. Appl. **36**, 11131–11144 (2009)
17. Martínez, L., Ruan, D., Herrera, F., Herrera-Viedma, E.: Linguistic decision making: Tools and applications Preface. Inf. Sci. **179**, 2297–2298 (2009)
18. Martínez, L., Rodríguez, R M., Herrera, F.: The 2-tuple linguistic model. Comput. Words Decis. Making 1–4 (2015)
19. Muñoza, E., Capón, E.: Operational, Tactical and strategical integration for enterprise decision-making. In: Proceedings of the 22nd European Symposium on Computer Aided Process Engineering, London (2012)
20. Paryani, K., Masoudi, A., Cudney, E.A.: QFD application in the hospitality industry: A hotel case study. Qual. Manage. J. **17**, 7–28 (2010)
21. Su, C.T., Lin, C.S.: A case study on the application of Fuzzy QFD in TRIZ for service quality improvement. Qual. Quant. **42**, 563–578 (2008)
22. Tang, J., Fung, R.Y.K., Xu, B.D., et al.: A new approach to quality function deployment planning with financial consideration. Comput. Oper. Res. **29**, 1447–1463 (2002)
23. Thomason, D.: Explains why all three levels of demand management are needed. Manufac. Eng. **83**, 34–37 (2014)
24. Yan, H.B., Ma, T.J., Li, Y.: A novel fuzzy linguistic model for prioritizing engineering design requirements in quality function deployment under uncertainties. Int. J. Prod. Res. **51**(21), 6336–6355 (2013)
25. Zadeh, L.A.: Fuzzy sets. Inf. Control **8**(3), 38–353 (1965)

Usages of Fuzzy Returns
on Markowitz's Portfolio Selection

Tanarat Rattanadamrongaksorn[✉], Jirakom Sirisrisakulchai,
and Songsak Sriboonjitta

Faculty of Economics, Chiang Mai University, Chiang Mai, Thailand
tanarat_ra@cmu.ac.th

Abstract. Given the unavailability of historical data, selecting portfolio by the Markowitz's becomes difficult, if not impossible. In this particular situation expert opinion is the inevitable option. To cope with the nature of subjective data and their different types of inherent risks, we have developed the fuzzy-set based approach following the direction of the modern theory. Instead of the deployment of traditional probability distribution, six possibilistic shapes of fuzzy numbers have been proposed in order to simplify the translation of linguistic terms into fuzzy returns. The returns on assets are scaled according to their allocation percentages and combined by operations on fuzzy restrictions while optimized on centroids and other indices in fuzzy set theory. The demonstration has been carried out on several asset types and solved by the general-purpose genetic algorithm. The structure of fuzzy number is still preserved throughout the process and, as a result, breeds the resultant portfolio distinct.

Keywords: Ambiguity · Fuzziness · Centroid of fuzzy number · Fuzzy arithmetic · Linguistic variable · Possibility distribution · Non-random uncertainty

1 Introduction

The applications in financial economics depend heavily on historical data and, therefore, modeling by possibility theory is arguably criticized to be less efficient than that by probability theory. There are, however, some particular aspects, i.e. re-evaluating existing assets with structural change, choosing Initial-Public-Offering (IPO) stocks, funding infrastructure projects, partnering in start-up businesses or patents, tendering for concessions and etc., that allow possibilistic uncertainties to blend naturally. In these situations the existing data are neither available nor applicable and, thus, leading to the utilization of expert opinion.

The mean-variance approach has been the widely accepted standard for portfolio selection since Markowitz's discovery [1]. Also known as the modern portfolio theory, MPT is the framework in allocating the proportion of assets within portfolio by maximizing expected return and minimizing risk in terms of mean

© Springer International Publishing AG 2016
V.-N. Huynh et al. (Eds.): IUKM 2016, LNAI 9978, pp. 124–135, 2016.
DOI: 10.1007/978-3-319-49046-5_11

and variance respectively. The key is the diversification resulting in the synergy of portfolio that must exceed the individual's sole performance. The continual work of MPT is to modify the specification of the risk as a semi-variance in order to emphasize the negative effects on unsymmetrical distributions [2].

Fuzzy set theory, on the other hand, is the generalization of set theory to represent the non-random uncertainties [3]. Considering the classification of object into subset - also called restriction, while that in classical set is said to be either completely in or out of the considered subset; the fuzzy set, on the other hand, allows the element partly being member of the subset. The formal definition of fuzzy number defined as $\mathcal{N} : \mathbb{R} \rightarrow [0,1]$ that is the mapping of the value of the element from real number to any number in the range of zero and one on the degree of membership. Fuzzy set is characterized by a pair - the value of element and its degree of membership, denoted by $[x, \mu_x]$. The element, x, is a universal object while its degree of membership or degree of compatibility, μ_x, is how much the element belongs to the subset. The degree of membership can be any real number from zero to one.

Although abundant and admitted as a state of the art in other disciplines, the applications of fuzzy set theory are still limited and scarce in the field of economics. Few samples exist still and their benefits shine. For example, fuzzy design in regression discontinuity [4] permits the soft criteria for policy evaluation. The fuzzy game enhances the empty core by relaxing the constraints gradually in order to ensure the coverage of solution in the equilibria [5]. In some recent works, the fuzzy random variable was introduced to describe a random phenomenon with fuzzy expected return [6] and later used with the newly proposed definition of risk that concerned the credibility of the intended return target [7]. Additionally the fuzzy random variable could be simplified by historical data for both short- and long-term observations constructed from maximum, minimum, and mean [8]. Fuzzy logic for underground economy took another level of advancement in predicting the hidden values of underground economies [9] and their tax burdens [10]. Because of the similar reasons as pointed out in the literature on different cultures of stochastic data modeling [11], it is unfortunate that the concept has been left neglected.

There are, however, several works that provide the mathematical proofs of the central limit theorem on the fuzzy random variables [12] and, in consequence, the possibilistic model can assume normal distribution. The fuzzy variable in this research is merely a specific case of fuzzy random variable [13] and, therefore, the asymptotic behavior could be claimed by such the same reason. In addition the fuzzy Delphi method [14] is suggested as a remedy approach for long term forecasting in which the expert's estimations are converge by statistical method.

The following discussions are divided into four sections. Section 2 contains the necessary concepts and the mapping of the entities from fuzzy set theory into portfolio selection problem. Section 3 states formally the mathematical descriptions of the portfolio selection problem. In Sect. 4 the demonstration of the proposed methodology is provided along with results and interpretations. Section 5 contains conclusions, advantages, disadvantages, and future topics of the research.

2   Proposed Methodology

The objective of this study is to propose the fuzzy set-based methodology as a complementary application to the modern portfolio theory into the realm of subjectivity. Where the past observations are absent, the method is aimed to transfer the knowledge from linguistic terms into mathematical expressions, manage the inherent uncertainties and produce the human-intuitive interpretation in a proper manner. This leads to the use of well-known fuzzy set theory for its abilities to deal with fuzzy data. The proposed methodology contains four steps of

Step 1: representing subjective data of the return on assets by fuzzy numbers,
Step 2: scaling the returns into fraction of investment by scalar multiplication,
Step 3: combining the proportions of investment into resultant portfolio, and
Step 4: evaluating and selecting portfolio by a set of indices.

The following subsections discuss on how to match project selection problem with fuzzy set theory. The necessary concepts include fuzzy returns in SubSect. 2.1, operations on fuzzy returns in SubSect. 2.2, centroids of fuzzy returns in SubSect. 2.3 and fuzzy risks in SubSect. 2.4. In the current stage of development, we strict our scope and direction similar to those of the modern portfolio theory [1] and shall briefly recommend the possible extensions in Sect. 5.

2.1   Fuzzy Returns

Knowledge about asset return is collected as percentage yields on investment. In the situation where observation is neither available nor usable, the best source of data is a decent reasoning from the qualified person in the field. However, human's judgment is always imprecise and uncertain, hence, fuzzy in nature. The return on asset in this study is therefore called fuzzy return for such a reason. For example, one may be familiar with investing in a type of asset, say real estate. He has worked extensively in this field for more than 20 years so that he could claim himself as an expert in the real estate investment. His opinion on the rate of return, after thorough assessing one of the incoming pre-sales of the commercial project on prime area, may be said as, for instance, "definitely at 12.5 percent", "between 10–15 percent", "most possible at 8.5 percent and least possible at 20 percent" and so on. Expert opinion is expressed verbally in this manner.

In the context of portfolio, fuzzy number should relate the asset return, x, to its possibility rather than its degree of membership. We interpret the degree of membership, $\mu_{(x)}$, as the level of possibility, $\pi_{(x)}$, in a sense that the range of zero and one on the degree of membership corresponds to the least (or not-possible) and the most possible respectively on the level of possibility. One asset, for a convincing example, with 12 percent return rate is approximated that its degree of compatibility to the moderately interesting asset is 0.7. In other words, this same asset which is moderately interesting is evaluated that the level of possibility at 12 percent rate of return of 0.7. Both statements have been defined equal [15] through the restriction. These justifications presume their equivalences at least in this work and, as a result, interchangeable even though not so in perpetual [16]. However, we stick to the μ symbol as our work is based mostly on the concept of fuzzy numbers, except for the better intuitive interpretation that has just been mentioned.

Fuzzy set theory naturally links to linguistic variables [17–19]. The translation and representation of linguistic terms into expression of fuzzy set utilizes the so-called fuzzy number. Fuzzy number demonstrates numerical quantities of fuzzy data in the form of relations between the values and its level of possibility. Theoretically fuzzy number satisfies the conditions [20] of (1) continuity that prevents jump or gap in a series of values, (2) convexity that maintains the validity of linear combination, (3) normality that normalizes the support of the level to the norm and ensures the existence of the apex at the upper bound, and (4) uniqueness that allows only single peak in each fuzzy number.

In Table 2 the common shapes of fuzzy number have been introduced for possibility distributions. Table 1(1) shows the case asset return can be determined precisely or no distribution. The shape in Table 2(2) is rectangle fuzzy number representing the uniform possibility distribution for the situation that the returns are certain but cannot be pinpointed into exact value.

It is the desired feature that the distribution could be derived either from linguistic terms or in probability-like style. The mean-variance using triangular shape as depicted in Table 2(3) could be constructed subjectively by the triple, namely the most pessimistic, the most optimistic, and the most possible. On the other hand, these shapes of fuzzy numbers could match with probability distributions such as normal distribution, in the case of triangle, which is so intuitive that migrating to the possibility side of uncertainties becomes more convenient. Fuzzy shape in Table 2(4) and 2(5) are aimed for the circumstances that the returns could be numerated at the most possible and the largest possible values. They could be viewed as the left and right skewed which are the variations of the the the shape of mean-variance. The trapezoidal fuzzy number or TFN in Table 2(6) is the general case and sufficient to represent every form of fuzzy number aforementioned. The mathematical representation of TFN is

$$\mathcal{N} = (a, b, c, d) \tag{1}$$

whose percentage of asset returns are a, b, c, and d and their corresponding possibilities are 0, 1, 1, and 0 respectively. That is the percentages of fuzzy return with $\mathcal{N}$ are the most possible between b and c and the largest possible between a and d. Any unknown in between would be calculated by linear interpolation.

Fuzzy number is usually decoded from human knowledge that is normally constructed linearly (of course there are many forms of nonlinear fuzzy numbers too) and, therefore, naturally constructs simple shapes such as triangle, trapezoid, rectangle, and etc. As a result, it is convenient and interesting to find the geometric properties such as positions of the points, lengths of the borders and areas of the shapes that lead further to versatile analysis on other properties such as operations on and centroids of fuzzy numbers that are used in this research.

Table 1. Fuzzy numbers for possibility distributions

#	Fuzzy Numbers	Linguistic Terms	Math. Expressions
1	**No distribution** 0 Possibility 1 — a Returns %	Definitely at a	$\mathcal{N} = (a,a,a,a)$ $\mathcal{L} = 0$ $\mathcal{A} = 0$ $\widetilde{x} = a$ $\widetilde{\mu} = 1$ $\mathcal{FZ} = 0$ $\mathcal{AG} = 0$
2	**Uniform** 0 Possibility 1 — b c % Returns	Equally possible- from b to c	$\mathcal{N} = (b,b,c,c)$ $\mathcal{L} = c-b$ $\mathcal{A} = c-b$ $\widetilde{x} = \frac{c+b}{2}$ $\widetilde{\mu} = 1$ $\mathcal{FZ} = 0$ $\mathcal{AG} = \frac{c-b}{2}$
3	**Mean-variance** 0 Possibility 1 — a b=c d % Returns	Most pessimistic at a, Most optimistic at d, Most possible at b or c	$\mathcal{N} = (a,b,b,d)$ $\mathcal{L} = d-a$ $\mathcal{A} = \frac{d-a}{2}$ $\widetilde{x} = \frac{d^2-a^2+cd-ab}{3d-3a}$ $\widetilde{\mu} = 0.5$ $\mathcal{FZ} = d-a$ $\mathcal{AG} = \frac{d-a}{6}$
4	**Negative-skewed** 0 Possibility 1 — a b % Returns	Least possible at a, Most possible at b	$\mathcal{N} = (a,b,b,b)$ $\mathcal{L} = b-a$ $\mathcal{A} = \frac{b-a}{2}$ $\widetilde{x} = 2b^2 + a$ $\widetilde{\mu} = 0.5$ $\mathcal{FZ} = b-a$ $\mathcal{AG} = \frac{b-a}{6}$
5	**Positive-skewed** 0 Possibility 1 — c d % Returns	Most possible at c, Least possible at d	$\mathcal{N} = (c,c,c,d)$ $\mathcal{L} = d-c$ $\mathcal{A} = \frac{d-c}{2}$ $\widetilde{x} = \frac{d+2c}{3}$ $\widetilde{\mu} = 0.5$ $\mathcal{FZ} = d-c$ $\mathcal{AG} = \frac{d-c}{6}$
6	**Interval** 0 Possibility 1 — a b c d % Returns	Largest possible- between a and d, Most possible- between b and c	$\mathcal{N} = (a,b,c,d)$ $\mathcal{L} = d-a$ $\mathcal{A} = \frac{d+c-b-a}{2}$ $\widetilde{x} = \frac{d^2+c^2-a^2-b^2+cd-ab}{3d+3a-3b-3a}$ $\widetilde{\mu} = \frac{d+c-b-a}{2d-2a}$ $\mathcal{FZ} = d-c+b-a$ $\mathcal{AG} = \frac{d+2c-2b-a}{6}$

2.2   Operations on Fuzzy Returns

Selecting portfolio requires allocating proportion of investment to each asset and combining them together into portfolio. The fractions of investment for each fuzzy returns can be weighed by scalar multiplication to fuzzy number,

$$\omega \mathcal{N} = (\omega a, \omega b, \omega c, \omega d) \tag{2}$$

where ω is weight or proportion of investment. The operations on fuzzy numbers require identification of the points with the same possibility usually at the ends or corners and then perform the calculation between each corresponding point in the normal style. After the allocations, the scaled fuzzy returns are aggregated by fuzzy addition,

$$\mathcal{N}_1 \oplus \mathcal{N}_2 = (a_1 + a_2, b_1 + b_2, b_1 + b_2, d_1 + d_2) \tag{3}$$

where $\oplus$ represents fuzzy addition and the subscripts specify the identifications of fuzzy returns.

The use of fuzzy addition could possibly be simply explained by an example. Supposedly there are two assets. At the end of the next maturity, asset A has the estimated prices of the most pessimistic at 1 million dollars, the most possible at 2 million dollars, and the most optimistic at 3 million dollars, or $R_A = (1, 2, 2, 3)$. Asset B is risk-free and will be definitely at 2 million dollars in value, or $R_B = (2, 2, 2, 2)$ – that is special case of crisp value of fuzzy return. Which portfolio is consisted of these two assets can be approximated to be the most pessimistic at 3 million dollars, the most possible at 4 million dollars, and the most optimistic at 5 million dollars respectively or, mathematically in the form of fuzzy addition, $(1, 2, 2, 3) \oplus (2, 2, 2, 2) = (3, 4, 4, 5)$. Geometrically, the shapes of fuzzy returns for asset A and B are triangle and a point respectively while the resultant is triangle and the linguistic interpretation has been reserved accordingly.

2.3   Centroids of Fuzzy Return

The expectation of fuzzy return is obtained by defuzzification. In this research the method of centroid has been selected [21]. The length and the area of fuzzy number represent the overall and the weighed possibility that constitutes different kinds of moments on the considered element. By the method of moment, the expectation of return is calculated by the centroid of fuzzy return,

$$\widetilde{x}_{(x)} = \frac{\int_x x \mu_{(x)} dx}{\mathcal{A}} \tag{4}$$

where x is percentage of asset return and $\mathcal{A}$ is the area of fuzzy number. In addition to the traditional method, we also consider the moments around the $x-$axis. The expectation of possibility is then computed by the centroid of possibility,

$$\widetilde{\mu}_{(x)} = \frac{\int_x \mu_{(x)} dx}{\mathcal{L}} \tag{5}$$

where $\widetilde{\mu}$ represents centre of length and $\mathcal{L}$ is the length of the fuzzy number's border. Normally centre of area for triangle is one-third accounted from the bigger end and two-thirds accounted from the smaller end. It is one-half for rectangle in perpendicular coordinates while the center of line is always one-half.

The centroid of possibility is the representation of overall possibility for each fuzzy return. For instance, comparing the asset with uniform fuzzy return to another asset with mean-variance, the former is more possible or more certain assuming equal expected return. It has also been shown in Table 2 that the more complex shape of fuzzy number, the more varied the centroid of possibility. The examination, probably for some purpose, can be made across the centroids because both are based on the same scale, i.e. zero to one. However, while the expectation of possibility can take value only within the unit interval, the expectation of return has no boundary i.e. it is possible for either profit that return exceeds a hundred percent in the case of extreme profit or negative return for loss.

2.4 Risks of Fuzzy Returns

The definition of risk is officially given as (1) the possibility of suffering harm or loss and (2) the variability of returns from an investment [22] that corresponds to the uncertainty and imprecision in this research respectively. Uncertainty means something not having sure knowledge and imprecision means the quality of lacking the number of significant digits to which a value has been reliably measured [22]. Fuzzy number takes care of the uncertainty by the level of possibility and the imprecision by the interval data as meant by the preceding definitions. Recalling SubSect. 2.1, the uncertainty is reflected in verbals such as "definitely", "the most possible", "the least possible" and while the imprecision refers to the verbals such as "at", "between", "from-to". These risks have been incorporated within each fuzzy number in Fig. 1. To quantify uncertainty and imprecision, two indices [13] have been utilized.

Fuzziness is a kind of uncertainty belonging to fuzzy number which means not clear [22]. Fuzziness index is the measurement of the degree that the arbitrary element belongs to the subset by how much the set differs from its complement. In our context the interpretation is a little different because the subset is interpreted as a possibility in defining the return on asset. The numerical value of fuzziness is calculated by fuzziness index,

$$\mathcal{FZ}_{(x)} = \int_x \left(1 - |2\mu_{(x)} - 1|\right) dx \tag{6}$$

and is probably called fuzziness measure in other literature [13]. Considering the extreme cases, the fuzziness is lowest when the possibility is either zero or one and highest at one-half. In other words, the risk is minimum in case of the most and the least possible and vice versa in the case of half-possible or ill-defined.

Ambiguity, on the other hand, is the measurement of the difficulty in making precision of the element by weighing all possible values of the element with its level of possibility by ambiguity index,

$$AG_{(\mu)} = \int_{\mu} \mu \left[\overset{*}{x}_{(\mu)} - \underset{*}{x}_{(\mu)} \right] d\mu \qquad . \qquad (7)$$

where $\overset{*}{x}_{(\mu)}$ and $\underset{*}{x}_{(\mu)}$ are maximum and minimum of returns for the particular level of possibility. The zero value of ambiguity means the most precise or the value of return can definitely be assigned and the more the ambiguity the wider the interval.

Simply, weighing of the fuzziness index and ambiguity index is sufficient to combine both effects into single determination of uncertainties of fuzzy number, namely fuzzy number quality index, $\mathcal{FNQI}_{(N)} = \omega_{\mathcal{FZ}}\mathcal{FZ}_{(N)} + \omega_{\mathcal{AG}}\mathcal{AG}_{(N)}$ where $\omega_{\mathcal{FZ}} + \omega_{\mathcal{AG}} = 1$ and $\omega_{\mathcal{FZ}}, \omega_{\mathcal{AG}} \in [0,1]$. In normal situation the weight of fuzziness and that of ambiguity index are the same, $\omega_{\mathcal{FZ}} = \omega_{\mathcal{AG}} = 0.5$, as we tend to evaluate them with equal importance.

3   Problem Statement

The optimization problem is defined by maximizing expectations and minimizing risks corresponding to the classical mean-variance approach. However with fuzzy set theory the problem is more diversified. It has been derived by maximizing the centroids of return, maximizing centroid of possibility, minimizing fuzziness, and minimizing ambiguity. Mathematically the portfolio selection problem could be rearranged into optimization by Eqs. 2–7 sequentially resulted as

$$\max \; \frac{\sum_i \omega_i^2 d_i^2 + \sum_i \omega_i^2 c_i^2 - \sum_i \omega_i^2 a_i^2 - \sum_i \omega_i^2 b_i^2 + \sum_i \omega_i c_i \sum_i \omega_i d_i - \sum_i \omega_i a_i \sum_i \omega_i b_i}{3 \left(\sum_i \omega_i d_i + \sum_i \omega_i a_i - \sum_i \omega_i b_i - \sum_i \omega_i a_i \right)}, \quad (8)$$

$$\max \quad \sum_i \omega_i d_i + \sum_i \omega_i c_i - \sum_i \omega_i b_i - \sum_i \omega_i a_i, \qquad (9)$$

$$\min \quad \sum_i \omega_i d_i - \sum_i \omega_i c_i + \sum_i \omega_i b_i - \sum_i \omega_i a_i, \qquad (10)$$

$$\min \quad \frac{1}{6} \left(\sum_i \omega_i d_i + 2 \sum_i \omega_i c_i - 2 \sum_i \omega_i b_i - \sum_i \omega_i a_i \right), \qquad (11)$$

$$s.t. \; \sum_i \omega_i = 1, \qquad (12)$$

$$and \; \omega_i \geq 0. \qquad (13)$$

We suggest the non-deterministic approach as a problem solving tool because of its capability and flexibility in a wide-range of problem configuration.

4   Examples

Seven assets i.e. three stocks, one concession, two projects, and one startup have been experimented for the demonstration of the proposed framework. The sources of data are various. The data of the stocks have been from the financial analysis reports. The returns for the startup have been from dissertation conducting on the actual business. The others have been summarized from public sources of consultants' views on their clients.

The general-purpose genetic algorithm [23] has been chosen for solving the optimization because it could be modified for reuses with the more complex specification in the future extensions. The engine has been configured on the population size, probability of crossover, and probability of mutation rate to 50, 0.75, and 0.25 respectively. The solver has run on several seeds in order to monitor the convergences but only number 403 will be reported. Each test has been repeated for 10,000 iterations and converged at around the six hundredth. The slice of the first 2000 iterations is shown in Fig. 1.

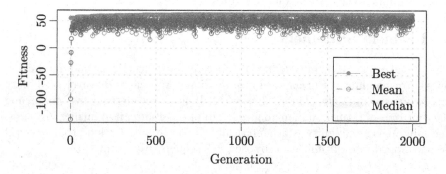

Fig. 1. Convergence of genetic algorithm on fuzzy returns of seven assets

Figure 2 depicts the fuzzy returns for all but one asset, dizP, since its extreme percentages are out of range. The final allocation for the portfolio is also plotted in the same figure and the structure of a fuzzy return is still retained. Table 2 is equipped with the numerical values of the necessary indices for using with Fig. 2. From the figure, it indicates that the most possible return of the optimized portfolio is between 19.2–20.6 percent while the most pessimistic return is 13.8 percent and the most optimistic return is 26.3 percent. Exception on the startup, dizP, the obtained returns seem to be compromised between the second highest and lowest in all levels of possibility. The centroids of return and possibility are 20.0 percent and 0.56 respectively. On the risk side, the indices of fuzziness are 11.2 and 2.55 on the ambiguity. This is quite satisfactory that the obtained portfolio yields high expectations and low risks. The expected return still stands in the range of the most possible. On the other hand, the overall possibility is also traded off between all assets. It could easily be seen some relation to ambiguity as they incline the same change.

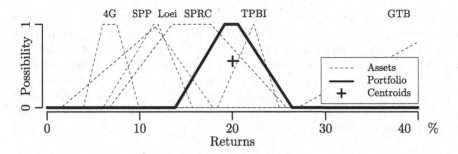

Fig. 2. Graphical illustration of fuzzy returns on assets and resultant portfolio

Table 2. Fuzzy numbers for possibility distributions

Asset Name	Asset Type	Asset Returns		Indices				
		Largest Possible (%)	Most Possible (%)	$\tilde{x}$ (%)	$\tilde{\mu}$	$\mathcal{FZ}$	$\mathcal{AG}$	ω (%)
TPBI	I	18.4–25.0	22.3	21.9	0.50[7]	6.6	1.1	61.2[1]
GTB	I	27.0[2]–72.8[2]	42.4[2]	47.4[2]	0.50[7]	45.8[2]	7.6[2]	0.1[7]
SPRC	S	6.7–25.8	13.5–17.7	16.0	0.61[2]	14.9	4.6	31.8[2]
4G	C	3.9–9.9[7]	6.0–7.5	6.8[7]	0.62[1]	4.5[7]	0.8[7]	6.3
Loei	P	1.5[7]–15.7	12.0	9.7	0.50[7]	14.2	2.4	0.2
SPP	P	6.0–17.9	11.5[7]	11.8	0.50[7]	11.9	2.0	0.3
dizP	B	146[1]–2128[1]	796[1]	1011[1]	0.50[7]	1981[1]	330[1]	0.1[7]
Portfolio		13.8–26.3	19.2–20.6	20.0	0.56	11.2	2.55	100.0

Note:
[1],[2],[7] indicates the highest, second highest, lowest in each index respectively.
Abbreviations:
I = Initial public offering,   TPBI = TPBI Public Co. Ltd.,
S = Structural change,   GTB = GETABEC Public Co. Ltd.,
C = Concession,   SPRC = Star Petroleum Refining Plc. Co. Ltd.,
P = Project,   4G = 4G Concession Auction,
B = startup Business,   Loei = Loei Water Development Project,
                                 SPP = Small Power Producer Project,
                                 dizP = diz P Company Limited

The proportions of portfolio for TPBI, GTB, SPRC, 4G, Loei, SPP, and dizP are 61.2, 0.1, 31.8, 6.3, 0.2, 0.3, and 0.1 percent respectively and are filled in the last column. The major investment or 93.0 percent in total is on TPBI and SPRC. Considering TPBI portion of investment, the returns on asset have been allocated by 61.2 percent resulting from 18.4 percent return on asset to 11.3 percent return on portfolio in the most pessimistic return, from 22.3 percent return on asset to 13.6 percent return on portfolio in the most possible return, and from 25.0 percent return on asset to 15.3 percent return on portfolio in the most optimistic return.

It is quite interesting on GTB and dizP that the computed proportion of investment is so low as 0.1 percent. These two assets are the top two unfavorable even though the returns are very high. Especially in the case of startup the lowest is 146 percent much higher than the return from all other assets and the highest is as high as 2128 percent. The reason is the uncertainties that are also extremely high, 1981 and 330 for fuzziness and ambiguity indices respectively. It is noted that in this case the estimation tends to be over estimated intentionally in order to draw attention of investors.

5   Conclusions

The methodology for selecting portfolio utilizing fuzzy set theory has been proposed and demonstrated. The contributions of this study include (1) the representations of fuzzy returns either from linguistic variables or via probability distribution-like formation, (2) the construction of portfolio using fuzzy set theory, and (3) the alternate interpretation and inference of portfolio.

The proposed methodology for portfolio selection has several advantages in that (1) no observation is required and only small set of subjective data is needed, (2) the representation is more comprehensive because the information is translated from common language which is intuitive to users, (3) varieties of indices are mobilized and, as a result, our understanding in the characteristics of portfolio has been raised, and (4) the outcome from the proposed methodology is comprehensively interpretable. However the disadvantages also exist. It is mainly due to the prejudices on the subjective data i.e. (1) subjective data is conservative because the expert tends to stay on the safe side, (2) the result is dependent on the reliability of information provider.

The fuzzy set theory is so sophisticated and useful that the unlimited number of discoveries could be made for economic applications. The suggestions for future researches may be on that (1) setting and solving fuzzy returns could be performed in more deterministic fashion, (2) some properties of fuzzy number could be relaxed so that the possibility distribution could be available in every scenario, (3) budget constraint could be relaxed in order to take borrowing into account, (4) time dimension is one of the interesting factors in portfolio analysis, and (5) more indices could be added in order to expand the definition of portfolio.

Acknowledgements. The authors would like to express much of our appreciations to Ms. Duangthip Sirikanchanarak, Senior Economist - Bank of Thailand (BOT), for her contributions on the data used in the examples and to the Faculty of Economics, Chiang Mai University for financial support.

References

1. Markowitz, H.: Portfolio selection. J. Finance **7**(1), 77–91 (1952)
2. Markowitz, H.: Portfolio Selection: Efficient Diversification of Investments. Cowles Foundation for Research in Economics at Yale University, Wiley (1959). Monograph
3. Zadeh, L.A.: Fuzzy sets. Inf. Control **8**(3), 338–353 (1965)
4. Hahn, J., Todd, P., Van der Klaauw, W.: Identification and estimation of treatment effects with a regression-discontinuity design. Econometrica **69**(1), 201–209 (2001)
5. Aubin, J.P.: Cooperative fuzzy games. Math. Oper. Res. **6**(1), 1–13 (1981)
6. Liu, Y., Liu, B.: Expected value operator of random fuzzy variable and random fuzzy expected value models. Int. J. Uncertainty Fuzziness Knowl. Based Syst. **11**(2), 195–215 (2003)
7. Huang, X.: A new perspective for optimal portfolio selection with random fuzzy returns. Inf. Sci. **177**(23), 5404–5414 (2007)
8. Sirisrisakulchai, J., Autchariyapanitkul, K., Harnpornchai, N., Sriboonchitta, S.: Portfolio optimization of financial returns using fuzzy approach with NSGA-II algorithm. JACIII **19**(5), 619–623 (2015)
9. Draeseke, R., Giles, D.E.: A fuzzy logic approach to modelling the New Zealand underground economy. Math. Comput. Simul. **59**(1), 115–123 (2002)
10. Autchariyapanitkul, K., Hanpornchai, N., Sriboonchitta, S.: On tax management for unobserved utility using fuzzy logic: An investigation from Thailand underground economy. In: The Eleventh International Symposium on Management Engineering, pp. 36–42, July 2014
11. Breiman, L., et al.: Statistical modeling: The two cultures (with comments and a rejoinder by the author). Stat. Sci. **16**(3), 199–231 (2001)
12. Boswell, S.B., Taylor, M.S.: A central limit theorem for fuzzy random variables. Fuzzy Sets Syst. **24**(3), 331–344 (1987)
13. Shaheen, A.A., Fayek, A.R., AbouRizk, S.: Fuzzy numbers in cost range estimating. J. Constr. Eng. Manag. **133**(4), 325–334 (2007)
14. Kaufmann, A., Gupta, M.M.: Fuzzy Mathematical Models in Engineering and Management Science. Elsevier Science Inc., New York (1988)
15. Zadeh, L.: Fuzzy sets as a basis for a theory of possibility. Fuzzy Sets Syst. **1**(1), 3–28 (1978)
16. Puri, M.L., Ralescu, D.: A possibility measure is not a fuzzy measure. Fuzzy Sets Syst. **7**(3), 311–313 (1982)
17. Zadeh, L.: The concept of a linguistic variable and its application to approximate reasoning–I. Inf. Sci. **8**(3), 199–249 (1975)
18. Zadeh, L.: The concept of a linguistic variable and its application to approximate reasoning–II. Inf. Sci. **8**(4), 301–357 (1975)
19. Zadeh, L.: The concept of a linguistic variable and its application to approximate reasoning–III. Inf. Sci. **9**(1), 43–80 (1975)
20. Hanss, M.: Applied Fuzzy Arithmetic: An Introduction with Engineering Applications. Springer, Heidelberg (2005)
21. Meriam, J.L., Kraige, L.G.: Engineering Mechanics: Statics, 7th edn. Wiley, New York (2011)
22. American Heritage Dictionary of the English Language. 5 edn. Houghton Mifflin Harcourt Publishing Company, New York (2016)
23. Scrucca, L.: GA: A package for genetic algorithms in R. J. Stat. Softw. **53**(4), 1–37 (2013)

A Flood Risk Assessment Based on Maximum Flow Capacity of Canal System

Jirakom Sirisrisakulchai[1]([⊠]), Napat Harnpornchai[1],
Kittawit Autchariyapanitkul[2], and Songsak Sriboonchitta[1]

[1] Faculty of Economics, Chiang Mai University, Chiang Mai, Thailand
jirakom.s@cmu.ac.th
[2] Faculty of Economics, Maejo University, Chiang Mai, Thailand

Abstract. System analysis of network of flows of water is essential for assessing risk of flooding. The flood risk management generally focuses on the meteorological forecasting together with the operation of hydraulic structure while overlooks the flood incurred by inefficient performance of natural instrument in flood mitigation, namely of canal systems. A new methodology for the risk assessment of flood from the prospect of the capacity of canal system is proposed in this paper. The methodology comprises the modeling of a canal system by a flow network in the graph theory, the formulation for the determination of the system capacity in terms of the maximum flow problem, the treatment of uncertainty using copula couple with maximum entropy models, the definition of flood risk event, and the method of risk assessment. The application of the proposed methodology is illustrated through a numerical example.

Keywords: Flood risk assessment · Canal system · Maximum flow problem · Systems of network flows · Copula with maximum entropy

1 Introduction

Evidence have shown the disastrous consequences of flood in terms of safety and economy. Such a recent evidence is the Thailand mega-flood in 2011 (see, [15]). The flood affected 12.8 million people, caused 728 deaths, and damaged 16,668.55 km^2 of agricultural area and 9,859 factories. The total damage and loss amounted to USD 46.5 billion. The World Bank estimates that recovery and reconstruction would cost about USD 50 billion. Four factors are identified as significant factors to the flood including the highest rainfall and five consecutive tropical storms, the water runoff from major rivers, the unsuitable land use in the flood plains, and the flood mismanagement (see, [19]).

The uncertainty in the nature results in inevitable flood risk. The flood risk management have always been an issue in disaster management. The flood risk management generally focuses on the meteorological forecasting together with the operation of hydraulic structures and overlooks natural devices of flood mitigation. With respect to the natural flood mitigation, canal networks play an

V.-N. Huynh et al. (Eds.): IUKM 2016, LNAI 9978, pp. 136–148, 2016.
DOI: 10.1007/978-3-319-49046-5_12

important role in transferring overwhelming water and alleviating possible inundations at the network upstream. When it is inevitably to mitigate the flood by transferring the water via a network, the maximum transfer capacity is the first and critical concern. Incapacitate networks can lead to the flood problems both at the network upstream and in the canals comprising the network. Consequently, the consideration of canal network management is vital for the efficient and effective flood risk management.

This paper proposes a methodology for the flood risk assessment from the viewpoint of the maximum transfer capacity which will be referred to as the system capacity hereinafter. The system capacity can be estimated under the framework of maximum flow problem. The maximum flow problem finds the largest amount of material that can be transported from sources to sinks [18]. The largest amount of transported material is referred to as the maximum flow in a flow network. In this context, the transported material is the water which is distributed through the canal system. The methodology comprises the modeling of a canal system by a flow network in the graph theory, the formulation for system capacity determination in terms of the maximum flow problem, the treatment of uncertainty, and the definition of flood risk event. The results from the assessment give the information for further decision and action in dealing with floods.

The remaining of the paper starts with modeling of flood risk based on system capacity in Sect. 2. This section includes all relevant definitions and mathematical modeling. Section 3 is the method of flood risk assessment. Section 4 shows the applicability of the proposed methodology through a numerical example. Section 5 addresses the conclusion.

2   Modeling of Flood Risk Based on System Capacity

2.1   Modeling of Canal System by Flow Network

A flow network in the graph theory is employed to model a canal system for the purpose of analyzing the flow characteristics in the canal system. A graph is formed from two components [21]. The first component is a set of points, called nodes or vertices. The second component is a set of lines which join pairs of these points. These lines are known as edges or arcs. A network is a graph on which one has defined additional information, usually numerical data, associated with the edges and sometimes with the nodes [21]. The data can be, for examples, distances, costs, times, limits on the flow of material through the node or edge, or a combination of these properties.

A flow network is a directed graph $G = (V, E)$ in which V is a non-empty set of nodes and a set $E \subseteq V \times V$ of directed edges. Each edge has a start node i and an end node j. Let n be the total number of nodes and m be the total number of edges in the flow network, respectively. The flow capacity of the edge k is denoted by $c_k = c_{ij}$. A source node is the node at which the flow starts or arrives at the network. A sink node is the node at which the flow leaves the network.

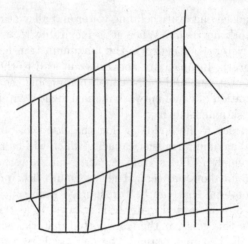

Fig. 1. Schematic diagram of the rangsit canal network in Thailand.

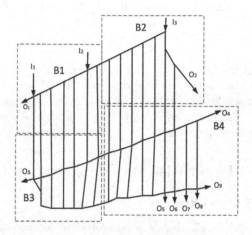

Fig. 2. Modeling of the canal network in Fig. 1 by the flow network.

Analogously, the nodes and edges in the flow network represent the canal confluences and canals. The transfer capacity of each respective canal is the flow capacity of the corresponding edge. Figure 1 shows a schematic diagram of the Rangsit canal network in Thailand, including its inlets and outlets. The canal system is modeled by the flow network in which the inlet flows are I_1 to I_3 and the outlet flows are O_1 to O_9, as shown in Fig. 2.

The flow network in Fig. 2 is divided into 4 blocks from B1 to B4. The flow network in each block is illustrated in Fig. 3 including its node and edge numbers. Each node number is given within a circle whereas each edge number is illustrated within a square. The arrows indicate the flow directions of respective canals.

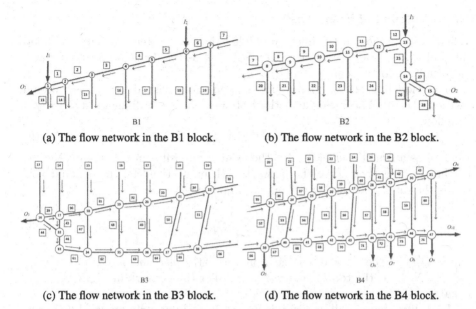

(a) The flow network in the B1 block.   (b) The flow network in the B2 block.

(c) The flow network in the B3 block.   (d) The flow network in the B4 block.

Fig. 3. The flow network for each block in Fig. 2

2.2   Formulation for System Capacity Determination

Upon modeling a canal network by a corresponding flow network, for example the network in Fig. 2, the system capacity can be estimated from the maximum flow problem. Based on the maximum flow problem, the maximum flow is taken as the system capacity. The maximum flow problem for a flow network is mathematically defined as follows.

$$\text{Maximize } q = \sum_{i=1}^{n} f_{it} \tag{1}$$

subject to

$$\sum_{j=1}^{n} f_{ij} - \sum_{l=1}^{n} f_{li} = \begin{cases} q_{max} & ; \quad i = 1 \\ 0 & ; \quad i = 2, \cdots, n-1 \\ -q_{max} & ; \quad i = n \end{cases}$$

$$0 \le f_{ij} \le c_{ij}; \quad i,j = 1, \cdots, n \ , i \ne j \ , \text{and } q \ge 0,$$

where f_{ij} is the flow magnitude in the kth edge from the node i to the node j and can also be denoted by f_k. t indicates the sink node. The solution of the problem as defined by (1) is the maximum flow. The maximum flow is denoted by q_{max}. It should be noted that q_{max} is the function of c_k, i.e. $q_{max}(c)$. c is the vector of c_k ($k = 1, \cdots, m$). According to the notion of network model, q_{max} is the system capacity.

2.3   Modeling of Flood Risk

When the flow rates at the inlets are higher than the system capacity, the flood takes place. Such a flow rate will be referred to as the demand flow Q_D. Mathematically, a flood occurs when $Q_D > q_{max}$.

The modeling of Q_D in practice can be accomplished using the rational method [11,13]. The rational method obtains the peak discharge Q_p (m^3/s) as follows:

$$Q_P = 0.278CIA, \tag{2}$$

where C is a runoff coefficient, I is the rainfall intensity (mm/hr), and A is the catchment area (km^2). For the network with multiple inlets, the demand flow Q_D is the total peak discharge from each inlet, i.e.

$$Q_D = \sum_{r=1}^{N_{inlet}} Q_{P,r}, \tag{3}$$

where $Q_{P,r}$ is the peak discharge from the r-th inlet and is computed from Eq. (2). N_{inlet} is the total number of inlets. For the network in Fig. 2, N_{inlet} is equal to (3).

In reality, the rainfall intensity is uncertain and normally modeled by a random variable. Moreover, the rainfall intensities that generate the inlet flows to a network can be correlated. There is a joint probability density function (JPDF) of the rainfall intensities, i.e. $p_I(\cdot)$, where $I = [I_1...I_r...I_{N_{inlet}}]^T$ and I_r is the r_{th} rainfall intensity that generates $Q_{P,r}$. When a marginal distribution of I_r is given, $p_I(\cdot)$ can be constructed using the copula concept which will be discussed in the next section.

2.4   Modeling Rainfall Intensities Using Copula with Maximum Entropy

The main purpose in this section is to build the model for rainfall intensities at the upstream sites from the canal network by taking into account the spatial dependence of multi-sites. [1,16,17] proposed the model to construct a joint probability distribution for monthly rainfall data that match a given set of grade correlation coefficients by using copula with maximum entropy. We apply the method proposed by [1,16,17] to model the spatial dependence of multi-sites rainfall intensities. A copula is a function which joins a multivariate distribution function to its marginal distributions. Copula function can be used to model a multivariate joint distribution with flexible dependence structures. For the entropy concept, which has a long history dating back to Boltzman in 1870s following by the works of Maxwell, Gibbs and Shannon, [8,9] proposed the method using entropy concepts [8,9] called it the maximum entropy principle in selecting the unknown probability distributions.

By Sklar's theorem [20], we can define an m-dimensional copula function $C : 0,1]^m \to [0,1]$ on the unit m-dimensional hyper-cube with uniform marginal probability distribution as a continuous, m-increasing, grounded probability distribution function. The readers are referred to [10,14] for more details discussion. An m-dimensional checkerboard copula is a probability distribution function with the density function defined by a step function on an m-uniform subdivision of the hypercube $[0,1]^m$ [16]. We can use a checkerboard copula to approximate any continuous copula function.

For simplicity, we will discuss a two-dimensional checkerboard copula. Note that the generalization to m-dimensions can be found in [16,17]. For n equal subintervals within $[0,1]$, we can define a checkerboard copula density function by the step function $h(u,v)$ on $[0,1] \times [0,1]$ as follows [16]:

$$h(u,v) = n h_{ij}, (u,v) \in (u_i, u_{i+1}) \times (v_j, v_{j+1}), \tag{4}$$

where h_{ij} is the element of n-dimensional square matrix H on $[0,1] \times [0,1]$. The constraints $\sum_{i=1}^{n} h_{ij} = 1$, $\sum_{j=1}^{n} h_{ij} = 1$ for all $h_{ij} \geq 0$ have to be specified to meet the condition of uniform margins. Then, the grade correlation can be computed by using the following formula:

$$\frac{1}{n^3} \sum h_{ij}(i - 1/2)(j - 1/2) = \frac{\rho + 3}{12}. \tag{5}$$

The entropy of the checkerboard copula can be defined as follows:

$$J(h) = -\frac{1}{n} \left[\sum_{i=1}^{n} \sum_{j=1}^{n} h_{ij} \ln(h_{ij}) + \ln(n) \right]. \tag{6}$$

To model the rainfall intensities by copula with maximum entropy, the aim is to find n-dimensional square matrix $H = [h_{ij}]$ matching with the sample grade correlation of the rainfall data. This can be done by maximizing the entropy subject to the constraints given in (5), and uniform margins constraints. We can write these constraints in matrix form of $AX = b$, where X is an $n^2 \times 1$ vector of the unknown parameter φ, and $A = a_{ij}$ is $n^2 \times n^2$ constant matrix. [16] showed how to solve this problem by using the Fenchel duality theory. The constrained optimization problem can be transformed to unconstrained optimization as follows:

$$\sum_{i=1}^{n^2} a_{ki} \exp \left(n \sum_{j=1}^{n^2} a_{ji} \varphi_j \right) = b_k, k = 1, 2, ..., n^2. \tag{7}$$

The solution of the unconstrained optimization problem can be calculated numerically by the Newton iteration as follows:

$$\varphi^{(l+1)} = \varphi^{(l)} - G^{-1}(\varphi^{(l)}) Q_k(\varphi^{(l)}), \tag{8}$$

where

$$Q_k(\varphi) = \sum_{i=1}^{n^2} a_{ki} \exp \left(n \sum_{j=1}^{n^2} a_{ji} \varphi_j \right) - b_k, \tag{9}$$

and $G(.)$ is the gradient of Q_k. Finally, we can obtain the vector of matrix element h by computing $h = \exp(nA^T\varphi)$.

Since the $Q_{P,r}$ is uncertain, Q_D becomes naturally uncertain and a function of random variables I, i.e. $Q_D(I)$. The uncertainty in $Q_D(I)$ leads to a flood risk. The flood risk can be quantitatively assessed due to the application of the probability measure to the uncertain rainfall intensities. The water flood risk is then quantitatively measured as the probability that $Q_D(I) > q_{max}$, i.e. $P[Q_D(I) > q_{max}]$. Here, $P[E]$ is the occurrence probability of the event E. Mathematically, the flood risk due to insufficient network capacity is then determined from

$$P[Q_D(I) > q_{max}] = \int_R p_I(i)di, \tag{10}$$

where

$$R = \{I|g(Q_D(I)) > 0\}, \tag{11}$$

$$g(Q_D(I)) = Q_D(I) - q_{max}, \tag{12}$$

q_{max} in the expression (12) is obtained from solving the maximum flow problem defined by Eq. (1).

3   Flood Risk Assessment

The flood risk as defined by $P[Q_D(I) > q_{max}]$ is generally not possible to be obtained in an explicit form. Consequently, a computational method is necessary. The Monte Carlo simulation (MCS) is employed here for such a purpose. When applying the MCS, Eq. (10) is rewritten as follows:

$$P[Q_D(I) > q_{max}] = \int_R I_F(i)p_I(i)di, \tag{13}$$

where I_F is the indicator function and defined as

$$I_F = \begin{cases} 1, & \text{if } g(Q_D(I)) > 0 \\ 0, & \text{otherwise} \end{cases}. \tag{14}$$

Using the MCS, the flood risk is approximately computed as

$$P[Q_D(I) > q_{max}] \approx \frac{1}{N_{sim}} \sum_{is=1}^{N_{sim}} I_F(I_{is}), \tag{15}$$

where I_{is} is the is-th simulation of the I and is generated by the checkerboard copula. N_{sim} is the total number of simulations. [17] discussed a procedure to simulate rainfall from a checkerboard copula defined by a hyper-matrix $h = [h_i] \in \mathbb{R}^l$, where $l = n^3$ and $i = (i,j,k) \in 1,2,\ldots,n^3$ on a uniform partition $\{I_i\}$ of the unit cube $(0,1)^3$ as follows. Firstly, define an order $(i,j,k) \prec (i_0,j_0,k_0)$ if $i < i_0$ or if $i = i_0$ and $j < j_0$ or if $i = i_0$ and $j = j_0$ and $k < k_0$. Then select the interval $I_{i_0 j_0 k_0} = (a(i_0), a(i_0 + 1)) \times (a(j_0), a(j_0 + 1)) \times (a(k_0), a(k_0 + 1))$ if

$$\sum_{(i,j,k) \prec (i_0,j_0,k_0)} h_{ijk} < nr < \left[\sum_{(i,j,k) \prec (i_0,j_0,k_0)} h_{ijk} \right] + h_{i_0 j_0 k_0}, \qquad (16)$$

for each standard uniform random number r.

The position of the random point $(u_r, v_r, w_r) \in I_{i_0 j_0 k_0}$ is fixed by generating three independent standard uniform random number (q_r, s_r, t_r) and setting

$$(u_r, v_r, w_r) = \left(\frac{(i_0 - 1) + q_r}{n}, \frac{(j_0 - 1) + s_r}{n}, \frac{(k_0 - 1) + t_r}{n} \right). \qquad (17)$$

Finally, the corresponding rainfall simulations is generated by

$$(x_r, y_r, z_r) = \left(F_x^{-1}(u_r), F_y^{-1}(v_r), F_z^{-1}(w_r) \right), \qquad (18)$$

where F_x, F_y, and F_z are the given marginal distributions.

Equations (13) and (15) imply that it is necessary to determine q_{max} first in order to compute the flood risk. Regarding the solution of the maximum flow problem, a number of algorithms have been proposed ([5, 6, 12]) to name a few. Among the aforementioned algorithms, a recent pseudo-flow algorithm introduced by Hochbaum [7] is faster than the best-known implementation of the highest-level push-relabel in most instances [2] for the present date. Therefore, the determination of q_{max} will use the pseudo-flow algorithm.

The pseudo-flow algorithm works with a tree structure called a normalized tree. This tree preserves some information about residual paths in the graph. The pseudo-flow allows excesses and deficits. Source and sink nodes play no different role. All edges adjacent to source and sink are kept saturated. The algorithm searches for a partition of the set of nodes to subsets some of which have excesses and some with deficits. The union of excess subsets forms a super optimal solution. Consequently, all edges going from excess subsets to deficit ones are saturated. If there are no unsaturated edges between excess subsets and deficit subsets, then the union of the excess subsets forms an optimal solution. The partition with this property provides a maximum flow in the graph.

An iteration of the algorithm consists of seeking a residual edge from a strong node to a weak node-a merger edge. If such an edge does not exist, the normalized tree is optimal. Otherwise, the selected merger edge will be included to the tree. The excess edge of the strong merger branch is then removed and the strong branch is merged with the weak branch. The entire excess of the respective strong branch is then pushed along the unique path from the root of the strong branch to the root of the weak branch. Any edge along this path that does not have sufficient residual capacity to accommodate the amount pushed will be split and the tail node of that edge becomes a root of a new strong branch with excess equal to the difference between the amount pushed and the residual capacity. The process of pushing excess and splitting is called normalization. The residual capacity of the split edge is pushed further until it either reaches another edge to split or the deficit edge adjacent to the root of the weak branch. The implementation of the algorithm in a source code is also available (see [3, 4]).

4 Numerical Example

The Rangsit canal network as shown in Figs. 1, 2 and 3 are considered here. The information of all canals in the system is given in Table 1. The information includes the start node number, end node number, edge number, and transfer capacity.

Table 1. Information of all canals in the Rangsit network.

Start	End	Edge k	C_k	Start	End	Edge k	C_k	Start	End	Edge k	C_k
2	1	1	17	14	28	26	29	22	38	51	68
3	2	2	17	14	15	27	41	23	39	52	68
4	3	3	17	15	29	28	29	24	40	53	68
5	4	4	17	16	17	29	41	25	41	54	68
6	5	5	17	17	18	30	41	26	42	55	68
7	6	6	17	18	19	31	41	27	43	56	68
8	7	7	17	19	20	32	41	28	44	57	68
9	8	8	17	20	21	33	41	29	45	58	68
10	9	9	17	21	22	34	41	30	46	59	68
11	10	10	17	22	23	35	41	31	47	60	68
12	11	11	17	23	24	36	41	33	34	61	24
13	12	12	17	24	25	37	41	34	35	62	24
1	16	13	48	25	26	38	41	35	36	63	24
2	17	14	47	26	27	39	41	36	37	64	24
3	18	15	36	27	28	40	41	37	38	65	24
4	19	16	38	28	29	41	41	38	39	66	24
5	20	17	44	29	30	42	41	39	40	67	24
6	21	18	43	30	31	43	41	40	41	68	24
7	22	19	35	16	32	44	35	41	42	69	24
8	23	20	31	17	32	45	27	42	43	70	24
9	24	21	29	32	33	46	27	43	44	71	24
10	25	22	29	18	34	47	59	44	45	72	24
11	16	23	29	19	35	48	70	45	46	73	24
12	27	24	29	20	36	49	68	46	47	74	24
13	14	25	29	21	37	50	68				

C_k is Transfer Capacity (m^3/s)

The rainfall data for 24-year period from 1990–2013 at three sites that are corresponding to the inlet I_1, I_2, and I_3 are used for the example. The descriptive statistics are shown in Table 2. Table 3 shows grade correlation coefficients for each pair of site. [1] have shown that the gamma distribution is the maximum

Table 2. Descriptive statistics for three sites in September

Site	1	2	3
Mean (mm/hr)	2.94	5.66	3.56
Standard deviation (mm/hr)	1.28	2.22	2.05

Table 3. Grade correlation coefficients for multi-site pairs in September

	1	2	3
1	1.00	−0.03	0.06
2	−0.03	1.00	−0.22
3	0.06	−0.22	1.00

entropy model for a random variable defined only by a finite number of strictly positive observations. In the three sites, there are no zero observations of rainfall intensities, thus the gamma distribution is the reasonable choice in modeling the marginal distribution. The gamma distribution defined on $(0, \infty)$ can be expressed by the following formula:

$$F_{\alpha,\beta}(x) = \int_0^x \frac{\xi^{\alpha-1}}{\beta^\alpha \Gamma(\alpha)} \exp(-\xi/\beta)d\xi, \tag{19}$$

where $\alpha > 0$, and $\beta > 0$ are parameters. We estimate the parameters α and β for the three sites by the maximum likelihood estimations. The results are $\alpha = (5.395, 6.296, 3.007)$, and $\beta = (23.15, 38.46, 49.75)$.

The copula with maximum entropy discussed previously are followed. Here, we set $n = 4$, $\rho_{12} = -0.03$, $\rho_{13} = 0.06$, and $\rho_{23} = -0.22$. The results of $\boldsymbol{h}_i = [h_{ijk}]$ are

$$\boldsymbol{h}_1 = \begin{bmatrix} 0.0408 & 0.0529 & 0.0653 & 0.0767 \\ 0.0583 & 0.0623 & 0.0633 & 0.0612 \\ 0.0796 & 0.0700 & 0.0585 & 0.0466 \\ 0.1038 & 0.0751 & 0.0517 & 0.0339 \end{bmatrix}$$

$$\boldsymbol{h}_2 = \begin{bmatrix} 0.0385 & 0.0527 & 0.0687 & 0.0852 \\ 0.0544 & 0.0613 & 0.0657 & 0.0671 \\ 0.0732 & 0.0680 & 0.0600 & 0.0504 \\ 0.0942 & 0.0720 & 0.0523 & 0.0362 \end{bmatrix}$$

$$\boldsymbol{h}_3 = \begin{bmatrix} 0.0362 & 0.0523 & 0.0720 & 0.0942 \\ 0.0504 & 0.0600 & 0.0680 & 0.0732 \\ 0.0671 & 0.0657 & 0.0613 & 0.0544 \\ 0.0852 & 0.0687 & 0.0527 & 0.0385 \end{bmatrix}$$

$$h_4 = \begin{bmatrix} 0.0339 \ 0.0517 \ 0.0751 \ 0.1038 \\ 0.0466 \ 0.0585 \ 0.0700 \ 0.0796 \\ 0.0612 \ 0.0633 \ 0.0623 \ 0.0583 \\ 0.0767 \ 0.0653 \ 0.0529 \ 0.0408 \end{bmatrix} .$$

The sample of $I_r(r = 1, 2, 3)$ are generated according to the procedure described in Sect. 3. The number of samples is equal to 5,000. The parameters for the rational method are given in Table 4.

Table 4. Parameters for the rational method.

Site index (r)	Runoff coefficient C_r	Catchment area A_r (km^2)
1	0.17	142
2	0.25	134
3	0.30	151

Based on the generated samples together with the site parameters in Table 4, the samples of Q_D are obtained using Eqs. (2) and (3).

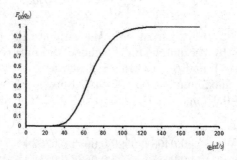

Fig. 4. CDF of Q_D

The maximum flow q_{max} for the given canal system is obtained from the pseudoflow algorithms as 171 m^3/s. Since the number of Q_D samples that is greater than 171 m^3/s is equal to 1, the flood risk according to the MCS as described in Sect. 2.1 is thus

$$P[Q_D > 171] = \frac{1}{5000} = 2.00 \times 10^{-4}. \tag{20}$$

It should be noted that the computed flood risk is of extremely low order. This can also be correspondingly observed from the resulting CDF in Fig. 4 that $P[Q_D \leq 171] \cong 1$.

Based on the numerical example, the flood risk assessment from the viewpoint of system capacity can be realized using a simple mathematical model. Since the cause of risk is attributed to the uncertainty, the rationale modeling of the uncertainty is thus crucial. The over-simplified treatment of the uncertainty like the assumption of statistical independence can lead to erroneous results. The dependence structure of random variables needs to be taken into account using such a concept of copula as shown in the present paper. This implies that the quality of the assessment largely depends on the data used in the mathematical model, including those of canal system and demand flow.

5 Conclusions

A methodology for flood risk assessment from the viewpoint of the capacity of canal system is proposed. The methodology comprises the modeling of a canal system by a flow network in the graph theory, the definition of system capacity by the maximum flow in the flow network, the quantification of uncertainty by probability distributions, and the mathematical definition of flood risk. The determination of the system capacity is accomplished by the computational algorithms for the maximum flow problem. A numerical example shows that the flood risk assessment can be realized by the proposed methodology. For the methodology implementation, it is necessary to acquire data for modeling the canal capacities and the demand flow. Since the present methodology is applied for the time-invariant risk assessment, the time-variant risk assessment is a subject of future research.

Acknowledgement(s). The authors thank Prof. Dr. Hung T. Nguyen for his helpful comments and suggestions.

References

1. Borwein, J., Howlett, P., Piantadosi, J.: Modelling and simulation of seasonal rainfall using the principle of maximum entropy. Entropy **16**, 747–769 (2014)
2. Chandran, B.G., Hochbaum, D.S.: A computational study of the pseudoflow and push-relabel algorithms for the maximum flow problem. Oper. Res. **57**(2), 358–376 (2009)
3. Fishbain, B., Hochbaum, D.S., Mueller, S.: Competitive Analysis of Minimum-Cut Maximum Flow Algorithms in Vision Problems (2010). arXiv preprint arXiv: 1007.4531
4. Fishbain, B., Hochbaum, D.S.: Hochbaum's Pseudo-flow Matlab implementation (2012). http://riot.ieor.berkeley.edu/riot/Applications/Pseudoflow/maxflow.html. Accessed 15 June 2014
5. Ford Jr., L.R., Fulkerson, D.R.: A simple algorithm for finding maximal network flows and an application to the Hitchcock problem. Can. J. Math. **9**, 210–218 (1957)
6. Goldberg, A.V., Tarjan, R.E.: A new approach to the maximum flow problem. J. ACM **35**, 921–940 (1988)
7. Hochbaum, D.S.: The pseudoflow algorithm: A new algorithm for the maximum flow problem. Oper. Res. **56**(4), 992–1009 (2008)

8. Jaynes, E.T.: Information theory and statistical mechanics. Phys. Rev. **106**, 620–630 (1957a)
9. Jaynes, E.T.: Information theory and statistical mechanics II. Phys. Rev. **108**, 171–190 (1957b)
10. Joe, H.: Multivariate Models and Dependence Concepts. Chapman & Hall, London (1997)
11. Kuichling, E.: The relation between the rainfall and the discharge of sewers in populous districts. Trans. Am. Soc. Civ. Eng. **20**, 1–56 (1889)
12. King, V., Rao, S., Tarjan, R.: A faster deterministic maximum flow algorithm. J. Algorithms **17**(3), 447–474 (1994)
13. McCuen, R.: Hydrologic Analysis and Design, 2nd edn. New Jersey, Prentice-Hall Inc. (1998)
14. Nelsen, R.B.: An Introduction to Copulas, 2nd edn. Springer, New York (2006)
15. Poapongsakorn, N., Meethom, P.: Impact of the 2011 Floods, and Flood Management in Thailand. No. DP-2013-34 (2013)
16. Piantadosi, J., Howlett, P.G., Borwein, J.M.: Copulas with maximum entropy. Optim. Lett. **6**, 99–125 (2012)
17. Piantadosi, J., Howlett, P.G., Borwein, J.M., Henstridge, J.: Maximum entropy methods for generating simulated rainfall. Numer. Algebra Control Optim. **2**, 233–256 (2012)
18. Schrijver, A.: On the history of the transportation and maximum flow problem. Math. Program. **91**, 437–445 (2002)
19. Suppaisarn, C.: Medium and heavy flood management in chao phraya river basin following the royal initiative. In: TRF Seminar (2011)
20. Sklar, A.: Fonctions de repartition a n dimensions et leurs marges. Publications de lInstitut de Statistique de lUniversite de Paris **8**, 229–231 (1959)
21. Smith, D.K.: Networks and Graphs: Techniques and Computational Methods. Woodhead Publishing Limited, Sawston (2011)

Soft Clustering and Classification

Generalizations of Fuzzy c-Means and Fuzzy Classifiers

Sadaaki Miyamoto$^{(\boxtimes)}$, Yoshiyuki Komazaki, and Yasunori Endo

Department of Risk Engineering, University of Tsukuba, Tsukuba, Japan
miyamoto@risk.tsukuba.ac.jp

Abstract. Different methods of generalized fuzzy c-means having cluster size variables and cluster covariance variables are compared, which include Gustafson-Kessel's method, Ichihashi's method of KL-information, and Yang's method of fuzzified maximum likelihood. Theoretical properties using fuzzy classifier functions as well as results of numerical experiments are shown.

Keywords: Fuzzy c-means · Fuzzy classifier · Fuzzy covariance

1 Introduction

Different types of extended fuzzy c-means have been proposed which include Gustafson-Kessel's method [3], Ichihashi's method of KL-information. [5,6] and Yang's method of fuzzified maximum likelihood [12]. They have been, however, independently proposed and have not been compared. This means that possibility of further considerations maybe overlooked. For example, some of the proposals includes consideration of a variable controlling cluster sizes (sometimes called cluster volumes), while other methods do not have that variable. We thus compare them using the both types of cluster size variables and cluster covariance variables. As a result we propose a new method that combines the two types of variables. We compare theoretical properties of them using the concept of fuzzy classifier functions. Moreover we show results of applying these methods to a set of numerical examples.

The rest of this paper is organized as follows. Section 2 discusses fuzzy c-means with an additional variable controlling cluster sizes. Section 3 then introduces cluster covariance variables and gives the whole combinations of the two types of variables for controlling cluster sizes and cluster covariances. Section 4 is devoted to investigation of theoretical properties, where the proofs of the propositions are omitted to save space, as they are not difficult. Section 5 shows results of numerical experiments, and finally Sect. 6 concludes the paper.

Yoshiyuki Komazaki is now with Forcia Inc., Shinjuku, Tokyo, Japan.

© Springer International Publishing AG 2016
V.-N. Huynh et al. (Eds.): IUKM 2016, LNAI 9978, pp. 151–162, 2016.
DOI: 10.1007/978-3-319-49046-5_13

2 Fuzzy c-Means with an Additional Variable

To begin with, let $X = \{x_1, \ldots, x_n\}$ be a set of objects for clustering, and let $V = \{v_1, \ldots, v_c\}$ be a set of *centroids* which mean representatives of clusters. They are vectors of real p-dimensional Euclidean space $\boldsymbol{R}^p$, that is, $x_k = (x_k^1, \ldots, x_k^p)^T \in \boldsymbol{R}^p$ and $v_i = (v_i^1, \ldots, v_i^p)^T \in \boldsymbol{R}^p$. Let $U = (u_{ki})$ be an $c \times n$ matrix of fuzzy membership of x_k to cluster i. $d(x_k, v_i)$ is *dissimilarity* between object x_k and centroid v_i. Unless stated otherwise, $d(x_k, v_i)$ is assumed to be the squared Euclidean norm:

$$d(x_k, v_i) = \|x_k - v_i\|^2 = \sum_{l=1}^{p} (x_k^l - v_i^l)^2.$$

We use the objective functions proposed in [5,9,12] with the additional variable A for controlling cluster sizes, which are as follows,

$$J_{\text{fcma}} = \sum_{i=1}^{c} \sum_{k=1}^{n} (\alpha_i)^{1-m} (u_{ki})^m d(x_k, v_i)$$

$$J_{\text{kfcm}} = \sum_{i=1}^{c} \sum_{k=1}^{n} u_{ki} d(x_k, v_i) + \nu \sum_{i=1}^{c} \sum_{k=1}^{n} u_{ki} \log \left(\frac{u_{ki}}{\alpha_i} \right)$$

$$J_{\text{pfcm}} = \sum_{i=1}^{c} \sum_{k=1}^{n} (u_{ki})^m d(x_k, v_i) + \nu \sum_{i=1}^{c} \sum_{k=1}^{n} (u_{ki})^m \log \left(\frac{1}{\alpha_i} \right),$$

where $A = (\alpha_1, \ldots, \alpha_c)$ is a variable for controlling cluster sizes, and ν is a positive parameter. The constraint for A is assumed to be

$$\mathcal{A} = \left\{ A = (\alpha_1, \ldots, \alpha_i) : \sum_{j=1}^{c} \alpha_j = 1; \alpha \geq 0, 1 \leq i \leq c \right\}.$$

and the constraint of U is given by

$$\mathcal{U}_{fcm} = \left\{ (u_{ki}) : u_{ki} \in [0, 1], \sum_{i=1}^{c} u_{ki} = 1, \forall k \right\}.$$

We call these FCMA, KFCM, and PFCM respectively. They use the next alternative minimization algorithm.

Step 1. Generate c initial values for V and A.
Step 2. Calculate optimal U that minimizes J.
Step 3. Calculate optimal V that minimizes J.
Step 4. Calculate optimal A that minimizes J.
Step 5. If (U, V, A) is convergent, stop; else return to **Step 2**.

where we put either $J = J_{\text{fcma}}$, $J = J_{\text{kfcm}}$, or $J = J_{\text{pfcm}}$.

Moreover we call basic fuzzy c-means SFCM (standard fuzzy c-means) which does not have the variable A, and call entropy-based fuzzy c-means EFCM without variable A. We can derive objective functions of SFCM and EFCM by putting $\alpha_i = 1$ in J_{fcma} and J_{kfcm}, respectively.

3 Generalized Fuzzy c-means with Two Additional Variables

The shape of each cluster have been supposed to be hypersphere up to now, as data set is partitioned using the direction-independent Euclidean distance. Thus, every method discussed before may fail to have good clusters when clusters are prolonged in some directions. To solve this problem, objective functions with covariance matrices have been considered. One method has been proposed by Gustafson and Kessel [3], and the other has been proposed in fuzzy c-means regularized by KL information in [5]. Note that the alternative minimization algorithm is used for (U, V, A, S) when the four variables are used with the additional step to minimize J with respect to $S = (S_1, \cdots, S_c)$:

Step 1. Generate initial values for V, A, and S.
Step 2. Calculate optimal U that minimizes J.
Step 3. Calculate optimal V that minimizes J.
Step 4. Calculate optimal A that minimizes J.
Step 5. Calculate optimal S that minimizes J.
Step 6. If (U, V, A, S) is convergent, stop; else return to **Step 2**.

Note that J is one of the objective function described below.

3.1 Gustafson-Kessel Method

A method to solve the above problem is to use clusterwise Mahalanobis distance instead of Euclidean distance, and thus the case of standard fuzzy c-means is:

$$J_{\text{sfcm}} = \sum_{i=1}^{c} \sum_{k=1}^{n} (u_{ki})^m d_{S_i}(x_k, v_i), \tag{1}$$

where

$$d_{S_i}(x, v_i) = (x - v_i)^T S_i^{-1}(x - v_i)$$

and S_i $(i = 1, \ldots, c)$ is positive definite matrix to minimize the objective function. Gustafson and Kessel [3] add the following constraint, since the optimal solution of S_i^{-1} obtained by differentiating 1 is zero matrix without a constraint:

$$|S_i| = \rho_i \quad (\rho_i > 0)$$

Therefore optimal S_i is derived from the Lagrangian function:

$$L = J_{\text{sfcm}} + \sum_{i=1}^{c} \lambda_i \log \frac{|S_i|}{\rho_i}.$$

Then, optimal solution of S_i is

$$S_i = \frac{\rho_i^{\frac{1}{p}}}{|\hat{S}_i|^{\frac{1}{p}}} \hat{S}_i$$

where

$$\hat{S}_i = \sum_{k=1}^{n} u_{ki}^m (x_k - v_i)(x_k - v_i)^T.$$

Thus S_i is a fixed-size fuzzy covariance matrix. We can also apply Gustafson-Kessel method to other methods in the previous section, substituting $d(x_k, v_i)$ by the Mahalanobis distance $d_{S_i}(x_k, v_i)$ in J_{fcma}, J_{kfcm}, and J_{pfcm}.

3.2   Logarithm of Covariance

KFCM is a special case where covariances are all identity matrices. The original KL information-based fuzzy c-means [5] uses

$$J_{\text{kfcm}} = \sum_{i=1}^{c} \sum_{k=1}^{n} \{u_{ki} d_{S_i}(x_k, v_i) + \nu u_{ki} \log\left(\frac{u_{ki}}{\alpha_i}\right) + u_{ki} \log |S_i|\}$$

The optimal solutions are as follows.

$$u_{ki} = \frac{\alpha_i |S_i|^{-\frac{1}{\nu}} \exp\left(-\frac{d_{S_i}(x_k, v_i)}{\nu}\right)}{\sum_{j=1}^{c} \alpha_j |S_j|^{-\frac{1}{\nu}} \exp\left(-\frac{d_{S_j}(x_k, v_j)}{\nu}\right)}$$

$$v_i = \frac{\sum_{k=1}^{n} u_{ki} x_k}{\sum_{k=1}^{n} u_{ki}}$$

$$\alpha_i = \frac{\sum_{k=1}^{n} u_{ki}}{n}$$

$$S_i = \frac{\sum_{k=1}^{n} u_{ki}(x_k - v_i)(x_k - v_i)^T}{\sum_{k=1}^{n} u_{ki}}$$

We can introduce covariance into the other methods in a similar way. We call here these *the log methods* to distinguish them from Gustafson-Kessel methods. The objective functions and the respective optimal solutions are listed in the following.

$$J_{\text{sfcm}} = \sum_{i=1}^{c} \sum_{k=1}^{n} (u_{ki})^m (d_{S_i}(x_k, v_i) + \log |S_i|)$$

$$J_{\text{efcm}} = \sum_{i=1}^{c} \sum_{k=1}^{n} u_{ki}(d_{S_i}(x_k, v_i) + \nu \log u_{ki} + \log |S_i|)$$

$$J_{\text{fcma}} = \sum_{i=1}^{c} \sum_{k=1}^{n} (\alpha_i)^{1-m}(u_{ki})^m (d_{S_i}(x_k, v_i) + \log |S_i|)$$

$$J_{\text{pfcm}} = \sum_{i=1}^{c} \sum_{k=1}^{n} \{(u_{ki})^m (d_{S_i}(x_k, v_i) + \log |S_i|) + \nu(u_{ki})^m \log\left(\frac{1}{\alpha_i}\right)\}$$

$$J_{\text{kfcm}} = \sum_{i=1}^{c} \sum_{k=1}^{n} \{ u_{ki} d_{S_i}(x_k, v_i) + \nu u_{ki} \log \left(\frac{u_{ki}}{\alpha_i} \right)$$
$$+ u_{ki} \log |S_i| \}$$

Optimal Solutions of J_{sfcm}:

$$u_{ki} = \frac{\left(\frac{1}{d_{S_i}(x_k, v_i) + \log |S_i|} \right)^{\frac{1}{m-1}}}{\sum_{j=1}^{c} \left(\frac{1}{d_{S_i}(x_k, v_j) + \log |S_i|} \right)^{\frac{1}{m-1}}}$$

$$v_i = \frac{\sum_{k=1}^{n} (u_{ki})^m x_k}{\sum_{k=1}^{n} (u_{ki})^m}$$

$$S_i = \frac{\sum_{k=1}^{n} u_{ki}^m (x_k - v_i)(x_k - v_i)^T}{\sum_{k=1}^{n} u_{ki}^m}$$

Optimal Solutions of J_{efcm}:

$$u_{ki} = \frac{|S_i|^{-\frac{1}{\nu}} \exp \left(\frac{-d(x_k, v_i)}{\nu} \right)}{\sum_{j=1}^{c} |S_j|^{-\frac{1}{\nu}} \exp \left(\frac{-d(x_k, v_j)}{\nu} \right)}$$

$$v_i = \frac{\sum_{k=1}^{n} u_{ki} x_k}{\sum_{k=1}^{n} u_{ki}}$$

$$S_i = \frac{\sum_{k=1}^{n} u_{ki}(x_k - v_i)(x_k - v_i)^T}{\sum_{k=1}^{n} u_{ki}}$$

Optimal Solutions of J_{fcma}:

$$u_{ki} = \frac{\alpha_i \left(\frac{1}{d_{S_i}(x_k, v_i) + \log |S_i|} \right)^{\frac{1}{m-1}}}{\sum_{j=1}^{c} \alpha_j \left(\frac{1}{d_{S_i}(x_k, v_j) + \log |S_i|} \right)^{\frac{1}{m-1}}}$$

$$v_i = \frac{\sum_{k=1}^{n} (u_{ki})^m x_k}{\sum_{k=1}^{n} (u_{ki})^m}$$

$$\alpha_i = \frac{(\sum_{k=1}^{n} (u_{ki})^m (d_{S_i}(x_k, v_i) + \log |S_i|))^{\frac{1}{m}}}{\sum_{i=1}^{c} (\sum_{k=1}^{n} (u_{ki})^m (d_{S_i}(x_k, v_i) + \log |S_i|))^{\frac{1}{m}}}$$

$$S_i = \frac{\sum_{k=1}^{n} u_{ki}^m (x_k - v_i)(x_k - v_i)^T}{\sum_{k=1}^{n} u_{ki}^m}$$

Optimal Solutions of J_{pfcm}:

$$u_{ki} = \frac{\left(\frac{1}{d_{S_i}(x_k, v_i) - \nu \log \alpha_i + \log |S_i|}\right)^{\frac{1}{m-1}}}{\sum_{j=1}^{c} \left(\frac{1}{d_{S_i}(x_k, v_j) - \nu \log \alpha_j + \log |S_i|}\right)^{\frac{1}{m-1}}}$$

$$v_i = \frac{\sum_{k=1}^{n} (u_{ki})^m x_k}{\sum_{k=1}^{n} (u_{ki})^m}$$

$$\alpha_i = \frac{\sum_{k=1}^{n} (u_{ki})^m}{\sum_{i=1}^{c} \sum_{k=1}^{n} (u_{ki})^m}$$

$$S_i = \frac{\sum_{k=1}^{n} u_{ki}^m (x_k - v_i)(x_k - v_i)^T}{\sum_{k=1}^{n} u_{ki}^m}$$

4 Classifier Functions

After finishing clustering, we are able to set membership value to a new object x using a classifier function. In the case of SFCM, the following has been considered [10].

$$U_i^s(x) = \frac{\left(\frac{1}{d_{S_i}(x, v_i)}\right)^{\frac{1}{m-1}}}{\sum_{j=1}^{c} \left(\frac{1}{d_{S_i}(x, v_j)}\right)^{\frac{1}{m-1}}}.$$

This function is simply derived from the optimal solution of u_{ki} by putting $x = x_k$, where v_i $(i = 1, \ldots, c)$ are the converged centroids. A classifier function helps us to consider theoretical properties of clusters since it is defined in the whole input space. The classifier functions of EFCM and three other methods using variables controlling the size of clusters are as follows.

$$U_i^e(x) = \frac{\exp\left(\frac{-d(x, v_i)}{\nu}\right)}{\sum_{j=1}^{c} \exp\left(\frac{-d(x, v_j)}{\nu}\right)}$$

$$U_i^a(x) = \frac{\alpha_i \left(\frac{1}{d(x, v_i)}\right)^{\frac{1}{m-1}}}{\sum_{j=1}^{c} \alpha_j \left(\frac{1}{d(x, v_j)}\right)^{\frac{1}{m-1}}}$$

$$U_i^p(x) = \frac{\left(\frac{1}{d(x, v_i) - \nu \log \alpha_i}\right)^{\frac{1}{m-1}}}{\sum_{j=1}^{c} \left(\frac{1}{d(x, v_j) - \nu \log \alpha_j}\right)^{\frac{1}{m-1}}}$$

$$U_i^k(x) = \frac{\alpha_i \exp\left(-\frac{d(x, v_i)}{\nu}\right)}{\sum_{j=1}^{c} \alpha_j \exp\left(-\frac{d(x, v_j)}{\nu}\right)}$$

4.1 Theoretial Properties of Classifier Functions

We can prove a number of properties of clusters using the classifier functions. The region of cluster i [10] in the following means

$$\mathcal{R}_i = \{x \in \mathbf{R}^p : U_i(x) \geq U_j(x), \quad j \neq i \; j = 1, 2, \ldots, c \}.$$

$\mathcal{R}_i$ reduces to the Voronoi regions with the center of cluster centroids when SFCM and EFCM are used. When crisp K-means are used, we also have the Voronoi regions. We have either a bounded or unbounded Voronoi region for cluster i.

Proposition 1. *Assume that the covariance variable is not used. As $\|x\| \to \infty$ in an unbounded $\mathcal{R}_i$, we obtain*

$$\lim_{\|x\| \to \infty} U_i^s(x) = \frac{1}{c}$$

$$\lim_{\|x\| \to \infty} U_i^e(x) = 1$$

$$\lim_{\|x\| \to \infty} U_i^a(x) = \alpha_i$$

$$\lim_{\|x\| \to \infty} U_i^p(x) = \frac{1}{c}$$

$$\lim_{\|x\| \to \infty} U_i^k(x) = 1$$

This shows robustness to outliers for each method. If far objects have high memberships, centroids more easily move to the objects. Hence, EFCM and KFCM are more sensitive to outliers than the others.

Proposition 2. *Assume that the covariance variable is used. We then have the following.*

$$\lim_{\|x\| \to \infty} U_i^s(x) = \frac{1}{1 + \sum_{j=1, j \neq i}^{c} \left(\frac{\sigma_j}{\sigma_i} \right)^{\frac{1}{m-1}}}$$

$$\lim_{\|x\| \to \infty} U_i^e(x) = \begin{cases} 1 & (i = \arg\max_j |S_j|) \\ 0 & (otherwise) \end{cases}$$

$$\lim_{\|x\| \to \infty} U_i^a(x) = \frac{1}{1 + \sum_{j=1, j \neq i}^{c} \frac{\alpha_j}{\alpha_i} \left(\frac{\sigma_j}{\sigma_i} \right)^{\frac{1}{m-1}}}$$

$$\lim_{\|x\| \to \infty} U_i^p(x) = \frac{1}{1 + \sum_{j=1, j \neq i}^{c} \left(\frac{\sigma_j}{\sigma_i} \right)^{\frac{1}{m-1}}}$$

$$\lim_{\|x\| \to \infty} U_i^k(x) = \begin{cases} 1 & (i = \arg\max_j |S_j|) \\ 0 & (otherwise) \end{cases},$$

where

$$\sigma_i = \frac{(x - v_i)^T S_i (x - v_i)}{\|x - v_i\|^2}.$$

This proposition shows again that EFCM and KFCM are sensitive to outliers, even when they use Mahalanobis distances. Other classifier functions have constant values as $\|x\| \to \infty$, where the constants are dependent on σ_i. In the case of $U_i^a(x)$, the constant also depends on α_i.

Proposition 3. *Assume that the covariance variables are not used. As x approaches v_i, the classifier function approaches unity in FCMA, whereas it doesn't approach unity in PFCM and KFCM:*

$$\lim_{x \to v_i} U_i^s(x) = 1$$

$$\lim_{x \to v_i} U_i^e(x) = \frac{1}{1 + C_e} < 1$$

$$\lim_{x \to v_i} U_i^a(x) = 1$$

$$\lim_{x \to v_i} U_i^p(x) = \frac{1}{1 + C_p} < 1$$

$$\lim_{x \to v_i} U_i^k(x) = \frac{1}{1 + C_k} < 1,$$

where

$$C_e = \sum_{j=1, j \neq i}^{c} \exp\left(-\frac{d(v_i, v_j)}{\nu}\right)$$

$$C_p = \sum_{j=1, j \neq i}^{c} \left(\frac{\nu \log \alpha_i}{d(v_i, v_j) - \nu \log \alpha_j}\right)^{\frac{1}{m-1}}$$

$$C_k = \alpha_i^{-1} \sum_{j=1, j \neq i}^{c} \alpha_j \exp\left(-\frac{d(v_j, x)}{\nu}\right).$$

Each centroid doesn't have maximum membership in EFCM, EFCA and KFCM although it is representative of its cluster.

Proposition 4. *Assume that the Gustafson-Kessel method is used. We then have:*

$$\lim_{x \to v_i} U_i^s(x) = 1$$

$$\lim_{x \to v_i} U_i^e(x) = \frac{1}{1 + C_e} < 1$$

$$\lim_{x \to v_i} U_i^a(x) = 1$$

$$\lim_{x \to v_i} U_i^p(x) = \frac{1}{1 + C_p} < 1$$

$$\lim_{x \to v_i} U_i^k(x) = \frac{1}{1 + C_k} < 1,$$

where

$$C_e = \sum_{j=1,j\neq i}^{c} \exp\left(-\frac{d_{S_j}(v_i, v_j)}{\nu}\right)$$

$$C_p = \sum_{j=1,j\neq i}^{c} \left(\frac{\nu \log \alpha_i}{d_{S_j}(v_i, v_j) - \nu \log \alpha_j}\right)^{\frac{1}{m-1}}$$

$$C_k = \sum_{j=1,j\neq i}^{c} \frac{\alpha_j}{\alpha_j} \exp\left(-\frac{d_{S_j}(v_j, x)}{\nu}\right).$$

Proposition 5. *Assume that the log method is used. We then have the following.*

$$\lim_{x \to v_i} U_i^s(x) = \frac{1}{1 + C_s} < 1$$

$$\lim_{x \to v_i} U_i^e(x) = \frac{1}{1 + C_e} < 1$$

$$\lim_{x \to v_i} U_i^a(x) = \frac{1}{1 + C_a} < 1$$

$$\lim_{x \to v_i} U_i^p(x) = \frac{1}{1 + C_p} < 1$$

$$\lim_{x \to v_i} U_i^k(x) = \frac{1}{1 + C_k} < 1,$$

$$C_s = \sum_{j=1,j\neq i}^{c} \left(\frac{\log |S_i|}{d_{S_j}(v_i, v_j) + \log |S_j|}\right)^{\frac{1}{m-1}}$$

$$C_e = \sum_{j=1,j\neq i}^{c} \exp\left(-\frac{d_{S_j}(v_i, v_j)}{\nu}\right)$$

$$C_a = \sum_{j=1,j\neq i}^{c} \frac{\alpha_j}{\alpha_i} \left(\frac{\log |S_i|}{d_{S_j}(v_i, v_j) + \log |S_j|}\right)^{\frac{1}{m-1}}$$

$$C_p = \sum_{j=1,j\neq i}^{c} \left(\frac{\nu \log \alpha_i + \log |S_i|}{d_{S_j}(v_i, v_j) - \nu \log \alpha_j + \log |S_j|}\right)^{\frac{1}{m-1}}$$

$$C_k = \sum_{j=1,j\neq i}^{c} \frac{\alpha_j}{|S_j|} \exp\left(-\frac{d_{S_j}(v_i, v_j)}{\nu}\right)$$

Thus the Gustafson-Kessel method behaves similarly to those in Proposition 3, while the log methods do not, as shown in the next proposition.

5   Experimental Results

We experimented using SFCM, EFCM, FCMA, PFCM, KFCM on three UCI data sets: Iris Data Set, Wisconsin Breast Cancer Data Set, and Wine Data Set with numeric attributes.

We tested all methods, SFCM, EFCM, FCMA, PFCM and KFCM, and these methods with covariance by Gustafson-Kessel method and the log method.

We chose the case when the value of objective function is minimum from the following: for each combination of parameter values 50 times trial have been made to reduce the effects of initial values: the parameters were as follows: initial centroids were selected from samples randomly, parameter m was in $[1.1, 4.0]$ with the increment 0.1, parameter ν was in $[0.1, 100.0]$ with the increment 0.1, and $\rho_i = 1$ for all trials in Gustafson-Kessel method.

The results are shown in Table 1. The table shows the Adjusted Rand Index [4] (see also [11]) when the converged value of objective function is minimum.

Table 1. Adjusted Rand Index of each result. (GK) and (Log) means Gustafson-Kessel method and the log method, respectively.

	Iris	BCW	Wine
SFCM	0.757	0.863	0.799
SFCM(GK)	0.868	0.798	0.814
SFCM(Log)	0.904	0.010	0.880
EFCM	0.782	0.852	0.785
EFCM(GK)	0.868	0.830	0.833
EFCM(Log)	0.960	0.114	0.964
FCMA	0.746	0.891	0.814
FCMA(GK)	0.868	0.810	0.740
FCMA(Log)	0.361	0.010	0.933
PFCM	0.747	0.852	0.799
PFCM(GK)	0.886	0.625	0.964
PFCM(Log)	0.732	0.783	0.842
KFCM	0.732	0.847	0.770
KFCM(GK)	0.868	0.830	0.830
KFCM(Log)	0.904	0.093	0.981

In the case of Iris and Wine data set, covariance methods appears to perform better than the same methods without covariance. However, covariance method didn't work well in BCW data set. Moreover, the performances of the log method in BCW experiments were much poorer than the GK method. In contrast, the

GK and the log methods perform equally well in other two examples, except FCMA(Log) for Iris.

Note that the results of KFCM(Log) is not the best, while KFCM(Log) is known to be a generalization of the Gaussian mixture with the EM algorithm [10], which means that the standard method of the Gaussian mixture may be improved to have better performances.

To select the best method seems difficult in view of these results, but to compare the results of these different methods will be useful in other applications. **Note:** We have omitted a number of experimental results. For the details of other results, readers may refer to [7].

6 Conclusion

We compared different methods of generalized fuzzy c-means having cluster size variables and covariance variables, in which basic ideas have already been proposed but all the combinations have not been tested; moreover they have not been compared using common examples. Numerical examples show that the Gustafson-Kessel method worked more stably than the log method, which implies that the Gaussian mixture model has rooms of further improvement, as it is a special case of KFCM with the log method herein.

Future studies should include more extensive comparisons and evaluations in both theoretical properties and larger-scale experiments. The validity indices [1] of clusters should also be studied in near future.

Acknowledgment. The authors deeply appreciate useful comments by reviewers. This study has partly been supported by the Grant-in-Aid for Scientific Research (KAKENHI), JSPS, Japan, No. 26330270.

References

1. Bezdek, J.C.: Pattern Recognition with Fuzzy Objective Function Algorithms. Kluwer, New York (1981)
2. Dunn, J.C.: A fuzzy relative of the ISODATA process and its use in detecting compact well-separated clusters. Cybern. Syst. **3**, 32–57 (1973)
3. Gustafson, D.E., Kessel, W.C.: Fuzzy clustering with a fuzzy covariance matrix. In: Proceedings of 1978 IEEE Conference on Decision and Control, pp. 761–766 (1978)
4. Hubert, L., Arabie, P.: Comparing partitions. J. Classifications **2**, 193–218 (1985)
5. Ichihashi, H., Honda, K., Tani, N.: Gaussian mixture PDF approximation and fuzzy c-means clustering with entropy regularization. In: Proceedings of the 4th Asian Fuzzy System Symposium, pp. 217–221 (2000)
6. Ichihashi, H., Miyagishi, K., Honda, K.: Fuzzy c-means clustering with regularization by KL information. In: Proceedings of the 10th IEEE International Conference on Fuzzy System, pp. 924–927 (2001)
7. Komazaki, Y.: Extensions of Fuzzy c-Means for Imbalanced Clusters, Master's thesis, Graduate School of Systems, Information Engineering (2013). http://www.soft.risk.tsukuba.ac.jp/miyamoto/

8. Li, R.P., Mukaidono, M.: A maximum-entropy approach to fuzzy clustering. In: Proceedings of International Joint Conference of the Fourth IEEE International Conference on Fuzzy Systems and The Second International Fuzzy Engineering Symposium, pp. 2227–2232 (1995)
9. Miyamoto, S., Kurosawa, N.: Controlling cluster volume sizes in fuzzy c-means clustering. In: Proceedings of SCIS and ISIS 2004, Yokohama, Japan
10. Miyamoto, S., Ichihashi, H., Honda, K.: Algorithms for Fuzzy Clustering. Springer, Heidelberg (2008)
11. Vinh, N.X., Epps, J., Bailey, J.: Information Theoretic Measures for Clustering Comparison: Is a Correction for Chance Necessary? In: ICML 2009, Proceedings of the 26th Annual International Conference on Machine Learning, pp. 1073–1080. ACM (2009)
12. Yang, M.S.: On a class of fuzzy classification maximum likelihood procedures. Fuzzy Sets Syst. **57**(3), 365–375 (1993)

Partial Data Querying Through Racing Algorithms

Vu-Linh Nguyen[1]([⊠]), Sébastien Destercke[1], and Marie-Hélène Masson[1,2]

[1] UMR CNRS 7253 Heudiasyc, Sorbonne Université, Université de Technologie de Compiègne CS 60319, 60203 Compiègne Cedex, France
{linh.nguyen,sebastien.destercke,mylene.masson}@hds.utc.fr
[2] Université de Picardie Jules Verne, Amiens, France

Abstract. This paper studies the problem of learning from instances characterized by imprecise features or imprecise class labels. Our work is in the line of active learning, since we consider that the precise value of some partial data can be queried to reduce the uncertainty in the learning process. Our work is based on the concept of racing algorithms in which several models are competing. The idea is to identify the query that will help the most to quickly decide the winning model in the competition. After discussing and formalizing the general ideas of our approach, we study the particular case of binary SVM and give the results of some preliminary experiments.

Keywords: Partial data · Data querying · Active learning · Racing algorithms

1 Introduction

Although classical learning schemes assume that every instance is fully specified, there are many cases where such an assumption is unlikely to hold, and where some features or the label (class) of an instance may be only partially known. The problem of learning from imprecise data has gained an increasing interest with applications in different fields such as image or natural language processing [2–4]. The imprecision of the data leads to uncertainties in the learning process and in the decision making.

This work explores an issue related to partially specified data: if we have the possibility to gain more information on some of the partial instances, which instance and what feature of this instance should we query? In the case of a completely missing label (and to a lesser extent of missing features), this problem is known as active learning and has already been largely treated [6]. However, we are not aware of such works for partial data. The present proposal is based on the concept of racing algorithms [5], initially used to select an optimal configuration of a given lazy learning model, and since then applied to other settings such as multi-armed bandits. The idea of such racing algorithms is to oppose a (finite) set of alternatives in a race, and to progressively discard losing ones as the race goes

© Springer International Publishing AG 2016
V.-N. Huynh et al. (Eds.): IUKM 2016, LNAI 9978, pp. 163–174, 2016.
DOI: 10.1007/978-3-319-49046-5_14

along. In our case, the set of alternatives will be composed of different possible models. As the data are partial, the performance of each model is uncertain (i.e. interval-valued) and several candidate models can be optimal. The race will consist in iteratively making queries, i.e., in asking to an oracle the precise value of a partial data. The key question is then to identify those queries that will help the most to reduce the set of possible winners in the race and to converge quickly to the optimal model. We illustrate this general approach using binary SVM classifiers.

The rest of this paper is organized as follows: we present in Sect. 2 the basic notations used in this paper. Section 3 introduces the general principles of racing algorithms and formalizes the problem of quantifying the influence of a query on the race. Section 4 is focused on the particular case of binary SVM. Finally, some experiments in Sect. 5 demonstrate the effectiveness of our proposals.

2   Preliminaries

In classical supervised setting, the goal of the learning approach is to find a model $m : \mathcal{X} \to \mathcal{Y}$ within a set $\mathcal{M}$ of models using n input/output samples $(\mathbf{x}_i, y_i) \in \mathcal{X} \times \mathcal{Y}$, where $\mathcal{X}$ and $\mathcal{Y}$ are respectively the input and the output spaces[1]. The empirical risk $R(m)$ associated to a model m is then evaluated as

$$R(m) = \sum_{i=1}^{n} \ell(y_i, m(\mathbf{x}_i)) \tag{1}$$

where $\ell : \mathcal{Y} \times \mathcal{Y} \to \mathbb{R}$ is the loss function, and $\ell(y, m(\mathbf{x}))$ is the loss of predicting $m(\mathbf{x})$ when observing y. The selected model is then the one minimizing (1), that is

$$m^* = \arg \min_{m \in \mathcal{M}} R(m). \tag{2}$$

Another way to see the model selection problem is to assume that a model m_j is said to be better than m_k (denoted $m_j \succ m_k$) if

$$R(m_k) - R(m_j) > 0, \tag{3}$$

or in other words if the risk of m_j is lower than the risk of m_k.

In this work, we are however interested in the case where data are partial, that is where general samples are of the kind $(\mathbf{X}_i, Y_i) \subseteq \mathcal{X} \times \mathcal{Y}$. In such a case, Eqs. (1), (2) and (3) are no longer well-defined, and there are different ways to extend them. Two of the most common ways to extend them is either to use a minimin (optimistic) or a maximin (pessimistic) approach [7]. That is, if we extend Eq. (1) to a lower bound

---

[1] As $\mathcal{X}$ is often multi-dimensional, we will denote its elements and subsets by bold letters.

$$\underline{R}(m) = \inf_{(\mathbf{x}_i, y_i) \in (\mathbf{X}_i, Y_i)} \sum_{i=1}^{n} \ell(y_i, m(\mathbf{x}_i)) \tag{4}$$

$$= \sum_{i=1}^{n} \inf_{(\mathbf{x}_i, y_i) \in (\mathbf{X}_i, Y_i)} \ell(y_i, m(\mathbf{x}_i)) := \sum_{i=1}^{n} \underline{\ell}(Y_i, m(\mathbf{X}_i))$$

and an upper bound

$$\overline{R}(m) = \sup_{(\mathbf{x}_i, y_i) \in (\mathbf{X}_i, Y_i)} \sum_{i=1}^{n} \ell(y_i, m(\mathbf{x}_i)) \tag{5}$$

$$= \sum_{i=1}^{n} \sup_{(\mathbf{x}_i, y_i) \in (\mathbf{X}_i, Y_i)} \ell(y_i, m(\mathbf{x}_i)) := \sum_{i=1}^{n} \overline{\ell}(Y_i, m(\mathbf{X}_i))$$

then the optimal minimin m_{mm}^* and maximin m_{Mm}^* models are

$$m_{mm}^* = \arg\min_{m \in \mathcal{M}} \underline{R}(m) \quad \text{and} \quad m_{Mm}^* = \arg\min_{m \in \mathcal{M}} \overline{R}(m).$$

The minimin approach usually assumes that data are distributed according to the model, and tries to find the best data replacement (or disambiguation) combined with the best possible model. Conversely, the maximin approach assumes that data are distributed in the worst possible way, and select the model performing the best in the worst situation, thus guaranteeing a minimal performance of the model. However, such an approach, due to its conservative nature, may lead to sub-optimal model.

In this paper, we are interested into another kind of approach, where we do not search for a unique optimal model but rather consider sets of potentially optimal models. In this case, we can say that a model m_j is better than m_k (still denoted $m_j \succ m_k$) if

$$\underline{R}(m_{k-j}) = \inf_{(\mathbf{x}_i, y_i) \in (\mathbf{X}_i, Y_i)} R(m_k) - R(m_j) > 0, \tag{6}$$

which is a direct extension of Eq. (3). That is, $m_j \succ m_k$ if and only if it is better under every possible precise instances $(\mathbf{x}_i, y_i)$ consistent with the partial instances $(\mathbf{X}_i, Y_i)$. We can then denote by

$$\mathcal{M}^* = \{m \in \mathcal{M} : \not\exists m' \in \mathcal{M} \ s.t. \ m' \succ m\} \tag{7}$$

the set of undominated models within $\mathcal{M}$, that is the set of models that are maximal with respect to the partial order $\succ$.

Example 1. Figure 1 illustrates a situation where $\mathcal{Y}$ consists of two different classes (gray and white), and $\mathcal{X}$ of two dimensions. Only imprecise data are numbered. Squares are assumed to have precise features. Stripped squares have unknown labels. Assuming that $\mathcal{M} = \{m_1, m_2\}$ (the models could be decision stumps, or one-level decision trees), we would have that $m_2 = m_{Mm}^*$ is the maximin model and $m_1 = m_{mm}^*$ the minimin one. The two models would however be incomparable according to (6), hence $\mathcal{M}^* = \mathcal{M}$ in this case.

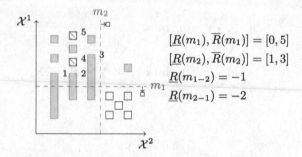

$$[\underline{R}(m_1), \overline{R}(m_1)] = [0, 5]$$
$$[\underline{R}(m_2), \overline{R}(m_2)] = [1, 3]$$
$$\underline{R}(m_{1-2}) = -1$$
$$\underline{R}(m_{2-1}) = -2$$

Fig. 1. Illustration of partial data and competing models

3   Partial Data Querying: A Racing Approach

Both the minimin and maximin approaches have the same goal: obtaining a unique model from partially specified data. The idea we consider in this paper is different. We want to identify and query those data that will help the most to reduce the set $\mathcal{M}^*$. Whether an information is useful for the race is formalized in what follows. Let us assume that $\mathcal{X} = \mathcal{X}^1 \times \ldots \times \mathcal{X}^P$ is a Cartesian product of P spaces, and that a partial data $(\mathbf{X}_i, Y_i)$ can be expressed as $(\times_{j=1}^{P} X_i^j, Y_i)$, and furthermore that if $\mathcal{X}^j \subseteq \mathbb{R}$ is a subset of the real line, then X_i^j is an interval.

A query on a partial data $(\times_{j=1}^{P} X_i^j, Y_i)$ consists in transforming one of its dimension X_i^j or Y_i into the true precise value x_i^j or y_i, provided by an oracle. More precisely, Q_i^j denotes the query made on X_i^j or Y_i, with $j = p + 1$ for Y_i. Given a model m_k and a data $(\times_{j=1}^{P} X_i^j, Y_i)$, the result of a query can have an effect on the interval $[\underline{R}(m_k), \overline{R}(m_k)]$, depending on whether it changes the interval $[\underline{\ell}(Y_i, m_k(\mathbf{X}_i)), \overline{\ell}(Y_i, m_k(\mathbf{X}_i))]$. Similarly, when assessing whether the model m_k is preferred to m_ℓ, the query can have an influence on the value $\underline{R}(m_{\ell-k})$ or not. This can be formalized by two functions, $E_{Q_i^j} : \mathcal{M} \to \{0, 1\}$ and $J_{Q_i^j} : \mathcal{M} \times \mathcal{M} \to \{0, 1\}$ such that:

$$E_{Q_i^j}(m_k) = \begin{cases} 1 \text{ if } \exists x_j^i \in X_i^j \text{ that reduces } [\underline{R}(m_k), \overline{R}(m_k)] \\ 0 \text{ else} \end{cases} \quad (8)$$

and

$$J_{Q_i^j}(m_k, m_\ell) = \begin{cases} 1 \text{ if } \exists x_j^i \in X_i^j \text{ that increases } \underline{R}(m_{\ell-k}) \\ 0 \text{ else.} \end{cases} \quad (9)$$

Of course, when $j = p + 1$, X_i^j is to be replaced by Y_i. $E_{Q_i^j}$ simply tells us whether or not the query can affect our evaluation of m_k performances, while $J_{Q_i^j}(m_k, m_\ell)$ informs us whether the query can help to differentiate m_k and m_ℓ.

Example 2. In Fig. 1, questions related to partial classes (points 4 and 5) and to partial features (points 1, 2 and 3) have respectively the same potential effect, so we can restrict our attention to Q_4^3 (the class of point 3) and to Q_1^1 (the first feature of point 3). For these two questions, we have

- $E_{Q_4^3}(m_1) = E_{Q_4^3}(m_2) = 1$ and $J_{Q_4^3}(m_1, m_2) = J_{Q_4^3}(m_1, m_2) = 0$.
- $E_{Q_1^1}(m_1) = 1$, $E_{Q_1^1}(m_2) = 0$ and $J_{Q_1^1}(m_1, m_2) = J_{Q_1^1}(m_2, m_1) = 1$.

This example shows that while some questions may reduce our uncertainty about many model risks (Q_4^3 reduce risk intervals for both models), they may be less useful than other questions to tell two models apart (Q_1^1 can actually lead to declare m_2 better than m_1).

The effect of a query being now formalized, we can propose a method inspired by racing algorithms to select the best query. An initial set of models can be created by sampling several times a precise data set $(\mathbf{x}_i, y_i) \in (\mathbf{X}_i, Y_i)$ and then learning several optimal models according to this selection. Algorithm 1 summarises the general procedure applied to find the best query and to update the race. This algorithm simply searches the query that will have the biggest impact on the minimin model and its competitors, adopting the optimistic attitude of racing algorithms. Once a query has been made, the data set as well as the set of competitors are updated, so that only potentially optimal models remain.

In the next sections, we illustrate our proposed setting with the popular SVM algorithm.

Algorithm 1. One iteration of the racing algorithm to query data.

Input: data (X_i, Y_i), set $\{m_1, \ldots, m_R\}$ of models
Output: updated data and set of models

1   $k^* = \arg\min_{k \in \{1,\ldots,R\}} \underline{R}(m_i)$;

2   **foreach** *query* Q_i^j **do**

3     $\lfloor$ $Value(Q_i^j) = E_{Q_i^j}(m_{k^*}) + \sum_{k \neq k^*} J_{Q_i^j}(m_{k^*}, m_k)$;

4   $Q_{i*}^{j^*} = \arg\max_{Q_i^j} Value(Q_i^j)$;

5   Get value $x_{i*}^{j^*}$ of $X_{i*}^{j^*}$;

6   **foreach** $k, \ell \in \{1, \ldots, R\} \times \{1, \ldots, R\}$, $k \neq \ell$ **do**

7     $\lfloor$ Compute $\underline{R}(m_{\ell-k})$;

8   **if** $\underline{R}(m_{\ell-k}) > 0$ **then** remove m_ℓ from $\{m_1, \ldots, m_R\}$;

---

4   Application to Binary SVM

In the binary SVM setting [1], the input space $\mathcal{X} = \mathbb{R}^p$ is the real space and the binary output space is $\mathcal{Y} = \{-1, 1\}$, where $-1, 1$ encode the two possible classes. The model $m_k = (\mathbf{w}_k, c_k)$ corresponds to the "maximum-margin" hyperplane $\mathbf{w}_k \mathbf{x} + c_k$ with $\mathbf{w}_k \in \mathbb{R}^p$ and $c_k \in \mathbb{R}$. For convenience sake, we will use $(\mathbf{w}_k, c_k)$ and m_k interchangeably from now on. We will also focus on the case of imprecise features but precise labels, and will denote y_i the label of training instances for short, instead of Y_i. We will also focus on the classical $0-1$ loss function defined as follows for an instance $(\mathbf{x}_i, y_i)$:

$$\ell(y_i, m_k(\mathbf{x}_i)) = \begin{cases} 0 \text{ if } y_i \cdot m_k(\mathbf{x}_i) \geq 0 \\ 1 \text{ if } y_i \cdot m_k(\mathbf{x}_i) < 0, \end{cases} \tag{10}$$

where $m_k(\mathbf{x}_i) = \mathbf{w}_k \mathbf{x}_i + c_k$.

4.1   Instances Inducing Imprecision in Empirical Risk

Before entering into the details of how risk bounds (4)–(6) and query effects (8)–(9) can be estimated in practice, we will first investigate under which conditions an instance $(\mathbf{X}_i, y_i)$ induces imprecision in the empirical risk. Such instances are the only ones of interest here, since if $\underline{\ell}(y_i, m_k(\mathbf{X}_i)) = \overline{\ell}(y_i, m_k(\mathbf{X}_i)) = \ell(y_i, m_k(\mathbf{X}_i))$, then $E_{Q_i^j}(m_k) = J_{Q_i^j}(m_k, m_l) = 0$ for all $j = 1, \ldots, P$

Definition 1. Given a SVM model m_k, an instance $(\mathbf{X}_i, y_i)$ is called an imprecise instance w.r.t m_k (or shortly, imprecise instance when m_k is fixed) if and only if

$$\exists \mathbf{x}_i', \mathbf{x}_i'' \in \mathbf{X}_i \ s.t \ m_k(\mathbf{x}_i') \geq 0 \ and \ m_k(\mathbf{x}_i'') < 0. \tag{11}$$

Instances that do not satisfy Definition 1 will be called precise instances (w.r.t m_k). Being precise means that the sign of $m_k(x_i)$ is the same for all $x_i \in \mathbf{X}_i$, which implies that the loss $\underline{\ell}(y_i, m_k(\mathbf{X}_i)) = \overline{\ell}(y_i, m_k(\mathbf{X}_i))$ is precisely known. The next example illustrates the notion of (im)precise instances.

Example 3. Figure 2 illustrates a situation with two models and where the two different classes are represented by grey ($y = +1$) and white ($y = -1$) colours. From the figure, we can say that $(\mathbf{X}_1, y_1)$ is precise w.r.t both m_1 and m_2, $(\mathbf{X}_2, y_2)$ is precise w.r.t m_1 and imprecise w.r.t m_2, $(\mathbf{X}_3, y_3)$ is imprecise w.r.t both m_1 and m_2 and $(\mathbf{X}_4, y_4)$ is imprecise w.r.t m_1 and precise w.r.t m_2.

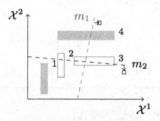

Fig. 2. Illustration of interval-valued instances

Determining whether an instance is imprecise w.r.t. m_k is actually very easy in practice. Let us denote by

$$\underline{m}_k(\mathbf{X}_i) := \inf_{\mathbf{x}_i \in \mathbf{X}_i} m_k(\mathbf{x}_i) \ \text{and} \ \overline{m}_k(\mathbf{X}_i) := \sup_{\mathbf{x}_i \in \mathbf{X}_i} m_k(\mathbf{x}_i) \tag{12}$$

the lower and upper bounds reached by model m_k over the space $\mathbf{X}_i$. The following result characterizing imprecise instances, as well as a hyperplane $m_k(\mathbf{x}_i) = 0$ intersects with a region $\mathbf{X}_i$, follows from the fact that the image of a compact and connected set by a continuous function is also compact and connected.

Proposition 1. *Given* $m_k(\mathbf{x}_i) = \mathbf{w}_k\mathbf{x}_i + c_k$ *and the set* $\mathbf{X}_i$, *then* $(\mathbf{X}_i, y_i)$ *is imprecise w.r.t.* m_k *if and only if*

$$\underline{m}_k(\mathbf{X}_i) < 0 \text{ and } \overline{m}_k(\mathbf{X}_i) \geq 0. \tag{13}$$

Furthermore, we have that the hyperplane $m_k(\mathbf{x}_i) = 0$ *intersects with the region* $\mathbf{X}_i$ *if and only if* (13) *holds. In other words,* $\exists \mathbf{x}_i \in \mathbf{X}_i$ *s.t.* $m_k(\mathbf{x}_i) = 0$.

This proposition means that to determine whether an instance $(\mathbf{X}_i, y_i)$ is imprecise, we only need to compute values $\underline{m}_k(\mathbf{X}_i)$ and $\overline{m}_k(\mathbf{X}_i)$, which can be easily done using Proposition 2. Note that, due to a lack of space, the proofs of proposition are omitted in this conference version.

Proposition 2. *Given* $(\mathbf{X}_i, y_i)$ *with* $X_i^j = \left[a_i^j, b_i^j\right]$ *and SVM model* $(\mathbf{w}_k, c_k)$, *we have*

$$\overline{m}_k(\mathbf{X}_i) = \sum_{w_k^j \geq 0} w_k^j b_i^j + \sum_{w_k^j < 0} w_k^j a_i^j + c_k$$

$$\underline{m}_k(\mathbf{X}_i) = \sum_{w_k^j \geq 0} w_k^j a_i^j + \sum_{w_k^j < 0} w_k^j b_i^j + c_k.$$

Again, it should be noted that only imprecise instances are of interest here, as only those can result in an increase of the lower empirical risk bounds. We will therefore focus on those in the next sections.

4.2 Empirical Risk Bounds and Single Effect

We are now going to investigate the practical computation of $\underline{R}(m_k)$, $\overline{R}(m_k)$ for $k = 1, \ldots, M$, as well as the value $E_{Q_i^j}(m_k)$ of a query on a model m_k. Equation (4) (resp. (5)) implies that the computation of $\underline{R}(m_k)$ (resp. $\overline{R}(m_k)$) can be done by computing $\underline{\ell}(y_i, m_k(\mathbf{x}_i))$ (resp. $\overline{\ell}(y_i, m_k(\mathbf{x}_i))$) for $i = 1, \ldots, n$ and then by summing the obtained values, therefore we can focus on computing $\underline{\ell}(y_i, m_k(\mathbf{x}_i))$ and $\overline{\ell}(y_i, m_k(\mathbf{x}_i))$ for a single instance. Similarly, $E_{Q_i^j}(m_k) = 1$ only when the interval $[\underline{\ell}(y_i, m_k(\mathbf{x}_i)), \overline{\ell}(y_i, m_k(\mathbf{x}_i))]$ can be modified by querying X_i^j, therefore we can also focus on a single instance to evaluate it. When $\mathbf{X}_i$ is imprecise w.r.t. m_k, we have $\underline{\ell}(y_i, m_k(\mathbf{X}_i)) = 0$ and $\overline{\ell}(y_i, m_k(\mathbf{X}_i)) = 1$. Let us now consider the problem of computing, for a query Q_i^j, the effect $E_{Q_i^j}(m_k)$ it can have on the empirical risk bounds. In the case of 0–1 loss, the only case where $E_{Q_i^j}(m_k) = 1$ is the one where $[\underline{\ell}(y_i, m_k(\mathbf{x}_i)), \overline{\ell}(y_i, m_k(\mathbf{x}_i))]$ goes from $[0, 1]$ before the query to a precise value after it, or in other words if there is $x_i^j \in X_i^j$ such that $\mathbf{X}_i' = \times_{j' \neq j} X_i^j \times \{x_i^j\}$ is precise w.r.t. m_k. According to Proposition 1, this means that either $\underline{m}_k(\mathbf{X}_i')$ should become positive, or $\overline{m}_k(\mathbf{X}_i')$ should become negative after query Q_i^j. This is formalised in the next proposition.

Proposition 3. *Given $(\mathbf{X}_i, y_i)$ with $X_i^j = \left[a_i^j, b_i^j\right]$ and model $m_k(\mathbf{x}_i)$ s.t. $\mathbf{X}_i$ is imprecise, then $E_{Q_i^j}(m) = 1$ if and only if one of the following conditions holds*

$$\underline{m}_k(\mathbf{X}_i) \geq -|w_k^j|(b_i^j - a_i^j) \tag{14}$$

or

$$\overline{m}_k(\mathbf{X}_i) < |w_k^j|(b_i^j - a_i^j). \tag{15}$$

$\underline{R}(m_k), \overline{R}(m_k)$, needed in the line 1 of Algorithm 1 to identify the most promising model k^*, are computed easily using the values of $\underline{\ell}(y_i, m_k(\mathbf{X}_i)) = 0$ and $\overline{\ell}(y_i, m_k(\mathbf{X}_i)) = 1$, while Eqs. (14) and (15) provide us easy ways to estimate the values of $E_{Q_i^j}(m_{k^*})$, needed in line 3 of Algorithm 1.

4.3   Pairwise Risk Bounds and Effect

Let us now focus on how to compute, for a pair of models m_k and m_l, whether a query Q_i^j will have an effect on the value $\underline{R}(m_{k-l})$. For this, we will have to compute $\underline{R}(m_{k-l})$, which is a necessary step to estimate the indicator $J_{Q_i^j}(m_l, m_k)$ of a possible effect of Q_i^j. To do that, note that $\underline{R}(m_{k-l})$ can be rewritten as

$$\underline{R}(m_{k-l}) = \inf_{\mathbf{x}_i \in \mathbf{X}_i, i=1,\ldots,n} (R(m_k) - R(m_l)) = \sum_{i=1}^{n} \underline{\ell}_{k-l}(\mathbf{x}_i, y_i) \tag{16}$$

with

$$\underline{\ell}_{k-l}(y_i, \mathbf{X}_i) = \inf_{\mathbf{x}_i \in \mathbf{X}_i} \left(\ell(y_i, m_k(\mathbf{x}_i)) - \ell(y_i, m_l(\mathbf{x}_i)) \right), \tag{17}$$

meaning that computing $\underline{R}(m_{k-l})$ can be done by summing up $\underline{\ell}_{k-l}(y_i, \mathbf{X}_i)$ over all $\mathbf{X}_i$, similarly to $\underline{R}(m_k)$ and $\overline{R}(m_k)$. Also, $J_{Q_i^j}(m_l, m_k) = 1$ if and only if Q_i^j can increase $\underline{R}(m_{k-l})$. We can therefore focus on the computation of $\underline{\ell}_{k-l}(y_i, \mathbf{X}_i)$ and its possible changes. First note that if $\mathbf{X}_i$ is precise w.r.t. both m_k and m_l, then $\ell(y_i, m_k(\mathbf{X}_i)) - \ell(y_i, m_l(\mathbf{X}_i))$ is a well-defined value, as each loss is precise, and in this case $J_{Q_i^j}(m_l, m_k) = 0$. Therefore, the only cases of interest are those where $\mathbf{X}_i$ is imprecise w.r.t. to at least one model. We will first treat the case where it is imprecise for only one, and then will proceed to the more complex case where it is imprecise w.r.t. both. Note that imprecision with respect to each model can be easily established using Proposition 1.

Imprecision with Respect to One Model. Let us consider the case where $\mathbf{X}_i$ is imprecise w.r.t. either m_k or m_l. In each of these two cases, the loss induced by $(\mathbf{X}_i, y_i)$ on the model for which it is precise is fixed. Hence, to estimate the lower loss $\underline{\ell}_{k-l}(y_i, \mathbf{X}_i)$, as well as the effect of a possible query Q_i^j, we only have to look at the model for which $(\mathbf{X}_i, y_i)$ is imprecise. The next proposition establishes the lower bound $\underline{\ell}_{k-l}(y_i, \mathbf{X}_i)$, necessary to compute $\underline{R}(m_{k-l})$.

Proposition 4. *Given* $(\mathbf{X}_i, y_i)$ *with* $X_i^j = [a_i^j, b_i^j]$ *and two models* $m_k(\mathbf{X}_i)$ *and* $m_l(\mathbf{X}_i)$ *s.t* $(\mathbf{X}_i, y_i)$ *is imprecise w.r.t. one and only one model, then we have*

$$\underline{\ell}_{k-l}(\mathbf{X}_i) = \ell(y_i, m_k(\mathbf{X}_i)) - 1 \qquad \text{if } \mathbf{X}_i \text{ is imprecise w.r.t. } m_l \tag{18}$$

$$\underline{\ell}_{k-l}(\mathbf{X}_i) = -\ell(y_i, m_l(\mathbf{X}_i)) \qquad \text{if } \mathbf{X}_i \text{ is imprecise w.r.t. } m_k. \tag{19}$$

Let us now study under which conditions a query Q_i^j can increase $\underline{\ell}_{k-l}(\mathbf{X}_i)$, hence under which conditions $J_{Q_i^j}(m_l, m_k) = 1$. The two next propositions respectively address the case of imprecision w.r.t. m_l and m_k. Given a possible query Q_i^j on $\mathbf{X}_i$, the only possible way to increase $\underline{\ell}_{k-l}(\mathbf{X}_i)$ is for the updated $\mathbf{X}_i'$ to become precise w.r.t. to the model for which $\mathbf{X}_i$ was imprecise, and moreover to be so that $\ell(y_i, m_l(\mathbf{X}_i')) = 0$ ($\ell(y_i, m_k(\mathbf{X}_i')) = 1$) if $\mathbf{X}_i$ is imprecise w.r.t. m_l (m_k).

Proposition 5. *Given* $(\mathbf{X}_i, y_i)$ *with* $X_i^j = [a_i^j, b_i^j]$ *and two models* $m_k(\mathbf{x}_i)$ *and* $m_l(\mathbf{x}_i)$ *s.t.* $(\mathbf{X}_i, y_i)$ *is imprecise w.r.t.* m_l, *the question* Q_i^j *is such that* $J_{Q_i^j}(m_l, m_k) = 1$ *if and only if one of the two following condition holds*

$$y_i = 1 \text{ and } \underline{m}_l(\mathbf{X}_i) \geq -|w_l^j|(b_i^j - a_i^j) \tag{20}$$

$$or$$

$$y_i = -1 \text{ and } \overline{m}_l(\mathbf{X}_i) < |w_k^j|(b_i^j - a_i^j). \tag{21}$$

Proposition 6. *Given* $(\mathbf{X}_i, y_i)$ *with* $X_i^j = [a_i^j, b_i^j]$ *and two models* $m_k(\mathbf{x}_i)$ *and* $m_l(\mathbf{x}_i)$ *s.t.* $(\mathbf{X}_i, y_i)$ *is imprecise w.r.t.* m_k, *the question* Q_i^j *is such that* $J_{Q_i^j}(m_l, m_k) = 1$ *if and only if one of the two following condition holds*

$$y_i = 1 \text{ and } \overline{m}_k(\mathbf{X}_i) < |w_l^j|(b_i^j - a_i^j) \tag{22}$$

$$or$$

$$y_i = -1 \text{ and } \underline{m}_k(\mathbf{X}_i) \geq -|w_k^j|(b_i^j - a_i^j). \tag{23}$$

In summary, if $\mathbf{X}_i$ is imprecise w.r.t. only one model, estimating $J_{Q_i^j}(m_l, m_k)$ comes down to identify whether the $\mathbf{X}_i$ can become precise with respect to such a model, in such a way that the lower bound is possibly increased. Propositions 5 and 6 show that this can be done easily using our previous results of Sect. 4.1 concerning the empirical risk.

Imprecision with Respect to both Models. Given $\mathbf{X}_i$ and two models m_k, m_l, we will adopt the following notations:

$$m_{k-l}(\mathbf{X_i}) > 0 \text{ if } m_k(\mathbf{x}_i) - m_l(\mathbf{x}_i) > 0 \quad \forall \mathbf{x}_i \in \mathbf{X}_i \tag{24}$$

$$m_{k-l}(\mathbf{X_i}) < 0 \text{ if } m_k(\mathbf{x}_i) - m_l(\mathbf{x}_i) < 0 \quad \forall \mathbf{x}_i \in \mathbf{X}_i. \tag{25}$$

In the other cases, this means that there are $x_i', x_i'' \in \mathbf{X}_i$ for which the model difference have different signs. The reason for introducing such differences is

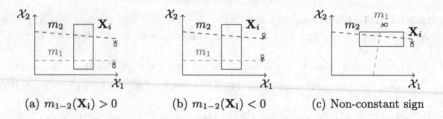

(a) $m_{1-2}(\mathbf{X_i}) > 0$ (b) $m_{1-2}(\mathbf{X_i}) < 0$ (c) Non-constant sign

Fig. 3. Illustrations for the different possible cases corresponding to the difference $m_1(x) - m_2(x)$

that, if $m_{k-l}(\mathbf{X_i}) > 0$ or $m_{k-l}(\mathbf{X_i}) < 0$, then not all combinations in $\{0,1\}^2$ are possible for the pair $(\ell(y_i, m_k(\mathbf{x}_i)), \ell(y_i, m_l(\mathbf{x}_i)))$, while they are in the other case. These various situations are depicted in Fig. 3, where the white class is again the negative one ($y_i = -1$).

Since $m_k(\mathbf{x}_i) - m_l(\mathbf{x}_i)$ is also of linear form (with weights $w_k^j - w_l^j$), we can easily determine whether the sign of $m_{k-l}(\mathbf{X_i})$ is constant: it is sufficient to compute the interval

$$\left[\inf_{\mathbf{x}_i \in \mathbf{X}_i} (m_k(\mathbf{x}_i) - m_l(\mathbf{x}_i)), \sup_{\mathbf{x}_i \in \mathbf{X}_i} (m_k(\mathbf{x}_i) - m_l(\mathbf{x}_i)) \right]$$

that can be computed similarly to $[\underline{m}_k(\mathbf{X}_i), \overline{m}_k(\mathbf{X}_i)]$ in Sect. 4.1. If zero is not within this interval, then $m_{k-l}(\mathbf{X}_i) > 0$ if the lower bound is positive, otherwise $m_{k-l}(\mathbf{X}_i) < 0$ if the upper bound is negative. The next proposition indicates how to easily compute the lower bound $\underline{\ell}_{k-l}(\mathbf{X}_i)$ for the different possible situations.

Proposition 7. *Given* $(\mathbf{X}_i, y_i)$ *with* $X_i^j = [a_i^j, b_i^j]$ *and two models* m_k, m_l *s.t.* $(\mathbf{X}_i, y_i)$ *is imprecise w.r.t. both the given models, then the minimal difference value is*

$$\underline{\ell}_{k-l}(\mathbf{X}_i) = \begin{cases} \min(0, -y_i) & \text{if } m_{k-l}(\mathbf{X}_i) > 0 \\ \min(0, y_i) & \text{if } m_{k-l}(\mathbf{X}_i) < 0 \\ -1 & \text{else} \end{cases} \tag{26}$$

The next question is to know under which conditions a query Q_i^j can increase $\underline{\ell}_{k-l}(\mathbf{X}_i)$ (or equivalently $\underline{R}(m_{k-l})$), or in other words to determine pair (i, j) s.t $J_{Q_i^j}(m_l, m_k) = 1$. Proposition 7 tells us that $\underline{\ell}_{k-l}(X_i)$ can be either 0 or -1 if $m_{k-l}(\mathbf{X}_i) > 0$ or $m_{k-l}(\mathbf{X}_i) > 0$, and is always -1 if $m_{k-l}(\mathbf{X}_i)$ can take both signs. The next proposition establishes conditions under which $\underline{\ell}_{k-l}(\mathbf{X}_i)$ can increase.

Proposition 8. *Given* $(\mathbf{X}_i, y_i)$ *with* $X_i^j = [a_i^j, b_i^j]$ *and two models* $m_k(x_i)$ *and* $m_l(x_i)$ *s.t* $(\mathbf{X}_i, y_i)$ *is imprecise w.r.t both of the given models, then* $J_{Q_i^j}(m_l, m_k) = 1$ *if the following conditions hold*

if $\underline{\ell}_{k-l}(X_i) = -1$ *and* $y_i = 1$:

$$\overline{m}_k(\mathbf{X}_i) < |w_l^j|(b_i^j - a_i^j) \text{ or } \underline{m}_l(\mathbf{X}_i) \geq -|w_l^j|(b_i^j - a_i^j) \tag{27}$$

if $\underline{\ell}_{k-l}(X_i) = -1$ *and* $y_i = -1$:

$$\underline{m}_k(\mathbf{X}_i) \geq -|w_k^j|(b_i^j - a_i^j) \text{ or } \overline{m}_l(\mathbf{X}_i) < |w_l^j|(b_i^j - a_i^j). \tag{28}$$

if $\underline{\ell}_{k-l}(X_i) = 0$ *and* $m_{k-l}(\mathbf{X}_i) < 0$:

$$\overline{m}_k(\mathbf{X}_i) < |w_l^j|(b_i^j - a_i^j) \text{ and } \underline{m}_l(\mathbf{X}_i) \geq -|w_l^j|(b_i^j - a_i^j) \tag{29}$$

if $\underline{\ell}_{k-l}(X_i) = 0$ *and* $m_{k-l}(\mathbf{X}_i) > 0$:

$$\underline{m}_k(\mathbf{X}_i) \geq -|w_k^j|(b_i^j - a_i^j) \text{ and } \overline{m}_l(\mathbf{X}_i) < |w_l^j|(b_i^j - a_i^j). \tag{30}$$

For instance, in Fig. 3(a) and (b), $J_{Q_i^1}(m_1, m_2) = 0$ for both cases, while $J_{Q_i^2}(m_1, m_2) = 0$ for 3(a) and $J_{Q_i^2}(m_1, m_2) = 1$ for 3(a)

5 Experiments

To demonstrate the usefullness of our approach, we run experiments on a "contaminated" version of a standard benchmark data set, namely Parkinson (195 instances, 22 features, binary labels) which contains precise features and labels. 10 % of data have been used for training and 90 % for testing, since querying partial data is most useful when only few data are available. For each feature x_i^j in the training set, a biased coin is flipped in order to decide whether or not this example will be contaminated; the probability of contamination is $p = 0.4$. In case x_i^j is contaminated, a width q_i^j will be generated from a uniform distribution. Then the generated interval valued data is $X_i^j = [x_i^j + q_i^j(\underline{D}^j - x_i^j), x_i^j + q_i^j(\overline{D}^j - x_i^j)]$ where $\underline{D}^j = \min_i(x_i^j)$ and $\overline{D}^j = \max_i(x_i^j)$. To evaluate the efficiency of our proposal, we query interval data using three approaches: our racing algorithm, a random querying strategy (each time, interval examples will be chosen randomly) and the most partial querying strategy (each time, examples with the largest imprecision will be queried). Firstly, we randomly generate 100 completions of interval-valued data. From each completion, one linear SVM model is trained and the set of such SVM models is considered as the initial set of undominated models. To limit the computational cost, at each iteration of the racing algorithm, we choose to perform 2 queries (batch query) instead of only one. After each batch, we discard the dominated models and determine the best potential model. In case of multiple minimal risk models, the one with minimum value of $\overline{R}_m$ will be chosen as the best potential model. The accuracy of the best potential model is computed on the test set. The learning process is repeated 10 times and the average size of the sets of models and the average accuracy of the best potential model are given in Fig. 4(a) and (b), respectively.

The experimental results show that, using our approach, the size of the undominated set can be quickly reduced and that the accuracy of the best potential model converges very fast to the one obtained when knowing all precise data, while the reduction of the size of the set and the convergence of the accuracy is slower and less stable for other querying strategies.

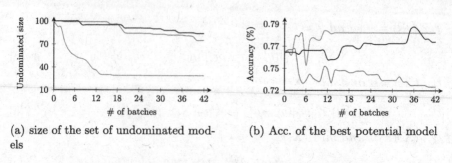

(a) size of the set of undominated models

(b) Acc. of the best potential model

—— Racing —— Most partial —— **Random**

Fig. 4. Results of the experiments

6   Conclusion

This paper has explored an issue related to partially specified data: what is the best information to query so that an optimal model can be quickly determined. We have proposed to use a racing algorithms approach in which several models are competing and some of them are discarded as long as new precise information become available. These general concepts have been illustrated in the case of binary SVM and the first experiments have shown the interest of the method. Future works will focus on the case of decision trees with set-valued labels.

Acknowledgement. This work was carried out in the framework of Labex MS2T, which was funded by the French National Agency for Research (Reference ANR-11-IDEX-0004-02).

References

1. Burges, C.J.: A tutorial on support vector machines for pattern recognition. Data Min. Knowl. Discov. **2**(2), 121–167 (1998)
2. Cour, T., Sapp, B., Jordan, C., Taskar, B.: Learning from ambiguously labeled images. In: CVPR 2009. IEEE Conference on Computer Vision and Pattern Recognition, pp. 919–926. IEEE (2009)
3. Cour, T., Sapp, B., Taskar, B.: Learning from partial labels. J. Mach. Learn. Res. **12**, 1501–1536 (2011)
4. Hüllermeier, E.: Learning from imprecise and fuzzy observations: data disambiguation through generalized loss minimization. Int. J. Approx. Reason. **55**(7), 1519–1534 (2014)
5. Maron, O., Moore, A.W.: The racing algorithm: Model selection for lazy learners. In: Aha, D.W. (ed.) Lazy Learning, pp. 193–225. Springer, Heidelberg (1997)
6. Settles, B.: Active learning literature survey. Computer Sciences Technical report 1648, University of Wisconsin-Madison (2009)
7. Troffaes, M.C.: Decision making under uncertainty using imprecise probabilities. Int. J. Approx. Reason. **45**(1), 17–29 (2007)

Fuzzy DA Clustering-Based Improvement of Probabilistic Latent Semantic Analysis

Takafumi Goshima, Katsuhiro Honda[⊠], Seiki Ubukata, and Akira Notsu

Osaka Prefecture University, Sakai, Osaka 599-8531, Japan
{honda,subukata,notsu}@cs.osakafu-u.ac.jp

Abstract. Probabilistic latent semantic analysis (pLSA) can be interpreted as a soft co-clustering model with an intrinsic fuzzification penalty
and the partition quality was shown to be improved by tuning the degree
of intrinsic partition fuzziness while the model is not supported by probabilistic constraints. In this paper, the mechanism of intrinsic fuzziness
tuning is utilized for improving the partition quality of pLSA under
the strict probabilistic constraints. The proposed deterministic annealing approach first initializes a co-cluster partition with a slightly fuzzier
penalty weight and then gradually reduces the intrinsic fuzziness until
it reaches the strict probabilistic constraints. Supported by the robust
feature of fuzzier models against random initialization, the derived pLSA
partition is demonstrated to be more stable in several numerical experiments.

Keywords: Probabilistic latent semantic analysis · Fuzzy
co-clustering · Deterministic annealing

1 Introduction

Co-cluster structure extraction is a fundamental step in such cooccurrence information analysis as document-keyword analysis and customer-products purchase
history data analysis, where mutual cooccurrence degree among objects and
items are available instead of multivariate vector-form observations of objects.
Probabilistic latent semantic analysis (pLSA) [1,2] is a statistical co-clustering
model, which was shown to have the ability of outperforming the conventional co-
clustering models in many document analysis tasks. In contract to multinomial
mixture models (MMMs) [3] and Dirichlet mixtures [4,5], which estimate the
document clusters under the assumption that each document consists of a single topic, pLSA constructs the document-keyword generative models under the
assumption of soft partition natures of both documents and keywords. Besides
each topic is characterized by several keywords, each document can also consist of multiple topics while each keyword is associated with a single topic in
each document. Then, the soft partition of both documents and keywords is
constructed by estimating the conditional probability of each keyword in each
document to be updated with the EM framework [6].

© Springer International Publishing AG 2016
V.-N. Huynh et al. (Eds.): IUKM 2016, LNAI 9978, pp. 175–184, 2016.
DOI: 10.1007/978-3-319-49046-5_15

Recently, another interpretation of pLSA was introduced considering the fuzzy clustering nature of the EM framework in pLSA. Fuzzy c-means (FCM) [7, 8] is a fuzzy variant of crisp k-means clustering, where fuzzy c-partition is achieved by adopting nonlinear features into the linear k-means objective function. Following the fuzzy partition concept, the pseudo-log-likelihood function of Gaussian mixture models (GMMs) was regarded as the combination of the k-means objective function and the fuzzification penalty [9], and a fuzzy counterpart of GMMs was proposed in conjunction with an adjustable fuzziness weight [10,11]. In the similar manner, Honda *et al.* [12] gave a fuzzy co-clustering interpretation of the pseudo-log-likelihood function of pLSA considering the intrinsic fuzziness penalty varied in the pLSA model. Introducing an additional adjustable fuzziness weight, the partition quality of pLSA was shown to be improved with slightly fuzzier situations although the model does not strictly obey the probabilistic constraints.

In order to improve the partition quality of pLSA obeying strict probabilistic constraints, this paper considers utilization of a deterministic annealing (DA) approach [13]. In the fuzzy clustering context, it was shown that very fuzzy models are often more robust to noise or random initialization rather than crisper models and we can get stable solutions by starting with very fuzzy model and by gradually decreasing the fuzziness degree. The DA model imitates the physical annealing process by identifying the degree of the model fuzziness as the temperature parameter and tries to converge the process to plausible solutions. The DA approach was demonstrated to be useful not only with FCM-type object clustering [13] but also with single topic-based co-clustering [14,15]. In this paper, the DA approach is further utilized in the pLSA context supported by the mechanism of tuning the intrinsic fuzziness of pLSA-based co-clustering.

The remaining parts of this paper are organized as follows: Sect. 2 presents a brief review on FCM-type and pLSA-based clusterings. In Sect. 3, the DA model for FCM-type clustering is introduced following a novel proposal of adopting a DA process in pLSA-based co-clustering in order to derive stable solutions in pLSA under strict probabilistic constraints. The characteristic features are demonstrated through numerical experiments in Sect. 4 and a summary conclusion is given in Sect. 5.

2   FCM-Type Clustering and Extension to pLSA-Based Co-clustering

2.1   FCM Clustering and GMMs

The goal of clustering is to partition objects into several homogeneous clusters such that intra-cluster objects are mutually similar while inter-cluster objects are dissimilar. Assume that we have n objects with their m-dimensional numerical observations $\boldsymbol{x}_i$, $i = 1, \ldots, n$ and try to partition them into C clusters. k-means clustering [16] adopts within-cluster mean vectors $\boldsymbol{b}_c$, $c = 1, \ldots, C$ as the prototypes of clusters and assigns objects into clusters so that the within-cluster errors are minimized.

$$L_{km} = \sum_{c=1}^{C} \sum_{i=1}^{n} u_{ci} \|\boldsymbol{x}_i - \boldsymbol{b}_c\|^2,$$ (1)

where u_{ci} ($u_{ci} \in \{0, 1\}$) is the membership of sample i to cluster c, and is constrained as $\sum_{c=1}^{C} u_{ci} = 1$. The algorithm iterates the two phases of assignment of objects to nearest prototype and estimation of within-cluster mean vectors after random initialization of prototypes.

FCM is a fuzzy extension of k-means, where u_{ci} is relaxed so that it represents the relative degree of object i belonging to cluster c such that $u_{ci} \in [0, 1]$ and $\sum_{c=1}^{C} u_{ci} = 1$. Even if the membership constraint is relaxed to $u_{ci} \in [0, 1]$, k-means model cannot be fuzzified because of the linear nature of k-means objective function. Then, fuzzy partition was achieved supported by two directions. First, Bezdek [7] introduced non-linearity into Eq. (1) by replacing u_{ci} with u_{ci}^θ ($\theta > 1$) and proposed the following objective function:

$$L_{fcm1} = \sum_{c=1}^{C} \sum_{i=1}^{n} u_{ci}^\theta \|\boldsymbol{x}_i - \boldsymbol{b}_c\|^2.$$ (2)

θ can tune the degree of fuzziness of c-partition such that a larger θ brings fuzzier memberships and $\theta \to \infty$ implies $u_{ci} \to 1/C$.

Second, Miyamoto and Mukaidono [17] proposed a regularization approach by introducing an additional penalty term into Eq. (1) as:

$$L_{fcm2} = \sum_{c=1}^{C} \sum_{i=1}^{n} u_{ci} \|\boldsymbol{x}_i - \boldsymbol{b}_c\|^2 + \lambda \sum_{c=1}^{C} \sum_{i=1}^{n} u_{ci} \log u_{ci}.$$ (3)

The negative entropy-like penalty works for *fuzzifier* such that a larger λ brings fuzzier memberships and $\lambda \to \infty$ implies $u_{ci} \to 1/C$.

When λ is the double variance of $\boldsymbol{x}$, Eq. (3) can be identified with a kind of the pseudo-log-likelihood of GMMs, which is a combination of k-means objective function and a soft partition penalty [9]. Although the regularized FCM model does not obey the strict probabilistic constraints, i.e., it has no corresponding probabilistic distributions, the partition quality and the interpretability of partition were shown to be improved rather than GMMs by careful tuning of fuzziness degrees [11].

2.2   pLSA and Fuzzy Co-clustering

Assume that we have an $n \times m$ co-occurrence information matrix $R = \{r_{ij}\}$ composed of the frequency of item j, $j = 1, \ldots, m$ in object i, $i = 1, \ldots, n$, and the goal is to extract C co-clusters of familiar objects and items. In contrast to the mixture of unigrams models such as MMMs and Dirichlet mixtures, pLSA [1, 2] captures the possibility that an object may contain multiple topics while each item is assumed to be generated from a single topic in each document. Let u_{ci} and w_{cj} be the generative probability of object i and item j with component topic c, and obey the sum-to-one conditions of $\sum_{c=1}^{C} u_{ci} = 1$, $\forall i$ and $\sum_{j=1}^{m} w_{cj} = 1$, $\forall c$, respectively. The probability p_{ij} of item j appeared in object i is defined as:

$$p_{ij} = \sum_{c=1}^{C} u_{ci} w_{cj}. \tag{4}$$

Then, the log-likelihood to be maximized is given as:

$$L_{plsa} = \sum_{i=1}^{n} \sum_{j=1}^{m} r_{ij} \log \sum_{c=1}^{C} u_{ci} w_{cj}. \tag{5}$$

Considering Jensen's inequality [18], the lower bound of L_{plsa} is reduced to:

$$L_{plsa'} = \sum_{c=1}^{C} \sum_{i=1}^{n} \sum_{j=1}^{m} \phi_{cij} r_{ij} \log(u_{ci} w_{cj})$$

$$- \sum_{c=1}^{C} \sum_{i=1}^{n} \sum_{j=1}^{m} \phi_{cij} \log \phi_{cij}, \tag{6}$$

where ϕ_{cij} is a latent variable representing the posteriori probability of component c given u_{ci} and w_{cj}. In the EM framework, E-step is responsible for estimating latent variable ϕ_{cij} while M-step estimates generative probabilities u_{ci} and w_{cj}. It was shown that the maximum value of L_{plsa} can be achieved by maximizing its lower bound $L_{plsa'}$.

In fuzzy co-clustering context [19, 20], the goal is to extract the pair-wise clusters of familiar objects and items by estimating the two types of fuzzy memberships of u_{ci} and w_{cj}, which represent the degree of belongingness of object i and item j into cluster c, respectively. Honda et al. [12] gave another interpretation of Eq. (6) such that it can be decomposed into two components of the aggregation measure of object-item pairs and the entropy-based fuzzification penalty in the same manner with the entropy-based FCM. The aggregation measure $\phi_{cij} r_{ij} \log(u_{ci} w_{cj})$ implies that the mutually familiar pairs of object i and item j having large r_{ij} tend to take large memberships u_{ci} and w_{cj} in the same cluster c. On the other hand, the entropy-based penalty $\phi_{cij} \log \phi_{cij}$ works for tuning the intrinsic fuzziness of the co-cluster partition.

Following the regularization concept of the entropy-based FCM, Honda et al. [12] introduced an additional adjustable weight for arbitrarily tuning the intrinsic fuzziness and proposed the following objective function:

$$L = \sum_{c=1}^{C} \sum_{i=1}^{n} \sum_{j=1}^{m} \phi_{cij} r_{ij} \log(u_{ci} w_{cj})$$

$$- \lambda_{\phi} \sum_{c=1}^{C} \sum_{i=1}^{n} \sum_{j=1}^{m} \phi_{cij} \log \phi_{cij}, \tag{7}$$

where λ_ϕ tunes the degree of intrinsic fuzziness of latent variable ϕ_{cij}. If $\lambda_\phi < 1$, the intrinsic fuzziness of the model is lower than that of pLSA. On the other hand, if $\lambda_\phi > 1$, higher than that of pLSA.

In [12], it was demonstrated that the partition quality of pLSA was improved by adopting a slightly fuzzier situation such as $\lambda_\phi = 1.2$ rather than pLSA with $\lambda_\phi = 1$ although the co-clustering model is not strictly obey the probabilistic constraints. In the following sections, the fuzzy co-clustering model is further extended so that we can derive better pLSA solutions under strict probabilistic constrains by introducing a DA scheme.

3   A Deterministic Annealing Approach for Improving pLSA Co-cluster Partition

3.1   DA Clustering in FCM Context

Caused by random initialization, FCM-type clustering models often brings several different solutions and can be easily trapped into local optima especially in noisy situations. DA clustering [13] is a process of avoiding local optima, which starts with a very fuzzy situation and gradually decreases the fuzziness of partition imitating the physical annealing process. Because very fuzzy models estimate cluster prototypes under a macroscopic view, they may often bring more robust partition against bad initialization by weakly reflecting whole objects rather than the microscopic view of crisp clustering models, where each cluster prototype is estimated using within-cluster objects only.

Rose et al. [13] proposed a DA clustering model originally based on a probabilistic framework but the updating formula is almost equivalent to that of the entropy-based FCM [17] except for the dynamic tuning of the penalty weight. In the DA clustering model, the fuzzification penalty λ is regarded as the temperature parameter and the FCM cost function is deterministically optimized at each temperature sequentially, starting at high temperature and going down. The DA approach is shown to be useful not only in FCM-type clustering but also in the co-clustering context [14]. Then, in the next subsection, the DA approach is utilized with the pLSA-induced fuzzy co-clustering model.

3.2   Introduction of DA Process into pLSA-Induced Fuzzy Co-clustering

The goal of this paper is to improve the partition quality of pLSA supported by the stable feature of slightly fuzzier models. So, a DA process is designed such that it starts from slightly fuzzier situation than the intended fuzziness degrees of pLSA ($\lambda_\phi = 1$) and is degraded until the model is reduced to the intended $\lambda_\phi = 1$. In general simulated annealing approaches, a practical way for decreasing the temperature parameter T_k with iteration index k is

$$T_{k+1} = \gamma T_k \ (0.8 \leq \gamma < 1), \tag{8}$$

where γ is the depletion rate. Based on the same concept, the fuzzification parameters are adjusted. Because the model fuzzifier λ_ϕ is directly identifiable with the temperature parameter of the conventional DA clustering model, it can be degraded as:

$$\lambda_{\phi,k+1} = \max\{\gamma\lambda_{\phi,k}, 1\}, \tag{9}$$

where $0 < \gamma < 1$, and $\lambda_\phi = 1$ corresponds to pLSA.

A sample procedure is written as:

[Algorithm: A pLSA Implementation with Deterministic Annealing of Intrinsic Fuzziness]

Step 1. Initialize fuzzy memberships u_{ci}, $c = 1,\ldots,C$, $i = 1,\ldots,n$ and w_{cj}, $c = 1,\ldots,C$, $j = 1,\ldots,m$ such that they satisfy $\sum_{c=1}^{C} u_{ci} = 1$, $\forall i$ and $\sum_{j=1}^{m} w_{cj} = 1$, $\forall c$. Choose the depletion rate γ, the initial fuzziness penalty weight λ_ϕ and termination criterion ε.

Step 2. Update latent variable ϕ_{cij}, $c = 1,\ldots,C$, $i = 1,\ldots,n$, $j = 1,\ldots,m$ by

$$\phi_{cij} = \frac{(u_{ci}w_{cj})^{(r_{ij}/\lambda_\phi)}}{\sum_{\ell=1}^{C}(u_{\ell i}w_{\ell j})^{(r_{ij}/\lambda_\phi)}}. \tag{10}$$

Step 3. Update w_{cj}, $c = 1,\ldots,C$, $j = 1,\ldots,m$ by

$$w_{cj} = \frac{\sum_{i=1}^{n} r_{ij}\phi_{cij}}{\sum_{\ell=1}^{m}\sum_{i=1}^{n} r_{i\ell}\phi_{ci\ell}}. \tag{11}$$

Step 4. Update u_{ci}, $c = 1,\ldots,C$, $i = 1,\ldots,n$ by

$$u_{ci} = \frac{\sum_{j=1}^{m} r_{ij}\phi_{cij}}{\sum_{\ell=1}^{C}\sum_{j=1}^{m} r_{ij}\phi_{\ell ij}}. \tag{12}$$

Step 5. Update fuzziness penalty by $\lambda_\phi = \max\{\gamma\lambda_\phi, 1\}$.

Step 6. If $\max_{c,i} | u_{ci}^{NEW} - u_{ci}^{OLD} | < \varepsilon$, then stop. Otherwise, return to Step 2.

4   Numerical Experiments

The characteristic features of the proposed DA-based scheme are demonstrated through numerical experiments with two real world benchmark data sets, which are available at LINQS webpage of Statistical Relational Learning Group @ UMD (http://linqs.cs.umd.edu/projects//index.shtml). The statistics of the two cooccurrence information matrices are shown in Table 1. The two databases consists of scientific publications (objects) classified into one of several classes, and the cooccurrence information among each object and unique words (items) was described by a 0/1-valued word vector indicating the absence/presence of the corresponding word from the dictionary. The goal of unsupervised co-clustering

Table 1. Statistics of two cooccurrence matrices

Database	# object	# item	# class (cluster)
Citeseer	3312	3703	6
Cora	2708	1433	7

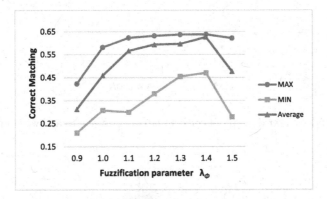

Fig. 1. Comparison of correct matching ratio of actual classes and estimated clusters with Citeseer [12].

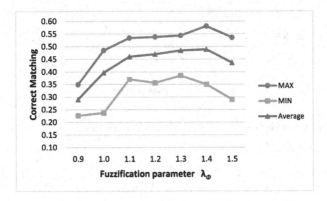

Fig. 2. Comparison of correct matching ratio of actual classes and estimated clusters with Cora [12].

is to reveal the intrinsic class structures withholding the class information of each object.

In [12], the partition quality was measured by the ratio of correct matching between the intrinsic (intended) classes and the estimated cluster labels after maximum membership classification. Figures 1 and 2 give the brief review of the experimental result in [12] comparing the maximum, minimum and average in 10 different random initialization with various fuzziness degrees, which show that the partition quality of pLSA ($\lambda_\phi = 1$) was improved in slightly fuzzier

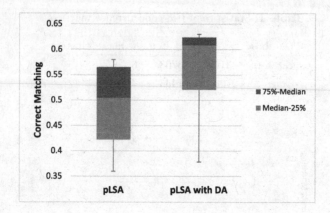

Fig. 3. Box-and-whisker diagram of correct matching ratio with and without annealing with Citeseer.

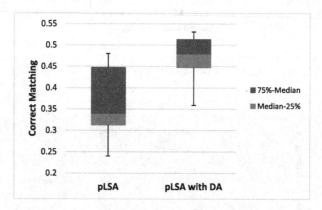

Fig. 4. Box-and-whisker diagram of correct matching ratio with and without annealing with Cora.

situations such as $\lambda_\phi = 1.2$ although the models of $\lambda_\phi \neq 1$ do not have their corresponding probabilistic distributions. Then, in this experiment, the goal is to improve the partition quality of pLSA ($\lambda_\phi = 1$) supported by the superior results of the model with $\lambda_\phi = 1.2$.

The annealing schedule of the proposed DA process was designed as $\lambda_{\phi,k+1} = \min\{0.99\lambda_{\phi,k}, 1\}$. Figures 3 and 4 compare the quality of pLSA partition with/without annealing scheme through the box-and-whisker diagrams. The figures show the box-and-whisker diagrams of the results of 10 trials with different random initializations. The figures indicate that the correct matching ratio was mostly improved by introducing the proposed DA scheme because it can reduce initialization sensitivity. Additionally, the dispersion of solutions with the DA scheme is smaller than that without DA (especially in Citeseer results). Therefore, fuzzy partition quality with the DA scheme were more stable to random initialization and more robust to noise than the conventional pLSA, which were brought with $\lambda_\phi = 1$.

5   Conclusion

In this paper, a novel deterministic annealing approach for improving the partition quality of pLSA was proposed considering the fuzzy co-clustering interpretation of pLSA. Starting from a slightly fuzzier situation and gradually decreasing the degree of intrinsic fuzziness, stable pLSA partition is expected to be derived under strict probabilistic constraints. The results of numerical experiments with two benchmark data sets demonstrated that the pLSA partition was improved so as to be robust against random initialization.

A possible future work includes the development of a better design of annealing schedules for achieving more effective operation of the DA framework. In this paper, with the goal of improving pLSA partition quality, the final fuzziness degree was fixed to $\lambda_\phi = 1$, which is equivalent to pLSA. However, the best co-cluster partition may be given with other fuzziness settings. So, the selection of the best degree of fuzziness should be also considered in the DA process. Another direction of future study is to investigate the influences of the DA schemes on the interpretability of co-cluster solutions.

Acknowledgment. This work was supported in part by JSPS KAKENHI Grant Number JP26330281.

References

1. Hofmann, T.: Probabilistic latent semantic analysis. In: Proceedings of 15th Conference on Uncertainty in Artificial Intelligence, pp. 289–296 (1999)
2. Hofmann, T.: Unsupervised learning by probabilistic latent semantic analysis. Mach. Learn. **42**(1–2), 177–196 (2001)
3. Rigouste, L., Cappé, O., Yvon, F.: Inference and evaluation of the multinomial mixture model for text clustering. Inf. Process. Manage. **43**(5), 1260–1280 (2007)
4. Sjölander, K., Karplus, K., Brown, M., Hughey, R., Krogh, A., Saira Mian, I., Haussler, D.: Dirichlet mixtures: a method for improved detection of weak but significant protein sequence homology. Comput. Appl. Biosci. **12**(4), 327–345 (1996)
5. Ye, X., Yu, Y.-K., Altschul, S.F.: Compositional adjustment of Dirichlet mixture priors. J. Comput. Biol. **17**(12), 1607–1620 (2010)
6. Dempster, A.P., Laird, N.M., Rubin, D.B.: Maximum likelihood from incomplete data via the EM algorithm. J. Roy. Stat. Soc. B **39**, 1–38 (1997)
7. Bezdek, J.C.: Pattern Recognition with Fuzzy Objective Function Algorithms. Plenum Press, New York (1981)
8. Miyamoto, S., Ichihashi, H., Honda, K.: Algorithms for Fuzzy Clustering. Springer, Heidelberg (2008)
9. Hathaway, R.J.: Another interpretation of the EM algorithm for mixture distributions. Stat. Probab. Lett. **4**, 53–56 (1986)
10. Ichihashi, H., Miyagishi, K., Honda, K.: Fuzzy c-means clustering with regularization by K-L information. In: Proceedings of 10th IEEE International Conference on Fuzzy Systems, vol. 2, pp. 924–927 (2001)
11. Honda, K., Ichihashi, H.: Regularized linear fuzzy clustering and probabilistic PCA mixture models. IEEE Trans. Fuzzy Syst. **13**(4), 508–516 (2005)

12. Honda, K., Goshima, T., Ubukata, S., Notsu, A.: A fuzzy co-clustering interpretation of probabilistic latent semantic analysis. In: Proceedings of the 2016 IEEE International Conference on Fuzzy Systems, pp. 718–723 (2016)
13. Rose, K., Gurewitz, E., Fox, G.: A deterministic annealing approach to clustering. Pattern Recogn. Lett. **11**, 589–594 (1990)
14. Oshio, S., Honda, K., Ubukata, S., Notsu, A.: A deterministic clustering framework in MMMs-induced fuzzy co-clustering. In: Integrated Uncertainty in Knowledge Modelling and Decision Making 2015. Lecture Notes in Artificial Intelligence, vol. 9376, pp. 204–213 (2015)
15. Honda, K., Oshio, S., Notsu, A.: Fuzzy co-clustering induced by multinomial mixture models. J. Adv. Comput. Intell. Intell. Inf. **19**, 717–726 (2015)
16. MacQueen, J.B.: Some methods of classification and analysis of multivariate observations. In: Proceedings of 5th Berkeley Symposium on Mathematical Statistics and Probability, pp. 281–297 (1967)
17. Miyamoto, S., Mukaidono, M.: Fuzzy c-means as a regularization and maximum entropy approach. In: Proceedings of the 7th International Fuzzy Systems Association World Congress, vol. 2, pp. 86–92 (1997)
18. Needham, T.: A visual explanation of Jensen's inequality. Am. Math. Mon. **100**(8), 768–771 (1993)
19. Oh, C.-H., Honda, K., Ichihashi, H.: Fuzzy clustering for categorical multivariate data. In: Proceedings of Joint 9th IFSA World Congress and 20th NAFIPS International Conference, pp. 2154–2159 (2001)
20. Kummamuru, K., Dhawale, A., Krishnapuram, R.: Fuzzy co-clustering of documents and keywords. In: Proceedings of 2003 IEEE International Conference on Fuzzy Systems, vol. 2, pp. 772–777 (2003)

Exclusive Item Partition with Fuzziness Tuning in MMMs-Induced Fuzzy Co-clustering

Takaya Nakano, Katsuhiro Honda$^{(\boxtimes)}$, Seiki Ubukata, and Akira Notsu

Osaka Prefecture University, Sakai, Osaka 599-8531, Japan
{honda,subukata,notsu}@cs.osakafu-u.ac.jp

Abstract. Fuzzy co-clustering achieves dual partition of object-item pairs by estimating fuzzy memberships of them. In the multinomial mixtures-induced model, object memberships present the exclusive assignment to clusters while item memberships describe relative typicality in each cluster. In order to improve the interpretability of item partition, exclusive penalty was adopted for item memberships in previous works, where item fuzzy memberships are estimated reflecting both fuzzification penalty and exclusive penalty. In this paper, the characteristics of exclusive item penalty are further studied considering the influences of the item fuzziness weight with different fuzziness degrees.

Keywords: Fuzzy clustering · Co-clustering · Exclusive partition

1 Introduction

Co-clustering is a fundamental technique in many web data analysis such as document-keyword analysis and customer-product market analysis, where co-occurrence information among objects and items are summarized into co-cluster structures. Each co-cluster is formed by mutually familiar objects in conjunction with their typical items. For example, in document analysis with bag-of-words data, similar documents are grouped into a cluster by associating with typical keywords, which are available in content summarization.

Fuzzy co-clustering [1,2] is a practical model of co-clustering, in which object and item assignment to co-clusters are represented by fuzzy memberships. Each co-cluster is extracted such that its aggregation degree is maximized, i.e., object-item pairs having large co-occurrence tend to have large memberships to a same cluster. In order to derive a unique but desirable solution, object and item memberships are estimated under different constraints. In the same manner with fuzzy c-means (FCM) [3,4], object memberships represent the exclusive assignment to clusters under the sum-to-one condition with respect to the cluster index such that each object belongs to at most one cluster with large memberships. On the other hand, item memberships represent the relative typicality in each cluster under the sum-to-one condition with respect to the item index such that the typicality of items is independently estimated in each cluster. Fuzzy co-clustering induced from multinomial mixture models (FCCMM) [5] is a fuzzy

© Springer International Publishing AG 2016
V.-N. Huynh et al. (Eds.): IUKM 2016, LNAI 9978, pp. 185–194, 2016.
DOI: 10.1007/978-3-319-49046-5_16

counterpart of multinomial mixture models (MMMs) [6], where the degrees of fuzziness of object and item partitions can be adjusted by introducing additional fuzzification penalties into the pseudo-log-likelihood function of MMMs.

Although object fuzzy memberships can be directly utilized for capturing the object belongingness to clusters, item assignment to clusters is not necessarily explicit because some items may belong to multiple clusters with high memberships or others may not belong to any clusters. In document co-clustering, some popular keywords may be shared by multiple clusters with large memberships while such severe sharing can conceal the cluster-wise peculiar features. Then, in order to improve the partition quality and interpretability, some previous works [7,8] introduced the exclusive nature into item partition, where an additional penalty for exclusive item partition was added to the FCCMM objective function. By assigning each item to at most one cluster with large memberships, typical items are utilized for emphasizing the peculiar feature of each cluster.

In this paper, the characteristics of the exclusive item penalty are further studied considering the influences of the item fuzziness weight. Some comparative experiments are performed for demonstrating that the exclusive item penalty still works well with different item fuzziness degrees. The remaining parts of this paper are organized as follows: Sect. 2 reviews the FCCMM algorithm and its extension with the exclusive item partition penalty. Some experimental results are presented in Sect. 3 and summary conclusions are given in Sect. 4.

2   MMMs-Induced Fuzzy Co-clustering and Exclusive Item Partition

2.1   MMMs-Induced Fuzzy Co-clustering

When we have co-occurrence information among objects and items such as document-keyword co-occurrence frequencies in document analysis, the intrinsic knowledge can be summarized into co-cluster structures, in which familiar objects are grouped into clusters in conjunction with their typical items. Assume that $R = \{r_{ij}\}$ is an $n \times m$ co-occurrence information matrix on n objects and m items, in which r_{ij} represents the co-occurrence degree among object i and item j.

In MMMs [6], each object is drawn from one of C component multinomial distributions with the generative probability u_{ci} of object i from component c under the probabilistic constraint of $\sum_{c=1}^{C} u_{ci} = 1$. Multinomial distribution is a multi-dimensional extension of binomial distribution, where each item occurs with the discrete probability w_{cj} of item j in component c under the constraint of $\sum_{j=1}^{m} w_{cj} = 1$. A mixture of multinomial distributions can be constructed based on the maximum likelihood principle. The pseudo-log-likelihood to be maximized is defined as:

$$L_{mmms} = \sum_{c=1}^{C} \sum_{i=1}^{n} \sum_{j=1}^{m} u_{ci} r_{ij} \log w_{cj} + \sum_{c=1}^{C} \sum_{i=1}^{n} u_{ci} \log \frac{\alpha_c}{u_{ci}}. \tag{1}$$

α_c represents the a priori probability of component c. Supported by the EM algorithm [9], α_c, u_{ci} and w_{cj} are iteratively updated until convergent.

From the fuzzy co-clustering context, Eq. (1) can be decomposed into two parts in the same manner with Gaussian mixture models [10]. The first part $\sum_{c=1}^{C} \sum_{i=1}^{n} \sum_{j=1}^{m} u_{ci} r_{ij} \log w_{cj}$ is the aggregation measure of co-clusters to be maximized so that familiar object-item pairs having large co-occurrence r_{ij} simultaneously have large memberships u_{ci} and w_{cj} in the peculiar cluster c. Here, because this first part is essentially a linear objective function with respect to u_{ci}, object memberships tend to have the solutions in the extremal values of $u_{ci} \in \{0, 1\}$. On the other hand, supported by the non-linear nature with respect to w_{cj}, item memberships are optimized in $w_{cj} \in [0, 1]$. Then, the second part $\sum_{c=1}^{C} \sum_{i=1}^{n} u_{ci} \log \frac{\alpha_c}{u_{ci}}$ plays a role for achieving soft partition of u_{ci}. α_c can be identified with the volume of cluster c such that $\sum_{c=1}^{C} \alpha_c = 1$ and u_{ci} are softened by reducing the deviation between u_{ci} and α_c.

The two soft partition principle for object and item probabilities u_{ci} and w_{cj} in MMMs have close connection with the fuzzification principles in FCM clustering. The K-L information-based soft partition penalty can be identified with the entropy-based or K-L information-based fuzzification schemes [11,12] while the log function-based objective function is a sort of non-linearized objective function of the standard FCM cost function [3].

Then, Honda et al. [5] proposed a novel fuzzy co-clustering algorithm, which is induced from MMMs so that the fuzziness degrees of both object and item memberships can be arbitrarily tuned with adjustable weights. The objective function of FCCMM clustering was defined as:

$$L_{fccmm} = \sum_{c=1}^{C} \sum_{i=1}^{n} \sum_{j=1}^{m} \frac{1}{\lambda_w} u_{ci} r_{ij} \left((w_{cj})^{\lambda_w} - 1 \right) + \lambda_u \sum_{c=1}^{C} \sum_{i=1}^{n} u_{ci} \log \frac{\alpha_c}{u_{ci}}. \quad (2)$$

The fuzzy memberships of u_{ci} for objects and w_{cj} for items are estimated under the same constraint with MMMs. In this objective function, the non-linear degree is tuned by two types of adjustable penalty weights. λ_u tunes the responsibility of the K-L information-based penalty, where $\lambda_u = 0$ implies a crisp object partition and a larger λ_u brings a fuzzier object memberships. On the other hand, λ_w tunes the non-linear degree of the aggregation criterion. Following the definition of log function:

$$\log w_{cj} = \lim_{t \to 0} \frac{1}{t} \left((w_{cj})^t - 1 \right), \quad (3)$$

the objective function of Eq. (2) is reduced to MMMs with $\lambda_w \to 0$ while $\lambda_w = 1$ implies a linear crisp objective function. Then, $\lambda_w < 0$ brings a fuzzier item memberships.

2.2 Exclusive Item Partition

From the dual partition concepts, we must note that the two types of fuzzy memberships of FCCMM play different roles and cannot be equally used in revealing cluster assignment of objects and items. Because item memberships are independently estimated in each cluster, some items may have large memberships in multiple clusters or may not belong to any clusters. So, the cluster-wise peculiar characteristics cannot be fairly compared only with large item memberships. In order to improve the partition quality and interpretability, some previous works [7,8] introduced the exclusive nature to item partition. Besides the previous works investigated the basic feature of the exclusive penalty only with the FCCMM model with $\lambda_w \to 0$, this paper extends the model with adjustable weight λ_w.

With the goal of forcing each item to belong to at most one cluster with large memberships, the co-occurrence degree of item should be degraded if the sum of degree of belongingness to other clusters is large. Then, the additional weight for evaluating the sharing degree of item j in cluster c is calculated as:

$$s_{cj} = \exp\left(-\beta \sum_{t \neq c} w_{tj}^*\right). \tag{4}$$

s_{cj} becomes small when item j belongs to other clusters. β tunes the sensitivity such that a large β causes a rapid decrease of s_{cj} while a small β brings a small change of s_{cj}. Indeed, $\beta \to 0$ reduces to all $s_{cj} \to 1$, i.e., the conventional FCCMM model without item exclusive penalty. A practical approach of setting β is to start with $\beta = 0$ and gradually increase to a prefixed maximum value β_{max} such that $\beta = \min\{0.1 \times (\tau - 1), \beta_{max}\}$ with iteration index τ [7,8]. This approach implies that the initial partition of the FCCMM result is gradually relaxed to an exclusive item partition model.

Multiplying the weight s_{cj} to cluster-wise aggregation, the objective function of FCCMM clustering is modified as:

$$L_{fccmm'} = \sum_{c=1}^{C} \sum_{i=1}^{n} \left(\sum_{j=1}^{m} \frac{1}{\lambda_w} u_{ci} r_{ij} s_{cj}\right) \left((w_{cj})^{\lambda_w} - 1\right) + \lambda_u \sum_{c=1}^{C} \sum_{i=1}^{n} u_{ci} \log \frac{\alpha_c}{u_{ci}}. \tag{5}$$

The clustering algorithm is composed of a four step iterative process of updating s_{cj}, α_c, u_{ci} and w_{cj}.

Here, we should note that item memberships w_{cj} are also included in s_{cj} and the optimal updating formula for w_{cj} cannot be rigorously derived. So, each s_{cj} can be calculated by Eq. (4) with w_{tj}^*, $t \neq c$ of the previous iteration, where the degrees of item sharing are temporally inherited. This type of trick is often utilized in relational clustering such as relational fuzzy c-means [13].

Then, the updating rule for α_c, u_{ci} and w_{cj} are given as:

$$\alpha_c = \frac{1}{n} \sum_{i=1}^{n} u_{ci}. \tag{6}$$

For $\lambda_w \neq 0$,

$$u_{ci} = \frac{\alpha_c \exp\left(\dfrac{1}{\lambda_u \lambda_w} \sum_{j=1}^{m} r_{ij} s_{cj}(w_{cj})^{\lambda_w}\right)}{\sum_{\ell=1}^{C} \alpha_\ell \exp\left(\dfrac{1}{\lambda_u \lambda_w} \sum_{j=1}^{m} r_{ij} s_{\ell j}(w_{\ell j})^{\lambda_w}\right)}. \tag{7}$$

For $\lambda_w = 0$, Eq. (3) brings

$$u_{ci} = \frac{\alpha_c \displaystyle\prod_{j=1}^{m} (w_{cj})^{(r_{ij} s_{cj})/\lambda_u}}{\sum_{\ell=1}^{C} \alpha_\ell \displaystyle\prod_{j=1}^{m} (w_{\ell j})^{(r_{ij} s_{\ell j})/\lambda_u}}. \tag{8}$$

$$w_{cj} = \left(\sum_{\ell=1}^{m} \left(\frac{\displaystyle\sum_{i=1}^{n} r_{ij} s_{cj} u_{ci}}{\displaystyle\sum_{i=1}^{n} r_{i\ell} s_{c\ell} u_{ci}}\right)^{\frac{1}{\lambda_w - 1}}\right)^{-1}$$

$$= \frac{\gamma_{cj}}{\displaystyle\sum_{\ell=1}^{m} \gamma_{c\ell}}, \tag{9}$$

where

$$\gamma_{cj} = \frac{1}{\left(\displaystyle\sum_{i=1}^{n} r_{ij} s_{cj} u_{ci}\right)^{\frac{1}{\lambda_w - 1}}}. \tag{10}$$

3 Numerical Experiments

In this section, the proposed model is applied to a document analysis task, and the results are shown to demonstrate the applicability of the exclusive item partition penalty in cases of tuning item fuzziness weight λ_w. The document-keyword co-occurrence matrix used in [14] was constructed from Japanese novel "Kokoro" written by Soseki Natsume, which can be downloaded from Aozora Bunko (http://www.aozora.gr.jp). The novel is composed of 3 chapters consisting of 36, 18 and 56 sections each, and the bag-of-words were generated after morphological analysis. The co-occurrence matrix R includes the frequencies of 83 most frequently used nouns and verbs ($m = 83$) in the 110 sections ($n = 110$),

Table 1. Chapter-cluster cross tabulation in non-exclusive partition

Chap.	$\lambda_w = 0$				$\lambda_w = -0.3$				$\lambda_w = 0.1$			
	$c=1$	$c=2$	$c=3$	$c=4$	$c=1$	$c=2$	$c=3$	$c=4$	$c=1$	$c=2$	$c=3$	$c=4$
1	30	5	1	0	30	5	1	0	31	5	0	0
2	0	18	0	0	0	18	0	0	0	18	0	0
3	1	1	13	41	1	0	16	39	14	1	2	39

Table 2. Chapter-cluster cross tabulation in exclusive partition

Chap.	$\lambda_w = 0$				$\lambda_w = -0.3$				$\lambda_w = 0.1$			
	$c=1$	$c=2$	$c=3$	$c=4$	$c=1$	$c=2$	$c=3$	$c=4$	$c=1$	$c=2$	$c=3$	$c=4$
1	31	5	0	0	30	5	1	0	32	4	0	0
2	0	18	0	0	0	18	0	0	0	18	0	0
3	2	1	13	40	1	0	16	39	2	1	15	38

in which each element r_{ij} corresponds to the tf-idf weight [15] of each section-keyword pair. Withholding chapter information of each section, unsupervised fuzzy co-clustering was performed with the goal of extracting section-keyword (object-item) co-cluster structures. Before application, cooccurrence information r_{ij} was normalized so that it has zero-minimum and one-maximum ($r_{ij} \in [0,1]$) for each item.

It was reported in a previous work [16] that the third chapter has two sub-co-clusters and the plausible cluster number is $C = 4$. Then, in this experiment, the cluster number was set as $C = 4$ in all trials. In order to study the influence of item fuzziness tuning, the FCCMM algorithm with/without exclusive penalty was applied in three fuzziness degrees of $\lambda_w \in \{-0.3, 0, 0.1\}$. Other parameters were set as $\lambda_u = 1$ and $\beta_{max} = 10$.

First, the partition quality is compared in Tables 1 and 2, in which chapter-cluster cross tabulation is shown after maximum membership classification. Table 1 implies that the FCCMM algorithm with $\lambda_w = 0$ could successfully reveal the four co-cluster structure reflecting the intrinsic chapter structure and still could work even when the item fuzziness degree was changes to slightly fuzzier ($\lambda = -0.3$) than MMMs ($\lambda_w = 0$). But, the partition quality was degraded with the crisper case of $\lambda_w = 0.1$. On the other hand, Table 2 indicates that the FCCMM algorithm could reveal the intrinsic chapter information in $\lambda_w \in \{-0.3, 0, 0.1\}$ with the exclusive item partition penalty. Then, the penalty may contribute to clarification of intrinsic chapter information under a proper weight settings.

Next, the interpretability of co-clusters is studied by comparing the cluster-wise typical items (keywords). Tables 3, 4, 5, 6, 7 and 8 show the top 5 keywords of each cluster, which have largest item memberships in the cluster. Each cell describe the Japanese word with its English translation in bracket. Tables 3, 4 and 5 imply that all clusters have some general common words such as *suru* (do)

Table 3. Top 5 keywords in non-exclusive clusters ($\lambda_w = 0$)

cluster	1st	2nd	3rd	4th	5th
$c = 1$	sensei (teacher)	suru (do)	sore (it)	naru (become)	watashi (I)
$c = 2$	suru (do)	chichi (father)	omou (think)	sore (it)	naru (become)
$c = 3$	suru (do)	watashi (I)	omou (think)	naru (become)	kokoro (heart)
$c = 4$	suru (do)	K (a name)	watashi (I)	naru (become)	sore (it)

Table 4. Top 5 keywords in non-exclusive clusters ($\lambda_w = -0.3$)

cluster	1st	2nd	3rd	4th	5th
$c = 1$	sensei (teacher)	suru (do)	sore (it)	naru (become)	watashi (I)
$c = 2$	suru (do)	chichi (father)	omou (think)	sore (it)	naru (become)
$c = 3$	suru (do)	watashi (I)	omou (think)	naru (become)	sore (it)
$c = 4$	suru (do)	K (a name)	watashi (I)	naru (become)	omou (think)

Table 5. Top 5 keywords in non-exclusive clusters ($\lambda_w = 0.1$)

cluster	1st	2nd	3rd	4th	5th
$c = 1$	suru (do)	watashi (I)	naru (become)	sore (it)	aru (be)
$c = 2$	suru (do)	omou (think)	sore (it)	naru (become)	chichi (father)
$c = 3$	oji (uncle)	suru (do)	sireru (be known)	watashi (I)	toru (take)
$c = 4$	suru (do)	watashi (I)	K (a name)	naru (become)	sore (it)

and *watashi* (I), which are popular in many documents but cannot be utilized for emphasizing differences among chapters. Then, the conventional FCCMM without exclusive item partition penalty is not necessarily useful for selecting cluster-wise peculiar keywords. On the other hand, Tables 6, 7 and 8 indicate that the exclusive item partition penalty can contribute to selection of cluster-wise peculiar keywords. The tables also show that the exclusive penalty still work well with slightly fuzzier and crisper item partitions.

Table 6. Top 5 keywords in all-exclusive clusters ($\lambda_w = 0$)

cluster	1st	2nd	3rd	4th	5th
$c = 1$	sensei (teacher)	sore (it)	naru (become)	kiku (listen)	aru (be)
$c = 2$	chichi (father)	haha (mother)	kaku (write)	tegami (letter)	kuru (come)
$c = 3$	oji (uncle)	watashi (I)	suru (do)	kokoro (heart)	sai (wife)
$c = 4$	K (a name)	kare (he)	ojosan (lady)	mukau (face)	watashi (I)

Table 7. Top 5 keywords in all-exclusive clusters ($\lambda_w = -0.3$)

cluster	1st	2nd	3rd	4th	5th
$c = 1$	sensei (teacher)	sore (it)	naru (become)	kiku (listen)	aru (be)
$c = 2$	chichi (father)	haha (mother)	suru (do)	kaku (write)	kuru (come)
$c = 3$	watashi (I)	suru (do)	oji (uncle)	sai (wife)	omou (think)
$c = 4$	K (a name)	kare (he)	ojosan (lady)	mukau (face)	watashi (I)

Table 8. Top 5 keywords in all-exclusive clusters ($\lambda_w = 0.1$)

cluster	1st	2nd	3rd	4th	5th
$c = 1$	sensei (teacher)	sore (it)	kiku (listen)	naru (become)	iu (say)
$c = 2$	chichi (father)	haha (mother)	kaku (write)	tegami (letter)	kuru (come)
$c = 3$	watashi (I)	suru (do)	oji (uncle)	kokoro (heart)	omou (think)
$c = 4$	K (a name)	kare (he)	ojosan (lady)	mukau (face)	shitsu (room)

These experimental results fairly support the applicability of the exclusive item partition penalty to the case of different item fuzziness degrees from the standard MMMs-like situation ($\lambda_w = 0$).

4   Conclusion

In this paper, the applicability of exclusive penalty in MMMs-induced fuzzy co-clustering was further studied with adjustable fuzziness tuning of item memberships. Some experimental results demonstrated that the exclusive penalty still works well with different fuzziness degrees from the standard fuzziness degree of MMMs.

A possible future work includes the investigation of utility in some real applications such as collaborative filtering based on fuzzy co-clustering [17]. Another direction of future study is to develop the mechanism of automatically tuning the fuzziness degrees in conjunction with the exclusive penalty weight following the intrinsic data fuzziness.

Acknowledgment. This work was supported in part by JSPS KAKENHI Grant Number JP26330281.

References

1. Oh, C.-H., Honda, K., Ichihashi, H.: Fuzzy clustering for categorical multivariate data. In: Proceedings of Joint 9th IFSA World Congress and 20th NAFIPS International Conference, pp. 2154–2159 (2001)
2. Kummamuru, K., Dhawale, A., Krishnapuram, R.: Fuzzy co-clustering of documents and keywords. In: Proceedings of the 2003 IEEE International Conference on Fuzzy Systems, vol. 2, pp. 772–777 (2003)
3. Bezdek, J.C.: Pattern Recognition with Fuzzy Objective Function Algorithms. Plenum Press, New York (1981)
4. Miyamoto, S., Ichihashi, H., Honda, K.: Algorithms for Fuzzy Clustering. Springer, Heidelberg (2008)
5. Honda, K., Oshio, S., Notsu, A.: Fuzzy co-clustering induced by multinomial mixture models. J. Adv. Comput. Intell. Intell. Inf. **19**(6), 717–726 (2015)
6. Rigouste, L., Cappé, O., Yvon, F.: Inference and evaluation of the multinomial mixture model for text clustering. Inf. Process. Manag. **43**(5), 1260–1280 (2007)
7. Honda, K., Nakano, T., Oh, C.-H., Ubukata, S., Notsu, A.: Partially exclusive item partition in MMMs-induced fuzzy co-clustering and its effects in collaborative filtering. J. Adv. Comput. Intell. Intell. Inf. **19**(6), 810–817 (2015)
8. Nakano, T., Honda, K., Ubukata, S., Notsu, A.: MMMs-Induced fuzzy co-clustering with exclusive partition penalty on selected items. In: Huynh, V.-N., Inuiguchi, M., Denoeux, T. (eds.) IUKM 2015. LNCS (LNAI), vol. 9376, pp. 226–235. Springer, Heidelberg (2015). doi:10.1007/978-3-319-25135-6_22
9. Dempster, A.P., Laird, N.M., Rubin, D.B.: Maximum likelihood from incomplete data via the EM algorithm. J. Roy. Stat. Soc. B **39**, 1–38 (1977)
10. Hathaway, R.J.: Another interpretation of the EM algorithm for mixture distributions. Stat. Probab. Lett. **4**, 53–56 (1986)
11. Miyamoto, S., Mukaidono, M.: Fuzzy c-means as a regularization and maximum entropy approach. In: Proceedings of the 7th International Fuzzy Systems Association World Congress, vol. 2, pp. 86–92 (1997)
12. Ichihashi, H., Miyagishi, K., Honda, K.: Fuzzy c-means clustering with regularization by K-L information. In: Proceedings of 10th IEEE International Conference on Fuzzy Systems, vol. 2, pp. 924–927 (2001)
13. Hathaway, R.J., Davenport, J.W., Bezdek, J.C.: Relational duals of the c-means clustering algorithms. Pattern Recogn. **22**(2), 205–212 (1989)
14. Honda, K., Notsu, A., Ichihashi, H.: Fuzzy PCA-guided robust k-means clustering. IEEE Trans. Fuzzy Syst. **18**(1), 67–79 (2010)

15. Salton, G., Buckley, C.: Term-weighting approaches in automatic text retrieval. Inf. Process. Manage. **24**(5), 513–523 (1988)
16. Honda, K., Oh, C.-H., Notsu, A.: Exclusive condition on item partition in fuzzy co-clustering based on K-L information regularization. In: Proceedings of the Joint 7th International Conference on Soft Computing and Intelligent Systems and 15th International Symposium on Advanced Intelligent Systems, pp. 1413–1417 (2014)
17. Honda, K., Muranishi, M., Notsu, A., Ichihashi, H.: FCM-type cluster validation in fuzzy co-clustering and collaborative filtering applicability. Int. J. Comput. Sci. Netw. Secur. **13**(1), 24–29 (2013)

A Hybrid Model of ARIMA, ANNs and *k*-Means Clustering for Time Series Forecasting

Warut Pannakkong[1(✉)], Van Hai Pham[2], and Van-Nam Huynh[1]

[1] Japan Advanced Institute of Science and Technology, Ishikawa, Japan
{warut,huynh}@jaist.ac.jp
[2] Pacific Ocean University, Nha Trang, Vietnam

Abstract. This paper proposed a new hybrid forecasting model of the combinations between autoregressive integrated moving average (ARIMA) models, artificial neural networks (ANNs) and *k*-means clustering to obtain the unique strength of the different forecasting models. The Wolf's sunspot time series (1700–1987), which is a well-known data set, is applied for the prediction capability testing. The experimental results imply that the proposed model can outperform the benchmark in short term prediction (37 points ahead) in term of MAE and MSE.

Keywords: ARIMA · ANN · *k*-means · Hybrid model · Time series forecasting

1 Introduction

Time series forecasting is an active research area for several decades with the purpose of improving the forecasting accuracy. The historical time series is used as the input of the time series forecasting models. This research area contributes to various practical applications: finance, agriculture, energy, environment, etc. [1].

Traditionally, the linear model such as autoregressive integrated moving average (ARIMA) models dominate the other methods because the ARIMA models are capable to deal with non-stationary time series. Nevertheless, the ARIMA models made prior assumption on relationship between the historical and the future series as the linear function which is very difficult to be satisfied in practical situations [2].

Artificial neural networks (ANNs), nonlinear models mimicking human brain neurons mechanism, become popular due to their advantages over the ARIMA models: the ANNs can fit the relationship between the historical and the future values without prior assumption, and they can be an universal approximator for any continuous function [3,4]. Comparative studies in several applications were done and the results implied that the ANNs usually outperformed the ARIMA models [5–10]. Recently, the ANNs have been improved by including the predicted values and the residuals of the ARIMA models as the inputs in order to obtain unique strength in linear and nonlinear modelling (so called

© Springer International Publishing AG 2016
V.-N. Huynh et al. (Eds.): IUKM 2016, LNAI 9978, pp. 195–206, 2016.
DOI: 10.1007/978-3-319-49046-5_17

ARIMA-ANN model) [11] but there is no guarantee that it can capture all the linear pattern as the ARIMA models because the nature the ANNs is the nonlinear model.

Recently, clustering techniques (e.g., self organizing map (SOM), k-means clustering) have been applied for the data preprocessing in purpose of grouping the time series that have the similar statistical distribution into the same cluster in order to capture non-stationary pattern before using the ANNs to forecast the future values for each cluster [12–14]. This approach can improve the forecasting accuracy because the ANNs focus on the pattern of the data in each cluster instead of considering the whole data set, however, this can lead to the overfitting problem. In addition, we cannot really know the cluster that belongs to each future value. In this case, the suitable way to use the predicted values from each cluster is also the problem. The recent work proposed to use the summation of the prediction values from every cluster [13], but in fact, it would be more logical if the future values are produced from the ANN dedicated to their cluster. For this reason, the ANN is selected based on the cluster prediction requiring the clustering technique that can provide the straight clear cut boundary between the clusters as the k-means clustering. Therefore, ARIMA-k-means-ANN model is developed.

This paper aims to develop a new hybrid forecasting model that combines unique strength of these three approaches (ARIMA, ARIMA-ANN and ARIMA-k-means-ANN models) to gain the more accuracy in time series forecasting. In addition, we also provide a new method for using the predicted value from each cluster from k-means clustering to predict the future value. The rest of the paper is organized as follows. In Sect. 2, the ARIMA models, the ANNs and k-means algorithm are introduced. Then, the proposed hybrid model is described in Sect. 3. In Sect. 4, the experiments are explained and the results are interpreted. Finally, the conclusions are given in Sect. 5.

2   Preliminaries

This section introduces the three techniques, which are involved in developing the new hybrid model, such as ARIMA models, ANNs and the k-means algorithm. Their brief history, concept, and mathematical formulation are described.

2.1   The Autoregressive Integrated Moving Average (ARIMA) Models

The autoregressive integrated moving average (ARIMA) models are a well-known linear model used in time series forecasting for decades [2]. The ARIMA models consist of three parts: autoregressive (AR), moving average (MA) and integration (I). The benefit of the ARIMA models is the capability in dealing with non-stationary data by differencing step corresponding to the integration (I) part.

The ARIMA models can be expressed generally as ARIMA(p,d,q) and mathematically as

$$\hat{Z}_t = c + \sum_{i=1}^{p} \phi_i Z_{t-i} + a_t - \sum_{j=1}^{q} \theta_j a_{t-j} \qquad (1)$$

where $\hat{Z}_t$ is the predicted value at time t, Z_t is the actual value at time t, c is the constant, a_t is the random error term at time t, ϕ_i and θ_j are autoregressive and moving average parameters respectively, p and q are the orders of the autoregressive and the moving average respectively, and d is the degree of differencing.

2.2 The Artificial Neural Networks (ANNs)

Artificial neural networks (ANNs) are a popular time series forecasting technique applied in broad range of the nonlinear problems because they can fit the relationship between the inputs and the future time series without prior assumption. The structure of the ANNs consists of input, hidden and output layers. Typically, the number of the input and the output layers is one but the number of the hidden layer can be more than one. However, the multilayer feedforward ANN with one hidden layer is capable to be the universal approximator for any continuous nonlinear function [4], and the relationship between the inputs and the outputs is shown in Eq. (2).

$$\hat{Z}_t = b_h + f\left(\sum_{h=1}^{R} w_h g\left(b_{i,h} + \sum_{i=1}^{Q} w_{i,h} p_i \right) \right) \qquad (2)$$

where $\hat{Z}_t$ is the predicted value at time t; $w_{i,h}$ and w_h are the connection weights between the layers; Q and R are the numbers of the input and the hidden nodes respectively; $b_{i,h}$ and b_h are the biases; and f and g are the transfer functions that usually are the linear and the nonlinear functions respectively.

2.3 The k-means Algorithm

The k-means algorithm is an unsupervised learning algorithm commonly used for grouping the data set into k clusters [15]. The algorithm (Fig. 1) starts with defining the number of cluster (k) and the centroids for each cluster. Second, each data is assigned to the nearest centroid using the Euclidian distance (d_{ij}) as below .

$$d_{ij} = \sqrt{ \sum_{m=1}^{M} (x_{im} - c_{jm})^2 } \qquad (3)$$

where x_{im} is the value of attribute m of the data i, and c_{jm} is the value of the attribute m of the centroid of the cluster j.

Third, the centroids are recalculated. Then, the second and the third steps are repeated until all controids do not move anymore.

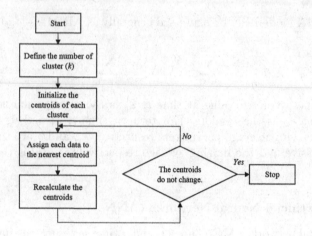

Fig. 1. k-means clustering algorithm

The problem of the k-means algorithm is the choosing of the optimal number of clusters (k). To find the optimal k, the Silhouette [16] can be the measure of the clustering quality, and it can be computed as

$$s_i = \frac{b_i - a_i}{\max(a_i, b_i)} \tag{4}$$

where a_i is the average dissimilarity of data i to all other data in the same cluster, b_i is the minimum dissimilarity of data i to all data in the other clusters. According to formula, the higher value of s_i implies the better matching between the data and its cluster. Thus, the number of k giving the highest average of s_i in all data can be considered as the optimal k.

3   Proposed Hybrid Model

In this section, the proposed hybrid model is presented (see Fig. 2). The objective of the proposed hybrid model is to take unique advantages from the different forecasting models: the ARIMA, the ARIMA-ANN and the ARIMA-k-means-ANN models. The detail of these three models are explained stage by stage, and the combination method in the final forecasting is described.

3.1   The ARIMA Model Stage

The ARIMA model is applied to the whole data set to obtain predicted values $(\hat{L}_t)$ and their residuals (e_t). The results are passed through the final forecasting and also used as the inputs for the ARIMA-ANN and the ARIMA-k-means-ANN models.

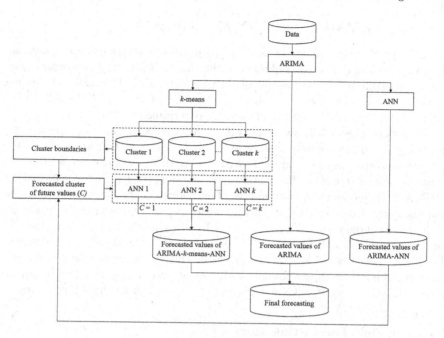

Fig. 2. The proposed hybrid model

3.2   The ARIMA-ANN Model Stage

The ARIMA-ANN model [11] is the ANN including the lagged values of the time series; and the predicted values and the residuals of the ARIMA model as the inputs, and fitting the relationship between the time series and the linear and nonlinear components as the function in Eq. (5).

$$\hat{Z}_t = f(\hat{L}_t, e_{t-1}, \ldots, e_{t-n}, Z_{t-1}, \ldots, Z_{t-m}) \tag{5}$$

where $\hat{Z}_t$ is the forecasted value at time t, f is the nonlinear function fitted by the ANN, $\hat{L}_t$ is the predicted value of the ARIMA model at time t, Z_t is the actual value at time t, e_t is the residual of the ARIMA model at time t, n and m are integers of the lag period identified in the design process.

The multilayer feedforward neural network with one hidden layer is applied. The transfer functions between the input and the hidden layers; and the hidden and the output layers are tan-sigmoid (Eq. (6)) and linear transfer functions respectively.

$$\text{Tan-sigmoid}(x) = \frac{2}{1 + e^{-2x}} - 1 \tag{6}$$

The number of hidden nodes, n and m are varied to find the best model. The forecasted values of the best model are moved to the final forecasting and used for the cluster forecasting in the ARIMA-k-means-ANN model as well.

3.3   The ARIMA-k-means-ANN Model Stage

In this part, the whole data set is classified into the clusters using the k-means algorithm finding the optimal number of the cluster (k) by trial and error experiments and computing the average of the Silhouette (s_i). Then, the whole data set is divided into the k clusters obtaining the highest average of the Silhouette, and the boundaries between the cluster are determined.

After that, the multilayer feedforward neural networks are constructed and trained for each cluster individually. In order to forecast the future values, the forecasted clusters of the future values are required. We propose a new approach to predict the cluster of the future values by mapping the forecasted values of the ARIMA-ANN model to the boundaries of each cluster. For example, we suppose that there are two clusters; the controid of the cluster 1 is lower than the cluster 2; and the boundary separating the cluster 1 and 2 is 20. If the forecasted value of the ARIMA-ANN model at time $t+1$ is 19, then, the forecasted value belongs to the cluster 1 because 19 is located in the cluster 1 which is below the boundary. Thus, the cluster 1 is chosen for the prediction at period $t+1$. Finally, the results of the ARIMA-k-means-ANN model are passed through the final forecasting.

3.4   The Final Forecasting Stage

In the final forecasting, the forecasting values of the three models are combined in all possible ways using the weights generated by discount mean square forecast error (DMSFE) combination method [19] expressed as

$$
w_i = \frac{\left[\sum_{t=1}^{T} \gamma^{T-t+1} \left(Z_t - \hat{Z}_t^i \right)^2 \right]^{-1}}{\sum_{i=1}^{m} \left[\sum_{t=1}^{T} \gamma^{T-t+1} \left(Z_t - \hat{Z}_t^i \right)^2 \right]^{-1}}
\tag{7}
$$

where w_i is the weight of the forecasting model i, Z_t is the actual value at time t, $\hat{Z}_t^i$ is the forecasted value at time t from method i, γ is the discount factor assumed 0.8, T is the total number of the training period, and m is the total number of the forecasting methods. In addition, in the testing period, the w_i is updated in every period by exponential smoothing with α assumed 0.2.

4   Experiments and Results

To test the capability of the proposed model, the Wolf's sunspot time series is applied and the forecasting performances are evaluated and compared. The detail of conducting the experiment is described and the interpreation of the results is also provided in this section.

4.1 The Wolf's Sunspot Time Series

The Wolf's sunspot time series is a well-known data set containing the historical annual record of the number of spots on the sun surface. In the experiments, the records from 1700 to 1987 (288 observations) are involved (Fig. 3) [21]. The data from 1700 to 1920 is used for model training. For the model testing including data from 1921 to 1987, there are two scenarios: short term (35 points ahead) and long term (67 point ahead).

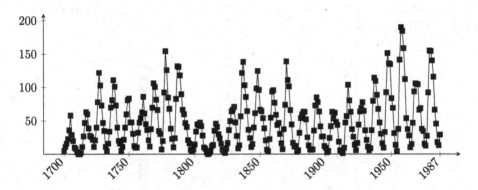

Fig. 3. The Wolf's sunspot time series (1700–1987)

4.2 Performance Measures

Two popular accuracy measures, mean absolute error (MAE) and mean square error (MSE), are considered as the measures for the forecasting performance evaluation. Their mathematical formulas can be expressed as follows:

$$\text{MSE} = \frac{1}{N} \sum_{t=1}^{N} (Z_t - \hat{Z}_t)^2 \tag{8}$$

$$\text{MAE} = \frac{1}{N} \sum_{t=1}^{N} |Z_t - \hat{Z}_t| \tag{9}$$

where Z_t denotes the actual time series at t, $\hat{Z}_t$ denotes forecasted time series at t, and N denotes number of total forecasting period.

Although both of them correspond to the error computed from the magnitude of the difference between Z_t and $\hat{Z}_t$, however, the meaning of their results are different. The model producing the lowest MAE has the lowest average of magnitude of error. In the other hand, the model generating the lowest MSE has the lowest maximum magnitude of error. In some situation, there is no model that can dominate in both MAE and MSE, therefore, in such case, the decision in selecting the suitable model depends on the user preferences.

4.3   Results

In the ARIMA model stage, the ARIMA(9,0,0) model is the best fitted model that is also chosen by several researchers [11,17,18]. The results of the ARIMA(9,0,0) are passed through the ARIMA-ANN model, the ARIMA-k-means-ANN model and the final forecasting.

In the ARIMA-ANN model stage, it includes the results of the ARIMA model and the lagged value of the time series as the inputs. The best model is determined by trial and error varying n and m (in Eq. (5)) from one to 11, and the number of hidden nodes from one to ten. The ARIMA-ANN(7-3-1) model (see Fig. 4) is the best model as in [11].

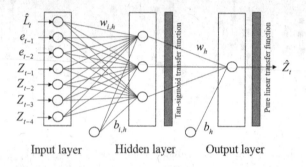

Input layer   Hidden layer   Output layer

Fig. 4. The ARIMA-ANN(7-3-1) model for the sunspot time series

In the ARIMA-k-means-ANN model stage, firstly, the sunspots time series is classified into the clusters. The number of cluster (k) is varried from two to five. After the trial and error, the optimal number of cluster (k) is two clusters showing the most average of Silhouettes. The two clusters are completely separated without overlapping (see Fig. 5) because the maximum value of the cluster 1 (52.2) is lower than the minimum value of the cluster 2 (53.8), in this case, we can compute the boundary separating the two clusters by taking average of these values (52.2 and 53.8) to get 53.0. Then, the two ARIMA-k-means-ANN models are dedicatedly constructed for each cluster. The ARIMA-k-means-ANN models fitted to cluster 1 and 2 are the ARIMA-k-means-ANN(4-3-1) model (see Fig. 6) and the ARIMA-k-means-ANN(5-2-1) model (see Fig. 7) respectively.

To forecast the future values, one of the ARIMA-ANN models has to be chosen based on the clusters that belong to the future values. However, practically, we cannot exaclty know the cluster of the future values, therefore, forecasting the cluster for the future value is required. In order to do so, the predicted future values from the ARIMA-ANN(7-3-1) model are assigned to the cluster by using the boundary between the clusters of training set as the criteria. For example, if the forecasted value in 1921 generated by the ARIMA-ANN(7-3-1) model is 21.1 which is below the boundary (53.0), the ARIMA-k-means-ANN(4-3-1) model of cluster 1 will be chosen as the prediction model for the sunspot in 1921.

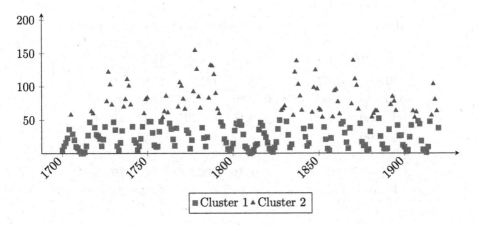

Fig. 5. Two clusters of the training set

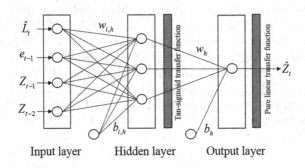

Fig. 6. The ARIMA-k-means-ANN(4-3-1) model for the cluster 1

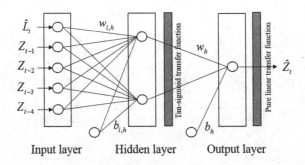

Fig. 7. The ARIMA-k-means-ANN(5-2-1) model for the cluster 2

Table 1. Performance comparison

Model	35 points ahead		67 points ahead	
	MAE	MSE	MAE	MSE
ARIMA (1)	10.10	174.30	12.55	276.35
ARIMA/ANN (2)	8.71	132.57	12.82	313.23
ARIMA/k-means/ANN (3)	9.89	189.18	13.40	368.29
(1) + (2)	8.48	127.20	11.47	234.80
(1) + (3)	8.64	138.39	11.87	265.29
(2) + (3)	8.29	135.09	12.31	302.17
(1) + (2) + (3)	**8.10**	**126.97**	11.67	250.08
Khashei and Bijari's model	8.85	129.43	**11.45**	**218.64**

In the final forecasting stage, the forecasted values from the ARIMA, the ARIMA-ANN and the ARIMA-k-means-ANN models are combined in all possible combinations using the weights from the discount mean square forecast error (DMSFE) method. The comparisons of MAE and MSE among the combinations are presented in Table 1 and the forecasted values of the best models are shown in Fig. 8. The Khashei and Bijari's model [11] is considered as the benchmark method. In short term prediction (35 points ahead), the best model is the combination of the ARIMA, the ARIMA/ANN and the ARIMA-k-means-ANN models ((1)+(2)+(3)) giving the lowest MAE (8.10) and MSE (126.97). However, in long term prediction (67 points ahead), the Khashei and Bijari's model [11], dominates the other models in term of MAE (11.45) and MSE (218.64).

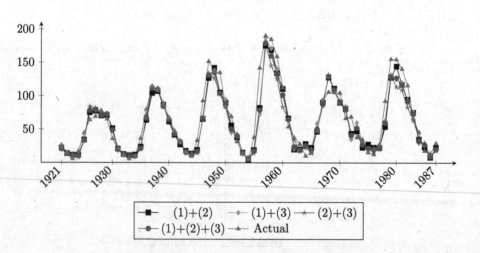

Fig. 8. Forecasted values of the best models

5 Conclusions

The hybird model of the ARIMA, the ARIMA-ANN and the ARIMA-k-means-ANN are proposed to obtain unique advantages among the different models in time series forecasting. The prediction capability of proposed model is tested with the Wolf's sunspot time series in the short term (35 points ahead) and the long term (67 points ahead) and the Khashei and Bijari's model [11] is used as the benchmark.

From the empirical results, the proposed model can give the best MAE and MSE in short term prediction with the combination of the ARIMA, the ARIMA-ANN and the ARIMA-k-means-ANN models. In long term prediction, the Khashei and Bijari's model is the best model in both MAE and MSE. In conclusion, clustering the data by k-means algorithm and combining the models that have different level of the complexity can improve the accuracy in time series forecasting as the proposed model can outperform the benchmarking in short term prediction. The limitations of the proposed model is that the forecasted clusters rely on the forecasted values from the ARIMA-ANN model that makes the error causing the wrong cluster forecasting especially when the forecasted values are closed to the cluster boundaries. Furthermore, the proposed model has been tested with only the sunspot data set.

In future work, the fuzzy clustering may be applied to overcome such limitation by assigning the clusters of the future values as probability to reduce the effect of the wrong cluster forecasting. Furthermore, the other practical applications will be included in the model testing and the ANNs may be used as the forecast combining method [20].

References

1. De Gooijer, J.G., Hyndman, R.J.: 25 years of time series forecasting. Int. J. Forecast. **22**(3), 443–473 (2006)
2. Box, G., Jenkins, G., Reinsel, G.: Time Series Analysis: Forecasting and Control. Wiley, New York (2008)
3. Zhang, G., Patuwo, B.E., Hu, M.Y.: Forecasting with artificial neural networks: the state of the art. Int. J. Forecast. **14**(1), 35–62 (1998)
4. Hornik, K., Stinchcombe, M., White, H.: Multilayer feedforward networks are universal approximators. Neural Netw. **2**(5), 359–366 (1989)
5. Zou, H.F., Xia, G.P., Yang, F.T., Wang, H.Y.: An investigation and comparison of artificial neural network and time series models for chinese food grain price forecasting. Neurocomputing **70**(16), 2913–2923 (2007)
6. Co, H.C., Boosarawongse, R.: Forecasting Thailand's rice export: statistical techniques vs. artificial neural networks. Comput. Ind. Eng. **53**(4), 610–627 (2007)
7. Prybutok, V.R., Yi, J., Mitchell, D.: Comparison of neural network models with ARIMA and regression models for prediction of Houston's daily maximum ozone concentrations. Eur. J. Oper. Res. **122**(1), 31–40 (2000)
8. Kohzadi, N., Boyd, M.S., Kermanshahi, B., Kaastra, I.: A comparison of artificial neural network and time series models for forecasting commodity prices. Neurocomputing **10**(2), 169–181 (1996)

9. Ho, S.L., Xie, M., Goh, T.N.: A comparative study of neural network and Box-Jenkins ARIMA modeling in time series prediction. Comput. Ind. Eng. **42**(2), 371–375 (2002)

10. Alon, I., Qi, M., Sadowski, R.J.: Forecasting aggregate retail sales: a comparison of artificial neural networks and traditional methods. J. Retail. Consum. Serv. **8**(3), 147–156 (2001)

11. Khashei, M., Bijari, M.: A novel hybridization of artificial neural networks and ARIMA models for time series forecasting. Appl. Soft Comput. **11**(2), 2664–2675 (2011)

12. Benmouiza, K., Cheknane, A.: Forecasting hourly global solar radiation using hybrid k-means and nonlinear autoregressive neural network models. Energy Convers. Manag. **75**, 561–569 (2013)

13. Ruiz-Aguilar, J.J., Turias, I.J., Jimnez-Come, M.J.: A novel three-step procedure to forecasting the inspection volume. Transp. Res. Part C **26**, 393–414 (2015)

14. Amin-Naseri, M.R., Gharacheh, E.A.: A hybrid artificial intelligence approach to monthly forecasting of crude oil price time series. In: The Proceedings of the 10th International Conference on Engineering Applications of Neural Networks, pp. 160–167 (2007)

15. MacQueen, J.: Some methods for classification and analysis of multivariate observations. Proc. Fifth Berkeley Symp. Math. Statist. Prob. 1(14), 281–297 (1967)

16. Rousseeuw, P.J.: Silhouettes: a graphical aid to the interpretation and validation of cluster analysis. J. Comput. Appl. Math. **20**, 53–65 (1987)

17. Zhang, G.P.: Time series forecasting using a hybrid ARIMA and neural network model. Neurocomputing **50**, 159–175 (2003)

18. Hipel, K.W., McLeod, A.I.: Time Series Modelling of Water Resources and Environmental Systems. Elsevier, Amsterdam (1994)

19. Winkler, R.L., Makridakis, S.: The combination of forecasts. J. R. Statist. Soc. A. **146**, 150–157 (1983)

20. Donaldson, R.G., Kamstra, M.: Forecasting combining with neural networks. J. Forecast. **15**, 49–61 (1996)

21. Tong, H.: Non-linear Time Series: A Dynamical System Approach. Oxford University Press, Oxford (1990)

The Rough Membership k-Means Clustering

Seiki Ubukata[✉], Akira Notsu, and Katsuhiro Honda

Osaka Prefecture University, Sakai, Osaka 599-8531, Japan
{subukata,notsu,honda}@cs.osakafu-u.ac.jp

Abstract. Fuzzy clustering approaches such as fuzzy c-means which is a fuzzified version of k-means method have been developed in order to deal with vague cluster memberships and used widely. Likewise, rough set approaches such as rough k-means and rough set k-means are also considered to be effective. In this paper, we propose the Rough Membership k-Means (RMKM) clustering in which values of the rough membership function are used as fuzzy cluster memberships. Furthermore, we carried out some numerical experiments in order to demonstrate the performance of the rough membership k-means clustering.

Keywords: k-means clustering · Rough set theory · Rough membership function

1 Introduction

Clustering such as k-means [6] is one of the key techniques in the field of data mining. Fuzzy clustering approaches such as Fuzzy c-Means (FCM) [1,2] which is a fuzzified version of k-means method have been developed in order to deal with vague cluster memberships.

As a way to deal with vague cluster memberships, rough set approaches [10, 11] are effective as well as fuzzy approaches. Lingras and West proposed the Rough k-Means (Lingras' RKM) clustering [5] introducing two regions which satisfy basic properties of the lower and upper approximations, respectively. Peters added some refinements to Lingras' rough k-means clustering [14]. Moreover, fuzzy extensions of rough k-means have been researched in a variety of ways. Mitra et al. proposed the Rough Fuzzy c-Means (Mitra's RFCM) clustering combining both rough and fuzzy sets [8,9]. In Mitra's RFCM, cluster memberships are calculated by the FCM strategy, and cluster centers are calculated by the weighted average of the lower area and the boundary. Maji et al. proposed Rough Fuzzy c-Means (Maji's RFCM) clustering [7] in order to represent the vagueness of the boundary by using fuzzy cluster memberships based on the FCM strategy. In Maji's RFCM, the objects in the lower area have crisp cluster memberships, on the other hand, the objects in the boundary have fuzzy cluster memberships.

These rough k-means approaches use analogous regions to the lower and upper approximations, but do not use original basic concepts of rough set theory,

© Springer International Publishing AG 2016
V.-N. Huynh et al. (Eds.): IUKM 2016, LNAI 9978, pp. 207–216, 2016.
DOI: 10.1007/978-3-319-49046-5_18

i.e., the lower and upper approximations based on binary relations are not used. Taking these facts, Ubukata *et al.* proposed the Rough Set k-Means (RSKM) clustering [16] using original definitions of the lower and upper approximations.

In our implementation of the RSKM clustering, the lower and upper approximations are calculated by the core (1-cut) and the support (strong 0-cut) of values of the rough membership function. We conceived an idea that we use values of the rough membership function directly from a fact that rough memberships have similar properties to fuzzy cluster memberships.

In this paper, we propose the Rough Membership k-Means (RMKM) clustering in which values of the rough membership function are used as fuzzy cluster memberships. In the proposed method, the objects in the boundary have a value in the open unit interval $(0, 1)$. Therefore, it can represent the vagueness of the boundary in analogy with Maji's RFCM by a context of rough set-based approximation. Furthermore, we carried out some numerical experiments in order to demonstrate the performance of the proposed method.

2   Preliminaries

2.1   Rough Set Theory

Pawlak Rough Set Model. Rough set theory [10] was proposed by Pawlak in 1982 and has received a lot of attention as a method which can deal with vagueness, inconsistency, uncertainty, and incompleteness.

Let U be a non-empty finite set of objects called the universe and $R \subseteq U \times U$ be a binary relation on U. Then, the pair

$$F = \langle U, R \rangle$$

is called a Pawlak approximation space. Rough set theory is a theory of approximations on approximation spaces. By using the relation R, the successor set of x is defined by:

$$U_R(x) = \{y \in U \mid xRy\}.$$

For any subset $X \subseteq U$, the lower and upper approximations of X with respect to R are defined respectively as follows:

$$\underline{apr}_R(X) = \{x \in U \mid U_R(x) \subseteq X\},$$
$$\overline{apr}_R(X) = \{x \in U \mid U_R(x) \cap X \neq \emptyset\}.$$

Then, the pair

$$(\underline{apr}_R(X), \overline{apr}_R(X))$$

is called the rough set of X with respect to R.

The rough R-membership function of X is defined as follows [12,13]:

$$\mu_X^R(x) = \frac{|U_R(x) \cap X|}{|U_R(x)|}.$$

Using the rough membership function, the definitions of the lower and upper approximation of X with respect to R are rewritten as:

$$\underline{apr}_R(X) = \{x \in U \mid \mu_X^R(x) = 1\},$$
$$\overline{apr}_R(X) = \{x \in U \mid \mu_X^R(x) > 0\}.$$

By the rough approximation, U is divided into three disjoint regions; the positive region, the boundary region, and the negative region:

$$POS_R(X) = \underline{apr}_R(X),$$
$$BND_R(X) = \overline{apr}_R(X) \setminus \underline{apr}_R(X),$$
$$NEG_R(X) = U \setminus \overline{apr}_R(X).$$

Considering a practical view of data mining, an information table IT is defined by:

$$IT = \langle U, A, V, \rho \rangle,$$

where A is a non-empty finite set of attributes, V is a non-empty set of attribute values, and $\rho : A \times U \to V$ is an information function which assigns an attribute value of an object with respect to a certain attribute. An indiscernibility relation with respect to $B \subseteq A$ is defined by:

$$IND(B) = \{(x,y) \in U \times U \mid \forall a \in B(\rho(a,x) = \rho(a,y))\}.$$

The indiscernibility relation is an equivalence relation.

Neighborhood Rough Set Model. The Pawlak rough set model aimed mainly at categorical data. If the indiscernibility relation is used for numerical data, it will often make the trivial discrete topology. Hu *et al.* introduced a neighborhood rough set model [3,4] to handle rough approximations for numerical data.

In neighborhood rough set model, the neighborhood $\delta(x)$ of $x \in U$ is defined by:

$$\delta(x) = \{y \in U \mid \Delta(x,y) \le \delta\},$$

where Δ is, for example, the Minkowski distance:

$$\Delta(x,y) = \left(\sum_{j=1}^{m} |x_j - y_j|^p \right)^{1/p},$$

where x and y are regarded as m-dimensional real vectors. In this paper, we choose $p = 2$, i.e., the Euclidean distance. Defining a binary relation named the neighborhood relation

$$R_\delta = \{(x,y) \in U \times U \mid \Delta(x,y) \le \delta\},$$

the lower and upper approximations of X with respect to R_δ are obtained in the same manner as before:

$$\underline{apr}_{R_\delta}(X) = \{x \in U \mid U_{R_\delta}(x) \subseteq X\}$$
$$= \{x \in U \mid \delta(x) \subseteq X\},$$
$$\overline{apr}_{R_\delta}(X) = \{x \in U \mid U_{R_\delta}(x) \cap X \neq \emptyset\}$$
$$= \{x \in U \mid \delta(x) \cap X \neq \emptyset\}.$$

2.2 k-Means Clustering and Rough k-Means Clustering

k-Means Clustering. The k-means clustering procedure is described as follows.

Step 1. Initialize crisp cluster membership u_{ci} by assigning each object to one cluster randomly.

Step 2. Calculate the cluster center $\boldsymbol{b}_c$ of each cluster c by the average of objects in the cluster:

$$\boldsymbol{b}_c = \frac{\sum_{x \in G_c} \boldsymbol{x}}{|G_c|},$$

where $G_c = \{\boldsymbol{x}_i \in U \mid 1 \leq i \leq n, u_{ci} = 1\}$ is a temporal cluster.

Step 3. Assign each object to the nearest cluster:

$$u_{ci} = \begin{cases} 1 & (c = \arg\min_{1 \leq l \leq k} \|\boldsymbol{x}_i - \boldsymbol{b}_l\|), \\ 0 & (otherwise). \end{cases}$$

Step 4. Repeat steps 2 and 3 until cluster assignments do not change.

Rough Set k-Means Clustering. Ubukata *et al.* proposed the Rough Set k-Means (RSKM) clustering [16]. The rough set k-means clustering procedure is described as follows.

Step 1. Determine a binary relation $R \subseteq U \times U$ on U. Set values of $w_{low}, w_{upp} \in [0, 1]$ such that they satisfy $w_{low} + w_{upp} = 1$.

Step 2. Initialize crisp cluster membership u_{ci} by assigning each object to one cluster randomly.

Step 3. Calculate the cluster center $\boldsymbol{b}_c$ of each cluster c by the lower and upper approximations:

$$\boldsymbol{b}_c = \begin{cases} w_{low} \frac{\sum_{x \in \underline{apr}_R(G_c)} \boldsymbol{x}}{|\underline{apr}_R(G_c)|} + w_{upp} \frac{\sum_{x \in \overline{apr}_R(G_c)} \boldsymbol{x}}{|\overline{apr}_R(G_c)|} & (\underline{apr}_R(G_c) \neq \emptyset), \\ \frac{\sum_{x \in \overline{apr}_R(G_c)} \boldsymbol{x}}{|\overline{apr}_R(G_c)|} & (\underline{apr}_R(G_c) = \emptyset, \overline{apr}_R(G_c) \neq \emptyset), \\ exception\ handling & (otherwise), \end{cases}$$

where $G_c = \{\boldsymbol{x}_i \in U \mid 1 \leq i \leq n, u_{ci} = 1\}$ is a temporal cluster.

Step 4. Assign each object to the nearest cluster:

$$u_{ci} = \begin{cases} 1 & (c = \arg\min_{1 \leq l \leq k} \|\boldsymbol{x}_i - \boldsymbol{b}_l\|), \\ 0 & (otherwise). \end{cases}$$

Step 5. Repeat steps 3 and 4 until cluster assignments do not change.

3   The Rough Membership k-Means Clustering

In this section, we propose the Rough Membership k-Means (RMKM) clustering. In iteration processes of k-means method, temporal clusters (denoted by G_c) and the temporal partition $P = \{G_1, G_2, ..., G_c, ..., G_k\}$ are generated. In our implementation of the RSKM clustering, the lower and upper approximations of G_c are calculated by the core (1-cut) and the support (strong 0-cut) of values of the rough membership function, respectively.

Now note that, as for the rough membership function, following hold similarly to fuzzy cluster membership, considering that G_c is an element of the partition:

$$\mu_{G_c}^R(\boldsymbol{x}_i) \in [0, 1],$$

$$\sum_{c=1}^{k} \mu_{G_c}^R(\boldsymbol{x}_i) = 1.$$

We use values of the rough membership function directly without α-cut from a fact that rough memberships have similar property to fuzzy cluster memberships.

The rough membership k-means clustering procedure can be described as follows.

Step 1. Determine a binary relation $R \subseteq U \times U$ on U.
Step 2. Initialize fuzzy cluster memberships u_{ci} such that they satisfy $\sum_{c=1}^{C} u_{ci} = 1, \forall c$.
Step 3. Calculate the cluster center of each cluster c by the weighted average of U:

$$b_c = \frac{\sum_{i=1}^{n} u_{ci} \boldsymbol{x}_i}{\sum_{i=1}^{n} u_{ci}}$$

Step 4. Generate each temporal cluster by nearest assign:

$$G_c = \{\boldsymbol{x}_i \in U \mid 1 \le i \le n, c = \arg\min_{1 \le l \le k} \|\boldsymbol{x}_i - \boldsymbol{b}_l\|\}.$$

Step 5. Calculate the membership of each object i to each cluster c by the rough membership function:

$$u_{ci} = \mu_{G_c}^R(\boldsymbol{x}_i) = \frac{|U_R(\boldsymbol{x}_i) \cap G_c|}{|U_R(\boldsymbol{x}_i)|}.$$

Step 6. Repeat steps 3 and 5 until cluster assignments do not change.

4   Numerical Experiments

We carried out some numerical experiments in order to demonstrate the performance of the Rough Membership k-Means (RMKM) clustering comparing with the k-means and the Rough Set k-Means (RSKM) clustering.

Table 1. An example of the cross-tabulation of k-means: class labels - clusters

k-means, Purity = 0.887				
Label\Cluster	0	1	2	sum
setosa	50	0	0	50
versicolor	0	47	3	50
virginica	0	14	36	50

4.1 Experimental Settings

We used the Iris dataset retrieved from the UCI Machine Learning Repository [15] as a benchmark dataset. The Iris dataset consists of $n = 150$ objects, a set of attributes $A = \{sepal\_length, sepal\_width, petal\_length, petal\_width\}$, and a set of class labels $\{setosa, versicolor, virginica\}$. A vague boundary is considered to exist between "*versicolor*" and "*virginica*." We tried to obtain $k = 3$ clusters detecting the boundary and positive regions of clusters by means of the proposed method. We determined the binary relation R_δ on U and used an approximation strategy in neighborhood rough set model. δ is determined based on the τ-percentile of the distance distribution between objects:

$$\delta = percentile(O_D, \tau),$$

where O_D is the sorted list of the value of distances $\{\|x_i - x_{i'}\| \mid x_i, x_{i'} \in U\}$ in ascending order. A subset $B \subseteq A$ of attributes is fixed to all four attributes, i.e., $B = A$.

The application was implemented using Python 3.5.1 and a desktop computer equipped with CPU Intel(R) Core(TM) i7-4790 @3.40GHz.

The average purity of the clusters is calculated by

$$\frac{1}{n} \sum_{c=1}^{k} \max_{1 \le t \le k^*} |G_c \cap G_t^*|,$$

where k^* is the original number of the partition and G_t^* is an element of the original partition.

4.2 The Examples of Cross-Tabulation

Tables 1, 2, 3, 4 and 5 are the examples of cross-tabulation between class labels and clustering results. Table 1 shows the cross-tabulation with respect to the partition by nearest assignment for all objects by k-means. "*setosa*" is classified certainly, but "*versicolor*" and "*virginica*" are not completely separated. There are 17 misclassified objects and the purity of is 0.887. Table 2 shows the cross-tabulation with respect to the partition by nearest assignment for all objects by RSKM($\tau = 15$). Misclassified objects are reduced to 16 and the purity is

Table 2. An example of the cross-tabulation of RSKM ($\tau = 15$): class labels - the crisp partition by nearest assignment.

RSKM $\tau = 15$, Purity = 0.893			
Label\Cluster	0	1	2	sum
setosa	50	0	0	50
versicolor	0	45	5	50
virginica	0	11	39	50

Table 3. An example of the cross-tabulation of RSKM ($\tau = 15$): class labels - the positive region and the boundary region.

RSKM $\tau = 15$, Purity = 0.991				
Label\Cluster	0	1	2	BND	sum
setosa	50	0	0	0	50
versicolor	0	29	0	21	50
virginica	0	1	29	20	50

Table 4. An example of the cross-tabulation of RMKM ($\tau = 15$): class label - the crisp partition by nearest assignment.

RMKM $\tau = 15$, Purity = 0.907			
Label\Cluster	0	1	2	sum
setosa	50	0	0	50
versicolor	0	50	0	50
virginica	0	14	36	50

Table 5. An example of the cross-tabulation of RMKM ($\tau = 15$): class label - the positive region and the boundary region.

RMKM $\tau = 15$, Purity = 0.958				
Label\Cluster	0	1	2	BND	sum
setosa	50	0	0	0	50
versicolor	0	36	0	14	50
virginica	0	5	28	17	50

improved to 0.893. Table 3 shows the cross-tabulation with respect to the positive and boundary region by RSKM ($\tau = 15$). 41 objects are detected as the boundary region. As for the positive region, only one object is misclassified and the purity is improved to 0.991. Hence, the certainty of clustering is considered to be improved. Table 4 shows the cross-tabulation with respect to the partition by nearest assignment for all objects by RMKM($\tau = 15$). There are 14 misclassified objects and the purity is 0.907. We can observe that the cluster 2 is composed only of "*versicolor*" and the certainty is improved. Table 5 shows the cross-tabulation with respect to the positive and boundary region by RMKM ($\tau = 15$). 31 objects are detected as the boundary region. As for the positive region, there are 5 misclassified objects and the purity is 0.958. This is worse than RSKM, but RMKM detects more objects as the positive region.

4.3   The Examples of Scatter Plots

Next, we show the examples of scatter plots. In these scatter plots, two attributes "*sepal_width*" and "*petal_length*" are used. Figure 1 shows a result of RSKM ($\tau = 15$). The positive region and boundary region are indicated by three primary colors and a neutral color, respectively. We can observe that the boundary region between "*versicolor*" and "*virginica*". Figure 2 shows a result of RMKM ($\tau = 15$). Likewise fuzzy cluster memberships, objects around the boundary

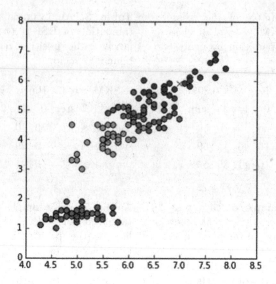

Fig. 1. An example of the scatter plot of the result of RSKM clustering: the positive region and the boundary region are indicated by three primary colors and an neutral color, respectively (Color figure online).

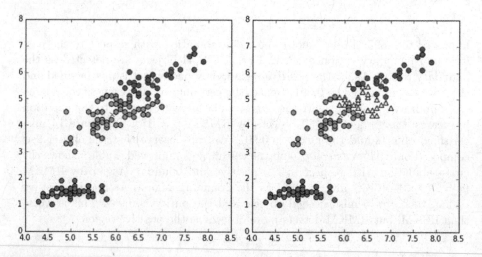

Fig. 2. An example of the scatter plot of the result of RMKM clustering: likewise fuzzy cluster memberships, gradation can be obserbed in the boundary region.

Fig. 3. The extracted positive region: the core (1-cut) of the values of the rough membership function. White triangles are the boundary whose rough membership values are in $(0, 1)$.

between "*versicolor*" and "*virginica*" have vague cluster membership values. Gradation can be observed in the boundary region. In RMKM, the positive region is detected by the core (1-cut) of the cluster membership u_{ci}. Figure 3 shows the extracted positive region in the same as Fig. 1. Objects whose cluster membership is $0 < u_{ci} < 1$ are detected as the boundary region.

Considering the results, we confirm that RSKM and RMKM can realize the detection of the boundary region and the certain clustering based on the positive region.

4.4 Performance Comparison

We measured the purity of the positive region by changing τ in $\{0, 1, 2, 3, ..., 100\}$ in order to compare the performance of RMKM with RSKM. These values are the average of the purity in 100 trials.

Figure 4 shows the changes of the purity of the positive region with τ-percentile in RSKM and RMKM. In cases of $2 < \tau < 15$, the performance of RSKM is better than RMKM. However, the performance of RSKM is significantly decrease with an increase in τ. On the other hand, in RMKM, we can obtain higher purity with an increase in τ.

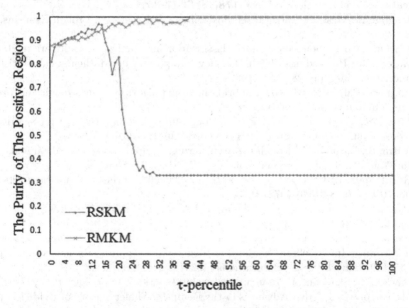

Fig. 4. The changes of the purity of the positive region with τ-percentile in RSKM and RMKM.

5 Concluding Remarks

In this paper, we proposed the rough membership k-means clustering in which values of the rough membership function are used as fuzzy cluster memberships.

We carried out some numerical experiments in order to demonstrate the characteristics and performance of the proposed method. By the experiments, we confirm that the proposed method can execute the detection of the boundary region and the certain clustering based on the positive region. More experiments using more datasets should be carried out in order to confirm characteristics of the method.

Acknowledgment. This work is supported by Program to Disseminate Tenure Tracking System, MEXT, Japan.

References

1. Bezdek, J.C.: Pattern Recognition with Fuzzy Objective Function Algorithms. Plenum Press, New York (1981)
2. Dunn, J.C.: A fuzzy relative of the ISODATA process and its use in detecting compact well-separated clusters. J. Cybern. **3**, 32–57 (1974)
3. Hu, Q., Yu, D., Xie, Z.: Neighborhood classifiers. Expert Syst. Appl. **34**(2), 866–876 (2008)
4. Hu, Q., Yu, D., Liu, J., Wu, C.: Neighborhood rough set based heterogeneous feature subset selection. Inf. Sci. **178**(18), 3577–3594 (2008)
5. Lingras, P., West, C.: Interval set clustering of web users with rough K-means. J. Intell. Inf. Syst. **23**(1), 5–16 (2004)
6. MacQueen, J.B.: Some methods of classification and analysis of multivariate observations. In: Proceedings of 5th Berkeley Symposium on Mathematical Statistics and Probability, pp. 281–297 (1967)
7. Maji, P., Pal, S.K.: RFCM: a hybrid clustering algorithm using rough and fuzzy sets. Fundamenta Informaticae **80**, 475–496 (2007)
8. Mitra, S., Banka, H., Pedrycz, W.: Rough-fuzzy collaborative clustering. IEEE Trans. Syst. Man Cybern. Part B (Cybern.) **36**(4), 795–805 (2006)
9. Mitra, S., Barman, B.: Rough-fuzzy clustering: an application to medical imagery. In: Wang, G., Li, T., Grzymala-Busse, J.W., Miao, D., Skowron, A., Yao, Y. (eds.) RSKT 2008. LNCS (LNAI), vol. 5009, pp. 300–307. Springer, Heidelberg (2008). doi:10.1007/978-3-540-79721-0_43
10. Pawlak, Z.: Rough sets. Int. J. Comput. Inf. Sci. **11**(5), 341–356 (1982)
11. Pawlak, Z.: Rough classification. Int. J. Man Mach. Stud. **20**(5), 469–483 (1984)
12. Pawlak, Z., Skowron, A.: Rough membership function: a tool for reasoning with uncertainty. In: Algebraic Methods in Logic and Computer Science, vol. 28, pp. 135–150. Banach Center Publications (1993)
13. Pawlak, Z., Skowron, A.: Rough membership functions. In: Yaeger, R.R., Fedrizzi, M., Kacprzyk, J. (eds.) Advances in the Dempster-Shafer Theory of Evidence, pp. 251–271. Wiley, New York (1994)
14. Peters, G.: Some refinements of rough k-means clustering. Pattern Recogn. **39**(8), 1481–1491 (2006)
15. UCI Machine Learning Repository. http://archive.ics.uci.edu/ml/
16. Ubukata, S., Notsu, A., Honda, K.: The rough Set k-means clustering. In: Proceedings of Joint 8th International Conference on Soft Computing and Intelligent Systems and 17th International Symposium on Advanced Intelligent Systems, pp. 189–193 (2016)

Instance Reduction for Time Series Classification by Exploiting Representative Characteristics using k-means

Vo Thanh Vinh, Hien T. Nguyen$^{(\boxtimes)}$, and Tin T. Tran

Faculty of Information Technology, Ton Duc Thang University,
Ho Chi Minh City, Vietnam
{vothanhvinh,hien,trantrungtin}@tdt.edu.vn

Abstract. 1-Nearest Neighbor has been endorsed an efficient method for time series classification as it outperforms more advanced classification algorithms in most cases. However, the time and space efficiency of this method depend on the number of instances in a training set. In order to improve its running time and space using, one can apply an approach called instance reduction. This approach reduces the training set size by choosing the best instances and using only them during classification of new instances. However, the high-dimensional characteristic of time series is a challenge for many mining tasks including instance reduction. Due to this challenge, only two instance reduction methods have been proposed: Naïve Rank Reduction and INSIGHT. In this work, we propose a new approach to instance reduction using K-means. Our method can select a reasonable set of instances so that it can preserve the representative characteristics and keep the generalization of the original training set. We conduct experiments to evaluate our method on 46 public datasets from the UCR Classification Archive. The experimental results show that the proposed method achieves state-of-the-art performance on these datasets.

Keywords: Time series · K-means · 1-Nearest Neighbor · Instance reduction

1 Introduction

Time series data can be found in many areas of our life such as finance, economy, or medicine. Mining data in these areas includes many different tasks such as motif discovery, anomaly detection, rule extraction, clustering, or classification. In which, classification is definitely important as it is the background of pattern recognition. For example, it can be applied in signature verification, handwriting recognition, analysis of brainwaves, or electrocardiograph signals. One of challenging properties of time series data is its high dimensionality. Some classification models such as 1-Nearest Neighbor (1-NN), Artificial Neural Network, or Bayesian Network have been applied successfully in time series. Among them, the simple approach 1-NN has been considered as very hard to be beaten [1, 3]. Nevertheless, time and space efficiency of 1-NN is still need to be improved. To this end, one can apply these directions:

© Springer International Publishing AG 2016
V.-N. Huynh et al. (Eds.): IUKM 2016, LNAI 9978, pp. 217–229, 2016.
DOI: 10.1007/978-3-319-49046-5_19

1. Using some indexing structures such as R-Tree [17], M-Tree [16], TS-Tree [18].
2. Reducing dimensions of time series [19, 20].
3. Speeding up computational time of distance measure [21, 22].
4. Reducing the number of instances in the training set [1, 2].

Following the fourth approach, Xi et al. [1] proposed Naïve Rank Reduction and Buza et al. [2] proposed Instance Selection based on Graph-coverage and Hubness for Time series (INSIGHT). Although these methods can help reduce instances, they have three drawbacks: (i) do not preserve the training set's distribution, (ii) may produce an imbalance training set, and (iii) must be provided the reduced training set size.

The contribution of our work can be summarized as follows. Firstly, we propose a new approach to instance reduction in time series classification using K-means. Secondly, our method can decide whenever to stop eliminating instances from the training set, in the meaning, if we continue to remove more instance, it would lead to poor results. The experiments were conducted over popular time series datasets from the UCR Classification Archive [7]. Experimental results show that the proposed method outperforms Naïve Rank Reduction and INSIGHT in most cases.

2   Backgrounds and Related Works

2.1   Time Series

A time series T is a sequence of real numbers collected at regular intervals over a period of time: $T = t_1, t_2, ..., t_n$. It can be considered as a n-dimensional object in a metric space.

2.2   1-Nearest Neighbor

The 1-Nearest Neighbor (1-NN) assigns a new data object to the same class of its nearest object in the training set. 1-NN has been considered hard to be beaten in classification of time series data among other methods such as Artificial Neural Network, Bayesian Network [1, 3]. In this work, we use 1-NN to classify time series data and evaluate the error rate of this classifier on the reduced training set obtained by our instance reduction method.

2.3   Instance Reduction

The main goal of instance reduction is: (i) to improve the running time of classification algorithms while losing accuracy as little as possible and (ii) to fit data, especially training data, into limited memory devices. Many instance reduction methods were proposed to speed up nearest neighbor classification of conventional data such as in [8–15]. However, it is hard to find an efficient classification method for time series data because of its high dimensionality characteristic. So far, there has been only two works related to instance reduction for time series data as above mentioned.

2.4 Naïve Rank Reduction

Xi et al. in 2006 [1] proposed a method called Naïve Rank Reduction for time series data which uses a ranking function developed from the ideas in the most referenced paper by Wilson and Martinez (1997) [6]. Naïve Rank Reduction method first assigns rank to each instance x in the training set by identifying its contribution to the classification of other instances. The instances are ranked by the following formula:

$$rank(x) = \sum_j \begin{cases} 1 & \text{if } class(x) = class(x_j) \\ -2 & \text{otherwise} \end{cases} \tag{1}$$

where x_j is the instance having x as its nearest neighbor. This ranking function attempts to keep the instances that contribute to correctly classifying other instances. The ranking function presented in Eq. (1) may produce the same rank for some instances. In order to break this tie, the authors used a second ranking function as presented in Eq. (2).

$$priority(x) = \sum_j \frac{1}{d(x, x_j)^2} \tag{2}$$

where x_j is the instance having x as its nearest neighbor and $d(x, x_j)$ is the distance between instances x and x_j. The idea behind Eq. (2) is that the farther the instance from its nearest neighbor, the lower priority it should have. Because this instance may be noisy or unrepresentative instance, so it should be eliminated first.

2.5 INSIGHT Method for Instances Selection

Buza et al. in 2011 [2] proposed a method called Instance Selection based on Graph-coverage and Hubness for Time series (INSIGHT), which ranks the time series in the training set by using one of their three score functions. These ranking functions attempt to exploit the hubness property of time series [2]. This property states that for data with high dimensionality, some objects tend to become nearest neighbors much more frequently than others. The term hubness is a phenomenon that the distribution of $f_N^k(x)$ significantly skewed to the right, where $f_N^k(x)$ denote the k-occurrence of an instance x in a training set D, which is the number of instances in D having x as their k nearest neighbor. The score functions of INSIGHT were defined as follows:

- $f_G(x) = f_G^1(x)$: Good 1-occurrence score.
- $f_R(x) = f_G^1(x) / (f_N^1(x) + 1)$: Relative score.
- $f_{Xi}(x) = f_G^1(x) - 2f_B^1(x)$: Xi's score (base on ranking criterion of Xi et al. [1]).

In which, $f_G^1(x)$ is the number of instances in the training set having x as their 1-nearest neighbor and x has the same label with them. $f_B^1(x)$ is the number of instances in the training set having x as their 1-nearest neighbor and label of x is different from their label.

2.6   Notes on Naïve Rank Reduction and INSIGHT

Naïve Rank Reduction was used in a method called Fast Time Series Classification Using Numerosity Reduction (FastAWARD) proposed by Xi et al. [1]. This method iteratively removes time series in a training set. In each iteration, ranks of all the time series in the training set is calculated and the lowest ranked time series is discarded. Thus, each iteration kept a particular number time series. Xi et al. claim that the best value of Sakoe-Chiba band [4, 5] for DTW distance depends on the number of time series in the reduced training set. Therefore, they include a final step in order to calculate the warping window size for each training set size. The INSIGHT method just ranks instances in the training set and reduces the training set only once. This method does not include the second step to find the value for Sakoe-Chiba band r as in FastAWARD. Instead, the warping-window size is set to 5 % of the time series length.

Our proposed method is different from the FastAWARD and INSIGHT in that we do not use a ranking function. Moreover, our method does not focus on finding Sakoe-Chiba band r since this additional step is not important but computationally expensive. In addition, our method can easily include this step. All things considered, we only compare our method with the Naïve Rank Reduction and INSIGHT.

2.7   Weaknesses of Naïve Rank Reduction and INSIGHT

The main idea of Naïve Rank Reduction and INSIGHT bases on ranking functions, which attempt to exploit the original training set to discover how much the instances impact in the classification process. Then, the instances are arranged from least to most impact. The instances contributing most to the classifier are selected first. This idea has three drawbacks as follows:

- It does not maintain the representative characteristics of the training set.
- It may lead to imbalance in the number of instances of each class in the reduced training set.
- It is parametric which means that we must provide the reduced training set size.

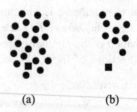

(a)   (b)

Fig. 1. Instance reduction using ranking strategy might lead to the representative characteristics of a training set are not maintained. *Circle*: training instances, *Square*: a new instance that need to be classified. (a) Original training set, (b) Reduced training set

With regards to the first drawback, because the instances that have most impact in the classification process are kept, the training set's distribution does not maintain as original. This means that representative characteristics of original training set are not preserved in reduced training set. As shown in Fig. 1, the original training set and the reduced training set have different distribution. Therefore, a new instance (square) in real world might be wrongly classified because the instances in the original training set that look like this new instance were removed.

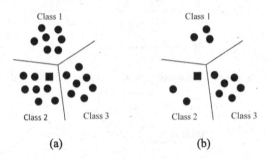

(a)   (b)

Fig. 2. There is some bias in the number of selected instances in each class. *Circle*: training instances, *Square*: a new instance that need to be classified. (a) Original training set with three classes has equal number of instances, (b) Reduced training set with more instances in class 1, meanwhile class 2 and class 3 has much less instances

For the second drawback, the ranking function may bias some classes while reducing instances. As a result, some classes may have much more instances than the others. We illustrate this issue in Fig. 2. The original training set has three classes; each contains 7 instances (circle). Meanwhile, the reduced training set is unbalanced, which has three instances for Class 1, two instance for Class 2, and six instances for Class 3. As a result, the new instance (square), which is classified as Class 2 with the original training set, will be classified as Class 3 with the reduced training set.

3 Proposed Method

3.1 The Framework

In order to overcome the drawbacks mentioned in Subsect. 2.7, we propose a framework to reduce training set size while maintaining distribution of each class. The outline of our method is described in Fig. 3. In order to keep $p\%$ instances in the training set, we run K-means on each class with the number of cluster $K = p$ x *number of instances in the class* to get K centroids. Then, the K centroids are used as training instances. All the instances kept in each class are combined together to get the reduced training set.

As we can see from Fig. 3, the framework uses K-means to partition n instances into K sets $C = \{C_1, C_2,..., C_K\}$ so as to minimize the sum of distance of each instance in the cluster to the K centroids as in Eq. (3).

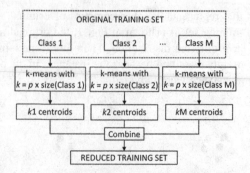

Fig. 3. The framework of instance reduction in a training set with M classes

$$\arg\min_{C} \sum_{i=1}^{K} \sum_{x \in C_i} dist(x, \mu_i) \qquad (3)$$

where μ_i is the centroids of data instances in C_i. As a result, the instances with similar characteristics are grouped into one cluster and they are all similar to their centroid. Therefore, we can keep the centroids only as they can represent the whole cluster. Figure 4 illustrates effect of our framework, which can help keep representative instances while the number of instances in each class is balance as original. While the component of the framework, K-means, is well-known, we emphasize that the framework as a whole is new for the instance reduction of time series classification.

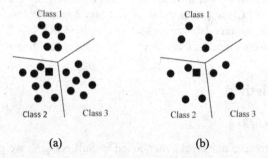

Fig. 4. *Circle*: training instances, *Square*: a new instance that need to be classified. (a) Original training set with three classes has equal number of instances, (b) Reduced training set with representative instances

3.2 Choosing Appropriate Percentage of Instances

One question in instance reduction problem is how many instances should be kept in the training set. In general, the bigger the training set size, the more accurate the classifier. Although accuracy rate should be considered as a major factor in deciding which training set is better, training set size should also be considered as a penalty.

If a training set has more instances than the others, we should penalize it more. For example, given a training set with 100 instances and accuracy rate for classification is 0.85. Now we remove this training set to 20 instances with its new accuracy rate for classification is 0.8. The accuracy rate decreases only 5 % but it can help to speedup 5 times. In this situation, it is more likely that we prefer the second choice. Akaike Information Criterion (AIC) and Bayesian Information Criterion (BIC) are the two popular model selections. The ideas behind these two methods are the same except that BIC gives more penalties to the model with more parameters. The formulas of AIC and BIC are as follows:

$$AIC = -2\log L + 2k$$
$$BIC = -2\log L + \log(N) \times k$$

(4)

where k is the number of parameters in the model, $\log L$ is a log-likelihood function, and N is the number of observed data. We can understand that AIC and BIC evaluate each model to find the best one bases on two factors: likelihood value and the number of parameters of the model. In other words, they attempt to balance likelihood and model's size. In this work, we adopt the above two formulas in Eq. (4) to select an appropriate point at which we should not remove anymore instances. Note that we only apply the idea of balancing between likelihood value and the number of parameters of a model. So we emphasize that our method is not AIC or BIC. The formulas are redefined as follows:

$$f1\_score = 2 \times percentage\_error + 2 \times k$$
$$f2\_score = 2 \times percentage\_error + \log(N) \times k$$

(5)

Where:

- *percentage_error*: the percentage of wrongly classified instances.
- k: the number used to penalize reduced training set, which is defined based on the percentage of selected instances as follows:

Percentage	10	15	20	25	30	35	40	45	50	55	60	65	70	75	80	85	90
k	2	3	4	5	6	7	8	9	10	11	12	13	14	15	16	17	18

- N: total number of instance in original training set.

In general, the percentage of selected instances with the lowest value of these two criterions is the best one. In Eq. (5), we map the percentage of selected instance to the number of parameters. For example, if we select 10 % instances as reduced training set, the value of k is 2. Now why do we need this mapping? The answer is that if we do not map but use the percentage of selected instance, then it will penalize a lot, i.e. the value of k will have a high weight in selecting appropriate training set, and thus the accuracy rate does not play an essential part in deciding the percentage at which we choose to reduce the training set to.

Figure 5 illustrates how to identify the number of instances for reduced training set. In which we run our framework in Subsect. 3.2 with different number of instances, i.e. 10 %, 15 %, 20 %,..., 90 % of original training set. Then, we use these reduced training sets to classify original one (100 % instances) to get the percentage of wrongly classified instances (*percentage_error*). Finally, Eq. (5) is applied to find out the best point to reduce the training set to.

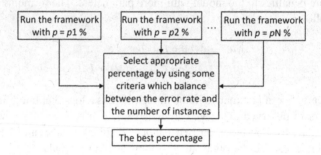

Fig. 5. Appropriate percentage selection

4 Experiments

This section shows two main experiments. Firstly, we compare accuracy rate of our method with those of Naïve Rank Reduction, and INSIGHT. Accuracy rate is the ratio of the number of correctly classified instances on tested data to the total number of tested instances. In general, the higher accuracy rate, the better method. Secondly, we find the percentage at which we should not remove any more instance. 10-fold cross-validation was applied over 46 datasets from the UCR Time Series Classification Archive [7]. Each datasets has from 2 to 50 classes. The length of each time series in each datasets is from 60 to 1882. Euclidean Distance was used as similarity measurement. The ranking function used in INSIGHT method is *Good 1-occurrence score* as this is the function used in the experiment of Buza et al. [2].

4.1 Comparing to Previous Methods

The two scatter graphs in Fig. 6 show comparison of the proposed method with those of Naïve Rank Reduction and INSIGHT. Each circle point in the scatter graphs represents a dataset. As we can see from the figures, most circle points deviate on the lower triangle, which means that the proposed method is better than the two previous ones. In detail, the proposed method outperforms INSIGHT in 41 datasets, draws in 1 datasets, and loses in 4 datasets. In comparison to Naïve Rank Reduction, the proposed method outperforms it in 42 datasets, draws in 1 datasets, and loses in only 3 datasets. Moreover, Table 1 shows accuracy rate of the three approaches on each dataset.

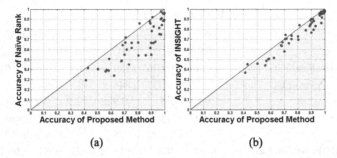

(a)   (b)

Fig. 6. Comparing the accuracy rates of the proposed method with those of Naïve Rank Reduction (a), INSIGHT (b) on 46 datasets

Table 1. Accuracy rate ± standard deviation in 46 datasets after reducing to 20 % of original training set

Datasets	Naïve Rank	INSIGHT	Proposed Method
50words	0.451 ± 0.07	0.66 ± 0.047	0.732 ± 0.043
Adiac	0.341 ± 0.052	0.515 ± 0.039	0.623 ± 0.041
Beef	0.38 ± 0.166	0.44 ± 0.12	0.56 ± 0.12
Car	0.6 ± 0.128	0.708 ± 0.085	0.717 ± 0.076
CBF	0.892 ± 0.036	0.983 ± 0.012	0.999 ± 0.003
ChlorineConcentration	0.858 ± 0.018	0.84 ± 0.009	0.813 ± 0.019
CinC_ECG_torso	0.961 ± 0.015	0.978 ± 0.016	0.997 ± 0.006
Coffee·	0.975 ± 0.075	0.95 ± 0.1	0.975 ± 0.075
Cricket_X	0.39 ± 0.053	0.501 ± 0.035	0.582 ± 0.054
Cricket_Y	0.403 ± 0.028	0.457 ± 0.063	0.503 ± 0.048
Cricket_Z	0.394 ± 0.041	0.485 ± 0.058	0.567 ± 0.072
DiatomSizeReduction	0.96 ± 0.036	0.973 ± 0.036	1 ± 0
ECG200	0.784 ± 0.076	0.868 ± 0.072	0.905 ± 0.046
ECGFiveDays	0.964 ± 0.017	0.957 ± 0.016	0.991 ± 0.012
FaceAll	0.665 ± 0.034	0.794 ± 0.041	0.917 ± 0.022
FaceFour	0.667 ± 0.131	0.767 ± 0.105	0.9 ± 0.105
FacesUCR	0.677 ± 0.045	0.831 ± 0.015	0.931 ± 0.012
FISH	0.669 ± 0.09	0.729 ± 0.05	0.803 ± 0.041
Gun_Point	0.86 ± 0.073	0.89 ± 0.066	0.895 ± 0.052
Haptics	0.416 ± 0.065	0.451 ± 0.072	0.429 ± 0.076
InlineSkate	0.295 ± 0.056	0.366 ± 0.059	0.413 ± 0.07
ItalyPowerDemand	0.902 ± 0.024	0.974 ± 0.017	0.972 ± 0.01
Lighting2	0.718 ± 0.143	0.745 ± 0.151	0.709 ± 0.134
Lighting7	0.55 ± 0.081	0.58 ± 0.108	0.7 ± 0.148
MALLAT	0.843 ± 0.075	0.955 ± 0.012	0.98 ± 0.01
MedicalImages	0.615 ± 0.044	0.704 ± 0.04	0.753 ± 0.034

(*continued*)

Table 1. (*continued*)

Datasets	Naïve Rank	INSIGHT	Proposed Method
MoteStrain	0.919 ± 0.025	0.902 ± 0.03	0.912 ± 0.028
NonInvasiveFatalECG_Thorax1	0.544 ± 0.046	0.789 ± 0.008	0.86 ± 0.01
NonInvasiveFatalECG_Thorax2	0.616 ± 0.037	0.848 ± 0.01	0.896 ± 0.011
OliveOil	0.55 ± 0.1	0.9 ± 0.122	0.9 ± 0.122
OSULeaf	0.507 ± 0.066	0.507 ± 0.068	0.585 ± 0.039
Plane	0.752 ± 0.092	0.938 ± 0.043	0.962 ± 0.029
SonyAIBORobotSurfaceII	0.908 ± 0.03	0.919 ± 0.031	0.978 ± 0.017
SonyAIBORobotSurface	0.884 ± 0.042	0.956 ± 0.023	0.99 ± 0.011
SwedishLeaf	0.54 ± 0.031	0.697 ± 0.027	0.805 ± 0.036
Symbols	0.813 ± 0.071	0.957 ± 0.016	0.961 ± 0.01
synthetic_control	0.88 ± 0.035	0.83 ± 0.018	0.888 ± 0.025
Trace	0.485 ± 0.098	0.64 ± 0.092	0.675 ± 0.098
TwoLeadECG	0.959 ± 0.016	0.973 ± 0.013	0.99 ± 0.01
Two_Patterns	0.751 ± 0.028	0.863 ± 0.016	0.974 ± 0.006
uWaveGestureLibrary_X	0.665 ± 0.027	0.735 ± 0.011	0.753 ± 0.02
uWaveGestureLibrary_Y	0.668 ± 0.018	0.704 ± 0.019	0.706 ± 0.017
uWaveGestureLibrary_Z	0.644 ± 0.03	0.67 ± 0.014	0.686 ± 0.015
wafer	0.955 ± 0.01	0.991 ± 0.005	0.998 ± 0.002
WordsSynonyms	0.49 ± 0.056	0.635 ± 0.045	0.68 ± 0.046
yoga	0.829 ± 0.026	0.881 ± 0.021	0.92 ± 0.017

4.2   Experiments on Choosing Appropriate Percentage

Figure 7(a) and (b) illustrate $f1\_score$ and $f2\_score$ respectively on Haptics dataset. As we can see, $f1\_score$ selects 70 % as the percentage that we should reduce the training set to and $f2\_score$ selects at 45 %. Comparing these points to the figures on Fig. 7(c), which shows the error rates when we use the reduced training sets to classify instances on testing set, these reductions are acceptable because the error rates at these points do not increase too much. In detail, the training set with 70 % chosen by $f1\_score$ has the

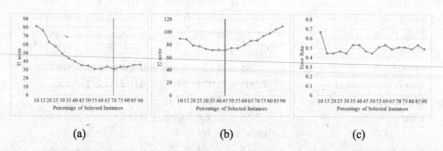

Fig. 7. Percentage of instances selected in Haptics, (a) $f1\_score$ selects at 35 %, (b) $f2\_score$ selects at 15 %, (c) error rates at each percentage of instances kept as training set

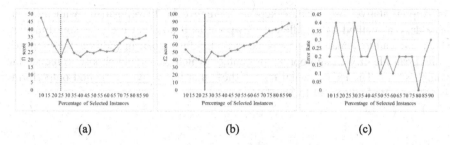

(a)   (b)   (c)

Fig. 8. Percentage of instances selected in Lightning 7 dataset, (a) *f1_score* selects at 25 %, (b) *f2_score* selects at 25 %, (c) error rates at each percentage of instances kept as training set

error rate of 0.51. The training set with 45 % chosen by *f2_score* has the error rate of 0.46. Meanwhile, the training set of 90 % has the error rate of 0.48.

Figure 8(a) and (b) illustrate *f1_score* and *f2_score* respectively on Lightning 7 dataset. Both *f1_score* and *f2_score* select at 25 % as the percentage that we should reduce the training set to. Compare these points to the corresponding error rates on Fig. 8(c), which use the reduced training sets to classify instances on testing set, these reductions are acceptable because the error rates at these points even decreased. In detail, the training set with 25 % instances has the error rate of 0.1. Meanwhile, the training set with 90 % instances has the error rate of 0.3.

Figure 9(a) and (b) illustrate *f1_score* and *f2_score* respectively on Gun Point dataset. *f1_score* selects at 25 % as the percentage that we should reduce the training set to and *f2_score* selects at 10 %. Comparing these points to the error rate on Fig. 9 (c) when we use the reduced training sets to classify instances on testing set, these reductions are reasonable because the error rate is 0.0 at all the percentage of reduction.

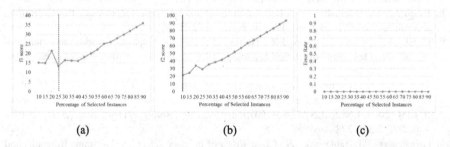

(a)   (b)   (c)

Fig. 9. Percentage of instances selected in Gun Point dataset, (a) *f1_score* selects at 25 %, (b) *f2_score* selects at 10 %, (c) error rates at each percentage of instances kept as training set

5   Conclusions

Reducing training set size is vitally important in improving speed and space using of 1-NN, an efficient method for time series classification. In this work, our method reduces training set size by using K-means to determine its representative characteristics. Then, only representative instances are kept as training data. In comparison with previous methods, our method has a different viewpoint as it does not use a ranking function, which can lead to some weaknesses mentioned in this paper. In addition, the proposed method can identify an appropriate percentage which we should choose to reduce the training set to. The experimental results show that the proposed method outperforms Naïve Rank Reduction and INSIGHT.

References

1. Xi, X., Keogh, E., Shelton, C., Wei, L., Ratanamahatana, C.A.: Fast time series classification using numerosity reduction. In: Proceedings of the 23rd International Conference on Machine learning, ICML 2006, pp. 1033–1040 (2006)
2. Buza, K., Nanopoulos, A., Schmidt-Thieme, L.: INSIGHT: efficient and effective instance selection for time-series classification. In: Huang, J.Z., Cao, L., Srivastava, J. (eds.) PAKDD 2011. LNCS (LNAI), vol. 6635, pp. 149–160. Springer, Heidelberg (2011). doi:10.1007/978-3-642-20847-8_13
3. Ding, H., Trajcevski, G., Scheuermann, P., Wang, X., Keogh, E.: Querying and mining of time series data: experimental comparison of representations and distance measures. Proc. VLDB Endowment 1(2), 1542–1552 (2008)
4. Keogh, E., Ratanamahatana, C.A.: Exact indexing of dynamic time warping. Knowl. Inf. Syst. 7(3), 358–386 (2005)
5. Berndt, D., Clifford, J.: Using dynamic time warping to find patterns in time series. In: Proceedings of AAAI Workshop on Knowledge Discovery in Databases, KDD 1994, Seattle, Washington, USA, pp. 359–370 (1994)
6. Wilson, D.R., Martinez, T.R.: Instance pruning techniques. In: Proceedings of ICML 1997, Morgan Kaufmann, pp. 403–411 (1997)
7. Chen, Y., Keogh, E., Hu, B., Begum, N., Bagnall, A., Mueen, A. and Batista, G.: The UCR Time Series Classification Archive (2015). www.cs.ucr.edu/~eamonn/time_series_data/
8. Brighton, H., Mellish, C.: Advances in instance selection for instance-based learning algorithms. Data Min. Knowl. Discov. 6, 153–172 (2002)
9. Garcia, S., Derrac, J., Cano, J.R., Herrera, F.: Prototype selection for nearest neighbor classification taxonomy and empirical study. IEEE Trans. Pattern Anal. Mach. Intell. 34(3), 417–435 (2012)
10. Jankowski, N., Grochowski, M.: Comparison of instances seletion algorithms I. algorithms survey. In: Rutkowski, L., Siekmann, Jörg, H., Tadeusiewicz, R., Zadeh, Lotfi, A. (eds.) ICAISC 2004. LNCS (LNAI), vol. 3070, pp. 598–603. Springer, Heidelberg (2004). doi:10.1007/978-3-540-24844-6_90
11. Grochowski, M., Jankowski, N.: Comparison of instance selection algorithms II. results and comments. In: Rutkowski, L., Siekmann, J.H., Tadeusiewicz, R., Zadeh, L.A. (eds.) ICAISC 2004. LNCS (LNAI), vol. 3070, pp. 580–585. Springer, Heidelberg (2004). doi:10.1007/978-3-540-24844-6_87
12. Triguero, I., Derrac, J., Garcia, S., Herrera, F.: A Taxonomy and experimental study on prototype generation for nearest neighbor classification. IEEE Trans. Syst. Man Cybern. Part C Appl. Rev. 41(1), 86–100 (2012)
13. Wilson, D.L.: Asymptotic properties of nearest neighbor rules using edited data. IEEE Trans. Syst. Man Cybern. 2(3), 408–421 (1972)

14. Wilson, D.R., Martinez, T.R.: Instance pruning techniques. In: Proceedings of ICML1997, Morgan Kaufmann, pp. 403–411 (1997)
15. Wilson, D.R., Martinez, T.R.: Reduction techniques for instance-based learning algorithms. Mach. Learn. **38**, 257–286 (2000)
16. Ciaccia, P., Patella, M., Zezula, P.: M-tree: an efficient access method for similarity search in metric spaces. In: VLDB 1997 Proceedings of the 23rd International Conference on Very Large Data Bases, pp. 426–435 (1997)
17. Guttman, A.: R-trees: a dynamic index structure for spatial searching. In: SIGMOD 1984 Proceedings of ACM SIGMOD International Conference on Management of Data, pp. 47–57 (1984)
18. Assent, I., Krieger, R., Afschari, F., Seidl, T.: The TS-tree: efficient time series search and retrieval. In: Proceedings of the 11th International Conference on Extending database technology: Advances in database technology (EDBT 2008), pp. 252–263 (2008)
19. Keogh, E., Chakrabarti, K., Pazzani, P., Mehrotra, S.: Dimensionality reduction for fast similarity search in large time series databases. Knowl. Inf. Syst. **3**, 263–286 (2001)
20. Keogh, E., Pazzani, M.: An enhanced representation of time series which allows fast and accurate classification, clustering and relevance feedback. In: Proceedings of the 4th International Conference on Knowledge Discovery and Data Mining, pp. 239–241 (1998)
21. Rakthanmanon, T., Campana, B., Mueen, A., Batista, G., Westover, B., Zhu, Q., Zakaria, J., Keogh, E.: Searching and mining trillions of time series subsequences under dynamic time warping. In: Proceedings of the 18th ACM SIGKDD International Conference on Knowledge Discovery and Data Mining (KDD 2012), pp. 262–270 (2012)
22. Keogh, E., Ratanamahatana, C.A.: Exact indexing of dynamic time warping. Knowl. Inf. Syst. **7**, 358–386 (2005)

A New Fault Classification Scheme Using Vibration Signal Signatures and the Mahalanobis Distance

Jaeyoung Kim, Hung Nguyen Ngoc, and Jongmyon Kim[✉]

School of Electrical Engineering,
University of Ulsan, Ulsan 680-749, South Korea
kjy7097@gmail.com, ngochung212@gmail.com,
jongmyon.kim@gmail.com

Abstract. Condition monitoring is a vital task in the maintenance of factory automation. Many feature extraction and selection algorithms have been studied to derive distinctive feature vectors. The common extraction methods for all fault classes can be ineffective in separating a new class out of existing classes or dealing with input signals under severely noisy environments. Thus, extraction and selection algorithms might need to be redesigned. Therefore, we propose a new approach to accurately identify fault classes from vibration signals even under severely noise conditions; our approach can also easily add a new group to the classification system. The proposed algorithm, a viable alternative to detect induction motor defects online, uses the differences in the fault-related harmonics of vibration signals to generate good feature vectors. This approach discriminates the harmonics for one specific fault to generate features and then classifies faults using a modified minimum distance classifier, which improves classification accuracy. In our experiments, the proposed technique shows a clear advantage over existing methods in classification performance in both noiseless and additive white Gaussian noise circumstances and demonstrates the capability to learn new signatures from unknown motor conditions.

Keywords: Fault classification · Spectral envelope · Mahalanobis distance · Fault signature · Vibration harmonics

1 Introduction

Induction motors are complex electromechanical devices used in most industries in applications such as industrial fans, blowers and pumps, and conveyors and compressors. However, they are susceptible to many types of faults, and malfunctioning motors can cause large economic losses. Although the costs to repair or refurbish the motors themselves might not be substantial, the costs associated with downtime can be enormous [1–7]. Thus, condition monitoring systems for electrical machines play important roles in the safe operation of machinery and expand the motors' lifetimes by providing adequate warning of imminent failures, diagnosing present maintenance needs, and allowing workers to schedule future preventive maintenance and repair work, thereby avoiding heavy production losses and maximizing uptime and optimum

© Springer International Publishing AG 2016
V.-N. Huynh et al. (Eds.): IUKM 2016, LNAI 9978, pp. 230–241, 2016.
DOI: 10.1007/978-3-319-49046-5_20

maintenance schedules. Early fault diagnosis techniques allow machine operators to have the necessary spare parts on hand before the machine is stripped down, thereby reducing outage times. Furthermore, fault prognosis systems can be integrated into maintenance policies, allowing usual maintenance at specific intervals to be replaced by condition-based maintenance [2, 3, 6–8].

Many papers in the literature address the problem of identifying faults in induction motors by monitoring the motor status. Researchers have mainly used stator current, voltage, and vibration signals as the inputs for fault detection systems. Among them, vibration signals are preferable because they link directly to most machines' operational and maintenance stages. Vibrations are consequences of defects in a machine's structure; therefore, identifiable fault signatures can be found in vibration signals. Axial and radial acceleration sensors are used to detect mechanical problems, and tangential acceleration sensors are used to detect electrical problems. Thus, vibration monitoring is the most effective and reliable method for identifying faults in induction motors [9–11].

In general, condition-monitoring schemes have been widely used to identify specific failure modes in one of the three main induction motor components: stator, rotor, and bearings. Several conventional vibration and current analysis techniques exist by which certain faults in rotating machinery can be identified for repair. Many researchers have applied advanced signal processing techniques over the measured physical magnitude to obtain good feature vectors for use as reliable fault indicators.

The existing techniques can be classified into three domains: time-domain analysis (mean, peak, peak-to-peak interval, standard deviation, crest factor, high-order statistics: RMS (root mean square), skewness, kurtosis, etc.) [11–17]; frequency-domain analysis (e.g., FFT, envelope analysis, spectral analysis) [9, 11]; and time-frequency domain analysis (e.g., wavelet transform) [10]. Based on the three domains, many features can be generated from vibration data. The original feature set from each of the three domains generally has high dimensionality, which under low variability from healthy and unhealthy states can complicate defect detection and degradation propagation prediction. Therefore, it is important to devise a systematic approach that can extract the most useful information about machine health states. Several methods have been introduced to reduce or select optimal features: linear and nonlinear principal component analysis [18], independent component analysis [11], and singular value decompositions [19], for example.

To the best of our knowledge, most methods follow the same rules. So-called "universal" feature generation and reduction methods are applied to all classes to derive distinctive condition signatures. That is a time consuming and difficult way to find a proper extraction method because the signature of a defective motor occurs within a wide frequency band and can be masked by noise. Instead of looking for highly significant features, this study aims to design and develop an effective, efficient, and simple method to generate different features to enhance the performance of a diagnosis system that will be able to learn new class signatures.

We select vibration signals, which are inexpensive and easy to measure, as the fault diagnosis system input for detection and classification. We then use spectral envelopes (spectral masks) and fast Fourier transforms to specify the harmonic components of a vibration signal. These signals model a particular motor failure signature such that the system can detect and classify comprehensive fault conditions in induction motors, as indicated in [20]. With subsequent vibration readings taken under identical conditions, we can determine whether any deterioration that affects the vibration signatures has taken place in the machine's condition. In the classification task, we use the Mahalanobis distances between a test sample and known faulty signatures to measure the similarity between the sample and the signatures, so that the minimum distance indicates the name of the fault.

The remainder of the paper is organized as follows. Section 2 surveys the fault-related frequencies and harmonics of vibration signals from faulty induction motors. Section 3 presents the proposed vibration signature production method and diagnosis model. Section 4 presents a discussion of our experimental results, including comparison with several published algorithms. Section 5 concludes this paper.

2   Fault-Related Harmonics in Vibration Signals

We consider seven different motor conditions in this paper, six mechanical faults: rotor unbalance (RU), broken rotor bar (BR), bowed rotor shaft (BS), faulty bearing (FB), angular misalignment (AM), and parallel misalignment (PM); and one healthy condition (NO).

2.1   Rotor Unbalance (RU)

The motor speed can be identified by a peak in the spectral range, which is determined by the number of poles in the rotor. If a motor is out of balance or misaligned, the fault signature normally takes the form of increased amplitude of the rotating frequency and its harmonics. Vibration analysis can provide a quick and relatively easy way to extract information about a rotor unbalance fault in an induction motor.

2.2   Faulty Bearings (FB)

The characteristic fault frequencies for FB depend on which bearing surface contains the fault, and they are predictable. In addition, bearing vibrations often occur at very high frequencies. From the geometry of the bearing, various theoretical frequencies can be calculated, such as the inner and outer race element pass frequencies, cage rotational frequency, and rolling-element spin frequency. A defect on the outer race will cause an impulse each time the rolling elements contact the defect. The theoretical ball pass outer raceway frequency (f_{BPOF}) is easily determined as

$$f_{BPOF} = \frac{n}{2}\frac{\omega_r}{60}\left[1 - \frac{d}{D}\cos\phi\right], \tag{1}$$

where ϕ is the ball contact angle with the races; D is the ball pitch diameter, measured from one ball center to the opposite ball center; d is the ball diameter; n is the number of balls in the bearing; and ω_r is rotational speed in revolutions per minute (*rpm*). This vibration frequency (f_{BPOF}) reflects itself in the current spectrum as follows:

$$f_{BRN} = (f_S \pm mf_{BPOF}), \tag{2}$$

where $m = 1, 2, 3, \ldots,$ and f_s is the electrical supply frequency.

2.3 Broken Rotor Bar (BR)

The space harmonics (f_{BB}) components indicate a BR defect in an induction motor:

$$f_{BB} = \left[k(\frac{1-s}{p}) \pm s \right] f_s, \tag{3}$$

where s is the per-unit motor slip, p is the number of poles in the motor, and $k/p = 1, 3, 5, \ldots$ denotes the characteristic values of the motor. The sideband components around the supply current frequency f_s are as follows:

$$f_{BRB} = (1 \pm 2ks)f_s, \tag{4}$$

where f_{BRB} denotes the sideband frequencies associated with the BR. The slip s is defined as the relative mechanical speed of the motor n_m with respect to the motor synchronous speed n_s as follows:

$$s = \frac{n_s - n_m}{n_s}, \tag{5}$$

The motor synchronous speed n_s is related to the electrical supply frequency f_s as follows:

$$n_s = \frac{120f_s}{P}, \tag{6}$$

where P is the number of poles in the motor, and the constant 120 expresses the motor's synchronous speed n_s in *rpm* units.

2.4 Bowed Rotor Shaft (BS)

When a motor runs with a bowed shaft, the vibration spectrum appears quite similar to that in an imbalance condition. A bowed shaft can be detected by measuring the separated shaft with a dial indicator. A bowed rotor shaft is an out-of-balance condition and shows up as a 1X-*rpm* and 2X-*rpm* vibration.

2.5   Rotor Misalignment (Parallel (PM) and Angular (AM))

Rotor misalignment results in a non-uniform air gap. Misalignment-related faults commonly occur as a result of FB. Higher orders of misalignment can cause rotor-to-stator rub, resulting in damage to the rotor or stator windings or cores.

3   Proposed Method

Our proposed fault classification system generates different fault-related features to enhance the accuracy of its classification decisions. The system is depicted in Fig. 1. In Sect. 2, we discussed the fault-related theoretical harmonics. The proposed approach exploits those different harmonics for online identification of induction motor states. First, we generate the signatures of the condition vibration signals in a training stage, and then we use those signatures as motor condition indicators in online detection and classification of induction motor defects.

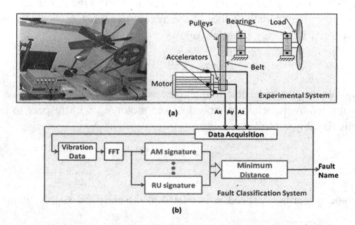

Fig. 1. Proposed online fault classification system

3.1   Production of Fault Signatures in Vibration Signals

Signature production contains two phases, as depicted in Fig. 2. The first phase specifies the set of relevant frequency bands and weights, where the frequency bands are the harmonic-related frequencies in each kind of vibration signal, and the weights are the values of the spectral envelope at each frequency band. In this paper, we use the all-pole model to define the spectral envelope because it directly describes a power spectrum and hence can be easily combined with our harmonics structure model, which is defined on a linear frequency axis.

Figure 2(a) shows the three concise steps we use to generate the frequency bands and weights phase.

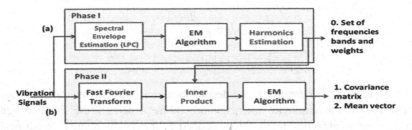

Fig. 2. Fault signatures in the vibration signal generation process

Phase I:

– *Step 1:* Use one-second faulty vibration signals to calculate linear predictive coding (LPC) coefficients $\{a_k\}$, where $k=1,2,...,$ and P is the LPC order. The spectral envelope of the vibration signal is then defined by the frequency response of the all-pole filter, $H(z)$:

$$\widehat{x}[n] = \sum_{k=1}^{P} a_k x[n-k], \tag{7}$$

$$H(z) = \frac{1}{A(z)} = \frac{1}{1 - \sum_{k=1}^{P} a_k z^{-k}}, \tag{8}$$

– *Step 2:* Apply the expectation and maximization algorithm to estimate the mean spectral envelope from all the training samples for a specific motor condition.
  • Expectation step: Calculate the expected value of the log likelihood function with respect to the conditional distribution of Z given X under the current estimate of parameter $\theta(t)$

$$Q(\theta|\theta(t)) = E_{Z||X,\theta(t)}[\log L(\theta; X, Z)], \tag{9}$$

where $\theta(t)$ is the current parameter estimation, and θ is the unknown actual parameter X.
  • Maximization step: Find the parameters that maximize this quantity

$$\theta(t+1) = \arg\max_{\theta} Q(\theta|\theta(t)), \tag{10}$$

– *Step 3:* Determine the frequency bands and weights. The peaks, or local maxima of the mean envelope of a specific motor condition, first defined in step 2 are found by first-order optimization techniques (gradient ascent method) on the frequency axis in steps proportional to the positive of the envelope gradient at the current point.

$$f_{n+1} = f_n + \gamma_n \frac{\partial F(f)}{\partial f}, \, n \geq 0, \tag{11}$$

$$F(f_0) \leq F(f_1) \leq F(f_2) \leq \cdots, \tag{12}$$

The sequence $\{f_n\}$ converges to the desired local maxima. The interested frequency bands contain the harmonics and neighborhood frequencies of the vibration signals of the specific induction motor faults, defined by the valley frequencies adjacent to the peak frequencies. Weights at the frequencies are values of the spectral envelope if the frequencies are in the band and are zeros if the frequencies are outside the band.

Therefore:
Phase II:
The second phase, shown in Fig. 2(b), uses the set of frequency bands and weights calculated in phase I to calculate the covariance matrix and mean vector of the fault signature. This phase contains three steps.

- *Step 1:* Determine the fast Fourier transform (FFT) of the vibration signal to obtain the power spectrum.

$$Y(m) = \frac{1}{N} \sum_{n=0}^{N-1} x(n)\omega(n)e^{-j\frac{2mn}{F}n}, \tag{13}$$

where N is the number of points used to calculate the discrete Fourier transform (DFT), $0 \leq n \leq N-1$, and $\omega(n)$ is the Hamming window function given by:

$$\omega(n) = \beta(0.5 - 0.5\cos\frac{2\pi n}{N-1}), \tag{14}$$

- *Step 2:* Find the inner product of the power spectrum for each frequency band and weight determined in phase I.
- *Step 3:* Apply the expectation and maximization algorithm to find the covariance matrix and mean vector of all training samples.

This procedure applies to each fault to generate its own signature. Thus, for our seven motor conditions, we need to run the process seven times to create seven signatures. Table 1 shows the mean value components of the fault signatures in the vibration signal. The vector dimensions are unequal because the number of harmonics in the vibration signals varies from fault to fault, as explained in Sect. 2.

3.2   Online Detection and Classification System

Figure 1(b) shows the schema of our online classification system based on the fault signatures in the vibration signal. The acquisition system receives the vibration signals from sensors attached to the induction motor. The vibration signal in the time domain is

Table 1. Fault signatures in vibration signals (mean vector)

Signatures						
AM	BR	BS	FB	NO	PM	RU
45242.74	280.05	1085.41	956.07	8890.90	70.48	148998.18
23853.37	41899.99	126.99	375.00	43450.87	3431.19	1229.45
1831.38	1326.22	4869.66	2069.77	1582.01	7481.50	2719.97
206.37	2651.83	120.97	749.67	3923.58	556.29	100.82
784.58	104.29	236.63	–	454.47	325.39	431.03
–	1051.46	–	–	229.27	46.95	316.14
–	–	–	–	191.23	282.15	–
–	–	–	–	–	45.76	–

transformed to the frequency domain by fast Fourier transform. The power spectrum of the vibration signal then simultaneously feeds to the 7 fault models, which each have 3 components: set of frequency bands and weights, covariance matrix, and mean vector. The set of harmonics and its neighbor frequencies vary from signature to signature in number of bands and weight values within each band. For each model, the power spectrum is divided into a vector by an inner product operation between the spectrum and weight values in each band. The dimension of the output vector is the number of bands, which are defined in the signature production progress. We use the Mahalanobis distance to measure the similarity between the generated vector, represented by its mean value with the covariance matrix, and the fault signatures. In total, we calculate 7 Mahalanobis distances to classify the vibration signal to the corresponding fault category using the reference signature that gives the minimum distance to the vibration signal. Each signature model works as a binary classifier that can detect its own conditions against other motor conditions.

Mahalanobis distance: Given unknown x, it is assigned to class ω_i if

$$d_i = \sqrt{(x - m_i)^T S_i^{-1}(x - m_i)}, \qquad (15)$$

where S_i is the covariance matrix, and m_i is mean point of the ith group. Table 2 shows, for example, the Mahalanobis distance from RU vibration signals to all predefined signatures.

Table 2. Mahalanobis distances from test samples to the RU signature

	Signatures						
Samples	AM	BR	BS	FB	NO	PM	RU
16	17.44	34.94	197.91	274.85	14.16	135.65	**4.07**
19	17.36	35.39	198.01	274.71	14.68	135.38	**5.75**
20	15.95	36.52	195.32	276.58	15.89	138.04	**5.15**
27	16.25	36.48	198.31	275.11	14.82	135.27	**4.96**
38	17.14	35.77	197.13	275.71	14.10	136.25	**4.45**

(*continued*)

Table 2. (*continued*)

Samples	AM	BR	BS	FB	NO	PM	RU
	Signatures						
39	16.04	36.47	195.41	275.25	15.40	135.27	**5.48**
40	16.68	36.52	195.66	274.94	15.62	134.40	**4.76**
44	17.31	34.84	196.34	274.09	14.16	136.21	**3.99**
54	17.39	36.20	196.37	274.13	14.39	137.49	**4.43**
56	18.17	35.23	198.66	273.30	14.57	137.56	**4.91**
61	17.48	35.61	197.31	275.66	14.20	137.83	**4.32**
68	16.41	37.03	195.07	275.61	15.65	135.17	**4.52**
74	17.20	36.24	196.70	273.88	14.33	138.26	**4.67**
75	16.24	34.53	195.95	275.49	14.52	133.13	**4.35**
88	16.30	37.49	196.43	274.27	15.56	138.38	**4.61**
91	17.07	36.05	195.86	274.09	14.48	134.01	**5.03**
93	17.12	36.28	195.93	274.17	15.33	137.54	**5.00**
94	17.86	35.71	197.59	273.44	13.82	135.71	**4.65**
99	17.44	35.32	195.18	274.62	14.23	134.69	**4.55**
102	18.04	35.93	198.32	275.08	14.55	136.23	**4.78**

4 Experimental Results

4.1 Experiment and Simulation Setup

To evaluate the performance of our proposed method, we set up an experimental test-bed to acquire several different faulty vibration signals, as shown in Fig. 1(a). This setup consists of motors, pulleys, a belt, a shaft, and a fan with changeable blade pitch angle. We use 6 0.5 kW, 60 Hz, 4-pole induction motors to generate data under full load conditions. We collect one normal (NO) and 6 fault signals: AM, PM, BR, BS, FB, and RU. Table 3 describes the different faulty conditions when the acquired vibration signals are sampled at 8 kHz. We use 105 one-second-long vibration signals for each faulty condition. More detailed information about this experiment is available in [11].

Table 3. Details of fault conditions

Fault condition	Fault description
angular and parallel misalignment	rearing diameter: 40 mm housing maximum diameter: 40.7 mm
broken rotor bar	12 of 34 rotor bars were broken by holes; depth: 15 mm, diameter: 5 mm
bowed rotor shaft	shaft deflection in mid-span: 0.075 mm, air gap: 0.025 mm
faulty bearing	spalling on the outer race: #6203
rotor imbalance	unbalanced mass on the rotor: 8.4 g, distance 40.2 mm, position: 0°, 36°, 72°
phase imbalance	adding resistance to one phase

In our experiment, we implement a minimum distance classifier for faulty vibration signals from induction machines. We carry out 10 trials to obtain the average result. For each trial, we use 34 randomly selected signals for each fault to generate the vibration signatures and total data for validation. Therefore, the test data contain 735 vibration signals for seven motor condition signal categories (AM, BR, BS, FB, NO, PM, and RU), 105 samples per fault, to validate the approach. A vibration signal in the test data is used to simultaneously generate 7 vectors following steps 1 and 2 from phase II of signature production. Then we calculate 7 Mahalanobis distances to measure the similarity of the vibration signal to all fault categories. The fault is classified into the category with the minimum Mahalanobis distance.

We consider the acquired vibration signals to be noiseless. To evaluate the efficiency and robustness of our work, we use simulated data with additive white Gaussian noise at SNR = 15 or 20 dB to the vibration signals.

4.2 Experimental Results

Overall, the results presented below show that our proposed algorithm outperforms the approaches in [12, 15] correctly classifying all induction motor conditions in both noiseless and noisy circumstances. Figure 3(a) illustrates the accuracy of classification when we select data for training or generating signatures from our whole original dataset. Figures 3(b) and (c) show the accuracy of the system when, after selecting training data from the original samples, we contaminate all samples with additive white Gaussian noise to a SNR of 15 or 20 dB, respectively.

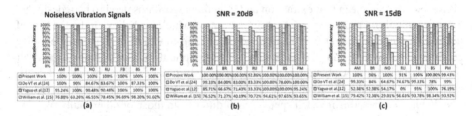

Fig. 3. Performance under different conditions

The accuracy values from our proposed approach are the highest for all kinds of conditions at 100 %, compared with the other 3 algorithms, which achieves 100 % correct classification only for certain faults: AM, FB, PM for the algorithm and BR, FB, BS, and PM in [12] (Fig. 3(a)). The general classification accuracy of our algorithm and the approaches in [12, 15] are 100 %, 92 %, 97 %, and 79 %, respectively. Our approach outperforms the algorithms in [12, 15] detecting NO and RU conditions with 100 % and 91 % precise condition indications, whereas the published algorithms have less than 60 % and 75 % accuracy.

When we test the systems with noisy vibration signals, the accuracy of our approach decreases slightly to 98.18 %, whereas the algorithms in [12, 15] show dramatic declines to 85.87 %, 61.45 %, and 74.78 %, respectively.

The strength of our algorithm over the others can be explained by two main reasons. First, our approach can precisely detect the characteristic harmonics of a vibration signal, which helps us detect the motor condition. Second, we extract different features for each condition, represented by the frequency bands and weights, which is far better than the existing extraction methods that apply to all kinds of motor fault categories.

In addition to classification accuracy, the ability to detect an unknown state is an important requirement in an induction motor state monitoring system. The algorithm in [12] is originally designed to detect only bearing faults. With our dataset, we introduce other mechanical faults, such as rotor unbalance and angular misalignment. The approach in [12] seems unable to recognize the RU fault, having the very low accuracy value of 33.33 % and 0 % when the SNR (Signal-to-Noise Ratio) is 20 and 15 dB, respectively, even though this kind of fault is used to train the diagnosis system. Our approach automatically detects the distinctive fault-related frequencies and then focuses on those frequency bands to generate more discriminant vectors for the introduced classes.

5 Conclusion

Fault detection and classification are essential processes for manufactory automation, used mainly for condition-based maintenance. Generally, researchers have tried to find "universal" feature generation and feature reduction methods to derive distinctive fault features. The features then apply to a classification system that recognizes the fault name. This approach has problems. For example, if a new class is taken into consideration, the extracted features can be ineffective at separating it out from existing classes, which can mean that extraction and selection algorithms need to be redesigned. To overcome those problems, we have found a way, which we believe to be the first of its kind, to accurately identify induction motor faults from vibration signals under noisy conditions; our system can also cope with a new group in the classification system. Our proposed algorithm, a viable alternative to online detection of induction motor defects, takes into account the differences in the fault-related frequencies in vibration signals. The features extracted from each specific fault differ from those of other faults. We can then use a modified minimum distance classifier to improve classification performance. Our experiments revealed that our technique shows a clear advantage over existing methods in classification accuracy in both noiseless and additive white Gaussian noise circumstances.

Acknowledgements. This work was supported by the Korea Institute of Energy Technology Evaluation and Planning (KETEP) and the Ministry of Trade, Industry, & Energy (MOTIE) of the Republic of Korea (No. 20162220100050); in part by The Leading Human Resource Training Program of Regional Neo Industry through the National Research Foundation of Korea (NRF), Ministry of Science, ICT, and Future Planning (NRF-2016H1D5A1910564); in part by the Business for Cooperative R&D between Industry, Academy, and Research Institute funded by the Korea Small and Medium Business Administration in 2016 (Grants No. C0395147); and in part by the "Leaders INdustry-university Cooperation" Project supported by the Ministry of Education (MOE).

References

1. Isermann, R.: Supervision, fault detection and diagnosis methods – an introduction. Control Eng. Pract. CEP, **5**(5), 639–652
2. Isermann, R.: Fault diagnosis system: An introduction from fault detection to fault tolerance. In: Plastics, 2nd ed., vol. 3, J. Peters, Ed. New York: McGraw-Hill, 1964, pp. 15–64
3. Isermann, R.: Fault diagnosis system: An introduction from fault detection to fault tolerance, pp. 13–43. Springer, Berlin (2006)
4. Kimmich, F., Schwarte, A., Isermann, R.: Fault detection for modern diesel engines using signal-and process model based methods. Cont. Eng. Prac. **13**, 189–203 (2005)
5. Kimmich, F., Schwarte, A., Isermann, R.: Model based fault detection for diesel engines. Aachen Colloquium, Aachen (2011)
6. Isermann, R.: Process fault detection on modeling and estimation methods – a survey. Automatica **20**, 387–404 (1984)
7. Jardin, A.K.S., Lin, D., Banjevic, D.: A review on machinery diagnostics and prognostics, implementing condition-based maintenance. Mech. Syst. Signal Process. **20**(7), 1483–1510 (2006)
8. Benbouzid, M.E.H.: A review of induction motors signature analysis as a medium for faults detection. IEEE Trans. Ind. Electron. **47**(5), 984–993 (2000)
9. Inerny, S.A., Dai, Y.: Basic vibration signal processing for bearing fault detection. IEEE Trans. Educ. **46**(1), 149–156 (2003)
10. Li, F., Meng, G., Ye, L., Chen, P.: Wavelet transform-based higher-order statistics for fault diagnosis in rolling element bearings. J. Vib. Control **14**(11), 1691–1709 (2008)
11. Widodo, A., Yang, B.S., Han, T.: Combination of independent component analysis and support vector machines for intelligent faults diagnosis of induction motors. Expert Syst. Appl. **32**, 299–312 (2007)
12. Lei, Y., He, Z., Zi, Y.: Application of an intelligent classification method to mechanical fault diagnosis. Expert Syst. Appl. **36**(6), 9941–9948 (2009)
13. Samanta, B., Al-Balushi, K.R., Al-Araimi, S.A.: Artificial neural networks and support vector machines with genetic algorithm for bearing fault detection. Eng. Appl. Artif. Intell. **16**(7), 665–687 (2003)
14. Delgado, M., Cirrincione, G., Espinosa, A.G., Ortega, J.A., Henao, H.: Bearing faults detection by a novel condition monitoring scheme based on statistical-time features and neural networks. IEEE Trans. Ind. Electron. **99**, Early Access
15. William, P.E., Hoffman, M.W.: Identification of bearing faults using time domain zero-crossings. Mech. Syst. Signal Process. **25**(8), 3078–3088 (2011)
16. Zhou, J.H., Yang, X.: Reinforced morlet wavelet transform for bearing fault diagnosis. In: Proceedings IECON, Glendale, AZ, pp. 1179 – 1184 (2010)
17. Chow, T.W.S., Hai, S.: Induction machine fault diagnostic analysis with wavelet technique. IEEE Trans. Ind. Electron **51**(3), 558–565 (2004)
18. Wang, X., Kruger, U., Irwin, G.W., McCullough, G., McDowell, N.: Nonlinear PCA with the local approach for diesel engine fault detection and diagnosis. IEEE Trans. Cont. Syst. Tech. **16**(1), 122–129 (2008)
19. Baranyi, P., Yam, Y., Kóczy, A.R.V., Patton, R.J.: SVD-based reduction to MISO TS models. IEEE Trans. Educ. **50**(1), 232–242 (2003)
20. Do, V.T., Chong, U.P.: Signal model-based fault detection and diagnosis for induction motors using features of vibration signal in two- dimension domain. Strojniški vestnik-J. Mech. Eng. **57**(9), 655–666 (2011)

Machine Learning for Social Media Analytics

Estimating Asymmetric Product Attribute Weights in Review Mining

Wei Ou[1(✉)], Anh-Cuong Le[2], and Van-Nam Huynh[1]

[1] Japan Advanced Institute of Science and Technology,
Asahidai 1-1, Nomi, Ishikawa, Japan
{ouwei,huynh}@jaist.ac.jp
[2] Faculty of Information Technology,
Ton Duc Thang University, Ho Chi Minh City, Vietnam

Abstract. In this paper we propose a probabilistic graph model to estimate the importance weights of product attributes from customer reviews. In this model, each product aspect has two weights: one weight indicates its importance level when customers' opinions about the product are generally positive; the other one indicates its importance level when the opinions are negative. Those weights provide on-line retailers with insight into the advantages and disadvantages of their products and allow them to devise effective methods to increase on-line sales.

Keywords: Aspect discovery · Sentiment classification · Aspect weight estimation · Asymmetric aspect weights

1 Introduction

Textual product reviews posted by previous shoppers have been serving as an important source of information that helps on-line retailers or shoppers to make their correct decisions. Customers usually read reviews to decide whether to place their orders; on-line retailers read those reviews to identify the aspects (or attributes, 'attribute' and 'aspect' will be used interchangeably throughout the following sections of this paper) their customers like or hate about. However, reading through all the reviews of a product is usually a time-demanding and frustrating task, especially when those reviews deliver conflicting information. Therefore, it is of great practical value to develop techniques to automatically generate brief but accurate summaries for the numerous reviews on a product.

There are existing a number of techniques aiming to generate product review summaries. Most of them focus on aspect discovery and sentiment analysis, that are to detect the aspects described in each review and classify the general sentiment orientations of them, respectively [3–5,7,8,11,12]. A few techniques that focus on estimating aspect weights of each product from product reviews [9,13,14]. Quantizing the importance level of each product aspect carries very good practical value, however, this area has drawn much less attention than aspect discovery and sentiment analysis from the research communities. We focus on this particular area in this paper.

© Springer International Publishing AG 2016
V.-N. Huynh et al. (Eds.): IUKM 2016, LNAI 9978, pp. 245–254, 2016.
DOI: 10.1007/978-3-319-49046-5_21

Matrix factorization is one of the most influential techniques in the weight estimation [9]. Matrix factorization assumes the rating of a review on a product is the dot product between the product's aspect weights vector and the customer's preferences. Aspect weights, and customer preferences can be estimated by minimizing the errors between actual ratings and predicted ratings. However, it is very difficult, if not impossible, to interpret those inferred weights since there is no natural mapping between the 'aspect' in the algorithm and the 'aspect' in our real world.

Exploiting both textual content and ratings of reviews, rather than using only the ratings, can add semantic information to those weights. One representative work is the rating regression model proposed by Wang et al. [13]. In this model, the authors assume each aspect has a latent rating and the total rating of a review is the weighted summation of those latent ratings. Therefore the weight estimation can be approached as a regression problem. Based on Wang's assumption, Yu et al. [14] propose an aspect ranking algorithm that is similar as Wang's work except they assume the aspect weights are further drawn from some certain Gaussian distributions. Julie et al. [9] first use matrix factorization to compute the aforementioned weights based on ratings then use topic models to map those weights into a semantic space in which the weights can be easily interpreted and utilized.

The weights obtained by the aforementioned methods can indicate which aspects are important and which are not. However, this information is gross grained and it doesn't take into account the fact that weights may vary by contexts of opinions: the contributions of an aspect to the ratings of positive reviews may be different from its contributions to the ratings of negative reviews. For example, a good giveaway may have a big contribution to the rating of a positive review where the customer is satisfied with the target product itself. However, it may have a trivial contribution to the rating of a negative review when the customer is totally dissatisfied with the target product.

In this paper we take into account the asymmetric characteristics of weights and propose a probabilistic model, Asymmetric Aspect Weight Model (AAWM), to estimate weights in different opinion contexts. In this model, each aspect has two weights that indicate its importance levels in low-rating reviews and high-rating reviews, respectively. This weights provide on-line retailers with insight into what their customers like most about their products when customers' opinions are in the positive side and what they hate most when the opinions are in the negative side. This information can be conducive for retailers to devise effective ways to improve sales. As far as we know, no previous work considering the asymmetric characteristics has been proposed and this is the main contribution of this work. In the remaining of this paper, we will first introduce the proposed model then present our experiment results. At the end of this paper, we will have a brief discussion about related work and our future work.

2   Asymmetric Aspect Weight Model

In this work, we first use Latent Dirichlet Allocation (LDA) [1] to discover aspects in each review sentences, then use the learnt results as the input for the proposed model to estimate aspect weights. We begin by introducing the LDA model before presenting our own model.

2.1   Latent Dirichlet Allocation

LDA was originally developed for topic discovery in text mining. It clusters words that frequently co-occur in the same documents into the same classes. Each of those classes represents a 'topic'. LDA has been applied in aspect discovery since the concept 'aspect' in review mining bears high resemblance to the 'topic' in text mining.

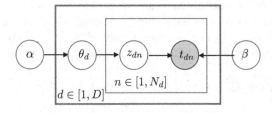

Fig. 1. The Bayesian network of LDA [1]

To use LDA for aspect discovery, a common practice is to treat each review sentence as a document. In this work we also treat each sentence as a document. Besides, we keep only nouns and verbs and remove other parts of speech (such as adjectives, adverbs, etc.) from those sentences for aspect discovery because the same adjectives, adverbs can be in conjunction with different aspect words in different sentences, they are usually very noisy in the learning process of LDA.

Let d denote a document, θ denote the document's topic mixture that is drawn from a Dirichlet distribution $Dir(\alpha)$ and represents its membership in each topic. Let β denote each topic's distribution over the vocabulary. According to LDA's Bayesian network shown in Fig. 1, we can get the following equation:

$$p(d|\alpha, \beta, t_d) = \int p(\theta_d|\alpha) \prod_{n=1}^{N_d} \sum_{z=1}^{K} p(z|\theta_d)p(t_{dn}|z, \beta)\, d\theta_d \tag{1}$$

where t_d represents the words in the document, N_d is the total number of words in the document, K is the number of topics.

Therefore the likelihood of the corpus can be expressed by the following equation:

$$p(D|\alpha, \beta, t) = \prod_d p(d|\alpha, \beta, t_d) \tag{2}$$

The topic mixture of each document θ and the word distribution of each topic β can be estimated by Gibbs sampling. Given θ, it is easy to determine which aspect is most probably described in a sentence d by the following equation:

$$z_d = \operatorname*{argmax}_{k} \theta_{dk} \tag{3}$$

2.2 The AAWM Model

We begin this subsection by presenting the Bayesian network of the proposed model in Fig. 2. In the network, r_d denotes the rating of sentence d, z_d denotes the aspect of the sentence, $s_d (s_d \in [e_+, e_-]$, where e_+ represents the positive sentiment and e_- represents the negative sentiment) denotes the sentiment polarity of the sentence. In this paper, the sentiment polarity serves as an intermediary that links the desirable weights with other observable variables. We assume the possibility for s_d to be positive or negative is determined by the aspect of the sentence z_d (for simplicity, we assume each sentence describes only one aspect), and the rating of the sentence r_d. Specifically, if $z_d = k$ (sentence d describes about aspect k), we define $\pi_{ds} = p(s|z_d, r_d)$ as follows:

$$\pi_{d+} = \begin{cases} \frac{\exp(w_{k+})}{\exp(w_{k+})+1} & r_d > 0.5 \\[2ex] 1 - \pi_{d-} & r_d \leq 0.5 \end{cases} \qquad \pi_{d-} = \begin{cases} 1 - \pi_+ & r_d > 0.5 \\[2ex] \frac{\exp(w_{k-})}{\exp(w_{k-})+1} & r_d \leq 0.5 \end{cases} \tag{4}$$

where w_{k+} is the weight of aspect k when the review rating is in the higher end and w_{k-} is its weight when the rating is in the lower end. π indicates that the sentiment polarity of a review sentence is inclined to be positive when the rating is greater than one-half of the full scale rating and to be negative when it is less than one-half of the full scale rating. Furthermore, it also implies that the inclinations on sentences describing more important aspects would be more clear than that on the sentences describing less important aspects.

a_d represents the sentimental words in document d. For simplicity, we use only adjectives in each sentence as the sentimental words of the sentence because

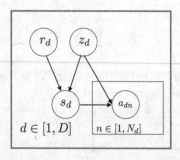

Fig. 2. The Bayesian network of Asymmetric Aspect Weight Model

adjectives are usually more sentimental than other parts of speech. We assume an adjective a_{dn} in a sentence d is drawn from the conditional probability distribution $p(a_{dn}|s_d, z_d)$. Let ϕ denote the parameters of the CPD. Therefore the likelihood of the adjectives in a sentence can be expressed by the following equation:

$$p(a_d|r_d, z_d) = \sum_s p(s|r_d, z_d) \prod_{n=1}^{N_d} p(a_{dn}|s, z_d) \tag{5}$$

$$= \sum_s \pi_{ds} \times \prod_{n=1}^{N_d} \phi_{s z_d a_{dn}} \tag{6}$$

The likelihood of all the adjectives can be expressed by the following equation:

$$p(a|r, d) = \prod_{d=1}^{D} p(a_d|r_d, z_d) \tag{7}$$

2.3 Model Inference

In this paper we use the EM (Expectation Maximization) algorithm to infer the aspect weights. In the proposed model, the sentiment polarities of sentences are latent variables. We initialize aspect weights w and distributions of adjectives ϕ with random values. In the E-step, we 'guess' the sentiment polarity of each sentence based on the given w and ϕ; in the M-step, we use the 'guess' to update w and ϕ. We iterate between the two steps until the algorithm is convergent.

In the E-step, we compute $\gamma_d = p(s_d|z_d, r_d, a_d)$ as follows:

$$\gamma_d = p(s_d|a_d, r_d, z_d) = \frac{p(s_d|r_d, z_d) \prod_{n=1}^{N_d} p(a_{dn}|z_d, s_d)}{\sum_s p(s_d|r_d, z_d) \prod_{n=1}^{N_d} p(a_{dn}|z_d, s_d)} \tag{8}$$

In the M-step, given the possibility of each sentiment polarity s of each sentence d, we update w and ϕ by maximizing the following equation:

$$L = \sum_{d=1}^{D} (\log p(s_d|r_d, z_d) + \sum_{n=1}^{N_d} \log p(a_{dn}|z_d, s_d)) \tag{9}$$

Based on γ_d obtained from the E-step, we maximize L with respect to w and ϕ and can get the following results:

$$w_{k+} = \log \frac{\sum_{d=1}^{D} \gamma_{d+} * 1\{z_d = k, r_d > 0.5\}}{\sum_{d=1}^{D} \gamma_{d-} * 1\{z_d = k, r_d > 0.5\}} \tag{10}$$

$$w_{k-} = \log \frac{\sum_{d=1}^{D} \gamma_{d-} * 1\{z_d = k, r_d \leq 0.5\}}{\sum_{d=1}^{D} \gamma_{d+} * 1\{z_d = k, r_d \leq 0.5\}} \tag{11}$$

$$\phi_{skt} = \frac{\sum_{d=1}^{D} \sum_{n=1}^{N_d} \gamma_{ds} * 1\{a_{dn} = t, z_d = k\}}{\sum_{d=1}^{D} \sum_{n=1}^{N_d} \gamma_{ds} * 1\{z_d = k\}} \tag{12}$$

We summarize the EM inference algorithm in Algorithm 1.

Algorithm 1. EM Inference algorithm

---

procedure INFERENCE(a,r,z)
    Initialize w and ϕ with random values
    while not convergent **do**
        compute $p(s_d|z_d, r_d, a_d)$ according to (8)
        update w and ϕ according to (10),(11),(12)
    end while
end procedure

---

3 Evaluation

In this section we evaluate the performance of the proposed model on a set of product reviews crawled from amazon.com [10]. The following 5 items are included in the dataset: #8288862993 (phone charger), #B0000WZWSI (cellphone), #B001F2Q0BU (phone charger), #B00001W0EQ (ear phone), #B00499DUC8 (phone charger). Each item contains around 2000 sentences in the training set and 250 sentences in the test set. For preprocessing, we use Part-of-Speech (POS) taggers to identify nouns and verbs for the aspect discovery process and identify adjectives for the weight inference process.

As mentioned in Sect. 2, we first use Latent Dirichlet Allocation for aspect discovery. In this paper, we set the number of topics K to be 4 that gives LDA the best performances on our dataset.

It is currently out of the question to survey on-line shoppers or retailers to verify the correctness of the proposed asymmetric weights. Instead, we evaluate the sentiment model π that plays an 'intermediary' role in the proposed probabilistic graph model. According to π, the most important aspects should have the highest fraction of positive sentences in high-rating reviews and have the highest fraction of negative sentences in low-rating reviews. Therefore, to evaluate π, we first manually label the sentiment polarity of each review sentence in the test set and compute the fraction of positive sentences f_{k+} for each aspect k in high-rating reviews and the fraction of negative sentences f_{k-} in low-rating reviews. At least the following condition should be satisfied to justify π.

$$\operatorname*{argmax}_{k} w_{k+} = \operatorname*{argmax}_{k} f_{k+} \tag{13}$$

$$\operatorname*{argmax}_{k} w_{k-} = \operatorname*{argmax}_{k} f_{k-} \tag{14}$$

where w_{k+} and w_{k-} represent the importance weight of each aspect k in high-rating reviews and low-rating reviews, respectively. In the experiment, we judge whether the condition is met on each item in the test set and get 7 (out of 10) matches. For performance comparisons, we also use the unsupervised ASUM model [5], and the supervised Standford sentiment analysis package to classify the sentiment polarity of each sentence and check whether the aspects with the highest fractions match that of the ground truth. We report the results in Table 1 and Figs. 3, 4.

Table 1. The most important aspect for each item

Item ID	Ground truth	AAWM	ASUM	Standford NPL Package
8288862993 (charger)	Portability	Portability	Power	Power
	Shipping	Shipping	Price	Shipping
B0000WZWSI (cell phone)	Voice	Voice	Software	Camera
	Shipping	Shipping	Shipping	Shipping
B001F2Q0BU (charger)	Price	Durability	Charging speed	Price
	Charging speed	Charging speed	Charging speed	Cable connection
B00001W0EQ (ear phone)	Material	Material	Sound	Sound
	Sound	Sound	Sound	Material
B00499DUC8 (charger)	Price	Charging	Price	Price
	Charging	Cable connection	Charging	Charging
Total matches		7 out of 10	5 out of 10	5 out of 10

The upper row in each item is populated by the most important aspect in high rating reviews and the lower row is for that in low rating reviews, respectively.

The results in Table 1 indicate that the important aspects obtained by the proposed model are basically aligned with the ground truth and perform better than ASUM and Standford package. We attribute the better performance to that both aspects and ratings are taken into account in the sentiment model.

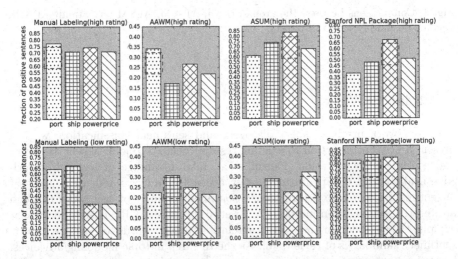

Fig. 3. Comparisons between the ground truth, AAWM, AUSM and Standford NPL Package on item No. 8288862993. The most important aspect of each model is marked by the dashline box

In Fig. 3, it is worth noting that ASUM gives very low fractions of negative sentences in low-rating reviews. This may be due to that the model doesn't
take into account the 'implication' of ratings on sentiment polarities. The supervised Stand NPL package has very high accuracies of sentiment classification in
both Figs. 3 and 4, however, its results don't reflect the relations between aspect
weights and sentiment polarities. In Fig. 4, the difference between each pair of
negative fractions of the proposed model is very slight, which may due to that
the high-rating reviews greatly outnumber the low-rating reviews in the training
data for the item.

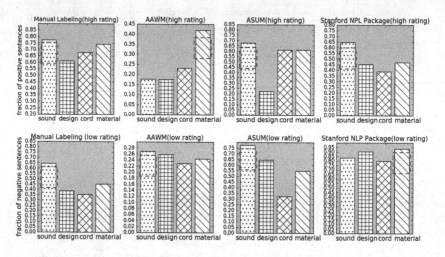

Fig. 4. Comparisons between the ground truth, AAWM, AUSM and Standford NPL
Package on item No. B00001W0EQ

4   Related Work

Since our work involves aspect discovery and sentiment classification, we will
present a brief introduction to existing research related to joint aspect discovery
and sentiment classification. Fahrni et al. take advantage of the rich knowledge available on Wikipedia to identify aspects in reviews and use a seed set
of sentimental words to determine the sentiment polarity of each word through
an iterative propagation method [2]. Li et al. propose a CRF-based model that
exploits the conjunction and syntactic relations among words to generate aspect-
sentiment summaries of reviews [6]. As the research in topic modeling advances,
a number researchers start to focus on topic models to perform the task. Oh
et al. [5] propose Aspect-Sentiment Unification Model (ASUM) based on LDA
in which the authors introduce the sentiment polarity of each word as another
latent factor and infer the aspect label and sentiment label simultaneously under

framework of LDA. Lin et al. [8] propose the Joint Sentiment and Topic (JST) model which shares similar latent components as ASUM but differentiate itself from ASUM by assuming a different generative processes. Mukherjee et al. collect a set of seed words with different sentiments then use LDA to learn the association between the seed words and unlabeled words [11].

However, none of those methods mentioned above takes into account ratings which usually reflect the overall sentiment orientations of reviews. In our work, we exploit the straightforward relation between ratings and sentiments and integrate it into aspect weight estimation. This is the main contribution of this work.

5 Conclusion

In this paper we propose Asymmetric Aspect Weight Model for aspect weight estimation. In this model we assume the aspect weights are asymmetric in different opinion contexts and verify the validity of the proposed model through a preliminary experiment. In the near future, we will devise a much more comprehensive experiment setting to evaluate the proposed model. Also, the relations between distributions of adjectives under different aspects and sentiment polarities are not taken into account in this paper. We will explore the relations in the near future in order to improve the effectiveness of the proposed model.

Acknowledgement. This paper is partly supported by The Vietnam National Foundation for Science and Technology Development (NAFOSTED) under grant number 102.01-2014.22.

References

1. Blei, D.M., Ng, A.Y., Jordan, M.I.: Latent dirichlet allocation. J. Mach. Learn. Res. **3**, 993–1022 (2003)
2. Fahrni, A., Klenner, M.: Old wine or warm beer: target-specific sentiment analysis of adjectives. In: Proceedings of the Symposium on Affective Language in Human and Machine, AISB, pp. 60–63 (2008)
3. Fletcher, J., Patrick, J.: Multi-attribute sentiment classification using topics
4. Guo, H., Zhu, H., Guo, Z., Zhang, X., Su, Z.: Product feature categorization with multilevel latent semantic association. In: Proceedings of the 18th ACM Conference on Information and Knowledge Management, CIKM 2009, pp. 1087–1096. ACM, New York (2009)
5. Jo, Y., Oh, A.H.: Aspect and sentiment unification model for online review analysis. In: Proceedings of the Fourth ACM International Conference on Web Search and Data Mining, pp. 815–824. ACM (2011)
6. Li, F., Han, C., Huang, M., Zhu, X., Xia, Y.J., Zhang, S., Yu, H.: Structure-aware review mining and summarization. In: Proceedings of the 23rd International Conference on Computational Linguistics, pp. 653–661. Association for Computational Linguistics (2010)
7. Li, F., Huang, M., Zhu, X.: Sentiment analysis with global topics and local dependency. In: AAAI, vol. 10, pp. 1371–1376 (2010)

8. Lin, C., He, Y.: Joint sentiment/topic model for sentiment analysis. In: Proceedings of the 18th ACM Conference on Information and Knowledge Management, pp. 375–384. ACM (2009)

9. McAuley, J., Leskovec, J.: Hidden factors and hidden topics: understanding rating dimensions with review text. In: pp. 165–172. ACM Press (2013)

10. McAuley, J., Pandey, R., Leskovec, J.: Inferring networks of substitutable and complementary products. In: Proceedings of the 21th ACM SIGKDD International Conference on Knowledge Discovery and Data Mining, pp. 785–794. ACM (2015)

11. Mukherjee, A., Liu, B.: Aspect extraction through semi-supervised modeling. In: Proceedings of the 50th Annual Meeting of the Association for Computational Linguistics: Long Papers, vol. 1, pp. 339–348. Association for Computational Linguistics (2012)

12. Titov, I., McDonald, R.: Modeling online reviews with multi-grain topic models. In: Proceedings of the 17th International Conference on World Wide Web, pp. 111–120. ACM (2008)

13. Wang, H., Lu, Y., Zhai, C.: Latent aspect rating analysis on review text data: a rating regression approach. In: Proceedings of the 16th ACM SIGKDD International Conference on Knowledge Discovery and Data Mining, pp. 783–792. ACM (2010)

14. Yu, J., Zha, Z.J., Wang, M., Chua, T.S.: Aspect ranking: identifying important product aspects from online consumer reviews. In: Proceedings of the 49th Annual Meeting of the Association for Computational Linguistics: Human Language Technologies, vol. 1, pp. 1496–1505. Association for Computational Linguistics (2011)

Deep Bi-directional Long Short-Term Memory Neural Networks for Sentiment Analysis of Social Data

Ngoc Khuong Nguyen[1], Anh-Cuong Le[2]([⊠]), and Hong Thai Pham[1]

[1] VNU University of Engineering and Technology,
144 Xuan Thuy, Cau Giay, Ha Noi, Vietnam
khuongnn@dhhp.edu.vn, thaiph@vnu.edu.vn
[2] Faculty of Information and Technology, Ton Duc Thang University, 19 Nguyen
Huu Tho Street, Tan Phong Ward, District 7, Ho Chi Minh City, Vietnam
leanhcuong@tdt.edu.vn

Abstract. Sentiment analysis (SA) has been attracting a lot of studies in the field of natural language processing and text mining. Recently, there are many algorithm's enhancements in various SA applications are investigated and introduced. Deep Convolutional Neural Networks (DCNNs) have recently been shown to give the state-of-the-art performance on sentiment classification of social data. Although, these solutions effectively address issues of multi-levels features presentation but having some limitations of temporal modeling. In addition, the Bidirectional Long Short-Term Memory (BLTSM) conventional models have encountered some limitations in presentation with multi-level features but can keep track of the temporal information while enabling deep representations in the data. In this paper, we propose to use Deep Bi-directional Long Short-Term Memory (DBLSTM) architecture with multi-levels feature presentation for sentiment polarity classification (SPC) on social data. By using DBLSTM, we can exploit more level features than BLTSM and inherit temporal modeling in BLTSM. Moreover, the language of social data is very informal with misspellings and abbreviations. One word can be appeared in multiple formalities, which is a challenge in word-level models. We use *character-level* as input of DBLSTM neural network (called Character DBLSTM - CDBLSTM) for learning sentence level presentation. The experimental results show that the performance of our model is competitive with state-of-the-art of SPC on Twitter's data. Our model achieves **85.86%** accuracy on Stanford Twitter Sentiment corpus (STS) and **84.82%** accuracy on the subtasks B of SemEval-2016 Task 4 corpus.

Keywords: Sentiment polarity classification · Deep learning · Multi-levels features presentation · Character deep bi-directional long short-term memory · Text classification

© Springer International Publishing AG 2016
V.-N. Huynh et al. (Eds.): IUKM 2016, LNAI 9978, pp. 255–268, 2016.
DOI: 10.1007/978-3-319-49046-5_22

1   Introduction

In the recent years, the social network has become a very popular and convenient communication environment in our life. Twitter and Facebook are most popular social network platforms which allow users to share their opinions on a wide variety of topics. Everyday, there are enormous amount of information covering a wide range of topics on various brands, products, politics and social events which are generated by the user of these platforms.

Sentiment analysis on social network's corpora has become an important area of research across the fields of NLP, and social science research. Through social networks, users usually share comments or behaviors which contain the their attitudes to some topics or products. Exploit sentiment information in these comments or behaviors will provide convincing information of users' interest or feedback on the topic help and build up a lot of practical NLP applications.

In this paper, we focus on sentiment polarity analysis on twitter messages (called tweet), in which we will classify each tweet into positive or negative classes. In twitter platform, the short nature of tweet is one of the most striking features. The tweet may contain multimedia information but our research only limits on text data and each tweet has a maximum of 140 characters. The user often do a behavior (like, follow,...) or post the tweet for sharing and expressing their views about topics.

Sentiment polarity classification on tweet has been facing many challenges because of the abstraction of feature levels of elements in tweets. There has been a lot of techniques Unlike other sentiment data types, Twitter messages have some unique characteristics which present special challenges for sentiment analysis: Tweets are short in length. There is a limitation of 140 words for each tweet which makes analyzing them challenging [16]; The language used in tweets is very informal with misspellings and abbreviations (often intentional, like different spellings of "cool": coool, coolll, coooolll!!), new words, slangs, and URLs; The number of tweets increases very fast and there are many new words, new trends in using abbreviations, which lead to a frequent problem of out-of-vocabulary words; Special symbols and their combinations are often used, like emoticons and hashtags.

Keeping in view the challenges posed by the nature of tweets. In this paper, we propose a model which is inspired from DBLSTM neural networks [5]. Deep BLSTM can be understood as stacking multiple BLSTM layers on top of each other, for example, each layer attempt to model high-level abstractions of the sentence. The first layer operates at *character-level input* and the last layer makes predictions at the Tweet-level whether it is of positive or negative sentiment. Such networks naturally handle the problem of very large vocabulary sizes and the presence of sub-word information, without having to keep many trained embedding which would be required for word-level models with very large vocabularies. We find out that the character-level model, interestingly, performs **better** than equivalent DBLSTM model which operates at the word level (even when initialized with a variety of pre-trained embeddings). We also compare our results with Deep Convolutional Neural Network (DCNN) [3] which has shown the state of the art performance on Twitter sentiment polarity classification. In addition, we also explore the properties of the CDBLSTM model and find out that it learns to represent the composition of characters and can effectively relate sub-word information (like the presence of period instead of exclamation marks) to the sentiment polarity of the tweet.

This paper is laid out as follows: Sect. 2 gives brief related works which relate to our research. In Sect. 3, we describe background knowledge on sentiment polarity classification using deep learning. The Character DBLSTM neural network architecture is presented in Sect. 4. In Sect. 5, we describe the datasets used in our experiments and show the results. We conclude with an analysis of our results, as well as describe areas of future research in the last section.

2 Related Works

Sentiment analysis on social data is increasingly drawing the attention of researchers in recent years. Given the tweets, sentiment classification on tweets is often considered as the problem of sentence-level sentiment analysis [14]. However, the approaches based on phrase or sentence levels can hardly define the sentiment of some specific topics. Considering opinions adhering on different topics, Wang et al. [24] proposed a hashtag-level sentiment classification method to generate the overall sentiment polarity for a given hashtag. Recently, following the work of [17] some researchers used neural networks to implement sentiment classification. For example, Kim [12] adopted convolutional neural networks to learn sentiment-bearing sentence vectors; Mikolov et al. [18] proposed Paragraph vector which outperformed bag-of-words model for sentiment analysis; and Tang et al. [22] used ConvNets to learn sentiment specific word embedding (SSWE), which encodes sentiment information in the continuous representation of words. Furthermore, Kalchbrenner [10] proposed a Dynamic Convolutional Neural Network (DyCNN) which uses dynamic k-max pooling, a global pooling operation over linear sequences. Instead of directly applying ConvNets to word embeddings, very recently [25] has applied the network only on characters. They showed that the deep ConvNets does not require knowledge of words and thus can work for different languages. LSTM [21] is the another state-of-the-art semantic composition model for sentiment classification. Similar to DCNN, it also learns fixed-length vectors for sentences of varying length of word-level input, captures words order in a sentence and does not depend on external dependency or constituency parse results. In addition, moreover deep bi-directional LSTM (DBLSTM) recurrent neural networks have recently been shown to give state-of-the-art performance on many problems such as: sentiment classification [12]; speech recognition [5]; Semantic Labeling [26]; Large Scale Acoustic Modeling [20],...

Recently, there has been a surge of research interest in exploring character-input models for a variety of natural language processing tasks such as language modeling [13], POS tagging [15], dependency parsing [1] and machine translation [2]. Specially, Santos et al. [3] uses Deep Convolutional Neural Network for sentiment analysis of short texts by exploiting from character- to sentence-level information. This research has achieved state-of-the-art results for single sentence sentiment prediction in both binary positive/negative classification with 85.7 % accuracy on the Stanford Sentiment Treebank (SSTb) and **86.4 %** accuracy on the Stanford Twitter Sentiment corpus (STS). Our exploration is in a similar direction, focused on social data which is naturally morphologically rich, and on using a variant of LSTM (DBLSTM) which are powerful models of sequence data at the character level and hence a natural choice for text modeling.

3 Background

Before explaining some aspects of the networks motivate DBLSTM model recurrent neural network, we briefly describe the Conventional LSTM and review BLTSM neural networks.

3.1 Long Short-Term Memory

Recurrent Neural Networks (RNNs) are a class of artificial neural networks used for modeling sequences. RNNs are highly flexible in their use of context information as they can learn what part of the input sequence to store to memory and what parts to ignore. They also allow modeling of various regimes of sequence modeling as shown in Fig. 1.

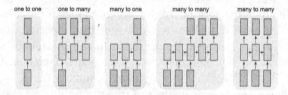

Fig. 1. RNNs allow modeling of multiple types of input and output sequences

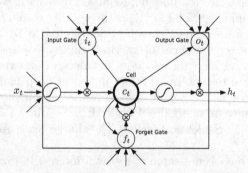

Fig. 2. The long short-term memory cell. (Source: [9])

One of the limitations of RNN is that it is very difficult to store information over long sequences because of problems due to vanishing and exploding gradients as explained in [6,7]. *Long Short-Term Memory (LSTM)* models [8] are designed to remedy this and store information over larger input sequences. They achieve this using special "memory cell" units as illustrated in Fig. 2. This cell is composed of five main elements: an input gate i, a forget gate f, an output gate o, a recurring cell state c and hidden state output h. These values are estimated through the equations Eqs. (1), (2), (3), (4), (5), where σ is the logistic sigmoid function. Refer to neural network[1] for a gentle introduction to LSTM and to [6] for a more comprehensive review and applications. ·

– The LSTM's final output: Output of the LSTM scaled by a tank transformation of the current state

$$h_t = o_t * tanh(c_t) \tag{1}$$

– Output gate: Scale the output from the cell

$$o_t = tanh(W_{x0}x_t + W_{h0}h_{t-1} + W_{c0}c_t + b_o) \tag{2}$$

– Cell state update step: computes the next timesteps state using the gated previous state and the gated input. In other hand, transforms the input and previous state to be taken into account into the current state.

$$c_t = f_t * c_{t-1} + i_t * tanh(W_{xc}x_t + W_{hc}h_{t-1} + b_c) \tag{3}$$

– Forget (reset) gate: Decides whether to erase (set to zero) or keep individual components of the memory

$$f_t = \sigma(W_{xf}x_t + W_{hf}h_{t-1} + W_{cf}c_{t-1} + b_f) \tag{4}$$

– Input gate: Controls how much of the current input x_t and the previous output h_{t1} will enter into the new cell

$$i_t = \sigma(W_{xi}x_t + W_{hi}h_{t-1} + W_{ci}c_{t-1} + b_i) \tag{5}$$

However, this variant LSTM model can only capture context sequence on one input direction. In order to access long-range context in both input directions, bi-directional LSTM is used through forward and backward sequence to two separate recurrent hidden layers.

3.2   Bi-directional Long Short-Term Memory

A BLSTM consists of two LSTMs that are run in parallel: the first LSTM processes the input sequence from left to right and the second LSTM does the input from right to left. At each time step, the hidden state of the BLSTM is the concatenation of the forward and backward hidden states. This initialization allows the hidden state to capture both past and future information that

---

[1] http://colah.github.io/posts/2015-08-Understanding-LSTMs.

exploits information follows both directions. The Fig. 3 shows the structure of a Bi-directional LSTM, in which y_t is one of elements in output vector sequence y and computed by equation Eq. (6).

$$y_t = W_{hy}h_t + b_y \tag{6}$$

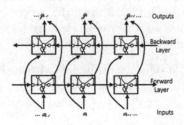

Fig. 3. The bi-directional LSTM (Source: [23])

A crucial element of the recent success of some natural language processing (NLP) systems is the use of deep architectures, which are able to build up progressively higher level representations of data. In the next section, we will describe deep bi-directional long short-term memory model for some applications in natural language processing.

3.3   Deep Bi-directional Long Short-Term Memory

In [5], the basic architecture of the DBLSTM neural network is shown in Fig. 4. In terms of nature, each DBLSTM architecture includes multiple BLSTM layers which are stacked each other. Besides, BLSTMs are already deep architectures in the sense that they consist two LSTM neural networks unrolled in time where each layer shares the same model parameters. Similar to the principle in DCNNs, the inputs to the model go through multiple DBLSTM layers. However, the features from a given time instant are only processed by a single layer before contributing the output for that time instant. Therefore, the depth concept in DBLSTM neural network has an additional meaning. The input to the neural network at a given time step goes through multiple BLSTM layers in addition to propagation through time and BLSTM layers. It has been argued that deep layers in RNNs allow the network to learn at different time scales over the input [5]. DBLSTM offer another benefit over standard BLSTM: They can make better use of parameters by distributing them over the space through multiple layers.

4   Character DBLSTM for Sentiment Analysis

This section describes how to use the DBLSTM neural network in learning tweet level encoding from character level.

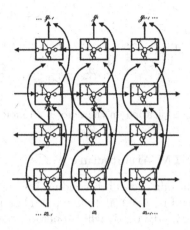

Fig. 4. The deep bi-directional LSTM

4.1   Character Level Model

The typical problem of sentiment analysis is described in Fig. 5 as follows: Given a text, for an example a twitter message we need to figure out if the message is positive (1) or negative (0). Generally, lets denote s_i the text input, which consists sequence of words $w_1, w_2, ..., w_n$, and y_i the corresponding sentiment, so we create a neural network corresponding $f(s_i)$ function that will predict the label of the sample. In such settings a typical approach is to split s_i into a sequence of words, and then learn some fixed length embedding of the sequence that will be used to classify it.

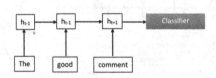

Fig. 5. Sentiment classification on word level model (Source: [11])

Like any typical model that uses a word as its smallest input entity, a character level model will use the character as the smallest entity.

This model is showed in Fig. 6 which reads characters one by one from the first character to the last character, to create an embedding of the given word or sentence. As such our model will try to learn word presentation from sequences character and sentence presentation form words separated by spaces or other punctuation points.

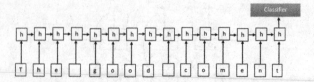

Fig. 6. Sentiment classification on character level model (Source: [11])

4.2 Character DBLSTM Architecture

In this section we present our deep bi-directional LSTM architecture for sentiment analysis of tweets as Fig. 7. We also explored different architectures neural networks (LSTM, BLSTM, DBSLTM) which operate character-level input.

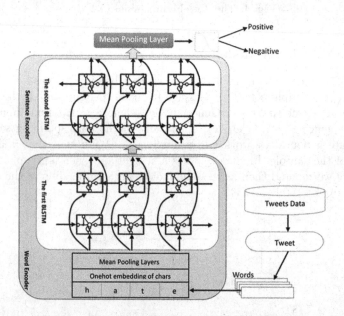

Fig. 7. The character deep bi-directional LSTM

Our goal is to encode tweet from character level, so tweet will be split into the list of words. After using the BLSTM for encoding each word from characters to getting fixed length encoding that be used as input encoding tweet from word level with other BLSTM. To illustrate this idea let's considering the modeling of sentiment classification problem as follows: Given a tweet as a sequence of characters or words $X = \{x_1, \ldots, x_n\}$, the task is to predict the sentiment of Tweet as being *positive* (1) or *negative* (0). Each x_i is a one-hot encoding of the character by indexing in given encoding table. The DBLSTM models then take this one-hot encoding and convert them into a *character embedding*. The embeddings can be randomly initialized and learned jointly with other model parameters. The model is trained using categorical cross-entropy (CCE). Note that it is possible to have multiple levels of granularity in predictions, like positive, "natural" and negative, or even finer but we restrict ourselves to two classes to be able to compare with the state of the art published results [10].

Intuitively, the first BLSTM operated at character level is expected to get information about the characters of a word from the first LSTM which is then composed into a word-level meaning by the second LSTM layer. This LSTM layer then encodes the information of the entire tweet in a r ($= 256$ in experiments) dimensional space. We also expect to get information of words from the second BLSTM which are then composed into a sentence-level meaning which can be used to classify the tweet.

Furthermore, all (LSTM, BLSTM, DBLSTM) model can be also used with either character input data or word input data. Note that when used with word input, it is common to initialize the models with pre-trained word embeddings [18]. We explore multiple such initializations in our experiments. For the character input model, the embeddings are always initialized randomly, since pre-trained character embeddings are not yet available.

In the same way, using character level model in [11] will exploit LSTM, BLSTM models in operating at the character level. Therefore, in this work we will focus on the model described in Fig. 7 with the character as inputs.

5   Experiments

5.1   Datasets

We conduct our experiments on two datasets: the latest benchmark dataset for SemEval 2016 and the dataset provided by [4]. In the latter dataset, the training set consists of 1.6 million tweets collected during 2009. In the experiments presented below, we first train our models on the Go dataset [4], then re-train the model parameters on the smaller, fully-supervised SemEval dataset (Table 1).

5.2   Results

We compare the performance of character-level models against word-level models. For the former, we compare across character sets - CSs ("utf8" and "ascii"),

Table 1. Data size and label distribution

	Go et al. (1.6 M)		SemEval2016	
	Neg	Pos	Neg	Pos
Train	800000	800000	781	2805
Dev	-	-	358	766
Test	177	182	286	886

Table 2. Accuracy across LSTMs with character level

Model	CSs	PIs	EDs	1.6 M (acc)	SemEval (acc)
DBLSTM	ascii	rnd	50	**85.86**	83.08
DBLSTM	ascii	rnd	200	83.45	**84.82**
DBLSTM	ascii	eye	50	81.44	82.18
DBLSTM	utf8			81.34	82.34
DCNN				**86.4** [3]	81.3

Table 3. Results for subtask B at SemEval [19]

#	System	Acc	Rank in SemEval
1	CDBLSTM	**84.82**	3–4
2	Tweester	86.20	3
4	INSIGHT-1	86.40	2
5	UNIMELB	87.00	1
7	Finki	84.80	4

parameter initializations - PIs ("rnd" and "eye"), and embedding dimension sizes - EDs (50 and 200). The vocabulary for the "utf8" setting consisted of 1949 characters, and for "ascii" consisted of 93 characters. We initialize model parameters randomly at all levels in the "rnd" setting whereas in the "eye" setting we initialize the second-level LSTM cell of both BLSTM layers and gate parameters using the identity matrix.

We compare word-level models under two settings: initializing the word embeddings - IWEs using random vectors ("word/rnd") and initialization using "sentiment-specific" word embeddings provided in [22] ("word/sswe").

Our results are shown in Tables 2, 4. We train all models using the implementation of the Adam algorithm provided in Lasagne[2] and a learning rate set to 0.1[3]. To learn word and character embeddings, we use two bi-directional LSTMs with 256 hidden units on each, followed by a mean pooling and a dropout layer ($p = 0.5$), a second forward-directional LSTM (again with 256 hidden units) and a final dropout layer ($p = 0.6$). We obtain final predictions using the logistic regression model.

Table 4. Accuracy across LSTMs with word level

Model	IWEs	1.6 M (acc)	SemEval (acc)
DBLSTM	rnd	83.85	80.07
DBLSTM	sswe	**85.24**	81.27
DCNN		**85.5** [3]	80.3

The character-level models using the "ascii" character set outperform the other models on the SemEval dataset, due to the highly-productive nature of the Twitter "lexicon", users' predisposition toward using slang dialects, and the constraint of the 140-character limit. This result makes sense that character-level models is better character-level ones in this task.

---

[2] https://github.com/Lasagne/Lasagne.

[3] We retrained the ascii/rnd/200 on SemEval using AdaGrad and a learning rate of 0.01 to achieve 84.13; using Adam and 0.1 learning rate, the result was 83.21.

It is worth noting that the "utf8" model performed comparably despite having a much larger vocabulary size. It is not surprising that the "utf8" model performed worse than the "ascii" model, as the test data consisted of only English tweets; however, since the "utf8" model is implicitly multi-lingual, our results suggest that the same model may perform well across multiple languages. We leave such experiments for future work.

As Table 4 shows, word/dcnn [3] outperform our DBLSTM on the Go dataset. Actually our character and word-level models achieve competitive performance on the Go dataset without tuning hyper-parameters such as the size of embeddings, the number of hidden units and learning rate.

We also compare the effectiveness of our proposed method with top 8 best systems on the SemEval competition. The performance comparison between various methods from Table 3 shows that the CDBLSTM achieves higher classification accuracy than almost of another systems on subtasks B of SemEval-2016 Task 4 corpus.

Figures 8, 9 show that the performance of DBLSTMs are almost better than two models (LSTMs, BLSTMs) for both character-level and word-level. The result on GO dataset (Fig. 10) also shows that performance of variant LSTMs with the character-level model is almost better than word-level model.

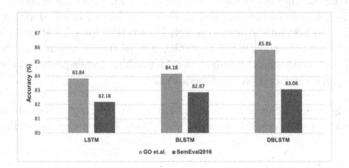

Fig. 8. Comparison of performance between variant LSTMs with character level model

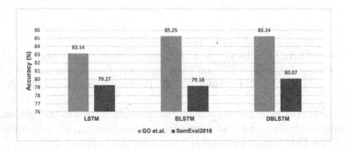

Fig. 9. Comparison of performance between variant LSTMs with word level model

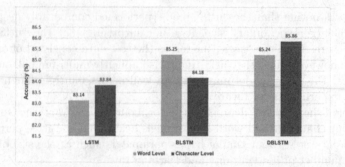

Fig. 10. Comparison of performance of variant LSTMs with character and word level model

6   Conclusion

In this paper, we have focused on the problem of polarity classification for sentiment analysis on the tweets. We have applied the DBLSTM model to solve this problem. Because the tweets are noisy and usually contain new words and new acronyms then in this paper we have proposed a character-based modeling for this DBLSTM model. Through various experiments we have shown that our approach is efficient for the problem, and have achieved the best result in comparison with all previous studies, on well-known benchmark datasets. In future work, we will explore other variants on the architectures presented in this paper, for example: compare mean pooling vs max pooling; hyper-parameter optimization; learn joint character-word embeddings.

Acknowledgement. This paper is supported by The Vietnam National Foundation for Science and Technology Development (NAFOSTED) under grant number 102.01-2014.22.

References

1. Ballesteros, M., Dyer, C., Smith, N.A.: Improved transition-based parsing by modeling characters instead of words with lstms. In: EMNLP (2015)
2. Costa-Jussà, M.R., Fonollosa, J.A.R.: Character-based neural machine translation. In: Proceedings of the 54th Annual Meeting of the Association for Computational Linguistics, ACL 2016, 7–12 August 2016, Berlin, Germany, vol. 2: Short Papers (2016)
3. dos Santos, C., Gatti, M.: Deep convolutional neural networks for sentiment analysis of short texts. In :The 25th International Conference on Computational Linguistics (COLING 2014), pp. 69–74. ACM (2014)
4. Go, A., Bhayani, R., Huang, L.: Twitter sentiment classification using distant supervision. CS224N Project Report, Stanford, 1: 12 (2009)
5. Graves, A., Jaitly, N., Mohamed, A.R.: Hybrid speech recognition with deep bidirectional lstm. In: 2013 IEEE Workshop on Automatic Speech Recognition and Understanding (ASRU), pp. 273–278, December 2013

6. Graves, A.: Supervised Sequence Labelling with Recurrent Neural Networks. SCI, vol. 385. Springer, Heidelberg (2012)
7. Hochreiter, S., Bengio, Y., Frasconi, P., Schmidhuber, J.: Gradient flow in recurrent nets: the difficulty of learning long-term dependencies. In: A Field Guide to Dynamical Recurrent Neural Networks. IEEE Press (2001)
8. Hochreiter, S., Schmidhuber, J.: Long short-term memory. Neural Comput. 9(8), 1735–1780 (1997)
9. Hong, J., Fang, M.: Sentiment analysis with deeply learned distributed representations of variable length texts. Technical report, Stanford University (2015). CS224d: Deep Learning for Natural Language Processing
10. Kalchbrenner, N., Grefenstette, E., Blunsom, P.: A convolutional neural network for modelling sentences. In: Proceedings of the 52nd Annual Meeting of the Association for Computational Linguistics, ACL 2014, 22–27 June 2014, Baltimore, MD, USA, vol. 1: Long Papers, pp. 655–665 (2014)
11. Karpathy, A., Johnson, J., Li, F.-F.: Visualizing and understanding recurrent networks. CoRR, abs/1506.02078 (2015)
12. Kim, Y.: Convolutional neural networks for sentence classification. In: Proceedings of the 2014 Conference on Empirical Methods in Natural Language Processing, EMNLP 2014, 25–29 October 2014, Doha, Qatar, A meeting of SIGDAT, a Special Interest Group of the ACL, pp. 1746–1751 (2014)
13. Kim, Y., Jernite, Y., Sontag, D., Rush, A.M.: Character-aware neural language models. In: Proceedings of the Thirtieth AAAI Conference on Artificial Intelligence, 12–17 February 2016, Phoenix, Arizona, USA, pp. 2741–2749 (2016)
14. Kouloumpis, E., Wilson, T., Moore, J.: Twitter sentiment analysis: the good the bad and the omg!. Icwsm 11, 538–541 (2011)
15. Ling, W., Dyer, C., Black, A.W., Trancoso, I., Fermandez, R., Amir, S., Marujo, L., Luís, T.: Finding function in form: compositional character models for open vocabulary word representation. In: Proceedings of the 2015 Conference on Empirical Methods in Natural Language Processing, EMNLP 2015, Lisbon, Portugal, 17–21 September 2015, pp. 1520–1530 (2015)
16. Mehrotra, R., Sanner, S., Buntine, W., Xie, L.: Improving lda topic models for microblogs via tweet pooling and automatic labeling. In: Proceedings of the 36th International ACM SIGIR Conference on Research and Development in Information Retrieval, pp. 889–892. ACM (2013)
17. Mikolov, T., Chen, K., Corrado, G., Dean, J.: Efficient estimation of word representations in vector space. CoRR, abs/1301.3781 (2013)
18. Mikolov, T., Sutskever, I., Chen, K., Corrado, G.S., Dean, J.: Distributed representations of words and phrases and their compositionality. In: Advances in Neural Information Processing Systems, pp. 3111–3119 (2013)
19. Nakov, P., Ritter, A., Rosenthal, S., Stoyanov, V., Sebastiani, F.: SemEval-2016 task 4: sentiment analysis in Twitter. In: Proceedings of the 10th International Workshop on Semantic Evaluation, SemEval 2016, San Diego, California. Association for Computational Linguistics, June 2016
20. Sak, H., Senior, A.W., Beaufays, F.: Long short-term memory based recurrent neural network architectures for large vocabulary speech recognition. CoRR, abs/1402.1128 (2014)
21. Tai, K.S., Socher, R., Manning, C.D.: Improved semantic representations from tree-structured long short-term memory networks. CoRR, abs/1503.00075 (2015)

22. Tang, D., Wei, F., Yang, N., Zhou, M., Liu, T., Qin, B.: Learning sentiment-specific word embedding for twitter sentiment classification. In: Proceedings of the 52nd Annual Meeting of the Association for Computational Linguistics, vol. 1, pp. 1555–1565 (2014)
23. Thireou, T., Reczko, M.: Bidirectional long short-term memory networks for predicting the subcellular localization of eukaryotic proteins. IEEE/ACM Trans. Comput. Biol. Bioinf. **4**(3), 441–446 (2007)
24. Wang, X., Wei, F., Liu, X., Zhou, M., Zhang, M.: Topic sentiment analysis in twitter: a graph-based hashtag sentiment classification approach. In: Proceedings of the 20th ACM International Conference on Information and Knowledge Management, pp. 1031–1040. ACM (2011)
25. Zhang, X., Zhao, J., LeCun, Y.: Character-level convolutional networks for text classification. In: Advances in Neural Information Processing Systems, pp. 649–657 (2015)
26. Zhou, J., Xu, W.: End-to-end learning of semantic role labeling using recurrent neural networks. In: Proceedings of the 53rd Annual Meeting of the Association for Computational Linguistics, the 7th International Joint Conference on Natural Language Processing of the Asian Federation of Natural Language Processing, ACL 2015, 26–31 July 2015, Beijing, China, vol. 1: Long Papers, pp. 1127–1137 (2015)

Linguistic Features and Learning to Rank Methods for Shopping Advice

Xuan-Huy Nguyen[✉] and Le-Minh Nguyen

Japan Advanced Institute of Science and Technology, Nomi, Japan
{s1420203,nguyenml}@jaist.ac.jp

Abstract. We present a recommendation system (RS) which helps users in buying products based on summarizing all the customer reviews of product attributes. Our recommendation system extracts users' opinions for products through a decision tree. Each node of the tree is a question to the users. From each node, our RS gives a ranked list of products which is matches the opinions of users. We explain (a) a learning tree structure, for instance, at each node which questions can be asked; and (b) producing a suitably ranked list at each node. Firstly, we use a top-down strategy to build a decision tree in order to select the best user attributes corresponding to a question which is asked at each node. Secondly, we use a learning-to-rank method to learn a ranked list of products for each node of the tree. In experimentation, we use amazon datasets for computer products. We evaluate our RS by using mean reciprocal rank (MRR). Experimental results show that RankBoost achieves better quality than RankSVM.

Keywords: Recommendation system · Information retrieval · Learning to rank · Feature extraction

1 Introduction

The traditional way for buying a new laptop is that customers would walk into a shop. A shopping assistant will help them select the best laptop for their needs. A good shopping assistant will ask intelligible questions to non-expert users, for example, "Do you intend to use the laptop for playing 3D games?", while mapping the answers to technical product specification, such as, "The customer will need at least 8GB of RAM and 2GB of monitor card". In the case of shopping online, customers access a website from their homes without any human help. Moreover, catalogs contain a large number of products information which are updated every day. Therefore, it is hard to find the suitable products for non-expert users. Consequently, the customers are leave without any guidance.

In this research, we proposed a recommendation system to assist users in shopping online. The task is performed as follows:

1. We extract the product features (technical information e.g., the monitor card of a laptop), user information (lifestyle preference, interests, etc.) and review

© Springer International Publishing AG 2016
V.-N. Huynh et al. (Eds.): IUKM 2016, LNAI 9978, pp. 269–279, 2016.
DOI: 10.1007/978-3-319-49046-5_23

{"Reviews": [
    {"Title": "Great", "Author": "Johnnie", "Job": "Student"
    "Hobby": "Playing 3D game, Coding", "ReviewID": "R0001",
    "Rate": "5.0", "Reviews": "The CPU speed is very height.", "ProductID": "P0001"}}

Fig. 1. An example of customer reviews product

scores of products from users' comments. In this step, we identify product features that customers have expressed opinions on (called *opinion features*). We illustrate a simple example in Fig. 1. We can see the reviews of a laptop on amazon website: The *CPU speed* is opinion feature, so it is a feature of this laptop. From this example, we also see that the user has three attributes (*student, Playing 3D game, and coding*), and the review score is *5.0*.

2. We address the problem in automatically designing a decision tree that helps customers in selecting the best product for their needs. Our recommendation system generates a decision tree, in which each node of the tree is a question addressed to customers. The question refers to only attributes from the user's information, which non-expert customers can understand easily. An example of our recommendation system is showed in Fig. 2.

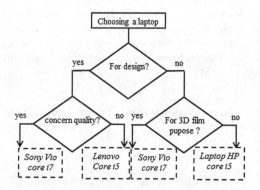

Fig. 2. An example of RS tree

Customers start from the root node, answer questions, follow the control flow, and descend towards a leaf node, where they receive a suitable product from a ranked list of products recommendations. In fact, a customer can stop answering at any internal node in the tree, and he/she also receives a ranked list of products recommendations at that node.

We build a RS tree structure like a traditional decision tree method. However, our system differs from the traditional decision tree method in the following aspects. Firstly, our system induces a ranked list of all products based on the products' attributes. From this, we learn a structure of the tree in which each node outputs as a ranked list of products, while each node outputs of the traditional decision tree method is a class label. Secondly, our approach is different from other traditional decision trees in the splitting criterion and ranking method. In fact, our RS tree splits the set of users based on users' attributes with the condition that they will prefer the similar products.

In Sect. 2 we briefly discuss other related work. Section 3 describes our problem definitions. We then discuss our algorithm in Sect. 4. Next, we present our experiments and results in Sect. 5. Finally, we discuss conclusions and future work in Sect. 6.

2 Related Work

Firstly, about the feature extraction, the customer always addresses comments or states their opinion on product's features. Therefore, the extracting product features and analyzing positive/negative points of product features are necessary for recommendation system. There are two kinds of product features including explicit and implicit features. Currently, there are many methods proposed like Hu et al., (Hu and Liu [4,5]) proposed a method to extract product features and classify customer opinions associated with these features from online product review text. This system uses association rules to extract feature mentions for a specific product and extracts opinion words that appear with these feature mentions. Our research is different. In order to extract product features, we use patterns of opinion words/phrases about the features of product from the review text through *adjective, adverb, verb* and *noun*.

Secondly, in order to generate a decision tree, our research is motivated by the idea proposed in [2]. But, our recommendation system is different. We extract product features from opinions of users about products and apply some learning-to-rank methods to analyze the recommendation quality. Our recommend system uses both interactive eliciting of user's preferences and learning-to-rank as a mixture of collaborative filtering. The traditional recommendation is formulated as the problem of estimating rating for products that have not been seen by a user [8]. When the ratings of products are estimated, a recommendation is generated by picking the highest rating products. However, we use the ranking of products rather than the estimated rating. Another formulation has made this assumption explicit and casts the recommendation task as a ranking problem [1]. In this research, the techniques attributes from the learning-to-rank method can be used to learn personalized ranking functions. Our system applies a learning-to-rank method to produce recommendations.

3 Problem Definition

There are three tables in our input data (cited from [2]): user table U, product table P, and review table R. The first, the table U: $(n_U \times m_U)$ stores information of n_U users, and each user has m_U attributes. The notation n_i is the ith user. The second, the table P: $(n_P \times (f_p, sc_p))$ stores information of n_P products in which each product is described with f_p features or opinions phrases, and each feature/opinion has sentiment score sc_p. The notation p_j is the jth product. The third, the review table R: $(n_c \times 3)$ stores reviewing actions (u_i, p_j, s_{ij}) of user u_i for product p_j with score s_{ij} (s_{ij} as a number in an interval).

We have notations: $|P| = n_P$; $|U| = n_U$; and $|A| = m_U$.

The main problems of our research are defined as follows:

Problem 1: (Feature extraction). We collect all customer reviews for products. The task is to extract product features, customer information and reviews score.

Problem 2: (Recommendation solution: cited from [2]). The input of the task includes reviews database (described above) and an integer k. Our work is to build a RS tree T. Each internal node of the tree T is a question corresponding to user's attribute $\alpha \in A$. Each node contains a top-k ranked list of the products in the set P.

When a RS tree T is built, each internal node of the T contains a user attribute $\alpha \in A$. The attribute α corresponds to a question to a customer. The leaf node of T is an ordering of the top-k ranked list of the products in P.

A customer starts from the root, follows the internal node of the tree, answers questions until reaching a leaf node, where he/she receives the top-k ranked list of the products at that leaf node.

The quality of the tree T reflects the quality of the recommendations made to customers. To evaluate our RS, we used 5-fold cross-validation. Users in the evaluation fold have reviewed products, which their scores are stored in the review table R and thus we can evaluate the quality of T.

4   Algorithm

4.1   Feature Extraction

We have proposed the feature extraction model using patterns of opinion words or phrases from the review text through adjectives, adverbs, verbs, and nouns. The feature extraction is useful for generating a meaningful summary that can provide the important information resource to help the users or merchants to select the most suitable choice of product. The feature extraction is the pre-process of our model. The product feature in this model is the explicit feature. For the first step, we use Standford CoreNLP (Manning et al. 2014 [7]) to parse and convert sentences with the POS tags. The second step is to extract features and opinions associated with these features based on the predefining patterns of words. The third step is to calculate sentiment score based on sentiment Treebank [9] in order to specify the features/opinions which are negative or positive. After that, we store information in the reviews database. Our feature extraction method is described in Fig. 3.

In order to extract product features and opinion words from the customer text reviews product, we represent the part-of-speech tagging to identify noun(s) phrases from the user reviews that can be product features. We use Standford CoreNLP to parse each sentence and yield the part of speech tag of the word (whether the word is a noun, adjective, verb, adverb, etc.). For instance, from the sentence:

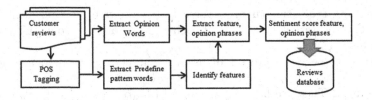

Fig. 3. Product feature extraction

This is a great indoor camera for stores.
We yield phrases:
(N P (DT a)(J J great)(J J indoor)(N N camera))
After POST tagging is done, we extract features that are noun(s) phrases. In the example, *camera* is a feature and we also have phrases *JJ-JJ-NN*. Therefore, *JJ-JJ-NN* is a pattern for feature extraction.

For opinion words extraction step, we saw that adjective/adverb is the opinion words/phrases that are used to qualify product features with noun(s) phrases. Therefore, we can extract the nearby adjective/adverb as opinion words if the sentences contain any features. For the above example, the feature, *camera*, is near the opinion words *greatindoor (JJ-JJ)*. Therefore, we need to extract phrases containing the adjective, adverb, verb, and noun(s) that imply opinion. Moreover, we also consider some verbs such as *like, dislike, appreciate, andlove* as opinion words, or some adverbs like *not, overall, absolutely, really, andwell* are also considered. Therefore, we collect all opinion words of mostly two or three consecutive words such as *(adjective, noun), (adverb, adjective), (verb, noun)* from POS-tagged review if their tag conforms to any of the feature patterns.

4.2   Recommendation Solution

Firstly, to find the best user attribute α ($\alpha \in A$) at each any node of the tree T, we use the *payoff* function (based on [2]). Secondly, to determine the ranked list of the products recommended to customers, we use a rank function (based on [2]).

When a decision tree T is built, each node of the tree divides the training samples into two sub-sets. Each internal node represents the set of users whose attributes match the attributes at all internal nodes on the path from the root to that node. The target is to find an attribute of the users so that it can separate the set of users into two sub-sets.

Each node of the tree T contains a top-k ranked list of products. We consider a customer who has followed until a node q of the tree T. If q is a leaf node, the customer receives a top-k ranked list of product recommendations. If q is an internal node, the customer is asked the question that corresponds to q. Depending on his/her answer, it leads him/her to one of the two sub-trees of node q. The customer also does not answer the question and receive the current recommendation at node q.

The learning tree structure: The structure of the tree T is equivalent to find a question (user's attribute $\alpha \in A$) at each node. In order to solve this task, we formulate as below:

Given a node q and the set of users U_q at node q, we find a candidate $\alpha \in A$ that splits set of U_q into two subsets: $U_q(\alpha)$ contains α and $U_q(\overline{\alpha})$ does not contain α. Thus, we have $U_q = U_q(\alpha) \bigcup U_q(\overline{\alpha})$ and $U_q(\alpha) \bigcap U_q(\overline{\alpha}) = \emptyset$. In order to find the best user attribute α to divide U_q at node q, we need to evaluate the *payoff* function as below.

$$payoff(q, \alpha) = combine(payoff(U_q(\alpha)), payoff(U_q(\overline{\alpha})), |U_q(\alpha)|, |U_q(\overline{\alpha})|, |U_q|) \tag{1}$$

Where $payoff(U)$ function evaluates the quality of ranking for the set of users U. We consider all possible user attributes $\alpha \in A$, and choose the maximum of payoff as a splitter criterion to partition the set of users into two groups. Thus, we have the equation as below:

$$splitter(q) = arg \max_{\alpha \in A} payoff(q, \alpha) \tag{2}$$

Learning to rank: We solve the problem of learning-to-rank for the products in P at node q. Learning to rank can be used in a wide variety of applications in information retrieval (IR), natural language processing (NLP) and data mining (DM). The input of this task are (based on [2]): the set of users U_q at node q; product table P; and the review table R. In fact, we only consider the set of records (denoted by R_q) of the table R corresponding to set of U_q.

The task is to learn a rank function: $P \rightarrow R$, which each record p ($p \in P$), specifies its features and returns a $rank(p)$. We use a linear function to learn weight coefficients for the product features. To handle this, we need to represent the categorical features as the boolean features (*0* or *1* values).

The target is to find a weight vector $w = \{w_1, w_2, .., w_{mp}\}$; m_p is the number of features of the product p. Thus, we can apply learning-to-rank method to node q. Node q contains a set P_q of training samples which are the results of joining the table P with the review sub-table R_q. In order to learn a ranking function, we use a pairwise RankSVM [6] and RankBoost method [3]. When the weight vector w is learned, the ranking function is calculated by $rank(p) = w^T p$. Thus, we have a ranking of the whole set of P by $rank(p_1) \geq ... \geq rank(p_{np})$.

The ranking of product at q is learned based on the set of users U_q who have ratings on the products of q. We denote that $rank_U$ is the ranking function for a set of users U. The quality of ranking is evaluated by the *eval* function such as the mean reciprocal ranking (*MRR*) method. We denote $eval(rank_U)$ as the quality of $rank_U$. Therefore, the quality of the ranking should be reflected in the *payoff* function.

$$payoff(U) = eval(rank_U) \tag{3}$$

Where:

$$eval(rank) = \frac{2|\{(p_i, p_j) \in P_q^2 | rank(p_i) > rank(p_j)\}|}{n(n-1)} \tag{4}$$

Discussion: Our recommendation solution is based on the method in [2] to build the structure of decision tree. However, our RS is different from the previous one. Specifically, we have built the feature extraction model to extract product features which are the input of our model; we used patterns of opinion words/phrases from the real review text through adjectives, adverbs, verbs, and nouns. From this task, we yielded product features and opinion words of users. This step generates very important information to make the summarization for products. We also applied and made the comparison for some of the learning-to-rank methods such as RankBoost and RankSVM that are described in the next section.

5 Experiment

5.1 Dataset

We implemented our RS on amazon datasets for the computer product. We used 35 products, 372 users, and 6300 reviews. Each review has an average of 7 sentences and each sentence has an average of 20 words.

5.2 Preparing

The First Step. To extract product features and user opinions, we used the feature extraction described in Sect. 4. We conducted our implementation on the PC (X201s core I7, Memory 8G), Java environment (JDK 1.8), and Eclipse. We focused on extracting product features and opinion words from customer review products. We extracted 22 product features. Our extraction feature model is described through examples in Fig. 4 and Table 1.

A fantatic ultra portable laptop. I bought the 12" Powerbok G4 1Ghz, and have been blown away by it. It has a good metallic design and its size is very easy to carry around, this is the major disadvantage of larger bulkier notebooks. I have had mine upgraded to 1.25 GB Ram, and the performance of applications like Microsoft Office and Adobe Photoshop is great. You can even play games like Halo on it pretty well, if that is your thing. You can have up to 80GB of HD, and my advice is to buy the largest you can affort. I also have an Airport (802.11g) card installed, and it allow wireless connectivity with a pretty good range. In addition, Bluetooth built-in allowed me to use a very nice wireless mouse with it. The keyboard is of good quality, an important feature if you have to write long documents in Word! The LCD screen is of good quality too. All in all it made a pleasant switch from my previous bulky Sony Vaio Lap. I would recommend this computer very highly.

Fig. 4. Customer review product

From our dataset, we also yielded the frequency of each feature in Fig. 5.

The Second Step. We built a decision tree denoted by T that guides the user in making their shopping decisions. The internal node of the tree T corresponds to user attributes such as user opinions on product features. A user starts from the root of the tree and answers the questions until reaching the leaf. Our decision is described in Fig. 6.

Table 1. Pattern phrases and Product Features.

Pattern phrases	Product features	Opinion phrases
JJ:metallic, NN:design	design	metallic design
JJ:major, NN:disadvantage	disadvantage	major disadvantage
JJ:good, NN:quality	quality	good quality
JJ:important, NN:feature	feature	important feature
JJ:pleasant, NN:switch	switch	pleasant switch
JJ:wireless, NN:connectivity, mouse	connectivity, mouse	

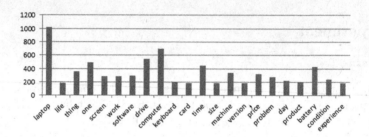

Fig. 5. Frequency of each feature

5.3   Setup and Metrics Evaluation

In order to evaluate our RS, we partitioned each of our datasets into training and test sets via 5 folds cross-validation. The main evaluation metric that we use in our experiments to measure the quality of recommendations is mean reciprocal rank (MRR), which is a meaningful measure of single-item retrieval.

Mean reciprocal rank (MRR). Reciprocal Rank (RR) is very important. It calculates the reciprocal of the rank at which the first relevant document was retrieved. The mean reciprocal rank (MRR) is calculated by the average of the reciprocal ranks of results for a sample of queries q. With one relevant item per test user, we measured the recommendation quality by finding out the top of the list the relevant items.

$$MRR = \frac{1}{q} \sum_{i=1}^{q} \frac{1}{rank_i} \tag{5}$$

where $rank_i$ is the rank position of the ith test user's relevant item in the ranked list of items returned. In our measure, we separated our data into the train and test sets in the way that the test set contains users who have created items high. Thus, the MRR is more meaningful.

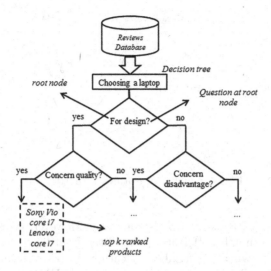

Fig. 6. Our recommendation system

5.4   Experimental Results

RankSVM. RankSVM [6] generates recommendations based on learning item feature weights. In our implementation, we set the training error regularization parameter c to 0.0001. RankSVM consists of a learning module (svm_rank_learn) and a module for making a prediction (svm_rank_classify). RankSVM uses the same input and output as SVM-light which can be called as follow.

   svm_rank_learn -c 0.0001 train.dat model.dat

which trains a RankSVM on the training set train.dat and outputs the learned rule to model.dat using the regularization parameter c set to 0.0001. When model.dat is showed, we can calculate weight vector $w=\{w_1, w_2, .., w_{mp}\}$ and rank function $rank(p) = w^T p$.

RankBoost. We used the RankLib[1] for calculating document ranking with boosting. The number of round [-round <T>] is set to be 10 for our training datasets, which can be called as follow.

   java -jar RankLib.jar -round 10 -ranker 2 -train train.dat -save model.dat

The input train.dat is the same as for SVM-rank.[2] When model.dat is showed, we can calculate RakBoost by using the command line below.

   java -jar RankLib.jar -load model.dat -rank train.dat -score rankBoost.dat

Figure 7 showed a comparison of recommendation quality measured by the MRR over 5 folds for RankSVM, RankBoost. We see that when the number of

---

[1] http://people.cs.umass.edu/vdang/ranklib.html.
[2] http://www.cs.cornell.edu/People/tj/svmlight/svmrank.html.

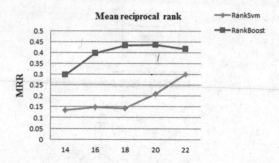

Fig. 7. Number of features with MRR

features increases, MRR of our RS also increases. However, RankBoost slightly decrease when the number of features increases from 20 to 22. RankSVM is gradually increased when the number of features increases from 18 to 22. The results showed that Rankboost is better than RankSVM when the number of features increases from 14 to 22. Nevertheless, if we increase the number of features to a threshold, RankSVM is considered to be better.

We measured the errors of RankSVM and RankBoost. Figure 8 shows a comparison of errors for each fold.

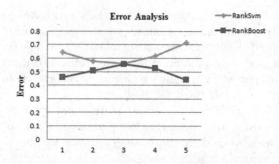

Fig. 8. Error analysis

It can be observed that the average error of RankBoost is 49.71 % while the average error of RankSVM is 62.26 %.

6   Conclusion

In this paper, we proposed some techniques for extracting opinion features from products reviews based on Standford CoreNLP and feature pattern methods. After that, we built a RS that helps users in shopping online based on opinions of other users for product attributes.

Our system represented a learning tree structure used to guide shopper in their purchases. Each node of the tree is a question that is easy to understand for non-expert users. The users will follow a path from the root node by answering the question at each internal node and reach to the leaf node where they can receive the top-k ranked list of products. Our RS also produces a suitable recommendation at each internal node of the tree so that the user may stop answering at any time. We used the learning-to-rank method to provide a list of top-k products recommendations for the user given the current answers. We implemented our RS with some learning to rank methods including RankSVM and RankBoost.

In our future work, we plan to further improve techniques for extracting features and add more data in our experiment. We try to extract which features customers strongly like and dislike. This will further improve the feature extraction.

References

1. Balakrishnan, S., Chopra, S.: Collaborative ranking. In: Proceedings of the Fifth ACM International Conference on Web Search and Data Mining, pp. 143–152. ACM (2012)
2. Das, M., De Francisci Morales, G., Gionis, A., Weber, I.: Learning to question: leveraging user preferences for shopping advice. In: Proceedings of the 19th ACM SIGKDD International Conference on Knowledge Discovery and Data Mining, pp. 203–211. ACM (2013)
3. Freund, Y., Iyer, R., Schapire, R.E., Singer, Y.: An efficient boosting algorithm for combining preferences. J. Mach. Learn. Res. 4, 933–969 (2003)
4. Hu, M., Liu, B.: Mining and summarizing customer reviews. In: Proceedings of the Tenth ACM
5. Hu, M., Liu, B.: Mining opinion features in customer reviews. In: Proceedings of the 19th National Conference on Artificial Intelligence, pp. 755–760
6. Joachims, T.: Training linear SVMS in linear time. In: Proceedings of the 12th ACM SIGKDD International Conference on Knowledge Discovery and Data Mining, pp. 217–226. ACM (2006)
7. Manning, C.D., Surdeanu, M., Bauer, J., Finkel, J.R., Bethard, S., McClosky, D.: The stanford corenlp natural language processing toolkit. In: ACL (System Demonstrations), pp. 55–60 (2014)
8. Resnick, P., Iacovou, N., Suchak, M., Bergstrom, P., Riedl, J.: Grouplens: an open architecture for collaborative filtering of netnews. In: Proceedings of the 1994 ACM Conference on Computer Supported Cooperative Work, pp. 175–186. ACM (1994)
9. Socher, R., Perelygin, A., Wu, J.Y., Chuang, J., Manning, C.D., Ng, A.Y., Potts, C.: Recursive deep models for semantic compositionality over a sentiment treebank. In: Proceedings of the Conference on Empirical Methods in Natural Language Processing (EMNLP), vol. 1631, p. 1642. Citeseer (2013)

An Evidential Method for Multi-relational Link Prediction in Uncertain Social Networks

Sabrine Mallek[1,2(✉)], Imen Boukhris[1], Zied Elouedi[1], and Eric Lefevre[2]

[1] LARODEC, Institut Supérieur de Gestion de Tunis,
Université de Tunis, Tunis, Tunisia
sabrinemallek@yahoo.fr, imen.boukhris@hotmail.com,zied.elouedi@gmx.fr
[2] Univ. Artois, EA 3926, Laboratoire de Génie Informatique Et D'Automatique de
L'Artois (LGI2A), 62400 Béthune, France
eric.lefevre@univ-artois.fr

Abstract. Link prediction is an important problem that permits to analyze networks' evolution. The task is to estimate the likelihood of the existence of future links. Yet, social networks relate individuals via several
types of relationships. Thus, it is more interesting to predict the existence
and the type of a future link. We focus in this paper on predicting links in
multiplex social networks since these latter allow simultaneous relationships with several types. Furthermore, we take into account the uncertainty characterizing the prediction process and social networks noisy and
missing data. To this end, we firstly propose an uncertain graph-based
model for multiplex social networks that encodes the uncertainty degrees
at the edges level using the belief function framework. Furthermore, a
novel link prediction approach is subsequently introduced to estimate
both the existence and the type of a new link while taking uncertainty
into account. Empirical evaluation on two preprocessed real world social
networks that support our proposals is provided.

Keywords: Social network analysis · Link prediction · Uncertain social
network · Multiplex social networks · Belief function theory

1 Introduction

During the last years, the World Wide Web has linked tens of thousands to
millions of individuals through the Social Web. A great deal of information has
become accessible, social network analysis has arisen as a tool to extract and
study the patterning of such data. Social networks became the main focus of
researchers and analysts from various domains. They are generally conceptual-
ized as graphs where the nodes represent the actors linked by social relationships.
One of the major problems handled in network mining and social network analysis is the study of social networks evolving including the prediction of future or
hidden links, which is known as the link prediction problem. The task is to
evaluate the likelihood of the establishment of a new link between two nodes
according to an observed snapshot of the network.

© Springer International Publishing AG 2016
V.-N. Huynh et al. (Eds.): IUKM 2016, LNAI 9978, pp. 280–292, 2016.
DOI: 10.1007/978-3-319-49046-5_24

The structure of social networks depends on the quality of the entities and the ties under consideration. They can be homogenous/heterogeneous, uniplex/ multiplex, weighted/unweighted, directed/undirected, etc. For instance, if there is only one type of a relationship between two actors i.e., two friends, two co-workers or two collaborated authors then the tie is called uniplex. On the other hand, if several relationships are shared then it is called multiplex (multi-relational, multi-layered, multi-dimensional) i.e., if two people are friends, co-workers and live in the same building, their association is a multiplex tie (a three-way one). The link prediction has to take into account the topological structure of the social network, especially the characteristics of the links. Most of the traditional methods ignore the relationships labels between the entities, they only treat the existence of the links. However, the likelihood and the type of a connection are frequently interrelated [2,4].

Yet, most of the state of the art link prediction methods consider links with binary values i.e., 1 (exists) or 0 (¬ exists). Conversely, social networks structure highly rely on the precise nature of the data. Sparse bias alter considerably the analysis results. In contrast, as discussed in [1,11], social networks data are often exposed to observation errors and are frequently noisy. Uncertainty can be cast from lack of information for determining the correctness of a statement then quantified on a numerical scale. For example, we can quantify imprecision about the responses that we get when constructing networks from surveys and encapsulate these uncertainties in the structure processing to get an uncertain network. According to [20], errors about the components of multiplex social networks are expected to be larger as a result of inaccurate experimental settings or technical issues. In particular, the complexity of multiplex network may affect their properties more adversely. For instance, the extra links can possibly be just duplicates that were generated erroneously by the tools used for the construction of the social network. Besides, real world collections are often missing or have a number of incorrect labels and links. Consequently, one has either to remove a probable valuable information or to take into account all the uncertain information from the data [11]. This imprecision impacts directly the network structure and therefore the outcomes of the analysis. Indeed, we show in previous works [14,15] the relevance of handling uncertainty whether within the structure or throughout the link prediction process. However, we treated uniplex social networks where there are only uni-relational links between the actors.

Accordingly, we embrace the belief function theory [5,19] to deal with imperfect data and manage uncertain knowledge as it is a general framework for reasoning under uncertainty. We first introduce a new graph model for multiplex social network graph that encodes uncertainty at the edges level. Subsequently, a novel approach for the prediction of new links along with their types in multiplex social networks is proposed. It is inspired from node neighborhood methods and uses exclusively the belief function tools. The common neighbors are considered as independent sources of evidence, information is transferred and combined and is revised afterwords to get a closer picture about the existence of a future link. This paper is organized as follows. Section 2 presents briefly some related works on link prediction. In Sect. 3, essential belief function notations and concepts are re-called. Sections 4 and 5 detail our proposals where a new model for uncertain multiplex social network is presented along with a method for the prediction of new connections and their type. Section 6 reports the experimental results. Section 7 concludes the paper.

2 Related Work on Link Prediction

Link prediction has a great applicability in a wide variety of domains as it plays a key role in network analysis. Namely, it is applied to recommend new friends or items in social networks, detect criminals in dark networks, explore missing links in biological networks, etc. The objective is to evaluate the likelihood of a new association between two unlinked nodes given a snapshot of the network.

Most traditional methods consider simple networks allowing only one type of relations. However, a relevant aspect is not treated which is the types of the links. Actually, prediction of link existence and link type are often considered as two independent problems. In the first case, one predicts the future linkage between two nodes, conversely, in the second case, one assumes that the link exists and tries to predict its type. Yet, these two problems are interrelated [2,4].

As a matter of fact, multiplex social networks highlight the diversity of the links' types and allow simultaneous relationships as it is an aspect of real social life. Besides, multiplex information is sometimes more useful which is why many works join networks from several platforms to get a more informative multi-relational network [7,22]. Formally, a multiplex social network can be defined as a graph $G(V, E_1, \ldots, E_n)$ where V is the set of entities and $E_1, \ldots, E_n$ are the sets of edges each belonging to specific relationship (layer, dimension). In the following, we present traditional methods for link prediction for both uniplex and multiplex structures and we introduce the intuition of our proposed approach.

2.1 Link Prediction in Uniplex Networks

There are two groups of link prediction methods depending on the considered information of the network. The first group of methods uses the local structural properties of the nodes i.e., common neighbors, common circles, or the nodes' attributes i.e., age, interests, gender. Yet, the nodes' attributes are usually not available or hidden due to privacy anonymization constraints. Thus, most of the local information based methods use structural similarities metrics. Namely, the most popular ones are "Common Neighbors", "Jaccard's Coefficient" and "Adamic/Adar". For instance, the common neighbors measure, denoted by CN, counts the jointly connected neighbors of a pair of nodes (u, v). It can be defined as $CN_{uv} = |\tau(u) \cap \tau(v)|$ where $\tau(u)$ and $\tau(v)$ are respectively the set of neighbors of u and v. The common neighbors metric predicts links to nodes with many common neighbors. It is based on the intuition that the more two persons share mutual friends, the likely to become friends which has been demonstrated by Kossinets' and Watts analysis [10] made on a large scale social network of student friendships. Furthermore, it is fast and yields to very well results in practice. The second group of methods uses global information based on global topological properties of the network. Popular approaches include Hitting time, SimRank or the shortest path to reach a node. However, these latter suffer from high complexity since they inquire for global topological properties. Besides, global information is not always available. Also, the additional complexity does not always pay off since local methods can give great performance as well [12].

2.2 Link Prediction in Multiplex Networks

Correspondingly, when dealing with multiplex social networks, two options are available. Treat each layer independently using uniplex graph measures or handle directly the network using multiplex measures [2]. For instance, the neighborhood of a node may be considered in different ways in a multiplex i.e., the union of all the neighbors in all the dimensions or more restrictively, the intersection of the neighbors set across all the layers [9]. Few attempts have been made to address link prediction in multiplex social networks due to manipulation difficulties and the lack of available data of such networks. Some methods tackled it via supervised and unsupervised learning [18]. There is another category of methods that treated link prediction in multiplex networks as a matrix factorization problem [3,8]. For instance, the authors in [8] proposed a latent factor model to predict multiple links using tensor factorization. However, this category of methods suffer from high complexity issues especially when the number of links' types increases. Other methods applied measures based on local or global information. For instance, the authors in [4] proposed a probabilistic weighted version of the common neighbors method and computed prediction scores for each link type. They subsequently tested the new score under unsupervised and supervised learning. Actually, structural similarity measures are frequently used as features for classification or as similarity scores for unsupervised learning.

In this paper, the intuition of the structural local measures is adopted. More precisely, we draw on the method of the common neighbors as it is simple and shows great results in many previous works [10,17]. Yet, we take both information across local layers and the overall information about the global network into consideration in the prediction task. We adopt the belief function theory to handle uncertainty as it permits to represent and manage imperfect knowledge.

3   Basics of Belief Function Theory

Mathematical notations and definitions of the belief function theory [5,19] essential for the understanding of our proposals are given in this section.

Let Θ be the frame of discernment. It is a finite set including exhaustive and mutually exclusive events associated to the problem. Let 2^{Θ} be the power set of Θ. A basic belief assignment (bba), denoted by m, is the mass attached to an event given a piece of evidence. It is defined as:

$$m : 2^{\Theta} \to [0, 1]$$

$$\sum_{A \subseteq \Theta} m(A) = 1. \tag{1}$$

When an element $A \subseteq \Theta$ of a mass function m such that $m(A) > 0$, it is called a focal element.

The fusion of two masses m_1 and m_2 derived from two reliable and distinct sources of evidence is ensured using the conjunctive rule of combination [21] denoted by $\bigcirc$. It is defined by:

$$m_1 \bigcirc m_2(A) = \sum_{B,C \subseteq \Theta : B \cap C = A} m_1(B) \cdot m_2(C). \tag{2}$$

On the other hand, when at least one of the sources is reliable but we do not know which one it is, the disjunctive rule of combination denoted by $\bigcirc$ is used [21]. It is defined by:

$$m_1 \bigcirc m_2(A) = \sum_{B,C \subseteq \Theta : B \cup C = A} m_1(B) \cdot m_2(C). \tag{3}$$

The reliability of the source can be evaluated by a coefficient $\alpha \in [0, 1]$. A discounting mechanism [19] could therefore be performed on m. The discounted mass function is denoted by $^{\alpha}m$ and we have:

$$\begin{cases} ^{\alpha}m(A) = (1 - \alpha) \cdot m(A), \forall A \subset \Theta \\ ^{\alpha}m(\Theta) = \alpha + (1 - \alpha) \cdot m(\Theta). \end{cases} \tag{4}$$

where α is called the discounting rate. $m(\Theta)$ symbolizes the state of ignorance, it is equal to 1 when $\alpha = 1$. It corresponds to a vacuous mass function. In other terms, the source is fully unreliable. On the other hand, when $\alpha = 0$, the source is considered to be fully reliable.

In some cases, one may want to revise a mass function m by reinforcing an element A of the frame. This can be done using the reinforcement correction mechanism [16], which is similar to the discounting operation but unlike the later the masses of the focal elements are recovered and redistributed to the element A instead of Θ. Let $\beta \in [0, 1]$ be the reinforcement rate, the reinforcement towards the element A is defined by:

$$\begin{cases} ^{\beta}m(A) = (1 - \beta)m(A) + \beta \\ ^{\beta}m(B) = (1 - \beta)m(B), \forall B \subseteq \Theta \text{ and } B \neq A. \end{cases} \tag{5}$$

To set up the relation between two disjoint frames Θ and Ω, we can use a multi-valued mapping operation [5]. Actually, a multi-valued mapping function denoted by τ allows to assign the subsets $B \subseteq \Omega$ that can possibly accord a subset $A \subseteq \Theta$:

$$m_{\tau}(A) = \sum_{\tau(B) = A} m(B). \tag{6}$$

One of the solutions for decision making in the belief function framework is the transformation of the mass functions into pignistic probabilities denoted by $BetP$. It is defined by [21]:

$$BetP(A) = \sum_{B \subseteq \Theta} \frac{|A \cap B|}{|B|} \frac{m(B)}{(1 - m(\emptyset))}, \forall A \in \Theta. \tag{7}$$

4   Evidential Multiplex Social Network

In previous works [14,15], we developed an evidential graph-based versions of social networks that handle uncertainty at the edges level. However, the latter model only supports uni-relational homogenous connections between the nodes. Consequently, this paper extends it the multi-relational case. The proposed evidential multiplex social network graph is defined as $G(V, E_1, \ldots, E_n)$ where $V = \{v_1, \ldots, v_{|V|}\}$ is the set of nodes, and $E_1, \ldots, E_n$ are the sets of edges with n being the number of types. Each edge $uv \in E_i$ has assigned a bba defined on the frame of discernment $\Theta_i^{uv} = \{E_{uv}, \neg E_{uv}\}$ denoted by m_i^{uv}. The event E_{uv} means that uv exists and $\neg E_{uv}$ means that it is absent. The bba m_i^{uv} quantifies the degree of uncertainty regarding the existence of a link of type i between (u, v).

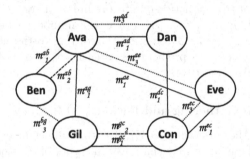

Fig. 1. A multiplex social network graph with bba's weighted edges

Figure 1 gives an illustration of such a graph structure. The links are weighted with bba's rather than binary values (either 1 or 0) to quantify the uncertainty regarding their existence. The nodes may share three different types of relationships namely m_1, m_2 and m_3. These latter are schematized differently where each type represents a specific association. Hence, there are three layers with the same number of nodes. For the sake of lucidity, a link uv of type i is schematized if its pignistic probability $BetP_i^{uv}(E_{uv}) > 0.5$. In other words, its likelihood to exist is greater than 50%.

5   Link Prediction in Evidential Multiplex Social Networks (LPEM)

Our proposed approach is based on the intuition of the common neighbors technique. This latter has proved its effectiveness in various real networks and usually got the best performances with respect to other local measures [10,17]. We have proposed, in previous works [14,15], methods for future links' existence in uncertain social networks. However, we worked on uniplex social network graphs where only a single type of a relationship is allowed. Application of these methods to the multiplex case is possible by carrying out the same process for each layer separately. However, information regarding the whole structure of the multiplex is not treated. In this work, we take into account the multiplicity of the relations between the nodes. Information about the links' types is treated in order to predict the likelihood of the existence of a link in a specific layer. We draw on the methods based on local structural properties by considering the common neighbors present throughout the global graph.

At first, evidence from the neighboring nodes is gathered from all the layers of the network where each one is considered as an independent source of evidence. Then, the evidence collected from each layer is evaluated according to its reliability. Subsequently, the resulting beliefs are revised according to the distribution of simultaneous links of specific types in the multiplex. Indeed, global information is mandatory for a successful overall link prediction. From this point of view, the steps for the prediction of a future link uv in an evidential multiplex social network $G(V, E_1, \ldots, E_n)$, where n is the number of possible types, are as outlined below. Each step is illustrated according to the network presented in Fig. 1 where the aim is to predict the potential existence of one or multiple relations between Ava and Con.

5.1   Information Gathering and Fusion

Firstly, we extract the subgraph $G_C(V_C, E_C)$ containing the common neighbors of u and v. Since these latter may share various types of relationships with their common neighbors, we decompose G_C into n graphs where each graph $G_i(V_i, E_i)$ includes the links belonging to the type of associations $i \in \{1, \ldots, n\}$ such that

$$V_C = V_i \text{ and } E_C = \bigcup_{i=1}^{n} E_i.$$

Then, for each graph G_i, the masses of each neighboring link xy are transferred to the frame of uv_i using a multivalued mapping operation (Eq. 6) $\tau : \Theta_i^{xy} \to 2^{\Theta_i^{uv}}$ in order to get the mass in G_i collected from xy denoted $m_{xy,i}^{uv}$ as follows:

- The mass $m_i^{xy}(\{E_{xy}\})$ is transferred to $m_{xy,i}^{uv}(\{E_{uv}\})$;
- The mass $m_i^{xy}(\{\neg E_{xy}\})$ is transferred to $m_{xy,i}^{uv}(\{\neg E_{uv}\})$;
- The mass $m_i^{xy}(\Theta_i^{xy})$ is transferred to $m_{xy,i}^{uv}(\Theta_i^{uv})$.

Thereafter, we combine the masses that we get from the neighboring links according to the presence of the common neighbors in G_i to get the overall mass of uv_i denoted m_i^{uv}. For instance, if all the common neighbors of the pair (u, v) are in G_i, the masses are fused using the conjunctive rule (Eq. 2). That is, all the common neighbors are considered as reliable sources. In contrast, if there are common and uncommon neighbors of (u, v) in G_i, the transferred masses are fused using the disjunctive rule of combination (Eq. 3). Hence, we consider at least one of the common neighbors as a reliable source of evidence. G_i is ignored if (u, v) does not share any common neighbor in it and we get a vacuous mass function.

Accordingly, if considering *Ava* and *Con* from Fig. 1 as the query nodes, we catch the subgraph containing their common neighbors which are *Dan*, *Eve* and *Gil* (the node *Ben* is disregarded as it is not a common neighbor). The common neighbors subgraph is subsequently decomposed into three graphs where each one includes the links belonging to a specific type as shown in Fig. 2. G_1 contains the shared links of type 1, G_2 contains the shared links of type 2 and G_3 contains the shared links of type 3.

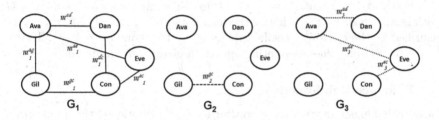

Fig. 2. Decomposition into common neighbors' subgraphs of each type

The masses of the neighboring links in each subgraph G_i are transferred to the mass m^{ac} of the query link ac connecting *Ava* and *Con*. Hence, for each subgraph G_i of type i we get the following bba's:

$$\begin{cases} G_1 : m_{ad,1}^{ac}, m_{ae,1}^{ac}, m_{ag,1}^{ac}, m_{gc,1}^{ac}, m_{dc,1}^{ac} \text{ and } m_{ec,1}^{ac} \\ G_2 : m_{gc,2}^{ac} \\ G_3 : m_{ad,3}^{ac}, m_{ae,3}^{ac} \text{ and } m_{ec,3}^{ac} \end{cases}$$

In order to get the overall mass function for each type, we combine the transferred masses according to the presence of the common neighbors in the subgraphs. For instance, we fuse the obtained masses in G_1 using the conjunctive rule of combination (Eq. 2) since all the common neighbors are present. Hence, we get:

$$m_1^{ac} = m_{ad,1}^{ac} \textcircled{\cap} m_{ae,1}^{ac} \textcircled{\cap} m_{ag,1}^{ac} \textcircled{\cap} m_{gc,1}^{ac} \textcircled{\cap} m_{dc,1}^{ac} \textcircled{\cap} m_{ec,1}^{ac}$$

However, the masses from G_3 are combined using the disjunctive rule (Eq. 3) since there are common and uncommon neighbors i.e., only *Eve* is a common neighbor, *Gil* and *Dan* are not common neighbors anymore. That is, the bba's are combined using the disjunctive rule:

$$m_3^{ac} = m_{ad,3}^{ac} \textcircled{U} m_{ae,3}^{ac} \textcircled{U} m_{ec,3}^{ac}$$

On the other hand, G_2 is discarded since there are no common neighbors on it and we get a final bba $(m_2^{ac}(\Theta_2^{ac}) = 1)$.

5.2 Reliability Evaluation

Next, we focus on the overall reliability of the sub-graphs G_i with respect to the global graph G_C. We compute the distribution of the common neighbors across all the subgraphs G_i defined by $\lambda_i = \frac{|CN_{uv_i}|}{|CN_{uv}|}$. We use $\alpha_i = 1 - \lambda_i$ as a discounting rate to discount m_i^{uv}. In doing so, we get a discounted mass function $^{\alpha_i}m_i^{uv}$ as follows:

$$\begin{cases} ^{\alpha_i}m_i^{uv}(\{E_{uv}\}) = (1 - \alpha_i) \cdot m_i^{uv}(\{E_{uv}\}) \\ ^{\alpha_i}m_i^{uv}(\{\neg E_{uv}\}) = (1 - \alpha_i) \cdot m_i^{uv}(\{\neg E_{uv}\}) \\ ^{\alpha_i}m_i^{uv}(\Theta_i^{uv}) = \alpha_i + (1 - \alpha_i) \cdot m_i^{uv}(\Theta_i^{uv}) \end{cases} \qquad (8)$$

For example, in Fig. 2, Ava and Con share three common neighbors in G_1 (Gil, Eve and Dan) thus, $\lambda_1 = \frac{3}{3} = 1$. They share one common neighbor in G_3 which is Eve thus, $\lambda_3 = \frac{1}{3}$. That is, the reliabilities of G_1 and G_3 are respectively quantified by the reliability coefficients $\alpha_1 = 1 - 1 = 0$ and $\alpha_3 = 1 - \frac{1}{3}$. That is, G_1 is fully reliable. The discounting operation gives $^{\alpha_1}m_1^{ac}$ and $^{\alpha_3}m_3^{ac}$.

5.3 Evidence Reinforcement

The obtained masses are revised according to the distribution of the simultaneous links of more than two types in the multiplex $G(V, E)$. For instance, if we consider a sub-graph G_i and it already exists exactly one link of a type $j \neq i$ between u and v. We compute the distribution of simultaneous 2-relational associations of types i and j denoted by S_{ij}^2 with respect to all the simultaneous relations of exactly two types in G denoted by S_G^2. Generally, when there are $m \leq n - 1$ simultaneous links between (u, v), we seek to the distribution S_{*j}^{m+1} where $* = \{1, \ldots, m\}$ are the types of the shared links. If $S_{*j}^{m+1} \neq 0$, we reinforce the mass on the element "exist" using $\beta = \frac{S_{*j}^m}{S_G^m}$ as a reinforcement rate (Eq. 5) and obtain the mass $^{\beta}m_i^{uv}$.

As illustrated in Fig. 1, Ava and Con do not share any links. Thus, the obtained bba's after the discounting operation are final. However, if they had a link, for example of type 1. One goes through the reinforcement step (Subsect. 5.3) by considering the overall simultaneous 2-relational connections in the global graph. As illustrated in Fig. 1, there are five 2-relational associations i.e., between (Ava, Ben), (Ava, Eve), (Ava, Dan), (Eve, Con), and (Gil, Con). Hence, we compute the distribution of simultaneous 2-relational associations of types 1 and 2 i.e., $S_{12}^2 = 2$ and the types 1 and 3 i.e., $S_{13}^2 = 3$ with respect to all the simultaneous relations of exactly two types in G i.e., $S_G^2 = 5$. That is, $^{\alpha_3}m_2^{ac}(\{E\})$ and $^{\alpha_3}m_3^{ac}(\{E\})$ are reinforced using respectively $\beta_2 = \frac{S_{12}^2}{S_G^2} = \frac{2}{5}$ and $\beta_3 = \frac{S_{13}^2}{S_G^2} = \frac{3}{5}$ as reinforcement rates. We get the final bba's $^{\beta_2}m_2^{ac}$ and $^{\beta_3}m_3^{ac}$.

5.4   Links Selection

Most of the methods use the ranking of the similarity scores and consider the L highest ones as the predicted links. In contrast, we first compute the pignistic probability $BetP_i^{uv}$ of the query link uv. Decision about its existence is made later according to the ranking of the pignistic probabilities on the event "exists" of all the analyzed links.

6   Experimental Evaluation

In order to evaluate our proposals, we test our approach on two real word networks. A network of 185 students cooperation linked by 362 edges according to 3 relations [6] and a relationships network of 21 K links between 84 persons with 5 links' types collected from the social evolution dataset [13]. To the best of our knowledge, there are no uncertain multiplex social network datasets available. Thus, we build our networks artificially by simulating mass functions on the links regarding their existence. We proposed, in previous works [14,15], methods for the pre-processing of a social network to transform it into an uncertain one where the links are valued with bba's. That is, we simulate mass functions according to the technique proposed in [14]. It is based on a widely applied procedure of graph sampling used in link prediction literature [23].

A comparative study with the method proposed in [15] is conducted for the two datasets. Actually, the latter method called belief link prediction (BLP), predicts new links under the belief function theory framework. It is stimulated from the common neighbor method. However, it only predicts new links of single types. Therefore, it is applied for each set of links of a particular type separately.

Precision and recall are used for performance evaluation. The precision depicts the number of relevant links ϵ according to the number of analyzed links δ. It is defined as follows:

$$\text{Precision} = \frac{\epsilon}{\delta}. \tag{9}$$

The recall expresses the correctly predicted existing links ϵ with respect to the number of correctly and falsely predicted existing ones γ. It is defined as follows:

$$\text{Recall} = \frac{\epsilon}{\gamma}. \tag{10}$$

Table 1 reports the obtained results in terms of precision and recall for the two networks. For the BLP method, we average the values of the precision and recall obtained solely in each layer in the networks. As it can be seen, the proposed algorithms outperforms the BLP method in terms of both precision and recall among the two datasets. The LPEM approach gives higher prediction quality measured by precision 80 % for the students cooperation dataset compared to 46 % for the BLP and 72 % for the relationships network compared to 53 % for the BLP. That is, our method is able to predict both links' existence and types efficiently. The same applies to recall, the LPEM gives much higher values

Table 1. Results measured by precision and recall

Method	Evaluation metric	Students cooperation	Relationships network
LPEM	Precision	0.80	0.72
	Recall	0.65	0.59
BLP	Precision	0.46	0.53
	Recall	0.35	0.41

than the BLP method i.e., 65 % for the students cooperation compared to 35 % given by the BLP method. A possible reason for having such results by the BLP method is the sparsity of the homogeneous relationships. This implies that global information about the network contributes to the prediction task. Besides, our method is more convenient as it does not inquire its application as many times as there are layers. Yet, results given by the precision are higher than those of the recall measure for both methods. This points out that we are getting more incorrect existing links than incorrect non existing links. Still, the new approach has proved validity and performance empirically. To the best of our knowledge, this is the first approach proposed to predict future links' existence and types in multiplex social network while dealing with uncertainty.

7   Conclusion

In this paper, we have presented an uncertain multi-relational graph model for multiplex social networks that encapsulates the uncertainty degrees using mass functions encoding the likelihood of the edges' existence. In addition, we have given a new link prediction method that permits to predict both the existence and the type of new associations in multiplex social networks and deals with uncertainty at the same time. Specifically, the proposed method fully operated using the tools of the belief function framework. Evidence from the common neighbors of the global graph is revised and transferred to the query link frame of discernment. It is subsequently updated and fused with the overall information gathered from all the layers to predict both existence and type of the link. A future direction is to take the actors attributes into consideration. It is clear that some attributes relate directly to the type of the associations between the nodes. Therefore, we intend to investigate the impact of the nodes' attributes on the links' types and consequently link prediction.

References

1. Adar, E., Ré, C.: Managing uncertainty in social networks. Data. Eng. Bull. **30**(2), 23–31 (2007)
2. Battiston, F., Nicosia, V., Latora, V.: Metrics for the analysis of multiplex networks. CoRR abs/1308.3182 (2013)

3. Cao, B., Liu, N.N., Yang, Q.: Transfer learning for collective link prediction in multiple heterogeneous domains. In: Proceedings of the 27th International Conference on Machine Learning, pp. 159–166 (2010)
4. Davis, D., Lichtenwalter, R., Chawla, N.V.: Multi-relational link prediction in heterogeneous information networks. In: Proceedings of the 2011 International Conference on Advances in Social Networks Analysis and Mining, pp. 281–288 (2011)
5. Dempster, A.P.: Upper and lower probabilities induced by a multivalued mapping. Ann. Math. Stat. **38**, 325–339 (1967)
6. Fire, M., Katz, G., Elovici, Y., Shapira, B., Rokach, L.: Predicting student exam's scores by analyzing social network data. In: Proceedings of the 8th International Conference on Active Media Technology, pp. 584–595 (2012)
7. Hristova, D., Noulas, A., Brown, C., Musolesi, M., Mascolo, C.: A multilayer approach to multiplexity and link prediction in online geo-social networks. CoRR abs/1508.07876 (2015)
8. Jenatton, R., Roux, N.L., Bordes, A., Obozinski, G.: A latent factor model for highly multi-relational data. In: Proceedings of the 26th Annual Conference on Neural Information Processing Systems, pp. 3176–3184 (2012)
9. Kanawati, R.: Multiplex network mining: a brief survey. IEEE Int. Info. Bull. **16**(1), 24–27 (2015)
10. Kossinets, G., Watts, D.: Empirical analysis of an evolving social network. Science **311**(5757), 88–90 (2006)
11. Kossinets, G.: Effects of missing data in social networks. Soc. Net. **28**, 247–268 (2003)
12. Liben-Nowell, D., Kleinberg, J.: The link-prediction problem for social networks. J. Am. Soc. Inf. Sci. Technol. **58**(7), 1019–1031 (2007)
13. Madan, A., Cebrian, M., Moturu, S., Farrahi, K., Pentland, A.: Sensing the health state of a community. Pervasive Comput. **11**(4), 36–45 (2012)
14. Mallek, S., Boukhris, I., Elouedi, Z., Lefevre, E.: Evidential link prediction based on group information. In: Prasath, R., Vuppala, A.K., Kathirvalavakumar, T. (eds.) MIKE 2015. LNCS (LNAI), vol. 9468, pp. 482–492. Springer, Heidelberg (2015). doi:10.1007/978-3-319-26832-3_45
15. Mallek, S., Boukhris, I., Elouedi, Z., Lefevre, E.: The link prediction problem under a belief function framework. In: Proceedings of the IEEE 27th International Conference on the Tools with Artificial Intelligence, pp. 1013–1020 (2015)
16. Mercier, D., Denœux, T., Masson, M.H.: Belief function correction mechanisms. In: Bouchon-Meunier, B., Magdalena, L., Ojeda-Aciego, M., Verdegay, J.-L., Yager, R.R. (eds.) Foundations of Reasoning Under Uncertainty. STUDFUZZ, vol. 249, pp. 203–222. Springer, Heidelberg (2010). doi:10.1007/978-3-642-10728-3_11
17. Newman, M.E.J.: Clustering and preferential attachment in growing networks. Phys. Rev. E **64**(2), 025102 (2001)
18. Pujaril, M., Kanawati, R.: Link prediction in multiplex bibliographical networks. Int. J. Complex Syst. Sci. **3**(1), 77–82 (2013)
19. Shafer, G.: A Mathematical Theory of Evidence. Princeton University Press, Princeton (1976)
20. Sharma, R., Magnani, M., Montesi, D.: Missing data in multiplex networks: a preliminary study. In: 2014 Tenth International Conference on Signal-Image Technology and Internet-Based Systems (SITIS), pp. 401–407, November 2014
21. Smets, P.: Application of the transferable belief model to diagnostic problems. Int. J Intell. Syst. **13**(2–3), 127–157 (1998)

22. Zhang, J., Yu, P.S., Zhou, Z.H.: Meta-path based multi-network collective link prediction. In: Proceedings of the 20th ACM International Conference on Knowledge Discovery and Data Mining, pp. 1286–1295 (2014)
23. Zhang, Q.M., Lü, L., Wang, W.Q., Zhu, Y.X., Zhou, T.: Potential theory for directed networks. PLoS ONE 8(2), e55437 (2013)

Detecting Thai Messages Leading to Deception on Facebook

Panida Songram[1(✉)], Atchara Choompol[2], Paitoon Thipsanthia[3], and Veera Boonjing[4]

[1] Polar Lab, Department of Computer Science, Faculty of Informatics, Mahasarakham University, Mahasarakham, Thailand
panida.s@msu.ac.th
[2] Department of Computer Science, Faculty of Agro-Industial Technology, Kalasin University, Kalasin, Thailand
atchara.pu@ksu.ac.th
[3] Department of Computer Education, Faculty of Liberal Arts and Sciences, Kalasin University, Kalasin, Thailand
paitoon@ksu.ac.th
[4] International College, King Mongkut's Institute of Technology Ladkrabang, Bangkok, Thailand
kbveera@kmitl.ac.th

Abstract. Social network has become a very popular communication for Thai people, especially Facebook. Unfortunately, this popularity also attracts deceiver spreading malicious messages to other users. Some messages lead to deception. This paper studies Thai messages posted on Facebook that lead to deception. We try to investigate different approaches to detect deceptive messages and find dominant words. To detect deceptive messages, the dataset is retrieved from Facebook pages. Next, content-based and context-based features are extracted from the dataset. Two algorithms, i.e. SVM and KNN, are applied to perform a prediction. We construct the experiments to investigate context-based and content-based features for detecting deceptive messages. The experimental results show that the context-based features gives the best performance and the F-measure for predicting deceptive messages achieves 99 % when using SVM classifier. In addition, dominant words in deceptive messages and truthful messages are reported in our work.

Keywords: Deception · Social network · Text mining · Machine learning · Thai messages

1 Introduction

Social networks are widely used in daily life of Thai people and have been growing at exponential rates. Currently, many people mainly use social media for communication. Facebook is one of the most popular social media in Thailand due to free of charge and sharing capability. In 2016, the amount of Facebook users in Thailand was expected to reach 18.9 million users [1]. Due to numerous users and sharing nature of Facebook, deceivers easily spread messages with malicious intention to other users. This may

© Springer International Publishing AG 2016
V.-N. Huynh et al. (Eds.): IUKM 2016, LNAI 9978, pp. 293–304, 2016.
DOI: 10.1007/978-3-319-49046-5_25

increase an opportunity of scams to use this channel to deceive people online. As reported in May 2016, many Thai people were victims of online deception and Thailand was on the first rank of the country with high rate of deception in Asia, especially job scam [2].

This paper focuses on Thai messages on Facebook relating to job application that possibly leading to deception. We concentrate deception on messages on social network because messages are first considered to be truth or deceptive and deceivers always post some deceptive messages on wall. Moreover, messages on Facebook are short and unstructure so studying about deception on Facebook messages is a challenge work. In addition, Thai messages are less structured, compared to English messages, which make deception detection is even more challenging. Furthermore, content-based features are investigated to detect deceptive messages in our work.

The rest of this paper is organized as follows. Section 2 summarizes related words. Section 3 explains our methodology for detecting deceptive messages on Facebook. Section 4 shows the experimental results and gives discussion. Finally, conclusion and direction of future work are given in Sect. 5.

2   Related Work

Currently, various approaches are proposed to detect online deception. Some researchers studied deception on messages. The use of linguistic cues on messages has been studied to indicate potentially deceptive communication. Zhou and Sung [3] proposed an empirical study on cues to detect deception on Mafia game message logs. Nine cues are selected to investigate deception behavior in Chinese community. The contents of messages are transformed to numerical values for each cue and then statistical measures are used to evaluate the cues. From the evaluation, it shows that the deceiver tended to communicate less and low complexity and high diversity in their messages. Next, Zhou [4] extensively studied deception behavior in Yahoo messenger. Eight variables, productivity, participation, initiation, spontaneous correction, word diversity, effect, cognitive complexity, non-immediacy, are investigated to compare between deceivers and truth tellers. The statistical measures are used to analyze the deceivers and truth-tellers. From the analysis, it shows that participation, initiation, cognitive complexity, non-immediacy and spontaneous corrections significantly differentiate deceivers from truth tellers.

Briscoe et al. [5] proposed linguistic cues to detect deception in social media communication and classified deception using machine learning on the proposed features. Deceptive communication was investigated to identify which cues are related to deception and dataset is created from the communication. Five cues are investigated: sentence length, sentence complexity, sentiment, emoticon usage and informality. From the investigation, it was found that higher sentence complexity, emoticon usage, and more extreme sentiment result in a lower perception of deception. Then five linguistic cues are used as features for classification. Various classifiers: Random Forest, Gradient Boosting, Support Vector Machines (SVM) and Perceptron are built on the proposed features and SVM achieves average 91 % accuracy in cross fold validation. Next, Appling et al. [6] proposed an method to evaluate the use of textual cues to discriminate between deception strategies. Participants are set to conversation on FaceFriend and asked to label one of four deception strategies; falsification, exaggeration, omission, and misleading. Then two classifiers, Random Forest and SVM, are developed to classify the deception strategies on structural data (e.g. psycholinguistic and trigram). Precision and recall are used to evaluate the performance of both classifiers. From the evaluation, Random Forest gives the best accuracy but still low accuracy because it is difficult to predict certain kinds of deception.

Moreover, some of researchers tried to detect deception from opinion messages. Ott et al. [7] studied deceptive opinion spam on hotel reviews and proposed several approaches for extracting features. The features are created from psychological deception, n-gram-based text categorization, and a combination of both n-gram and psychological deception. Naïve Bayes and SVM are used to perform the classification. From experimental results, they demonstrate that the features extracted from psychological deception and bigram with SVM gives accuracy to 89.8 % for detecting deceptive opinion spam. They proved that accuracy of text categorization is higher than psycholinguistic deception detection, genre identification, and human. Rayana and Akoglu [8] proposed opinion spam detection from behavior and content of messages. They proposed a new holistic approach called SPEAGLE to detect opinion spam and used a very small set of review message features as prior information. SPEAGLE is evaluated on three datasets YelpChi, YelpNYc and YelpZip that contain reviews for a set of restaurants and hotels. From the evaluation, it shows that SPEAGLE outperforms RANDOM, FRAUDEAGLE, PRIOR and Graph-based approach. Yafeng et al. [9] proposed an approach to correct the mislabel instances to find deceptive opinion spam from message reviews. A dataset is divided into several subsets and classifiers are generated for each subset and then the best classifier is selected to evaluate the whole dataset. Features are divided into four groups; lexicon, part of speech, deep syntactic information of text, and psycholinguistic features. The classifiers, Naïve Bayes, SVM, and Logistic regression, are tested on the features and selected the best classifier.

Detection of spam is another approach that tries to protect online deception due to deception comes with spam message [10, 11]. Bhat et al. [12] use community-based features for detecting spammer. Single and ensemble classifiers are compared for spammer classification. Decision tree, Naïve Bayes and KNN are evaluated as a single classifier. Three algorithms are also combined as ensemble classifiers by using Bagging, Boosting and Stacking. Zheng et al. [13] detect Spammer on Sina Weibo social network in China. A crawler is implemented for collecting 30,116 users and more than 16 million messages. Message content and user behavior are studied to extract a set of most important features. The features are used to classify spammer and non-spammer by using SVM. Gupta and Kaushal [14] proposed to classify spammer and non-spammer on Twitter. Features are extracted from user content and message content. Then Naïve Bayes, Clustering and Decision Tree are applied to detect Spammers.

A study of fake profile on social network is an approach for detecting deception. Fire et al. [15] proposed a method for detection of malicious profiles by using the social network's own topological features. First, fake profiles are simulated into a social network and real profiles are collected from social network. Then all profiles are extracted to four features: (1) the degree of the user, (2) the number of communities the user is connected to, (3) the number of connections exists among the friends of the user, and (4) the average number of friends inside each of the user's connected communities. Naïve Bayes and Decision Tree are applied to classify fake profile. Alowibdi et al. [16] proposed a frame work for detecting deception about gender information in online social networks. Gender indicators are obtained from first name, user name, and layout color. The indicators are used as features and applied Bayesian classifier for detecting deception profile.

From the most previous works, online deception was studied on content of messages. Actually, context of messages is an important data to detect deception as shown in [7]. Now, there are few works studying deception on context of messages. For example, Mihalcea and Strapparava [17] proposed to recognize deceptive language on statement datasets. More and Kalkundri [10, 11] proposed to evaluate deceptive mails using messages in the email as dataset. In our works, we propose an approach to detect deceptive messages on Facebook which are short and lack of structure, which are different from messages in E-mail or statements that are more formal. In addition, we first study deceptive messages by considering context of complex Thai messages on Facebook.

3   Deception Detection on Facebook

In our work, deceptive messages are detected based on text categorization manner. The overview of deception detection is shown in Fig. 1.

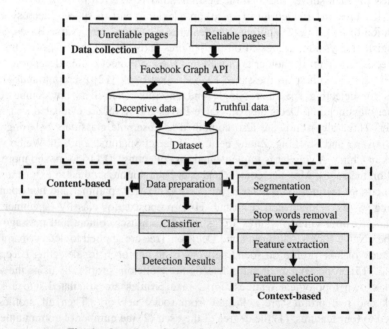

Fig. 1. Overview of deception detection on Facebook

3.1 Dataset Collection and Analysis

In this work, we concentrate on job application deception on Facebook due to Thai people are mostly trapped online job scams [2] and Facebook is the most popular social media in Thailand. To study deceptive and truthful messages on Facebook, first, we manually select 21 unreliable pages and 12 reliable pages about job application with our observations as follows:

- The number of page likes: The number of likes of unreliable pages is very small. While the number of likes of reliable pages are very large because people always share useful information on reliable pages to other people and then the others always likes the pages.
- Page information: Unreliable pages do not give details of page information because the owners do not want to be tracked. The owners may give mobile phone number or email that can be changed latter. If deception is occurred, victims cannot contact them. For reliable pages, the owners give all details of page information and put the office phone number in page information.
- The number of likes of message: The number of likes of messages on unreliable pages is very small because people aware that the messages are unreliable and potentially be deceptive messages so they do not like the messages. While the messages on reliable pages give useful information and reliable so many people like the messages.
- The number of shares: if people see useful information they always share the information to others so that the number of shares of messages on reliable pages is high. On the other hand, messages on unreliable pages do not be shared because people have known that the messages are untruthful.
- Photo: Unreliable pages always show money, travel, and eating photos to trap people work with them.

To confirm the reliable and unreliable pages, Cohen's kappa statistic is used to support the observations. Five experts evaluate 33 pages to determine whether they are reliable or unreliable. The kappa value is 0.85 which indicates the agreement between the five experts. Therefore, the pages are capable for experiments and evaluation.

We develop a program to retrieve dataset on reliable and unreliable pages by using Facebook Graph API [18]. The dataset consists of post messages, a number of likes, a number of shares, a number of comments and a number of pictures. The data on unreliable pages are collected as a set of candidate deceptive data. The candidate truthful data are collected from reliable pages. Then we manually select 1,189 truthful data and 1,189 deceptive data as the dataset. All selected messages relate to job application and the average length of messages is 926 characters.

3.2 Data Preparation

For context-based features, they are extracted from messages. Since the messages are unstructured data, they are transformed to vector space model before classification as follows:

First, the messages are segmented to find a collection of features. Since Thai texts do not have word boundaries as English texts, dictionary-based approaches are adopted to segment Thai messages. In our work, we adopt the longest matching and maximal matching algorithms to segment Thai messages. Both algorithms are dictionary-based and available in SWATH [19, 20] program. The longest matching algorithm (LM) is a greedy algorithm. It scans an input message from left to right and selects a longest word that is matched a dictionary entry. The matched longest word is segmented from the message and then scans the rest of message to match a longest one in the dictionary and so on. If the rest of message cannot match with any dictionary entry, the algorithm will backtrack to find the next longest one [21]. Maximal Matching algorithm (MM) applies a dynamic programming technique to segment Thai message. It produces all possible segmentation of message and then selects the one that contain the fewest words [22].

Second, all of non-redundant words from the segmentation are collected. The stop words, numbers, and signs are removed from the collection. We also remove numbers and signs because they are not necessary for deceptive detection and the number of features will be large because of a large amount of numbers and signs in dataset. Moreover, there are words in messages that specify what the numbers are. The rest of collection of words is selected as unigram feature sets. In addition, we try to consider a bigram feature set where n is a contiguous sequence of n words. The bigram feature set is created based on maximal matching method.

Third, All messages are transformed to vector space model. Each message is a vector. Each vector consists of feature weights. In our work, we investigate three features weighting methods, Boolean weighting, TF weighting, and TF $\times$ IDF weighting. The Boolean weighting is simple but effective. It gives a weight of feature to 1 if the feature occurs in the message, else 0. TF weighting gives the weight of feature to frequency of features appeared in the message. TF $\times$ IDF weighting combines the term frequency with the inverted message frequency to reflect the importance of feature in a dataset.

Due to a large number of features always generate from text dataset, we also perform feature selection to improve the prediction performance of classifier and to reduce computation cost of classifiers. In our work, we perform feature selection by using linear classifier weights. We adopt feature weights learned by SVM classifier to select the most relevant features. The SVM classifier was improved that it is effective on text dataset or sparse data [23] with linear kernel output predictions in form:

$$prediction(x) = sgn[b + w^T x] \text{ for } w = \sum_i \alpha_i x_i \tag{1}$$

where $x_i = (x_{i1}, x_{i2}, \ldots, x_{id})$ is a vector and d is the number of distinct features in the model. The vector of weights $w = (w_1, w_2, \ldots, w_d)$ can be computed and access directly. For feature selection, the absolute value $|w_j|$ is used as the weight of feature j. The feature j with weight close to 0 has a smaller effect on the classification than features with high value of weight. Features with small value of weight are removed because they are not important for classification.

For content-based features, they consist of seven features; a number of likes, a number of shares, a number of comments, a number of pictures, a number of hashtags, a number of URLs, and the length of message. A number of hastags, a number of URLs and the length of message are extracted from the messages.

3.3   Classifiers

In our work, we test dataset with SVM, and KNN to find the best classifier for detecting deceptive messages. Both classifiers perform well in text classification which is the significant manner for detecting deceptive messages. SVM [24] uses linear or non-linear delineations between classes to divide the data space. The method tries to determine the optimal boundaries that maximally split the different classes. SVM is a popular classifier adopted in text classification [25–28]. In our work, we use linear kernel in SVM for learning because liner kernel performs well when there is a lot of features and document vectors are sparse that are characteristic of our dataset. A new message x can be classified in form as shown in (1).

Also, the KNN [29] method is often used in text classification [30–32]. KNN is a non-parametric classification method with effective computation for text classification which is a manner used in our work. It classifies a new document by using the class label of the k most similar neighbor in training documents. The similarity is measured by Euclidean distance or cosine similarity between two documents [33]. In our work, K is set to 3 and we use Euclidean distance as shown in (2), where x_i is weight of feature i in a testing document and y_i is weight of feature i in a training document, n is the number of distinct features.

$$d = \sqrt{\sum_{i=1}^{n} (x_i - y_i)^2} \tag{2}$$

3.4   Evaluation Metrics

For evaluation of classifiers, deceptive and truthful messages are predicted in a confusion matrix as shown in Table 1, where a represents the number of truthful messages correctly classified, b is the number of truthful messages misclassified as deceptive messages, c refers to the number of deceptive messages misclassified as truthful messages, and d expresses the number of deceptive messages correctly classified. Accuracy, precision, recall, F-measure are metrics evaluated in our work. Accuracy evaluates the overall predictions of classifier that were correct. It is defined as Acc = $(a + d)/(a + b + c + d)$. Precision is the number of correct predictions from all predictions for each class. Precision for deceptive class is defined as $P_D = d/(b + d)$. Precision for truthful class is defined as $P_T = a/(a + c)$. Recall is the number of correct prediction from all true data for each class. Recall for deceptive class is defined as $R_D = d/(c + d)$ and recall for truthful class is defined as $R_T = a/(a + b)$. F-measure is the harmonic mean between precision and recall, F-measure for deceptive class is defined as $F_D = 2P_D R_D/(P_D + R_D)$ and F-measure for truthful class is defined as $F_T = 2P_T R_T/(P_T + R_T)$.

Table 1. Confustion matrix

true	predicted	
	truthful	deceptive
truthful	a	b
deceptive	c	d

4 Identifying Dominant Words

In this work, we are also interested to find characteristics of deceptive messages. We proposed a simple method to find dominant words for each class. Given a collection of words $W = \{w_1, w_2, ..., w_d\}$ be a set of distinct words appearing in deceptive and truthful messages, the potential of word w_i appearing in truthful messages is calculated by $P(w_i|T) = n(w_i, T)/n(T)$, where $n(w_i, T)$ is the number of occurrences of word w_i in truthful messages and $n(T)$ is the number of occurrences of words in truthful messages. The potential of word w_i appearing in deceptive messages is calculated by $P(w_i|D) = n(w_i, D)/n(D)$, where $n(w_i, D)$ is the number of occurrences of word w_i in deceptive messages and $n(D)$ is the number of occurrences of words in deceptive messages. The dominant score of word w_i is calculated as in $P(w_i|D) - P(w_i|T)$. If the score is high positive value, the word is very dominant in deceptive messages. If the score is high negative value, the word is very dominant in truthful messages.

5 Experiments and Discussion

After dataset preparation phase, two unigram feature sets are constructed. The first unigram feature set (unigram$_{LM}$) consists of 4,256 features which is the collection of words segmented by longest matching method and the second unigram feature set (unigram$_{MM}$) consists of 4,276 features which is the collection of words segmented by maximal matching method. A bigram feature set consists of 4,275 features. First we tried to train classifiers from all features with different extraction methods. Then SVM and KNN are applied for deceptive detection. We conduct 10-fold cross validation to divide training and testing sets. The performance was evaluated in average accuracy, average precision, average recall, and average F-measure of each class. From Table 2, we can see that Boolean weighting gives highest accuracy. Particularly, the unigram$_{MM}$ with Boolean weighting gives 98.28 % accuracy and 98.28 % F-measure for deceptive class when using SVM. In addition, we found that the unigram feature sets give higher accuracy than bigram feature set in most cases. Since deceptive messages often contain exaggerated language [7], the most accuracies, precision, recall, F-measure are more than 90 % when using unigram feature sets.

Next, we tried to select important features to improve the deceptive detection performance. Features were selected on vector space generated by the unigram$_{max}$ with Boolean weighting. The selected features were learned by the best SVM classifier from 10-fold cross validation. From Table 3, we eliminate 122 features with weight 0 and then get 4,154 features with 98.11 % F-measure when using SVM classifier. Then we tried to find a sufficient number of features that give the best F-measure. We observe that SVM classier gives at least 90 % F-measure when selecting features from top 100 highest weighted features. The F-measure increases to 99.12 % when selecting the top 400 highest weighted features to detect deceptive messages.

In addition, we present the top 10 highest dominant words in deceptive messages and truthful messages in Table 4. The dominant words in deceptive messages are about income and work at home because deceivers want to trap other people join with them. The dominant words in truthful messages are more specific about job application and position.

Table 2. Classifer performance with context-based features

Feature extraction	Classifier	Acc	Truthful			Deceptive		
			P_T	R_T	F_T	P_D	R_D	F_D
unigram$_{LM}$ +Boolean	SVM	97.86	97.71	98.32	97.86	98.09	99.15	97.84
	KNN	97.86	99.22	98.32	97.82	96.58	100.00	97.89
unigram$_{LM}$ +TF	SVM	97.52	97.50	100.00	97.52	97.59	98.32	97.52
	KNN	96.76	96.97	96.64	96.76	96.58	97.48	96.77
unigram$_{LM}$ + TF × IDF	SVM	97.44	97.98	97.64	97.42	96.96	97.48	97.45
	KNN	92.64	98.47	86.55	92.13	87.94	99.16	93.09
unigram$_{MM}$ +Boolean	SVM	**98.28**	98.41	96.61	**98.27**	98.17	99.16	**98.28**
	KNN	97.31	98.47	94.96	97.27	96.25	100.00	97.34
unigram$_{MM}$ + TF	SVM	97.52	97.93	97.48	97.51	97.19	95.76	97.52
	KNN	96.93	97.07	95.80	96.93	96.83	99.16	96.93
unigram$_{MM}$ + TF × IDF	SVM	97.48	97.91	97.48	97.47	97.10	99.16	97.48
	KNN	92.56	98.66	80.67	92.02	87.91	99.16	93.02
Bigram + Boolean	SVM	93.92	94.50	96.77	96.77	94.21	54.55	90.39
	KNN	95.39	94.91	98.39	95.52	96.61	84.85	93.20
Bigram + TF	SVM	89.04	87.95	100.00	92.22	95.11	73.53	81.23
	KNN	80.15	77.04	98.39	86.74	98.60	38.24	60.19
Bigram + TF × IDF	SVM	93.00	92.57	98.39	94.94	96.03	67.65	88.51
	KNN	95.09	95.29	98.39	96.24	94.92	94.12	92.89

Table 3. Classifer performance with feature selection

No. of Features	F_D	
	SVM	KNN
100	90.26	97.38
200	97.23	98.10
300	98.36	98.61
400	**99.12**	98.53
500	98.95	98.49

(continued)

Table 3. (*continued*)

No. of Features	F_D	
	SVM	KNN
600	98.21	98.91
700	98.70	97.66
800	98.99	97.49
900	**99.21**	97.84
1000	99.04	97.87
4,154	98.11	97.39
All	98.28	97.34

Table 4. Top 10 dominant words

Deceptive		Truthful	
รายได้ (income)	ติด (contact)	เปิดรับสมัคร (call for)	ตำแหน่ง (position)
สนใจ (interest)	ข้อมูล (information)	สมัครงาน (application)	สมัคร (apply)
ที่บ้าน (at home)	คน (human)	อัตรา (position)	รับสมัคร (openings)
บ้าน (home)	ติดต่อ (contact)	จำกัด (limited)	เปิดรับ (openings)
รายได้เสริม (extra income)	ทำที่ (location)	รายละเอียด (details)	งาน (job)

We also investigate the experiment by using content-based features and the combination of the content-based features with the top 400 highest weighted features in the context-based features. The experimental result is shown in Table 5. From Tables 3 and 5, they show that using context-based features for detecting deceptive messages provides the highest F-measure when using SVM classifier.

Table 5. Classifer performance with content-based features and the combination of content-based and context-based features

Features	Classifier	Acc	Truthful			Deceptive		
			P_T	R_T	F_T	P_D	R_D	F_D
content-based	SVM	81.45	89.72	82.35	79.22	76.10	95.80	83.25
	KNN	**91.93**	95.67	88.24	**91.57**	88.87	96.64	**92.25**
content-based +context-based	SVM	**98.82**	99.08	99.16	**98.82**	98.58	100.00	**98.83**
	KNN	92.30	95.92	86.55	91.91	89.27	94.12	92.60

6   Conclusion

In this work, we propose a methodology for detecting Thai messages leading to deception on Facebook. We extract context-based and content-based features from the dataset. Then, we apply two classifiers for deception detection, SVM and KNN. The experimental results show that context-based features give higher performance than content-based features for detecting deceptive messages, especially, the unigram feature sets with Boolean weighting. The selected features learned from linear SVM classifier can improve F-measure to 99 %. In addition, we observe that dominant words of deceptive messages are about income and work at home. To extend the proposed method, a job for scam filtering on Thai messages is implemented on Facebook. Also, the deception detection in other domains on Facebook is designated as a future wrok.

Acknowledgements.   This work was supported by National Research Council of Thailand Grants.

References

1. Statistics. http://www.statista.com/statistics/490467/number-of-thailand-facebook-users/
2. Dailynews. http://www.dailynews.co.th/it/385782 (in Thai)
3. Zhou, L., Sung, Y.-W.: Cues to deception in online chinese groups. In: 41st Annual Hawaii International Conference on System Sciences, p. 146. IEEE Computer Society (2008)
4. Zhou, L.: An empirical investigation of deception behavior in instant messaging. IEEE Trans. Prof. Commun. **48**, 147–160 (2005)
5. Briscoe, E.J., Appling, D.S., Hayes, H.: Cues to deception in social media communications. In: 47th Hawaii International Conference on System Sciences, pp. 1435–1443 (2014)
6. Appling, D.S., Briscoe, E.J., Hutto, C.J.: Discriminative models for predicting deception strategies. In: 24th International Conference on World Wide Web, pp. 947–952. ACM. Florence, Italy (2015)
7. Ott, M., Choi, Y., Cardie, C., Hancock, J.T.: Finding deceptive opinion spam by any stretch of the imagination. In: 49th Annual Meeting of the Association for Computational Linguistics: Human Language Technologies. vol. 1, pp. 309–319. Association for Computational Linguistics, Portland, Oregon (2011)
8. Rayana, S., Akoglu, L.: Collective opinion spam detection: bridging review networks and metadata. In: 21th ACM SIGKDD International Conference on Knowledge Discovery and Data Mining, pp. 985–994. ACM, Sydney, NSW, Australia (2015)
9. Ren, Y.F., Ji, D.H., Yin, L., Zhang, H.B.: Finding deceptive opinion spam by correcting the mislabeled instances. Chin. J. Electron. **24**, 52–57 (2015)
10. More, S., Kalkundri, R.: Evaluation of deceptive mails using filtering & WEKA. In: 2015 International Conference on Innovations in Information, Embedded and Communication Systems (ICIIECS), pp. 1–4 (2015)
11. More, S., Kulkarni, S.A.: Data mining with machine learning applied for email deception. In: 2013 International Conference on Optical Imaging Sensor and Security (ICOSS), pp. 1–4 (2013)
12. Bhat, S.Y., Abulaish, M., Mirza, A.A.: Spammer classification using ensemble methods over structural social network features. In: 2014 IEEE/WIC/ACM International Joint Conferences on Web Intelligence (WI) and Intelligent Agent Technologies (IAT), pp. 454–458 (2014)

13. Zheng, X., Zeng, Z., Chen, Z., Yu, Y., Rong, C.: Detecting spammers on social networks. Neurocomputing **159**, 27–34 (2015)
14. Gupta, A., Kaushal, R.: Improving spam detection in online social networks. In: 2015 International Conference on Cognitive Computing and Information Processing (CCIP), pp. 1–6 (2015)
15. Fire, M., Katz, G., Elovici, Y.: Strangers Intrusion Detection - Detecting Spammers and Fake Profiles in Social Networks Based on Topology Anomalies (2012)
16. Alowibdi, J.S., Buy, U.A., Yu, P.S., Stenneth, L.: Detecting deception in online social networks. In: 2014 IEEE/ACM International Conference on Advances in Social Networks Analysis and Mining (ASONAM), pp. 383–390 (2014)
17. Mihalcea, R., Strapparava, C.: The lie detector: explorations in the automatic recognition of deceptive language. In: ACL-IJCNLP 2009 Conference Short Papers, pp. 309–312. Association for Computational Linguistics, Suntec, Singapore (2009)
18. Facebook. https://developers.facebook.com/docs/graph-api
19. Charoenpornsawat, P.: Feature-based Thai Word Segmentation (in Thai). Computer Engineering, Master. Chulalongkorn University, Bangkok (1999)
20. Surapant Meknavin, P.C., Boonserm, K.: Feature-based thai word segmentation. In: The Natural Language Processing Pacific Rim Symposium (1997)
21. Poowarawan, Y.: Dictionary-based Thai Syllable Separation (in Thai). In: Ninth Electronics Engineering Conference (1986)
22. Sornlertlamvanich, V.: Word Segmentation for Thai in a Machine Translation System (1993)
23. Mladeni, D., 0263, Brank, J., Grobelnik, M., Milic-Frayling, N.: Feature selection using linear classifier weights: interaction with classification models. In: 27th Annual International ACM SIGIR Conference on Research and Development in Information Retrieval, pp. 234–241. ACM, Sheffield, United Kingdom (2004)
24. Cortes, C., Vapnik, V.: Support-Vector Networks. Mach. Learn. **20**, 273–297 (1995)
25. Joachims, T.: Text categorization with support vector machines: learning with many relevant features. In: Nédellec, C., Rouveirol, C. (eds.) ECML 1998. LNCS, vol. 1398, pp. 137–142. Springer, Heidelberg (1998). doi:10.1007/BFb0026683
26. Chen, D., Liu, Z.: A new text categorization method based on HMM and SVM. In: 2010 2nd International Conference on Computer Engineering and Technology (ICCET), pp. V7-383-V387-386 (2010)
27. Yu-ping, Q., Xiu-Kun, W.: Study on multi-label text classification based on SVM. In: Sixth International Conference on Fuzzy Systems and Knowledge Discovery, pp. 300–304 (2009)
28. Ramaswamy, S.: Multiclass Text Classification a Decision Tree Based SVM Approach. CS294 Practical Machine Learning Project (2006)
29. Guo, G., Wang, H., Bell, D., Bi, Y., Greer, K.: KNN model-based approach in classification. In: Meersman, R., Tari, Z., Schmidt, Douglas, C. (eds.) OTM 2003. LNCS, vol. 2888, pp. 986–996. Springer, Heidelberg (2003). doi:10.1007/978-3-540-39964-3_62
30. Fang, L., Qingyuan, B.: A refined weighted K-Nearest neighbors algorithm for text categorization. In: 2010 International Conference on Intelligent Systems and Knowledge Engineering (ISKE), pp. 326–330 (2010)
31. Han, X., Liu, J., Shen, Z., Miao, C.: An optimized K-Nearest neighbor algorithm for large scale hierarchical text classification. In: ECML-PKDD PASCAL Workshop, pp. 2–12 (2011)
32. Jiang, S., Pang, G., Wu, M., Kuang, L.: An Improved K-nearest-neighbor Algorithm for Text Categorization. Expert Syst. Appl. **39**, 1503–1509 (2012)
33. Liao, Y., Vemuri, V.R.: Using text categorization techniques for intrusion detection. In: 11th USENIX Security Symposium, pp. 5–9 (2002)

Answer Validation for Question Answering Systems by Using External Resources

Van-Tu Nguyen[1] and Anh-Cuong Le[2(✉)]

[1] VNU University of Engineering and Technology, Ha Noi City, Vietnam
tuspttb@gmail.com
[2] Faculty of Information Technology, Ton Duc Thang University,
Ho Chi Minh City, Vietnam
leanhcuong@tdt.edu.vn

Abstract. This paper focuses on extracting question-answer pairs on the Internet which is an useful resource for building Automated Question Answering systems. Question-answer pairs from public resources usually contain noisy information, mostly in the answers. Therefore to obtain reliable question-answer pairs, the answers need to be validated. Previous studies usually handled this problem based on the relationship between a question and its corresponding answers. Differently, this paper proposes a new approach that uses external resources to validate the reliability of answers from question-answer pairs crawled from the Web. We will combine both kinds of information, one is the matching between question and its answers while the other is based on the supporting of external resources to the answers. The experiment conducted on the question-answer pairs extracted from Yahoo!Answer and StackOverflow shows the effectiveness of our proposed method.

Keywords: Answer validation · Using external resources · Definition question · Wikipedia · Question-Answering

1 Introduction

Automated Question Answering (QA) has been becoming an important research direction in natural language processing and artificial intelligence. In general, there are two kinds of QA systems. The first one builds a generative model and answer a question by knowledge encoded in this model. The second one is called Community Question Answering systems because it uses the answering information from communities (usually in the Web) where questions and answers are posted. This paper follows the second one, in which when receiving a question, the system will search its corresponding answers from somewhere in communities. Recent years have seen an explosion of question-answer information created by the user on the Internet such as Yahoo!Answers (https://answers.yahoo.com/) or StackOverflow (http://stackoverflow.com), where somebody post their questions and others give answers. These are rich resources containing a large

© Springer International Publishing AG 2016
V.-N. Huynh et al. (Eds.): IUKM 2016, LNAI 9978, pp. 305–316, 2016.
DOI: 10.1007/978-3-319-49046-5_26

number of question-answer pairs in many different topics. We can see many valuable answers here therefore they are important resources for any QA sytems to exploit.

The cQA sites offers a large and rich resource of question-answer information. Extracting question-answer pairs with high reliability is an important task for building the knowledge base, information retrieval and non-factiod QA. However, the question-answer pairs on cQA sites are usually very noisy, mostly in the answers. Therefore before using, the question-answer pairs needs to be validated of the answers. This validation helps us to build a repository of question-answer pairs with high reliability. Under our observation, previous studies just focus only on exploiting the relationship between questions and answers, or use users' votes.

QA aims to seek an accurate and concise answer to a free form question from a large collection of text data, rather than a full document, judged relevant as in standard information retrieval tasks. Nevertheless, most of the QA researches mainly focus on locating the exact answer to a given factoid question in the related documents. The most well known evaluation on the factoid QA task is the Text Retrieval Conference (TREC)[1] and Cross Language Evaluation Forum (CLEF)[2]. The annotated questions and answers released by TREC have become important resources for the researchers. However, when facing a non-factoid question such as *why, how, what about, however*, almost no automatic QA systems work very well [12]. There are more challenges for the researchers when developing QA systems for non-factoid questions. There is no training data available for this problem, most automated systems train either on a small hand-annotated corpus built in-house [4] or on a data set of question-answer pairs harvested from Frequently Asked Questions (FAQ) lists [10]. None of these situations is ideal because the cost of building a training data set is so high. In such systems, due to the cost of data construction then it just works in a specific domain. Therefore extracting knowledge from open resources (usually from Internet) is an appropriate approach for building community QA systems for open domains.

In this paper we propose a new approach for validating the answers extracted from cQA websites, in which we will use external resources as additional information to determine a answer is reliable or not. We limit our method on the definition question - a kind of non-factoid question. This kind of question appears frequently in the cQA websites, and it is easier to find the information from external resources for validation. Wikipedia - the open encyclopedia is chosen as the external resource to validate reliability of answers. We will utilize features extracted from a question and its answers, as well as new features by using documents from Wikipedia. We performed our experiments using Wikipedia English version with question-answer information source is the question-answer pairs taken from the cQA sites Yahoo!Answers and StrackOverflow.

---

[1] http://trec.nist.gov/.

[2] http://clef.isti.cnr.it/.

2   Related Work

The recent studies on evaluating the quality of the answers provided by cQA sites through its textual representation features [2], such as the length of a question, length of an answer, overlapped words between a question and its answer, length ratio between a question and its answer. Another common features used in quality analysis come from social interaction measures [2,7,9], such as the number of the best answers assigned by users, the votes of users, quality ratings, and answers acceptance ratio.

Another approach is to use the combination of features such as lexical, syntactical, information users and the similarity between questions and answers. To evaluate the similarity between questions and answers, some recent studies have used word representation [2,11,15]. To recognize the high-quality answers, Hu et al. [6] learned the joint representation for each question-answer pair by taking both of textual and non-textual features as the input of the model. To select high quality answers, Yu et al. [14] and Zhou et al. [17] proposed convolution neural networks based models to represent the question and answer sentences. For matching between a question and an answer, some methods have used deep learning techniques to learn the distributed representation of each question-answer pair and use it as the input. Instead of extracting a variety of features, these approaches learn the semantic features to represent questions and answers. However, these approaches only focus on modeling the semantic relevance between a question and its answer, there is no consideration for evaluating the reliability of this answer.

For the definition questions, i.e., questions like "What is atom?" or "Who is Aaron Copland?" have drawn much attention recently. H. Toba et al. [13] have used multiple feature kinds extracted from a pair of question-answer such as lexicon, statistical features (e.g. number of sentences in an answer, number of words in the question, etc.), the similarity between the question and its answer. And then, this study used a SVM classifier to label an answer with good or bad, and achieve the highest accuracy score of 81.56 % . There are many studies have addressed on finding the correct answers given a question, such as in [5,12,16]. These studies based on the target question to find corresponding answers using many different resources. Wesley Hildebrandt et al. [5] used a set of collected documents and online dictionaries. Zhang et al. [16] combined different resources such as a collection of documents and some well-known websites containing rich knowledge as Encyclopedia (www.encyclopedia.com), Wikipedia (www.wikipedia.com), Merriam-Webster dictionary (www.mw.com) and a dictionary biographies(www.s9.com).

Scanning over related studies, we found that they just used the relationship between a question and its answer candidates. There is no studies focusing on determining the reliability of the answer candidates. Our study is different, we not only use the features representing the relationship between a question and its answer candidates but also use the features representing the support from external resource to the answer candidate which validates the reliability of this answer candidate.

3 Classification Approach for Answer Validation

3.1 Problem Definition

Validation of answers can be considered as a binary classification problem. Given a set of Q questions, where each question $q_i \in Q$ has a set candidate answers $\{a_{i1}, a_{i2}, ..., a_{in}\}$ (n = 1, 2,...). The validation of answers to questions q_i is labeled for the answers $\{a_{i1}, a_{i2}, ..., a_{in}\}$ respectively labels $\{l_{i1}, l_{i2}, ..., l_{in}\}$ where $l_{ij} = 1$ if a_{ij} is the correct answer to the question q_i, the reverse $l_{ij} = 0$.

3.2 Classification Algorithms and Evaluation

Machine Learning Approach. Most studies in answer classification of questions in cQA follow supervised machine learning approach. There are many different classification methods used such as: Support Vector Machine, Maximum Entropy Models [15], Conditional Random Field (CRF) [17]. Among these methods, Support Vector Machine with linear kernel function is using recent studies [1,11]. Therefore, SVM is the machine learning method used in our system. We can easily search for a many documents introducing about SVM and its applications, thus it is not necessary for presenting in detail here.

Classification Evaluation. Performance in answer classification is evaluated by the global accuracy of the classifier.

$$Accuracy = \frac{\# \ of \ correct \ predictions}{\# \ of \ predictions}$$

In addition to evaluating the performance of an answer classifier on a specific class, we use the measure of precision, recall, F_1-measure is presented in [3].

3.3 Feature Extraction

There are many different features that were extracted in order to serve for validating the reliability of answers in the cQA. In this section we present our summarization of feature extraction from previous studies as well as some new features proposed by us.

Lexical features: Lexical features are usually the context words appearing in the question. Each question-answer pair can be represented as the a vector. These features are called bag-of-words features or n-gram features. To extract n-gram features, any n consecutive words in a question-answer pair is considered as a feature.

Word counts in the question and answer: To validate the reliability of the answers we can base on the features which is the number of words in the question, the number of words in the answer. These features have also been used in studies of J. Bian et al. [2] and H. Toba et al. [13].

The number of sentences for each answer: The answer has more sentences often contain more informative content. These features have also been used in studies of J. Bian et al. [2] and H. Toba et al. [13].

The number of votes for each answer [7,9]: This is an important feature to know whether an answer is good or not. If an answer has a large number of votes (in comparison with other answers) it means that this answer is more meaningful and reliable. In contrast, the answers which received a few votes or no vote may carry little information to the question [15].

Word matching [11]: This feature calculates the similarity between a question and an its answer which is used in many studies to recognize the good answers [11,15]. If an answer candidate has a high similarity with the question, the answer candidate may contain a lot of information for answering the question.

To estimate the similarity between a question and an answer, we use the cosine score as the following:

$$cosin\_sim = \frac{\sum_{i=1}^{n} u_i * v_i}{\sqrt{\sum_{i=1}^{n} (u_i)^2} * \sqrt{\sum_{i=1}^{n} (v_i)^2}}$$

where u and v are binary bags-of-words vectors of the question and answer respectively (note that the stop-words are removed); u_i is the ith dimension. This feature returns the cosine similarity between the question and the answer.

Word vector representation [15]: We use the word vector representation to model the relevance between a question and an answer. We choose the word2vec[3] model proposed by Mikolov et al. [8]. The general idea of word2vec is to represent each word as a real vector that captures the contexts of word occurrences in a corpus. For a given question-answer pair, we extract word2vec vectors from a pre-trained set for all tokens for which one is available. We compute the centroids of the question and the answer, then use the cosine between the two obtained centroids as a feature.

4   Wikipedia as the External Resource

We will propose new features which represent for the supporting from Wikipedia to the answers for the validation. If the information in Wikipedia documents cover the content of a considering answer then it is more reliable.

---

[3] https://code.google.com/p/word2vec.

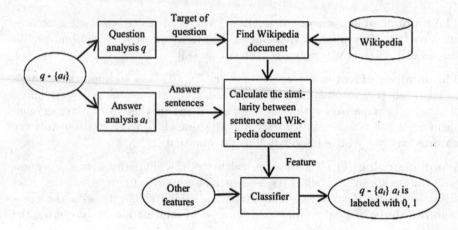

Fig. 1. The architecture of the answer validation system

4.1   Wikipedia

Wikipedia is today the largest encyclopedia in the world and surpasses other knowledge bases in its coverage of concepts, rich semantic knowledge and up-to-date content. Recently, Wikipedia has gained a wide interest in IR community and has been used for many different problems such as document classification, text clustering, and concept definition. Each article in Wikipedia describes a single topic: its title is a succinct, well-formed phrase that resembles a term in a conventional thesaurus.

4.2   Our Proposal of Using Wikipedia for Answer Validation

Non-factoid questions usually have forms such as *why* question, *definition* question or *how* question, etc. In this paper we focus on the answer validation of definition questions, which are often asked in QA systems. As mentioned before, in this paper we will use Wikipedia as the external resource to validation the reliability of the answers of the definition questions. The architecture of the answer validation system are described as Fig. 1.

Question Analysis. The question analysis module aims to find out what is the core information (the target of question) that the question to ask. To this end we implement the following steps:

1. Split each question q into a set of sentences $q = \{s_1, s_2, .., s_m\}$. ($m = 1, 2, ...$)
2. Determine whether a sentence is a definition sentence or not. We do that by using patterns described in Table 1. These patterns describe the templates of definition questions.
3. Each definition sentence is then analyzed to find the target of question. The target here includes the entities which need to be defined.

Table 1. The question and answer patterns used to definition question

Question pattern	Answer pattern
Who + tobe + <person name>?	<person name> + ...
What + tobe + <organization name>?	<organization name> +

For example, for the question: "What is atomic radius? What is atomic radius and why does it decrease when you go across the row (left to right) and increase down the group?", the definition sentence extracted is "What is atomic radius?" and the target of question is "atomic radius".

Extracting Related Wikipedia Documents. This module crawls Wikipedia documents which are relevant to the questions. Based on the target of questions, we will find the documents that their titles match the target. When a Wikipedia document is found, it will be used for the validating process.

Analyzing the Answer Candidates. The purpose of this module of analyzing the answer is to extract the best corresponding sentences for the question from a set of answer candidates. These sentences are determined by the Algorithm 1.

Algorithm 1. Algorithm determines the answer sentences

Input: question sentence q, the set of answer $\{a_i\}$.

Output: answer sentences for the question sentence q.

Step 1: Split each answer a_i into a set of sentences $a_i = \{s_1, s_2, ..., s_n\}$. (n = 1, 2,...). Initialize the score of each sentence $score(s_i) = 0$

Step 2: Calculate the similarity between each sentence s_j and question sentence q

$$score(s_j) = sim(s_j, q)$$

where $sim(s_j, q)$ is defined as the cosine similarity between s_j and q based on the vector of words of s_j and q

Step 3: If there exists s_j in a_i so that $score(s_j) > t$ (where t is the predefined threshold), then sentence s_j is chosen as an answer sentence to the question q

---

Extracting Features from Wikipedia. The similarity between an answer candidate and a Wikipedia document is estimated by the Algorithm 2.

Algorithm 2. Estimating the similarity between each answer and Wikipedia

Input: q - $\{a_1, a_2, \ldots, a_k\}$.
Output: $score(a_i)$

Step 1: Find corresponding Wikipedia documents whose titles match the target of question q. When each document is found, go to step 2
Step 2: Calculate the similarity between the answer sentence and the abstract of the selected Wikipedia document

---

The input for the Algorithm 2 includes a question q, its answer candidates $\{a_1, a_2, ..., a_k\}$, and the Wikipedia. At the first stage of this algorithm, we will find the corresponding Wikipedia documents given the answer candidate a_i. For the selected Wikipedia documents we just use their abstracts based on our assumption that they contain the most informative content of the definition required by the question q. The second stage will compute the similarity score between a_i and each selected Wikipedia document. For estimating the similarity, we propose to use two methods. The first one is based on word matching, and the second one use word representation (i.e. word2vec).

Denote s as the set of the answer sentences obtained by the Algorithm 1 and w is the abstract of the selected Wikipedia document, we have:

Method 1: We define the similarity $score(a_i)$ by:

$$score(a_i) = \sum_{t \in s} \frac{f(t, w)}{|s|}$$

where t is a word of s; $f(t, w)$ is the number of occurrences of the word t in the document w; $|s|$ is the number of words in s.

Method 2: We used the word vector representation to model the relevance between s and w. All the sentences in s and w are tokenized and then all words in s and w are transformed to vectors using a pre-trained word2vec model. Each word t_k in s will then be aligned to all words in w and select the highest vector cosine similarity, as follows:

$$score(t_k) = \max_{1 \leq h \leq m} (word2vec\_sim(t_k, b_h))$$

where:
m: the number of words in w.
t_k: word vector representation of the kth word in s.
b_h: word vector representation of the hth word in w.

$word2vec\_sim(t_k, b_h)$: is the cosine similarity of the two vector representations of t_k and b_h.

The similarity score between s and w is calculated as follows:

$$score(a_i) = \frac{\sum_{k=1}^{n} score(t_k)}{n}$$

where n is the number of words in s.

5   Experiments and Results

Data Preparation. In order to set up our experiment, we collected data from Yahoo!Answer and StackOverflow[4]. The dataset from Yahoo!Answers contains 5977 questions and 15895 answers respectively. The dataset from StackOverflow contains 1407 questions and 4013 answers respectively. All question-answer pairs are written in English. From these datasets, we select the definition questions which satisfy the patterns in Table 1. Then we received 653 questions and their corresponding answers. Among them we randomly select 523 tuples of question-answers for training and the remain for testing. In this corpus each answer is labeled manually with a label "good" or "bad". The label "good" means that it is the good answer for the corresponding question and the label "bad" means that it is an inappropriate answer. For the pre-processing step for analyzing the questions and the answers into words, sentences we used The Stanford CoreNLP Natural Language Processing Toolkit[5].

Wikipedia data can be obtained easily from https://dumps.wikimedia.org/enwiki/20150901/ for free research use. It is available in the form of database dumps that are released periodically. The version we used in our experiments was released on Sep. 1, 2015. The Wikipedia dump we use contains 6.4GB data include titles and abstract of articles.

We design different experiments as follows.

Experiment 1. For the first experiment we investigate different combinations of the single feature sets for the task of answer validation, that includes: unigram, bigram, the number of words in question, the number of words in answer, the number of sentences in answer, the number of votes, word matching feature, word vector representation based feature. Table 2 shows the results of the various combinations of feature kinds. From this experimental result we can see that the word matching feature plays a more important role in comparison with the word vector representation. In addition, combining all the features gives the highest accuracy.

Experiment 2. In this experiment we want to examine our proposal on the use of external resources in the problem of validating the reliability of the answers in question-answer pairs. We also want to examine the contribution of the feature generated based on Wikipedia by adding this feature to the best feature set from the Experiment 1. Note that we will use the two different features corresponding to the two methods as shown in Sect. 4. The obtained results are shown in Table 3.

The obtained results from Experiments 1 and 2 show that by using external resources (here, it means Wikipedia) we can achive the best accuracies. Moreover, Fig. 2 also shows that using word vector representation give better results in comparison with using the word matching method.

---

[4] http://stackoverflow.com/.
[5] http://nlp.stanford.edu/software/.

Table 2. The accuracy of using SVM classifier with combining the multiple feature kinds

Features	Accuracy	F_1-measure
Unigram, Bigram, the number of words in question, the number of words in answer, the number of sentences in answer, the number of votes	76.15 %	62.65 %
Unigram, Bigram, the number of words in question, the number of words in answer, the number of sentences in answer, the number of votes, word matching feature	78.46 %	65.85 %
Unigram, Bigram, the number of words in question, the number of words in answer, the number of sentences in answer, the number of votes, word vector representation based feature	76.92 %	64.29 %
All	79.23 %	66.67 %

Table 3. The accuracy of using SVM classifier when adding features from Wikipedia

Features	Accuracy	F_1-measure
All features in Experiment 1, Extracting features from Wikipedia. (Method 1: Based on words matching)	83.08 %	71.79 %
All features in Experiment 1, Extracting features from Wikipedia. (Method 2: Based on word vector representation)	84.62 %	75.00 %

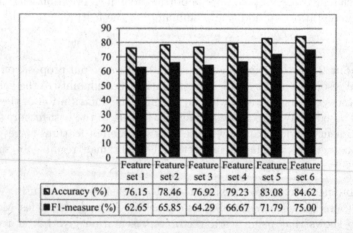

	Feature set 1	Feature set 2	Feature set 3	Feature set 4	Feature set 5	Feature set 6
Accuracy (%)	76.15	78.46	76.92	79.23	83.08	84.62
F1-measure (%)	62.65	65.85	64.29	66.67	71.79	75.00

Fig. 2. The comparisons of the Experiments 1 and 2 of results for definition question classes

Table 4. Comparison with previous studies using the same evaluation metrics and number of class classification

Study	Feature set	Question type	Accuracy
Toba et al. (2014) [13]	Features extracted from the question-answer pairs, user information	Definition question	81.56%
Toba et al. (2014) [13]	Features extracted from the question-answer pairs, user information	All question types	78.74%
Our work	Features extracted from the question-answer pairs, user information, Wikipedia	Definition question	84.62%

Comparison with previous studies. In addition, we also make a comparison with well-known previous studies of this task which also using the same SVM classifier. The Table 4 shows our result gives the much better accuracy.

6 Conclusion

In this paper we have presented our proposal of using external resources for the problem of validating answers from cQA websites, which is very useful for constructing a community based QA system. We have formulated this validation task as a binary classification problem and used a SVM classifier to determine each answer candidate is "good" or "bad". We investigated traditional kinds of features and then adding new features based on measuring the similarity between Wikipedia documents and answer candidates. We also used the word embedding models (i.e. word2vec representation) for this measurement. The experimental results shows that our proposal of using novel features gives the best accuracy in comparison with previous studies as well as with using the traditional feature set.

Acknowledgement. This paper is supported by The Vietnam National Foundation for Science and Technology Development (NAFOSTED) under grant number 102.01-2014.22.

References

1. Belinkov, Y., Mohtarami, M., Cyphers, S., Glass, J.: A continuous word vector approach to answer selection in community question answering systems. In: Proceedings of SemEval, pp. 282–287 (2015)
2. Bian, J., Liu, Y., Agichtein, E., Zha, H.: Factoid question answering over social media. In: Proceedings of WWW (2008)
3. Gunawardana, A., Shani, G.: A survey of accuracy evaluation metrics of recommendation tasks. J. Mach. Learn. Res. **10**, 2935–2962 (2009)

4. Higashinaka, R., Isozaki, H.: Corpus-based question answering for why-questions. In: Proceedings of the Third International Joint Conference on Natural Language Processing (IJCNLP), pp. 418–425 (2008)
5. Hildebrandt, W., Katz, B., Lin, J.: Answering definition questions using multiple knowledge sources. In: proceedings of HLT/NAANC (2004)
6. Hu, H., Liu, B., Wang, B., Liu, M., Wang, X.: Multimodal DBN for predicting high-quality answers in cQA portals. In: Proceedings of ACL, pp. 843–847 (2013)
7. Lou, J., Fang, Y., Lim, K.H., Peng, J.Z.: Contributing high quantity and quality knowledge to online Q&A communities. J. Am. Soc. Inf. Sci. Technol. **64**(2), 356–371 (2013)
8. Mikolov, T., Chen, K., Corrado, G., Dean, J.: Efficient estimation of word representations in vector space. CoRR, abs/1301.3781 (2013a)
9. Shah, C., Pomerantz, J.: Evaluating, predicting answer quality in community QA. In: Proceedings of SIGIR (2010)
10. Soricut, R., Brill, E.: Automatic question answering using the web: beyond the factoid. J. Inf. Retr. Web IR **9**(2), 191–206 (2006)
11. Tran, Q.H., Tran, V.D., Vu, T.T., Nguyen, M.L., Pham, S.B.: Combining multiple features for answer selection in community question answering. In: Proceedings of SemEval, pp. 215–219 (2015)
12. Wang, B., Wang, X., Sun, C., Liu, B., Sun, L.: Modeling semantic relevance for question-answer pairs in web social communities. In: Proceedings of the 48th Annual Meeting of the Association for Computational Linguistics, pp. 1230–1238 (2010)
13. Toba, H., Ming, Z.Y., Adriani, M., Chua, T.: Discovering high quality answers in community question answering archives using a hierarchy of classifiers. Inf. Sci. **261**, 101–115 (2014)
14. Yu, L., Hermann, K.M., Blunsom, P., Pulman, S.: Deep learning for answer sentence selection. In: Proceeding of Neural Information Processing Systems (NIPS): Deep Learning and Representation Learning Workshop, Montreal, Quebec, Canada (2014)
15. Zamanov, I., Hateva, N., Kraeva, M., Yovcheva, I., Nikolova, I., Angelova, G.: A hybrid system for answer validation based on lexical and distance features. In: Proceedings of SemEval, pp. 242–246 (2015)
16. Zhang, Z., Zhou, Y., Huang, X., Wu, L.: Answering definition questions using web knowledge bases. In: Dale, R., Wong, K.-F., Su, J., Kwong, O.Y. (eds.) IJCNLP 2005. LNCS (LNAI), vol. 3651, pp. 498–506. Springer, Heidelberg (2005). doi:10.1007/11562214_44
17. Zhou, X., Hu, B., Lin, J., Xiang, Y., Wang, X.: A deep learning based comment sequence labeling system for answer selection challenge. In: Proceedings of SemEval, pp. 210–214 (2015)

Optimizing Selection of PZMI Features Based on MMAS Algorithm for Face Recognition of the Online Video Contextual Advertisement User-Oriented System

Bao Nguyen Le[1], Dac-Nhuong Le[2], Gia Nhu Nguyen[1(✉)], and Do Nang Toan[3]

[1] Duytan University, Da Nang, Vietnam
{baole,nguyengianhu}@duytan.edu.vn
[2] Haiphong University, Hai Phong, Vietnam
nhuongld@hus.edu.vn
[3] Information Technology Institute, VNU University, Hanoi, Vietnam
toandt@vnu.edu.vn

Abstract. Presently, the advertising has been grown to focus on multimedia interactive model with through the Internet. The *Online Video Advertisement User-oriented* (OVAU) system is combined of the machine learning model for face recognition from camera, multimedia streaming protocols, and video meta-data storage technology. Face recognition is an importance phase which can improve the efficiency performance of the OVAU system. The *Feature Selection* (FS) for face recognition is solved by MMAS-FS algorithm used PZMI feature. The heuristic information extracted from the selected feature vector as ant's pheromone. The feature subset optimal is selected by the shortest length features and best presentation of classifier. The experiments were analyzed on face recognition show that our algorithm can be easily applied without the priori information of features. The performance evaluated of our algorithm is better than previous approaches for feature selection.

Keywords: Contextual advertising · Face recognition (FR) · Video-based face recognition (VbFR) · User-oriented system · Feature selection (FS) · PZMI feature · Max-Min Ant System (MMAS)

1 Introduction

Presently, the online-advertising has been grown to focus on multimedia interactive model with the high interoperability through the Internet, such as: *Google AdWords, Google AdSense*.... The OVAU system generate the content of advertising relevant, truly, and useful to customs in each explicit context over the object detected and recognized from the camera. The three phases of OVAUS system presented in Fig. 1 [24,25]. In the first phase, the objects have been identified and classify directly from the camera to get the features and characteristics. In the second phase, we can access videos based the objects classified using

© Springer International Publishing AG 2016
V.-N. Huynh et al. (Eds.): IUKM 2016, LNAI 9978, pp. 317–330, 2016.
DOI: 10.1007/978-3-319-49046-5_27

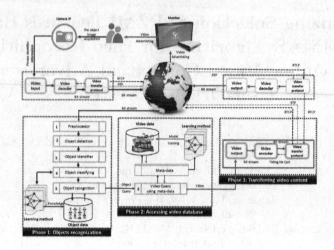

Fig. 1. The OVAUS system proposed

video meta-data storage technology. Finally, the content of advertising videos suitable will be transfer to customs by multimedia streaming protocols (RTP-*Real-time Transport Protocol* or RTP/RTCP-*Real-Time Control Protocol*). The media stream is sent as chunks of data, put into RTP/RTCP packets [25].

Our aiming is detect face of objects extracted from the video camera after we removed unwanted elements. There are two steps of face recognition process: (i) face detection and (ii) objects identifier automatically. The keys of face recognition is the distinguishing extraction methods from objects in images and performed standard for identification automatically (Fig. 2).

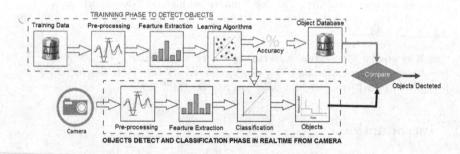

Fig. 2. Object recognition model from the camera in phase 1

The paper proposed a novel MMAS algorithm solve *face recognition* (FR) problem used *feature selection* (FS). Ant's pheromone presents the heuristic information of the selected feature vector extracted. The feature subset optimal is selected by the shortest feature length and best presentation of classifier.

The article is structured as follows: Sect. 2 introduces the related works of FS approaches for FR. Section 3 presents face recognition approaches for video used FS framework and Sect. 4 implemented our MMAS-FS algorithm. The experiments was analyzed and evaluated performance on face recognition presents in Sect. 5. Our conclusion and some approaches in the future mentioned in Sect. 6.

2 Related Works

The input of OVAUS system is real videos from cameras. So, video-based face recognition used feature selection for the real videos from cameras problem is a complex problem. Many approaches for VBFR problem proposed in [52]. The tracking algorithms are tracked faces in the videos. However, these approaches could not applied directly to video sequences [46]. Therefore, the challenges for researchers are unlimited orientations and positions of human faces can be detect in the video scenes [13]. Our motivation is finding the temporal information, an basic characteristic available in videos to improving face recognition in video. The local models or subspaces extract a variety of features from the complex face manifold in the video [49].

Two keys for the video based face detection are: How to extract features? and How to apply learning algorithm? Table 1 shows the summary of approaches and methods in face recognition.

Feature Selection (FS) [15,21,48] is a widespread areas. It contains, data mining, document classification, biometrics, object recognition, and computer vision. Usually, FS is algorithms include the random search strategies or heuristic search strategies to reduce the search space and computational complexity. It means that the optimal feature subset found was also restricted. Depending on the evaluation procedure, FS algorithms are classified into two categories. The first, FS algorithms implemented independently with learning algorithms. The second, the FS algorithms based on the approach wrapper. In which, the learning algorithm is used to evaluate. Five main methods mentioned approaches ares: Forward Selection, Backward Selection, Forward/Backward, Random Choice, and Instance. The majority of methods often begin with a random feature subset heuristically beforehand. The features will be added to or removed after each iteration in the Forward/Backward methods [41]. The FS approaches and methods are summarized in Table 2.

Recently, the famous approaches for feature selection are used on meta-heuristic algorithms, such as: GA [48], SA [37], ACO [1,5,11,12,41,54]. A hybrid ACO method for speech classication presented in [34]. The ACO algorithm employed mixture of shared information and ACO. The HACO used mutual information solve the forecaster FS [5]. The ACO-FDR has used the FDR heuristic information for FS in network intrusion introduced in [11]. The ACO solves the rough set reduces presented in [37]. The Ant-Miner algorithm employed a difficult pheromone updating strategy and state transition rule [1]. The combined ACO with SVM is called ACOSVM [54], the ACS and AS_{rank} are proposed in [12]. The combined ACOG and ACOFS algorithms introduced in [26,41].

Table 1. The approaches and methods for face recognition

Approaches and methods	Year	Reference
PCA	1991	[44]
LDA	1991, 2001	[14, 23, 28]
Bayesian Intra-personal/Extra-personal Classier (BIC)	1996	[9]
Combined PCA and LDA (PCA + LDA)	1997	[30]
SVM	2000	[2]
Iterative Dynamic Programming (DP)	2000	[53]
Boosted Cascade of Simple Features (BOOST)	2001	[31]
Isomap extended, KDF-Isomap	2002, 2005	[29, 36]
Kernel Principal Angles (KPA)	2006	[45]
Locality Preserving Projections (LPP)	2006	[17]
Statistical Local Feature Analysis (LFA)	2006	[10]
Discriminative Canonical Correlations (DCC)	2003, 2007	[7, 22]
Locally Linear Embedding (LLE)	2008	[51]
Local Graph Matching (LGM)	2007	[8]
KNN model and Gaussian mixture model (GMM)	2007	[38]
Hidden Markov Mode (HMM)	2008	[20]
3D model based approach	2007, 2008	[33, 47]
Feature Subspace Determination	2008	[39]
Learning Neighborhood Discriminative Manifolds (LNDM)	2011	[10]
MFA	2012	[55]
NPE	2012	[19]
Multi-dimensional scaling (MDS)	2012	[40]
Data Uncertainty in Face recognition	2014	[50]
Orthogonal Locality Preserving Projections (OLPP)	2015	[18]

All the objective function are calculated in accordance of a feature subset is created and compared with the best previous of candidates. Then characteristics of feature subset are compared, selection and replacement of the best features are selected. The algorithm will stop when the implementation of the iterators or no features is selected and replaced. Almost of the proposed ACO algorithms proposed for FS problem recommended use a complete graph to present features. The ants will find the way through the nodes (*feature*) on the graph that with the selection of features. The nodes of each path corresponding to a set of features selected as a solution. Therefore, we do not necessarily have to use a complete graph to represent all solutions. At a time, an ant is in a node, it will choose an edge to connected to the another nodes by the heuristic information of this edge. But in the FS problem, one feature to be selected is independent with the last feature added to the partial solution. So, ACO algorithms [1, 5, 11, 12, 41, 54]

Table 2. The approaches and methods for FS problem

Algorithm: Approaches and methods	Year	References
GA- Genetic algorithm	1989, 2008	[12,48]
FOCUS- Learning with many irrelevant features	1991	[15]
RELIEF	1992	[21]
LVW- A probabilistic wrapper approach	1996	[16]
Neural Network	1997	[34]
KFD- Invariant feature extraction and classication in feature spaces	2000	[42]
FDR- The fractal dimension	2000	[4]
EBR- A rough set	2001	[35]
SCRAP- Instance-based lter	2002	[3]
FGM- Feature Grouping Methods	2002	[32]
SA- Simulated annealing	2005	[37]
ACO- Ant colony optimization		
+ Ant-Miner find the rough set reduces	2004	[1]
+ ACOSVM- ACO with SVM	2004	[54]
+ Hybird-ACO (HACO)	2005	[5]
+ ACO- Fisher Discrimination Rate (FDR)	2005	[11]
+ ACO- Combining rough and fuzzy sets	2005	[37]
+ ACS- Ant Colony System	2008	[12]
+ AS_{rank}- Ant System Rank-Based	2008	[12]
+ ACOG- Ant colony optimization and Genetic Algorithm	2010	[41]
+ ACOFS	2011	[26]

have built a complete-graph G with $O(n^2)$ edges. Next section, we present our framework and algorithm based on MMAS to solve feature selection problem.

3   Video-Based Face Recognition Used Feature Selection

3.1   Video-Based Face Recognition Used Feature Selection Problem

Definition 1. *The sequence of N_c face images in the video can define as*

$$X_c = \{x_{c,1}, x_{c,2}, ..., x_{c,N_c}\} \tag{1}$$

Definition 2. *(The face recognition problem used feature selection)*

– *Given $F = \{f_1, ..., f_n\}$ (n is number of features) is a feature set.*
– *Find a optimal feature subset $S \subseteq F$ of minimal m features (m < n) while ensuring represented the original features.*

Definition 3. *(A combinatorial optimization of the feature selection problem)*

- *Given $F = \{f_1, ..., f_n\}$ is a feature set of basic components.*
- *A solution s of the problem is a subset of components.*
- *The number of feasible solutions is $S \subseteq 2^F$, It means that a solution S is feasible if $s \in S$.*
- *$f : 2^F \mapsto R$ is the cost function.*
- *Find a minimum cost feasible solution s^*, $s^* \in S$ and $f(s^*) \leq f(s)$, $\forall s \in S$.*

3.2 Framework of Face Recognition

Our algorithm framework proposed for feature selection shows in Fig. 3. In the first step, the input face image has converted to a binary before extract features. The face image is calculated from the centroid (X, Y) as follows:

$$X = \frac{\sum mx}{\sum m} \text{ and } Y = \frac{\sum my}{\sum m} \tag{2}$$

Here, x, y are the values of co-ordinate, and $m = f(x, y) = 0/1$. After that, only face has cropped and converted to the gray level and the features have been collected, the PZMI [12] features have been extracted. In the second step, we used the MMAS algorithm to select feature and reduce dimension. The final step, the images will be classification by Nearest Neighbor Classifier (NNC) [27] to make decision. The decisions will be feedback for MMAS algorithm.

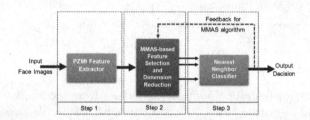

Fig. 3. Framework of face recognition used MMAS algorithm

4 MMAS Proposed for Feature Selection Problem

4.1 Construct Ant Solutions

To efficiently apply MMAS algorithm [6,43], we generated the construct graph $G(E, V)$ for the n features as $F = \{f_1, f_2, ..., f_n\}$. We used a digraph with $O(n) = 2n$ arcs to present the search space. The digraph $G(E, V)$ described in Fig. 4.

The features represent by nodes, and two features related to each other is represented by an edge connectivity, in which:

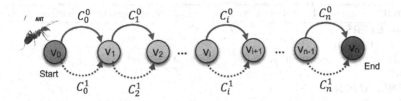

Fig. 4. The construct graph $G = (E, V)$ for feature set F

1. Each feature represent a node in the graph $G(E, V)$, node v_i represent feature $f_i \in F$ $(i = \overline{0..n})$.
2. Suppose there are m features in the solution, we add a source node (v_0) is the start node of the graph where the ants started.
3. For all nodes v_i and v_{i+1}, $i = \overline{0..n}$, we add two edges named C_i^0 and C_i^1. If the ant at v_i selects edge C_i^0 means the feature i^{th} is selected. Otherwise, the edge C_i^1 is choice means that the feature i^{th} is not selected.
4. The ants move on the graph $G(E, V)$ from v_0 to v_1, then v_2... The process of moving of each ant will stop when it moving to v_n, and the feature subset outputs.

When each ant move from v_0 to v_n, it mean that a solution s is constructed.

4.2 MMAS-FS Algorithm Implementation

4.3 Update Pheromones

Let $\tau_{\min}$ is lower and $\tau_{\max}$ is upper bounds of pheromone values. The pheromone value $\tau_{i,j}$ is the information feedback in the graph $G(E, V)$ when the ants searching. All pheromone trails initial to $\tau_{\max}$:

$$\tau_{i,j} = \tau_{\max} = 1, \forall i = \overline{1..n}, \forall j = \overline{0..1} \qquad (3)$$

For each candidate feature f_i, at time a new feature f_i is added to the solution s_k means that the ant moves on edge C_i^0. The pheromone factor τ_{s_k} incremented by $\tau_{i,0}$. The highest pheromone are calculated by

$$\tau_s = \sum_{i=1}^{n} \sum_{j=0}^{n} \tau_{i,j}, \forall f_i \in s, i = \overline{1..n} \qquad (4)$$

The ants start with an empty solution to construct a feature subset and move on the graph $G(E, V)$ to find the optimal solution. Assuming that, at the present time an ant is at v_{i-1} when constructing new solutions. It must decide to select the next v_i by moving over an edge $\{v_{i-1}, v_i\}$. After each iteration, the pheromone of the edge $\{v_{i-1}, v_i\}$ will increase when the features f_i and f_{i-1} in the best solution s^*. Each ant is initialized with a empty solution. In the first step, an ant chooses v_0, features are selected follow their related pheromone value.

Algorithm 1. MMAS-FS Algorithm proposed for feature selection problem

PARAMETERS:
  $\alpha = 1, \beta = 0.1, \rho = 0.5.$
  Size of Ant population: $K = 100$. Number of iteration: $N_{Max} = 500$;
BEGIN
  GENERATION:
  $\tau_{min} = 0.01; \tau_{max} = 1;$
  For each i=1..n do
    For each j=0..1 do
      $\tau_{ij} = \tau_{max};$
    end for
  end for
  $i = 1; s^{Best} \Leftarrow \emptyset;$
  Repeat
    For each ant_k $(k = 1...K)$ do
      Construct solution of Ant_k: s^k by choose features follows by (5) and (6);
      If $(s^{Best} = \emptyset)$ then $s^{Best} \Leftarrow s^k$;
      If $f(s_{Best}) < f(s^k)$ then
        $s_{Best} \Leftarrow s^k$;
        Update pheromone trails follows by (7);
        $\tau_{i,j}$ is set in $[\tau_{min}, \tau_{max}]$ by (10);
      end if
    end for
  Until $(i > N_{Max})$ or (s^* optimal solution found);
  Return Best subset s^*;
END

The probability equation based on local heuristic information of the remaining candidate selects of ant at v_{i-1} to select the edge C_i^j updated according to the formula:

$$P_{i,j}(t) = \frac{[\tau_{i,j}(t)]^{\alpha}[\eta_{i,j}]^{\beta}}{[\tau_{i,0}(t)]^{\alpha}[\eta_{i,0}]^{\beta} + [\tau_{i,1}(t)]^{\alpha}[\eta_{i,1}]^{\beta}}, \forall i = \overline{0..n}, \forall j = \overline{0,1} \qquad (5)$$

where, $\tau_{i,j}(t)$ is the pheromone factor in the edge C_i^j between nodes (v_{i-1}, v_i) at time t which reflects the potential tend for ants to follow edge $C_i^j (j = \overline{0,1})$. The influence of the pheromone concentration to the probability value is presented by the constant α, while constant β do the same for the desirability and control the relative importance of pheromone trail versus local heuristic value.

The knowledge-based information $\eta_{i,j}$ expected to choose the edge C_i^j can be defined as:

$$\eta_{i,j} = \frac{\sum_{k=1}^{m}(\overline{x_i^k} - \overline{x_i})}{\sum_{k=1}^{m}[\frac{1}{N_i^k-1}\sum_{j=1}^{N_i^k}(x_{i,j}^k - \overline{x_i^k})^2]}, \forall i = \overline{1..n} \qquad (6)$$

where, N_i^k denotes the number of feature samples $f_i(i = \overline{1..n}$ in class of image set $k(k = \overline{1..m})$; $x_{i,j}^k$ is the $j^{th}(j = \overline{1..N_i^k})$ training sample for the feature f_i of images in class k. $\overline{x}_i$ is value of the feature f_i of all images, $\overline{x_i^k}$ is the value of the feature f_i of all images in class k. $\eta_{i,1}$ implies the feature f_i has a greater discriminative ability, and a constant $\eta_{i,0} = \frac{\xi}{n} \sum\limits_{i=1}^{n} \eta_{i,1}$.

The ant's pheromone values is usually much better after each loop. The original pheromone trail will be updated follow the best solution by:

$$\tau_{i,j}(t+1) \leftarrow (1-\rho) \times \tau_{i,j}(t) + \Delta\tau_{i,j}(t)^{Best} + Q_{i,j}(t) \tag{7}$$

where, ρ is tuning parameter and the trail evaporation rate $(0 \leq \rho < 1)$; $\Delta\tau_{i,j}(t)^{Best} = \frac{1}{S_{i,j}(t)} \sum\limits_{s \in S_{i,j}(t)} f(s)$; $Q_{i,j}(t) = Q$ is a positive constant if $C_i^j \in s^{Best}$, otherwise $Q_{i,j}(t) = 0$; $S_{i,j}(t)$ is set of solution in the iteration t^{th} traverse over the edge C_i^j; The best solution found is s^{Best}, the extra pheromone increment on the edges included in s^{Best}.

The cost function $f(s)$ of solution s is calculated as follows:

$$f(s) = \frac{N_{Correct}}{1 + \lambda N_{Feat}} \tag{8}$$

in which, N_{Feat} is size of features selected in s, $N_{Correct}$ is the number of classified correctly examples, constant λ is the degree of accuracy for features selected.

The best solution s^* is given by:

$$f(s^*) < f(s), \forall s \in S_{i,j}(\forall i = \overline{1..n}, \forall j = \overline{0..1}) \tag{9}$$

Stop condition of the algorithm is either when the ant found a optimal solution, or when has executed the maximum number of iteration. After pheromone update, $\tau_{i,j}$ is set in $[\tau_{min}, \tau_{max}], \forall i = \overline{1..n}, j = \overline{0..1}$ defined by:

$$\tau_{i,j} = \begin{cases} \tau_{\max} & \text{if } \tau_{i,j} > \tau_{max} \\ \tau_{\max} & \text{if } \tau_{i,j} \in [\tau_{min}, \tau_{max}] \\ \tau_{\min} & \text{if } \tau_{i,j} < \tau_{min} \end{cases} \tag{10}$$

5   Experiment and Results

5.1   Experiment Implementation

The ORL gray scale database includes 400 facial images from 40 different states individuals with dimensions is 92×112 [56]. After pre-processing, we extracted the PZMI features with orders 1 to 20 from each face image. We used MMAS-FS algorithm proposed to select the optimal feature subsets. In this experiment, we used 40 classes and 10 image in each class. Some samples images are also included in Fig. 5.

Fig. 5. Sample face images of ORL Databases

We analyze, compare, and evaluate the effectiveness of MMAS-FS with other meta-heuristic approaches. The performance comparison between various methods from Table 3 shows that the MMAS-FS proposed produce much lower classification error rates 1.5 % and the execution times than others of the GA-based and ACO-based methods.

Table 3. The comparison of performance meta-heuristic algorithms

Methods	Mean square error (%)	Number of features selected	Time execution(s)
GA [12,48]	3.5	55	1080
ACO [1,37]	3.25	53	215
ACOSVM [54]	3	45	325
ACO [5]	4.5	55	255
ACO [11]	5	54	280
ACS [12]	3	49	780
AS_Rank [12]	1.5	42	300
ACOG [41]	3.5	43	315
ACOFS [26]	4	45	347
MMAS-FS	1.5	40	285

Finally, we compare the obtained recognition rates of PZMI feature subsets of each algorithm are shown in Fig. 6. Our algorithm proposed can achieve 98.83 % and the recognition rate is 98.57 % only with 40 features selected.

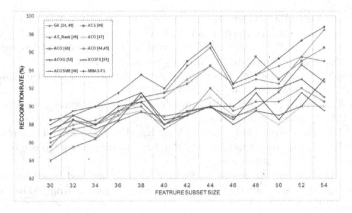

Fig. 6. The comparison of recognition rate of PZMI feature subsets of each algorithm

6 Conclusions

We proposes MMAS-FS algorithm used feature selection based on PZMI feature for face recognition. The heuristic information extracted from the selected feature vector as ant's pheromone. The feature subset optimal is selected by the minimal feature length and the best presentation of classifier. The experiments were analyzed on face recognition show that our algorithm can be easily applied without the priori information of features. The performance evaluated of MMAS-FS is more effective than other meta-heuristic approaches for FS. Our algorithm proposed can achieve 98.83 % and 98.57 % recognition rate only with 40 features selected on the ORL database.

References

1. Liu, B., Abbass, H.A., McKay, B.: Classication rule discovery with ant colony optimization. IEEE Comput. Intell. Bull. **3**(1), 31–35 (2004)
2. Heisele, B., et al.: Face detection in still gray images. AI Laboratory, MIT (2000)
3. Raman, B., Ioerger, T.R.: Instance-based filter for feature selection. J. Mach. Learn. Res. **1**, 1–23 (2002)
4. Traina, C., Traina, A., Wu, L., Faloutsos C.: Fast feature selection using the fractal dimension. In: Proceedings of the 15th Brazilian Symposium on Databases (SBBD), pp. 158–171 (2000)
5. Zhang, C.K., et al.: Feature selection using the hybrid of ant colony optimization, mutual information for the forecaster. In: Proceedings of the 4 International Conference on Machine Learning and Cybernetics (2005)
6. Le, D.-N.: Evaluation of pheromone update in min-max ant system algorithm to optimizing QoS for multimedia services in NGNs. In: Satapathy, S.C., Govardhan, A., Srujan Raju, K., Mandal, J.K. (eds.) Emerging ICT for Bridging the Future. Advances in Intelligent Systems and Computing, vol. 338, pp. 9–17. Springer, Switzerland (2014)

7. Dai, D.Q., Yuen, P.C.: Regularized discriminant analysis and its applications to face recognition. Pattern Recogn. **36**(3), 845–847 (2003)
8. Ersi, E.F., Zelek, J.S., Tsotsos, J.K.: Observation of strains: robust face recognition through local graph matching. J. Multimedia **2**, 31–37 (2011)
9. Etemad, K., Chellappa, R.: Discriminant analysis for recognition of human face images. J. Opt. Soc. Am. **14**, 1724–1733 (1997)
10. Ersi, E.F., Zelek, J.S.: Local feature matching for face recognition. In: Proceedings of the 3rd Canadian Conference on Computer and Robot Vision (2006)
11. Gao, H.H., et al.: Ant colony optimization based network intrusion feature selection and detection. In: Proceedings of the 4 International Conference on Machine Learning and Cybernetics (2005)
12. Kannan, H.R., et al.: An improved feature selection method based on ant colony optimization (ACO) evaluated on face recognition system. Appl. Math. Comput. **205**, 716–725 (2008)
13. Wang, H., Wang, Y., Cao, Y.: Video-based face recognition: a survey. World Acad. Sci. Eng. Technol. **3**, 273–283 (2009)
14. Yu, H., Yang, J.: A direct LDA algorithm for high-dimensional data-with application to face recognition. Pattern Recognit. **34**(10), 2067–2070 (2001)
15. Almuallim, H., Dietterich, T.G.: Learning with many irrelevant features. In: The 9th National Conference on Articial Intelligence. MIT Press, pp. 547–552 (1991)
16. Liu, H., Setiono, R.: Feature selection and classication - a probabilistic wrapper approach. In: Proceedings of the 9th ICIEAAIES, pp. 419–424 (1996)
17. Choi, J.Y., et al.: Feature subspace determination in video-based mismatched face recognition. In: 8th IEEE International Conference on Automatic Face and Gesture Recognition (2008)
18. Soldera, J., et al.: Customized orthogonal locality preserving projections with soft-margin maximization for face recognition. IEEE Trans. Instrume **64**(9), 2417–2426 (2015)
19. Gui, J., et al.: Discriminant sparse neighborhood preserving embedding for face recognition. Pattern Recogn. **45**(8), 2884–2893 (2012)
20. Kim, M., Kumar, S., Pavlovic, V., Rowley, H.A.: Face tracking and recognition with visual constraints in real-world videos. In: CVPR (2008)
21. Kira, K., Rendell, L.A.: The feature selection problem: traditional methods and a new algorithm. In: Proceedings of 9 National Conference on Articial Intelligence, pp. 129–134 (1992)
22. Kim, T.K., et al.: Discriminative learning and recognition of image set classes using canonical correlations. IEEE Trans. Pattern Anal. Mach. Intell. **29**(6), 1005–1018 (2007)
23. Chen, L.F., et al.: A new LDA-based face recognition system which can solve the small sample size problem. Pattern Recogn. **33**(10), 1713–1726 (2000)
24. Bao, L.N., Le, D.-N., Van Chung, L., Nguyen, G.N.: Performance evaluation of video-based face recognition approaches for online video contextual advertisement user-oriented system. In: Satapathy, S.C., Mandal, J.K., Udgata, S.K., Bhateja, V. (eds.) Information Systems Design and Intelligent Applications, vol. 435, pp. 287–295. Springer, India (2016)
25. Bao, L.N., Van Chung, L., Toan, D.N.: A proposed framework for the online video contextual advertisement user-oriented system using video-based face recognition. Int. J. Appl. Eng. Res. **11**(15), 8609–8617 (2016)
26. Chen, L., Chen, B., Chen, Y.: Image feature selection based on ant colony optimization. In: Wang, D., Reynolds, M. (eds.) AI 2011. LNCS (LNAI), vol. 7106, pp. 580–589. Springer, Heidelberg (2011). doi:10.1007/978-3-642-25832-9_59

27. Kozma, L.: k- Nearest Neighbours Algorithm, Helsinki University Technology (2008)
28. Moghaddam, B., Nastar, C., Pentland, A.: Bayesian face recognition using deformable intensity surfaces. In: Proceedings of Computer Vision and Pattern Recognition, pp. 638–645 (1996)
29. Yang, M.-H.: Face recognition using extended isomap. In: Proceedings of the 2002 International Conference on Image Processing 2012 (2002)
30. Belhumeur, P.N., et al.: Eigenfaces vs. sherfaces: recognition using class specic linear projection. IEEE Trans. Pattern Anal. Mach. Intell. **19**, 711–720 (1997)
31. Viola, P., Jones, M.: Rapid object detection using a boosted cascade of simple features. In: Proceedings of the Conference on Computer Vision and Pattern Recognition, pp. 511–518 (2001)
32. Paclík, P., Duin, R.P.W., Kempen, G.M.P., Kohlus, R.: On feature selection with measurement cost and grouped features. In: Caelli, T., Amin, A., Duin, R.P.W., Ridder, D., Kamel, M. (eds.) SSPR /SPR 2002. LNCS, vol. 2396, pp. 461–469. Springer, Heidelberg (2002). doi:10.1007/3-540-70659-3_48
33. Park, U., Jain, A.K.: 3D model-based face recognition in video. In: Lee, S.-W., Li, S.Z. (eds.) ICB 2007. LNCS, vol. 4642, pp. 1085–1094. Springer, Heidelberg (2007). doi:10.1007/978-3-540-74549-5_113
34. Setiono, R., Liu, H.: Neural network feature selector. IEEE Trans. Neural Networks **8**(3), 645–662 (1997)
35. Jensen, R., Shen, Q.: A rough set-aided system for sorting WWW bookmarks. In: Zhong, N., Yao, Y., Liu, J., Ohsuga, S. (eds.) WI 2001. LNCS (LNAI), vol. 2198, pp. 95–105. Springer, Heidelberg (2001). doi:10.1007/3-540-45490-X_10
36. Li, R.-F., et al.: Face recognition using KFD-Isomap. In: International Conference on Machine Learning and Cybernetics, pp. 4544–4548 (2005)
37. Jensen, R.: Combining rough and fuzzy sets for feature selection, Ph.D. thesis, University of Edinburgh (2005)
38. Stallkamp, J., Ekenel, H.K.: Video-based face recognition on real-world data (2007)
39. Chen, S., et al.: Face recognition from still images to video sequences: a local feature-based framework. EURASIP J. Image Video Process. (2010)
40. Biswas, S., et al.: Multidimensional scaling for matching low-resolution face images. IEEE Trans. Pattern **34**(10), 2019–2030 (2012)
41. Venkatesan, S., et al.: Face recognition system with genetic algorithm and ANT colony optimization. Int. J. Innov. Manage. Technol. **1**(5), 469–471 (2010)
42. Mika, S., et al.: Invariant feature extraction and classication in feature spaces. In: Solla, S.A., Leen, T.K., Müüller, K.-R. (eds.) Advances in Neural Information Processing Systems, pp. 526–532. MIT Press, Cambridge (2000)
43. Stutzle, T., Ibanez, M.L., Dorigo, M.: A Concise Overview of Application of Ant Colony Optimization. Wiley, New York (2010)
44. Turk, M., et al.: Eigenfaces for recognition. J. Cogn. Neuro-Sci. **3**, 71–86 (1991)
45. Chin, T.-J., et al.: Incremental Kernel SVD for face recognition with image sets. In: Proceedings of the 7th International Conference on Automatic Face and Gesture Recognition (2006)
46. Zhao, W.Y., Chellappa, R., Rosenfeld, A., Phillips, P.J.: Face recognition: a literature survey. ACM Comput. Surv. **35**, 399–458 (2003)
47. Wu, Y., et al.: Integrating illumination, motion and shape models for robust face recognition in video. EURASIP J. Adv. Signal Process. (2008)
48. Siedlecki, W., et al.: A note on genetic algorithms for large-scale feature selection. Pattern Recogn. Lett. **10**(5), 335–347 (1989)

49. Yan, S., Xu, D., Zhang, B., Zhang, H.J., Yang, Q., Lin, S.: Graph embedding: a general framework for dimensionality reduction. IEEE Trans. PAMI **29**(1), 40–51 (2007)

·50. Yong, X., et al.: Data uncertainty in face recoginition. IEEE Trans. Cybern. **44**(10), 1950–1961 (2014)

51. Pang, Y.H., et al: Supervised locally linear embedding in face recognition. In: Biometrics and Security Technologies, ISBAST 2008, pp. 1–6 (2008)

52. Zhang, Z., Wang, C., Wang, Y.: Video-based face recognition: state of the art. In: Sun, Z., Lai, J., Chen, X., Tan, T. (eds.) CCBR 2011. LNCS, vol. 7098, pp. 1–9. Springer, Heidelberg (2011). doi:10.1007/978-3-642-25449-9_1

53. Liu, Z., Wang, Y.: Face detection and tracking in video using dynamic programming. In: Proceedings of International Conference on Image Processing (2000)

54. Yan, Z., Yuan, C.: Ant colony optimization for feature selection in face recognition. In: Zhang, D., Jain, A.K. (eds.) ICBA 2004. LNCS, vol. 3072, pp. 221–226. Springer, Heidelberg (2004). doi:10.1007/978-3-540-25948-0_31

55. Wang, Z., et al.: Optimal Kernel marginal fisher analysis for face recognition. J. Comput. **7**(9), 2298–2305 (2012)

56. http://www.face-rec.org/databases/

Phrase-Based Compressive Summarization for English-Vietnamese

Tung Le[1(✉)], Le-Minh Nguyen[2], Akira Shimazu[2], and Dinh Dien[1]

[1] Faculty of Information Technology, University of Science,
Vietnam National Univeristy, Ho Chi Minh City, Vietnam
1212494@student.hcmus.edu.vn, ddien@fit.hcmus.edu.vn
[2] School of Information Science, Japan Advanced Institute of Science
and Technology, Nomi, Japan
{nguyenml,shimazu}@jaist.ac.jp

Abstract. Cross-language summarization is the novel topic which is extremely practical and necessary for capturing, tracing, and retrieving the huge data. Especially, for many low-resource languages as Vietnamese, Chinese, ..., there are not any previous works to solve this problem as well as datasets. Therefore we propose to apply Phrase-based Compressive Summarization for English-Vietnamese. This model takes advantages of the relation between translation and summarization phases to overcome the popular drawback in most antecedent researches. Besides, the bilingual corpus for English-Vietnamese summarization built manually on the dataset is extremely helpful for a lot of later works. In this dataset, our system achieves approximately 37 % in ROUGE-1 score which is equivalent to systems on other language pairs. This significant and encouraging result proves the effectiveness of our approach and the quality of our manual datasets in English-Vietnamese.

Keywords: Phrase-based compressive summarization · Cross-language summarization

1 Introduction

In the age of information explosion, there is more and more multilingual information. This is actually a difficult challenge for researchers to store, process and retrieve the knowledge from a ton of raw data. Especially, in the areas related to natural language processing as summarization, opinion mining, etc., the variety of languages requires the cross-language systems. The cross-language domain in summarization is extremely interesting and promising to put it into practice.

Cross-language summarization is the task to produce the summary in target language from document(s) in source language [1]. Recently, researchers have shown an increased interest in cross-language summarization which is categorized into two groups. The first one is to use the machine translation system to convert the input document in source language into target language and then apply

© Springer International Publishing AG 2016
V.-N. Huynh et al. (Eds.): IUKM 2016, LNAI 9978, pp. 331–342, 2016.
DOI: 10.1007/978-3-319-49046-5_28

summarization system into the translated data to produce the final summary. The second one is to swap two above steps.

In most researches, two above steps are done in turn independently. Researchers make an effort to increase the accuracy in each phase. The independence and lack of interaction in two phases are actually a drawback in many current systems. In addition, working at the word-based level that almost eliminates the context of words also reduces the quality of two phases.

To overcome the above disadvantages, we propose to put the knowledge in phrase-based machine translation system [3] as an important element in objective function for choosing sentences (phrases) for summarization. We also use the phrase-based algorithm to eliminate the redundancies of the translated sentences in compression step. After, the greedy algorithm is used to choose the compressive sentence to produce the summary. Additionally, the main challenge faced by many experiments for cross-language summarization is the lack of dataset. To deal with this drawback, we build the bilingual datasets on manual based on Opinosis 1.0 [2] for English-Vietnamese summarization.

The contributions of this paper include:

- Developing the cross-language compressive summarization system in English and Vietnamese at first as well as achieving the quite high accuracy (approximately 38.5 %- F1 score for ROUGE-1)
- Building reliable bilingual summarization dataset for English-Vietnamese which is extremely helpful for the later works.

The remainder of this paper is organized as follows. Section 2 provides a brief overview of the related works. Section 3 presents our method. Section 4 describes the experimental setup and results. Section 5 concludes the paper and points to avenues for future work.

2   Related Works

2.1   Cross-Language Summarization

The information explosion puts the challenges for researchers about dealing with multi-document and multi-language summarization. For the first problem, there are a lot of approaches as maximal marginal relevance (MMR), Submodular function, graph-based methods [8], etc. Recently, the development of Deep Learning gets much success in multi-document summarization [11].

For the bilingual summarization, it is extremely novel and difficult. The earlier methods often use knowledge from one side and the separation of two phase: translation and summarization [7], etc. Especially, for many low-resource languages as Asian one, there are no works for cross-language summarization model as well as the published datasets. Especially, in Vietnamese summarization, there is only a published dataset as Vietnamese MDS[1] for multi-document summarization which is, however, quite small and low-qualified. Therefore, together with

---

[1] https://github.com/lupanh/VietnameseMDS

building the first multi-language summarization for Vietnamese, the English-Vietnamese dataset for summarization is built on manual, which is extremely necessary and helpful for the later works.

2.2 Submodular Function in Summarization

Submodular function is firstly introduced in natural language processing task by Lin and Bilmes in 2010 [5] and become extremely suitable for summarization task. It shares a variety of properties in common with convex function as wide applicability, generality, closure, etc.

Mathematically, if the mapping $F : 2^V \to \mathbb{R}$ satisfies the Eq. 1, F is submodular function.

$$\forall A \subseteq B \subseteq V \setminus \{v\}, F(A + v) - F(A) \geq F(B + v) - F(B) \tag{1}$$

That means the incremental value of element v in the smaller set A is greater than the bigger set B. We have to also note that the function is close to some operations as mixtures, truncation, complementation, etc. The positive linear combination of submodular function set is submodular function $\mathcal{F} = \sum_i \alpha_i \mathcal{F}_i$.

In summarization, submodular function can be used as the objective function in selection phase yet finding the subset to maximize this function is an NP-hard problem [6]. However, the significant advantage of this kind of function is that the greedy solution is no much difference from the optimal one [9].

$$F(S^{greedy}) \geq \frac{1}{2}(1 - e^{-1})F(OPT) \tag{2}$$

2.3 Phrase-Based Machine Translation

From its appearance, phrase-based machine translation achieves the state-of-the-art result for the most pairs of language and become the best method in machine translation. The most significant improvement is that the system works with phrases instead of separated words. As we know, in inflection typology language as English, whitespace is the range of words, however, obviously, the meaning of words is based on their context. It means deciding the meaning of the phrases is better than the meaning of word. It is the reason that phrase-based method is more efficient than traditional word-based one in both translation and other natural language processing tasks, especially in summarization. We notice that phrases are the continuous sequence of words or n-grams as we know. They do not require any meaning of structure, grammar, etc.

In the context of phrase-based translation, we define y as a phrase-based derivation which can be done as the combination of two-side alignment by GIZA++ [3]. It means y is the set of phrases $\{p_1, p_2, ..., p_L\}$. The scoring function to estimate the quality of translation from source phrase-based sentence y to target one $e(y)$ as Eq. 3 whereas $e(.)$ is the translator.

$$f(y) = \sum_{k=1}^{L} g(p_k) + LM(e(y)) + \sum_{k=1}^{L-1} \eta|start(p_{k+1} - 1 - end(p_k))| \tag{3}$$

$LM(.)$ is the language model score which is simply determined as n-grams score (usually $n = 2$). $g(p_k)$ is the phrase' score of p_k. Besides, the distortion parameter $\eta < 0$ is to penalize for the distances among phrases. However, to find out the best solution phrase derivation for translation is extremely hard so the greedy algorithm is applied with some extra constraint [3].

3   Phrase-Based Summarization Model

In this section, our model is represented as the combination of two phases: compression and selection. Both of them are inspired by phrase-based machine translation model [3].

3.1   Phrase-Based Sentence Scoring

In summarization, most system tries to select the important sentences (phrases) from the input document(s). The importance of sentence can be calculated by the ranking systems. In our method, we assume that each sentence includes many phrases rather than are created from phrases. In compression step, the redundant phrases are eliminated to increase the amount of information. In selection step, the compressive sentence is put into summary to maximize the gain-cost ratio which is based on the phrase-based sentence scoring function as Eq. 4.

$$F(s) = \sum_{p \in s} d_0 g(p) + bg(s) + \eta dist(y(s)) \tag{4}$$

$g(p)$ is the score of phrase p which is simply calculated as the document frequency of phrase p in the corpus. The damping factor d_0 is a trade-off coefficient to balance between popular and rare phrases. The second term $bg(s)$ is bi-gram score of sentence s which is shown in Eq. 5 whereas w_i is the i'th word in sentence s and $P(w_i)$ can be estimated as frequency of word w_i in the corpus.

$$bg(s) = \sum_{n=1}^{N} P(w_n | w_{n-1}) \tag{5}$$

Bi-gram score $bg(s)$ is extremely important to simulate the effect of language model in phrase-based translation model [10]. It is the constrains to ensure the sentence's meaning. The last term $dist(y(s))$ is the distortion penalty term in phrase-based translation models. $dist(y(s))$ is determined as the sum of all pair phrase in sentence s whose distance is shown in Eq. 6.

$$dist(a, b) = |start(b) - 1 - end(a)| \tag{6}$$

Besides, $y(s)$ is the phrase derivation of sentence s. That process can be done as the combination of two-side alignments or result from translation system. The parameter $\eta < 0$ is the penalized one to discipline for their distance. We notice again that sentence s contains many phrases yet the total length of its phrases is smaller than the sentence one.

Based on the definition of sentence scoring, we expand into the summary scoring function as Eq. 7 where summary S is the set of sentence s.

$$F(S) = \sum_{p \in S} \sum_{i=1}^{count(p,S)} d^{i-1} g(p) + \sum_{s \in S} bg(s) + \eta \sum_{s \in S} dist\,(y(s)) \tag{7}$$

Equation 7 is simply calculated as the sum of each sentence scoring function where d^{i-1} is the hyperparameters which are defined by users and have the same meaning as d_0 in Eq. 4. The function $count(p, S)$ is the number of occurrences of the phrase p in summary S. This function is used to evaluate the quality of summary in both translation and summarization features. We notice that as the distortion term is ignored (means $\eta = 0$), the scoring function in Eq. 7 is obviously satisfied with submodular property mentioned in Sect. 2.2. However, Yao et al. proved that when $\eta < 0$, this setting does not affect the performance guarantee too much [10].

3.2   Phrase-Based Compression Function

In compression phase, the sentences are eliminated some redundant phrases into more meaningful one based on the scoring function mentioned above. First, the seed phrases which are the most meaningful are chosen by maximizing the gain-cost ratio in Eq. 8. These seed phrases are used to re-evaluate for previous and next phrases. The evaluation is based on the current summary and the context of phrases. When these phrases no meet the threshold, they are eliminated.

Re-evaluation is really significant as considering the meaning of phrases depended on their context (surrounding words/phrases). It helps the selection phase with more precision and guarantee. The detail of compression phase is represented in Algorithm 1.

$$gain(p, S) = \frac{scr\_p(p, S)}{cost(p)} = \frac{\sum_{i=1}^{count(p,S)} d^{i-1} g(p)}{cost(p)} \tag{8}$$

$gain\_p(.,.)$ is defined as Eq. 9 to determine the importance of previous and next phrases based on the context of the seed phrases. $bigr\_scr(*)$ is the bi-gram score of two phrase where the operator $*$ is the concatenating phrase a and phrase b. Bi-gram score takes an important role in compression phase. It is the key factor in order to maintain the sentence's fluency.

$$gain\_p(a, b, S) = \frac{scr\_p(a, S) + bigr\_scr(a * b) + \eta dist(a, b)}{cost(a) + cost(b)} \tag{9}$$

In addition, $cost(.)$ is the expense of phrase/sentence which is the number of sentences, words or character. It depends on the budget of summary that is predefined by users. In our research, we consider the budget as the average of the number of words in all references.

The significant difference to the original model in English is that *compress_threshold* is used as the predefined parameters to control the quality of compression phase. In particular, *compress_threshold* includes *select_thres*, *ppv_thres* and *pnx_thres*. In the previous works, they are fixed as the constant threshold, which is not suitable for low-source language due to the lack of data. The findings should make an important contribution to the field of Vietnamese cross-language summarization.

Algorithm 1. A Greedy Algorithm to compress the sentence

1: **function** COMPRESSION(s, S_{i-1})
2:     queue $Q \leftarrow \emptyset, kept \leftarrow \emptyset$
3:     **for** each phrase p in s.phrases **do**
4:         **if** $gain(p, S_{i-1}) > select\_thres$ **then**
5:             $kept \leftarrow kept \cup \{p\}$
6:             $Q.enqueue(p)$
7:         **end if**
8:     **end for**
9:     **while** $Q \neq \emptyset$ **do**
10:         $p \leftarrow Q.deque()$
11:         $ppv \leftarrow p.previous\_phrase$
12:         $pnx \leftarrow p.next\_phrase$
13:         **if** $gain\_p(ppv, p, S_{i-1}) > ppv\_thres$ **then**
14:             $Q.enqueue(ppv, p)$
15:             $kept \leftarrow kept \cup \{ppv\}$
16:         **end if**
17:         **if** $gain\_p(p, pnx, S_{i-1}) > pnx\_thres$ **then**
18:             $Q.enqueue(pnx)$
19:             $kept \leftarrow kept \cup \{pnx\}$
20:         **end if**
21:     **end while**
22:     **return** $\tilde{s} = kept$
23: **end function**

3.3   Phrase-Based Compressive Summarization Model

Phrase-based compressive summarization model is the sequential combination of two phases: compression and selection. As being mentioned in the Sect. 2.2, the greedy algorithm is an appropriate solution combining between the complexity and accuracy. First, the input sentence is translated into target language. Secondly, in our model, the phrase are derived from the result of translation phase. Thirdly, the translated sentences are compressed in Algorithm 1 by removing their redundant phrases. Finally, the point-wise ranking algorithm helps us to select the important sentence for the summary as Algorithm 2.

Algorithm 2. Phrase-based compressive summarization algorithm

1: $S \leftarrow \emptyset$
2: $i \leftarrow 1$
3: **while** $U \neq \emptyset$ **do**
4:     $max\_score = 0, max\_idx = -1$
5:     **for** each sentence s in U **do**
6:         $\tilde{s} = COMPRESSION(s, S_{i-1})$
7:         $score\_add = \frac{F(S_{i-1} \cup \tilde{s}) - F(S_{i-1})}{cost(\tilde{s})}$
8:         **if** $score\_add > max\_score$ **then**
9:             $selected\_sen = \tilde{s}$
10:             $max\_score = score\_add, max\_idx = i$
11:         **end if**
12:     **end for**
13:     **if** $C(S_{i-1} \cup \{selected\_sen\}) \leq B$ **then**
14:         $S_i \leftarrow S_{i-1} \cup \{selected\_sen\}$
15:         $i \leftarrow i + 1$
16:     **end if**
17:     $U \leftarrow U \setminus \{s_{max\_idx}\}$
18: **end while**
19: **return** S

Unlike some previous works no concern the effect of the length, our model tries to maximize the gain-cost ratio in the objective function between the score of the sentence and the cost to put it into the summary. It is extremely necessary and important in summarization due to the limitation of budgets. The greedy algorithm in Algorithm 2 looks quite simple, however, it is completely effective and reliable thanks to the bound in Eq. 2.

4 Experiments

4.1 Dataset

As we mentioned above, summarization especially cross-language one is too new to have any bilingual datasets in Vietnamese. Therefore, in the most previous works in bilingual summarization, the first important work is to build the datasets for evaluation step. However, creating the new corpus for summarization is the big challenge. Therefore, two alternative solutions are to translate the input of Vietnamese summarization dataset into English one or the gold summary of English corpus. The first one is too difficult to carry out due to the cost and the lack of Vietnamese summarization corpus. For that reason, we make a decision on translating the gold references of English corpus (Opinosis ver 1.0 [2]) into Vietnamese one.

Opinosis 1.0 dataset [2] is extremely popular and well-qualified due to being confident to be used in many works. In this corpus, there are 51 topics in many areas as short users' comments or reviews. Each topic has from 50 to more 550 sentences which are not processed. Together with each topic, there are between 4

Table 1. Our bilingual datasets

		English	Vietnamese
Input document	No. Sentences (sen)	7086	7086
	No. Ave. Tokens (token/topic)	2617.82	3538.94
References	No. Sentences (sen)	1190	1190
	No. Ave. Tokens (token/topic)	16.14	23.33

and 5 references which are created by human authors. For each reference, there are around 2 sentences with 16 tokens. However, their tokens are processed perfectly as also existing some mistakes. It is the reason that the translated summary we create manually is longer than the root one in English. It is about 23 tokens which are only separated by spaces, not word segmentation tools. Some statistics are represented in Table 1.

The quality of our dataset is ensured by manual process and evaluation. It proves that our dataset is extremely confident and reliable to evaluate for this problems. Especially, when there are not any published dataset for cross-language summarization, this corpus is an important contribution to the later works.

4.2 Phrase Derivation

In our approach, the quality of machine translation model is one of the main factors which affect the final accuracy of our system. We use the Google Translator API which is provided by the TextBlob Library[2]. Together with the translated output, we also determine the phrases as the result of TextBlob Translation.

The Fig. 1 shows an example of translation and phrase derivation of two sentences. Words are marked by bold character and underline as they are in the same phase. However, in some cases like the last example, the phrase derivation of TextBlob is so extremely bad that we can not use it to compress the sentence, which reduces our accuracy in all model.

4.3 Evaluation Metrics

We use ROUGE (Recall-Oriented Understudy for Gisting Evaluation) [4] to evaluate our model. However, we also focus on ROUGE-1, ROUGE-2, and ROUGE-SU4. All of them are based on the phrase overlap between automatic summary and manual references.

4.4 Baseline

Due to the novelty of cross-language summarization in Vietnamese, we also set up the popular and simple model to compare with our model. This baseline is

---

[2] http://textblob.readthedocs.io/en/dev/.

E2: 0 out of 5 stars **GPS Navigator** **doesn't navigate accurately on a straight road** .

V2: 0 trong 5 sao **GPS Navigator** **không điều hướng chính xác trên một đường thẳng**.

E3: 0 out of 5 stars **Honest**, **accurate review**, , **PLEASE** READ !

V3: 0 trong 5 sao **Trung** thực, **xem xét chính xác**,, **XIN** ĐỌC!

Fig. 1. The example of phrase derivation via TextBlob

divided into two independent phases: summarization and translation. First, input documents are summarized by open-source library[3]. Then, English summary is translated to Vietnamese one by Bing Translator[4]. It is simpler and more feasible than the opposite model. The reason is that it is too difficult to find an available and efficient Vietnamese summarizer.

4.5 Results

Together with our contribution in the dataset, our first model for cross-language summarization in Vietnamese is a significant baseline for many later works. In spite of its first occurrences, our model also achieves the relevant results to many previous works in the other languages. The dil of our result is shown in Table 2. Another experiment conducted in Opinosis demonstrates that our model outperforms the baseline (Table 3).

Table 2. Our model results

	ROUGE-1	ROUGE-2	ROUGE-SU4
Recall (%)	36.81	11.54	16.70
Precision (%)	41.74	13.79	20.73
F1-score (%)	38.52	12.34	17.94

Besides, parameter *compress_threshold* also proves that our compression component is extremely effective. To evaluate the effect of *compress_threshold*, some parameters are adjusted to increase the fluctuation in F1-score of ROUGE-1. When *compress_threshold* increases, much information is eliminated (Fig. 3). However, when we can choose the appropriate threshold, the compressive summary is better than general one which is not processed by Algorithm 1.

[3] https://github.com/despawnerer/summarize.
[4] https://github.com/dookgulliver/BingTranslator.

Table 3. Comparison between our model and baseline

	ROUGE-1 (%)	ROUGE-2 (%)	ROUGE-SU4 (%)
Baseline	34.9	**12.4**	17.2
Our model	**38.5**	12.3	**17.9**

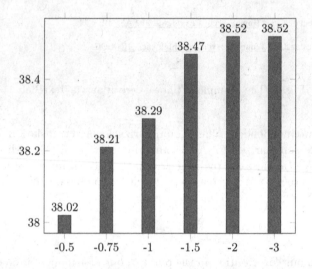

Fig. 2. The effect of eta with ROUGE-1

Obviously, the fixed threshold as Yao et al.'s work is not suitable for low-resource language. Their thresholds are too tight to maintain the necessary information in the sentences. Our predefined thresholds are considered as the new findings for low-resource language, (Fig. 2).

Although we eliminate the redundant phrases in each sentence, our summary also maintains its meaning and fluency thanks to the bi-gram component in scoring function (Eq. 7). In Fig. 4, we represent an example between references and our model summary. With the manual reference, our summary is also fluent and meaningful.

5   Conclusion

In this paper, we build manually the first bilingual dataset based on Opinosis ver 1.0 for cross-language summarization task. Our corpus is extremely effective and reliable to use for evaluation in the later works. It is our important contribution due to the lack of dataset in this task especially for the low-resource language as Vietnamese. Together with our dataset, we also firstly apply the compressive summarization model in Vietnamese. Our model also achieves the significant and encouraging results as the first cornerstone of this task in Vietnamese. Although it is the first work in English-Vietnamese summarization, our

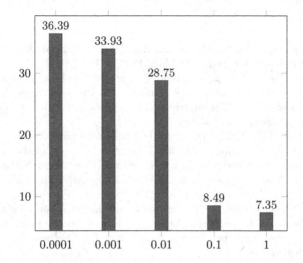

Fig. 3. The effect of *compress_threshold* with ROUGE-1 ($d = 0.001, \eta = 0.1$)

References:
E1: This unit is generally quite accurate.
V1: Đơn vị này nhìn chung khá chính xác .
E2: Set-up and usage are considered to be very easy.
V2: Thiết lập và cách sử dụng được đánh giá là rất dễ dàng .
E3: The maps can be updated, and tend to be reliable.
V3: Bản đồ có thể được cập nhật , có xu hướng đáng tin cậy .

Our summary:
Dễ dàng sử dụng và ngạc nhiên trước cách chính xác mặt hàng này .
có vẻ là khá chính xác .

Fig. 4. An example for topic *accuracy_garmin_nuvi_255W_gps*

model overcomes the drawback in many previous works in cross-language summarization. By adding the knowledge from translation phase, the scoring function is improved significantly. Besides, we also maintain the meaning and fluency through adding the bigram score in the objective function. All our development helps to make the selection phase better and better.

In future works, we intend to build the bigger datasets that are suitable and effective for training in many Machine Learning or Deep Learning methods. It is our grand ambition as it is too hard to find the huge and qualified dataset for Vietnamese summarization. However, if we can reach our ambition, it is extremely outstanding for the later works or the practical application in industry. In addition, to overcome the lack of dataset, we also put more knowledge from translation phase into summarization one and otherwise. Besides, together with the development of Deep Learning and its significant results in many tasks, we are going to take advantage of its power and effectiveness. It is obviously worth expecting if we had the huge and qualified dataset in Vietnamese which is not available now.

References

1. Boudin, F., Huet, S., Torres-Moreno, J.: A graph-based approach to cross-language multi-document summarization. Polibits **43**, 113–118 (2011)
2. Ganesan, K., Zhai, C., Han, J.: Opinosis: a graph-based approach to abstractive summarization of highly redundant opinions. In: Proceedings of the 23rd International Conference on Computational Linguistics, pp. 340–348. Association for Computational Linguistics (2010)
3. Koehn, P., Och, F.J., Marcu, D.: Statistical phrase-based translation. In: Proceedings of the 2003 Conference of the North American Chapter of the Association for Computational Linguistics on Human Language Technology, NAACL 2003, vol. 1, pp. 48–54. Association for Computational Linguistics, Stroudsburg, PA, USA (2003)
4. Lin, C.Y.: Rouge: a package for automatic evaluation of summaries. In: Marie-Francine Moens, S.S. (ed.) Text Summarization Branches Out: Proceedings of the ACL-04 Workshop, pp. 74–81. Association for Computational Linguistics, Barcelona, Spain (2004)
5. Lin, H., Bilmes, J.: Multi-document summarization via budgeted maximization of submodular functions. In: Human Language Technologies: The 2010 Annual Conference of the North American Chapter of the Association for Computational Linguistics, HLT 2010, pp. 912–920 (2010)
6. Lin, H., Bilmes, J.: A class of submodular functions for document summarization. In: Proceedings of the 49th Annual Meeting of the Association for Computational Linguistics: Human Language Technologies, HLT 2011, vol. 1, pp. 510–520. Association for Computational Linguistics, Stroudsburg, PA, USA (2011)
7. Litvak, M., Last, M., Friedman, M.: A new approach to improving multilingual summarization using a genetic algorithm. In: Proceedings of the 48th Annual Meeting of the Association for Computational Linguistics, ACL 2010, pp. 927–936. Association for Computational Linguistics, Stroudsburg, PA, USA (2010)
8. Nenkova, A., McKeown, K.: Automatic summarization. Found. Trends Inf. Retr. **5**(2–3), 103–233 (2011)
9. Sviridenko, M.: A note on maximizing a submodular set function subject to a knapsack constraint. Oper. Res. Lett. **32**(1), 41–43 (2004)
10. Yao, J.g., Wan, X., Xiao, J.: Phrase-based compressive cross-language summarization. In: Proceedings of the 2015 Conference on Empirical Methods in Natural Language Processing, pp. 118–127. Association for Computational Linguistics, Lisbon, Portugal, September 2015
11. Zhong, S., Liu, Y., Li, B., Long, J.: Query-oriented unsupervised multi-document summarization via deep learning model. Expert Syst. Appl. **42**(21), 8146–8155 (2015)

Improve the Performance of Mobile Applications Based on Code Optimization Techniques Using PMD and Android Lint

Man D. Nguyen[1]($\boxtimes$), Thang Q. Huynh[2], and T. Hung Nguyen[2]

[1] International School, Duy Tan University, Da Nang, Vietnam
mannd@duytan.edu.vn
[2] School of Information and Communication Technology of HUST,
Hanoi, Vietnam
{thanghq,hungnt}@soict.hust.edu.vn

Abstract. Analyzing, testing, and optimizing source code are techniques that improve software quality and the performance of features and energy consumption of systems. Source code analysis includes analyzing the source code of an application and checking aspects to detect potential problems based on previous experience. In this paper, we investigate the rules and techniques of analyzing and optimizing Java source code by using PMD and Android lint. An automatic code-analyzing and code refactoring tool is developed with a set of rules based on the Eclipse Refactoring API (plug-in) to get optimized code that consumes less energy and improves performance for Android applications. The optimized code was tested in real environments with positive results. It reveals that programmer could use these techniques and support tool for developing Android applications with high quality source code and reliable and performance.

Keywords: Static testing · Code analyzing · Optimized code · Code refactoring · Android testing · PMD · Android lint

1 Introduction

Analyzing and optimizing source code are techniques that improve the software quality and increase the performance of a system. Source code inspection and review are static testing techniques used for enhancing quality and reliability of applications. Code analysis includes reviewing and analyzing the source code in order to uncover the defects and potential problems within an application based on previous experience [1–3]. Static testing is a form of software testing that can be applied in each software development phase where the software has not been executed yet. It is generally not a detailed test because it mainly checks for the sanity of the code, algorithms, or documents. A developer who wrote the code, in isolation, can use this type of testing to improve the quality of the code and product [17].

Energy efficiency can have a significant influence on user experience of mobile devices. Although energy is consumed by hardware, software optimization plays an important role in saving energy, so that the developers have to participate in the

© Springer International Publishing AG 2016
V.-N. Huynh et al. (Eds.): IUKM 2016, LNAI 9978, pp. 343–356, 2016.
DOI: 10.1007/978-3-319-49046-5_29

optimization process. The source code is the interface between the developer and hardware resources. According to Manotas and et al. [20] developers do not understand how the software engineering decisions they make affect the energy consumption of their applications. Developing energy efficient mobile applications is an important goal for the developers as energy usage can directly affect the usability of a mobile device. However, there is a lack of guidance for developers to improve the energy efficiency of their implementation and to choose the most useful practices. There are existing studies and tools that can help developers to analyze, improve, and optimize their applications, such as, Shuai Hao and et al. proposed program analyzes [19], and statistical based measurement techniques [21]. These techniques enable developers to understand where energy is consumed within an application; they do not provide guidance to improve the application's energy consumption. Xueliang Li and John P. Gallagher [18, 22] proposed an energy optimization framework guided by a source code energy model. This technique enables developers to be aware of energy usage induced by the code and to apply very targeted source-level refactoring strategies. In previous our work [5], we presented the techniques to optimize the source code for Android mobile applications that are optimizing the process, shifting bits, using intermediate variables, removing redundant mathematical operations, processing loop, reusing objects, and refactoring source code, in order to increase the performance. At present, there are few techniques automated and supported by a variety of tools [6, 10] that increase developers' performance and lighten their burden. However, in fact, these techniques are not enough to address the gap between understanding where energy is consumed and how the code can be changed to reduce the energy consumed.

In this research, we mainly utilize PMD (Programming Mistake Detector), and Android lint to create a new set of rules and Eclipse plugins, which are used to analyze, optimize, and refactor Java source code, and thus, increase the performance as well as decrease the energy consumption of Android applications. PMD analyzes Java source-code based on the evaluative rules that have been enabled during a given execution. The tool comes with a default set of rules, which can be used to unearth common development mistakes, such as having empty try-catch blocks, unused variables, objects that are unnecessary, etc. PMD also allows users to execute custom analyzes by allowing them to develop new evaluative rules in a convenient manner [4, 11, 14]. The Android lint is a static code analysis tool that checks Android project source files for potential bugs and optimization improvements on correctness, security, performance, usability, accessibility, and internationalization [4, 12, 13].

The paper is organized as follows: In Sect. 2, we discuss about the optimizing and analyzing techniques for Java source code, the strategy for building and enforcing the inspection rules and the introduction of a new set of rules. In Sect. 3, the techniques along with all functions of Eclipse IDE's plugins are presented. Section 4 presents test results on real devices. The discussion of related works is presented in Sect. 5. Finally, the conclusion remarks, limitation, and future works are presented in Sect. 6.

2 Java Source-Code Analyzing and Optimizing Techniques for Android Applications

PMD and Android lint are used to carry out the rules for code optimization and static code analysis. There are general rules built for multiple languages and specific rules for Java only. Therefore, we realized that the richness of the optimized source code may cause conflicts and misunderstandings between the rules, such as optimizing speed may increase the size of memory. So that, what we need to solve in this paper is to optimize code in terms of energy consumption and performance improvement of the system. The contributions of this paper includes a set of new rules that are built to reduce energy consumption and improve performance of the system based on a set of rules of PMD and Android lint; a tool to optimize source code for Android applications, which can automatically change source code through Eclipse Refactoring API [9, 10].

2.1 Source Code Analyzing Rules

According to [7] and [11], different rules have an impact on different source code's elements, which are the result of transforming the original code into an Abstract Syntax Tree (AST) for Java and Document Object Model (DOM) for XML. Therefore, we use the following strategy for discovering the potential elements:

- Step 1: Analyze Java (*.java) and XML (*.xml) source code of project into AST and DOM, respectively.
- Step 2: With each achieved AST and DOM, we categorize the elements into groups with a common type. (E.g., element method calls, variable declaration)
- Step 3: Each rule in the ruleset is applied to the elements that it associates with. If an element fails to meet the requirement of any rules, we consider it as a potential element.
- Step 4: We collect the necessary information of potential elements and put in a list, which will be delivered to users later.

The set of rules may be dramatically various and the application of the rules in a particular case, sometimes, depends on the semantics of the program. Therefore, the potential elements should be explicitly listed so that programmers can decide the application of the rules.

2.2 Source Code Altering Rules

After analyzing the source code to find the potential elements, altering source code without violating the rules is the next issue. Proposing source-code altering and fixing methods are among the required functions in the application that we built. However, there is one thing to be noticed is that not all the rules can easily change and fix the source code because of the semantics of the program. Therefore, fixing source code is developers' duty. The steps to fix the source code are as follows:

- Step 1: Pick a potential element from the list and all necessary information about the rules that it has violated.
- Step 2: Check if the element still violates the rule (using previously mentioned analyzing techniques such as AST). If the violation remains, or to be removed, there are two possible situations: (a) if the rule does not depend on the program's semantics, the source code will be changed automatically; (b) if the rule depends on the program's semantics, the programmer will change the code manually based on the basic of the information provided by the rule.
- Step 3: Apply the source code alteration

The rules themselves have their own violation testing and re-adjusting process. By using this tool, we propose a general strategy that either automatically simplifies the re-adjustment, or suggests a solution to the developers.

2.3 Strategy for Applying the Rules

According to the rules of code analysis and alteration, the strategy is as follows:

- Classify AST's elements into different groups from beginning to increase the performance of the program. Sometimes different rules can affect the same elements and therefore, these rules can be reused to classify groups.
- The benefit of having AST's elements classified from the beginning is that each element goes through only one inspection process. Otherwise, if AST's elements classification takes place every time, the rules are applied to one particular group; each element will go through another inspection process. It means that the initial classification helps achieve better performance.
- Higher compatibility after extension: When the number of Java source-code files or rules increases, the proposed strategy will still take effect without any changes. Especially, the larger amounts of rules, the higher probability of AST's elements are used. That strengthens the effectiveness of the initial classification.
- Always make sure that the potential elements still violate the rules prior to source code alternation.

2.4 The Establishment of Source Code Analyzing and Altering Rules

The rules are built based on [7, 8] and [11]. The rules are also included useful information into these rules for future needs, such as suggesting examples and methods of using them without any violation. These details provide users a deeper understanding of the rules so that they can avoid breaking them without using other help from the application. The set of rules built based on groups of (1) decrease battery consumption (e.g. lower CPU Usage, brightness, color display and network processes), (2) general rules (which can be applied to different programming languages), (3) specification rules built only for Java and XML files for layout display, (4) rules that clarify the source code and simplify the maintenance process, and (5) rules that help decrease other costs.

In this paper, the set of 49 rules was built for optimizing source code and decreasing battery consumption. These rules are listed in Table 1:

Table 1. Rules for optimizing source code and decreasing battery consumption

No	Name of rule	Description
1	FinalVariable	Add "final" attribute to the variable that you do not want to change its value.
2	FieldShouldBeStatic	Variable with "final" attribute should be "static"
3	AlwaysReuseImmutable-ConstantObjectsFor Better-MemoryUtilization	Reuse constant objects if possible
4	UnusedVariable	Unused variable should be removed
5	UncalledPrivateMethod	Unused private method should be removed
6	ShouldBeStaticInnerClass	Inner class should be static if possible
7	AvoidUsingMathClass-MethodsOnConstant	Avoid using Math class methods on constant
8	ConstantExpression	Expression with constant value should be immediately calculated.
9	UseShiftOperators	The use of bit shifting operation is more recommended than multiplication and division of the power of 2.
10	UseShortCircuit-BooleanOperators	Use "||", "&&" instead of "|", "&"
11	UseStringLength-CompareEmptyString	Use "String.length()" instead of "String.equal()" when checking empty string
12	UseStringEqualsIgnoreCase	Use "String.equalsIgnoreCase ()" instead of "String.equals()" when comparing string if possible
13	AvoidUnnecessarySubstring	Avoid using "String.substring ()" if unnecessary
14	AvoidUsingStringCharAt	Avoid using "String.charAt ()"if unnecessary
15	AvoidConcatenatingString-UsingPlusOperatorInLoop	Avoid using operator "+" of class "String" to concatenate string in loop
16	UseSingleQuotesWhen-ConcatenatingCharacter-ToString	Using single quote instead of a double quote when concatenating a character to the string.

(continued)

Table 1. (*continued*)

No	Name of rule	Description
17	AvoidUsingStringTokenizer	Avoid using "StringTokenizer"
18	BoxedPrimitiveJust-CallToString	A boxed primitive is allocated just to call toString(). It is more effective to just use the static form of toString which takes the primitive value.
19	AvoidConstructing-PrimitiveType	Avoid initializing objects of the box of the primitive data types via the operator "new"
20	AvoidInstantiation-ForGetClass	Avoid initializing the object just to call the method "getClass()"
21	AvoidBooleanArray	Avoid using an array of elements of "bool" type
22	AvoidEmptyLoops	Avoid using empty loops
23	AvoidEmptyStaticInitializer	static initialization block
24	AvoidEmptyTryBlocks	Avoid using empty "try" blocks
25	AvoidEmptyFinallyBlock	Avoid using empty "finally" blocks
26	AvoidEmptyCatchBlocks	Avoid using empty "catch" block
27	AvoidEmptySynchronized-Block	Avoid using empty "synchronized" blocks
28	AvoidEmptyIf	Avoid using empty If
29	AvoidUnnecessaryIf	Eliminate condition statement always or never occur
30	AvoidWriteByteMethod-InLoop	Avoid using "DataOutputStream.writeByte()" in loop
31	AvoidReadByteMethod-InLoop	Avoid using "DataInputStream.readByte()" in loop
32	AvoidMethodCallsInLoop	Avoid calls methods in the loop if possible
33	AvoidObjectInstantiation-InLoops	Avoid initializing objects in the loop
34	PlaceTryCatchOutOfLoop	Place "try/catch" outside of loop.
35	AvoidVectorElementAt-InsideLoop	Avoid using "Vector.elementAt()" inner of loop.

(*continued*)

Table 1. (*continued*)

No	Name of rule	Description
36	EnsureEfficientRemoval-OfElementsInCollection	Delete Elements of "Collection" inner loop efficiently
37	EnsureEfficientRemoval-OfMapEntries	Delete Elements of "Map" inner loop efficiently
38	EnsureEfficientIteration-OverMapEntries	Iterate the Elements of "Map" efficiently
39	UseEntrySetInsteadOf-KeySet	Using "Map.entrySet()" instead of "Map.keySet()"
40	AvoidPollingLoops	Avoid using "Thread.sleep()" inner loop
41	AvoidUsingThreadYield	Avoid using "Thread.yield()"
42	AvoidCreatingThread-WithoutRunMethod	Avoid initializing "Thread" if have no "implement" object and "run" method
43	AvoidSynchronized-BlocksInLoop	Avoid using "synchronized" block inner loop
44	AvoidSynchronized-MethodsInLoop	Avoid using "synchronized" method inner loop.
45	AvoidSynchronized-ModifierInMethod	Avoid using "synchronized" in the method
46	AvoidLinkedList	Avoid using "LinkedList" in necessary.
47	DefineInitialCapacities	Should predefine the size of "AbstractList"
48	UseToArrayWithArray-AsParameter	Using "Collection.toArray (Object[])" instead of "Collection.toArray()"
49	UseDataSourceInstead-OfDriverManager	Using "DataSource" instead of "DriverManager" to get "Connection"

In generally, from the source code of the program, PMD creates the AST of the checked source file and executes each rule against that tree. AST is Abstract Syntax Tree which is a finite, labeled tree where the nodes represent the operators and the edges represent the operands of the operators. The violations are collected and presented in a report. PMD executes the following steps when being invoked from Eclipse:

1. The Eclipse PMD plugin passes a file name, a directory or a project to the core PMD engine. This engine then uses the rules as defined in package rule (the rules are configured in XML file of config folder of a project) to check the file(s) for violations.
2. PMD uses JavaCC to obtain a Java language parser and passes an Input Stream of the source file to the parser.

3. The parser returns a reference of an Abstract Syntax Tree back to the PMD
4. PMD hands the AST off to the symbol table layer which builds scopes, finds declarations, and finds usages
5. Each rule in the RuleSet gets to traverse the AST and checks for violations
6. A list of RuleViolations is recorded in Problem List and Refactoring Unit for a report and refactors source code.

3 Develop a Plug-in Tool for Eclipse IDE

Code inspection techniques for Android plug-in tool have (1) Android applications' project analysis feature, which enables the analysis of a project containing Java source code to figure out the potential problems and violated rules, (2) the re-adjustment of problematic elements that allow developers to fix and to solve the potential problems to achieve optimal source code, (3) information on violated rules that allows developers to select one potential element and review the violation details, (4) the elimination of the potential element on the list, and (5) the complement of new rules that allow developers to add new rules based on their experience.

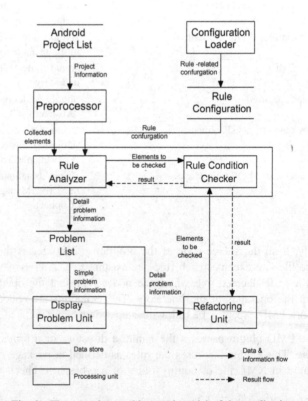

Fig. 1. The overview architectural model of the application

The architectural model of the application is shown in Fig. 1. The application has five main components: (1) *Configuration Loader* enables loading and saving the configuration of optimizing rules. Further information about the rules is also included. The loading must be done before process takes place because this configuration will be applied in most of the functions; (2) *Preprocessor* collects useful information from Java source code, analyzes source file(s) from Eclipse project, transforms source file(s) into AST, and finally, classifies AST's elements into groups with the same type; (3) *Analyzer* detects the potential elements in Java source code, retrieves the data from the preprocessor that has the same type of node group and manipulates them; (4) *Display Problem Unit* displays all potential elements with problems found by the Analyzer. All information about these elements will be fully displayed so that developers can set the overview of the source code that they are writing; (5) *Refactoring Unit* provides users a tool that supports source code alteration to solve and to eliminate all potential problems; the result is a much better and is the optimal source code.

Analyzer component is one of the most important modules of the tool. It plays a vital role in identifying potential elements in Java source code. Each analysis will be based on a number of specific rules to analyze and detect errors. Each rule has an impact on a group or a node obtained from the certain preprocessor. The class diagram of Analyzer component is shown in Fig. 2(a). The rule's information is described by RuleDescriptor and is stored in RuleDescriptionStore components.

The rule includes the following information: *Name of rule, Rule Identifying, Categorizing, Critical Levels of rule, Rule description, The reason for forming the rule, Examples and solutions, and Other information.*

ProblemItem is one of the potential elements detected by RuleAnalyzer. It contains a suite of RuleDescriptor, representing the violated rule and other information such as potential elements (ASTNode's derivative), locations (containing file, line...), and related information. Display Problem Unit is in charge of listing all problems detected in the source code. Refactoring Unit is responsible for refactoring the source code and supporting users in repairing and solving problems in order to achieve optimal source code. The class diagram of Refactoring Unit is shown in Fig. 2(b). When a user chooses a problem from MarkingItem on Eclipse's user interface, all information of the

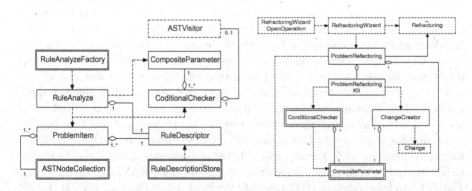

Fig. 2. Class diagram of (a) Analyzer component, (b) Refactoring Unit

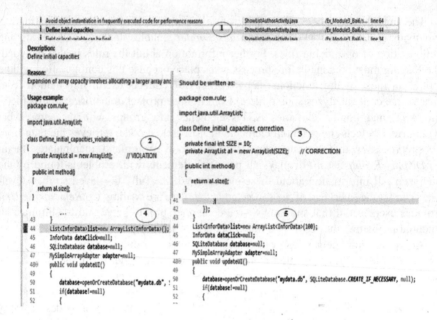

Fig. 3. Problem information

problem will be obtained immediately. All information (Fig. 3 is an example) needed for the refactoring process will be included in CompositeParameter.

The dashed rectangle notations in Figs. 1, 2(a) and (b) are eclipse's components. The single border rectangles are classes or interfaces. The double border rectangles are significant classes that the tool built. The dashed arrows indicate the reply of the information.

4   Testing Result

In the scope of the paper, we set the goal of improving the quality of source code, decreasing the energy consumption of applications as well as enhancing the system performance. The set of rules are mentioned in Table 1 helps us optimize source code and decrease battery consumption for Android phones. Testing result is collected from "Show CPU Usage" within Android's Developer Options. During the test, we collect the average percentage of CPU being used in the last 1-minute, 5 min and 10 min before and after the optimization. Also, we conducted this test on 5 different applications and on the same devices with different settings. Figures 4 and 5 display test results from two different devices respectively (*A and B in* Figs. 4 and 5 *mean After and Before the optimization*). One is Motorola moto X: Android 4.4.4, OpenGL ES 2.0, Kernel Architecture i686, Kernel version 3.10.0-genymotion, screen size 4.59 inches, resolution 720 × 1280 pixels, screen density 320dpi, total RAM 2020 MB, storage 4.84 GB. And the other is HTC One XL: Android 4.2.2, OpenGL ES 2.0, Kernel

Architecture i686, kernel 3.4.67-qenu + , screen size 4.59 inches, screen resolution 720 × 1280 pixels, screen density 320dpi, total RAM 1006 MB, storage 4.78 GB.

According to Figs. 4 and 5, the use of source-code optimizing and refactoring techniques obviously enhances the performance of not only the application but also the device (the percentage of CPU used has dropped). The percentage of CPU used in 1 min has dropped significantly after the optimal source code. This value will stabilize after 10 min, and in close proximity to the value before the optimization.

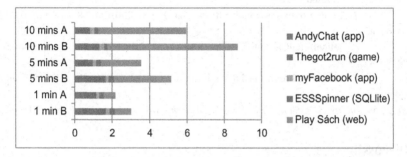

Fig. 4. Statistics of CPU Usage (%) testing on Motorola Moto X

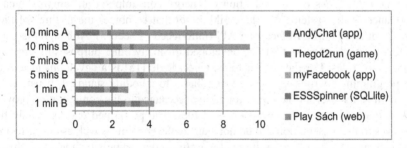

Fig. 5. Statistics of CPU Usage (%) testing on HTC One XL

Table 2 indicates the CPU is used for each rule to be invoked. Depending on the specific rule that the reduction rate after 1 min of CPU usage is very high, such as AvoidMethodCallsInLoop, UseStringLength CompareEmptyString, DefineInitialCapacities, and so on. Corresponding to each program, all the rules in Table 1 are implemented to find the violations of the program that can help programmers optimize source code to improve performance for applications. Furthermore, if more than one installed applications are optimized, the decrease in battery consumption is much more significant.

Table 2. - Statistics of CPU Usage (%) testing on HTC One XL for each rule

Application	Rule	Before optimization	After optimization
Estate	AvoidEmptyIf	0.28/0.22/0.14	0.27/0.19/0.14
Estate	AvoidMethodCallsInLoop	0.34/0.13/0.12	0.08/0.07/0.10
Who is millionaire	UseStringLength-CompareEmptyString	0.51/0.17/0.08	0.09/0.10/0.08
Estate	Avoid object instantiation in frequently executed code	0.38/0.12/0.08	0.32/0.11/0.07
Book management	A constant expression can be evaluated	0.43/0.12/0.08	0.11/0.09/0.07
Book management	DefineInitialCapacities	0.14/0.09/0.06	0.0/0.04/0.05
Who is millionaire	AvoidSynchronized-MethodsInLoop	0.22/0.07/0.06	0.01/0.08/0.07

(0.39/0.5/0.59: the average % of CPU being used in the last 1-minute, 5 min and 10 min)

5 Discussion

In summary, the contributions of this paper are as following. We take the approach to check the quality of code and build the set of rules based on PMD and Android lint rules. The rules focus on the reduction of energy consumption and improvement of performance of the system. We also build a tool that combines the features of PMD, Android lint and Eclipse Refactoring API (introduced in Sect. 3). Although this tool is built for specific purposes, it can be developed for many other purposes by creating a custom set of rules. For the similar tools, CheckStyle [15] is a development tool that helps programmers writes Java code that adheres to a coding standard. It parses Java code and informs user all the errors found in accordance to the configuration provided in checkstyle.xml and the suppressions.xml files. The performed checks mainly limit themselves to the presentation and do not analyze the content, as well as the correctness or completeness of the program. It may be useful to determine which level of check is needed for a certain type of program. FindBugs [16] uses static analysis to inspect Java bytecode for occurrences of bug patterns. It detects common errors such as wrong Boolean operator. It also detects errors due to the misunderstanding of language features, such as reassignment of parameters in Java. The goal is globally the same CheckStyle, to find the patterns, which can lead to bugs using static analysis. However, it can be much more controversial than FindBugs. The basic ruleset is always executed alongside with the custom ruleset. Android Lint provides great piece of advice to improve the quality of code. The user can use a separate lint file to define which rules to use and to activate all the exacted rules. Android Lint always tests all the rules except the one of which the "severity" level is "ignore". So if new rules are released with new version of ADT, they will be examined without ignorance. Although each tool has its own advantages and disadvantages, our tool is built by combining the features of PMD and Android lint for a specific purpose. Our future work is to research on combining into a framework for improving the quality of code and increasing the performance of mobile applications for many different purposes.

6 Conclusion

In this paper, we studied several types of research and utilized plenty of Java source-code optimization techniques. We studied and built a set of rules and developed a plug-in, which automates the analysis, finds the potential problems and refactors the code. The quality of code is sharply improved and the battery consumption is dramatically decreased. The result is very useful for developers in agile development process, which supposedly requires developers' acquisition of software tester's skills. The tool helps developers analyze and reveal problems quickly, effectively by doing the refactoring process, and easily share the result in the group. Moreover, with our own method, engineers are able to extend and enrich the rules (based on their demand) to fulfill different needs in a very simple, effective, and reliable way.

The future work includes adding more rules to enrich the capabilities of the tool, as well as applying more rules for refactoring and supporting automatically. Currently, the tool can refactor potential problems on each component through the Eclipse interface. In the future, the tool can support automated refactoring potential problems across multiple components simultaneously. Further evaluation of usages and application for developing different types of mobile applications are also necessary. The long-term future work includes making the tool into a framework by combining the advanced features of Checkstyle, FindBugs, PMD, and Android lint to check the quality of code and performance improvement of the system.

References

1. Fagan, M.: Design and code inspections to reduce errors in program development. In: Broy, M., Denert, E. (eds.) Software pioneers, pp. 575–607. Springer, Heidelberg (2002)
2. Ackerman, A.F., Fowler, P.J., Ebenau, R.G.: Software inspections and the industrial production of software. In: Proceedings of a Symposium on Software Validation: Inspection-Testing-Verification-Alternatives. Elsevier North-Holland, Inc. (1984)
3. Ackerman, A.F., Buchwald, L.S., Lewski, F.H.: Software inspections: an effective verification process. IEEE Softw. 3, 31–36 (1989)
4. Gomes, I., et al.: An overview on the static code analysis approach in software development. Faculdade de Engenharia da Universidade do Porto, Portugal (2009)
5. Thang, H.Q., Duc-Man, N., Nam, D.L.: Some optimization techniques and performance testing for android application development. In: Proceedings of the 7th National Conference FAIR'7), ISBN:978-604-913-300-8, pp. 365–375 (2014)
6. Wedyan, F., Alrmuny, D., Bieman, J.M.: The effectiveness of automated static analysis tools for fault detection and refactoring prediction. In: International Conference on Software Testing Verification and Validation. ICST 2009. IEEE (2009)
7. Bauer, L., et al.: Mobile SCALe: Rules and Analysis for Secure Java and Android Coding. No. CMU/SEI-2013-TR-015. Carnegie-Mellon Univ Pittsburgh Pa SEI (2013)

8. Sharkey, J.: Coding for life–battery life, that is. In: Google IO Developer Conference (2009)
9. Petito, Michael.: Eclipse refactoring, http://people.clarkson.edu/ ∼ dhou/courses/EE564-s07/Eclipse-Refactoring.pdf
10. Frenzel, L.: The Language Toolkit: an API for automated refactorings in Eclipse-based IDEs. Eclipse Mag. 5 (2006). https://eclipse.org/articles/Article-LTK/ltk.html
11. How to write a PMD rule. http://pmd.sourceforge.net/snapshot/-howtowritearule.html
12. Android Lint. http://tools.android.com/tips/lint
13. Lint. http://developer.android.com/tools/help/lint.html
14. Christopher, C.N.: Evaluating static analysis frameworks. Anal. Softw. Artif., 1–17 (2006)
15. Checkstyle tool. http://checkstyle.sourceforge.net/
16. FindBugs tool FindBugs™ - Find Bugs in Java Programs. http://findbugs.sourceforge.net/
17. Bourque, P., Fairley, R.E.: Guide to the software engineering body of knowledge(SWEBOK (R)): Version3.0. Chap.10. IEEE Computer Society Press, Los Alamitos (2014). ISBN: 0769551661 9780769551661
18. Li, X., Gallagher, J.P.: A Source-level Energy Optimization Framework for Mobile Applications (2016). arXiv preprint arXiv:1608.05248
19. Hao, S., Li, D., Halfond, W.G.J., Govindan, R.: Estimating mobile application energy consumption using program analysis. In: 2013 35th International Conference on Software Engineering (ICSE), pp. 92–101. IEEE (2013)
20. Manotas, I., Pollock, L., Clause, J.: SEEDS: a software engineer's energy-optimization decision support framework. In: Proceedings of the 36th International Conference on Software Engineering, pp. 503–514. ACM (2014)
21. Li, D., Hao, S., Halfond, W.G.J., Govindan, R.: Calculating source line level energy information for android applications. In: Proceedings of the 2013 International Symposium on Software Testing and Analysis, pp. 78–89. ACM (2013)
22. Li, X., Gallagher, J.P.: A top-to-bottom view: Energy analysis for mobile application source code (2015). arXiv preprint arXiv:1510.04165

Biomedical and Image Applications

Clustering of Children with Cerebral Palsy with Prior Biomechanical Knowledge Fused from Multiple Data Sources

Tuan Nha Hoang[1,2], Tien Tuan Dao[1(✉)], and Marie-Christine Ho Ba Tho[1]

[1] UTC CNRS UMR 7338 Biomechanics and Bioengineering,
University of Technology of Compiegne, BP 20529, 60205 Compigne cedex, France
tien-tuan.dao@utc.fr
[2] UTC CNRS UMR 7253 Heuristics and Diagnostics for Complex Systems,
University of Technology of Compiegne, BP 20529, 60205 Compigne cedex, France

Abstract. Clustering models have been used to automatically identify groups in children with cerebral palsy using their clinical observable data. Current models used a limited data (e.g. gait data) and without integrating the underlying data uncertainty. Thus, this could be lead to unreliable clustering results. The objectives of this present work were to: (1) develop a new clustering approach to integrate multimodal biomechanical data; (2) incorporate prior biomechanical knowledge extracted from multiple data sources to improve the reliability of the clustering results. Thus, a new variant of classical evidential C-means (ECM) method called US-ECM (Uncertainty-Space Evidential C-means) was developed and implemented. We tested the performance and robustness of the proposed method on a synthetic database of children with cerebral palsy. Using Davies−Bouldin index, computational results showed that the use of multimodal data leads to a better clustering result. Moreover, the integration of prior knowledge about the uncertainty space of each parameter of interest allowed the clustering algorithm to be converged quickly as well as the clustering performance to be improved significantly. In fact, our new clustering method could be used to assist the clinicians in their decision making process (diagnosis, evaluation of medical intervention outcomes, communications with medical professionals) in a more reliable manner.

Keywords: Data clustering · Evidential C-means (ECM) · Uncertainty-Space Evidential C-means (US-ECM) · Belief functions · Data uncertainty space · Data fusion · Children with cerebral palsy

1 Introduction

Cerebral palsy (CP) is a chronic orthopedic pediatric disorder of the human musculoskeletal system [1]. Children with cerebral palsy have abnormal but recoverable locomotion functions. At the present time, clinicians have used the palpation

© Springer International Publishing AG 2016
V.-N. Huynh et al. (Eds.): IUKM 2016, LNAI 9978, pp. 359–370, 2016.
DOI: 10.1007/978-3-319-49046-5_30

and video-based observation techniques to obtain the patient gait data and analyze them to choose a suitable method of treatment [2]. Despite their simplicity and ease-to-use characters, these techniques allow a limited patient data (e.g. spatiotemporal data) to be used in the clinical decision making process. For the future clinical routine practice, more patient data such as morphological, mechanical, and kinetic and electromyography (EMG) ones need to be used for choosing a more appropriate treatment prescription as well as for performing an objective evaluation of the treatment outcomes (functional rehabilitation or surgical treatments). However, when dealing with these multimodal and multidimensional data, which are highly heterogeneous and subject to the data uncertainties, a new predictive approach needs to be developed because of the lack of available methods to tackle this problem.

Data uncertainties are classified as random uncertainty (e.g. experimental error due to the systematic intrinsic variations of the experimental observations) [3,4] or epistemic uncertainty (e.g. error due to lack of information or knowledge about the observations of interest) [5,6]. Biomechanical data are influenced not only by the subject specific intrinsic variations [7,8] but also by experimental operator [9–11] or by instrumentation used [12]. These sources lead to the intra-subject error (e.g. error expressed by the mean standard deviation values) from repeated trials on the same subject and the inter-subject error (e.g. a range of values measured for N subjects of interest) coming from different subject origins/races or from different laboratories [13]. In fact, these uncertainties need to be considered into our clustering method.

To represent data uncertainties, a probability uniform distribution has been commonly used [14–16]. However, this classical representation may be used only under some limited conditions such as minimum limits or 100 % containment criterion. Moreover, it deals only with random uncertainty. Advanced uncertainty representations such as box-encoding probability (i.e. distribution function accumulating upper and lower bounds) [17–19] or the possibility distribution (or fuzzy interval) [20–22] or the Dempster-Shafer belief function [23–25] have been recently used to represent both random and epistemic uncertainties. Furthermore, belief theory was successfully applied in many applications domains ranging from postal recognition [26] to knee function classification [27]. Regarding the available clustering methods using belief function, an evidential clustering (EVCLUS) algorithm [28] for relational data and an Evidential C-Means (ECM) [29] with partial knowledge for object data have been proposed recently. Numerical results showed the performance of these approaches comparing to traditional clustering ones. Furthermore, these approaches provided flexible mathematical structures to be adapted for incorporating prior knowledge into a clustering model. Consequently, the objectives of this present work were to: (1) develop a new clustering approach to integrate multimodal biomechanical data; (2) incorporate prior biomechanical knowledge extracted from multiple data sources to improve the reliability of the clustering results.

The rest of the paper is organized as follows. The background on belief functions will first be recalled in Sect. 2. The new clustering approach will be described and evaluated, respectively, in Sects. 3 and 4. Finally, Sect. 5 will conclude the paper.

2   Background on the Theory of Belief Functions

In this section, we summary briefly some basic concepts and elements of Dempster-Shafer theory used in the development of our predictive clustering model. The theory of belief functions was introduced by Dempster and Shafer [23,24] and then developed by Smets [25].

2.1   Mass Function

A frame of discernment (Ω) is defined as a set of mutually exclusive elementary propositions from available data sets. Considering a variable x, the partial knowledge regarding the actual value taken by x can be represented by a mass function m from 2^{Ω} to $[0,1]$ with:

$$\sum_{A \subseteq \Omega} m(A) = 1 \tag{1}$$

From mathematical point of view, m can be considered as a generated probabilistic distribution (i.e. mass distributes on the 2^{Ω} instead of Ω).

2.2   Decision Making

Using bba, we have several methods to make decision on choosing a singleton in 2^{Ω}. One of them is to use the highest plausibility [30]. In this case, an interval of belief $([Bel(A), Pl(A)])$ is defined as follows:

$$Bel(A) = \sum_{A,B \in 2^{\Omega}, B \subseteq A} m(B) \tag{2}$$

$$Pl(A) = \sum_{A,B \in 2^{\Omega}, B \cap A \neq \emptyset} m(B) \tag{3}$$

3   New Clustering Approach

3.1   Workflow

To extract prior knowledge from literature as well as to integrate multimodal biomechanical data into clustering model, a workflow was developed and illustrated in Fig. 1.

The first step concerns the literature-based data collection aiming to establish an uncertainty space considered as prior knowledge, which represents fused data from multiple sources. The second step relates to the development of a new variant of classical evidential C-means (ECM) method called uncertainty-space evidential C-means (US-ECM), which was proposed to incorporate prior knowledge into the clustering process. Then, the third step is the generation of a synthetic patient database. The final step relates to the performance and sensitivity analyses of the proposed clustering model on the generated database.

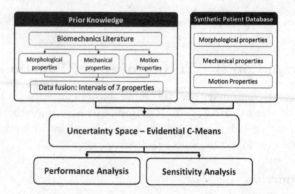

Fig. 1. Flow diagram of different components of our study.

3.2   Building Prior Knowledge

Seven morphological, mechanical, and motion properties of the normal children and those with cerebral palsy (CP) were acquired from biomechanics literature. The selection criterion is based on the availability of data for constructing the uncertainty space of each parameter of interest. In fact, multiple data sources (i.e. research papers from Science Direct and Pubmed) were collected and analyzed. Each retrieved paper was scanned and reviewed by a biomechanical expert to ensure the pertinence of experimental and processed data. Then, data were extracted from each paper to define an uncertainty space (US) for each property of interest such as muscle volume or muscle shear modulus. An example of uncertainty space is the shear modulus of soleus muscle ranging from 50 to 54 kPa.

3.3   Clustering Models

Evidential C-means (ECM) method [29] is based on the framework of belief functions. Credal partition is used in ECM to define the class membership of an object i by a bba m_i on the set $\Omega = \{\omega_1, \ldots, \omega_c\}$ where $|\Omega| = c$ is number of singleton classes. To incorporate prior knowledge considered as an uncertainty space (denoted as DOM - value DOMain) into the clustering model, the objective function is modified by using the dissimilarity from each group (whose center is C_i) to its uncertainty space (whose center is G_i) (Fig. 2). In the next subsections, a new variant of this method called Uncertainty-Space Evidential C-means (US-ECM) is addressed.

Evidential C-Means (ECM). Each object has p parameters $x_i(x_{i_1}, \ldots, x_{i_p})$. Its corresponding membership is presented by a bba $m_i(.)$ on $\Omega = \{\omega_1, \ldots, \omega_c\}$. With each element $A_j \in 2^{\Omega}$, m_{ij} is determined by the distance d_{ij} between x_i and center of A_j: $m_{ij} \triangleq m_{x_i}(A_j)$ in such a way that $(m_{ij} \sim \frac{1}{d_{ij}})$. Each subset

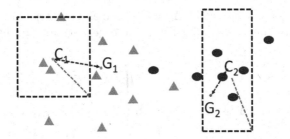

Fig. 2. Illustration of the integration of prior knowledge into the bi-classes (triangle and circle) clustering model.

A_j is represented by a center vector $\overline{v_j}$ computed as follows:

$$\overline{v_j} = \frac{1}{c_j} \sum_{k=1}^{c} s_{kj} v_k \qquad (4)$$

with $c_j = |A_j|$ is the cardinality of A_j and $s_{kj} = \begin{cases} 1 \text{ if } \omega_k \in A_j \\ 0 \text{ if other.} \end{cases}$

Uncertainty-Space Evidential C-Means (US-ECM). Prior knowledge was considered as an uncertainty space for each singleton class of interest (e.g. normal or CP cases). The uncertainty space of all p properties for each class is represented by the following p-dimension matrix:

$$\begin{bmatrix} min_{1_{\omega_k}} & \dots & min_{p_{\omega_k}} \\ max_{1_{\omega_k}} & \dots & max_{p_{\omega_k}} \end{bmatrix} \qquad (5)$$

Then, the fused uncertainty space Z_j of each element $A_j \in \Omega$ is computed using the following combination rule:

$$\begin{cases} min_{i_{A_j}} = \min_{\omega_k \in A_j} \{min_{i_{\omega_k}}\} \\ max_{i_{A_j}} = \max_{\omega_k \in A_j} \{max_{i_{\omega_k}}\} \end{cases} \qquad (6)$$

It is important to note that the units of morphological, mechanical and motion properties are different from one to another. To keep the same meaning for the use of the Euclidean distance metric, each parameter was normalized by its DOM range ($M_i, \forall i = \overline{1, p}$). Thus, the distance is redefined as follows:

$$d_{ij}^2 = \frac{(x_i^1 - \overline{v_j^1})^2}{M_1^2} + \dots + \frac{(x_i^p - \overline{v_j^p})^2}{M_p^2} \qquad (7)$$

And the center-vertex distance of $Z_j, \forall j = \overline{1, 2^c - 1}$ is calculated as follows:

$$l_j^2 = \frac{\sum_{i=1}^{p} \left\| max_{i_{A_j}} - min_{i_{A_j}} \right\|^2}{4} \qquad (8)$$

In addition, a new component Δ_3, considered as a constraint of each group to their uncertainty space, was added into the objective function of US-ECM as follows:

$$J_{US-ECM}(M,V) = \Delta_1 + \Delta_2 + \Delta_3, \text{ with } \begin{cases} \Delta_1 = \sum_{i=1}^{n} \sum_{A_j \in \Omega, A_j \neq \varnothing} c_j^\alpha m_{ij}^\beta d_{ij}^2 \\ \Delta_2 = \sum_{i=1}^{n} \delta^2 m_{i\varnothing}^\beta \\ \Delta_3 = \sum_{i=1}^{n} \sum_{A_j \in \Omega, A_j \neq \varnothing} c_j^\alpha m_{ij}^\beta D_j^2 \end{cases} \tag{9}$$

with

$$D_j^2 = \gamma \sum_{\omega_k \in A_j} \frac{\|v_k - \Gamma_j\|^2}{c_j} \times \frac{l_j^2}{l_{min}^2} \tag{10}$$

where $l_{min}^2 = \min_{A_j \in \Omega, A_j \neq \varnothing} l_j^2$ and Γ_j is the center of Z_j. γ represents the degree of belief on the uncertainty space.

In case of fixing V, the method n Lagrange multipliers λ_i is used to minimize J_{US-ECM} with respect to M:

$$\mathcal{L}(M, \lambda_1, ..., \lambda_n) = J_{US-ECM}(M,V) - \sum_{i=1}^{n} \lambda_i \left(\sum_{j/A_j \in \Omega, A_i \neq \varnothing} m_{ij} + m_{i\varnothing} - 1 \right) \tag{11}$$

By differentiating the Lagrangian with respect to the m_{ij}, $m_{i\varnothing}$, λ_i and setting the derivatives to zero, we have:

$$\frac{\partial \mathcal{L}}{\partial m_{ij}} = c_j^\alpha \beta m_{ij}^{\beta-1} \overline{d_{ij}^2} - \lambda_i = 0 \tag{12}$$

$$\frac{\partial \mathcal{L}}{\partial m_{i\varnothing}} = \beta m_{i\varnothing}^{\beta-1} \delta^2 - \lambda_i = 0 \tag{13}$$

$$\frac{\partial \mathcal{L}}{\partial \lambda_i} = \sum_{j/A_j \in \Omega, A_i \neq \varnothing} m_{ij} + m_{i\varnothing} - 1 = 0 \tag{14}$$

$$\overline{d_{ij}^2} = d_{ij}^2 + D_j^2 \tag{15}$$

After solving this equations system, we have the necessary condition of optimality for M as follows:

$$\Rightarrow \begin{cases} m_{ij} = \dfrac{c_j^{-\alpha/(\beta-1)} \overline{d_{ij}^{-2/(\beta-1)}}}{\sum\limits_{A_k \neq \varnothing} \overline{d_{ik}^{-2/(\beta-1)}} + \delta^{-2/(\beta-1)}} \\ m_{i\varnothing} = 1 - \sum\limits_{A_k \neq \varnothing} m_{ij}, \forall i = \overline{1,n} \end{cases} \tag{16}$$

To minimize J_{US-ECM} in case of fixing m, the partial derivatives of J_{US-ECM} with respect to the centers are used, as results we have:

$$\frac{\partial J_{UCM}}{\partial v_l} = \sum_{i=1}^{n} \sum_{A_j \neq \varnothing} c_j^\alpha m_{ij}^\beta \frac{\partial \overline{d_{ij}^2}}{\partial v_l} = 0, \forall l = \overline{1,c} \tag{17}$$

With:

$$\frac{\partial \overline{d_{ij}^2}}{\partial v_l} = \frac{1}{c_j}(\frac{1}{c_j}\sum_k s_{kj}v_k - x_i) + \frac{\gamma l_j^2}{c_j l_m^2}(v_l - \Gamma_j) \tag{18}$$

$$\Rightarrow \sum_{i=1}^n \sum_{A_j \neq \varnothing} c_j^{\alpha-1} m_{ij}^\beta s_{lj}(x_i + \gamma\Gamma_j \frac{l_j^2}{l_m^2}) = \sum_{i=1}^n \sum_{A_j \neq \varnothing} c_j^{\alpha-1} m_{ij}^\beta s_{lj}(\frac{1}{c_j}\sum_k s_{kj}v_k + \frac{\gamma l_j^2}{c_j l_m^2}v_l) \tag{19}$$

This leads us to a linear equation as follows:

$$B_{c\times p} + C_{c\times p} = H_{c\times c}V_{c\times p} \tag{20}$$

where:

$$B_{lq} = \sum_{i=1}^n \sum_{A_j \neq \varnothing} c_j^{\alpha-1} m_{ij}^\beta s_{lj} x_{iq} = \sum_{i=1}^n \sum_{\omega_l \in A_j} c_j^{\alpha-1} m_{ij}^\beta x_{iq} \tag{21}$$

$$C_{lq} = \sum_{i=1}^n \sum_{A_j \neq \varnothing} c_j^{\alpha-1} m_{ij}^\beta s_{lj} \gamma \Gamma_{jq} \frac{l_j^2}{l_m^2} = \sum_{i=1}^n \sum_{\omega_l \in A_j} c_j^{\alpha-1} m_{ij}^\beta \gamma \Gamma_{jq} \frac{l_j^2}{l_m^2} \ \forall l = \overline{1,c} \ \forall p = \overline{1,p} \tag{22}$$

V is the solution of above equations which can be solved by a standard linear system solver. Then the maxtrix of mass functions was used to compute the $Bel(A)$ for decision making purpose.

4   Results

4.1   Uncertainty Space

The ranges of values of 7 morphological, mechanical and motion properties (medialis gastrocnemius (MG) muscle volume [32–34], medialis gastrocnemius (MG) muscle physiological cross-sectional area (pCSA) [32,35], the shear modulus of soleus muscle [36], the shear modulus of lateralis gastrocnemius muscle [36], the cadence [37], the stance time and the maximal dorsiflexion angle [37]) for the normal children and those with cerebral palsy were depicted in Table 1.

4.2   Systematic Performance Analysis

To quantify the performance and the robustness of our new clustering algorithm according to the classical evidential C-means, the Davies−Bouldin (DB) index [31] was used. In our bi-class clustering problem, 2-credal partition (normal and ceberal palsy classes) was imposed. Thus, four focal elements (ω_1 , ω_2 , Ω, and the empty set) were considered in the optimization process. The impact of parameter type on the clustering result was shown in Fig. 3. The Davies−Bouldin (DB) index showed that the use of muscle mechanical property gives the best clustering performance according to the use of other properties as well as the use of all properties.

Table 1. Multi-dimensional Uncertainty Space (US) of 7 selected parameters

Parameter	Normal	CP	All
Maximal dorsiflexion angle (ankle)(∘)	[20–32]	[5–15]	[5–32]
Normalized MG muscle volume (cm^3/kg)	[31–35]	[23–27]	[23–35]
MG pCSA (cm^2)	[41–61]	[31–37]	[31–61]
Cadence(step/min)	[64–115]	[58–99]	[58–115]
Stance time(s)	[0.6–0.7]	[0.7–1.6]	[0.6–1.6]
Shear Modulus of Soleus (kPA)	[16.5–17]	[50–54]	[16.5–54]
Shear Modulus of LG (kPA)	[16–16.4]	[37.3–38.8]	[16–38.8]

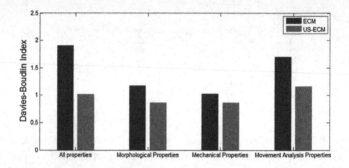

Fig. 3. Impact of parameter type on the clustering results.

4.3   Evaluation with a Variety of Data Sets

In order to demonstrate the effeciency of our method, several data sets with different distributional assumptions were used. It is impractical to have experimental data for all selected parameters. Consequently, a synthetic data was used for the performance analysis of our clustering approach. Related data of 1000 subjects were generated. Each subject data includes a 7-values vector. Data are

Fig. 4. Evaluation with a variety of data sets.

generated using their literature-based value rang. Note that a noise-control parameter was used to ensure the mixture of two classes (i.e. normal and cerebral palsy). We have generated many datasets for testing. Two dataset with related density for each generated property is used for evaluation. The results (Fig. 4) show that the the the DB Index is reduced with US-ECM.

5 Discussion

This present study proposed a powerful clustering model of children with cerebral palsy using multimodal biomechanical data (morphological, mechanical and motion properties) and prior biomechanical knowledge fused from multiple data sources. This allows grouping a set of children in such a way that children in the cerebral palsy group are more similar to each other than to those in the normal group. The clinical benefit of such useful model is the automatic determination of a cluster of the children before and after a treatment prescription (e.g. functional rehabilitation or surgery) to evaluate it. Thus, if the treatment is efficient, the children under investigation may be clustered in CP group before and in normal group after the treatment. To improve the performance and the robustness of our clustering model, a prior knowledge base was established from literature-based multiple data sources (i.e. scientific research papers). The idea of building such knowledge base comes from the fact that biomechanical data uncertainties exist and the use of one-value property for clustering task may lead to unreliable results, especially for a clinical application [11]. In fact, by using this prior knowledge base considered as a multi-dimensional uncertainty space, the clustering performance of our model was improved.

The use of belief function to represent this piece of knowledge allows the knowledge combination and propagation to be performed in a flexible and robust manner. In this present study, this method was extended with the incorporation of a multi-dimensional uncertainty space (US) for a credal partition considered as an interval-based prior knowledge on the cluster parameters leading to a better clustering result. In the literature, clustering model using interval data is commonly used to improve the clustering results [36–44]. However, each parameter in these studies is represented by an interval value while each parameter in our model is represented by one value and a prior knowledge represented by an extendable interval value. The use of prior knowledge in semi-supervised clustering and classification applications is a common approach [45,46]. In our present study, a prior knowledge base was established on the random uncertainty information of the upper and lower bound of each parameter. In fact, this can help the clusters to be converged rapidly and the search space is reduced considerably. In perspectives, our methodology will be extrapolated for others supervised data mining models such as artificial neural networks [47] and decision tree, or evidential clustering models such as EK-NNclus [48] and EVCLUS [49]. Moreover, data sets with different sizes as well as mutli-class clustering problem also need to be tested and analysed.

References

1. Blair, E., Watson, L.: Epidemiology of cerebral palsy. Semin Fetal Neonatal Med. **8**, 1–9 (2005)
2. Grunt, S., van Kampen, P.J., van der Krogt, M.M., Brehm, M.A., Doorenbosch, C.A.M., Becher, J.G.: Reproducibility and validity of video screen measurements of gait in children with spastic cerebral palsy. Gait Posture. **31**, 489–494 (2010)
3. Tonon, F.: Using random set theory to propagate epistemic uncertainty through a me-chanical system. Reliab. Eng. Syst. Saf. **85**(1–3), 169–181 (2004)
4. Cozman, F.G.: Concentration inequalities and laws of large numbers under epistemic and regular irrelevance. Int. J. Approximate Reasoning **51**(9), 1069–1084 (2010)
5. Dubois, D., Prade, H.: On several representations of an uncertain body of evidence. Fuzzy Inf. Decis. Process. **205**, 167–181 (1982)
6. Zadeh, L.A.: The concept of a linguistic variable and its application to approximate reasoning. Inf. Sci. **8**(3), 199–249 (1975)
7. Steinwender, G., Saraph, V., Scheiber, S., Zwick, E.B., Uitz, C., Hackl, K.: Intrasubject repeatability of gait analysis data in normal and spastic children. Clin. Biomech. **15**(2), 134–139 (2000)
8. Yavuzer, G., Oken, O., Elhan, A., Stam, H.J.: Repeatability of lower limb three-dimensional kinematics in patients with stroke. Gait Posture **27**(1), 1–5 (2008)
9. Leardini, A., Chiari, L., Della Croce, U.: Human movement analysis using stereophotogrammetry. Part 3. Soft tissue artifact assessment and compensation. Gait Posture **21**(2), 212–225 (2005)
10. Della Croce, U., Leardini, A., Chiari, L., Cappozzo, A.: Human movement analysis using stereophotogrammetry. Part 4: assessment of anatomical landmark misplacement and its effects on joint kinematics. Gait Posture **21**(2), 226–237 (2005)
11. Dao, T.T., Marin, F., Pouletaut, P., Aufaure, P., Charleux, F., Ho Ba Tho, M.C.: Estimation of accuracy of patient specific musculoskeletal modeling: case study on a post-polio residual paralysis subject. Comput. Method Biomech. Biomed. Eng. **15**(7), 745–751 (2012)
12. Chiari, L., Della Croce, U., Leardini, A., Cappozzo, A.: Human movement analysis using stereophotogrammetry. Part 2: instrumental errors. Gait Posture **21**(2), 197–211 (2005)
13. Gorton, G.E., Hebert, D., Gannotti, M.E.: Assessment of the kinematic variability among 12 motion analysis laboratories. Gait Posture **29**(3), 398–402 (2009)
14. Janak, S.L., Lin, X., Floudas, C.A.: A new robust optimization approach for scheduling under uncertainty: II. Uncertainty with known probability distribution. Comput. Chem. Eng. **31**(3), 171–195 (2007)
15. Horlek, V.: Analysis of basic probability distributions, their properties and use in determining type B evaluation of measurement uncertainties. Measurement **46**(1), 16–23 (2013)
16. Karmeshu, F., Rosano, L.: Modelling data uncertainty in growth forecasts. Appl. Math. Model. **11**(1), 62–68 (1987)
17. Ferson, S., Kreinovich, V., Ginzburg, L.R., Myers, D.S., Sentz, K.: Constructing Probability Boxes and Dempster-Shafer structures. Technical Report SAND 2002-4015, Sandia National Laboratories, Albuquerque, New Mexico, pp. 1–143 (2003)
18. Mehl, C.H.: P-boxes for cost uncertainty analysis. Mech. Syst. Sig. Process. **37**(1–2), 253–263 (2013)

19. Destercke, S., Dubois, D., Chojnacki, E.: Unifying practical uncertainty representations I: Generalized p-boxes. Int. J. Approximate Reasoning **49**(3), 649–663 (2008)
20. Dubois, D.: Possibility theory and statistical reasoning. Comput. Stat. Data Anal. **51**(1), 47–69 (2006)
21. Mohamed, S., McCowan, A.K.: Modelling project investment decisions under uncer-tainty using possibility theory. Int. J. Project Manage. **19**(4), 231–241 (2001)
22. Henn, V., Ottomanelli, M.: Handling uncertainty in route choice models: From proba-bilistic to possibilistic approaches. Eur. J. Oper. Res. **175**(3), 1526–1538 (2006)
23. Dempster, A.: Upper and lower probabilities induced by multivalued mapping. Ann. Math. Stat. **38**, 325–339 (1967)
24. Shafer, G.: A mathematical theory of evidence. Princeton University Press, Princeton (1976)
25. Smets, P.: The normative representation of quantified beliefs by belief functions. Artif. Intell. **92**, 229–242 (1997)
26. Mercier, D., Cron, G., Denoeux, T., Masson, M.H.: Decision fusion for postal address recognition using belief functions. Expert Syst. Appl. **36**, 5643–5653 (2009)
27. Jones, L., Beynon, M.J., Holt, C.A., Roy, S.: An application of the Dempster-Shafer theory of evidence to the classification of knee function and detection of improvement due to total knee replacement surgery. J. Biomech. **39**, 2512–2520 (2006)
28. Denoeux, T., Masson, M.H.: EVCLUS: evidential clustering of proximity data. IEEE transactions on systems, man, and cybernetics. Part B, Cybern. Publ. IEEE Syst. Man Cybern. Soci. **34**(1), 95–109 (2004)
29. Masson, M., Denoeux, T.: ECM: An evidential version of the fuzzy c-means algorithm. Pattern Recogn. **41**(4), 1384–1397 (2008)
30. Cobb, B.R., Shenoy, P.P.: On the plausibility transformation method for translating belief function models to probability models. Int. J. Approximate Reasoning **41**(3), 314–330 (2006)
31. Davies, D., Bouldin, D.: A cluster separation measure. Pattern Anal. Mach. **2**, 224–227 (1979)
32. Barber, L., Hastings-Ison, T., Baker, R., Barrett, R., Lichtwark, G.: Medial gastrocnemius muscle volume and fascicle length in children aged 2 to 5 years with cerebral palsy. Dev. Med. Child Neurol. **53**(6), 543–548 (2011)
33. Malaiya, R., McNee, A.E., Fry, N.R., Eve, L.C., Gough, M., Shortland, A.P.: The morphology of the medial gastrocnemius in typically developing children and children with spas-tic hemiplegic cerebral palsy. J. Electromyogr. Kinesiol. **17**(6), 657–663 (2007)
34. Oberhofer, K., Stott, N.S., Mithraratne, K., Anderson, I.: Subject-specific modelling of lower limb muscles in children with cerebral palsy. Clin. Biomech. **25**(1), 88–94 (2010)
35. Barber, L., Barrett, R., Lichtwark, G.: Passive muscle mechanical properties of the me-dial gastrocnemius in young adults with spastic cerebral palsy. J. Biomech. **44**(13), 2496–2500 (2011)
36. Basford, J.R., Jenkyn, T.R., An, K.N., Ehman, R.L., Heers, G., Kaufman, K.R.: Evaluation of healthy and diseased muscle with magnetic resonance elastography. Arch. Phys. Med. Rehabil. **83**(11), 1530–1536 (2002)
37. Sekiguchi, Y., Muraki, T., Kuramatsu, Y., Furusawa, Y., Izumi, S.I.: The contribution of quasi-joint stiffness of the ankle joint to gait in patients with hemiparesis. Clin. Biomech. **27**(5), 495–499 (2012)

38. De Carvalho, F.A.T., Lechevallier, Y.: Partitional clustering algorithms for symbolic in-terval data based on single adaptive distances. Pattern Recogn. **42**(7), 1223–1236 (2009)
39. De Carvalho, F.A.T.: Fuzzy c-means clustering methods for symbolic interval data. Pattern Recogn. Lett. **28**(4), 423–437 (2007)
40. De Almeida, C.W.D., de Souza, R.M.C.R., Candeias, A.L.B.: Fuzzy Kohonen clustering networks for interval data. Neurocomputing **99**(1), 65–75 (2013)
41. De Carvalho, F.A.T., Tenrio, C.P.: Fuzzy K-means clustering algorithms for interval-valued data based on adaptive quadratic distances. Fuzzy Sets Syst. **161**(23), 2978–2999 (2010)
42. Masson, M.H., Denux, T.: Clustering interval-valued proximity data using belief functions. Pattern Recogn. Lett. **25**(2), 163–171 (2004)
43. Chen, M., Miao, D.: Interval set clustering. Expert Syst. Appl. **38**(4), 2923–2932 (2011)
44. Masson, M.H., Denoeux, T.: Ensemble clustering in the belief functions framework. Int. J. Approximate Reasoning **52**(1), 92–109 (2011)
45. Lauer, F., Bloch, G.: Incorporating prior knowledge in support vector machines for classification: A review. Neurocomputing **71**(7–9), 1578–1594 (2008)
46. Tari, L., Baral, C., Kim, S.: Fuzzy c-means clustering with prior biological knowledge. J. Biomed. Inform. **42**(1), 74–81 (2009)
47. Denoeux, T.: A neural network classifier based on Dempster-Shafer theory. IEEE Trans. Syst. Man Cybern. Part A: Syst. Hum. **30**(2), 131–150 (2000)
48. Denoeux, T., Kanjanatarakul, O., Sriboonchitta, S.: EK-NNclus: a clustering procedure based on the evidential K-nearest neighbor rule. Knowl. Based Syst. **88**, 5769 (2015)
49. Denoeux, T., Masson, M.-H.: EVCLUS: evidential clustering of proximity data. IEEE Trans. Syst. Man Cybern. B **34**(1), 95–109 (2004)

Co-Simulation of Electrical and Mechanical Models of the Uterine Muscle

Maxime Yochum[(✉)], Jérémy Laforêt, and Catherine Marque

Sorbonne University, Université de Technologie de Compiègne, CNRS,
UMR 7338 Biomechanics and Bioengineering,
Centre de Recherche Royallieu-CS 60319, 60203 Compiègne Cedex, France
maxime.yochum@utc.fr

Abstract. Currently, one of the most studied signals to detect preterm labor is the electrohysterogram. Indeed, its components can be interpreted to quantify preterm labor risk. However, these kind of analyses do not give information about what is going on the uterine muscle itself. The literature already offers models that describe the electrical or the mechanical behavior of the uterine smooth muscle. But usually, these models are not related to each other. A model that could relate the electrical and the mechanical uterus behaviors, in order to understand the global electro-mechanical uterine activity, could be useful for preterm labor detection. It will permit to model the links existing during uterine contraction between the measurable component (the electrical activity) and the effective non measurable one (the mechanical activity). The study proposed here presents a co-simulation of three different models of the uterine smooth muscle related to its electromechanical components. The results show the feasibility to combine those three models in order to generate a global contraction on a 3D realistic uterine muscle mesh.

Keywords: Uterine activity · Modeling · Multiscale · Multi-physic · Co-simulation

1 Introduction

Preterm birth is one of the world largest public health problems [1] leading to high morbidity and mortality of newborns [2]. To prevent it, it is necessary to detect preterm labor symptoms as soon as possible via some biochemical or biophysical markers of preterm labor threats. One of the most promising tools is the analysis of the electrohysterogram (EHG, uterine electrical activity recorded on woman's abdomen) that gives an electrical representation of the contractile activity of the uterus, in a non-invasive way [3]. But the links existing between the EHG characteristics and the physiological events leading to preterm labor remain complex and unknown. EHG has been proved to be representative of the uterine electrical activity [3]. It has also been used to estimate the intrauterine pressure [4], However, EHG does not give information about the mechanical activity of the smooth muscle tissue, such as the force generated or the tissue

© Springer International Publishing AG 2016
V.-N. Huynh et al. (Eds.): IUKM 2016, LNAI 9978, pp. 371–380, 2016.
DOI: 10.1007/978-3-319-49046-5_31

deformation during a contraction. These mechanical features are out of reach of any measurement. But they could be extracted from a model linking the electro-mechanical phenomena involved in uterine contraction. In addition, it is well known that a correlation between the electrical activity and the contraction of the uterine muscle exists. But the electrical activity on its own cannot explain the whole uterine contraction process. The action potential propagation on the uterine tissue cannot induce the global uterine contraction occurring during labor. Therefore, other effects than the electrical diffusion should also exist to generate the uterine synchronization observed during labor contractions [5].

A big challenge is thus to develop a realistic model of the uterine muscle activity (electrical and mechanical) during pregnancy, integrating the physiological phenomena of interest in order to simulate preterm labor process. A multiscale model, going from the electrical activity of uterine cell to the EHG has already been developed [6].

We introduce here an electrical and mechanical model of the uterine smooth muscle through co-simulation. The objective is then to combine three models of the uterine activity, from the cell to the tissue levels, in order to model the electrical as well as the mechanical activities at the whole uterine muscle level. The electrical model describes the ionic currents generated during a uterine smooth muscle cell activity (SMC) and their propagation over the tissue through spatial electrical diffusion (cable theory). The second, force generation, models the mechanical contraction related to the electrical activity, from the calcium concentration output of the first model. The third one is a deformation model which describes the shape modification of the uterine muscle resulting from the forces generated by the active cells, outputs of the second model. We use a multiscale (from SMCs to the whole uterine muscle) and a multi-physics (electrical and mechanical properties) approach. We present results of the co-simulation of these 3 models, on a matrix of cells as well as on a realistic 3D surface mesh of the uterine muscle extracted from MRI images. Results show the feasibility to co-simulate the uterine electrical, mechanical and deformation behaviors. Forces generated on the tissue seem to have a slower dynamics than the electrical component of the model, leading to a synchronization of the mechanical uterine contraction.

2   Method

Three models (one electrical, 2 mechanical) are associated to make a multi-scale and multi-physic co-simulation of the uterine muscle contractile activity. The first is an electrical model, the second is a force generation model and the third one a deformation model.

The electrical model is a physiological model of the ionic currents of the uterus cells [6]. It uses only three variables and contains the calcium concentration which is used to link the electrical model to the force generation model. The three variables of the electrical model are described by the following equations:

$$\frac{dV_m}{dt} = \frac{1}{C_m} \left(I_{stim} - I_{Ca} - I_K - I_{KCa} - I_{leak} \right), \tag{1}$$

$$\frac{dn_K}{dt} = \frac{h_{K_\infty} - n_K}{\tau_{n_K}}, \tag{2}$$

$$\frac{d[Ca^{2+}]_i}{dt} = f_c \left(-\alpha I_{Ca} - K_{Ca}[Ca^{2+}]_i \right), \tag{3}$$

where V_m is the transmembrane potential, n_K is the potassium activation variable, K_{Ca} is the Calcium extraction factor and $[Ca^{2+}]_i$ the intracellular calcium concentration. The ionic currents are I_{Ca} for the voltage dependent calcium channels, I_K for the voltage dependent potassium channels, I_{KCa} for the calcium dependent potassium channels and I_{leak} for the leakage current. .

The diffusion process between neighboring cells is described by an electrical coupling simulating the gap junction effect (electrical propagation) [7]. A coupling term is thus added to Eq. 1:

$$\frac{dV_m}{dt} = \frac{\sum I_x}{C_m} - \bigtriangledown \cdot D \bigtriangledown (V_m), \tag{4}$$

where C_m is the membrane capacitance, I_x represents the different ionic currents and D is the diffusion tensor. The latter is dependent on the distance d_k that exists between neighboring cells. Therefore:

$$D_k = \frac{1}{RC_m d_k^2}, \tag{5}$$

where R is the membrane resistance and k a pair of neighboring cells.

The second used model, the force generation model, is the one proposed by Maggio et al. [8], who modified the original Hai-Murphy's model [9] to better correspond to uterine smooth muscle cells. Briefly, the force generation is based on four-state kinetic cross-bridges (M, Mp, AMp and AM) where the distribution of cross-bridges in each state can be computed by:

$$\frac{dM}{dt} = -k_1 M + k_2 M_p + k7 AM \tag{6}$$

$$\frac{dM_p}{dt} = k_1 M - (k_2 + k_3) Mp + k_4 AM_p \tag{7}$$

$$\frac{dAM_p}{dt} = k_3 M_p - (k_4 + k_5) AM_p + k_6 AM \tag{8}$$

$$\frac{dAM}{dt} = k_5 AM_p - (k_6 + k_7) AM \tag{9}$$

where initial conditions are M(0)=1 and Mp(0)=AMp(0)=AM(0)=0. k1 to k7 are the transition rates from one state to another which can be found in [8].

The link between the electrical and the force generation models is mediated through the calcium concentration, which is an output of the electrical model. Burzstyn et al. [10] introduced this link on the k1 parameter which corresponds to the phosphorylated rate of cross-bridges. k1 is then computed as:

$$k1(t) = \frac{[Ca^{2+}]_i^{n_M}}{[Ca^{2+}]_i^{n_M} + [Ca^{2+}_{1/2MLCK}]^{n_M}},\tag{10}$$

where $[Ca^{2+}_{1/2MLCK}]$ is the concentration of intracellular calcium that corresponds to the half activation of the myosin light-chain kinase, and n_M is the Hill coefficient of this activation. The force generated by the SMC is the sum of both attached states (AM and AMp) with a proportional coefficient K (given in [8]).

$$F(t) = K(AMp(t) + AM(t))\tag{11}$$

The last model is a simple Kelvin Voigt model approach, frequently used to describe the viscoelastic biological materials [11]. The equations are given by:

$$\varepsilon = \frac{1}{E}\sigma_1,\tag{12}$$

$$\frac{d\varepsilon}{dt} = \frac{1}{\eta}\sigma_2,\tag{13}$$

$$\sigma = \sigma_1 + \sigma_2,\tag{14}$$

where σ_1 is the spring stress, σ_2 is the dash-pot stress and σ is the total model stress, E is the stiffness coefficient and η the viscosity of the material. The stretch between two cells is then computed by ε (Eq. 15).

$$\sigma = E\varepsilon + \eta\frac{d\varepsilon}{dt}\tag{15}$$

The forces generated from the second model (force generation), are linked to the deformation model by means of the external forces (F_{ext}) computed for each cell. The F_{ext} of a given cell is obtained by a tensor according to the force generated by the given cell and the forces generated by its neighbors. F_{ext} is computed as follows

$$F_{ext} = \sum_{i=1}^{N}(C_i - C_0) \cdot (F_0 + F_i),\tag{16}$$

where, N is the number of neighbors for the concerned cell, C is the coordinate vector, F is the force generated from the force generation model for each cell, and F_0 the force of the cell for which the F_{ext} is computed. For further information, all details about the model can be found in [12].

3   Results

The previous sections introduced the model used to represent the different components of the uterine muscle activity. In this section we first co-simulated the different model components at the tissue level, by applying the models to a matrix of cells. Then, we co-simulated the 3 models on a realistic 3D mesh of the uterine muscle, in order to see the integration of electrical and mechanical effects of the contraction at the whole organ level.

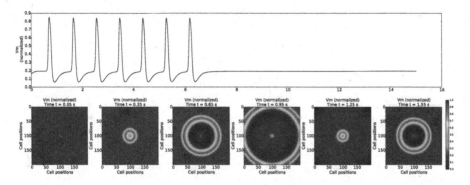

Fig. 1. Example of Vm for a co-simulation with 2D tissue. Top: the temporal signal of the cell (position [50 ; 50]) potential Vm. Bottom: six snapshots of the two dimensional Vm where we can see the APs spreading over the tissue.

3.1   Tissue Level

A two dimensional tissue is created as a matrix of cells. Each point of a Cartesian matrix represents a uterine SMC, which is associated with an electrical and a force model. Each cell is connected to its direct neighbors (top, bottom, left and right) by the diffusion process (Eq. 4) simulating gap junctions. First, the co-simulation is done without any deformation of the tissue, in order to simulate only the electrical coupling between cells' activity and the resultant force. Then, the deformation model is added in order to highlight that the generated force affects the cell positions.

Without Deformation. Figures 1 and 2 present a result of the co-simulation of both electrical and force models for a 2D tissue (200×200 SMCs). A pacemaker area is defined at the center of the tissue. The top panels present the temporal signals for this pacemaker cell located at [50:50]. The bottom panels show a series of snapshots of the signal over the whole tissue. We can notice on the temporal signals the difference of dynamics between both models and the cumulative effect of the force generated. On Fig. 1, we see APs spreading, from the pacemaker area toward the edges, due to the electrical diffusion between cells. Figure 2 shows the force generated according to the calcium level generated from the electrical model. We can notice here that SMCs are contracted on the whole tissue, while APs are not present on the entire tissue. This is induced by the slower dynamic of the mechanical model. This behavior is similar to real uterine contraction, where a burst of EHG (containing lots of APs) is associated with a unique increase in intrauterine pressure [13]. A video is available online at: http://www.utc.fr/~jlaforet/Suppl/2delecmeca.gif.

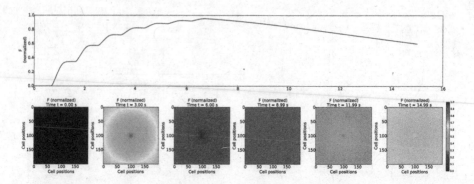

Fig. 2. Example of Force for a co-simulation with two dimensional tissue. Top: the temporal signal of the cell (position [50 ; 50]) force F. Bottom: six snapshots of the two dimensional force view where we can see the SMCs being contracted on the whole tissue.

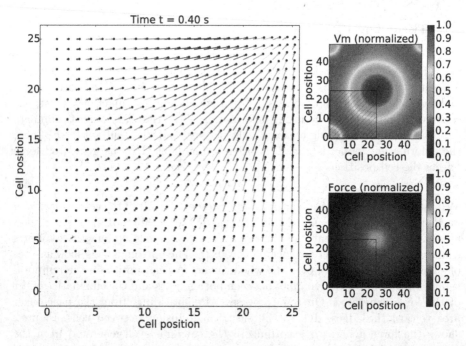

Fig. 3. Co-simulation example with a two dimensional tissue (50 × 50 SMCs). Top: Vm normalized (top), and the force generated (bottom). Left: the positions of the 625 cells with their [x:y] coordinates (blue dots) corresponding to the lower left quarter of the simulation matrix. Quiver plot: displacement of each cell. (Color figure online)

With Deformation. This result has been done by co-simulating the electrical, the force and the deformation models for a two dimensional tissue (50 × 50 SMCs). A pacemaker area is defined at the center of the tissue (cell coordinates

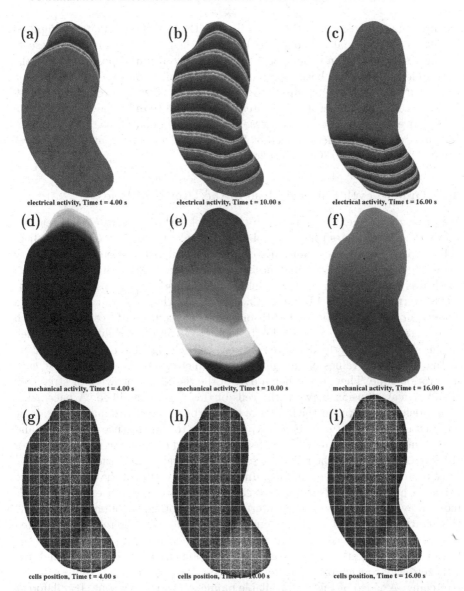

Fig. 4. Result of electrical, force and deformation models applied to a realistic 3D uterine mesh with a view from left mother's side. Top row: electrical activity of the uterus; Middle row: mechanical activity of the uterus; Bottom row: mechanical deformation of the uterus (a grid was added for a better visualization).

[25:25]). Figure 3 top right presents a snapshot of Vm, where we can see an AP spreading all over the tissue. Figure 3 bottom right presents the force generated at the same time, where we can notice the presence of a force gradient from the pacemaker area toward the edges, due to the AP spreading. Figure 3 left presents the position of the 625 cells with their [x:y] coordinates, computed thanks to the deformation model. For each cell, the head of the arrows represents the current position and the base represents the cell initial position. Only the bottom left quarter has been represented for readability purpose (corresponding to the hatched part of the snapshots). Due to the gradient of force, the distances between cells are smaller at the center of the tissue than at the edges. We can thus notice a contraction of the tissue. In this co-simulation, we keep constant the surface of the tissue. A video of Fig. 3 is available online at: http://www.utc.fr/~jlaforet/Suppl/2delecmecadeformation.gif

3.2   Organ Level: Surfacic Mesh of the Uterus Muscle

We present here the co-simulation of the model with deformation on a realistic uterine mesh. The mesh has been obtained thanks to the FEMONUM project [14] (http://femonum.telecom-paristech.fr/) that offers the scientific community 3D fetal, uterine and abdominal meshes extracted from MRI images. The initial mesh was resampled for spatial resolution enhancement purpose and contains 99,084 vertices and 198,146 faces. Each vertex of the mesh is considered as a smooth muscle cell associated with one electrical and one force models. Each edge is associated to the electrical diffusion process and to a Kelvin Voigt model. In this simulation, as the distances between cells change due to the deformation of the tissue, the values of the diffusion parameters D_k are computed at each integration time. Therefore, when a part of the tissue is contracted or stretched, the electrical diffusion between the cells of this part is modified. A pacemaker area is defined, located at the fundus of the uterus. In order to simulate a realistic contraction situation (volume quasi constant due to the presence of the fetus and of the amniotic fluid), the external force is directed outside the uterus wherever the force is lower than the mean force of all cells, and to the inside wherever the force is greater than the mean force. Indeed, the curvature of the uterus tends to bring the contracted cells toward the center of the uterus and the stretched cells move away from the center of the uterus. The percentage of volume change using this strategy is low (0.1 %) permitting thus to validate it. The result is presented in Fig. 4. We can notice here that the cells become closer on the part where the force is the strongest, representing the contraction effect. On the other hand, the cells get away to each other on the part where the force is weak, representing a stretching. A video of Fig. 4 is available online at: http://www.utc.fr/~jlaforet/Suppl/model_uterus3D_elec_meca_stretch.gif

4   Conclusion-Perpectives

We presented in this paper a co-simulated model of the uterine muscle developed with a multi-scale and multi-physics approach. The uterine muscle is modeled in different aspects by using 3 models: one electrical at the cell level, and 2 mechanical: force (cell level) and deformation (tissue level). The result of the co-simulation of these models show that the uterine muscle is able to be fully contracted by combining the electromechanical aspect of the contraction. Several electrical APs are needed to induce a maximal force generation on the whole organ. In order to enhance the performance of the model, we plan to add a biofeedback that exists between the stretching part of the uterus and the electrical model. Indeed, it has been proved that specific stretch activated channels, evidenced on the cell membrane, control some ionic current [15]. In addition, for the co-simulation on a realistic 3D mesh, we let the action potentials spread uniformly to the whole uterine muscle. This phenomenon is not consistent with the real uterine muscle structure that is neither homogeneous nor isotropic [16]. We will take into account this inhomogeneity in our future work.

Acknowledgment. This work was carried out and funded in the framework of the Labex MS2T. It was supported by the French Government, through the program "Investments for the future" managed by the National Agency for Research (Reference ANR-11-IDEX-0004-02).

References

1. Beck, S., Wojdyla, D., Say, L., Betran, A.P., Merialdi, M., Requejo, J.H., Rubens, C., Menon, R., Van Look, P.F.: The worldwide incidence of preterm birth: a systematic review of maternal mortality and morbidity. Bull. World Health Organ. **88**(1), 31–38 (2010)
2. Huddy, C., Johnson, A., Hope, P.: Educational and behavioural problems in babies of 32–35 weeks gestation. Arch. Dis. Child. Fetal Neonatal Ed. **85**(1), F23–F28 (2001)
3. Devedeux, D., Marque, C., Mansour, S., Germain, G., Duchêne, J.: Uterine electromyography: a critical review. Am. J. Obstet. Gynecol. **169**(6), 1636–1653 (1993)
4. Rabotti, C., Mischi, M., van Laar, J.O., Oei, G.S., Bergmans, J.W.: Estimation of internal uterine pressure by joint amplitude and frequency analysis of electrohysterographic signals. Physiol. Meas. **29**(7), 829 (2008)
5. Young, R.C.: Myocytes, myometrium, and uterine contractions. Ann. N.Y. Acad. Sci. **1101**(1), 72–84 (2007)
6. Laforet, J., Rabotti, C., Terrien, J., Mischi, M., Marque, C.: Toward a multiscale model of the uterine electrical activity. IEEE Trans. Biomed. Eng. **58**(12), 3487–3490 (2011)
7. Koenigsberger, M., Sauser, R., Lamboley, M., Bény, J.-L., Meister, J.-J.: Ca 2+ dynamics in a population of smooth muscle cells: modeling the recruitment and synchronization. Biophys. J. **87**(1), 92–104 (2004)
8. Maggio, C.D., Jennings, S.R., Robichaux, J.L., Stapor, P.C., Hyman, J.M.: A modified hai murphy model of uterine smooth muscle contraction. Bull. Math. Biol. **74**(1), 143–158 (2012)
9. Hai, C.-M., Murphy, R.A.: Cross-bridge phosphorylation and regulation of latch state in smooth muscle. Am. J. Physiol. Cell Physiol. **254**(1), C99–C106 (1988)
10. Bursztyn, L., Eytan, O., Jaffa, A.J., Elad, D.: Mathematical model of excitation-contraction in a uterine smooth muscle cell. Am. J. Physiol. Cell Physiol. **292**(5), C1816–C1829 (2007)
11. Özkaya, N., Nordin, M., Goldsheyder, D., Leger, D.: Mechanical properties of biological tissues. In: Özkaya, N., Nordin, M., Goldsheyder, D., Leger, D. (eds.) Fundamentals of Biomechanics: Equilibrium, Motion, and Deformation, pp. 221–236. Springer, New York (2012)

12. Yochum, M., Laforêt, J., Marque, C.: An electro-mechanical multiscale model of uterine pregnancy contraction. Comput. Biol. Med. **77**, 182–194 (2016)
13. Garfield, R.E., Maner, W.L.: Physiology and electrical activity of uterine contractions. Semin. Cell Dev. Biol. **18**(3), 289–295 (2007). Elsevier
14. Bibin, L., Anquez, J., de la Plata Alcalde, J.P., Boubekeur, T., Angelini, E.D., Bloch, I.: Whole-body pregnant woman modeling by digital geometry processing with detailed uterofetal unit based on medical images. IEEE Transactions. Biomed. Eng. 57(10), 2346–2358 (2010)
15. Sachs, F.: Stretch-activated ion channels: what are they? Physiology **25**(1), 50–56 (2010)
16. Young, R.C., Hession, R.O.: Three-dimensional structure of the smooth muscle in the term-pregnant human uterus. Obstet. Gynecol. **93**(1), 94–99 (1999)

Computing EHG Signals from a Realistic 3D Uterus Model: A Method to Adapt a Planar Volume Conductor

Maxime Yochum$^{(\boxtimes)}$, Pamela Riahi, Jérémy Laforêt, and Catherine Marque

CNRS, UMR 7338 Biomechanics and Bioengineering,
Centre de Recherche Royallieu, Sorbonne University,
Université de Technologie de Compiègne, CS 60319,
60203 Compiègne cedex, France
maxime.yochum@utc.fr

Abstract. Modelling the uterus is a suitable way to explore the links existing between its electrical activity and the contractile synchronization during labor. In order to compare simulated and real electrical uterine activities, it is necessary to obtain the same signals in both cases. The only real signal related to uterine activity, that can be recorded is the electrohysterogram. This paper presents a method to obtain electrohysterograms from a uterine model applied on a realistic uterine mesh, by using a volume conductor, which contains abdominal muscle, fat and skin layers.

Keywords: Uterine model · Multiscale · Multi-physic · Electrohysterogram

1 Introduction

Modelling the uterus enables us to gain an understanding of the phenomena happening in this complex organ, especially during the critical time of labor. The synchronization of the uterine contraction leading to a global efficient contraction of the uterus during term and preterm labor is not well known yet [1]. A comprehensive uterine model could help to predict the risk of preterm labor in order to prevent it. But, we do not have access to real in-vivo muscle cell information at the organ level such as the ionic currents for each muscle cell or between cells, the force generated by each cell, the mechanical interaction due to the muscle contraction. However, the validation of such a model is challenging. Indeed, to validate a model, it is necessary to simulate quantities that can be measured experimentally in a real situation similar to the simulated one. In the case of pregnancy monitoring, one of the measurements that is possible and of potential clinical interest is the electrohysterogram (EHG) [2]. It is the measure of the electrical activity of the uterus at the skin surface, by means of abdominal electrodes. During this measure, the electrical activity originating from the uterus level passes through the abdominal muscles, the fat and the skin (the

© Springer International Publishing AG 2016
V.-N. Huynh et al. (Eds.): IUKM 2016, LNAI 9978, pp. 381–388, 2016.
DOI: 10.1007/978-3-319-49046-5_32

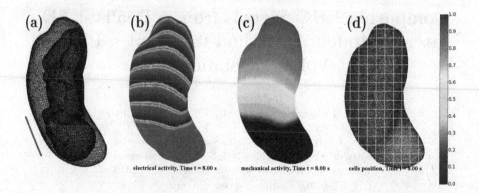

Fig. 1. Example of model co-simulation. (a) 3D realistic mesh, the purple line corresponds to the electrode grid plane location. (b) potential Vm, notice the action potential propagation front going downward. (c) generated force. (d) 3D cell position.

volume conductor) prior to being recorded [3]. Therefore, the electrical activity simulated by the model at the uterine muscle (myometrium) level needs to be filtered and integrated (as done by the volume conductor) in order to simulate a realistic EHG signal. Then this simulated EHG could be compared to real data sets for model validation purpose.

In this paper, we propose a method to compute the EHG from the co-simulation of a uterine muscle electrical and mechanical activity model applied on a realistic surfacic 3D mesh. All details about the model can be found in [4], Although, the model is still briefly explained in this section. We then use a method designed to model a 2D flat pregnant woman's volume conductor that we modified to adapt to our uterine 3D muscle model. The method is composed of 4 steps that include a principal component analysis (PCA) for dimension reduction purpose, an interpolation, a spatial frequency domain filtering and an electrode grid integration. This method could permit us to link the effect of some physiological parameters used in the model (i.e. application of clinical agent, like calcium channel blockers, used to inhibit uterine contraction) to the EHG characteristics in order to monitor properly the uterine contractile activity during pregnancy, by means of non-invasive EHG.

2   Method

2.1   Uterus Model

We modeled 3 different aspects of the uterine activity: electrical activity, force generation and tissue deformation. These three kinds of activity are associated in a multi-scale and multi-physic co-simulation of the uterine muscle behavior. Our multi-scale approach ranges from the cell level to the complete organ level. The ionic currents through the cell membrane are modeled at the cell level with a physiological model based on the Hodgkin-Huxley formalism [5]. It computes

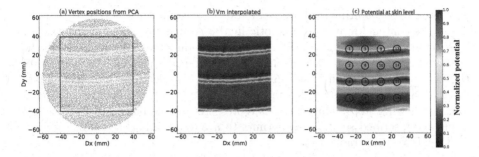

Fig. 2. (a) Position of vertex in 2D from the PCA dimension reduction, the square represents the vertices that are considered for the interpolation process. (b) Potentials interpolated on the myometrium surface. (c) Potentials on the skin surface after volume conductor filtering: the circles represent the electrode locations and the numbers represent the electrode indexes.

two important variables, which are the cell potential (Vm) and the calcium concentration. Vm is important because it permits the computation of the EHG with the method exposed later. Calcium concentration is important because it permits to link the electrical activity to the force generated, also computed at the cell level, as proposed by Burzstyn et al. [6]. The force generation model is the one proposed by Maggio et al. [7] who modified the original Hai-Murphy model [8] to better correspond to uterine smooth muscle cells. The organ/tissue levels contain two components, which are the electrical propagation between neighboring cells, by using a gap junction model from [9], and a deformation model with a Kelvin Voigt model approach [10]. This uterine model is applied to a 3D realistic mesh of a uterus obtained from the FEMONUM project [11]. To obtain a better spatial resolution the mesh was oversampled. It contains 99 084 vertices and 198 146 faces. An example of co-simulation is shown in Fig. 1.

2.2 Principal Component Analysis (PCA)

In order to compute the EHG from Vm, output of the uterine electrical model, only vertices which are underneath the electrode grid are considered [12]. During our recording of real EHGs, we use a matrix of 16 electrodes (grid 6.05 cm × 6.05 cm) located on the woman's abdomen, under the belly button [13] where the uterus is rather flat. The matrix location is indicated by the purple line Fig. 1(a). For simulation purpose, the vertex located closest to the center of the electrode grid is defined as the center of a disk that will contain all the vertices to be used as inputs to the next step. The disk is 60 mm in radius in order to have a surface larger than the square electrode grid. This selection is done for each time step as the tissue moves during the contraction.

As the volume conductor method assumes that the interface between two layers is a plane, in a first simplified approach, we chose to transform the positions of our input vertices into a 2D plane. Indeed, the original vertices (each one corresponding to a urine cell) are located on the uterine muscle that is rounded due to its content (fetus and amniotic fluid). To find the plane fitting at best with these vertices, we apply a PCA dimension reduction technique [14]. All the used vertices are thus projected onto the resulting PCA plane. An example of this dimension reduction is shown Fig. 2(a). As the electrode grid is a square matrix, only the vertices inside the square inscribed in the circle are considered for the following steps. This square is represented by the black square Fig. 2(a). Using the inscribed square also permits to avoid any missing values, specially for the 2D Fourier transform step.

2.3   Interpolation

The volume conductor is a 2D spatial filter, defined in the frequency domain. We thus need to apply a spatial 2D Fourier transform to our uterine surface data. However, the vertex positions obtained after the PCA reduction are non-uniformly located in the PCA plane. To solve this problem, we chose to interpolate the non-uniform Vm values on a Cartesian grid with a spatial sampling step of 0.5 mm in both directions. This step corresponds approximately to the size of a uterine muscle cell [15]. We use the cubic interpolation method provided by the Scipy Python library to get enough accuracy. A result is presented in Fig. 2(b), which corresponds to the interpolation of the data contained in the square Fig. 2(a).

2.4   Volume Conductor Model

To simulate the effect of the tissues present between the myometrium and the recording surface, we adapted the model proposed in [3]. This model permits to model the surface EHG in the spatial frequency domain as the product between an electrical source, defined at the myometrium level and an analytical expression representing the effect of the volume conductor.

The volume conductor is considered as made of parallel interfaces separating the different abdominal tissues, namely, the myometrium, where the source is placed at a depth $z = z_0$, the abdominal muscle, the fat, and the skin. The volume conductor effect depends on the tissue thicknesses, their conductivities, and the source depth, z_0. This assumption of parallel interfaces is close enough to the real anatomy when considering the size and the location of our electrode grid. All the tissue are assumed to be isotropic with the exception of the abdominal muscle.

We implemented the frequency filter (G_{skin}) for all these layers. It is possible with this method to compute the potential at each layer (vector X components, a : myometrium, b : abdominal muscle, c : fat and d : skin), but we considered here only the potentials at the skin level (A_d and B_d). X is computed as:

$$X = A^{-1}B, \tag{1}$$

$$A = \begin{bmatrix} 1 & -1 & -1 & 0 & 0 & 0 & 0 \\ \sigma_a k_y & -\sigma_{yb} k_{yb} & \sigma_{yb} k_{yb} & 0 & 0 & 0 & 0 \\ 0 & e^{h_1 k_{yb}} & e^{-h_1 k_{yb}} & -e^{h_1 k_y} & -e^{-h_1 k_y} & & \\ 0 & \beta e^{h_1 k_{yb}} & -\beta e^{-h_1 k_{yb}} & -\gamma e^{h_1 k_y} & \gamma e^{-h_1 k_y} & 0 & 0 \\ 0 & 0 & 0 & e^{h_2 k_y} & e^{-h_2 k_y} & -e^{h_2 k_y} & -e^{-h_2 k_y} \\ 0 & 0 & 0 & \gamma e^{h_2 k_y} & -\gamma e^{-h_2 k_y} & -\eta e^{h_2 k_y} & \eta e^{-h_2 k_y} \\ 0 & 0 & 0 & 0 & 0 & \eta e^{h_3 k_y} & -\eta e^{-h_3 k_y} \end{bmatrix} \quad (2)$$

with

$$\beta = \sigma_{yb} k_{yb}, \gamma = \sigma_c k_y, \text{ and } \eta = \sigma_d k_y. \quad (3)$$

where A and B are expressed in Eqs. (2) and (4).

$$X = \begin{bmatrix} A_a \\ A_b \\ B_b \\ A_c \\ B_c \\ A_d \\ B_d \end{bmatrix}, \quad B = \begin{bmatrix} \frac{1}{\sigma_a k^2} \\ 0 \\ 0 \\ 0 \\ 0 \\ 0 \\ 0 \end{bmatrix}. \quad (4)$$

The σ_i are the different conductivities, the h_i the different layer thicknesses, and k the angular frequency according to [3]. The frequency filter is then defined by:

$$G_{skin} = A_d + B_d. \quad (5)$$

The Vm values, obtained at the uterine level after the interpolation process, are then filtered with G_{skin} in the frequency domain giving thus a representation of the uterine electrical potential on the skin surface, after application of an inverse 2D FFT. An example of the obtained filtered potential distribution is depicted Fig. 2(c), which corresponds to the filtering of the Vm values of Fig. 2(b).

2.5   Electrodes Grid Model

We included in our simulation process a simple model of the electrode grid to simulate EHG signals similar to the ones recorded experimentally [13]. This model is flexible and has been designed to be able to reproduce the different kinds of array that may be used in human or animal experiments. In addition, the size and placement of electrodes can easily be changed as well as their shape (circle, ellipse, concentric or rectangular). We simulated here the same grid of electrodes as the one used in [13] for EHGs recording (4 × 4 electrodes, 8 mm diameter, and 17.5 mm electrode spacing). This electrode grid is shown in Fig. 2(c) as black circles with each electrode (each circle) numbered from 1 to 16.

3   Results

The methodology presented previously was used to co-simulate the model in order to obtain EHG signals simulated during 35 s. A pacemaker area was activated at the fundus of the uterus during half the co-simulation time, leading

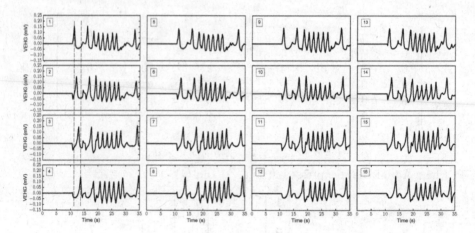

Fig. 3. Example of EHG signals ($VEHG$) simulated for a 35 s co-simulation of the models. The number in each plot corresponds to the electrode number in Fig. 2(c). The red dashed lines evidence the propagating of the electrical front from electrode 1 to electrode 4.

to the generation and propagation of 9 action potentials over the whole uterus. A snapshot of the simulation result is shown in Fig. 1 where the potential Vm at the uterus surface is represented in Fig. 1(b). We can notice action potential waves propagating over the whole uterus from top to bottom. EHG signals were computed from these electrical potential by using the method presented in this paper. Figure 3 presents the EHGs obtained by the 4 by 4 electrodes grid previously described (the 16 electrodes are numbered according to their position on the grid in Fig. 2(c). This result clearly evidences the spatial propagation of the wave front. The time of appearance of one front under electrodes 1 and 4 is highlighted by the two red dashed lines present in panels 1 to 4. The propagation velocity is then computed as the distance between the electrodes 1 and 4 (5.25 cm) divided by the time delay of the front propagation (2.42 s), which gives 2.17 cm/s. This propagation velocity corresponds to the value obtained from experimental EHG recorded on women: 2.18 ($\pm$0.68) cm/s [16].

Figure 4(a) to (e) shows the Power Spectral Density of the EHG signals computed with different thickness of the fat layer, range classically noticed for women (from 0 mm in panel a -no fat- to 40 mm in panel e). The Y axis represents the electrode number and the X axis the frequency content. The color represents the PSD energy. For visual comparison purpose, the same representation has been generated from real EHG signals [17] and is presented Fig. 4(f) for a thin woman (BMI=23 at 38 weeks of pregnancy). The frequency contents of both simulated and real EHGs are similar. It is also worth noting in Fig. 4 that the attenuation of the signal increases with fat tissue thickness, as expected.

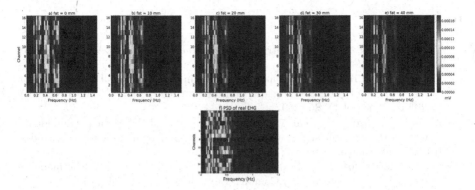

Fig. 4. Power Spectral Density of the modeled EHG for different fat thickness. (a) 0 mm (no fat), (b) 10 mm, (c) 20 mm, (d) 30 mm, (e) 40 mm. (f) Power Spectral Density of experimental EHGs from [17]. Channel numbers represent the corresponding electrode number

4   Conclusion

This paper presents a methodology to adapt a classical planar volume conductor to a 3D mesh source in order to compute the EHGs recorded on the abdominal skin surface. Sources are first restricted to a 2D plane from a portion of the 3D mesh by PCA and projection. In order to obtain the potential over a Cartesian grid we then apply a cubic interpolation. The Cartesian interpolated sources are then represented in the frequency domain by a spatial 2D FFT and spatially filtered by a 3 layer volume conductor. Finally EHG signals are computed by simulating an electrode grid similar to the one used for real EHG recording. An example of simulated EHG signals is given where it is possible to evidence a propagating front. Results show that the frequencies contained in the simulated and real EHGs are similar. Our simulations also demonstrated an increased in the signal attenuation with increasing fat thicknesses. This result is expected given the insulating properties of fat. This method will be further used to compare simulated and real specific EHG signals, first for model validation in order to correctly identify model parameters and so decrease model errors, then for model aided diagnosis to predict preterm labor.

Acknowledgment. This work was carried out and funded in the framework of the Labex MS2T. It was supported by the French Government, through the program "Investments for the future" managed by the National Agency for Research (Reference ANR-11-IDEX-0004-02).

References

1. Young, R.C.: Myocytes, myometrium, and uterine contractions. Ann. N.Y. Acad. Sci. **1101**(1), 72–84 (2007)
2. Marque, C.K., Terrien, J., Rihana, S., Germain, G.: Preterm labour detection by use of a biophysical marker: the uterine electrical activity. BMC Pregnancy Childbirth **7**(Suppl 1), S5 (2007)
3. Rabotti, C., Mischi, M., Beulen, L., Oei, G., Bergmans, J.W.: Modeling and identification of the electrohysterographic volume conductor by high-density electrodes. IEEE Trans. Biomed. Eng. **57**(3), 519–527 (2010)
4. Yochum, M., Laforêt, J., Marque, C.: An electro-mechanical multiscale model of uterine pregnancy contraction. Comput. Biol. Med. **77**, 182–194 (2016)
5. Laforet, J., Rabotti, C., Terrien, J., Mischi, M., Marque, C.: Toward a multiscale model of the uterine electrical activity. IEEE Trans. Biomed. Eng. **58**(12), 3487–3490 (2011)
6. Bursztyn, L., Eytan, O., Jaffa, A.J., Elad, D.: Mathematical model of excitation-contraction in a uterine smooth muscle cell. Am. J. Physiol. Cell Physiol. **292**(5), C1816–C1829 (2007)
7. Maggio, C.D., Jennings, S.R., Robichaux, J.L., Stapor, P.C., Hyman, J.M.: A modified hai murphy model of uterine smooth muscle contraction. Bull. Math. Biol. **74**(1), 143–158 (2012)
8. Hai, C.-M., Murphy, R.A.: Cross-bridge phosphorylation and regulation of latch state in smooth muscle. Am. J. Physiol. Cell Physiol. **254**(1), C99–C106 (1988)
9. Koenigsberger, M., Sauser, R., Lamboley, M., Bény, J.-L., Meister, J.-J.: Ca2+ dynamics in a population of smooth muscle cells: modeling the recruitment and synchronization. Biophys. J. **87**(1), 92–104 (2004)
10. Özkaya, N., Nordin, M., Goldsheyder, D., Leger, D.: Mechanical properties of biological tissues. In: Özkaya, N., Nordin, M., Goldsheyder, D., Leger, D. (eds.) Fundamentals of Biomechanics, pp. 221–236. Springer, New York (2012)
11. Bibin, L., Anquez, J., de la Plata Alcalde, J.P., Boubekeur, T., Angelini, E.D., Bloch, I.: Whole-body pregnant woman modeling by digital geometry processing with detailed uterofetal unit based on medical images. IEEE Trans. Biomed. Eng. **57**(10), 2346–2358 (2010). http://femonum.telecom-paristech.fr/
12. Carriou, V., Boudaoud, S., Laforet, J., Ayachi, F.S.: Fast generation model of high density surface emg signals in a cylindrical conductor volume. Comput. Biol. Med. **74**, 54–68 (2016)
13. Alexandersson, A., Steingrimsdottir, T., Terrien, J., Marque, C., Karlsson, B.: The icelandic 16-electrode electrohysterogram database. Sci. Data **2**, Article ID: 150017 (2015). doi:10.1038/sdata.2015.17
14. Fodor, I.K.: A survey of dimension reduction techniques (2002)
15. Grudzinskas, J.G.: The Uterus. Cambridge University Press, New York (1994)
16. Mikkelsen, E., Johansen, P., Fuglsang-Frederiksen, A., Uldbjerg, N.: Electrohysterography of labor contractions: propagation velocity and direction. Acta obstetricia et gynecologica Scandinavica **92**(9), 1070–1078 (2013)
17. Diab, A.: Study of the nonlinear properties and propagation characteristics of the uterine electrical activity during pregnancy and labor. Ph.D. dissertation, Université de Technologie de Compiègne (2014)

Ant Colony Optimization Based Anisotropic Diffusion Approach for Despeckling of SAR Images

Vikrant Bhateja[1(✉)], Abhishek Tripathi[1], Aditi Sharma[1],
Bao Nguyen Le[2], Suresh Chandra Satapathy[3], Gia Nhu Nguyen[2],
and Dac-Nhuong Le[4]

[1] Department of Electronics and Communication Engineering,
Shri Ramswaroop Memorial Group of Professional Colleges (SRMGPC),
Lucknow, Uttar Pradesh, India
bhateja.vikrant@gmail.com, abhishekl.
srmcem@gmail.com, aditiii065@gmail.com
[2] Duytan University, Danang, Vietnam
{baole,nguyengianhu}@duytan.edu.vn
[3] Department of Computer Science and Engineering,
ANITS, Visakhapatnam, Andhra Pradesh, India
sureshsatapathy@gmail.com
[4] Haiphong University, Haiphong, Vietnam
nhuongld@dhhp.edu.vn

Abstract. Synthetic Aperture Radar (SAR) images are known to be corrupted by granular noise known as speckle. This noise is inherently present in these images owing to acquisition constraints and is a major cause of visual quality degradation. The anisotropic diffusion approaches for despeckling are constrained in terms exercising control over the non-homogeneous regions. This paper proposes to improve the non-linear Anisotropic Diffusion (AD) filter for despeckling using Ant Colony Optimization (ACO) algorithm. The main essence of this work is to suppress speckle and preserve the structural content. The issue of residual speckle content has been minimized by optimal selection of AD parameter(s) using ACO algorithm. Experimental results advocate the performance improvement achieved and has been validated using objective measures of image quality evaluation.

Keywords: Anisotropic Diffusion (AD) · Ant Colony Optimization (ACO) · Despeckling · Iterations

1 Introduction

SAR is a coherent imaging radar system that generates high resolution remote sensing images via antenna(s) mounted on an aircraft or spacecraft. The received signals by SAR antenna are complex in nature; as it yields outcomes from incoherent sum of several backscattered waves. Owing to its all-weather, day and night imaging capabilities, SAR imaging systems are widely deployed in applications like:

© Springer International Publishing AG 2016
V.-N. Huynh et al. (Eds.): IUKM 2016, LNAI 9978, pp. 389–396, 2016.
DOI: 10.1007/978-3-319-49046-5_33

reconnaissance, surveillance, navigation, ground penetration, etc. [1, 2]. As an outcome of de-phased but coherent signals in SAR imaging; these images are plagued inherently with a granular noise called 'Speckle'. Perceptually, speckle affects the brighter regions of the image with an increase in the mean gray-level values of the local region. This very aspect limits the ability to discern and extract textural and radiometric information. Speckle is modeled as a multiplicative noise and thus despeckling approaches employing incoherent averaging serves to minimize speckle at the expense of resolution [3, 4]. Bayesian approaches employing local statistics are simple in approach but generally limited in functionality to distinguish homogeneous and non-homogeneous regions during speckle filtering. With this, the despeckled images consist of smoothened edges and blurring [5–7]. Non-Bayesian approaches operate based on Anisotropic Diffusion (AD) filtering [7, 8]; where continuous and monotonically decreasing conduction function(s) are employed as decision criteria to classify the speckled image into heterogeneous and homogeneous regions. Also, the usage of gradient factor helps in discrimination of true and false edges [9]. The traditional Perona-Malik Anisotropic Diffusion (PMAD) filtering [8] deals faithfully with additive noises but there yield visually unpleasant results with speckle (being a multiplicative noise). OSRAD filter [10] has emerged as an improvement over PMAD; extending its applicability towards speckle suppression. However, the performance deteriorates at higher degree of speckle variances. The literature review and comparison of such approaches has been detailed in works of Bhateja et al. [4, 11, 12]. In this paper, the performance of AD filtering algorithm has been enhanced by employing optimal selection of parameters using ACO algorithm. The parameter tuning via ACO has served to minimize the residual speckle in noisy images; yielding better restoration responses with increasing levels of speckle variances. In the remaining part of the manuscript, Sect. 2 presents the ACO modified AD filtering approach. This is followed by result analysis in Sect. 3 and concluding the work in Sect. 4.

2 Proposed Despeckling Approach

2.1 Anisotropic Diffusion

Traditionally, filtering an image using Gaussian low-pass filter is equivalent to processing the image according to the isotropic heat equation. On the other hand, PMAD algorithm modifies the image via Partial Differential Equations (PDE) to suppress noise. The classical isotropic diffusion equation was replaced with the Anisotropic Diffusion (AD) equation in continuous form as in Eq. (1).

$$\frac{\partial I(x, y, t)}{\partial t} = div[g(||\nabla I||)\nabla I] \tag{1}$$

Where: $g(||\nabla I||)$ is an edge stopping function and $||\nabla I||$ denotes the gradient magnitude of the image (I). The discrete version of Eq. (1) has been presented by Perona and Malik [8] as given in Eq. (2).

$$I_s^{t+1} = I_s^t + \frac{\lambda}{|\eta_s|} \sum_{p \in \eta_s} g\left(\nabla I_{s,p}\right) \nabla I_{s,p} \tag{2}$$

Where: t denotes the discrete time steps, s is the pixel position and I_s^t is the discretely sampled image. The rate of diffusion is determined by λ, ($\lambda \in IR^+$), η_s represents spatial neighborhood of pixel s and $|\eta_s|$ gives the number of neighbors. The image gradient in the particular direction is approximated as in Eq. (3) which yields the

$$\nabla I_{s,p} = I_p - I_s^+ \text{ ,where } p \in \eta_s \tag{3}$$

image intensity difference between pixel s and its neighboring pixel p. At constant image regions, these neighbor differences will be normally distributed and small in magnitude. However, for an image that includes an edge, the neighbor differences will not be normally distributed. The edge stopping function (herein, also referred to as the conductance function) is chosen in such a manner that when x $\rightarrow$ ∞, $g(x)$ $\rightarrow$ 0 so as to stop diffusion across the edges. On the other hand, when x $\rightarrow$ 0, $g(x)$ $\rightarrow$ 1 so that diffusion should take place in homogenous region. The gradient parameter therefore provides discrimination between true and false edges.

2.2 Ant Colony Optimization

Ants are insects which live in colony and they focus on survival of colony rather than individual. The behavior that provides inspiration for Ant Colony Optimization (ACO) is the ant forging which means their search to find food within shortest path. While exploring food in a random manner near their nest, these ants leaves a chemical known as pheromone. Ants choose the way where there is high concentration of pheromone and find the food source. They evaluate the quality and quantity of food and carry some of the food back to the nest. While returning they leave pheromone on the basis of evaluation of food present in the source [13]. A general overview of the same has been presented as Algorithm 1 below. However, the detailed description of ACO algorithm and its variants may be referred to in literature [14, 15].

Algorithm 1. The Ant Colony Optimization Meta-heuristic

      Set parameter, initialize pheromone trails
      While termination condition is not met **do**
        Construct Ant Solution
        Update Pheromones
  End while

2.3 Ant Colony Optimized AD Filtering Algorithm

PMAD filtering algorithm modifies the conduction coefficient in the nonlinear diffusion equation as in Eq. (4).

$$\frac{\partial}{\partial t} I(x,y,t) = \nabla.(c(x,y,t)\nabla I) \tag{4}$$

Here in, $c(x,y,t)$ is the monotonically decreasing function of the image gradient which is defined as the conduction function as defined in Eq. (2). As it has been uses this gradient for preservation of edges; it is referred to as edge stopping function.

$$c(x,y,t) = \frac{1}{1 + \left(\frac{\|\nabla I\|}{k}\right)^2} \tag{5}$$

In Eq. (5) k is referred to as diffusion constant. If $c(x,y,t)$ is equal to 1, linear isotropic diffusion is achieved while when $c(x,y,t)$ tends to 0, non linear anisotropic diffusion takes place. The transformed pixel is estimated within a spatial window of using the local gradient, which is calculated using nearest-neighbor differences as:

$$\begin{cases} \nabla_N I_{i,j} = I_{i-1,j} - I_{i,j} \\ \nabla_S I_{i,j} = I_{i+1,j} - I_{i,j} \\ \nabla_E I_{i,j} = I_{i,j+1} - I_{i,j} \\ \nabla_W I_{i,j} = I_{i,j-1} - I_{i,j} \end{cases} \tag{6}$$

Subscripts N, S, E, and W (North, South, East, and West) describe the direction of computation of the local gradients within the aforesaid spatial window. The equation for the transformed pixel is given by:

$$I_t(i,j) = I(i,j) + \lambda(\nabla_E I \bullet c_E + \nabla_W I \bullet c_W + \nabla_N I \bullet c_N + \nabla_S I \bullet c_S) \tag{7}$$

Where: the parameter $\lambda \in \left[0, \frac{1}{4}\right]$; and c denotes the diffusion coefficient computed in each direction. It is known that the factor λ exercise control over the smoothing of the speckle along with preservation of structural content. In this work, the selection of value for parameter λ has been optimized using ACO. This has been applied based on the notion that usage of constant value of λ during the entire course of filtering may saturate the despeckling performance of filter. Hence, an optimized value of λ has been determined during each stage of iteration of AD filtering (adaptively depending upon the residual speckle content) using ACO algorithm. Hence, the factor λ in Eq. (7) has been updated as per the movement of ants and pheromone paradigm. Within a spatial window of size (w) processing the speckled SAR image, the movement of ants according to pheromone trail probability can be calculated as:

$$p_{ij} = \frac{w(i,j) \bullet n_{ij}^{0.1}}{\sum\limits_{i=1}^{m} \sum\limits_{j=1}^{n} w(i,j) \bullet n_{ij}^{0.1}} \tag{8}$$

Where: p_{ij} ranges within [0, 1]. Herein, n_{ij} denotes the quantity of pheromone as per ACO algorithm. This parameter (n_{ij}) has been mapped to local average gradient computed in a particular direction within w and can be mathematically expressed as:

$$n_{ij} = \frac{[((w(i-1,j) - w(i,j)) + (w(i,j-1) - w(i,j)]}{\sum_i w}$$ (9)

Based on the value of the probability $p_{ij} > 0$; the fitness function of ACO in Eq. (10) is executed otherwise, the window is passed to next center pixel. Now, the value of parameter λ has been optimized based on the following fitness function; with t_i being the random value.

$$\lambda_0 = \lambda + t_i(0.01)$$ (10)

The update function for the same using ACO is given by:

$$t_i = (1-\rho)t_i + \rho(\Delta t_{i,j})$$ (11)

Where: ρ is tuning parameter and the $\Delta t_{i,j}$ denotes the initial pheromone value. The new value of λ is used in Eq. (7) for generation of the restored image. Henceforth, Mean Squared Error (MSE) has been computed taking last processed image as reference-x_{ij} and y_{ij} being the currently restored image with new value of λ.

$$mse = \frac{1}{MN} \sum_{i=1}^{M} \sum_{j=1}^{N} \left(x_{i,j} - y_{i,j}\right)^2$$ (12)

The process is repeated unless the last iteration mse_{i-1} is greater than the value of current iteration, mse_i.

3   Results and Discussions

SAR images of 'East Coast of India' (Test image#1) and 'Moon Crater' (Test image #2) are used in the simulations carried out in this paper. The simulation setup is initiated by normalizing the input SAR image. The normalized image is then simulated speckle noise of different levels of variances ranging from low speckle variance of 0.01 to high magnitude speckle of 0.1. The parameters of the proposed algorithm that are initially selected during simulation are: spatial window size ($w = 3 \times 3$), ρ is selected as 0.05 and $\Delta t_{i,j}$ is set to 0.001.The obtained results on Test image#1 with proposed despeckling algorithm are depicted in Fig. 1. It shows the outcomes of the proposed algorithm on low (0.01), medium (0.04) and high (0.1) magnitude of speckle variances each. It is evident here that there is smoothening in homogenous regions and edges are preserved to a reasonable extent. With optimal selection of λ, minimizing the residual speckle content; the applicability of proposed algorithm is depicted on high speckle

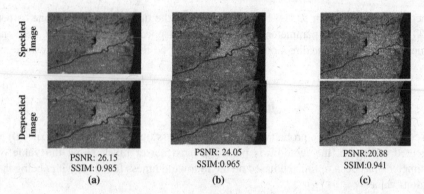

Fig. 1. Simulation results obtained using proposed despeckling algorithm on test image#1 at (a) low (0.01), (b) medium (0.04) and (c) high (0.1) levels of speckle variances.

variances without blurring edges. The quantitative assessment of image quality has been also carried out using Peak Signal-to-Noise Ratio (PSNR) and Structural Similarity (SSIM) as quality metrics. The computed values of PSNR & SSIM for different variance levels of Test image#1 are also included in Fig. 1.

In order to benchmark the performance of proposed despeckling algorithm; a comparison has been made with other speckle filtering techniques like: Kuan Filter [7], PMAD [8] and OSRAD [10]. The obtained results for visual inspection are shown in Fig. 2.

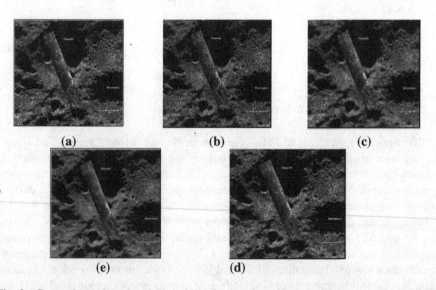

Fig. 2. Comparison of results of despeckling from various filtering techniques. (a) Noisy SAR (Test Image#2) with variance 0.1. Results using (b) Kuan Filter [7], (c) PMAD [8], (d) OSRAD [10], (e) Proposed Despeckling Algorithm.

The objective evaluation of image quality for the sake of benchmarking using PSNR and SSIM are shown in Tables 1 and 2 respectively. It can be interpreted from Fig. 2(b) that Kuan filter, with a Bayesian type filtering approach enhances the visual perception to a limited extent only. In addition, the same is obtained at the cost of over smoothing the edges with ample of persistent residual noise. Although, there are considerate improvements with PMAD and OSRAD; yet the performance of the filter saturates with increasing levels of speckle variances. Further, the performance deteriorates at a high speckle variance of 0.1 as shown in Figs. 2(c) and (d) and in Tables 1 and 2 respectively. Finally, Fig. 2(e) shows a significant improvement in performance with relative increment in the values of PSNR and SSIM. It can be observed the tremendous amount of residual speckle has been minimized by incorporating proposed ACO based AD filtering algorithm. There is better smoothing in homogeneous regions of the image along with edge preservation in non-homogeneous regions. In this paper, ACO plays a prominent role in minimizing the residual speckle in subsequent iterations by optimizing the value of parameter λ of Eq. (7).

Table 1. Performance comparison of proposed algorithm with other despeckling techniques using psnr(dB) as quality metric for test image#2.

Speckle variance	Kuan filter [7]	PMAD [8]	OSRAD [10]	Proposed despeckling algorithm
0.01	20.1494	23.6846	25.5522	27.3848
0.04	19.2013	22.5342	25.4721	26.1722
0.06	18.1531	22.4858	24.2660	25.5192
0.08	17.3262	21.4178	23.2599	24.3414
0.1	16.5589	19.4756	21.1538	23.4352

Table 2. Performance comparison of proposed algorithm with other despeckling techniques using SSIM as quality metric for test image#2.

Speckle variance	Kuan filter [7]	PMAD [8]	OSRAD [10]	Proposed despeckling algorithm
0.01	0.9226	0.9666	0.9838	0.9852
0.04	0.9088	0.9685	0.9826	0.9843
0.06	0.8891	0.9699	0.9772	0.9849
0.08	0.8721	0.9510	0.9707	0.9801
0.1	0.8534	0.9426	0.9644	0.9705

4 Conclusion

SAR images when processed for speckle suppression, poses a threat during in terms of tradeoff between detail preservation and residual speckle minimization. On account of this AD algorithm leads to under-filtering of the edges and at high variance the effect become more variant and visible. In the proposed work, the optimization of PMAD has been carried out via ACO so as to provide better filtering performance at higher noise variances. This has been achieved by optimal selection of parameter(s) of AD algorithm. The reduction in speckle content at higher variances is achieved without much computational complexity. This preliminary work can be further extended by analyzing the performance optimization based on multiple parameters of AD algorithm. This would provide further enhancement in edge preservation and restoration of image content for better interpretation or data extraction.

References

1. ChanY, K., KooV, C.: An introduction to Synthetic Aperture Radar (SAR). J. Prog. Electromag. Res. B **2**, 27–60 (2008)
2. Griffo, G., Piper, L., Lay-Ekuakille, A., Pellicano, D., De Franchis, E.: Modelling a buoy for sea pollution monitoring using fiber optics sensors. In: 4th Imeko TC19 Symposium. Lecce, Italy, pp. 182–186, June 2013
3. Lopes, A., Touzi, R., Nezry, E.: Adaptive speckle filter and scene heterogeneity. IEEE Trans. Geosci. Remote Sens. **28**(6), 992–1000 (1990)
4. Bhateja, V., Tripathi, A., Gupta, A., Lay-Ekuakille, A.: Speckle suppression in SAR images employing modified anisotropic diffusion filtering in wavelet domain for environment monitoring. Measurement **74**, 246–254 (2015)
5. Sharma, A., Bhateja, V., Tripathi, A.: An improved Kuan algorithm for despeckling SAR images. Inf. Syst. Des. Intell. Appl. **434**, 663–672 (2016)
6. Bhateja, V., Tripathi, A., Gupta, A.: An improved local statistics filter for denoising of SAR images. In: Proceedings of (Springer) 2nd International Symposiumon Intelligent Informatics (ISI 2013). Maysore, India, vol. 235, pp. 23–29 (2013)
7. Argenti, F., Lapini, A., Bianchi, T., Alparone, L.: A tutorial on synthetic aperture radar images. IEEE Geosci. Remote Sens. Mag. **1**(3), 6–35 (2013)
8. Perona, P., Malik, J.: Scale space and edge detection using anisotropic diffusion. IEEE Trans. Pattern Anal. Mach. Intell. **12**(7), 629–639 (1990)
9. Bhateja, V., Singh, G., Srivastava, A., Singh, J.: Despeckling of ultrasound images using non-linear conductance function. In: Proceedings of the (IEEE) International Conference of Signal Processing and Integrated Networks (SPIN-2014). Noida (U.P.), India, pp. 722–726 (2014)
10. Krissian, K., Karl, et al.: Oriented speckle reducing anisotropic diffusion. IEEE Trans. Image Process. **16**(5), 1412–1424 (2007)
11. Singh, S., Jain, A., Bhateja, V.: A comparative evaluation of various despeckling algorithms for medical images. In: Proceedings of (ACMICPS) CUBE International Information Technology Conference & Exhibition. Pune, India, pp. 32–37 (2012)
12. Bhateja, V., Misra, M., Urooj, S., Lay-Ekuakille, A.: Bilateral despeckling filter in homogeneity domain for breast ultrasound images. In: Proceedings of the 3rd (IEEE) International ConferenceonAdvamce in Computing. Communication and Informatics (ICACCI-2014). Greater Noida (U.P.), India, pp. 1027–1032 (2015)
13. Blum, C.: Ant colony optimization: introduction and recent trends. Elsevier Phys. Life Rev. **2**, 353–373 (2005)
14. Dorigo, M., Birattari, M., Stützle, T.: Ant colony optimization. IEEE Comput. Intell. Mag. **1**(4), 28–39 (2006)
15. Gao, L., Gao, J., Li, J., Plaza, A., Zhuang, L., Sun, X., Zhang, B.: Multiple algorithm integration based on ant colony optimization for endmember extraction from hyperspectral imagery. IEEE J. Sel. Top. Appl. Earth Obs. Remote Sens. **8**(6), 2569–2582 (2015)

A Fusion of Bag of Word Model and Hierarchical K-Means++ in Image Retrieval

My Kieu[1($\boxtimes$)], Khai Dinh Lai[2], Tam Duc Tran[3], and Thai Hoang Le[1]

[1] Faculty of Information Technology, VNUHCM - University of Science,
Ho Chi Minh City, Vietnam
mrkieumy@gmail.com, lhthai@fit.hcmus.edu.vn
[2] Saigon University, Ho Chi Minh City, Vietnam
khai.ld@cb.sgu.edu.vn
[3] Ho Chi Minh City University of Pedagogy, Ho Chi Minh City, Vietnam
tamtd@hcmup.edu.vn

Abstract. This paper proposes an Image Retrieval method using the Bag of Word model combining with Hierarchical K-Means++ algorithm to store and arrange databases, called BoW-HKM++ approach. A Hierarchical tree is proposed to improve the processing speed in retrieval data. In this research: (1) we first build a bag of word model based on SIFT feature extraction on images; (2) then show its improvement by using inverted index technical for whole feature databases; (3) the Hierarchical K-Means++ algorithm is also proposed in the bag of word model to optimize retrieval system. Experiment results were evaluated with k-fold cross-validation method on 2 datasets of James Z. Wang Research Group. Finally, the obtained results will be compared with other method and discussed to prove the effectiveness of proposed framework in Image Retrieval problems.

Keywords: HKM++ · Bag of word · Image retrieval · Inverted index

1 Introduction

Content-Based Image Retrieval (CBIR) has been an interesting research field, which has been attracting many global researchers. The purpose is to find an image retrieval method which is more efficient and simple. There are two important tasks in Content-Based Image Retrieval: the feature extraction method on the image and arrange feature databases. Until now, Image Retrieval based on Semantic Content is very challenging because there are many existing factors affecting the system performance such as illumination variation, resolution or occluded objects. Especially, global features are known to be limited in the face of these difficulties, which is very difficult to describe the image. Fortunately, Scale Invariant Feature Transforms (SIFT) provided the way to describe local features like a salient point or interest point of the image which is demonstrated great discriminative power on the image. SIFT was introduced by David Lowe in 1994. Until now, SIFT was cited more than 31,000 times [1]. In recent years, SIFT has been showed advantages with successful applications in various

© Springer International Publishing AG 2016
V.-N. Huynh et al. (Eds.): IUKM 2016, LNAI 9978, pp. 397–408, 2016.
DOI: 10.1007/978-3-319-49046-5_34

models because SIFT is invariant to scale, rotation and translation as well as partially invariant to affine distortion and illumination changes. Local features (local descriptors) like SIFT illustrates clearly interest points on the image. There were several experiments and applications of SIFT on Image Retrieval System [3, 9, 10] like Medical Images [14], Speed classification on spectrogram image [12].

In recent years, one of the most popular and simple models for Image Retrieval is Bag of Word (BoW) or Bag of Feature model [5]. It was an idea from Information Retrieval field, but it has been very useful for the image database, especially when the number of the image is very large and the number of the interest point of each image is different. Bag of Word model applied so early in the beginning of 21st century. Until now, it has been becoming to arrange database model, which is very common and efficient on Image Retrieval [6]. Furthermore, the combination of BoW model and Support Vector Machine classification was integrated effectively in a Microsoft SQL server database [4]. When the number of interest point in every image large and all image in the database are very large, the combination between SIFT feature extraction and Bag of Word model illustrated the structure of a model, which is very close and logical in the arrangement of image database. It is due to the fact that the number of interest point on the image was extracted by SIFT are very different, therefore the application Bag of Word model is suitable [7].

Another important challenge is the size of databases. It is difficult to select or compare one out of a large number of objects in acceptable time. One of the most popular methods is building Vocabulary tree or Hierarchical K-Means tree. Hierarchical K-Means tree has been applying to optimize searching and comparing local feature descriptors in the database by hierarchical clustering. Moreover, it contributes to decrease the process speed. Also, HKM was proposed to build Vocabulary tree of Scalable Recognition by David Nistér and Henrik Stewénius [2], which achieves a high result for image retrieval in a large database. Furthermore, K-Means plus plus (K-Means++) substantially outperformed standard K-Means in term of speed and accuracy [8, 13]. In [2] one of the contributions of the author is indexing mechanism that enables efficient retrieval. An Inverted index is a simple technique and reference from Information Retrieval model, every interest point will be assigned a label to count the frequency of appearance when we compare.

In this paper, we propose the BoW-HKM++ method for image retrieval, which is the fusion of bag of word model and hierarchical k-means++ in image retrieval in order to optimize the storage and arrangement the database. Specifically, we used K-Means++ to hasten both speed and accuracy, inverted index technique to improve the efficient retrieval, hierarchical tree to improve retrieval process and build the database. The proposed approach was verified by Cross-validation method before was evaluated experiment results on 2 public datasets about 11,000 images of prof. Wang, PSU [16].

2   Background

In this section, we will illustrate some brief rows about theory, which is used in order to build a retrieval image system, like SIFT, Bag of Word model, Hierarchical K-Means algorithm, Inverted Index and K-Means++ algorithm.

2.1   SIFT Feature

In recent years, some researches have illustrated that SIFT is one of the best descriptors for keypoint on image [6]. The way how to extract SIFT feature from the image was explained very detail in many papers, such as in [1, 10]. Therefore, we will not repeat those explanations in this paper. In general, there are four steps in the process: (1) Detection of Scale - space extrema; (2) Keypoint localization; (3) Orientation assignment; (4) Local feature vector.

After we extract SIFT feature on the image, we have several descriptors (or interest point). Each descriptor is a 128-dimensional vector. The number of descriptors is different from every image. It depends on the complicatedness of the image. For example, the image mono color (every image has only one color) has zero descriptor, which means it does not have any SIFT descriptor. But the common image has from a hundred to several thousand descriptors. Figure 1 shows an example of feature extraction on a sample image in the dataset [16]. On the left hand, it is an input image. On the right hand, it is the image after SIFT extraction. The red circle is the interest point or descriptor 128-dimensions vector.

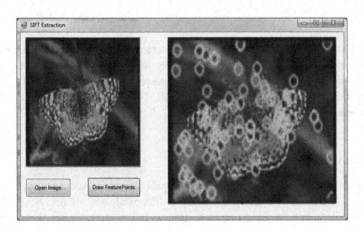

Fig. 1. SIFT extraction on an image.

2.2   Bag of Word Model

The principal problem of Retrieving images depends on how to store features of images. A number of images are large and the image features rise dramatically causing the extreme expansion of databases, which leads to some difficulties in retrieving, comparing and searching the necessary information in acceptable time.

BoW model, in this case, could be understood as a database of words and every interest point is a visual word. In the last decade, BoW model has been proved to be the leading strategy and remains a very competitive representation [5]. The main idea of BoW model is to quantize local feature (interest point) into visual word. Every SIFT interest point is a 128-dimensional descriptor.

2.3   Inverted Index

Depending on each image, the number of descriptors, which extracted by SIFT is very different. Nevertheless, a descriptor can appear in several images. On the other hand, descriptors were sorted and arranged by HKM++, so that the original order was lost. Inverted Index will use descriptors as an index, and a set of images having that descriptor is a value of the index. When the query image is used in retrieving, descriptors will be extracted from the image. Their descriptors will be compared with other descriptors in the database by searching algorithms in HKM++. The result will choose images, which have a high ranking when we count in the histogram. In this way, we will choose images, which have a high similarity with retrieval image.

Table 1. Inverted index

Descriptor	Images
1	2, 5, 8
2	1, 3, 6, 8, 10, 12
3	1, 2, 5, 10
4	2, 6, 8, 12
...	

Table 1 shows the result after we apply inverted index technique on the database. The SIFT descriptor is the index. Images having this SIFT descriptor are the value of this index. For example, images 2, image 5, image 8 have descriptor 1. Descriptor 3 appears in image 1, image 2, image 5, image 10. This process improves the efficiency of the retrieval task.

2.4   Hierarchical K-Means

Hierarchical tree proposed in [2] to build Vocabulary tree. Accordingly, in the first level, they will cluster all database by using K-Means algorithms (ex: K = 10), which means every element belongs to one of K groups. In the second level, they continue to cluster for each group (K groups) by using K-Means algorithms again. In other words, each group (in K groups) will have K small groups, and the third level continues, so on, this work continues until each element is a group. They call it: a leaf node. Thus, with N is the number of elements in the database, K = 10, we will have H level calculated by (1) formula:

$$H = \log_K N. \tag{1}$$

The most advantageous point of this model is that when we want to add or search new data into the hierarchical tree, the number of comparing operator is K*H (with K is the number of groups on each level and H is the level). This number is very small comparing with the number of elements in the database is very large. We used Euclid distance to compare between two elements (two descriptors) [2] for build vocabulary tree and retrieval process.

2.5   K-Means++

K-Means++ was proved to be better than standard K-Means about the speed to cluster and the quality of element in groups [8]. Because the initial step of standard K-Means is to choose random K centroids, it leads to several disadvantages such as local optimizing, quantum error and wasting time in loops to clustering later. But in K-Means++ we replace the initial step by random under control. Figure 2 indicates the pseudo code of the initialization of K-Means++. Only the first step (the initial step) is different with standard K-Means. The other steps are the same with standard K-Means.

```
Input: a set of descriptors
Output: k centroids
Process:
        C ← Choose random 1 centroid
        while ( |C| < k )
        {
                Choose x    X with probability
                            distance(x,C)maximum
                C ← C    x
        }
```

Fig. 2. An example of the pseudo code of the initial step in K-Means++.

Applying standard K-Means continues. After using K-means++, we achieve a set of descriptors, which assigned labels belong to a group.

K-Means++ will repair some disadvantages of standard K-Means, K-Means++ will help global optimize and make clustering operator better. The good experiment of K-Means++ is illustrated in [13], especially when the number of elements is very large, K-Means++ shows advantages both processing speed and quality of clustering.

For that reason, we used K-Means++ instead of standard K-Means in Hierarchical, thus we have Hierarchical K-Means++.

3   The Proposal Method

In this section, we proposed an effective method for image retrieval based on Bag of Word model combining with Hierarchical K-Means++. Following is the whole process of the model. It consists of 2 stages. The first stage is to build vocabulary tree, which means to build feature database (Fig. 3). The second stage is the retrieval process, which returns a set of result image when a query image input into the system (Fig. 4). We borrowed a few images from [2].

Our method could be described in Fig. 3. Detail Fig. 3 is building vocabulary tree: (1) we first extract SIFT feature on each image in the dataset and push all feature descriptors into the database (BoW). Thus, BoW is a feature database, each row is an

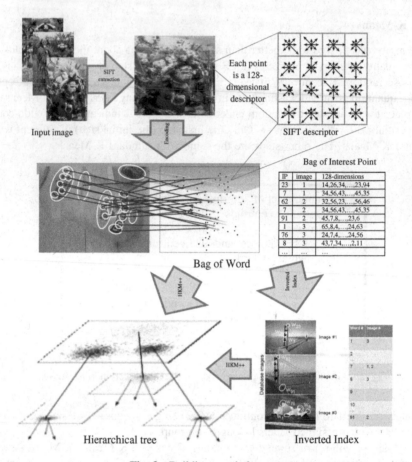

Fig. 3. Building vocabulary tree.

interest point (SIFT descriptor) 128-dimensional vector; (2) Inverted index algorithms was applied on feature database to combine coincident descriptors.

In this step, the index is the descriptor. Images belonged to this descriptor is the value of that index; (3) Feature database was applied by HKM++ algorithms to build vocabulary tree. That means clustering on each level by using k-means++. After 3 steps, we have a new feature database, which is sorted and arranged in the hierarchical tree.

Figure 4 shows retrieval tasks: the first step is to extract SIFT feature of a query image. Secondly, for each descriptor of a query image, we search and compare in the hierarchical tree. In this step, we count the appearance frequency of the images in the histogram. Thirdly, the process returns the highest ranking image in frequency histogram, which we count.

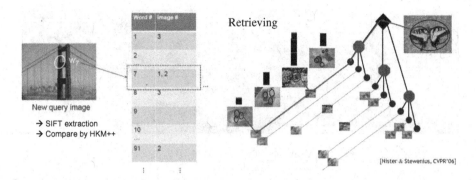

Fig. 4. Retrieval task.

4 Experiment and Evaluation

4.1 Databases

To evaluate the proposed method, we set up our method on 2 public datasets from a website of Professor James Z. Wang, Penn State University, the United State [16]. There are 2 datasets: The first one includes 1000 images (have folder name image.orig), we called D1K dataset. The second one includes 9909 images (have folder name image.vary), called D10K dataset (Fig. 5).

The D1K dataset consists of 10 different types of objects, such as indigenes, beach, heritage, bus, dinosaur, elephant, rose, horse, mountain, food. Each type has 100 images. Each image has resolution 256 × 384 (or 384 × 256).

The D10K dataset has many types of object. The number of each type of object is not clearly. For example, butterfly images have 100 images, but background images have more than 300 images, the earth images have 40 images. Each image has low resolution 128 × 96 (or 96 × 128).

Table 2. Two public datasets of prof. Wang from PSU.

Dataset name	Number of images	Number of features	Size
D1K dataset	1000	1.126.948	256 × 384
D10K dataset	9909	1.573.935	128 × 96

Table 2 expresses the number of images, the number of features and the size of 2 datasets from Penn State University, the United State.

To evaluate the performance of our model, firstly, we use the K–fold Cross-validation method to check the realizable of the method (Sect. 4.2). Secondly, we calculate the Precision and Recall proportion to evaluate the retrieval result (Sect. 4.3).

Fig. 5. A sample set of images in D1K dataset.

Fig. 6. A sample set of images in D10K dataset.

4.2 Verification by K-Fold Cross-Validation Method

In K-fold Cross-Validation, we choose K = 10, thus we run 10 times on each dataset. Each time, our approach pick out 1/10 of the number of descriptors in the dataset to test. In this step, we erase the label of 1/10 of the database, called a set of test data. To verify our model, we execute a set of test data by searching algorithm on Hierarchical tree to find the new label. The descriptor is correct if the new label is the same as the old label (classification is correct).

Table 3 illustrated the percentage of K-fold Cross-validation method running on 2 datasets in 10 times (k = 10 in Cross-validation method).

Table 3. The result tested by K-fold Cross-validation on 2 datasets in 10 times.

Times	1	2	3	4	5	6	7	8	9	10
D1K	78.5	82.4	92.7	87.6	83	79.5	86	96.3	86	74.6
D10K	84.5	78.7	90.4	87.1	89.8	77.5	89.2	90.5	93.5	77.6

From the result of 10 times in Table 3, we calculated the average of 10 times on each database, we have the result of each dataset. The D1K dataset result is 84.66 %, the D10K dataset result is 85.88 %. The final result of proposed method by K-fold Cross-validation (85.27 %) is the average of 2 datasets. Thus, the accuracy of our method is 85.27 % (Fig. 6).

4.3 Evaluated Results

Following formulas describe the way to calculate precision rate and recall rate.

$$Precision\ rate = \frac{the\ number\ of\ correct\ received\ image}{the\ number\ of\ the\ total\ received\ image}. \tag{2}$$

$$Recall\ rate = \frac{the\ number\ of\ correct\ received\ image}{the\ number\ of\ the\ total\ correct\ image\ in\ the\ database}. \tag{3}$$

We retrieve 10 times on the D1K to correspond with 10 objects. Each time we choose randomly 1 of 10 types of object to retrieval. The retrieval system returns 90 images for each retrieval task.

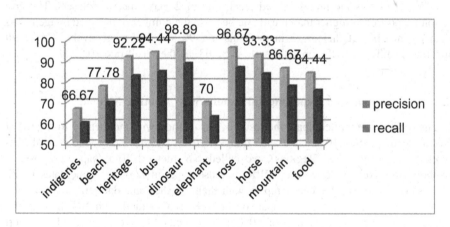

Fig. 7. The precision and recall rate for each group on the D1K dataset.

Figure 7 describes the precision and recall proportion on the D1K dataset for 10 objects. Because the number of images of each kind in the D10K dataset is not clearly, thus we chose 6 objects to retrieve. We retrieve 9 times on the D10K dataset.

Table 4. The number of image of some objects on the D10K dataset.

	Background	Butterfly	Earth	Sunset	Rose	Car
The number of image	385	100	40	100	99	100
The number of query	4	1	1	1	1	1

Table 4 shows the number of image and the number of query for 6 group objects. Figure 8 shows the precision and recall rate on the D10K dataset for 9 retrieval times on 6 groups object.

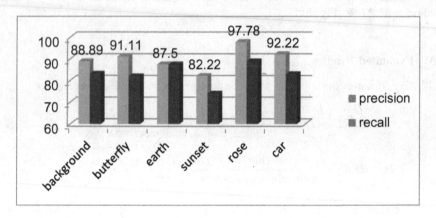

Fig. 8. The precision and recall rate retrieved on the D10K dataset.

Table 5 shows the precision and recall rate of 2 experimental datasets. The precision proportion is higher than recall rate because the number of received image is less than the number of the image in the database, 90 and 100 respectively. The result illustrated in Table 5, the precision rate is 88.03 %, the recall rate 80.29 %.

4.4   The Comparison Experimental Results with Another Method

In this section, we will compare the result of our method with the method from [15] of Rehan Ashraf author (Content Based Image Retrieval Using Embedded Neural Networks with Bandletized Regions), we called NN method. That paper proposed a method using Neural Network in CBIR, and also experiment on the same dataset with us. So that the reason we can compare with them on the same dataset.

The Fig. 9 illustrated the comparison between our method with NN method, the figure shows the precision rate of 10 objects on dataset of each method. The left column (blue color) is our method, and the right is NN method (their method). The number on the top column is of NN method.

Table 5. Precision and recall rate.

Dataset	Precision	Recall
D1K dataset	86.11 %	77.5 %
D10K dataset	89.95 %	83.08 %
Both 2 datasets	88.03 %	80.29 %

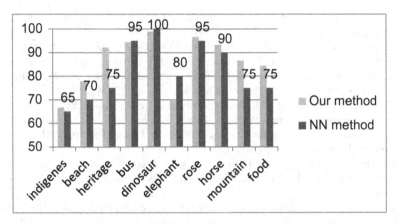

Fig. 9. The comparision of precision rate retrieved on the 2 methods. (Color figure online)

From the result of Fig. 9 illustrated the precision rate of NN method (their method) is 82 % (the average of 10 objects). While our method (86.11 %) is higher than NN method 4.11 %.

Our experimental results show that the proposed method using Bag of Word model combining with HKM++ technical is not only to produce excellent retrieval result for Image Retrieval but also for CBIR problem.

5 Conclusion

This paper proposed an image retrieval approach based on the bag of word model combining with advance hierarchical tree, called BoW-HKM++. Indeed, our method focuses on two main techniques: (1) arranging database with BoW model in order to optimize feature database, which is extracted by SIFT. It is potential that our method can be dealt with the large feature database, especially when the number of SIFT descriptors are extremely huge; (2) improving the performance of building database and retrieval data by HKM++ algorithms, which based on the combination HKM and K-Mean++. Inverted index method is also very useful for bag of word model in order to hasten retrieval process on database. The performance of our approach was evaluated with the public dataset [16] and got a better result when we compare with other method in CBIR problem. Our experimental results illustrated the feasibility of our proposed models.

References

1. Lowe, D.G.: Distinctive image features from scale-invariant keypoints. Int. J. Comput. Vis. **60**(2), 91–110 (2004)
2. Nister, D., Stewenius, H.: Scalable recognition with a vocabulary tree. In: Proceedings of the 2006 IEEE Computer Society Conference on Computer Vision and Pattern Recognition, CVPR 2006, vol. 2, pp. 2161–2168. IEEE Computer Society, Washington, DC, USA (2006)

3. Lopes, A.P.B., Avila, S.E.F., Peixoto, A.N.A., Oliveira, R.S., Araújo, A., de A.A.: A bag-of-features approach based on Hue-SIFT descriptor for nude detection. In: 17th European Signal Processing Conference (EUSIPCO 2009), pp. 1552–1556 (2009)
4. Korytkowski, M., Scherer, R., Staszewski, P., Woldan, P.: Bag-of-features image indexing and classification in Microsoft SQL server relational database. In: 2015 IEEE 2nd International Conference on Cybernetics (CYBCONF), pp. 478–482 (2015)
5. Law, M.T., Thome, N., Cord, M.: Bag-of-words image representation: key ideas and further insight. In: Ionescu, B., Benois-Pineau, J., Piatrik, T., Quénot, G. (eds.) Fusion in Computer Vision. Advances in Computer Vision and Pattern Recognition, pp. 29–52. Springer International Publishing, Switzerland (2014)
6. Alkhawlani, M., Elmogy, M., Elbakry, H.: Content-based image retrieval using local features descriptors and bag-of-visual words. Int. J. Adv. Comput. Sci. Appl. (IJACSA) 6(9) (2015)
7. Liu, J.: Image retrieval based on bag-of-words model. arXiv preprint arXiv:1304.5168
8. Arthur, D., Vassilvitskii, S.: k-means++: the advantages of careful seeding. In: SODA 2007 Proceedings of the Eighteenth Annual ACM-SIAM Symposium on Discrete Algorithms, pp. 1027–1035 (2007)
9. Ledwich, L., Williams, S.: Reduced SIFT features for image retrieval and indoor localisation. In: Australian Conference on Robotics and Automation (2004)
10. Almeida, J., Torres, R.S., Goldenstein, S.: SIFT applied to CBIR. Revista de Sistemas de Informacao da FSMA n. 4, pp. 41–48 (2009)
11. Philbin, J.: Scalable object retrieval in very large image collections, D. Phil thesis (2010)
12. Nguyen, Q.T., Bui, T.D.: Speech classification using SIFT features on pectrogram images. Vietnam J. Comput. Sci. (2016). doi:10.1007/s40595-016-0071-3
13. Bahmani, B., Moseley, B., Vattani, A., Kumar, R., Vassilvitskii, S.: Scalable K-Means++. Proc. VLDB Endowment 5(7), 622–633 (2012)
14. Sargent, D., Chen, C.-I., Tsai, C.-M., Wang, Y.-F., Koppel, D.: Feature detector and descriptor for medical images. In: SPIE Proceedings: Medical Imaging, vol. 7259, 27 March 2009. doi: 10.1117/12.811210
15. Ashraf, R., Bashir, K., Irtaza, A., Mahmood, M.T.: Content based image retrieval using embedded neural networks with bandletized regions. Entropy 17, 3552–3580 (2015)
16. http://wang.ist.psu.edu/docs/related/

Accelerating Envelope Analysis-Based Fault Diagnosis Using a General-Purpose Graphics Processing Unit

Viet Tra[1], Sharif Uddin[1], Jaeyoung Kim[1], Cheol-Hong Kim[2],
and Jongmyon Kim[1(✉)]

[1] School of Electrical Engineering,
University of Ulsan, Ulsan 680749, South Korea
traviet.vt@gmail.com, sharifruet@gmail.com,
kjy7097@gmail.com, jongmyon.kim@gmail.com
[2] School of Electronics and Computer Engineering,
Chonnam National University, Gwangju 61186, South Korea
cheolhong@gmail.com

Abstract. Reliable fault diagnosis in the bearings of an induction motor is of paramount importance for preventing unscheduled motor breakdowns and significant economic losses. This paper presents a fault diagnosis approach using a genetic algorithm and time-varying multi-resolution envelope analysis to select an optimal passband and the most discriminative fault components, respectively, in the acoustic emission signal from bearings. However, the computational complexity of the approach limits its use in real-time applications. To address that issue, this paper presents a general-purpose graphics processing unit (GPGPU)-based fault diagnosis methodology to accelerate the process via the optimal use of the GPGPU's global and shared memory resources and parallel computing abilities. Experimental results show that the proposed GPGPU implementation is approximately 19 times faster and uses 570 % less energy than CPU implementation.

Keywords: Fault diagnosis · Envelope analysis · Gaussian mixture model · Graphics processing unit · Genetic algorithm

1 Introduction

Bearings are vital for the continuous and efficient operation of industrial machines, but they are also a major cause of failure in rotational machinery [1]. The failure of a bearing can cause the shutdown of a critical machine, a stalled production line, or damage to the machine or personnel. Therefore, many researchers have proposed bearing condition monitoring techniques, which often use advanced signal processing to detect and diagnose faults. Information about bearing faults is often acquired in the form of a current, vibration, or acoustic emission (AE) signal. Those signals are then processed using time-frequency-based signal processing methods [2–7] to extract optimal sub-band signals for fault diagnosis.

© Springer International Publishing AG 2016
V.-N. Huynh et al. (Eds.): IUKM 2016, LNAI 9978, pp. 409–420, 2016.
DOI: 10.1007/978-3-319-49046-5_35

This paper presents an AE signal-based fault diagnosis method for bearings. AE signal processing is a highly effective technique for diagnosing bearing faults because of its ability to capture low-energy signals [8]. The AE signal captured from faulty bearings is both time-varying and non-stationary. This study uses a method that first finds the most informative sub-band from an AE signal using a time-varying multi-resolution envelope analysis (TVMREA) technique that evaluates each sub-band signal with a Gaussian mixture model-based residual-component-to-defect-component ratio (GMM-RDR) method [21]. In addition, a genetic algorithm (GA) is employed to select an optimum passband that maximizes the GMM-RDR value [9]. The techniques used are computationally intensive, making them impractical in real-time applications.

Therefore, this paper proposes a general-purpose graphics processing unit (GPGPU)-based implementation to accelerate the computational fault diagnosis method. A GPGPU has thousands of processing elements that execute an algorithm in parallel. CUDA was developed by NVidia for its GPGPUs to accelerate computationally intensive tasks in advanced signal processing, image processing, and computer vision [10]. In this study, we use CUDA to implement and accelerate TVMREA, the GMM-RDR, and a GA for fault diagnosis in rolling-element bearings.

The remainder of this paper is organized as follows. Section 2 describes the process of data acquisition for the experiment. Section 3 provides an overview of the fault diagnosis methodology. Section 4 describes the parallel implementation of the fault diagnosis methodology on a GPGPU, and Sect. 5 presents our experimental results and analyses. Finally, Sect. 6 concludes the paper.

2 Data Acquisition

The data for this study were collected using a machinery fault simulator that comprises a gearbox, cylindrical roller bearings (FAG NJ206-E-TVP2), a general-purpose wideband AE sensor to capture low-energy AEs and yield the AE fault signal, and a PCI-2-based data acquisition system to record the AE signal for further processing using the proposed technique in the GPGPU. The AE sensor is attached to the top of the bearing housing, as shown in Fig. 1. The AE signal is sampled at 250 kHz to capture fault information. The bearing faults are shown in Fig. 1 and are mostly cracks

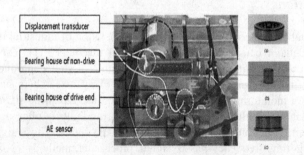

Fig. 1. Machine fault simulator and cracked bearings: (a) crack on the outer race, (b) crack on the roller, and (c) crack on the inner race.

on different parts of the bearing (i.e., the outer raceway shown in Fig. 1(a), the roller shown in Fig. 1(b), and the inner race shown in Fig. 1(c)). Faults are induced using a diamond cutter bit. Normal signals from defect-free bearings are also collected under different shaft speeds.

3 Fault Diagnosis Methodology

A comprehensive fault diagnosis methodology is implemented using a GPGPU, as illustrated in Fig. 2.

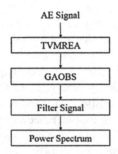

Fig. 2. Flow diagram of our fault diagnosis methodology.

The original AE signal is searched for the sub-band with the most distinctive fault features using TVMREA. This sub-band signal is then band-pass filtered in order to remove the low-frequency and high-amplitude frequency components. The optimal passband is determined using a GA with an objective function that maximizes the GMM-RDR value. Finally, the power spectrum of the filtered signal is obtained through envelope analysis and is then used to diagnose various fault conditions via inspection near different fault frequencies.

3.1   Time-Varying and Multi-resolution Envelope Analysis

TVMREA explores the input AE signal to determine its most informative sub-band using the sequence of steps depicted in Fig. 3. First, the AE signal is divided into several segments at different levels, as shown in Fig. 3(b), using a Hanning window with 50 % overlap. A candidate segment is then selected for frequency analysis. The frequency spectrum of each segment is evaluated using the GMM-RDR. The segment with the highest GMM-RDR value among all of the segments is selected as the most informative sub-band.

3.2   GMM-RDR Calculation

The GMM-RDR value, which is used to select the most informative sub-band during the TVMREA process, is calculated as follows.

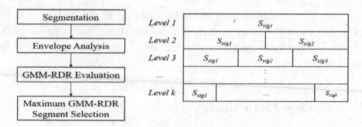

Fig. 3. Flow diagram of the fault diagnosis methodology.

Step 1: The GMM-RDR is calculated from the envelope power spectrum of the signal; therefore the first step is to perform envelope analysis [9].

Step 2: The GMM-RDR value is calculated from the power spectrum (Algorithm 1). This algorithm uses fault frequency f_i and frequency range f_{range}; where, f_i is the i-th harmonic of the bearing defect frequency, and f_{range} is the region around the defect frequency. The value of f_{range} is calculated for each fault according to its characteristic fault frequency, which is highly correlated to the ball pass frequency for the outer raceway (BPFO), ball pass frequency for the inner raceway (BPFI), twice the ball spin frequency (2 × BSF), and the fundamental train frequency (FTF) [9]. When the load conditions for each fault are varied, the frequencies for outer defects are sharp near the defect frequency, whereas those for roller and inner raceway faults are bell-shaped (less sharp) [9]. Therefore, a narrow ($f_{range} = \frac{1}{4}BPFO$) frequency range is selected for the outer raceway, and a wide ($f_{range} = \frac{1}{2}BPFI$ or BSF) frequency range is used for the inner raceway and roller defects (Table 1).

Table 1. Algorithm 1: GMM-RDR evaluation.

INPUT: Power spectrum
OUTPUT: GMM-RDR value

For harmonics, $i = 1$ to h

P_i = power spectrum values between $f_i - f_{range}$ and $f_i + f_{range}$

  for $k = 1$ to N_{rfreq}

$$w_{i,k} = \exp\left(-\frac{1}{2}\left(\beta\frac{\left(k-\frac{N_{rfreq}}{2}\right)}{\frac{N_{rfreq}}{2}}\right)^2\right)$$

$$D_{i,k} = w_{i,k} \cdot P_{i,k}$$
$$R_{i,k} = D_{i,k} - P_{i,k}$$

  end

$$GMM-RDR = 10 \cdot \log\left(\sum_{i=1}^{h}\frac{\sum_j^{N_{rfreq}} D_{i,j}^2}{\sum_j^{N_{rfreq}} R_{i,j}^2}+10\right)$$

3.3   Genetic Algorithm-Based Optimal Frequency Band Selection (GAOFBS)

After determining the three most informative segments, a GA-based approach is used for optimal frequency band selection (OFBS). Since the input AE fault signal is subject to modulation by the different rotating components, other frequencies affect the fault frequency response. To obtain a better fault frequency response, some of those frequency components that affect the fault frequency response are filtered before the envelope analysis.

GAOFBS uses a frequency band as a chromosome whose fitness depends on the GMM-RDR value of the filtered signal for that frequency band. In the fitness calculation, the AE signal is first filtered with band-pass filtering. That filtered signal is then used to calculate the GMM-RDR values. Then, the GAOFBS calculates the fitness, f, of each chromosome as follows:

$$f = \frac{1}{GMM - RDR} \tag{1}$$

4   Parallel Implementation

The fault diagnosis scheme just described is very computationally expensive and takes a substantial amount of time on a general-purpose computer. To reduce the execution time, the fault diagnosis scheme is accelerated using a GPGPU, which offers massive parallelization and thereby produces a better execution time than sequential execution. However, this speedup can only be achieved if the algorithms in the fault diagnosis scheme are effectively parallelized and efficiently implemented using shared memory for good coalesced memory access.

4.1   Segmentation

To increase parallelization and reduce memory demands, only the start and end indices of each signal segment are calculated instead of storing the whole signal segment in memory. The start and end indices of a segment are determined by finding the row and column numbers from the two-dimensional block in a CUDA kernel. However, the number of segments differs in each of the steps; therefore, many threads remain unused. This approach thus results in unaligned memory and threads and a lack of coalesced memory access. That problem can be alleviated by using a one-dimensional thread block and calculating the corresponding row and column values within the kernel as follows.

Code for calculating rows and columns from thread index tid.

```
segKernel(){
    . . . .
    int row = ceil( 0.5*sqrt((float)(16*tid+1))- 0.5);
    int col = tid - (row)*(row+1) ;
    . . . .

}
```

4.2   Envelope Analysis with GMM-RDR

The primary operations involved in envelope analysis are the Fourier and inverse Fourier transforms, which are implemented using the CUDA cuFFT library on a signal segment obtained using the Hanning window. Each thread is responsible for calculating one coefficient of the Hanning window. The magnitudes of the complex Fourier coefficients and their squares are calculated by a thread for each coefficient. Since GMM-RDR is not complex, it is implemented using only two thread blocks.

Code for computing Hanning window coefficients.

```
henningWindowCoef(int tid, int N){
    . . . .
    if(idx < N/2;
    return 0.5 * (1 - cos((2*M_PI*tid) / (N-1)));
    else
    return 0.5 * (1 - cos(2*M_PI*(N-1-tid) /(N-1)));
    }.
```

4.3   Filtering

This study uses finite impulse response (FIR) filtering to generate a band-pass filtered signal. FIR filters are parallel. Two kernels are implemented here: coefKernel, for calculating coefficients, and filterKernel, for the frequency response of the signal with calculated coefficients. The filterKernel performs convolution between the filter coefficients and the segmented signal. In convolution, both the signal and filter coefficients are used frequently; therefore, both the coefficients and the required signal segment are first copied to the shared memory.

4.4   Parallel Genetic Algorithm

In this study, the GA uses both envelope analysis and filtering when calculating the fitness value of a chromosome. Hence, those two processes present the primary design concerns for implementing the GA-based optimal passband selection on a GPGPU. The sequence of steps is shown in Fig. 4, and each step is discussed in detail in the ensuing

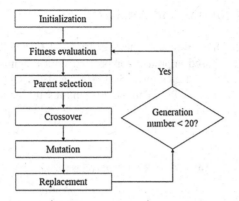

Fig. 4. Flow diagram of the genetic algorithm on the GPGPU.

text. The process starts by initializing the chromosome population and evaluating chromosome fitness in parallel. After the fitness evaluation, the selected parents engage in crossover and mutation to generate new offspring, which replace the existing population. This process continues for 20 generations.

In the initialization step, the first-generation chromosomes or population is initialized. Kernel `initializationKernel` creates chromosomes randomly. Each thread creates one chromosome, and in this way, the 256 threads in the thread block create the initial population in parallel.

The fitness calculation step determines the fitness of each chromosome in the population using a set of kernels. Computationally, this step is the most demanding task in GAOFBS. The `filterKernel` filters the selected segment of the time-domain AE signal using a band-pass filter for the lower and upper bands of the chromosome. Then, the envelope power spectrum of the filtered signal is generated using envelope analysis. Finally, the `rdrKernel` calculates the GMM-RDR value from the envelope power spectrum. The processes of filtering, envelope analysis, and GMM-RDR calculation are executed for each chromosome sequentially and in parallel for the entire population.

In the selection step, the GAOFBS uses a tournament selection algorithm in which three candidates are selected as the first parent, and three candidates are selected as the second parent. From each set of three candidate parents, the parent with the best fitness value is selected according to the tournament selection algorithm. The random selection of three parents results in un-coalesced memory access. To circumvent that problem, this study randomly selects one candidate parent and then selects the next two chromosomes as the other candidate parents.

In the crossover and mutation step, kernel `geneticOpsKernel` performs crossover, mutation, and replacement. Since is responsible for the crossover, mutation, and replacement of one pair of parents, which yields one new chromosome. The kernel generates the same number of offspring as the population size, i.e., 256. The new generation of chromosomes then replaces the old generation (generational GA).

5 Experimental Results and Analyses

This section analyzes and compares the performance of the GPU-based parallel approach with the CPU-based approach for bearing fault diagnosis. The GPU-based approach is tested on an Nvidia Geforce 560 GPU, and the CPU-based approach is executed on an Intel Core i5 CPU. This study explores the various implementation options and parameters, such as shared memory usage and the optimal size for thread blocks (Table 2).

Table 2. GPU system configuration.

Property	Value
GPU	nVidia GeForce 560
CUDA cores	336
Clock frequency	810 MHz
Number of SMs	6
Maximum number of threads per block	1024
Maximum number of threads per SM	1536
Total amount of shared memory per block	49152 bytes

5.1 Exploration of Thread-Block Configuration

An optimal thread-block configuration that ensures maximum parallelism depends on many parameters, such as coding style and the number of threads being used. Therefore, it is necessary to tune the thread-block size for each specific implementation.

Shared memory access is a lot faster than global memory access. When the number of thread blocks increases or decreases from its optimal value of two, shared memory is not used properly, which increases global memory access and, consequently, increases execution time. Figure 5 shows the execution time vs. block size for two different implementations: one of them uses global memory access, whereas the other uses shared memory access for both the coefficients and the required segment of the signal. It is evident that the shared memory implementation always takes less time irrespective of the block size because shared memory access is faster than global memory. The execution time for both implementations generally decreases with increasing block size until it reaches a constant value, which occurs at a block size of 768. This behavior is expected because using a higher number of threads delays the saturation, which is normally achieved for a thread-block size of 512 to 768. In subsequent experiments, the optimal thread-block size for filterKernel is 1024.

5.2 Performance Comparison with CPU-Based Implementation

This section compares the CPU-based sequential approach and GPU-based parallel approach of the fault diagnosis methodology in terms of execution time and energy consumption. The CPU-based approach is implemented in MATLAB R2013. Table 3 presents the execution environment for the CPU-based implementation.

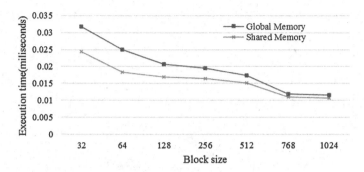

Fig. 5. Filtering with different block sizes.

Table 3. Specifications of the CPU environment.

Property	Value
Processor	Intel(R) Core (TM) i5-750
Clock speed	2.66 GHz
Number of cores	4
Number of threads	4
RAM	4 GB
Bus/core ratio	34
Operating system	Windows 8 (64 bit)

5.2.1 Execution Time

This section compares the execution time of the GPU- and CPU-based implementations of the two most expensive modules: envelope analysis with RDR evaluation and the FIR filter, and the GA.

Figure 6 shows the execution times for the module that carries out envelope analysis with RDR evaluation. In both implementations, the execution time is short when the number of samples is small, but it increases irregularly as the number of samples increases. The irregular increase in execution time is caused by the fast Fourier and inverse fast Fourier transforms, which are the main operations in the envelope analysis. Clearly, the execution time for the GPU implementation is always shorter than that for the CPU implementation. The GPU implementation is 3 to 10 times shorter than the CPU implementation for the sample range considered in this study.

The execution time for the FIR filter in the CPU increases almost linearly with an increasing number of samples, whereas in the GPU implementation, the execution time changes little except for a couple of spikes. The speedup that the GPU implementation achieves is significant: the CPU execution time is on the order of seconds, whereas the GPU execution time is on the order of milliseconds.

This study uses a GA to determine the optimal passband for the FIR band-pass filter. The objective or fitness function of the GA is the most computationally expensive part of the process because it involves band-pass FIR filtering and envelope analysis with RDR evaluation. Figure 7 shows the time required to create a single generation of

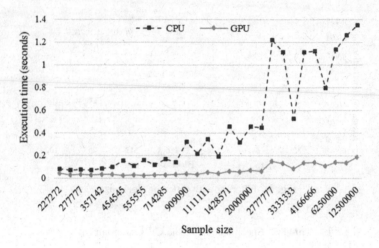

Fig. 6. Execution time for envelope analysis.

chromosomes using the CPU- and GPU-based implementations. The execution time increases with an increasing number of samples for the CPU, but it stays almost constant for the GPU-based implementation. The GPU achieves speeds for the GA-based optimal passband selection that are 15 to 40 times faster than those of the CPU-based implementation, as shown in Fig. 7.

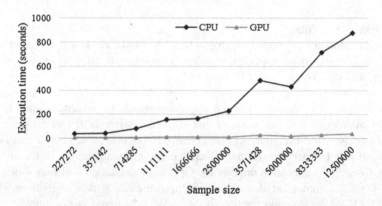

Fig. 7. Execution time of the genetic algorithm for a single generation.

5.2.2 Energy Consumption

In addition to performance improvement in terms of the execution time, reducing the energy consumption is also important. The energy consumed by each implementation can be expressed as

$$E_{consumption} = P_{avg} \times T_{exec}, \tag{2}$$

where P_{avg} is the average power consumed over the duration, and T_{exec} is the execution time required by each implementation. In this experiment, the GeForce GTX 560 is powered by an advanced technology extended power supply, which is connected to the power connector interface express (PCIe) slot via a PCIe riser card. The average instantaneous power from each power source for the Nvidia GTX 560 is calculated as

$$P = I_{inst} \times V_{inst}, \tag{3}$$

where V_{inst} is the instantaneous voltage, and I_{inst} is the instantaneous current, which is estimated by measuring the voltage drop across a 20 mΩ current-sensing resistor at a rate of one million samples per second. The average power consumed by the CPU is measured using the Intel® Power Gadget.

Table 4 shows the energy consumption of both the CPU and GPU implementations. Although the instantaneous power required by the GPU is more than that of the CPU, its significantly shorter execution time means that it consumes 570 % less energy than the CPU.

Table 4. Execution time and energy consumption.

Metrics	Measures
Execution time in CPU	6449 s
Execution time in GPU	340 s
Speedup	18.97×
Energy consumption CPU	210 kilojoules
Energy consumption GPU	36 Kilojoules
Energy consumption reduction	570 %

6   Conclusion

This paper presented an energy-efficient and high-performance bearing fault diagnosis methodology that uses acquired AE fault signals. The fault diagnosis scheme was accelerated using a GPGPU, and various aspects of this acceleration were studied. An optimal thread-block configuration that ensures maximum parallelization and reduces global memory access was developed in order to reduce the computational time required to execute the proposed fault diagnosis scheme in the GPGPU. The GPGPU-based implementation outperformed the CPU implementation, achieving an 18.97 times speedup and consuming 570 % less energy.

Acknowledgements. This work was supported by the Korea Institute of Energy Technology Evaluation and Planning (KETEP) and the Ministry of Trade, Industry & Energy (MOTIE) of the Republic of Korea (No. 20162220100050), funded in part by The Leading Human Resource Training Program of Regional Neo Industry through the National Research Foundation of Korea (NRF), Ministry of Science, ICT, and Future Planning (NRF-2016H1D5A1910564); in part by the Business for Startup R&D funded by the Korea Small and Medium Business Administration in 2016 (Grants S2381631); and in part by the "Leaders Industry-University Cooperation" Project supported by the Ministry of Education (MOE).

References

1. Cabal-Yepez, E., Valtierra-Rodriguez, M., Romero-Troncoso, R.J., Garcia-Perez, A., Osornio-Rios, R.A., Miranda-Vidales, H., AlvarezSalas, R.: FPGA-based entropy neural processor for online detection of multiple combined faults on induction motors. Mech. Syst. Signal Process. **30**, 123–130 (2012)
2. Cabal-Yepez, E., Garcia-Ramirez, A.G., Romero-Troncoso, R.J., Garcia-Perez, A., Osornio-Rios, R.A.: Reconfigurable monitoring system for time-frequency analysis on industrial equipment through STFT and DWT. IEEE Trans. Ind. Inf. **9**(2), 760–771 (2013)
3. Yan, R., Gao, R.X., Chen, X.: Wavelets for fault diagnosis of rotary machines: a review with applications. Signal Process. **96**, 1–15 (2014)
4. Seshadrinath, J., Singh, B., Panigrahi, B.K.: Investigation of vibration signatures for multiple fault diagnosis in variable frequency drives using complex wavelets. IEEE Trans. Power Electron. **29**(2), 936–945 (2014)
5. Lei, Y., Lin, J., He, Z., Zuo, M.J.: A review on empirical mode decomposition in fault diagnosis of rotating machinery. Mech. Syst. Signal Process. **35**(1–2), 108–126 (2013)
6. Zheng, J., Cheng, J., Yang, Y.: Generalized empirical mode decomposition and its applications to rolling element bearing fault diagnosis. Mech. Syst. Signal Process. **40**(1), 136–153 (2013)
7. Yan, J., Lu, L.: Improved Hilbert-Huang transform based weak signal detection methodology and its application on incipient fault diagnosis and ECG signal analysis. Signal Process. **98**, 74–87 (2014)
8. Widodo, A., Yang, B.-S., Kim, E.Y., Tan, A.C.C., Mathew, J.: Fault diagnosis of low speed bearing based on acoustic emission signal and multi-class relevance vector machine. Nondestruct. Test. Eval. **24**(4), 313–328 (2009)
9. Kang, M., Kim, J., Wills, L., Kim, J.-M.: Time-varying and multi-resolution envelope analysis and discriminative feature analysis for bearing fault diagnosis. IEEE Trans. Ind. Electr. **62**(12), 7749–7761 (2015)
10. Sanders, J., Kandrot, E.: CUDA by example: an introduction to general-purpose GPU programming (2010). http://www.amazon.com/CUDA-Example-Introduction-General-Purpose-Programming/dp/0131387685

The Marker Detection from Product Logo for Augmented Reality Technology

Thummarat Boonrod$^{(\boxtimes)}$, Phatthanaphong Chomphuwiset,
and Chatklaw Jareanpon

Polar Lab, Faculty of Informatics, Mahasarakham University,
Mahasarakham, Thailand
thummaratboon@gmail.com, {phatthanaphong.c,chatklaw.j}@msu.ac.th

Abstract. This paper proposed the development of an effective algorithm for marker detection from products for augmented reality by Speeded-Up Robust Features (SURF) algorithm that provided the efficiency in term of speed and accuracy. The SURF alorithm is consisted of 3 processes that are (1) feature extraction calculates the interested point and interested descriptions, (2) feature matching is that the correlation of all points is calculated from the distance of similarity of featuers, and (3) logo indentification is used to find the four corner point of the logo. This experiment is conducted from the recording video at 100 frames with resolution of 640×360 pixels and logo appeared all frames. Objects used in the experiment are consists of 3 shapes, cylindrical (can), rectangular (bag), and bottle. The logo template is divided into 5 sizes. The result of experiment found that the best detection accuracy of logo detection is from the size of 100×100 pixels. The accuracy of the region of marker detection compared with ground truth shows that the bag is equal to 94.96 %, can is equal to 93.99 %, and bottle is equal to 91.01 %, respectively. The difference of the logo is not affected with the computational time. However, the fast moving camera creates the blurred image and the reflection on the packaging creats a shiny surface which affects with the accuracy.

Keywords: Augmented reality (AR) · Speeded-Up Roust Features (SURF) · Reflection · Logo detection · 3D modeling

1 Introduction

The current computer vision technology has been applied in various fields such as image classification of vehicle logos [1]. Augmented reality (AR) is a one of computer vision technologies that is mixing between the world (Real environment) and the virtual (Virtual environment). The virtual reality displays the virtual through the computer monitors, projectors or other display devices on the real environment. Object detection is the technique for detecting the position and the movement of objects that is important for augmented reality to analyze the images [2]. The current augmented reality is applied in various applications

© Springer International Publishing AG 2016
V.-N. Huynh et al. (Eds.): IUKM 2016, LNAI 9978, pp. 421–432, 2016.
DOI: 10.1007/978-3-319-49046-5_36

such as sport, education, medical, entertainment and advertiments. For example, an applied augmented reality is used to show the distance and the border of the sailing competition of the America's Cup which shows the audience about the distance of each sailing to stay away from the finish line [3]. Advertising of T-Touch company shows the watches that the customers are able to download and put on the wrist when views through the screen, the system will display a virtual wristwatch [4].

Marker detection for augmented reality is an important step since it is used to find the position that 3D modeling will be displayed. Failed or missing marker detection is one of the problems that the 3D model cant show on a top of the marker correctly or the 3D rendering can't work in the some frame. The current design of the designed markers for solving a specific problems are not applied for using with the existing products on the market. The marker on the product will not need to print or design the new packaging. Moreover, some area of products are used to display an important information that is unable to display the marker. In addition, the problem of interference from the light directed to the surface of object is affected with the accuracy of the detection.

This paper proposed the development of an algorithm that is able to detect the marker from the products for augmented reality in terms of the effective speed and accuracy. This paper is organized as follows: Sect. 2 is the literature review, Sect. 3 explains the proposed method, Sect. 4 shows the experiment and results, Sect. 5 explains the discusses and conclusions, and future work.

2   Literature Review

This paper divides the literature reviews into 6 parts as follows: (1) augmented reality, (2) logo design for products, (3) logo detection, (4) Speeded-Up Robust Features (SURF), (5) 3D geometric transformations and (6) detection of specular reflection.

2.1   Augmented Reality (AR)

Augmented reality technology was emerged in the 1968, which showed the virtual system that can see-through display on the head (Head-mounted display) which was able to track the six movement directions (forward/backward, up/down, left/right) [5]. In the 1992, Caudell and Mizell definded the definition of "augmented reality" that a computer showed the superimposed object on top of the real world [6]. In the 1993, global positioning system (GPS) together with a compass was used for the development of prototype outdoor navigation system for the blurred vision people [7]. In the 1994, Milgram and Kishino proposed a taxonomy of the display by definition of a continuation of the real environment (RE) associated with the virtual environment (VE) or virtual reality (VR) [8,9]. In the 1997, Azuma proposed the augmented reality defined as follow: (1) combining real and virtual, (2) interacting in real time and (3) registering the 3D [10]. That is generally accepted for word definition of "Augmented reality" as shown in Fig. 1.

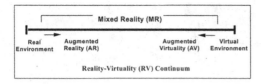

Fig. 1. The relationship between the augmented reality and augmented virtuality.

2.2 Logo Design for Product

Product logo created by human is the symbols or markers to represent the product. These symbols are easy to remember and simple. The logoes are divided into 4 types: (1) Letter mark, (2) Word mark, (3) Pictorial symbol, and (4) Combination mark [11].

(1) Letter marks are the logo designed from letters or acronyms as shown is Fig. 2a.
(2) Word marks are designed from texts or words as shown in Fig. 2b.
(3) Pictorial symbols are designed from picture or symbols as shown in Fig. 2c.
(4) Combination marks are mixed from the letters, text, words, picture or symbol as shown in Fig. 2d.

Fig. 2. Logo design, (a) Letter mark, (b) Word mark, (c) Pictorial symbol, and (d) Combination mark [12]. (Color figure online)

2.3 Logo Detection

Logo detection is the method for detecting the region of logo. The logo detection is usually used the object detection technique. Object detection technique captures the object by setting from the initial step or when an object appears. The popular object detections 4 techniques: (1) Point detectors, (2) Segmentation, (3) Background subtraction, and (4) Supervised learning [13].

(1) Point detection is the method to find the interest points used for detection within the image for the texture characteristics of the region as shown in Fig. 3a.
(2) Segmentation is the method to divide the image into the sections based on the similarities of the region as shown in Fig. 3b.
(3) Background subtraction is the method to find the difference of the scene called the background model. The background model represents the difference between frame, when the area has changed in the image that denotes a moving object as shown in Fig. 3c.

(a)   (b)   (c)

Fig. 3. Object detection technique, (a) Point detection, (b) Segmentation, and (c) Background subtraction [13].

(4) Supervised learning is the object detection method by learning the differences of the object automatically from a set of sample data.

2.4 Speeded-Up Robust Features (SURF)

Speeded-Up Robust Features (SURF) [14,15] is the feature extraction method that is considered in fastest process and robust to change in the rotation, size, blur and brightness. The SURF algorithm is consisted of 2 processes. (1) Interested point detection and (2) Interested point description.

Interested Point Detection is used to find the interested regions from images. This process is consisted of 4 processes as follows: (1) Integral images, (2) Hessian matrix, (3) Scale space and (4) Determine interesting point.

(1) Integral image process is calculated by the sum of pixels for reducing the computational time for large image calculated by the Eq. (1).

$$I_{\sum}(X) = \sum_{i=0}^{i \leq x} \sum_{i=0}^{j \leq y} I(i,j) \tag{1}$$

when
$I_{\sum}(X)$ is integral image of position $X = (x,y)$ which is the sum of pixels from input image I.
i is the position of row.
j is the position of column.
(2) The Hessian matrix is the region of interested point in the image calculated by the Eq. (2).

$$H(X,\sigma) = \begin{bmatrix} L_{xx}(X,\sigma) & L_{xy}(X,\sigma) \\ L_{xy}(X,\sigma) & L_{yy}(X,\sigma) \end{bmatrix} \tag{2}$$

When $L_{xx}(X,\sigma), L_{xy}(X,\sigma), L_{yy}(X,\sigma)$ are the second-order derivative of Gaussian image I at position X.
(3) Scale space is generated by the pyramid format of the convolution image with the multiple Gaussian kernel rounds of the box. The filters are applied in various sizes such as 9×9, 15×15, 21×21, 27×27, and they are used for each different sizes without reducing the size of images as shown in Fig. 4a. This helps to reduce the computational time and memory.

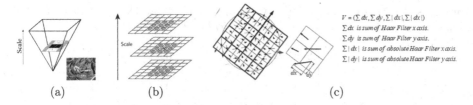

Fig. 4. (a) Pyramid format box filter for scale space, (b) The comparison of pixels with the scale adjacent, and (c) Orientation of key point.

(4) Determining the interested point is the points that are robust to change with in images using Non-maximal suppression. The selected-interested points are robust by comparison of the 26 adjacent pixels from 8 pixels at the same layers and 9 pixels from above and below layers. If the central point is greater than the all neighbor points, it will be define as an interested point as shown in Fig. 4b.

Interested Point Descriptions is using the orientation form the region of around interested points. This process consists of 2 processes.

(1) Orientation assignment, which uses is using Haar wavelets filter (robust of rotation and reduces the computational time). Haar wavelets filter is used as gradient of x axis and y axis.
(2) Discriptor based on sum of Haar wavelet processes the square window covering the key points and assigning the orientations as shown in Fig. 4c.

2.5   3D Geometric Transformations

Vector space is a mathematical structure determined by the number of vectors, which the size of vector space is in three dimensions. 3D model is created by the three dimensional system at working in the model space. Each point and the surface of a 3D model is related to the coordinate system. 3D Transformation is consisted of 3 types: translation, scale, and rotation [16].

Translation is the motion from the 3D coordinates of the origin to the new coordinates. Translation matrix can be calculated from the Eq. (3).

$$\begin{bmatrix} 1 & 0 & 0 & Translation\ x \\ 0 & 1 & 0 & Translation\ y \\ 0 & 0 & 1 & Translation\ z \\ 0 & 0 & 0 & 1 \end{bmatrix} \tag{3}$$

Scale is changing the ratio of the 3D model which it can be transformed to the small or large size. Scale matrix can be calculated from the Eq. (4).

$$\begin{bmatrix} Scale\ x & 0 & 0 & 0 \\ 0 & Scale\ y & 0 & 0 \\ 0 & 0 & Scale\ z & 0 \\ 0 & 0 & 0 & 1 \end{bmatrix} \tag{4}$$

Rotation is calculated by the rotation of the angle around the axis x, y and z.

(a) The x-axis rotation is can be calculated from the Eq. (5)

$$\begin{bmatrix} 1 & 0 & 0 & 0 \\ 0 & \cos(\theta) & -\sin(\theta) & 0 \\ 0 & \sin(\theta) & \cos(\theta) & 0 \\ 0 & 0 & 0 & 1 \end{bmatrix} \tag{5}$$

(b) The y-axis rotation is calculated from the Eq. (6)

$$\begin{bmatrix} \cos(\theta) & 0 & \sin(\theta) & 0 \\ 0 & 1 & 0 & 0 \\ -\sin(\theta) & 0 & \cos(\theta) & 0 \\ 0 & 0 & 0 & 1 \end{bmatrix} \tag{6}$$

(c) The z-axis rotation is calculated from the Eq. (7)

$$\begin{bmatrix} \cos(\theta) & -\sin(\theta) & 0 & 0 \\ \sin(\theta) & \cos(\theta) & 0 & 0 \\ 0 & 0 & 1 & 0 \\ 0 & 0 & 0 & 1 \end{bmatrix} \tag{7}$$

2.6   Detecting of Specular Reflection

Specular reflection is a problem causing from lighting effecting to the surface of objects. The reflected light is energized and increased as the source of light near objects [17]. The task of computer graphics, computer vision and image processing require the correct information. Therefore, the reflection will affect with the incorrect extraction information [18]. To solving this issue, the reflection must be detected and improved by various methods. Detection of specular reflection by the histogram decomposition and detection method can be achieved by using the criteria proposed by [17]. Specularities is defined by the intensity of the pixels of the images that are very high where the color matches with the source of the light. Creating a histogram of the image will help to identify these region and able to create a histogram that can separate the image color chanels (red, green, blue). Normally, the light source is white. Therefore, the region with high intensity of white assumes as the reflection area and a histogram of the colors red, green and blue will reach high intensity and that indicates the reflection area. Moreover, it can be used in a grayscale image for detection the region of specularities by using a simple thresholding, because the pixels in the region is high intensity, and independent from other regions. Specularities finds a component increased by saturation (S) of HSV Structure. Therefore, the detection area is a reflection achieved from gray scale images and saturation by creating a bi-dimensional histogram from the Eqs. (8) and (9).

$$M = \frac{1}{3}(r + g + b) \tag{8}$$

$$S = \begin{cases} \frac{1}{2}(2r - g - b) = \frac{3}{2}(r - m), if(b + r) \geq 2g \\ \frac{1}{2}(r + g - 2b) = \frac{3}{2}(m - b), if(b + r) \leq 2g \end{cases} \tag{9}$$

Specularities is used to identify by bi-dimensional histogram based on the maximum value of M and S of image. The identification of the pixels p as part of the reflected light can be calculated from the Eq. (10). The m_{max} and s_{max} are maximum of density of M and S From all pixels of image, from the Eq. (10)

$$\begin{cases} m_p \geq \frac{1}{2}m_{max} \\ s_p \leq \frac{1}{2}s_{max} \end{cases} \tag{10}$$

if the condition is true, this value will be masked as the refection and set it to 1.

3   Our Proposed Method

The proposed framework is divided into 3 parts: (1) Input video, (2) Logo detection and (3) 3D Rendering as shown in Fig. 5.

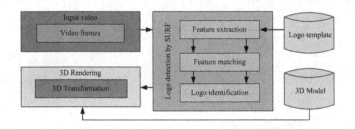

Fig. 5. The proposed framework diagram.

3.1   Input Video

This process inputs the video into the system for separating video frames that is recorded videos of the packaging of produces. The camera moves to the top, bottom, left, right, in and out. Recorded video resolution is 640 × 360 pixels and comprises with 100 of frames which logo appears in all frames. The packaging used in the experimental are 3 product shapes as follows: can, bag and bottle as shown in Fig. 6.

(a)   (b)   (c)

Fig. 6. The packaging product in used this experiment: (a) can, (b) bag, and (c) bottle. (Color figure online)

3.2   Logo Detection by SURF

The proposed logo detection is performed by using SURF algorithm and is divided into 3 parts: (1) feature extraction, (2) feature matching, and (3) logo identification.

(1) Feature extraction process takes input the images from the input video and extracts the logo using the logo templates (Logo template is divided into 5 sizes: 50×50, 70×70, 100×100, 130×130, and 150×150 pixels as shown in Fig. 7a.). After that, interested points and interested descriptions will be calculated and represented as shown in Fig. 7b.

(2) Feature matching is carried out by calculating the distance of the similarity of features. The correlation of all points are calculated for cutting the false matching when compares the distance with threshold values. After calculating the correlation of all points, there have a set of relationships of points as shown in Fig. 8a.

(3) Logo identification using estimation of geometric transforms is a set of the corresponding point that finds the position of the four corner points of the logo as shown in Fig. 8b.

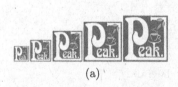 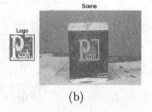

(a)   (b)

Fig. 7. (a) Sizes of logo: 50×50, 70×70, 100×100, 130×130, and 150×150 pixel, and (b) Interested point detection. (Color figure online)

3.3   3D Rendering

This process is divided into 3 steps: (1) 3D model loading, (2) 3D Geometrical transformating, and (3) 3D model placing.

(a)   (b)   (c)

Fig. 8. (a) Corresponding point, (b) Logo identification, and (c) The 3D model showed on the top of the marker. (Color figure online)

(1) 3D model loading is the process of loading the 3D model is from the stored area and replacing it on the marker. 3D model is consists of the vertex that coordinates in the axis of X, Y and Z. Edge is the index of the point connected to each point and polygon is the location of a surface of the 3D model.

(2) 3D geometrical transformating is a process to transform the 3D geometry using a transformation matrix that can be calculated from the Eq. (11).

$$R = \begin{bmatrix} 1 & 0 & 0 & 0 \\ 0 & \cos(\theta) & -\sin(\theta) & 0 \\ 0 & \sin(\theta) & \cos(\theta) & 0 \\ 0 & 0 & 0 & 1 \end{bmatrix} * \begin{bmatrix} \cos(\theta) & 0 & \sin(\theta) & 0 \\ 0 & 1 & 0 & 0 \\ -\sin(\theta) & 0 & \cos(\theta) & 0 \\ 0 & 0 & 0 & 1 \end{bmatrix} * \begin{bmatrix} \cos(\theta) & -\sin(\theta) & 0 & 0 \\ \sin(\theta) & \cos(\theta) & 0 & 0 \\ 0 & 0 & 1 & 0 \\ 0 & 0 & 0 & 1 \end{bmatrix}$$

(11)

(3) 3D model is a placement 3D of model into the centroid of 4 corner points as shown in Fig. 8c.

3.4 Measurement

The evaluation marker detection from the logo of the packaging of product is divided into 2 parts, accuracy and computational time.

(1) Accuacy of measurement of marker detection is divided into 3 part, accuracy of all marker detections which can be calculated from the Eq. (12), accuracy of the region of marker detection as shown in Fig. 9, and determining the region of marker on the packaging of products using ground truthas shown in Fig. 9.

$$Accuracy = \frac{Number\ of\ detetedcorrect}{Number\ of\ frames} * (100)$$

(12)

(2) Computational time is the evaluation of marker detection speed which will examine the logo detection process by SURF. The timer starts from feature extraction process to Logo identification process.

4   Experimental and Result

The experiment is set as a Real-time environment which records the video at 100 frames of a normal digital camera with 640×360 pixels of resolution which

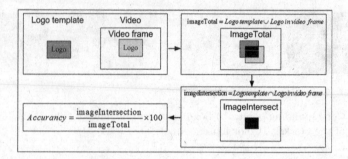

Fig. 9. Measurement accuracy.

logo appears in the all frames. Objects are used in the experiment consisted of 3 shapes: cylindrical, rectangular and bottle. Logo template is divided into 5 sizes: 50×50, 70×70, 100×100, 130×130, and 150×150 pixels. The result measurement is divided into 3 parts, (1) accuracy of marker detections as shown in Fig. 10a, (2) the average accuracy of the region of marker detection compared with ground truth found that the cylindrical(Can) is equal to 93.99 %, rectangular(Bag) is equal to 94.96 %, and bottle is equal to 91.01 %, and (3) computational time as shown in Fig. 10b. The experiment implemented using MatlabR2015b run on an Intel core i7 3.6 GHz processor with memory 4 GB.

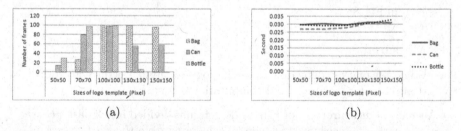

(a)   (b)

Fig. 10. The results of experiment, (a) Accuracy of marker detection from video frame and (b) Computational time of marker detection.

5   Conclusion and Future Work

5.1   Discussion

Biswas C. and Mukherjee J. [19] proposed the comparison technique for logo recognition between SIFT, SURF, and HOG descriptor and compared only in the accuracy. In that research, the accuracy of SURF is higher than SIFT and HOG algorithm. Moreover, the Boonrod T. et al. [13], proposed the comparison technique of logo classification between Template matching and SURF algorithm in term of accuracy and speed. This research found that the accuracy of SURF is higher and faster than template matching.

Therefore, this paper selected to use the SURF algorithm for logo detection process. The experiment as shown in Fig. 10a tested in the 5 different sizes of logo template: 50×50, 70×70, 100×100, 130×130, and 150×150 pixels. The high accuracy is from the 100×100 pixels, because size of logo template is similar to size of logo in the video frames. From the Fig. 10b shows the computational time of the logo detection which the result is similar to the Boonrod T. et al. [12]. The size of logo is not affected with the computational time.

5.2 Conclusion

This paper proposed the development of an algorithm for marker detection from products for augmented reality based on the SURF algorithm. From the result of experiment shows the efficacy in 3 parts: (1) the accuracy of marker detection. The result found that 100×100 pixels of the size of the logo template is the best accuracy of 3 product shapes as shown in Fig. 10a, and the size of logo template is affected with the accuracy of the marker detection., (2) The accuracy of the region of marker detection compared with the ground truth. The result found that the accuracy of bag is equal to 94.96 %, can is equal to 93.99 %, and bottle is equal to 91.01 %, respectively. Because the shape of bag is rectangular shape and plane which is clearly displayed the logo, the logo detection on the bag is the highest accuracy., and (3) the Computational time of the marker detection. The size of the logo template is not affected with the computational time as shown in Fig. 10b. The logo detection problem of frame is caused by the focus lossing, while the fast movement of camera creates the blurred image. Moreover, the problem of reflections on the packaging appears a shiny surface, which the logo is partial loss. The future work will try to recover the area caused by the reflection interference.

Acknowledgement. This research is supported by Mahasarakham University.

References

1. Zhang, B., Pan, H.: Reliable classification of vehicle logos by an improved local-mean based classifier. In: 2013 6th International Congress on Image and Signal Processing (CISP), vol. 1, pp. 176–180 (2013)
2. Augmented Reality. http://researchguides.dartmouth.edu/content.php
3. Americas Cup Updates as it Trawls for Viewers. http://www.nytimes.com/2012/06/28/sports/americas-cup-seeking-bigger-audience-changes-its-rules.html
4. Augmented Reality for Online Retail and Advertising in the Future. http://www.slashgear.com/augmented-reality-for-online-retail-and-advertising-in-the-future-28136588/
5. Sutherland, I.E.: A head-mounted three dimensional display. In: Proceedings of the Fall Joint Computer Conference, pp. 757–764 (1968)

6. Caudell, T., Mizell, D.: Augmented reality: an application of heads-up display technology to manual manufacturing processes. In: Proceedings of the Twenty-Fifth Hawaii International Conference on Systems Sciences, vol. 2, pp. 659–669 (1992)

7. Loomis, J., Golledge, R., Klatzky, R.: Personal guidance system for the visually impaired using GPS, GIS, and VR technologies. In: Proceedings of Conference on Virtual Reality and Persons with Disabilities, pp. 17–18 (1993)

8. Milgram, P., Kishino, F.: A taxonomy of mixed reality visual displays. IEICE Trans. Inf. Syst. **77**, 1321–1329 (1994)

9. Milgram, P., Takemura, H., Utsumi, A., Kishino, F.: Augmented reality: a class of displays on the reality-virtuality continuum. In: International Society for Optics and Photonics for Industrial Applications, pp. 282–292 (1995)

10. Azuma, R.T.: A survey of augmented reality. In: Presence Teleoperators and Virtual Environments, pp. 355–358 (1997)

11. Dimarco, P.: Logo Design for record company. In: Computer Graphics, pp. 1–5 (2014)

12. Yilmaz, A., Javed, O., Shah, M.: Object tracking: a survey. ACM Comput. Surv. (CSUR) **38**(4), 13 (2006)

13. Boonrod, T., Jareanpon, C., Chomphuwiset, P.: The comparison of template matching and SURF for logo classification on product. In: IIAE International Conference on Intelligent Systems and Image Processing, pp. 256–263 (2015)

14. Bay, H., Ess, A., Tuytelaars, T., Van Gool, L.: Speeded-up robust features (SURF). In: Computer Vision and Image Understanding, pp. 346–359 (2008)

15. Juan, L., Gwun, O.: A comparison of SIFT, PCA-SIFT and SURF. Int. J. Image Process. (JJIP) **3**, 143–152 (2009)

16. Foley, J.D., Van Dam, A.: Fundamentals of Interactive Computer Graphics. Addison-Wesley, Reading (1982)

17. Tchoulack, S., Langlois, J.P., Cheriet, F.: A video stream processor for real-time detection and correction of specular reflections in endoscopic images. In: Circuits and Systems and TAISA Conference, pp. 49–52 (2008)

18. Artusi, A., Banterle, F., Chetverikov, D.: A survey of specularity removal methods. In: Computer Graphics Forum 2011, vol. 30, pp. 2208–2230. Blackwell Publishing Ltd (2011)

19. Biswas, C., Mukherjee, J.: Logo recognition technique using sift descriptor, Surf descriptor and Hog descriptor. Int. J. Comput. Appl. **117**(22), 34–37 (2015)

Data Mining and Application

An Approach to Decrease Execution Time and Difference for Hiding High Utility Sequential Patterns

Minh Nguyen Quang[1], Ut Huynh[2], Tai Dinh[3], Nghia Hoai Le[4], and Bac Le[2(✉)]

[1] Academy of Cryptography, Techniques, Ho Chi Minh City, Vietnam
minhnq1962@yahoo.com.vn
[2] Department of Computer Science,
Ho Chi Minh City Industry and Trade College,
Ho Chi Minh City, Vietnam
ut.huynhvn@gmail.com, lhbac@fit.hcmus.edu.vn
[3] Department of Computer Science, University of Science, VNU-HCMC,
Ho Chi Minh City, Vietnam
duytai@fit-hitu.edu.vn
[4] University of Information Technology, VNU HCMC, Ho Chi Minh City, Vietnam
nghialh@uit.edu.vn

Abstract. Nowadays, databases are shared commonly in various types between companies and organizations. The essential requirement is to release *aggregate* information about the data, without leaking *individual* information about participants. So the Privacy Preserving Data Mining (PPDM) has become an important research topic in recent years. PPDM models are applied commonly on hiding association rule, hiding high utility itemsets mining and also on hiding High Utility Sequential Patterns (HUSPs) mining. The goal of hiding utility sequential patterns is to find the way to hide all HUSPs so that the adversaries cannot mine them from the sanitized database. The exiting researches hasn't considered in details about the difference ratio between the original database and the sanitized database after hiding all HUSPs. To address this issue, this paper presents two algorithms, which are HHUSP-D and HHUSP-A (Hiding High Utility Sequential Pattern by Descending and Ascending order of utility) to decrease the difference and also decrease execution time. In the proposed algorithms, a additional step is added to the exiting algorithm, HHUSP, to rearrange the hiding order of the HUSPs. Experimental results show that HHUSP-D is better performance than HHUSP [4] not only on the difference but also on the execution time.

Keywords: Data mining · Privacy preserving · Sensitive sequential pattern · High utility sequential patterns · High utility sequential patterns hiding

1 Introduction

In recent year, the volume of data has been increasing very quickly, that leading to increase demand of data mining and the PPDM has been taking an

© Springer International Publishing AG 2016
V.-N. Huynh et al. (Eds.): IUKM 2016, LNAI 9978, pp. 435–446, 2016.
DOI: 10.1007/978-3-319-49046-5_37

important role. Especially, there are some sensitive information areas such as health care, banking, marketing, etc., where the PPDM has to be considered carefully. There are many researches about PPDM on hiding association rule such as [5], hiding sequential pattern [2], hiding high utility itemset [3]. However, there are few researchs on hiding HUSPs. Two state-of-the-art algorithms for this purpose were proposed in [4]. The existing researches approach is changing value (support or utility) of items to hide sensitive items. There have been two types of privacy protection concerning data mining [5]. The first type of privacy, called output privacy protection, that is minimally altered so that the mining result will preserve certain privacy. The second type of privacy, input privacy, is that the data is manipulated so that mining result is not affected or minimally affected. In this study, a new proposed technique that is changing the hiding order of high sequential utility patterns. The experimental results showed that the hiding HUSPs by ascending order of utility makes the execution time and the difference increase. In contrast, the hiding by descending order of utility makes not only the execution time but also the difference decrease. The rest of this paper is organized as follows. Section 2 reviews the related works. Section 3 describes two proposed algorithms: HHUSP-A and HHUSP-D. Section 4 discusses the experimental results and evaluates the performance of the proposed algorithms on some datasets. Finally, Sect. 5 concludes the work.

2 Related Works

2.1 High Utility Sequential Patterns Mining

The purpose of the sequential pattern mining is to discover frequent sequential patterns from datasets in time orders of items. The traditional sequential pattern mining algorithms considers only binary frequency values. They are not applicable for numerous real-life scenarios, which should consider not only co-occurrence value but also quantity, utility (profit, cost or fee, etc.) and time, e.g. real-life application domains such as market basket data analysis, web log mining, biomedical gene data analysis, especially in the market basket data analysis, gold or diamond with low frequency purchase but high profit. HUSPs mining discovers all sequential patterns in sequence database whose utility values are equal to or greater than a given minimum utility. In recent years, the researchers [1,6,8] have proposed a new research direction called utility sequential pattern mining to solve HUSPs mining problem. After that, based on the comment *"main purpose of sequential pattern mining is to find most customers' purchase behavior, but not personal behavior"*, the research [6] proposed a new measure namely maximum utility measure and also developed a new method to calculate quickly maximum utility of sequential patterns in mining process. In 2012, USpan algorithm was proposed in [8] and also applied maximum utility measure model. USpan [8] based on Lexicographic Qsequence Tree to maintain pattern utility values on the tree. USpan [8] used two concatenated mechanisms (I-Concatenation and S-Concatenation) and two pruning strategies (Width Pruning and Depth Pruning) for the mining. In this study, the expansion algorithm of USpan is used to obtain all HUSPs and collect the input data for proposed hiding algorithms. The expansion was detailed in [4].

2.2 High Utility Itemset Hiding

One of the approaches for hiding high utility itemsets is reducing their utility in the original database so that the adversaries cannot mine them from the modified database by using mining algorithms. There are a few algorithms proposed for hiding high utility itemsets in transaction database. In 2013, a hiding high utility itemsets algorithm and an improvement of HHUIF [7] was proposed in [3] to hide all high utility itemsets. The algorithm reduced utility of each itemset in the set of high utility itemsets until their utilities were less than a given minimum utility by using a ratio α where α was the rate of decrease on quantity in each transaction. [3] outperformed HHUIF [7] both the runtime and the difference between the sanitized database and the original database. In 2011, two algorithms named DBSH and SBSH were proposed in [2] to hide sequences pattern by selecting a number of transactions to sanitize and then deciding which events needed to be deleted to perform the sanitization sequence database. The first algorithm sanitized data with minimal distortion, whereas the second focuses on reducing the side-effects [2]. But two algorithms could not support utility-based sequential pattern mining and just handled specific sequences with simple structures.

2.3 High Utility Sequential Patterns Hiding

In 2015, two novel algorithms HHUSP and MSPCF were proposed in [4] to hide all HUSPs. The research [4] proves that HHUSP outperformed more than MSPCF. However, the research [4] hasn't considered in details about the difference between the original database and the sanitized database after hiding phase. In the rest of this paper, the difference between the original database and the sanitized database will be called the missing cost. In this study, HHUSP-D is an improved algorithm of HHUSP [4] is proposed to decrease execution time and missing cost. Besides, we considered more details how to modify the missing cost by the simple way, which is changing the hiding order of the HUSPs.

3 Proposed Algorithms

3.1 Background

Assume a quantitative sequential database or be called q-sequence database, is given in Table 1, each q-sequence consists sequence identification (SID) and a sequential pattern with items quantity. Also, assume the profit value of each item is shown in Table 2.

Let $I = \{i_1 i_2 \ldots i_n\}$ is a set of all items in q-sequence database. An q-itemset $s_l = \{(x_1, q_1)(x_2, q_2) \ldots (x_n, q_n)\}$, where $(x_k, q_k)(1 \leq k \leq m)$ is a q-item, which

Table 1. Sequential database with internal utility

SID	Sequence
S_1	$e(5)$; $\{c(8), f(1)\}$; $b(10)$
S_2	$\{a(4), e(6)\}$; $\{a(2), b(5), c(8)\}$; $\{a(4), d(9), e(3)\}$
S_3	$c(4)$; $\{a(12), d(9), e(2)\}$
S_4	$\{b(10), e(2)\}$; $\{a(14), d(9)\}$; $\{a(8), b(5), e(2)\}$
S_5	$\{b(10), e(3)\}$; $\{a(12), e(3)\}$; $\{a(4), b(5)\}$
S_6	$g(1)$

Table 2. The external utility of the items

Item	a	b	c	d	e	f	g
Profit	3	5	1	8	4	6	7

consists an item, $x_k \in I$ and q_k is a positive number representing the quantity or the purchased number of i. If a q-itemset has only one q-item then the brackets are omitted. A q-sequence $S = \{s_1 s_2 \ldots s_l\}$ is a list of q-itemsets sorted in ascending order according to their occurrence time, where $s_j (1 \leq j \leq l)$ is an q-itemset. Each q-itemset s_j appears in q-sequence is also called an element of the q-sequence. A quantitative sequential pattern $p = \langle e_1 e_2 \ldots e_l \rangle$, where $e_j (1 \leq j \leq l)$ is an q-itemset, is called an r–subsequence if $|p| = r$, where $|p| = \sum_{e_j \in p} |e_j|$. Given two quantitative sequential pattern $\alpha = \langle a_1 a_2 \ldots a_m \rangle$ and $\beta = \langle b_1 b_2 \ldots b_n \rangle$ $(m \leq n)$, where a_i, b_i is q-itemsets. If there exists $j_1 < j_2 < \cdots < j_m \leq j_n$, such that $a_1 \subseteq b_{j_1}, a_2 \subseteq b_{j_2}, \ldots, a_m \subseteq b_{j_n}$, then α is subsequence of β and β is super-sequence of α, denoted by $\alpha \subseteq \beta$. Given a quantitative sequential database, denoted by SDB, is includes a list of quantitative sequential transactions, that is $SDB = \{TS_1, TS_2, \ldots, TS_z\}$ where $TS_y (1 \leq y \leq z)$ is the y-th q-sequence in SDB. Each q-sequence TS_y contains a tuple $\langle SID_y, S_y \rangle$ where SID_y is the q-sequence identity and S_y is q-sequence content of TS_y. TS_y is said to contain a pattern p, if p is subsequence of S_y.

Definition 1. *The internal utility value of an item i_j in the s_l q-itemset of q-sequence S_y, denoted by $iu(i_j, s_l, S_y)$, is the quantity or the purchased number of i_j in the s_l q-itemset of q-sequence S_y.*

For example, in Table 1 $iu(a, s_1, S_2) = 4$; $iu(e, s_1, S_2) = 6$

Definition 2. *The internal utility value of an item i_j in q-sequence S_y, denoted by $iu(i_j, S_y)$, is the sum of all the quantities of i_j in sequence S_y. The maximum internal utility value, denote by $iu_{max}(i_j, S_y)$ is maximum quantity of i_j in S_y. The external utility value of an item i_j, denoted by $eu(i_j)$, is utility value of item i_j in the profit table.*

For example, in Table 1 $iu(a, S_2) = 4 + 2 + 4 = 10$; $iu_{max}(a, S_2) = 4$; $eu(a) = 3$

Definition 3. *The utility value of an item i_j in the s_l q-itemset of q-sequence S_y, denoted by $u(i_j, s_l, S_y)$, is the external utility $eu(i_j)$ of item i_j multiplied* by the internal utility value $iu(i_j, s_l, S_y)$ of item i_j in the s_l q-itemset of the q-sequence S_y.

For example, in Table 1, item e two distinct occurrences in S_2 with internal utility is 6 and 3 in s_1 and s_3 respectively, $u(e, s_1, S_2) = iu(e, s_1, S_2) \times eu(e) = 6 \times 4 = 24$; $u(e, s_3, S_2) = iu(e, s_3, S_2) \times eu(e) = 3 \times 4 = 12$.

Definition 4. *The utility value of an item i_j in q-sequence S_y, denoted by $u(i_j, S_y)$, is the external utility $eu(i_j)$ of item i_j multiplied by the internal utility value $iu(i_j, S_y)$ of item i_j in the q-sequence S_y.*

Definition 5. *The maximum utility value of an item i_j in sequence S_y, denoted by $u_{max}(i_j, S_y)$, is the external utility $eu(i_j)$ of item i_j multiplied by the maximum internal utility value $iu_{max}(i_j, S_y)$ of item i_j in the sequence S_y.*

For example, in Table 1, item e two distinct occurrences in S_2 with internal utility is 6 and 3 respectively, $iu(e, S_2) = 6 + 3 = 9$; $u(e, S_2) = iu(e, S_2) \times eu(e) = 9 \times 4 = 36$, $iu_{max}(e, S_2) = 6$; $u_{max}(e, S_2) = iu_{max}(e, S_2) \times eu(e) = 6 \times 4 = 24$.

Definition 6. *The utility value of a quantitative sequential pattern p in a q-sequence S_y, denoted by $u(p, S_y)$, is maximum utility value of p in S_y.*

For example, in Table 1, sequential pattern $\langle (ae) \rangle$ appears two times in S_2 with utility values are 36 and 24 respectively, based on the maximum utility model, utility value of $\langle (ae) \rangle$, $u(\langle (ae) \rangle, S_2) = u_{max}(\langle (ae) \rangle, S_2) = 4 \times 3 + 6 \times 4 = 36$.

Definition 7. *The sequence utility of a quantitative sequential pattern p in SDB, denote by $u(p)$, is the sum of utility values of pattern p in all sequences of SDB.*

$$u(p) = \sum_{S_y \in SDB} u(p, S_y) \tag{1}$$

For example, in Table 1, $u(\langle (ae) \rangle) = u(\langle (ae) \rangle, S_2) + u(\langle (ae) \rangle, S_3) + u(\langle (ae) \rangle, S_4) + u(\langle (ae) \rangle, S_5) = 36 + 44 + 32 + 48 = 160$.

Definition 8. *Given a predefined minimum utility threshold, λ, the utility sequential pattern is HUSP if $u(p) \geq \lambda$.*

For example, in Table 1, given $\lambda = 228$. Then $\langle (ad) \rangle$ is HUSP because $u(\langle (ad) \rangle) = 306 > \lambda$.

Definition 9. *The sequence utility value of a q-itemsets s_q in sequence S_y, denoted by $su(s_q, S_y)$, is the sum of utility values of items are contained in s_q in sequence S_y.*

$$su(s_q, S_y) = \sum_{i_j \in s_q \wedge s_q \in S_y} u(i_j, s_q) \tag{2}$$

For example, in Table 1 $su(s_1, S_2) = u(a, s_1, S_2) + u(e, s_1, S_2) = 4 \times 3 + 6 \times 4 = 36$.

Definition 10. *The sequence utility value of a sequence S_y, denoted by $su\,(S_y)$, is the sum of utility values of q-itemsets are contained in S_y.*

$$su\,(S_y) = \sum_{s_q \in S_y} u\,(s_q, S_y) \tag{3}$$

For example, in Table 1 $su\,(S_1) = u\,(s_1, S_1) + u\,(s_2, S_1) + u\,(s_3, S_1) = 20 + 14 + 50 = 84$.

Definition 11. *The total utility value of SDB, denoted by $su(SDB)$, is the sum of sequence utility values of all sequences are included in SDB.*

$$su\,(SDB) = \sum_{S_y \in SDB} su\,(S_y) \tag{4}$$

Definition 12. *Given the sanitized database, denote by SDB' of the original database, SDB. The missing cost, denoted by MC, is the subtraction of the sequence utility values of SDB' from the sequence utility values of SDB.*

$$MC = su(SDB) - su(SDB') \tag{5}$$

Theorem 1. *The smallest MC is the best solution.*

Proof (1). The proposed HHUSP algorithm [4] modifies the quantity of the items in q-itemsets based on α ration, where α is the rate of decrease on quantity in each q-sequence S_y. If the ratio α lower than 1, the quantity of the item is modified a total q, where $q = [q(i_j) - q(i_j) \times \alpha] < q(i_j)$. If the ratio α greater than 1, the quantity of the item is modified to 1. In two cases above, the quantity of the item is always equal or greater than 1 after modifying. Therefore, There isn't any item will be removed from the original database and the number of missing patterns is smallest. That proves the smallest MC is the best solution.

3.2   HHUSP: Hiding High Utility Sequential Pattern [4]

In this section, the HHUSP algorithm [4] is considered in detail to know how it works. This is preparing step before new hiding algorithms are proposed. The algorithm performs in two steps. The first step is the expansion algorithm of USpan [8] to mining all HUSPs. The first step outputs two sets, which contain the input information for the second step. Two sets are respectively U for storing all the HUSPs during mining process and Q for storing all the q-sequences which each HUSP belongs to. The information is stored in the form of key-value pairs.

The second step is hiding step. The pseudo-code of the HHUSP algorithm is illustrated in Algorithm 1. The HHUSP algorithm runs on the HUSPs in set

U from mining step. First the algorithm calculates $diff$, the utility that needs to be decreased on each HUSP p_k, by subtracting minimum threshold utility from its utility. If $diff$ is still greater than zero, the algorithm selects an item i_j in p_k to reduce quantity, where total utility of i_j is the highest value. Because a HUSP can be appeared in many sequences, the algorithm calculates $\alpha = \lceil \frac{diff}{eu(i_j)} \rceil \times eu(i_j)/sum(i_j)$ where α is the rate of decrease on quantity in each q-sequence S_y and $\lceil \frac{diff}{eu(i_j)} \rceil$ is the total quantity of item i_j in all $S_y \in Q$ would be reduced. After calculating ratio α, the algorithm will modify the quantity of i_p base on the ratio α, where $i_j \in p_k \wedge p_k \subseteq s_l \wedge s_l \subseteq S_y \wedge S_y \in Q$. After the reduction is completed, the algorithm updates $diff$. If $diff$ is still greater than zero, the algorithm continues to find the other i_j which has the highest total value and reduce until $diff \leq 0$. Then the algorithm recalculates the utility of all HUSPs in U. For each pattern p_k, the algorithm scans all the sequences in Q which this pattern belongs to and update utility. Then the algorithm continues to modify other patterns in U until U is empty. Finally, the algorithm returns the sanitized database. The HHUSP performs hiding HUSPs by the default order of mining step. The hiding order takes significance effected to the running time, missing cost and failure cost of hiding phase. Therefore, our study focus on rearranging the hiding order to reach better performance and lower missing cost. We proposed two algorithms and described more details in next section.

3.3   Proposed Algorithm HHUSP-A: Hiding High Utility Sequential Pattern by Ascending Order of Utility

When hiding the HUSPs by ascending order of utility value, it means the HUSPs with the lowest utility value will be first considered. For example, considering the list of all HUSPs for the quantitative sequential dataset given above in Table 3, based on ascending order of utility value the pattern $\langle bab \rangle$ corresponding to the smallest utility value 228 will be hidden first. Because the pattern with the smallest utility value will be considered first, the $diff$ will be the smallest. Then, if we don't consider the dependency to item i_p, the ratio is α also reached to smallest (Line 6 - Algorithm 1). All above that mean the modified quantity of i_p is the smallest. In addition, when we modify quantity of items with small value, we must modify more HUSPs to hide all them. Therefore, we must spend more execution time. The pseudo-code of the HHUSP-A algorithm is illustrated in Algorithm 2

3.4   Proposed Algorithm HHUSP-D: Hiding High Utility Sequential Pattern by Descending Order of Utility

On the contrary, when hiding the HUSPs by descending order of utility value, it means the pattern with the highest utility value will be first considered. Therefore, the ratio α will be reached to the highest value and the modified quantity also is high value. Finally, the total time for hiding all the HUSPs as well as the count of modified patterns will decrease because we have modified a lot on

---

Algorithm 1. HHUSP ALGORITHM

Input: Two sets are U and Q from mining step
Output: The sanitized database SDB'

1 **for** $HUSP$ $p_k \in U$ **do**
2     $diff = u(p_k) - \lambda$
3   **while** $diff > 0$ **do**
4        $sum(i_j) = \sum_{i_j \in p_k \wedge p_k \subseteq s_l \wedge s_l \subseteq S_y} u(i_j, s_l, S_y)$ (where $S_y \in Q$)
5   Select i_j where $sum(i_j)$ is the maximum utility
6        $\alpha = \lceil \frac{diff}{eu(i_j)} \rceil \times eu(i_j)/sum(i_j)$
       `/* This iteration to modify the quantity of item      */`
7   **for** S_y where $i_j \in p_k \wedge p_k \subseteq s_l \wedge s_l \subseteq S_y \wedge S_y \in Q$ **do**
8   Modify the quantity of i_j $q(i_j) = \begin{cases} q(i_j) - q(i_j) \times \alpha \text{ if } \alpha < 1 \\ 1 \text{ if } \alpha \geq 1 \end{cases}$
9           $diff = \begin{cases} diff - u(i_j, s_l, S_y) \text{ if } diff > u(i_j, s_l, S_y) \\ 0 \text{ if } diff \leq u(i_j, s_l, S_y) \end{cases}$

    `/* This iteration to recalculate the utility of all patterns in`
    `U after modifying                                          */`
10   **for** S_y where $p_k \subseteq s_l \wedge s_l \subseteq S_y \wedge S_y \in Q$ **do**
11   Calculate the utility of all p_k in U

    `/* This iteration to update the utility of all patterns in U`
    `after recalculating                                        */`
12   **for** p_k in U **do**
13   **if** *new utility of* $p_k > \lambda$ **then**
14   Update utility of p_k
15   **else**
16   Remove p_k from U

17 **return** SDB'

Table 3. The list of all HUSPs for the quantitative sequential dataset given above

Pattern	Utility	Pattern	Utility
$\langle (ad) \rangle$	306	$\langle bab \rangle$	228
$\langle b(ad) \rangle$	273	$\langle b(ae) \rangle$	229
$\langle ba(ab) \rangle$	264	$\langle (be)aa \rangle$	234
$\langle (be)ab \rangle$	248	$\langle (be)a(ab) \rangle$	284
$\langle (be)(ad)(abe) \rangle$	229	$\langle e(ad) \rangle$	230

Algorithm 2. HHUSP-A ALGORITHM

Input: Two sets are U and Q from mining step
Output: The sanitized database SDB'

1 $U' = $ Rearrange U by ascending order of utility
2 $SDB' = $ HHUSP(U', Q)
3 **return** SDB'

Algorithm 3. HHUSP-D Algorithm

    Input: Two sets are U and Q from mining step
    Output: The sanitized database SDB'
1 U' = Rearrange U by descending order of utility
2 SDB' = HHUSP(U',Q)
3 **return** SDB'

---

each item. It seems that the missing cost will be higher, but the experiment results proven the missing cost was lower than when hiding the HUSPs follow the descending order approach. The pseudo-code of the HHUSP-D algorithm is illustrated in Algorithm 3

4   Experimental Results

4.1   Environment and Datasets

The algorithms was written in C# (Visual Studio 2013), and run on the Windows 7 operating system on an Intel Core 2 Duo 2.8 Ghz CPU with 3 GB main memory. The experiment datasets were obtained from the SPMF library website [9]. The way how to generate these datasets is described in [9]. We evaluated the performance of HHUSP, HHUSP-A and HHUSP-D algorithms on all datasets are showed in Table 4 in term of execution time, memory usage and missing cost on several times and calculated the average value of the results. The experiment results just perform only in hiding step.

Table 4. Datasets description

Dataset	Size	# Seq	Distinct item	Avg items/seq
SIGN	375 KB	800	310	51.99
BIBLE	8.56 MB	36, 369	13, 905	21.64
BMSWebView1	2.80 MB	59, 601	497	2.51
BMSWebView2	5.46 MB	77, 512	3, 340	4.62
Kosarak990k	57.2 MB	990, 000	41, 270	8.14

4.2   The Execution Time Evaluation

In the Fig. 1, the vertical axis of the following charts describes the execution time in millisecond and the horizontal axis describes the minimum utility thresholds. In general, all evaluation results show that HHUSP-D outperformed than the others and HHUSP-A is the worst performance. The HHUSP-A spends more time to finish hiding task and execution time increase more quickly when the minimum utility thresholds decrease. Specially, on SIGN datasets, the execution time evaluation shows that when the minimum utility thresholds decrease from 20 k to 15 k, the count of HUSPs only increase from 1,530 to 6,550 but the execution time of the HHUSP-A increase from 47,346 ms to 726,220 ms, the

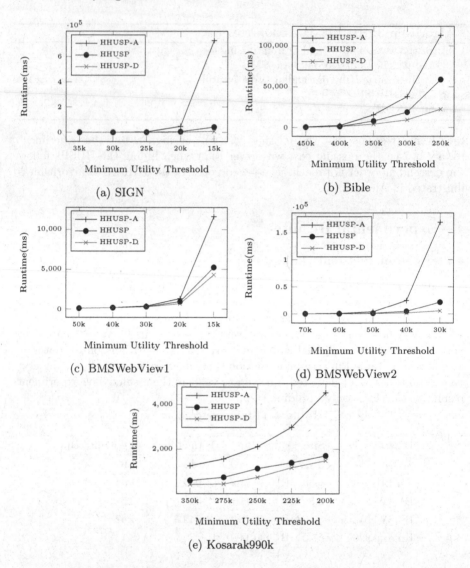

Fig. 1. Execution time evaluation on the datasets

ratio is approximately 15.3 times. In the same case, the ratio is approximately 8.2 times by 5,091 ms on 41,551 ms for the HHUSP and approximately 6.3 times by 1,290 ms on 8,089 ms for the HHUSP-D. In almost cases, the HHUSP-D is always faster than the HHUSP and HHUSP-A on the experimental datasets with various characteristics include very large dataset such as Kosarak990k.

4.3   The Memory Usage Evaluation

We performed the memory usage evaluation in all datasets but due to limitation of pages so we only show the results of the best case and the worst case.

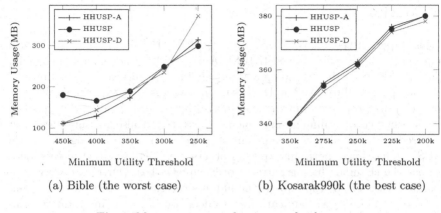

(a) Bible (the worst case)   (b) Kosarak990k (the best case)

Fig. 2. Memory usage evaluation on the datasets

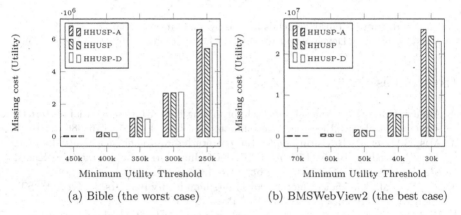

(a) Bible (the worst case)   (b) BMSWebView2 (the best case)

Fig. 3. Missing cost evaluation on the datasets

In the Fig. 2, the vertical axis of the following charts describes the memory usage in megabyte and the horizontal axis describes the minimum utility thresholds. In general, HHUSP, HHUSP-A and HHUSP-D are quite equal in memory usage, depending on each dataset. However, in almost cases, the memory usage of the HHUSP-D is always less than the HHUSP and HHUSP-A. Specially, in all evaluations, the HHUSP-D is the best performance on large dataset such as Kosarak990k

4.4   The Missing Cost Evaluation

The same as the memory usage evaluations, we perform the missing cost evaluations in all datasets but due to limitation of pages so we only show the results of the best case and the worst case in the Fig. 3. The missing cost is the reflection of the changing ratio between the original data and the sanitized database. In this study, the missing cost is figured out in the simplest way, which follows Definition 12. The evaluations results, which is the lowest missing cost, is the best case and the opposite.

5 Conclusion

In this paper, based on the novel algorithms [4] HUSPs for hiding we have proposed two algorithms named HHUSP-A and HHUSP-D, which are changing hiding HUSPs order by ascending and descending of utility value. The improvements are purposed to speed up HUSPs hiding process. The experimental results show that the proposed improvements make the HHUSP-D algorithm has better performance on running time, memory usage and also missing cost when compared with the HHUSP algorithm [4]. The other proposed algorithm, the HHUSP-A has lower performance than others but it is presented in this study to prove the changing hiding patterns order approaches. Our future work is on expanded research to study about the hiding process take effect on the original database. After that, we can learn how to decrease the changing ratio not only the missing cost.

Acknowledgment. This research is funded by Vietnam National Foundation for Science and Technology Development (NAFOSTED) under grant number 102.05-2015.07.

References

1. Ahmed, C.F., et al.: A novel approach for mining high-utility sequential patterns in sequence databases. Electron. Telecommun. Res. Inst. J. **32**(5), 676–686 (2010)
2. Aris, G.D., et al.: Revisiting sequential pattern hiding to enhance utility. In: KDD 2011 Proceedings of the 17th ACM SIGKDD International Conference on Knowledge Discovery and Data Mining, pp. 1316–1324 (2011)
3. Vo, B., et al.: An efficient method for hiding high utility itemsets. In: KES-AMSTA 2013, Hue, Vietnam, pp. 356–363. IOS-Press (2013)
4. Dinh, T., Nguyen, Q.M., Le, B.: A novel approach for hiding high utility sequential patterns. In: Proceedings of The Sixth International Symposium on Information and Communication Technology, Vietnam, pp. 121–128 (2015)
5. Jain, Y.K.: An efficient association rule hiding algorithm for privacy preserving data mining. Int. J. Comput. Sci. Eng. **3**(7), 2792–2798 (2011)
6. Lan, G.C., et al.: Applying the maximum utility measure in high utility sequential pattern mining. Expert Syst. Appl. **41**(11), 5071–5081 (2014)
7. Yeh, S., Hsu, P.C.: HHUIF and MSICF: novel algorithms for privacy preserving utility mining. Expert Syst. Appl. **37**(7), 4779–4786 (2010)
8. Yin, J., et al.: USpan: an efficient algorithm for mining high utility sequential patterns. In: The 18th ACM SIGKDD, pp. 660–668 (2012)
9. An Open-Source Data Mining Library. http://www.philippe-fournier-viger.com/spmf/index.php. Accessed July 1 2016

Modeling Global-scale Data Marts Based on Federated Data Warehousing Application Framework

Ngoc Sy Ngo[1(✉)] and Binh Thanh Nguyen[2,3]

[1] Hue University, Hue, Vietnam
ngongoc11906@gmail.com
[2] Duy Tan University, Danang, Vietnam
ttb_2001@gmail.com
[3] International Institute for Applied Systems Analysis (IIASA),
Schlossplatz 1, 2361 Laxenburg, Austria
nguyenb@iiasa.ac.at

Abstract. Data warehouses become large in size, dynamic, and physically distributed. In this context, federated data warehousing approach is a promising solution for specifying, designing, deploying as well as managing data marts and their data cubes. In this paper, the Federated Data warehousing Application (FDWA) framework has been used to specify global-scale data marts. To proof of our concepts, a global-scale analytical tool, namely GAINS-IAM (integrated assess model) will be presented as a case study to analyze recent trends and future world emission scenarios.

Keywords: FDWA (Federated Data Warehousing Application) · GAINS (Greenhouse Gas - Air Pollution Interactions and Synergies) · Global-scale data mart · Greenhouse Gases (GHGs) · IAM (Integrated Assess Model)

1 Introduction

Nowadays, data warehouses have become a core technology for competitive organizations in the globalized world [6]. Main challenges of those systems are size, speed and distributed operations. However, data warehouses always store big amount of data and their important requirement is that queries have to be processed quickly and efficiently, so parallel and distributed solutions are deployed to render the necessary efficiency [18]. As a result, federated data warehousing approach has been taken in account as a promising solution for different problems in the area of business decision support systems [9].

In our research [1, 2], regional federated data warehousing application framework [15–17] has been taken into our consideration. In this context, multi-scale regional data marts have been designed for specific regional analytical requirements and a data warehouse has been designed to store underline data organized in term of multidimensional data models.

First, concepts and requirements of global-scale data marts are specified in a formal manner as well as described in term of a motivation example. Hereafter, a global-scale

© Springer International Publishing AG 2016
V.-N. Huynh et al. (Eds.): IUKM 2016, LNAI 9978, pp. 447–456, 2016.
DOI: 10.1007/978-3-319-49046-5_38

multidimensional data model and its data mart data will be defined and generated by inheriting from our previous studies about FDWA framework [13–16]. Finally, the GAINS-IAM will be presented as a case study to illustrate how the just-defined data mart supports the development of global-scale science-driven policies.

The rest of this paper is organized as follows: Sect. 2 introduces some approaches and projects related to our work; the global-scale concepts will be introduced in Sect. 3. Afterwards, Sect. 4, our implementation results will be illustrated in context of a case study. And lastly, Sect. 5 gives a conclusion of what have been achieved and our future works.

2 Related Work

The features of our approach can be focused in the research area of federated data warehousing technologies [6, 8, 10, 11], especially in modeling global scale data marts.

In [18], a cloud-based federated data warehousing application framework has been proposed to improve the building and maintaining federated data warehousing systems, including cloud-based regional data marts specified from the data warehouse. Thus, a federated multidimensional data model has been specified in [15, 18].

In [12], the concepts of city scale win-win policy by policy requirements, which have been specified in [13, 14] as policy options and packages, have been used to develop city-scale data marts. The tool has been used by national policy makers to identify viable and efficient solutions at city-scale level.

The studies of global integrated assessment models in term of "representative concentration pathway" (RCP) have recently added future emission scenarios of air pollutants to their projections of long-lived GHGs (Greenhouse Gas) [21]. According to [12], these studies have been proposed to improve the ability of building, managing and analysis of global emission analytical tool.

3 The Concepts of Global-scale Data Marts

In this section, first we introduce the global-scale data mart concepts and requirements. Hereafter, system architecture of the FDWA framework [18] is customized to specify a global-scale data mart. In this context, two main operations of the FDWA framework are used to define multidimensional schema, then to generate data cubes of the global-scale data mart.

3.1 Motivation Example

According to [12], integrated assessment models, e.g. GAINS-IAM, are studied to identify portfolios of measures that improve air quality and reduce greenhouse gas emissions at least cost at global scales [5]. By using such models, scientific knowledge and quality-controlled data are brought together on future socio-economic driving forces of emissions, on the technical and economic features of the available emission control options, on the chemical transformation and dispersion of pollutants in the atmosphere, and the resulting impacts on human health and the environment [1, 4]. In this context, those computer simulation models have to meet requirements that other scientific models typically do not face [2], thus it will be useful for global experts and decision makers in a much broader international policy process. As a result, those models need to be:

- accessible, i.e. both the typical upstream and downstream workflows of the analysis need to be available online;
- transparent, i.e. rationale and mathematical relations of those models need to be comprehensible even to non-experts;
- participatory, i.e. global and national experts should have knowledge on and be involved in model design;
- able to store large amounts of data for global online emission analysis and process;
- able to deliver just-in-time results as required;
- able to interact and link with multiple regional modeling tools.

According to [12], each of the GAINS-IAM global-scale emission scenarios for future emissions contains three main building blocks: (a) activity projections; (b) control strategies; and (c) emission factors and cost factors. Those scenarios can be configured not only for different regions, but also certain classes of macro sectors individually. As a result, large numbers of scenarios can be specified and generated from a linked compact set of components. As shown in Fig. 1, a global scenario, namely *ECLIPSE_V5a_CLE_climate* is defined based on the the baseline case *ECLIPSE_V5a_CLE_base* scenario [12].

ECLIPSE_V5a_CLE_climate (zig)

Scenario relying on the IEA 2 degree (2DS) pathway (see Energy Technology Perspectives; IEA, 2012). For VOC and agriculture the same paths as for **ECLIPSE_V5a_CLE_base** are used. The emission vector and control strategies are the same as in the **ECLIPSE_V5a_CLE_base**.

Gridded data:
The global gridded data sets can be downloaded from here (Choose respective ECLIPSE set on the right side of the screen).

Created: *5 July, 2015*
Modified: *5 July, 2015*

internal scenario_id: ECLIPSE_V5a_CLE_climate

Scenario definition				
Country	**Emission vector**	**Control strategy**	**Activity type**	**Pathway**
Austria	JUN2015_EU	AUST_TS_Aug14	Agriculture	CAPRI_ref140613_C14_v2
			Energy	TSAP_PRIMES2012_DECARB
			FGAS sources	ETP_6DS_r
			Mobile	TSAP_PRIMES2012_DECARB
			Process	TSAP_P12_D40_flr_E
			VOC sources	GAINS_P13_C14

Fig. 1. Description of the global-scale scenario *ECLIPSE_V5a_CLE_climate*

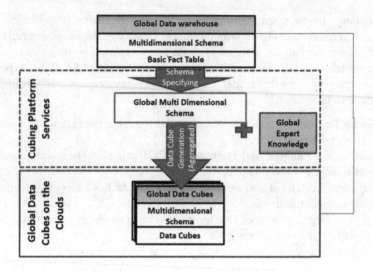

Fig. 2. Defining global data cubes based on the FDWA framework

3.2   Global-scale Data Mart Data Model

The global data mart M^G is specified based on the FDWA framework [18] as illustrated in Fig. 2 and defined as follows:

$$M^G = <D^G, L^G, V^G, C^G>$$

where:

- D^G is a set of dimensions of the global scale data mart, e.g. *IAM region, pollutant, IAM sector, IAM fuel_activity, IAM technology, IAM year, IAM scenario.*
- L^G is a set of dimensional levels of the global-scale data mart,e.g. *IAM-RegionGroup, IAM-Region*, etc.
- V^G is a set of decision variables of the global-scale data mart, e.g. *Activity, Emission* values.
- C^G is a set of global-scale data cubes, i.e. *IAM Activity Pathway* and *IAM Emission*.

3.3   Specifying Global-scale Data Cubes

According to [17], there are two main services, namely *MSchema* and *MData* used to specify the global-scale data mart M^G. Two services will be applied at the data warehouse to map from its schema to the global-scale data mart schema as well as to generate the global data cubes. The two services can be formulated as follows:

$$MSchema(M^G) = <D^G, L^G, V^G>$$

Hereafter, global data cube data could be generated by using the following service:

$$C^G = MData(M^G).$$

As illustrated in Fig. 2, in the global case, the *Global Multi Dimensional Schema* can be specified by using $MSchema(M^G)$ operation, afterwards, the *Global Data Cubes* are generated in the context of aggregation, or linking data from multiple regional data mart data.

4 GAINS-IAM (Integrated Assess Model): A Case Study of Global-scale Data Marts

The GAINS model [2, 3, 6, 18–20] has been proposed, developed and already used in various international policy fora as a tool to quantify costs and environmental benefits of reducing emissions of greenhouses gases and air pollutants. In the global-scale data mart context, basic concepts and requirements of the system are defined and illustrated how short-term co-benefits for local pollution can motivate mitigation greenhouse gases (GHGs) as a response of the otherwise intangible long-term and global risk of climate change [12]. So, we will present how the GAINS-IAM can be used as a common framework to make available and to compare the implications of the outputs of different energy system models working at different spatial and temporal scales in term of typical examples.

4.1 The IAM Dimensions

The following paragraphs shows descriptions of the version-specific dimensions, i.e. *IAM Region, IAM Sector*.

4.1.1 IAM Region Dimension
IAM *Region* dimension is defined by mapping of 162 GAINS regions [2, 3, 6] to three groups of regions worldwide [21], i.e. *Developed, Developing* and *Transition* groups. Figure 3 shows the list of IAM regions. Thus, the hierarchy of the *IAM Region* dimension can be seen as follows:

$$All-IAM-Regions- > IAM-RegionGroup- > IAM-Regions$$

4.1.2 IAM Sector Dimensions
Figure 4 shows the mapping between the GAINS and IAM *Sector* dimenisons. In this context, GAINS sectors are aggregated into IAM ones, e.g. *AUTO_P, DRY, FOOD,* etc. are grouped into *CHEM*. the hierarchies of IAM *Fuel Activity* and *Sector* dimensions can be specified as: $All\_IAM-Sectors- > IAM-Sector-Agg - > IAM-Sectors$.

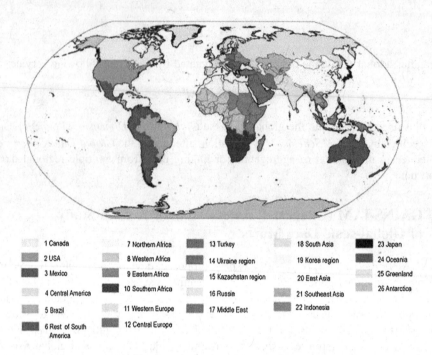

Fig. 3. The *IAM region* dimension

	GAINS_SECTORS	IAM_SECTORS
1	PR_ADIP	AACID
2	PR_CEM	CEMENT
3	PR_LIME	CEMENT
4	AUTO_P	CHEM
5	AUTO_P_NEW	CHEM
6	COIL	CHEM
7	DECO_P	CHEM
8	DEGR	CHEM
9	DEGR_NEW	CHEM
10	DOM_OS	CHEM
11	DRY	CHEM
12	DRY_NEW	CHEM
13	FATOIL	CHEM
14	FERTPRO	CHEM
15	FOOD	CHEM

Fig. 4. Mapping between GAINS and IAM *sector* dimensions

4.2   Specifying the IAM *Activity Pathway and Emission* Data Cubes

This section describes two steps: (1) to specify the *IAM Activity Pathway* data cube in the context of global-scale emission analysis; and (2) to calculate the *IAM Emission* data cube.

There are two main services, namely *MSchema* and *MData* used to specify the IAM data mart M^{IAM}.

$$MSchema(M^{IAM}) = <D^{IAM}, L^{IAM}, V^{IAM}>$$

Figure 5 shows multi dimensional schema of the *IAM Activity Pathway* data cube. In this context, *IAM fuel_activity*, *IAM sector*, and *IAM region* dimensions have global scale dimension levels as described in SubSect. 4.1. Afterwards, the *IAM Activity Pathway* and the *IAM Emission* data cube data could be generated by using the following service:

$$MData(M^{IAM}) = \{IAM\ Activity\ Pathway,\ IAM\ Emission\}$$

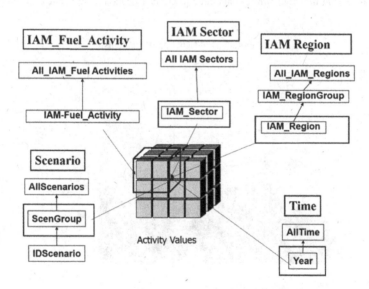

Fig. 5. IAM *activity pathway* data cube schema

China's activity pathway data of *ECLIPSE_V5a_CLE_climate* and *ECLIPSE_-V5a_CLE_base* global-scale scenario are compared and illustrated in Fig. 6. Afterwards, Fig. 7 as a typical example of the *IAM* application results shows the divergence of regional emission trends for different sectors in China and in India based on the current legislation between 1990 and 2050. China expects to alleviate the situation to some extent, although high emission densities have been expected to prevail over large areas. By contrast, given the lack of emission control legislation and the expected growth in power generation from coal, the Indian subcontinent will emerge as a new global hotspot of NOX emissions [12].

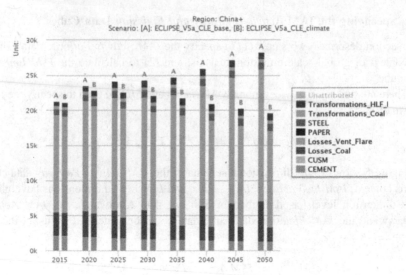

Fig. 6. Using *activity pathway* data cube to compare data among Chinese scenarios

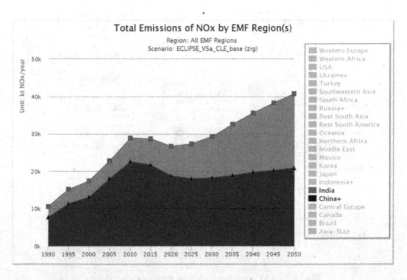

Fig. 7. Using *ECLIPSE-V5a_CLE_base* global-scale scenario to compare total NOX emission data between China and India

5 Conclusion

The concepts of the global-scale data mart has been introduced in this paper. We also introduced some additional global-scale requirements, e.g. mapping regional multidimensional schemas to global-scale aggregated ones, aggregating data from existing regional data marts to global levels of data granularities in term of linked data cubes [7, 15]. As a result, the global-scale multidimensional data model and its data mart data have been specified and generated by using two main services, namely *MSchema* and *MData* of the FDWA framework [13–16]. To proof of the proposed concepts and design, the GAINS IAM global data mart has been developed and presented in term of a case study.

In the context of the GAINS model, global cost analysis among linked emission data cubes will be taken into consideration of our future work.

References

1. Amann, M., Klimont, Z., Wagner, F.: Regional and global emissions of air pollutants: Recent trends and future scenarios. Annu. Rev. Environ. Resour. **38**, 31–55 (2013)
2. Amann, M., Bertok, I., Cofala, J., Heyes, C., Klimont, Z., Nguyen, T.B., Posch, M., Rafaj, P., Sandler, R., Schöpp, W., Wagner, F., Winiwarter, W.: Cost-effective control of air quality and greenhouse gases in Europe: modeling and policy applications. Environ. Model Softw. **26**(12), 1489–1501 (2011)
3. Bond, T.C., Streets, D.G., Yarber, K.F., Nelson, S.M., Woo, J.H., Klimont, Z.: A technology-based global inventory of black and organic carbon emissions from combustion. J. Geophys. Res. **109**(D14), 1–43 (2004)
4. Cofala, J., Amann, M., Klimont, Z., Kupiainen, K., Höglund-Isaksson, L.: Scenarios of global anthropogenic emissions of air pollutants and methane until 2030. Atmos. Environ. **41**, 8468–8499 (2007)
5. Dentener, F., Keating, T., Akimoto, H., (eds.): Hemispheric Transport of Air Pollution; Part A: Ozone and Particulate Matter. Air Pollution Studies No. 17. New York/Geneva: U.N
6. Furtado, P.: A survey of parallel and distributed data warehouses. Int. J. Data Warehouse. Min. **5**(2), 57–77 (2009)
7. Hoang, D.T.A., Ngo, N.S., Nguyen, B.T.: Collective cubing platform towards definition and analysis of warehouse cubes. In: Nguyen, N.-T., Hoang, K., Jędrzejowicz, P. (eds.) ICCCI 2012. LNCS (LNAI), vol. 7654, pp. 11–20. Springer, Heidelberg (2012). doi:10.1007/978-3-642-34707-8_2
8. Jindal, R., Acharya, A.: Federated data warehouse architecture, in White Paper
9. Kern, R., Ryk, K., Nguyen, N.T.: A framework for building logical schema and query decomposition in data warehouse federations. In: Jędrzejowicz, P., Nguyen, N.T., Hoang, K. (eds.) ICCCI 2011. LNCS (LNAI), vol. 6922, pp. 612–622. Springer, Heidelberg (2011). doi:10.1007/978-3-642-23935-9_60
10. Nguyen, N.T.: Model for knowledge integration. Advanced Methods for Inconsistent Knowledge Management. Advanced Information and Knowledge Processing, pp. 101–122. Springer, London (2008)
11. Nguyen, N.T.: Processing inconsistency of knowledge in determining knowledge of a collective. Cybern. Syst. **40**(8), 670–688 (2009)
12. Binh, N.T.: Integrated assessment model on global-scale emissions of air pollutants. In: Le Thi, H.A., Nguyen, N.T., Do, T.V. (eds.). AISC, vol. 358, pp. 345–354Springer, Heidelberg (2015). doi:10.1007/978-3-319-17996-4_31
13. Nguyen, T.B.: Policy by policy analytical approach to develop GAINS-city data marts based on regional federated data warehousing framework. In: Do, T., Thi, H.A.L., Nguyen, N.T. (eds.). AISC, vol. 282, pp. 243–253Springer, Heidelberg (2014). doi:10.1007/978-3-319-06569-4_18. ISBN 978-3-319-06568-7, 978-3-319-06569-4

14. Pitarch, Y., Favre, C., Laurent, A., Poncelet, P.: Context-aware generalization for cube measures. In: Proceedings of the ACM 13th International Workshop on Data warehousing and OLAP - DOLAP 2010, pp. 99–104. ACM Press (2010)
15. Nguyen, T.B., Ngo N.S.: Semantic cubing platform enabling interoperability analysis among cloud-based linked data cubes. In: Proceedings of the 8th International Conference on Research and Practical Issues of Enterprise Information Systems, CONFENIS 2014. ACM International Conference Proceedings Series (2014)
16. Nguyen, T.B., Tjoa, A.,Min, Wagner, R.: Conceptual multidimensional data model based on metacube. In: Yakhno, T. (ed.) ADVIS 2000. LNCS, vol. 1909, pp. 24–33. Springer, Heidelberg (2000). doi:10.1007/3-540-40888-6_3
17. Nguyen, T.B., Wagner, F.: Collective intelligent toolbox based on linked model framework. J. Intell. Fuzzy Syst. 27(2), 601–609 (2014)
18. Nguyen, T.B., Wagner, F., Schoepp, W.: Federated data warehousing application framework and platform-as-a-services to model virtual data marts in the clouds. Int. J. Intell. Inf. Database Syst. 8(3), 280 (2014). doi:10.1504/IJIIDS.2014.066635. http://www.inderscience.com/link.php?id=66635, ISSN 1751-5858
19. Nguyen, T.B., Wagner, F., Schoepp, W.: EC4MACS – An integrated assessment toolbox of well-established modeling tools to explore the synergies and interactions between climate change, air quality and other policy objectives. In: Auweter, A., Kranzlmüller, D., Tahamtan, A., Tjoa, A.M. (eds.) ICT-GLOW 2012. LNCS, vol. 7453, pp. 94–108. Springer, Heidelberg (2012). doi:10.1007/978-3-642-32606-6_8
20. Nguyen, T.B., Wagner F., Schoepp W.: GAINS-BI: Business intelligent approach for greenhouse gas and air pollution interactions and synergies information system. In: Proceedings of the International Organization for Information Integration and Web-based Application and Services IIWAS 2008, Linz (2008)
21. Van Vuuren, D.P., Edmonds, J.A., Kainuma, M., Riahi, K., Thomson, A.M., et al.: The representative concentration pathways: an overview. Clim. Change 109, 5–31 (2011)

How to Select an Appropriate Similarity Measure: Towards a Symmetry-Based Approach

Ildar Batyrshin[1], Thongchai Dumrongpokaphan[2], Vladik Kreinovich[3(✉)], and Olga Kosheleva[3]

[1] Centro de Investigaciń en Computación (CIC),
Instituto Politécnico Nacional (IPN), México, D.F., Mexico
batyr1@gmail.com
[2] Department of Mathematics, Chiang Mai University, Chiang Mai, Thailand
tcd43@hotmail.com
[3] University of Texas at El Paso, El Paso, USA
{vladik,olgak}@utep.edu

Abstract. When practitioners analyze the similarity between time series, they often use correlation to gauge this similarity. Sometimes this works, but sometimes this leads to counter-intuitive results, in which case other similarity measures are more appropriate. An important question is how to select an appropriate similarity measures. In this paper, we show, on simple examples, that the use of natural symmetries – scaling and shift – can help with such a selection.

1 Correlation and Other Similarity Measures: Formulation of the Problem

Practitioners routinely use correlation to detect similarities. When a practitioner is interested in gauging similarity between two sets of related data or between two time series, a natural idea seems to be to look for (sample) *correlation*; see, e.g., [5]: $\rho(a,b) = \dfrac{C_{a,b}}{\sigma_a \cdot \sigma_b}$, where

$$C_{a,b} \stackrel{\text{def}}{=} \frac{1}{n} \cdot \sum_{i=1}^{n} (a_i - \overline{a}) \cdot (b_i - \overline{b}), \quad \overline{a} \stackrel{\text{def}}{=} \frac{1}{n} \cdot \sum_{i=1}^{n} a_i, \quad \overline{b} \stackrel{\text{def}}{=} \frac{1}{n} \cdot \sum_{i=1}^{b} b_i,$$

$$\sigma_a \stackrel{\text{def}}{=} \sqrt{V_a}, \quad \sigma_b \stackrel{\text{def}}{=} \sqrt{V_b}, \quad V_a \stackrel{\text{def}}{=} \frac{1}{n} \cdot \sum_{i=1}^{n} (a_i - \overline{a})^2, \quad V_b \stackrel{\text{def}}{=} \frac{1}{n} \cdot \sum_{i=1}^{n} (b_i - \overline{b})^2.$$

Of course, correlation has its limitations. Practitioners understand that correlation only detects *linear* dependence. In some cases, the dependence is non-linear; in such cases, simple correlation does not work, and more complex methods are needed to detect dependence.

V.-N. Huynh et al. (Eds.): IUKM 2016, LNAI 9978, pp. 457–468, 2016.
DOI: 10.1007/978-3-319-49046-5_39

Also, correlation assumes that the value b_i is affected by the value of a_i at the same moment of time i – and only by this value. In real life, we may have a delayed effect – and the corresponding delay may depend on time.

However, in simple cases, when we do not expect nonlinear dependencies and/or delays, many practitioners expect correlation to be a perfect measure of similarity. And often it is. But sometimes, it is not. Let us give two examples.

Example of a simple case when correlation is not an adequate measure of similarity. Let us consider a simple case, when we ask people to evaluate several newly released movies on a scale from 0 to 5, and then we compare their evaluations a_i, b_i, ..., of different movies i to gauge how similar their tastes are; see, e.g., [1].

For simplicity, let us assume that for six movies, the first person gave them the following grades:

$$a_1 = 4, \quad a_2 = 5, \quad a_3 = 4, \quad a_4 = 5, \quad a_5 = 4, \quad a_6 = 5,$$

while the second person gave

$$b_1 = 5, \quad b_2 = 4, \quad b_3 = 5, \quad b_4 = 4, \quad b_5 = 5, \quad b_6 = 4.$$

From the common sense viewpoint, these two viewers have similar tastes – they seem to like all the movies very much. This similarity is especially clear if we compare them with the evaluations of a picky third person who does not like any new movies at all:

$$c_1 = 0, \quad c_2 = 1, \quad c_3 = 0, \quad c_4 = 1, \quad c_5 = 0, \quad c_6 = 1.$$

However, if we compute correlations, we will get exactly opposite conclusions:

- between a_i and c_i, there is a perfect correlation $\rho = 1$, while
- between a_i and b_i, there is a perfect *anti*-correlation $\rho = -1$.

In other words:

- a cheerful viewer a and a gloomy viewer c – who, from the commonsense viewpoint, are opposites – have a perfect positive correlation, while
- two cheerful viewers a and b – who, from the commonsense viewpoint, are almost Siamese twins – show perfect negative correlation.

This example clearly shows that we need to go beyond correlation to capture the commonsense meaning of similarity.

Second example. Let us have a somewhat less trivial example – based on the saying that when America sneezes, the world catches cold. Let us use a simplified example. Suppose that the US stock market shows periodic oscillations, with relative values

$$a_1 = 1.0, \quad a_2 = 0.9, \quad a_3 = 1.0, \quad a_4 = 0.9.$$

In line with the above saying, the stock market in a small country X shows similar relative changes, but with a much higher amplitude:

$$b_1 = 1.0, \quad b_2 = 0.5, \quad b_3 = 1.0, \quad b_4 = 0.5.$$

Are these sequences similar? Somewhat similar yet, but not exactly the same: while the US stock market has relatively small 10 % fluctuations, the stock market of the country X changes by a factor of two.

However, if we use correlation to gauge the similarity, we will see that these two stock markets have a perfect positive correlation $\rho = 1$. This example confirms that we need to go beyond correlation to capture the commonsense meaning of similarity.

Other similarity measures. The need to go beyond correlation to describe the intuitive idea of similarity is well known. Many effective similarity measures have proposed; see, e.g., [2,3,6] and references therein.

Most of these measures start either with correlation, or with the Euclidean distance $d(a,b) = \sqrt{\sum_{i=1}^{n}(a_i - b_i)^2}$ – or with a more general l^p-distance $\left(\sum_{i=1}^{n}|a_i - b_i|^p\right)^{1/p}$. Sometimes, a linear or nonlinear transformation is applied to the result, to make it more intuitive.

In other situations, modifications take care of the possible time lag in describing the dependence. For example, we may look for a correlation between b_i and the delayed series a_{i+c} for an appropriate constant delay. More generally, we can look for delay $c(i)$ that changes with time, i.e., for correlation between b_i and $a_{i+c(i)}$; an example of such a similarity measure is the move-split-merge metric described in [6].

An important problem: how to select the most appropriate similarity measure? The very fact that there exist many different effective similarity measures is an indication that in different practical situations, different similarity measures are appropriate. From the practical viewpoint, it is therefore important to be able to select the most appropriate similarity measure for each given situation.

There have been several papers comparing the effectiveness of different similarity measures in *clustering*, when we have several processes characterized by time series and we need to group them into clusters of similar ones; see, e.g., [2].

Another important practical case is when we simply have two time series and we are looking for the best measure to check if there is similarity between these two series.

What we do in this paper. In this paper, we show that natural symmetries – shifts and scalings – can help select the most appropriate similarity measure.

What we describe here are preliminary (and rather simple) results, results that – so far – only cover the cases when the time series are perfectly aligned in time, when there is no time lag, and when there are no non-linear effects.

The fact that the symmetry-based approach has helped to clarify the selection of a similarity measure in such simple cases makes us hope that this approach will be helpful in more complex situations as well.

A comment on intended audience. One of our major objectives is to provide a guidance for practitioners. Because of this practice-oriented goal, we tried our best to make our explanations and derivations as detailed as possible.

2 Natural Symmetries

Compared values come from measurements. To better understand why, for some time series a_i and b_i, there is sometimes such a discrepancy between commonsense meaning of similarity and correlation, let us recall how we get the values a_i and b_i. Usually, we get these values from measurements (see, e.g., [4]) – or, as in the example of evaluating movies, from expert estimates, which can also be considered as measurements, measurements performed by a human being as a measuring instrument.

Natural symmetries related to measurements. In the general measurement process, we transform actual physical quantities into numbers. For example, when we measure time, we transform an actual moment of time into a numerical value.

In general, to perform such a transformation, we need to select:

– a starting point and
– a measuring unit.

For example, if to measure time, we select the birth year of Jesus Christ as the starting point, and a usual calendar year as a measuring unit, we get the usual date in years. If instead we select the moment 2000.0 and use seconds as units, then we get astronomical time.

Similarly, we can measure temperature in the Fahrenheit (F) scale or in the Celsius (C) scale; these two scales have:

– different starting points: $0°C = 32°F$, and
– different units: a difference of 1 degree C is equal to the difference of 1.8 degrees Fahrenheit.

If we change a measuring unit to a new one which is u times smaller, then all numerical values get multiplied by this factor u: the same quantity that had the value x in the original units has the value $x' = u \cdot x$ in the new unit. For example, if we replace meters with centimeters, with $u = 100$, then a height of $x = 2\,\text{m}$ becomes $x' = 100 \cdot 2 = 200\,\text{cm}$ in the new units.

Similarly, if we change from the original the starting point to a new starting point which is s units earlier, then the original numerical value x is replaced by a new value $x' = x + s$.

In general, if we change both the measuring unit and the starting point, we get new values $x' = u \cdot x + s$.

In such general cases, correlation is a good description of similarity. For quantities for which we can arbitrarily select the measuring unit and the starting point, the same time series which is described by the numerical values x_i can be described by values $x_i' = u \cdot x_i + s$ in a different scale.

If we have a perfect correlation $\rho = 1$ between the two time series a_i and b_i, this means that after an appropriate linear transformation, we have $b_i = u \cdot a_i + s$. In other words, if we select an appropriate measuring unit and an appropriate starting point for measuring a, then the values $a_i' = u \cdot a_i + s$ of the quantity a as described in the new units will be identical to the values of the quantity b. And equality is, of course, a perfect case of what we intuitively understand by similarity.

This is why in many cases, correlation is indeed a perfect measure of similarity.

Not all quantities allow an arbitrary selection of measuring unit and starting point. The problem is that some quantities only allow some of the above symmetries – or none at all.

For example:

- while (as we have mentioned earlier) we can select different *units* for the distance between the points,
- we cannot select an arbitrary *starting point*: there is a natural starting point 0 that corresponds to the distance between the two identical points.

In this case, the fact that we can, e.g., obtain two series of distances a_i and b_i from one another by a shift does not make them similar: since this shift no longer has an intuitive sense.

This was exactly the case of two stock markets: for any price, 0 is a natural starting point, so:

- while scalings $x \to u \cdot x$ make sense,
- shifts $x \to x + s$ change the situation – often drastically.

For movie evaluations, the results are even less flexible: here, both the measuring unit and the starting point are fixed: any transformation will change the meaning. For the same physical distance, we can have two different values, e.g., 100 miles and 160 km, but for evaluations on a scale from 0 to 5, different numbers simply mean different evaluations.

In such cases, correlation – which is based on detecting a general linear dependence – is clearly not an adequate measure of similarity.

So how should we gauge similarity in such cases? Up to now, we showed that natural symmetries explain why correlation is not always a perfect measure of similarity. Let us now use natural symmetries to come up with measures of similarity which are adequate in such situations.

3   Starting Point: Case When No Scaling Is Possible

Description of the case. Let us start with the case when both measuring unit and starting point are fixed, all measurement results are absolute, and no scaling is possible – as in the case of viewers evaluating movies.

Natural idea. In this case, how can we gauge similarity between the two times series $(a_1, \ldots, a_n)$ and $(b_1, \ldots, b_n)$? The closer the two tuples, the more similar are these tuples. Thus, a natural measure of dissimilarity is simply the distance $d(a, b)$ between these two tuples: $d(a, b) = \sqrt{\sum_{i=1}^{n} (a_i - b_i)^2}$.

From the computational viewpoint, this idea can be slightly improved. The above formula is reasonable. However, our goal is not just to come up with a reasonable idea, but, ideally, to come with an idea to be applied in practice. From the practical viewpoint, the simpler the computations, the easier it is to apply the corresponding idea.

From this viewpoint, the above expression is not perfect; namely, in addition to:

- subtractions $b_i - a_i$ (which are easy to perform even by hand),
- multiplications $(b_i - a_i) \cdot (b_i - a_i)$ (which are also relatively easy to perform), and
- additions to compute the sum $(b_1 - a_1)^2 + (b_2 - a_2)^2 + \ldots$ (also easy),

we also need to compute the square root – which is not easy to perform by hand.

Good news is that the main purpose of gauging similarity is not so much to come up with some "absolute" number describing similarity, but rather to be able to compare the degree of similarity between different pairs (a_i, b_i). For example, in prediction, we can say that if a new situation a is sufficient similar to one of the past situations b – i.e., if the degree of similarity between them exceeds a certain threshold – then it is reasonable to predict that the situation a will come up with the same changes as were observed in the situation b in the past.

From this viewpoint, it does not matter how we assign numerical values to different degrees of similarity. We can change the numerical values of these degrees – as long as we preserve the order between them. In particular, when we square all the distances, then clearly larger distances become larger squares, and vice versa. Thus, instead of the original distances, we can as well consider their squares $d^2(a, b) = \sum_{i=1}^{n} (a_i - b_i)^2$.

Conclusion: in this case, distance is a natural measure of similarity. Summarizing, we can say that in situations when no scaling is possible – like in the case of movie evaluations – a reasonable idea is to use, as a reasonable measure of similarity,

- *not* the correlation (as practitioners are sometimes tempted to), but
- the *distance* (or squared distance) between the two series.

Comment. It should be emphasized once again, that, in contrast to correlation
– which attempts to describe *similarity* – distance describes *dissimilarity*:

– the larger correlation, the more similar the two time series, but
– the larger the distance, the less similar are the two time series.

4 When All Scalings Are Allowed, We Get Correlation

Description of the case. To see how good is the distance as the measure
of similarity, let us apply this idea to the generic case, when all scalings are
applicable. In other words, we consider the case when numerical values of both
quantities a_i and b_i are defined only modulo general linear transformations $a \to$
$u \cdot a + s$ and $b \to u' \cdot b + s'$, for any $u > 0$, $u' > 0$, s, and s'.

Starting point is distance: reminder. When the units and the starting points
are fixed, we get the usual distance – or, to be more precise, squared distance
$d^2(a, b) = \sum_{i=1}^{n} (a_i - b_i)^2$.

How to take care of possible re-scalings of the quantity a. If this distance
is small, this means that the time series a_i and b_i are similar. However, when
this distance is large, this does not necessarily mean that the time series a_i and
b_i are not similar – maybe we chose a wrong unit and/or a wrong starting point
for measuring a, and the distance will be much smaller if we use a different unit
and/or a different starting point. From this viewpoint, instead of considering
the distance $d(a, b)$ between the original numerical values a_i and b_i, it makes
more sense to consider the distance between b_i and re-scaled values $u \cdot a_i + s$ –
and consider the smallest possible value of this distance as a measure of this
dissimilarity:

$$D_g(a, b) = \min_{u,s} d^2(u \cdot a + s, b) = \min_{u,s} \sum_{i=1}^{n} (b_i - (u \cdot a_i + s))^2. \tag{1}$$

How to take care of re-scalings of the quantity b. The formula (1) takes
care of re-scaling the values a_i, but it may change if we re-scale the values b_i.
At first glance, it may seem that it we can solve this problem by also taking the
minimum also over all possible re-scalings of b as well. However, this will not
work: e.g., if we choose a very large measuring unit for measuring b, then the
numerical values of b_i can become very small – and thus, the value (1) can also
become arbitrarily small, and the minimum will always be 0.

To make the formula (1) scale-invariant, it is reasonable:

– instead of considering the *absolute* size of the discrepancies $b_i - (u \cdot a_i + s)$
  between the values b_i and the values $u \cdot a_i + s$ predicted by a_i,
– to consider *relative* size, relative to the size of how the values b_i themselves
  are different from 0,

i.e., the value

$$D'_g(a,b) = \frac{\min\limits_{u,s} \sum\limits_{i=1}^{n} (b_i - (u \cdot a_i + s))^2}{\sum\limits_{i=1}^{n} b_i^2} = \min\limits_{u,s} \frac{\sum\limits_{i=1}^{n} (b_i - (u \cdot a_i + s))^2}{\sum\limits_{i=1}^{n} b_i^2}. \qquad (2)$$

Need to take care of possible shifts in b. The formula (2) take care of scaling $b_i \to u' \cdot b_i$ – which, as one can check, does not change the value (2), but it still does change with the shift $b_i \to b_i + s'$.

Again, at first glance, it may seem reasonable to consider all possible shifts of b and to take the minimum – but then, after shifting b by a large amount, we do not change the numerator but we can make the denominator arbitrarily large. Thus, the result will be a meaningless 0.

Good news is that instead of taking the *minimum* over all possible shifts, we can get a meaningful result if we take the *maximum* overall all possible shifts. Thus, we arrive at the following definition.

Definition 1. *For every two tuples $a = (a_1, \ldots, a_n)$ and $b = (b_1, \ldots, b_n)$, we define a measure of dissimilarity as*

$$d_g(a,b) = \max\limits_{u',s'} \min\limits_{u,s} \frac{\sum\limits_{i=1}^{n} ((u' \cdot b_i + s') - (u \cdot a_i + s))^2}{\sum\limits_{i=1}^{n} (u' \cdot b_i + s')^2}. \qquad (3)$$

Discussion. From this expression, it is not even clear whether this expression is symmetric in terms of a and b, i.e., whether $d_g(a,b) = d_g(b,a)$. This is indeed true, and it is easy to see once we realize that $d_g(a,b)$ is directly related to the usual sample correlation:

Proposition 1. $d_g(a,b) = 1 - \rho^2(a,b)$.

Proofs. The proof of all the results form this paper is reasonably straightforward: to find the corresponding minima and/or maxima, we differentiate the corresponding expression and equate the resulting derivatives to 0.

For example, in this case, we first differentiate with respect to u and s and equate both partial derivatives to 0. We thus get a system of two linear equations with two unknowns u and s. Solving this system of equations, we conclude that $u = \dfrac{C_{a,b'}}{V_a}$ and $s = \overline{b'} - u \cdot \overline{a}$. Substituting these expressions for u and s into the minimized function, we get the expression $D'_g(a,b') = \dfrac{V_{b'}}{(b')^2} \cdot (1 - \rho^2(a,b'))$.

One can easily check that the correlation does not change under a linear transformation of one of the variables, so $\rho(a,b') = \rho(a,b)$. Thus, the above takes a simplified form $D'_g(a, b') = \dfrac{V_{b'}}{(b')^2} \cdot (1 - \rho^2(a, b))$. Here, $V_{b'} = \overline{(b')^2} - \left(\overline{b'}\right)^2 \le \overline{(b')^2}$

thus the ratio $\dfrac{V_{b'}}{(b')^2}$ is always smaller than or equal to 1. The largest possible value 1 of this ratio is attained when $\overline{b'} = 0$ – which we can always achieve by selecting an appropriate shift s' (namely, $s' = -\overline{b}$). In this case, the value $D'_g(a, b')$ is equal to $1 - \rho^2(a, b)$. Thus, $d_g(a, b) = \max\limits_{u', s'} D'_g(a, b') = 1 - \rho^2(a, b)$.

The proposition is proven.

Discussion. So, we confirmed that our approach makes sense – and it even leads to a non-statistical explanation of correlation. This enables us to use correlation beyond its usual Gaussian distribution case.

Let us now see what this approach results in in situations when only some of the natural symmetries are meaningful.

5   Case When Only Scaling Makes Sense – But Not Shift

Description of the case. Let us consider the case when a starting point is fixed, but we can choose an arbitrary measuring unit. (This is true, e.g., in the above the case of stock markets.)

In this case, we can have transformations $a_i \to a'_i = u \cdot a_i$ and $b_i \to b'_i = u' \cdot b_i$.

Analysis of this situation. In this case, instead of considering the distance $d(a, b)$ between the original numerical values a_i and b_i, it makes more sense to consider the distance between b_i and re-scaled values $u \cdot a_i$ – and consider the smallest possible value of this distance as a measure of this dissimilarity:

$$D_u(a, b) = \min_u d^2(u \cdot a, b) = \min_u \sum_{i=1}^{n} (b_i - u \cdot a_i)^2.$$

This takes care of re-scaling the values a_i. To take case of re-scalings of the values b_i, we can use the same idea as in the general case, and consider the ratio

$$D'_u(a, b) = \frac{\min\limits_u \sum\limits_{i=1}^{n} (b_i - u \cdot a_i)^2}{\sum\limits_{i=1}^{n} b_i^2} = \min_u \frac{\sum\limits_{i=1}^{n} (b_i - u \cdot a_i)^2}{\sum\limits_{i=1}^{n} b_i^2}.$$

It turns out that this ratio does not change if we re-scale b_i as well – this follows, e.g., from Proposition 2 proven below. So, we arrive at the following definition:

Definition 2. *For every two tuples* $a = (a_1, \ldots, a_n)$ *and* $b = (b_1, \ldots, b_n)$, *we define a measure of dissimilarity as* $d_u(a, b) = \min\limits_u \dfrac{\sum\limits_{i=1}^{n} (b_i - u \cdot a_i)^2}{\sum\limits_{i=1}^{n} (b_i)^2}.$

Comment. The following result provides an explicit formula for this measure of dissimilarity.

Proposition 2. $d_u(a, b) = 1 - \dfrac{\left(\overline{a \cdot b}\right)^2}{\overline{a^2} \cdot \overline{b^2}}.$

Comment. Here, in line with notations from Sect. 1, we denoted

$$\overline{a \cdot b} \overset{\text{def}}{=} \frac{1}{n} \cdot \sum_{i=1}^n a_i \cdot b_i, \quad \overline{a^2} \overset{\text{def}}{=} \frac{1}{n} \cdot \sum_{i=1}^n a_i^2, \quad \overline{b^2} \overset{\text{def}}{=} \frac{1}{n} \cdot \sum_{i=1}^n b_i^2.$$

Comment. In particular, when $\overline{a} = \overline{b} = 0$, we have $\overline{a \cdot b} = C_{a,b}$, $\overline{a^2} = V_a$, $\overline{b^2} = V_b$, and thus, this formula turns into the correlation-related formula $d_u = 1 - \rho^2(a, b)$.

In this case, correlation can be reconstructed as $\rho(a, b) = \sqrt{1 - d_u(a, b)}$. In general, we can therefore view the expression $\sqrt{1 - d_u(a, b)}$ as an analogue of correlation.

In the above example of two stock markets for which $\rho(a, b) = 1$ but common-sense similarity is not perfect, we have

$$\overline{a^2} = \frac{1^2 + 0.9^2 + 1^2 + 0.9^2}{4} = 0.905, \quad \overline{b^2} = \frac{1^2 + 0.5^2 + 1^2 + 0.5^2}{4} = 0.625, \text{and}$$

$$\overline{a \cdot b} = \frac{1 \cdot 1 + 0.9 \cdot 0.5 + 1 \cdot 1 + 0.9 \cdot 0.5}{4} = 0.725.$$

Thus, here, $d_u(a, b) = 1 - \dfrac{(0.725)^2}{0.905 \cdot 0.625} \approx 1 - 0.929 = 0.071 > 0$. Hence, the above equivalent of correlation $\sqrt{1 - d_u}$ is approximately equal to 0.96, which is smaller than 1 – as desired.

6   Case When Only Shift Makes Sense – But Not Scaling

Description of the case. Let us consider the case when a measuring unit is fixed, but we can choose an arbitrary starting point.

In this case, we can have transformations $a_i \to a_i' = a_i + s$ and $b_i \to b_i' = b_i + s'$.

Analysis of the situation. In this case, instead of considering the distance $d(a, b)$ between the original numerical values a_i and b_i, it makes more sense to consider the distance between b_i and shifted values $a_i + s$ – and consider the smallest possible value of this distance as a measure of this dissimilarity:

$$D_s(a, b) = \min_s d^2(a + s, b) = \min_s \sum_{i=1}^n (b_i - (a_i + s))^2.$$

This takes care of shifting the values a_i.

It turns out that this value does not change if we shift b_i as well – this follows, e.g., from Proposition 3 proven below. So, we arrive at the following definition:

Definition 3. *For every two tuples $a = (a_1, \ldots, a_n)$ and $b = (b_1, \ldots, b_n)$, we define a measure of dissimilarity as $D_s(a, b) = \min_s \sum_{i=1}^{n} (b_i - (a_i + s))^2$.*

Comment. The following result provides an explicit formula for this measure of dissimilarity.

Proposition 3. $D_s(a, b) = n \cdot (V_a + V_b - 2C_{a,b})$.

Comment. In the previous two cases, there was a possibility to re-scale b_i. To make the resulting measure of (dis)similarity independent on such re-scaling, we had to divide the squared distance by $\sum_{i=1}^{n} b_i^2$, i.e., consider *relative* discrepancy instead of the absolute one.

In the case when only shifts make physical sense, re-scaling of b_i is not possible, so there is no need for such a division.

What we *can* do is make sure that the value of dissimilarity does not depend on the sample size – in the sense that if we combine two identical samples, the dissimilarity will be the same. Such doubling does not change the sample variances and covariances V_a, V_b, and $C_{a,b}$; thus, after this doubling, the above expression $D_s(a, b)$ also doubles. To make it independent on such doubling, we can therefore divide the above expression $D_s(a, b)$ by the sample size and thus, get a new measure $d_s(a, b) \stackrel{\text{def}}{=} \dfrac{D_s(a, b)}{n} = V_a + V_b - 2C_{a,b}$. Such a division was not needed in the above two cases – since there, as we can see from Propositions 1 and 2, the division by the sum $\sum_{i=1}^{n} b_i^2$ automatically resulted in doubling-invariance.

7   Conclusions

The above analysis leads to the following recommendations for selecting an appropriate similarity measure in cases when we ignore time lag and non-linearities:

Situations when both a measuring unit and a starting point are fixed. In such situations, the most appropriate similarity-dissimilarity measure is the distance $\sqrt{\sum_{i=1}^{n} (a_i - b_i)^2}$.

Example: this is an appropriate similarity measure for movie evaluations.

Situations when neither a measuring unit nor a starting point are fixed. In such situations, the most appropriate similarity measure is the correlation $\rho = \dfrac{C_{a,b}}{\sigma_a \cdot \sigma_b}$.

Examples: there are many practical applications of this similarity measure.

Situations when a starting point is fixed, but we can choose an arbitrary measuring unit. In such situations, the most appropriate similarity measure is the ratio $\dfrac{\left(\overline{a \cdot b}\right)^2}{\overline{a^2} \cdot \overline{b^2}}$.

Example: this an appropriate measure for comparing the fluctuations of two stock markets.

Situations when a measuring unit is fixed, but we can choose an arbitrary starting point. In such situations, the most appropriate dissimilarity measure is $\sigma_a^2 + \sigma_b^2 - 2C_{a,b}$.

Example: this measure is appropriate for comparing two sequences of events that occurred in different time periods.

Acknowledgments. This work is supported by Chiang Mai University, Thailand. It was also supported in part:

   – by the National Science Foundation grants HRD-0734825 and HRD-1242122 (Cyber-ShARE Center of Excellence) and DUE-0926721,
   – by an award "UTEP and Prudential Actuarial Science Academy and Pipeline Initiative" from Prudential Foundation, and
   – by a grant Mexico's Instituto Politecnico Nacional.
   The authors are very thankful to the anonymous referees for valuable suggestions.

References

1. Batyrshin, I.: Fuzzy logic and non-statistical association measures. In: Proceesings of the 6th World Conference on Soft Computing, Berkeley, California, 22–25 May 2016
2. Iglesias, F., Kastner, W.: Analysis of similarity measures in times series clustering for the discovery of building energy patterns. Energies **6**, 579–597 (2013)
3. Liao, T.W.: Clustering of time series data - a survey. Pattern Recogn. **38**, 1857–1874 (2005)
4. Rabinovich, S.G.: Measurement Errors and Uncertainty. Theory and Practice. Springer, Berlin (2005)
5. Sheskin, D.J.: Handbook of Parametric and Nonparametric Statistical Procedures. Chapman and Hall/CRC, Boca Raton (2011)
6. Stefan, A., Athitsos, V., Das, G.: The move-split-merge metric for time series. IEEE Trans. Knowl. Data Eng. **25**(6), 1425–1438 (2013)

A Convex Combination Method for Linear Regression with Interval Data

Somsak Chanaim[1(✉)], Songsak Sriboonchitta[1], and Chongkolnee Rungruang[2]

[1] Faculty of Economics, Chiang Mai University, Chiang Mai, Thailand
somsak_ch@cmu.ac.th
[2] Faculty of Commerce and Management,
Prince of Songkla University Trang Campus, Trang, Thailand

Abstract. This paper introduces a new approach to fitting a linear regression model to interval-valued data by relaxing an assumption about using the center of interval data. We use convex combination between lower and upper values of the interval data as a parameter with value between [0,1]. Thus, the center method becomes a special case of this method. For the real application we use Capital Asset Pricing model (CAPM) and Autoregressive model (AR(p)) with interval-valued data to show that this method can provide a better result than the center method based on the Akaike information criterion (AIC).

Keywords: Convex combination method · Linear regression · Interval data · CAPM model · Autoregressive model

1 Introduction

When having to solve some problems by regression analysis, we cannot always get the sample point data. What we can get is perhaps only the interval data or the mix between point and interval. For example, the information about income and expense in questionnaire is very unlikely to be reported in exact number. Because the respondents may not want to tell the truth or the income and expense level are not stable every day, month, or year. It is easier to get answer close to the truth if the question is asked in terms of range, or lower and higher income and expense. Meanwhile, in financial market analysis, either open price or close price of stock is normally inputted in the regression model for stock price prediction. That means we lose a lot of information about prices between the day or during the week. With interval-valued data of a stock price, we can capture more realistic the movement or information of the stock price. All of the classical regression methods require only point data for parameters estimation (odinary least squares (OLS), maximum likelihood, maximum entropy method, Bayesian statisticsand fuzzy model [1]). Billard and Diday [3,6] defined regression analysis for interval data, putting the midpoint value of each interval in the OLS regression model to estimate the parameters. The method using the midpoint of interval data has gained popularity among many researchers who

© Springer International Publishing AG 2016
V.-N. Huynh et al. (Eds.): IUKM 2016, LNAI 9978, pp. 469–480, 2016.
DOI: 10.1007/978-3-319-49046-5_40

subsequently modified or extended this technique for application in many problems in econometrics and financial economics studies such as Neto et al. [7,8], Domingues et al. [10], Piamsuwannakit et al. [12].

In this paper, we use a new method for regression analysis with interval data called the convex combination method. It is a generalization of the center method because the parameters obtained from method with OLS or maximum likelihood estimation will be in the solution set of convex combination technique. We used the real interval data from financial market and put them into the CAPM model to explain the relationship between market index return and stock price return (for detail see William [2] or Lintner [11]).

2 Operation with Interval Arithmetics

Let $P_i = [\underline{P}_i, \overline{P}_i]$, be a lower and higher interval data i, $i = 1, 2, \cdots, n$. we can define arithmetic operations. For addition

$$P_i + P_j = [\underline{P}_i + \underline{P}_j, \overline{P}_i + \overline{P}_j]. \tag{1}$$

For Subtraction

$$P_i - P_j = [\underline{P}_i - \overline{P}_j, \overline{P}_i - \underline{P}_j]. \tag{2}$$

For Multiplication

$$P_i \cdot P_j = [\min \ A, \max \ A] \ , A = \{\underline{P}_i\underline{P}_j, \underline{P}_i\overline{P}_j, \overline{P}_i\underline{P}_j, \overline{P}_i\overline{P}_j\}. \tag{3}$$

For Division, we can define when $\underline{P}_i > 0$, we have

$$1/P_j = [1/\overline{P}_j, 1/\underline{P}_j] \tag{4}$$

$$P_i/P_j = P_i \cdot (1/P_j). \tag{5}$$

Additive and Multiplicative by scalar

$$P_i + a = [\underline{P}_i + a, \overline{P}_i + a] \tag{6}$$

$$a \cdot P_i = \begin{cases} [a \cdot \overline{P}_i, a \cdot \underline{P}_i], & a < 0 \\ 0, & a = 0 \\ [a \cdot \underline{P}_i, a \cdot \overline{P}_i], & a > 0. \end{cases} \tag{7}$$

For logarithm function we can define, if $\underline{P}_i > 0$ then

$$\log P_i = [\log \underline{P}_i, \log \overline{P}_i]. \tag{8}$$

For more detail about interval arithmetics see Moore et al.([4], Chap. 2) or Nguyen et al. [5].

3   Linear Regression Models with Interval Data

In this section, we introduce the past methods for linear regression models with interval data from many researchers and next, we will propose the new method.

3.1   The Center Method

This method was proposed by Billard and Diday [3] with the main idea to use the center of the interval data to make a prediction using regression equation in the simplest form:

$$Y_c = X_c \beta + \varepsilon_c, \tag{9}$$

where $Y_c = \dfrac{\underline{Y} + \overline{Y}}{2}, X_c = (1, \dfrac{\underline{X_1} + \overline{X_1}}{2}, \cdots, \dfrac{\underline{X_m} + \overline{X_m}}{2}), \beta = (\beta_0, \beta_1, \cdots, \beta_m)^T$.
This model can use OLS or maximum likelihood method to estimate parameter β and the coefficient of determination (R^2) is easy to define like an ordinary coefficient of determination (R^2) by

$$R_c^2 = \frac{\sum_{i=1}^{n}(\hat{Y}_{c,i} - \overline{Y}_c)^2}{\sum_{i=1}^{n}(Y_{c,i} - \overline{Y}_c)^2}, \quad \overline{Y}_c = \sum_{i=1}^{n} \frac{Y_{c,i}}{n}. \tag{10}$$

The center and range method was proposed by Lima Neto and De Cavalho [7] in 2008, by adding into the center method one more regression line by using radian for forecasting lower bound and upper bound defined by

$$Y_r = \frac{\overline{Y} - \underline{Y}}{2} \tag{11}$$

$$X_r = \frac{\overline{X} - \underline{X}}{2} \tag{12}$$

and the regression equation for radii is

$$Y_r = X_r \beta_r + \varepsilon_r,$$

and $\hat{Y}_c$ and $\hat{Y}_r$ are used to predict $\underline{Y}$ and $\overline{Y}$ by

$$\hat{\underline{Y}} = \hat{Y}_c - \hat{Y}_r$$
$$\hat{\overline{Y}} = \hat{Y}_c + \hat{Y}_r.$$

In 2010, Neto et al. [8] improved the center and range method by adding some constrain to parameter β for radian Y_r to ensure that $\hat{\underline{Y}} \leq \hat{\overline{Y}}$ and they called it the constrained linear regression model for interval-valued data. In 2015, Piamsuwanakit [12] modified the center method by adding more assumptions from only regression equation in the form $y = ax + b$ to be:

$$x_i = x_{c,i} + \delta_{x_i}, \quad \delta_{x_i} \sim N(0, (\sigma_0 x_{r,i})^2) \tag{13}$$
$$y_i = y_{c,i} + \delta_{y_i}, \quad \delta_{y_i} \sim N(0, (\sigma_0 y_{r,i})^2). \tag{14}$$

Where $(\sigma_0 x_{r,i})^2$ and $(\sigma_0 y_{r,i})^2$ are variances; and x_i, y_i are put into to the regression equation

$$y_{c,i} + \delta_{y_i} = a(x_{c,i} + \delta_{x_i}) + b$$
$$y_{c,i} = ax_{c,i} + b + (a\delta_{x_i} - \delta_{y_i})$$

So that $(a\delta_{x_i} - \delta_{y_i}) \sim N(0, \sigma_0^2(a^2 x_{r,i}^2 + y_{r,i}^2))$, $x_{r,i}$ and $y_{r,i}$ are independent, by maximum likelihood method they optimize

$$\max_{a,b,\sigma_0} L(a,b,\sigma^0|X,Y) = \max_{a,b,\sigma_0} \prod_{i=1}^{n} \left(\frac{1}{\sqrt{2\pi\sigma_0^2(a^2 x_{r,i}^2 + y_{r,i}^2)}} \exp\left[-\frac{(y_{c,i} - ax_{c,i} - b)^2}{2\sigma_0^2(a^2 x_{r,i}^2 + y_{r,i}^2)} \right] \right)$$

(15)

In 2009 Maia et al. [9] applied the center and range method to time series by using AR(p) model, ARMA(p,q) and artificial neural network (ANN) model.

4   Convex Combination Method for Linear Regression Model for Interval-Valued Data

In this model we make generalization of the center method using convex combination defined by

$$Y = \alpha_y \underline{Y} + (1 - \alpha_y)\overline{Y}, \; \alpha_y \in [0,1] \tag{16}$$

$$X_j = \alpha_j \underline{X}_j + (1 - \alpha_j)\overline{X}_j, \; \alpha_j \in [0,1] \; \forall j = 1, 2, \cdots, m \tag{17}$$

$$Y = \beta_0 + \sum_{j=1}^{m} \beta_j X_j + \varepsilon. \tag{18}$$

It is easy to see that if we choose α_y, α_j equal $\frac{1}{2}$, this is the center method. In this method we find parameters $\alpha_y, \; \alpha_j$ and β_j by using OLS

$$\min_{\alpha_y, \alpha_1, \cdots, \alpha_m, \beta_0, \beta_1, \cdots, \beta_m} \sum_{i=1}^{n} \left(Y_i - \beta_0 - \sum_{j=1}^{m} (\beta_j X_{ji}) \right)^2 \tag{19}$$

or maximum likelihood with given assumption about density function $f(\varepsilon; \theta)$ with parameter θ

$$\max_{\alpha_y, \alpha_1, \cdots, \alpha_m, \beta_0, \beta_1, \cdots, \beta_m, \theta} \prod_{i=1}^{n} f \left(Y_i - \beta_0 - \sum_{j=1}^{m} (\beta_j X_{ji}); \theta \right) \tag{20}$$

and we can define coefficient of determination (R^2) by

$$R^2 = \frac{\sum_{i=1}^{n}(\hat{Y}_i - \overline{Y}_t)^2}{\sum_{i=1}^{n}(Y_i - \overline{Y}_t)^2}, \; \overline{Y}_t = \frac{\sum_{i=1}^{n}(\alpha_y \underline{Y}_i + (1 - \alpha_y)\overline{Y}_i)}{n} \tag{21}$$

5   Convex Combination Method for Autoregressive Model for Interval-Valued Data

By the center method, we can write AR(p) model by

$$X_{c,t} = \phi_{c0} + \sum_{i=1}^{p} \phi_{ci} X_{c,t-i} + \varepsilon_{c,t} \tag{22}$$

where $X_{c,t}$, $t = 1, 2, \cdots, N$ are the time series data. In this model we generalize the center method using convex combination defined by

$$X_t = \alpha_0 \underline{X}_t + (1 - \alpha_0)\overline{X}_t, \qquad \alpha_0 \in [0,1],\ t = p+1, p+2, \cdots, N$$

$$X_{t-1} = \alpha_1 \underline{X}_{t-1} + (1 - \alpha_1)\overline{X}_{t-1}, \qquad \alpha_1 \in [0,1],\ t = p, p+1, \cdots, N-1$$

$$X_{t-2} = \alpha_2 \underline{X}_{t-2} + (1 - \alpha_2)\overline{X}_{t-2}, \qquad \alpha_2 \in [0,1],\ t = p-1, p, \cdots, N-2$$

$$\vdots$$

$$X_{t-p} = \alpha_p \underline{X}_{t-p} + (1 - \alpha_p)\overline{X}_{t-p}, \qquad \alpha_p \in [0,1],\ t = 1, 2, \cdots, p.$$

Thus, the convex combination method for AR(p) model is

$$\alpha_0 \underline{X}_t + (1 - \alpha_0)\overline{X}_t = \phi_0 + \sum_{i=1}^{p} \phi_i \left(\alpha_i \underline{X}_{t-i} + (1 - \alpha_i)\overline{X}_{t-i} \right) + \varepsilon_t \tag{23}$$

or

$$X_t = \phi_0 + \sum_{i=1}^{p} \phi_i X_{t-i} + \varepsilon_t \tag{24}$$

We use (23) or (24) to forecast by using expectation of X_t from the assumption of OLS or maximum likelihood about $\mathbb{E}(\varepsilon_t) = 0$

$$\mathbb{E}(X_t) = \phi_0 + \sum_{i=1}^{p} \phi_i X_{t-i} \tag{25}$$

It is easy to see that if we choose $\alpha_0 = \alpha_1 = \cdots = \alpha_p = \dfrac{1}{2}$, this is the center method. By this method we can find parameters $(\alpha_0, \alpha_1, \cdots, \alpha_p)$ and $(\phi_0, \phi_1, \cdots, \phi_p)$ by using OLS

$$\min_{\alpha_0, \alpha_1, \cdots, \alpha_p, \phi_0, \phi_1, \cdots, \phi_p} \sum_{t=p+1}^{N} \left(X_t - \phi_0 - \sum_{j=1}^{p} \phi_j X_{t-j} \right)^2 \tag{26}$$

or maximum likelihood by adding some assumptions about density function $f(\varepsilon; \theta)$ with parameter θ. For example $f(x; \theta)$ is normal with $\mu = 0$ and variance $= \sigma^2$.

$$\max_{\alpha_0, \alpha_1, \cdots, \alpha_p, \phi_0, \phi_1, \cdots, \phi_p, \theta} \prod_{t=p+1}^{N} f \left(X_t - \phi_0 - \sum_{j=1}^{p} \phi_j X_{t-j}; \theta \right) \tag{27}$$

6   Application to CAPM Model

Let P_i be an interval price of stock or interval of market index at week i, $i = 1, 2, \cdots, n$, so that we can define interval return of market index $r_{m,i}$ and interval return of stock $r_{s,i}$ and the risk free $r_{f,i}$ which is the point data by

$$r_{m,i} = \log \frac{P_{m,i}}{P_{m,i-1}} = \left[\log(\frac{\underline{P}_{m,i}}{\overline{P}_{m,i-1}}), \log(\frac{\overline{P}_{m,i}}{\underline{P}_{m,i-1}}) \right] = [\underline{r}_{m,i}, \overline{r}_{m,i}] \qquad (28)$$

$$r_{s,i} = \log \frac{P_{s,i}}{P_{s,i-1}} = \left[\log(\frac{\underline{P}_{s,i}}{\overline{P}_{s,i-1}}), \log(\frac{\overline{P}_{s,i}}{\underline{P}_{s,i-1}}) \right] = [\underline{r}_{s,i}, \overline{r}_{s,i}] \qquad (29)$$

$$r_{m,i} - r_{f,i} = [\underline{r}_{m,i} - r_{f,i}, \overline{r}_{m,i} - r_{f,i}] \qquad (30)$$

$$r_{s,i} - r_{f,i} = [\underline{r}_{s,i} - r_{f,i}, \overline{r}_{s,i} - r_{f,i}] \qquad (31)$$

We put 3 variables $r_{m,i}$, $r_{s,i}$ and $r_{f,i}$ into the CAPM model. In finance, we use this model to explain the relationship between return of market and return of stock price in terms of linear regression model defined by

$$r_s - r_f = \beta_0 + \beta_1(r_m - r_f) + \varepsilon \qquad (32)$$

$$\mathbb{E}(r_s - r_f) = \beta_0 + \beta_1 \mathbb{E}(r_m - r_f) \qquad (33)$$

where r_s is return of stock price, r_m return of market index, r_f risk free rate of return and assuming that ε_i is $N(0, \sigma^2)$, then by maximum likelihood method we have

$$\max_{\alpha_s, \alpha_m, \beta_0, \beta_1, \sigma^2} \prod_{i=1}^{n} f\left(r_{s,i} - \beta_0 - \beta_1 r_{m,i}; \sigma^2\right)$$

where $r_{s,i} = \alpha_s \underline{r}_{s,i} + (1 - \alpha_s)\overline{r}_{s,i} - r_f$, $r_{m,i} = \alpha_m \underline{r}_{m,i} + (1 - \alpha_m)\overline{r}_{m,i} - r_f$. For the real data we use weekly data stock price of Airports of Thailand Public Company Limited(AOT), CP ALL Public Company Limited (CPALL) and Thailand Stock market index by SET50 by Bloomberg database from May 5, 2006 to 29 April, 2015. For the risk free we used Thailand government bond 10 years at the same time. The summary statistics are shown in Table 1.

In Table 2, we show the values of parameter $\alpha_m, \alpha_s, \beta_0, \beta_1$ and σ from CAPM estimation with the use of interval data, obtained from Convex Combination Method (CCM) with maximum likelihood, of AOT, SET50 returns and the fixed risk free. The result can be written in equation form;

$$r_{AOT} = .7373 \cdot \underline{r}_{AOT} + (1 - .7373) \cdot \overline{r}_{AOT}$$

$$r_{set50} = .8784 \cdot \underline{r}_{set50} + (1 - .8784) \cdot \overline{r}_{set50}.$$

The CAPM for predicting return of AOT given market return and risk free is

$$r_{AOT} - r_f = .0044 + 1.0928(r_{set50} - r_f)$$

Table 1. Summary statistics

	AOT		CPALL		SET50		Risk free
	Low	High	Low	High	Low	High	
Mean	$-.0556$	.0695	$-.0539$	.0615	$-.0332$	.0360	.0007
Median	$-.0445$	.0615	$-.0435$	.0532	$-.0270$	.0306	.0007
Min	$-.2822$	$-.0028$	$-.2529$	.0057	$-.1769$	$-.0154$	.0005
Max	.0231	.2984	.0168	.2601	.0278	.1475	.0008
SD	.0450	.0464	.0401	.0393	.0290	.0236	.0001
Skewness	-1.7185	1.3727	-1.6011	1.2952	-1.6043	1.1072	$-.4704$
Kurtosis	7.2123	5.7115	6.7166	5.3323	6.7512	4.8441	2.5296
Observation	313						

Table 2. Estimated parameters

	AOT				CPALL			
	CM		CCM		CM		CCM	
	Value	SD	Value	SD	Value	SD	Value	SD
α_s	.5	-	0.7373	0.0201	.5	-	0.6847	0.0323
α_m	.5	-	0.8784	0.0529	.5	-	0.5670	0.0694
β_0	0.0054	0.0018	0.0044	0.0159	0.0025	0.0016	0.0049	0.0359
β_1	1.1207	0.0331	1.0928	0.0579	0.8200	0.0532	0.7880	0.0583
σ	0.0276	0.0012	0.0251	0.0016	0.0262	0.0010	.0259	0.0011
$LogLike$	679.96	-	709.21	-	695.69	-	699.35	-
AIC	$-1,353.9$	-	$-1,408.4$	-	$-1,385.4$	-	$-1,388.7$	-
R^2	0.4231	-	0.5682	-	0.3027	-	0.3289	-

and by Center Method(CM) $\alpha_s = \alpha_m = .5$, we have the CAPM as

$$r_{AOT} - r_f = .0054 + 1.1228(r_{set50} - r_f)$$

The CAPM by CCM for returns of CPALL, SET50 and risk free is

$$r_{CPALL} = .6847 \cdot \underline{r}_{CPALL} + (1 - .6847) \cdot \overline{r}_{CPALL}$$
$$r_{set50} = 0.5670 \cdot \underline{r}_{set50} + (1 - 0.5670) \cdot \overline{r}_{set50}.$$

The CAPM for predicting return of CPALL given market return and risk free is

$$r_{AOT} - r_f = .0049 + .7880(r_{set50} - r_f)$$

and by using the center method (CM), we have the CAPM as

$$r_{AOT} - r_f = .0025 + 0.8200(r_{set50} - r_f)$$

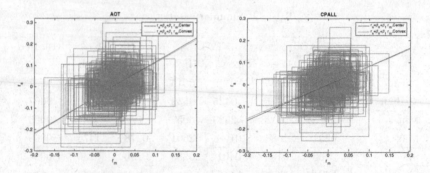

Fig. 1. Interval graph and regression line.

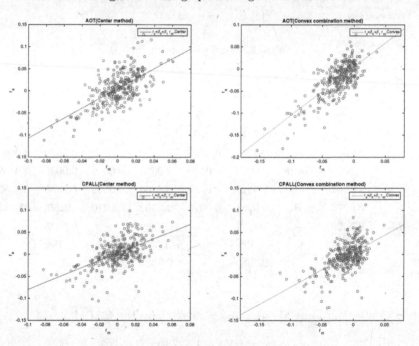

Fig. 2. Scatter plot from Center method and Convex combination method with regression line.

We can compare the performance of CAPM from CCM and CM data set, upon minimum AIC and R^2 value. From Table 2, AIC value from CCM is less than AIC value from CM and also R^2 from CCM is higher than R^2 from CM for both AOT and CPALL.This means that the Convex Combination Method is better than the Center Method by measure of AIC and R^2.

Figure 1 (left) is the graph of interval return data of SET50 and AOT with prediction line by CAPM with risk free being zero. From CCM and CM. and Fig. 1 (right) is the graph of interval return data of SET50 and CPALL with prediction line by CAPM with risk free being zero too.

Fig. 2 is the scatter plot of data point (SET50,AOT) and (SET50,CPALL). These data points are selected by CM and CCM and predicted by CAPM.

7 Application to AR(p) Model

Let P_t be an interval stock price at time t, $t = 1, 2, \cdots, n$, so that we can define interval return of stock r_t by

$$r_t = \log \frac{P_t}{P_{t-1}} = \left[\log(\frac{\underline{P_t}}{\overline{P}_{t-1}}), \log(\frac{\overline{P_t}}{\underline{P}_{t-1}}) \right] = [\underline{r}_t, \overline{r}_t] \qquad (34)$$

We put the variable r_i into the AR(p) model. For the real data we use weekly stock price data of PTT PUBLIC COMPANY LIMITED (PTT) and THE SIAM CEMENT PUBLIC COMPANY LIMITED (SCC) from Thailand stock market

Table 3. Summary statistics

	PTT		SCC	
	Low	High	Low	High
Mean	−.0492	.0491	−.0467	.0508
Median	−.0407	.0426	−.0406	.0434
Min	−.2380	−.0122	−.1922	−.0159
Max	.0087	.2299	.0196	.2299
SD	.0384	.0357	.0326	.0352
Skewness	−1.8453	1.3256	−1.3323	1.3711
Kurtosis	8.4195	6.0053	5.5891	6.0544
Observation	313			

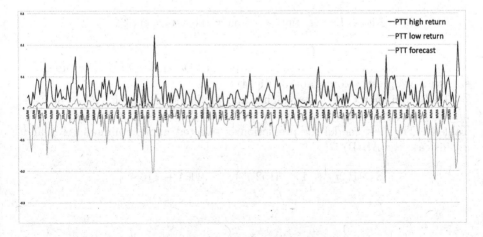

Fig. 3. Interval time series on weekly return of PTT.

Table 4. Estimation of parameter from PTT.

PTT												
	AR(1)			AR(2)			AR(3)			AR(4)		
	θ	SD		θ	SD		θ	SD		θ	SD
α_0	0.4152	0.0537	α_0	0.4116	0.4495	α_0	0.4144	0.0499	α_0	0.4222	0.6596
α_1	0.0000	0.3188	α_1	0.0000	0.3839	α_1	0.0000	0.4787	α_1	0.0000	0.3299
ϕ_0	-0.0004	0.0044	α_2	0.3698	2.6156	α_2	0.1962	0.9806	α_2	0.0002	1.0090
ϕ_1	0.1768	0.0525	ϕ_0	0.0003	3.3951	α_3	0.3896	2.3423	α_3	0.7894	4.6077
σ	0.0286	0.0012	ϕ_1	0.1806	0.1766	ϕ_0	0.0010	0.0136	α_4	0.0000	1.7376
			ϕ_2	-0.0377	0.2399	ϕ_1	0.1797	0.2517	ϕ_0	0.0023	0.0510
			σ	0.0286	0.0822	ϕ_2	-0.0327	0.1234	ϕ_1	0.1743	0.1844
						ϕ_3	-0.0300	0.1069	ϕ_2	-0.0307	0.2039
						σ	0.0285	0.0020	ϕ_3	-0.0230	0.1480
									ϕ_4	-0.0430	0.9369
									σ	0.0284	0.0557
LL	666.5895		LL	664.2163		LL	662.8765		LL	662.4693	
AIC	-1323.1790		AIC	-1314.4326		AIC	-1307.753		AIC	-1302.9386	

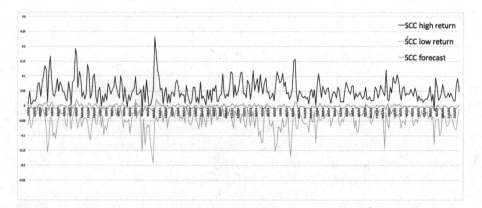

Fig. 4. Interval time series on weekly return of SCC.

by Bloomberg database from January 2010 until December 2015. The summary statistics are presented in Table 3 (Fig. 4).

The graph shows the lower, higher and predicted return of PTT. The estimated parameters are from AR(P) model by MLE and CCM method (see Table 4). The best model for prediction is selected upon minimum AIC, and in this case the AR(P) is

$$\mathbb{E}\left(.4152\underline{r}_t + (1 - .4152)\overline{r}_t\right) = -.0004 + .1768\overline{r}_{t-1}$$

Table 5. Estimation of parameter from SCC.

SCC

	AR(1)			AR(2)			AR(3)			AR(4)	
	θ	SD		θ	SD		θ	SD		θ	SD
α_1	0.4011	0.0313	α_1	0.4423	0.2581	α_1	0.4050	0.1917	α_1	0.3761	0.4466
ϕ_0	−0.0047	0.1067	α_2	0.5682	0.3623	α_2	0.3597	0.3202	α_2	0.3615	1.0382
ϕ_1	0.2146	0.0522	ϕ_0	−0.0042	0.0043	α_3	0.9999	0.2669	α_3	0.9998	1.3670
σ	0.0265	0.0011	ϕ_1	0.2370	0.0532	ϕ_0	−0.0040	0.0045	α_4	0.9986	1.0741
			ϕ_2	−0.0993	0.1659	ϕ_1	0.2349	0.0576	ϕ_0	−0.0031	0.0112
			σ	0.0264	0.0011	ϕ_2	−0.0972	0.0554	ϕ_1	0.2313	0.0577
						ϕ_3	−0.0362	0.0627	ϕ_2	−0.0899	0.0566
						σ	0.0263	0.0010	ϕ_3	−0.0527	0.0758
									ϕ_4	0.0402	0.0759
									σ	0.0263	0.0011
LL	690.5899		LL	689.5105		LL	688.2018		LL	686.0661	
AIC	−1371.1798		AIC	−1365.021		AIC	−1358.4036		AIC	−1350.1322	

The graph shows the lower, higher and predicted return of SCC. The estimated parameters are from AR(P) model by MLE and CCM method (see Table 5). The best model for prediction is selected upon minimum AIC, and in this case the AR(P) is

$$\mathbb{E}\left(.5444\underline{r}_t + (1 - .5444)\bar{r}_t\right) = -.0047 + .2146\left(.4011\underline{r}_{t-1} + (1 - .4011)\bar{r}_{t-1}\right)$$

8 Conclusions and Future Research

Convex combination method is a generalization of the center method. This method is easy to implement for solving regression problems with interval data. For the next future research, we will apply this method to interval data based on other econometric problems or use this method with Bayesian statistic.

Acknowledgement. We are grateful for financial support from Puey Ungphakorn Center of Excellence in Econometrics, Faculty of Economics and Graduate School, Chiang Mai University.

References

1. Tanaka, H., Uejima, S., Asai, K.: Linear regression analysis with fuzzy model. IEEE Trans. Syst. Man Cybern. **12**(6), 903–907 (1982)
2. William, F.: Capital asset prices a theory of market equilibrium under conditions of risk. J. Financ. **19**(3), 425–442 (1964)
3. Billard, L., Diday, E.: Regression analysis for interval-valued data. In: Kiers, H.A.L., Rasson, J.-P., Groenen, P.J.F., Schader, M. (eds.) Data Analysis, Classification, and the Related Methods. Studies in Classification, Data Analysis, and Knowledge Organization, pp. 369–374. Springer, Heidelberg (2000)

4. Moore R.E., Kearfott R.B., Cloud M.J.: Introduction to Interval Analysis, pp. 7–18. Siam, Philadelphia (2009)
5. Nguyen, H.T., Kreinovich, V., Wu, B., Xiang, G.: Computing Statistics Under Interval and Fuzzy Uncertainty. Studies in Computational Intelligence, vol. 393. Springer, Heidelberg (2012)
6. Billard, L., Diday, E.: Symbolic regression analysis. In: Jajuga, K., Sokołowski, A., Bock, H.-H. (eds.) Classification, Clustering, and Data Analysis. Studies in Classification, Data Analysis, and Knowledge Organization, pp. 281–288. Springer, Heidelberg (2002)
7. Neto, E.A.L., Carvalho, F.A.T.: Centre and range method for fitting a linear regression model to symbolic interval data. Comput. Stat. Data Anal. 52, 1500–1515 (2008)
8. Neto, E.A.L., Carvalho, F.A.T.: Constrained linear regression model for symbolic interval-valued. Comput. Stat. Data Anal. 54, 333–347 (2010)
9. Maia, A.L.S., Carvalho, F.D.A., Ludermir, T.B.: Forecasting models for interval-valued time series. Neurocomputing 71(16), 3344–3352 (2008)
10. Domingues, M.A.O., Souza, R.M.C.R., Cysneiros, F.J.A.: A robust method for linear regression of symbolic interval data. Pattern Recogn. Lett. 31, 1991–1996 (2010)
11. John, L.: The valuation of risk assets and selection of risky investments in stock. Rev. Econ. Stat. 47(1), 13–37 (1965)
12. Piamsuwannakit, S., Autchariyapanitkul, K., Sriboonchitta, S., Ouncharoen, R.: Capital asset pricing model with interval data. In: Huynh, V.-N., Inuiguchi, M., Denoeux, T. (eds.) IUKM 2015. LNCS (LNAI), vol. 9376, pp. 163–170. Springer, Heidelberg (2015). doi:10.1007/978-3-319-25135-6_16

A Copula-Based Markov Switching Seemingly Unrelated Regression Approach for Analysis the Demand and Supply on Sugar Market

Pathairat Pastpipatkul[(✉)], Nisit Panthamit, Woraphon Yamaka, and Songsak Sriboochitta

Faculty of Economics, Chiang Mai University, Chiang Mai, Thailand
ppthairat@hotmail.com, woraphon.econ@gmail.com

Abstract. This paper conducted a Markov switching seemingly unrelated regression without assuming a normal distribution of the error term. We proposed the use of both Archimedean and Elliptical copula classes to join the different marginal of the system equations. The results show that normal distribution for both demand and supply equations and joint distribution by Frank copulas present the lowest AIC and BIC. Moreover, the model is, then, applied for estimating the demand and supply in Thai sugar market. Thai export price and Brazil's export price were found to be the factors affecting the demand and supply of the Thai sugar market. Finally, the results on smoothed probabilities indicate the over-supply condition in Thai sugar market along our sample period.

Keywords: Markov switching seemingly unrelated regression · Copula · Demand and supply · Thai sugar market

1 Introduction

Thailand is the second largest exporter of sugar in the world, after Brazil and accounts for 11 % of the global market shares. Brazil's dominant position in the world sugar economy implies that domestic sugar production affects the global sugar market balance and prices as well as Thai sugar market. In the last decade, world sugar kept fluctuating due to either the oversupply or overdemand in the world market. Moreover, increasing in the demand of the mandated ethanol blend in gasoline substantial fuelled sugar prices. According to Foreign Agricultural Service (2015), Consumption is projected to reach a record 173 million tons and global imports are expected to grow to a record 52 million tons.

Considering Thai sugar market, the consumption trend is continue growth along the decade according to the growing demand of food and beverage industries. In addition, more than a half of annual sugarcane production in Thailand diverted to ethanol production. Thai sugar Exports are forecasted to be reach 8.8 million tons. These indicate that Thai sugar will become an important in the agriculture sector of Thailand.

Thai sugar price is tied to the world market price. Therefore, Thai government has to take measures both to maintain the high sugar producer price and to increase the export of sugar. There are more than 100,000 small holders who plant sugarcane in all regions of Thailand thus deriving the demand and supply functions of the Thai sugar market

© Springer International Publishing AG 2016
V.-N. Huynh et al. (Eds.): IUKM 2016, LNAI 9978, pp. 481–492, 2016.
DOI: 10.1007/978-3-319-49046-5_41

will benefit the government as the information will be pertinent for the design and implementation of appropriate policies. However, the agriculture market seems to present the non-linearity in the responsiveness of supply and demand to fluctuations in their price, as mention in Deaton and Laroque, [3], Balcombe and Rapsomanikis [2], and Fahmy [5], and Lin [10]. Therefore, the implementation of linear model might not appropriate to explain Thai sugar market behavior.

Thus, the principal efforts of this paper are first, extending the regime-switching model to the system equation, SUR, and secondly, providing an extension of MS-SUR model by relaxing typical correlated errors assumption, and thirdly, deriving the demand and supply configurations of Thai sugar market in two different regimes, including market upturn regime and market downturn regime.

Deriving the demand and supply equations is crucial for many economic market studies that intend to determine market equilibrium and elasticity of any products. Commonly, the multiple equations or system equations such as seemingly unrelated regression (SUR), Two-stage least squares (2SLS), Three-stage least squares (3SLS) are proposed to estimate all equations simultaneously [6]. Thus, in the past decade, these methods have been widely applied to derive demand and supply equations of several markets e.g. Pryce [17], Angrist ans Krueger [1], and Moghaddasi and Azizi (2011), Lin [12]. However, these studies seem failing to provide accurate results due to the strong assumption of normal distribution $\varepsilon_T \sim N(0, \Sigma \otimes I_T)$. To deal with this strong assumption, Pitt, Chan, Kohn [16], and Masarotto and Varin [11] proposed a flexible approach, Gaussian copula models for marginal regression model, to develop the traditional regression models with normal correlated errors. This approach becomes more flexible and useful since it allows us to model the non-normal correlated errors. Thus, in this study, we applied this method using both Archimedean and Elliptical classes to SUR model of Zellner [20] in order to enhance the model's efficiency and accuracy. However, the original Zellner model is not appropriate for capturing market that behave different over time. To capture more reality in the estimation, the Markov-switching approach of Hamilton [7] is, then, considered as the extension of copula based SUR model (see, [14]).

The Markov switching model has recently gained popularity in non-linear time series modeling since it can characterize the different behaviors in different time periods and can capture the dynamic change in the data series [9]. Thus, this study aimed to combine the system equation based SUR model with the Markov switching technique and proposed as Markov switching SUR model. Although, there are some studies which also work on the regime switching with system equations, such as the Markov switching simultaneous equations of Yoon [19] and Markov switching, Bayesian, Seemingly unrelated regression of Rotaru [18] and Pastpipatkul, Maneejuk, and Sriboonchitta [13], these papers still assume a normal distribution on the structure of the error. Hence, we developed a new estimation technique to implement in the likelihood of the MS-SUR model for achieving accuracy of the estimated parameters of the system equations with regime switching. Consequently, the model will become more flexible to use since it can capture the different behavior of the time series data as well as relax the assumption of the normal distribution in the structural disturbances of the multiple equations. Thus, in this paper, we proposed the copula based MS-SUR model to study and derive the demand and supply configurations in Thai sugar market.

The remainder of this paper is organized follows: Sect. 2 provides a methodology, Sect. 3 elaborate an estimation procedure; Sect. 4, simulation study and finally, the application of the model and conclusions are provided in Sects. 5 and 6.

2   Methodology

2.1   Markov Switching Seemingly Unrelated Regression (MS-SUR) Model

The SUR model that was introduced by Zellner [20], proposed the estimation of the multiple correlated equations to improve estimation efficiency. The study extends the Markov switching model of Hamilton [7] to SUR model and obtain the Markov switching seemingly unrelated regression (MS-SUR) model. Thus, the present general formula of two equations can be described as follows

$$
\begin{aligned}
y_{1,t} &= \beta_1(s_t) + \beta_2(S_t)X_{1,1t} + \ldots + \beta_j(S_t)X_{j,1t} + \in_{1,t}(S_t) \\
y_{2,t} &= \gamma_1(s_t) + \gamma_2(S_t)X_{1,2t} + \ldots + \gamma_j(S_t)X_{j,2t} + \in_{2,t}(S_t)
\end{aligned}
\tag{1}
$$

with

$$
\begin{aligned}
\in_{1,t} &\sim N(0, \sigma^2_{1,S_t}) \\
\in_{2,t} &\sim N(0, \sigma^2_{2,S_t})
\end{aligned}
\tag{2}
$$

$$
\Sigma_{S_t} = \begin{bmatrix}
\sigma^2_{1,S_t} & \sigma^2_{1,2S_t} & \cdots & \sigma^2_{1,k,S_t} \\
\sigma^2_{2,1S_t} & \ddots & & \vdots \\
& & \ddots & \vdots \\
\sigma^2_{k,1,S_t} & \cdots & \cdots & \sigma^2_{k,k_t}
\end{bmatrix}
\tag{3}
$$

where $S_t = \{1, \ldots, k\}$ is unobserved state variable with k regimes being governed by first order Markov process that is defined by the transition probabilities matrix P and $y_{i,t}$ and $X_{i,t}$ are denote as dependent and regressor variable i, respectively. $\in_i$ are the error term which assumed to have a normal distribution. To estimate the MS-SUR model, the maximum likelihood is employed. If the values of S_t are known, then the log likelihood of MS-SUR with k = 2 regimes can be written as

$$
\ln L = \sum_{t=1}^{T} [\sum_{k=1}^{K} \frac{MT}{2} \ln(2\pi) - \frac{T}{2} \ln |\Sigma_t| \{ -\frac{1}{2}(Y_t - (B_1 S_t + B_0(1 - S_t))X_t)' \Sigma_t^{-1}
$$
$$
(Y_t - (\beta_1 S_t + \beta_0(1 - S_t))X_t)\} p_{11}^{s_t s_{t-1}}(1 - p_{11})^{(1-s_t)s_{t-1}} p_{00}^{(1-s_t)(1-s_{t-1})}(1 - p_{00})^{s_t(1-s_{t-1})}]
\tag{4}
$$

But, if the values of S_t are unknown, let $\Theta =, \beta_i(S_t), \gamma_i(S_t), \sum(S_t)$ thus the full log-likelihood function can be rewritten as

$$
\ln L = \sum_{t=1}^{T} (\ln \sum_{j=1}^{2} (f(y_t|S_t = j, \Theta) \Pr(S_t = j))
\tag{5}
$$

Where the weight of likelihood function in each state (k) is given by the state's probabilities. However, these state's probabilities are unknown thus the Hamilton's filter is used to estimate the filter probabilities ($\Pr(S_t = j)$) of each state based on the available information set.

Considering φ_t as the matrix of available information at time t. The Hamilton's filter, as provided in Perlin [15], is determined using the following algorithm.

1. Given an initial guess of transition probabilities P_{ij} which are the probabilities of switching between regimes

$$P_{ij} = \Pr(S_{t+1} = j | S_t = i) \text{ and } \sum_{j=1}^{k} p_{ij} = 1 \quad i, j = 1, \dots, k \tag{6}$$

Thus the transition probabilities in the transition matrix Q is,

$$Q = \begin{bmatrix} p_{11} & p_{21} & \cdots & p_{k1} \\ p_{12} & p_{22} & \cdots & p_{k2} \\ \vdots & \vdots & \cdots & \vdots \\ p_{1k} & p_{2k} & \cdots & p_{kk} \end{bmatrix} \tag{7}$$

2. Updating the transition probabilities of each state with the past information including the parameters in the system equation, Θ_{t-1} and P_{ij}, for calculating the likelihood function in each state ($f(y_t | S_t = j, \varphi_{t-1})$) for time t. After that, the probability of each state is to be updated by the following formula

$$\Pr(S_t = j | \varphi_t) = \frac{f(y_t | S_{t=j}, \varphi_{t-1}) \Pr(S_{t=j} | \varphi_{t-1})}{\sum_{j=1}^{k} f(y_t | S_{t=j}, \varphi_{t-1}) \Pr(S_{t=j} | \varphi_{t-1})} \tag{8}$$

3. Iterating step 1 and 2 for $t = 1, \dots, T$.

2.2 Basic Copula Approach

Sklar [17] has proposed the link between the joint distributions and its margins which is possible to have different distributions but different dependence structure. The linkage between the marginal distributions can be called as a copula. Any d-dimensional joint distribution H for continuous random variables x_n with marginal distribution functions F_n can be decomposed to copula C. Hence, the d-copula C such that for all x_n is given as

$$H(x_1, \dots, x_d) = C(F_1(x_1), \dots, F_d(x_d)) \tag{9}$$

where C is copula distribution function of a d-dimensional random variables. If the marginals are continuous, C is unique. Equation 9 defines a multivariate distribution functions F_n. Thus, we can model the marginal distribution and joint dependence separately. If we have a continuous marginal distribution, the copula can be determined by

$$C(u_1, \ldots, u_d) = C(F_1^{-1}(u_1), \ldots, F_d^{-1}(u_d)) \tag{10}$$

where u is uniform $[0, 1]$. And a general $d-$ dimensional density copula can be expressed as

$$h(x_1, \ldots, x_d) = c(F_1(x_1), \ldots, F_d(x_d)) \tag{11}$$

There are two widely used copula classes namely Elliptical copulas and Archimedean copulas.

(1) Elliptical copulas are dependence structure with symmetric data. Gaussian or normal copula and student-t-copula are the copula families in this class. The different between these two families is that Gaussian copula cannot capture the tail dependence whereas student-t-copula can do.
(2) Archimedean copulas including Clayton, Gumbel, Joe, are used to obtain an asymmetric and allow us to model dependence in high dimensions with only one parameter.

Therefore, in this study, we consider falling into the six families of copulas; Gaussian, student-t, Frank, Clayton, Gumbel, and Joe. The density of these copulas as proposed in Embrechts, Lindskog, and McNeil [4] and Hofert, Machler, and McNeil [8].

3 Estimation of MS-SUR Based Copula

In this study, the MS-SUR model with two regimes, high growth market and low growth market, and two equations, demand and supply, is proposed to study the Thai sugar market. Thus, we can extend the MS-SUR model as:

$$
\begin{aligned}
Q_t^d(S_t = 1) &= \beta_1 + \beta_2 P_t^{export} + \beta_3 P_t^{Bra} + \in_{1,1t} \\
Q_t^d(S_t = 2) &= \alpha_1 + \alpha_2 P_t^{export} + \alpha_3 P_t^{Bra} + \in_{1,2t} \\
Q_t^s(S_t = 1) &= \gamma_1(s_t) + \gamma_2 P_t^{export} + \gamma_3 W_t + \in_{2,1t} \\
Q_t^s(S_t = 2) &= \delta_1(s_t) + \delta_2 P_t^{export} + \delta_3 W_t + \in_{2,2t}
\end{aligned}
\tag{12}
$$

The advantage of the copula is it can measure joint marginal distribution functions which are possible to have different distributions. Hence, in this study, we aim to eliminate the strong restriction of normal distribution in all equations by allowing the residuals ($\in$) of demand and supply equations to have a different distribution. To estimate the MS-SUR model based copula, we have to follow four steps. In the first step, the conventional MS-SUR model is estimated by the maximum likelihood method to obtain the initial values. In the second step, we construct the MS-SUR copula likelihood using the chain rule, we have

$$
\begin{aligned}
\frac{\partial^2}{\partial u_1(S_t)\partial u_2(S_t)} F(u_1(S_t), u_2(S_t)) &= \frac{\partial^2}{\partial u_1(S_t)\partial u_2(S_t)} C(F_1(u_1(S_t)), F_1(u_1(S_t))) \\
&= f_1(u_1(S_t))f_2(u_2(S_t))c(F_1(u_1), F_2(u_2))
\end{aligned}
\tag{13}
$$

where $u_1(S_t)$ and $u_2(S_t)$ are the marginals assumption of normal or student-t distribution for $S_t = 1, 2$ which are transformed form variable demand and supply and have uniform

$(0,1). f_1(u_1(S_t))$ and $f_2(u_2(S_t))$ are marginals of demand and supply. $c(F_1(u_1(S_t)), F_2(u_2(S_t)))$ are density functions of Gaussian, T, Frank, Clayton, Gumbel, and Joe copulas for constructing the jointly demand and supply equations.

By taking the logarithm in Eq. 13, and constructing the full likelihood function of copula based MS-SUR by multiplying the likelihood in Eq. 13 with Eq. 8, we get

$$\ln L = \sum_{t=1}^{T} (\ln \sum_{j=1}^{2} (f(y_t | S_t = j, \Theta) \Pr(S_t = j | \varphi_t)) \tag{14}$$

where $f(y_t | S_t = j, \Theta)$ is the joint density from Eq. 13 and $\Pr(S_t = j | \varphi_t))$ is the filter probabilities form Eq. 8. Θ denotes all parameters in copula based MS-SUR.

4   Simulation Study

In the simulation study, we apply elliptical copulas (Gaussian) and Archimedean copulas (Frank), to model the dependence structure of the MS-SUR. In this study, the simulation is the realization of MS-SUR with two equations, thus we generate random data from the following model specifications (Tables 1 and 2):

Table 1. Estimation results of bivariate elliptical gaussian copula MS-SUR

Case	True value	Gaussian N-N	Gaussian N-T	Gaussian T-N	Gaussian T-T
$\beta_{1,S=1}$	−1	−1.0175 (0.0541)	−1.0078 (0.0543)	−1.0115 (0.0684)	−0.9964 (0.0681)
$\beta_{2,S=1}$	2	1.9231 (0.0542)	1.9238 (0.0541)	1.9243 (0.0655)	1.9265 (0.0654)
$\beta_{1,S=2}$	1.5	1.5279 (0.0635)	1.5345 (0.0641)	1.5287 (0.0614)	1.5312 (0.0625)
$\beta_{2,S=2}$	3	3.0488 (0.0564)	3.0489 (0.0564)	3.0541 (0.0544)	3.0549 (0.0544)
$\alpha_{1,S=1}$	-2	−2.1228 (0.0794)	−2.1248 0.918	−2.1176 (0.0831)	−2.0618 (0.0781)
$\alpha_{2,S=1}$	2	1.8964 (0.0713)	0.916 (0.088)	1.8969 (0.0714)	1.8523 (0.0689)
$\alpha_{1,S=2}$	1	0.9414 (0.0.651)	0.9623 (0.0664)	0.9451 (0.0245)	0.9567 (0.0645)
$\alpha_{2,S=2}$	2	2.0444 (0.0588)	2.0386 (0.0579)	2.0422 (0.0589)	2.0377 (0.0581)
$\sigma^2_{1,S=1}$	0.8	0.7930 (0.0384)	07973 (0.0389)	1.0100 (0.0346)	1.0100 (0.0522)
$\sigma^2_{1,S=2}$	0.9	1.0698 (0.0476)	0.918 (0.0485)	1.0542 (0.0448)	1.0100 (0.0652)
$\sigma^2_{2,S=1}$	1	1.0483 (0.04518)	1.0611 (0.0461)	1.0100 (0.0511)	1.0100 (0.0541)
$\sigma^2_{2,S=2}$	2	2.1631 (0.0563)	2.0100 (0.0552)	2.2229 (0.0654)	2.0100 (0.0566)
$v_{1,S=1}$	5	n.a.	n.a.	4.9845 (0.0414)	5.3361 (1.1816)
$v_{1,S=2}$	5	n.a.	6.9638 (2.6606)	n.a.	5.3608 (1.8578)
$v_{2,S=1}$	10	n.a.	n.a.	10.1114 (0.0577)	12.4631 (1.9038)
$v_{2,S=2}$	10	n.a.	10.1364 (0.0480)	n.a.	11.2188 1.0379
$\theta_{c,S=1}$	0.5	0.5236 (0.0445)	0.5395 (0.0440)	0.5069 (0.0413)	0.5199 (0.0517)
$\theta_{c,S=2}$	0.5	0.4133 (0.0569)	0.4209 (0.0575)	0.4998 (0.0588)	0.5059 (0.0586)
P_{11}	0.95	0.9594 (0.0121)	0.9622 (0.0117)	0.9589 (0.0126)	0.9616 (0.0115)
P_{22}	0.95	0.9486 (0.0152)	0.9517 (0.0148)	0.9483 (0.0155)	0.9513 (0.0155)

Source: Calculation (Note: N = Normal margin, T = Student-t margin)

Table 2. Estimation results of bivariate archimedean frank copula MS-SUR

Case	True value	Gaussian N-N	Gaussian N-T	Gaussian T-N	Gaussian T-T
$\beta_{1,S=1}$	-1	-1.0492 (0.0549)	-1.0356 (0.0546)	-1.0540 (0.0685)	-1.0335 (0.0658)
$\beta_{2,S=1}$	2	1.9269 (0.0550)	1.9269 (0.0553)	1.9303 (0.0642)	1.9304 (0.0663)
$\beta_{1,S=2}$	1.5	1.5067 (0.0648)	1.5108 (0.0651)	1.5088 (0.0617)	1.5088 (0.0623)
$\beta_{2,S=2}$	3	3.0407 (0.0582)	3.0413 (0.0584)	3.0471 (0.0562)	3.0479 (0.0566)
$\alpha_{1,S=1}$	-2	-2.1337 (0.0783)	2.0855 (0.0750)	-2.1332 (0.0821)	-2.0830 (0.0769)
$\alpha_{2,S=1}$	2	1.8493 (0.0701)	1.8209 (0.0686)	1.8472 (0.0691)	1.8189 (0.0669)
$\alpha_{1,S=2}$	1	0.9630 (0.0646)	0.9765 (0.0653)	0.9659 (0.0640)	0.9794 (0.0651)
$\alpha_{2,S=2}$	2	2.0340 (0.0589)	2.0312 (0.0582)	2.0324 (0.0591)	2.0296 (0.0587)
$\sigma^2_{1,S=1}$	0.8	0.8045 (0.0395)	0.8008 (0.0389)	1.0100 (0.0344)	1.0100 (0.0160)
$\sigma^2_{1,S=2}$	0.9	0.0960 (0.0646)	0.0979 (0.0653)	1.0100 (0.588)	1.0001 (0.0485)
$\sigma^2_{2,S=1}$	1	1.0765 (0.0474)	1.0100 (0.0488)	1.0645 (0.0459)	1.0000 (0.0509)
$\sigma^2_{2,S=2}$	2	2.1669 (0.0564)	1.0010 (0.0488)	2.1967 (0.0592)	1.0010 (0.0562)
$v_{1,S=1}$	5	n.a.	n.a.	6.9330 (2.5145)	8.9875 (3.5384)
$v_{1,S=2}$	5	n.a.	6.9185 (2.6646)	n.a.	6.4099 (2.4753)
$v_{2,S=1}$	10	n.a.	n.a.	14.4770 (6.482)	9.5351 2.5839
$v_{2,S=2}$	10	n.a.	10.0605 (5.0789)	n.a.)	10.4885 (5.3318)
$\theta_{c,S=1}$	3	3.5581 (0.4385)	3.5449 (0.4295)	3.3968 (0.4062)	3.4133 (0.3978)
$\theta_{c,S=2}$	3	2.9543 (0.4809)	2.8300 (0.4295)	3.5048 (0.4876)	3.2684 (0.5000)
P_{11}	0.95	0.9584 (0.0123)	0.9606 (0.0121)	0.9576 (0.0125)	0.9601 (0.0123)
P_{22}	0.95	0.9474 (0.0154)	0.9499 (0.0152)	0.9467 (0.1561)	0.9495 (0.0154)

Source: Calculation Note: N = Normal margin, T = Student-t margin

Here, we consider two equations with two regimes, $S_t = 1, 2$, consisting of the high regime $S_t = 1$ and low regime $S_t = 2$. We simulate the value of the value of filtered probabilities from $U[0, 1]$, and the independent variable X are distributed normally with mean 0 and variance 1. The relationship between X and Y for two equations in each regime are specified as in Tables 1 and 2. The error terms are assumed to follow a normal distribution with $\varepsilon_{1,S=1} \sim N(0, 0.8)$ and $\varepsilon_{2,S=1} \sim N(0, 1)$ for regime 1 and $\varepsilon_{2,S=1} \sim N(0, 0.9)$ and $\varepsilon_{2,S=2} \sim N(0, 2)$ for regime 2. And student distribution with $\varepsilon_{1,S=1} \sim t(0, 0.8, 5)$ and $\varepsilon_{2,S=1} \sim t(0, 1, 10)$ for regime 1; and $\varepsilon_{2,S=1} \sim t(0, 0.9, 5)$ and $\varepsilon_{2,S=2} \sim t(0, 2, 10)$ for regime 2. For the dependence parameter for the copula function, we set the true value for the Gaussian correlation coefficient at 0.5 for Gaussian copula in both regimes while the true value for the Frank correlation coefficient is set at 3 for both regimes. In each simulation case, we assess the performance of our proposed MS-SUR model and compare the simulation results with the true values.

The estimation results from Tables 1 and 2 show the estimated mean form 8 cases of model specifications. We observe that our proposed model and method produce the unbiased parameter estimates when compare with their true values. The estimated parameter means are close to the true values in all cases and the standard error from the mean of each parameter is reasonable. Thus, we can conclude that our model is perform well in the simulation study. In the next section, we will apply our model to the real data analysis.

5 Application to Thai Sugar Market

The data sets include monthly data from January 2005 to May 2015 for demand for sugar (sum of quantity of export and domestic consumption:Q_t^d),output of sugar (Q_t^s), average export price of Thai sugar ($P^{exp\,ort}$), export price of Brazilian sugar (P^{Bra}_t), water storage (W_t) which were obtained from Thomson Reuter data stream, Faculty of Economics, Chiang Mai University, and Office of Agricultural Economics of Thailand. The data were converted into $(x_t - x_{t-1})/x_{t-1}$ form. To avoid the phenomenon of false regression caused by the regression analysis of non-stationary time series, an Augmented Dickey-Fuller unit root test for time series stationary was performed before estimating the MS-SUR model. We found that all variables are stationary at the level with 99 % confidence.

5.1 Model Selection

Table 3 reports the information criteria values including Akaiki information criteria and Bayesian information criteria. We observed that among the trial runs of several alternative marginal distribution and copula functions for MS-SUR model, the results provide

Table 3. Information criteria values

Model	Marginal Eq. 1	Marginal Eq. 2	Copula family	AIC	BIC	LL
1	**Gaussian**	**Gaussian**	**Frank**	**−714.588**	**−714.406**	44320.46
2	Gaussian	Student-t	Frank	−4.5	−4.31805	295
3	Student-t	Gaussian	Frank	−5.62974	−5.44778	364.998
4	Student-t	Student-t	Frank	−77.1421	−76.9601	4798.804
5	Gaussian	Gaussian	Joe	−4.19355	−4.0116	275.997
6	Gaussian	Student-t	Joe	−4.74194	−4.55998	309.997
7	Student-t	Gaussian	Joe	−5.62974	−5.44779	365.041
8	Student-t	Student-t	Joe	−5.70968	−5.52772	369.958
9	Gaussian	Gaussian	Gumbel	−31.018	−31.836	1939.116
10	Gaussian	Student-t	Gumbel	−94.1774	−93.8617	5854.99
11	Student-t	Gaussian	Gumbel	−341.485	−341.303	21188.07
12	Student-t	Student-t	Gumbel	−5.6129	−5.43095	363.999
13	Gaussian	Gaussian	Clayton	−14.4823	−14.3003	913.9026
14	Gaussian	Student-t	Clayton	−97.2587	−97.0768	6046.039
15	Student-t	Gaussian	Clayton	−17.4905	−17.3086	1100.38
16	Student-t	Student-t	Clayton	−4.83247	−4.64063	315.6088
17	Gaussian	Gaussian	Student-t	−3.70824	−3.52629	245.9084
18	Gaussian	Student-t	Student-t	−4.0655	−3.88355	268.061
19	Student-t	Gaussian	Student-t	−26.2223	−26.0404	1641.764
20	Student-t	Student-t	Student-t	−228.8	−228.618	268.061
21	Gaussian	Gaussian	Gaussian	−110.52	−110.3	6868.24
22	Gaussian	Student-t	Gaussian	−10.3057	−10.1237	6868.24
23	Student-t	Gaussian	Gaussian	−10.9478	−10.7478	654.95
24	Student-t	Student-t	Gaussian	−11.4201	−11.2381	694.7636
25	Conventional MS-SUR			−2.241	−2.059	122.942

Source: Calculation

evidence that the given Gaussian or normal distribution for both demand and supply equations and joint distribution by Frank copulas present the lowest AIC and BIC.

5.2 Estimation Result

5.2.1 Estimated Demand and Supply Equations from Copula Based MS-SUR Model

$$Q_t^d = -0.963^{***} - 0.899^{***}P_t^{export} + 0.272^{***}P_t^{Bra}, \quad (S_t = 1)$$
$$\quad\quad (0.001) \quad (0.223) \quad\quad\quad (0.100)$$
$$Q_t^d = 0.654^{***} - 3.933^{***}P_t^{export} + 9.158^{***}P_t^{Bra}, \quad (S_t = 2)$$
$$\quad\quad (0.001) \quad (0.005) \quad\quad\quad (0.001)$$
$$Q_t^s = 0.286 + 3.245P_t^{export} - 9.969^{***}W_t, \quad\quad (S_t = 1) \quad\quad\quad\quad (15)$$
$$\quad\quad (0.221) \quad (2.79) \quad\quad (2.796)$$
$$Q_t^s = 0.674^{***} + 0.685^{***}P_t^{export} + 1.295^{***}W_t, \quad (S_t = 2)$$
$$\quad\quad (0.003) \quad (0.063) \quad\quad (0.019)$$

Note: *** Significant at 1 % level, ** Significant at 5 % level, and Significant at 1 % level. () is Std.error

$$Q = \begin{bmatrix} 0.35 & 0.10 \\ 0.65 & 0.90 \end{bmatrix} \quad\quad (16)$$

The demand and supply functions estimated by the MS-SUR model with the assumption of two regimes are provided in Eq. 15. In this study we interpret regime 1 as market upturn regime, while regime 2 is interpreted as market downturn regime. According to the results on the demand for sugar (Q_t^d) with two equations for two regimes as presented in Eq. 15, we learned that Thai export price (P^{export}) and Brazilian export price (P_t^{Bra}) had significant effect on the demand for sugar in both regimes. Specifically an increase in Thai export price by 1 % will lead to a decrease in demand for Thai sugar by 0.9112 % (regime 1) and 3.803 %(regime 2) while Brazilian export price increase by 1 % will lead to increase the demand for Thai sugar by 0.272 % (regime 1) and 9.158 % (regime 2) for the reason of product substitutability. Normally, any goods is called the substitutes goods if the demand for goods is increase/decrease when the price of the other good increase/decrease, according to the substitute goods approach in Microeconomic approach. For sugar supply (Q_t^s), the function includes the average export price of Thailand (P^{export}) and water storage (W_t). The results show that export price (P_t^{export}) is positively, statistically significant at 10 % level, associated with the supply of sugar in market downturn regime. This means that export price is the factor affecting to the supply of sugar during the market downturn. The higher price of sugar will increase the willingness to plant of farmers, Water storage (W_t) showed negative significant effect on the supply of sugar in high market regime while a positive significant effect is obtained in market downturn regime. Naturally, abundant water supply should support the expansion of sugarcane cultivation or ensure the crop success. However, it is surprising to see that water storage has a negative effect on sugar supply in market upturn regime, We suggest that the higher water storage is reflected by the heavy rain fall which might disrupt and/or damage sugar production in Thailand. According to Better Management Practices and Agribusiness Commodities, the excessive for water irrigation can cause soil erosion and lead to lower sugar yield.

Figure 1 illustrates the smoothed probabilities of sugar market condition in upturn regime. This result indicates that the Thai sugar market is mostly operating in downturn regime rather than upturn. We found that the oversupply in world sugar market is the main factor that puts a high pressure on price of Thai sugar along our sample period. The lower price of sugar worldwide might lead to overstocking in large importing countries, including the United States, China, Indonesia, and the European Union. On the contrary, by focusing on the switching regime period, and where the high probability takes place for market upturn condition, we observed the similar pattern in each event along our study period to take place around October-January, namely events 1–7. Those periods apparently are the marketing year[1] of Thai sugar market. However, it is surprising to see the three sub periods of market downturn, including 2004/2005, 2007/2008, and 2008/2009, had taken place in the period of marketing year. According to Global Information and Early Warning System on food and agriculture (GIEWS), Thai sugar production decreased by 15 % in 2004/2005 and 30 % in 2007/2008, due to consecutive droughts. In the period of 2009/2010, the probabilities of staying in market upturn present a small increase. We found that production in 2009/10 pointed to 6 % growth because Thailand had a fine weather condition and farmer tended to use more fertilizers in the plantation process. In addition, we have learned from Eq. 16 that regime 1 is not persistent while regime 2 is persistent because the probabilities of staying in regime 1 and regime 2 are 35 and 90 %, respectively. This indicates that only an extreme event can switch the series to change from regime 2 to regime 1. Conversely, sugar market is easily to switch from regime 1 to regime 2 as presented by ($P_{12} = 0.65$). Moreover, the result also shows that the market upturn regime has a duration of approximately 1.53 months, while the market downturn regime has a duration of 10.54 months. This result indicates that Thai sugar markets have high volatility because the duration of each regime corresponds to a short period of time.

---

[1] **Marketing year** refers to the 12-month period, generally from the beginning of a new harvest, over which a crop is marketed (Report for Congress: Agriculture: A Glossary of Terms, Programs, and Laws, 2005 Edition).

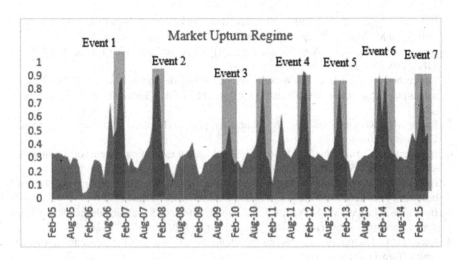

Fig. 1. Smoothed probabilities in high price market regime

6 Conclusion

This paper aims to relax the assumption of normal distribution on the structure of the error term in the system equation, conventional SUR model, and to combine the Markov Switching model with SUR to construct the Markov switching seemingly unrelated regression. In this study, we applied our model to derive the demand and supply in Thai sugar market. The model evidently can generate good estimation and outperform the MS-SUR model based on conventional likelihood function. We found that MS-SUR based Frank copula function show the best fit the estimate this model. Moreover, the model was applied to demand and supply equations with two regimes and it is found that water storage is the factor that effect supply of Thai sugar while Brazil's export price is the factor effect demand of Thai sugar. Thai export price is the factors that affect demand and supply of Thai sugar. The results of regime probabilities suggest that Thai sugar markets are mostly staying in market downturn regime rather than upturn. We found that Thai oversupply in the world sugar market is the main factor that puts a high pressure on price of Thai sugar along our sample period. Moreover, marketing year of Thailand which takes place around October-January is an event that leads the sugar market to stay in the market upturn regime.

Acknowledgement. The authors are grateful to Puay Ungphakorn Centre of Excellence in Econometrics, Faculty of Economics. Chiang Mai University for the financial support.

References

1. Angrist, J., Krueger, A.B.: Instrumental variables and the search for identification: From supply and demand to natural experiments. (No. w8456), National Bureau of Economic Research (2001)

2. Balcombe, K.G., Rapsomanikis, G.: Bayesian Estimation of Non-Linear Vector Error Correction Models: The Case of the Sugar-Ethanol-Oil Nexus in Brazil (2006)
3. Deaton, A., Laroque, G.: A model of commodity prices after Sir Arthur Lewis. J. Dev. Econ. **71**(2), 289–310 (2003)
4. Embrechts, P., Lindskog, F., McNeil, A.: Modelling dependence with copulas. Rapport technique, Département de mathématiques, Institut Fédéral de Technologie de Zurich, Zurich (2001)
5. Fahmy, H.: Regime switching in commodity prices. (Doctoral dissertation, Concordia University) (2011)
6. Henningsen, A., Hamann, J.D.: Systemfit: a package for estimating systems of simultaneous equations in R. J. Stat. Softw. **23**(4), 1–40 (2007)
7. Hamilton, J.D.: A new approach to the economic analysis of nonstationary time series and the business cycle. Econometrica: J. Econometric Soc. **57**(2), 357–384 (1989)
8. Hofert, M., Machler, M., McNeil, A.J.: Likelihood inference for archimedeancopulas in high dimensions under known margins. J. Multivar. Anal. **110**, 133–150 (2012)
9. Kuan, C.M.: Lecture on the Markov switching model. Institute of Economics, Academia Sinica, Taipei (2002)
10. Lin, C.Y.C.: Estimating supply and demand in the world oil market. J. Energy Dev. **34**(1), 1–32 (2011)
11. Masarotto, G., Varin, C.: Gaussian copula marginal regression. Electron. J. Stat. **6**, 1517–1549 (2012)
12. Moghaddasi, R., Azizi, K.: Measuring Welfare Effects of Importing sugar by Changing in Domestic Sugar Been Supply: the Case of Iran (2011)
13. Pastpipatkul, P., Maneejuk, P., Sriboonchitta, S.: welfare measurement on thai rice market: a markov switching bayesian seemingly unrelated regression. In: Huynh, V.-N., Inuiguchi, M., Denoeux, T. (eds.) IUKM 2015. LNCS (LNAI), vol. 9376, pp. 464–477. Springer, Heidelberg (2015). doi:10.1007/978-3-319-25135-6_42
14. Pastpipatkul, P., Maneejuk, P., Wiboonpongse, A., Sriboonchitta, S.: Seemingly unrelated regression based copula: an application on thai rice market. In: Huynh, V.-N., Kreinovich, V., Sriboonchitta, S. (eds.). SCI, vol. 622, pp. 437–450Springer, Heidelberg (2016). doi: 10.1007/978-3-319-27284-9_28
15. Perlin, M.: MS Regress - The MATLAB Package for Markov Regime Switching Models (2014)
16. Pitt, M., Chan, D., Kohn, R.: Efficient bayesian inference for gaussian copula regression models. Biometrika **93**(3), 537–554 (2006)
17. Pryce, G.: Construction elasticities and land availability: A two-stage least-squares model of housing supply using the variable elasticity approach. Urban Studies **36**(13), 2283–2304 (1999)
18. Rotaru, I.: A Bayesian MS-SUR Model for Forecasting Exchange Rates (2014)
19. Yoon, J.H.: Simultaneous equations in the markov-switching model. POSCO Research Institute working paper (2004)
20. Zellner, A.: An efficient method of estimating seemingly unrelated regressions and tests for aggregation bias. J. Am. Stat. Assoc. **57**(298), 348–368 (1962)

The Best Copula Modeling of Dependence Structure Among Gold, Oil Prices, and U.S. Currency

Pathairat Pastpipatkul[✉], Paravee Maneejuk,
and Songsak Sriboonchitt

Faculty of Economics, Chiang Mai University, Chiang Mai, Thailand
ppthairat@hotmail.com, mparavee@gmail.com

Abstract. As internationally traded commodities typically depend on the value of US dollar, this paper especially focuses on the most traded commodities, gold and crude oil, and tries to examine the dependence structures between these variables and the US currency. We employ various types of copulas i.e. the multivariate copula, vine copula, and the Markov switching copula and examine for the best-fit copula functions to model the dependency. Evidence from this study shows that gold and oil prices follow an inverse relationship with the value of US dollar but the relationship between gold and oil itself is strongly positive. However, the pair copulas given condition by another variable results in some attractive correlations.

Keywords: Multivariate copula · Vine copula · Markov switching · Dependence structures · Co-movement

1 Introduction

As described in a fundamental theory of international trade, the prices of commodities should be determined by the global supply and global demand for those internationally traded commodities. But, sometimes, a change in commodity price may be related nontrivially with a change of some specific factor. Due to the openness of the world economy, the U.S. dollar becomes a standard currency of international trade and widely available across the globe. Commodities are priced and traded primarily in dollars, which in turn seem to be associated with commodity prices. We can see often that when the dollar is strong, the internationally traded commodities including grains, food, metals, and energy, especially crude oil, seem to be more expensive in other non-dollar currencies. Just like other commodities, gold which is the most traded precious metal is also traded commonly in dollars. A weak dollar makes gold look cheaper for other nations to purchase this yellow metal. As gold is valuable and maintains its value over time, people tend to invest more in gold when a crisis is expected to force down the value of the national currency, which then drives up the gold price.

Some Observations on the Co-Movements of Gold, Oil, and the US Dollar. Motivated by these reasoning, we suspect that there is some economic relationship

© Springer International Publishing AG 2016
V.-N. Huynh et al. (Eds.): IUKM 2016, LNAI 9978, pp. 493–507, 2016.
DOI: 10.1007/978-3-319-49046-5_42

between gold, oil, and the U.S. currency. To prove this suspicion, the co-movements of these variables, we collected the monthly data of the real dollar index, gold and oil prices since April 1996 up to present and plotted the movements of these variables in Fig. 1. This Figure shows that oil and gold prices tend to move together while the U.S. currency seems to relate negatively to those prices of oil and gold. Shafiee and Topal [15] identified an extremely positive correlation between gold and oil prices similarly to what we can see from this picture. They found that during the run-up of oil prices happened in 2007, the gold prices also rose. After the peak of oil prices in 2007, there was a stunning fall in oil prices. Economists pointed that the drop in 2008–09 was almost entirely due to a collapse in demand or so-called demand destruction which happened around the world, especially in emerging markets. However, the prices of oil aren't set by demand alone. Some economists argued that this stunning fall in oil prices was related to the futures exchange. The drop in oil prices caused many companies including hedge funds, banks and pension funds bankruptcy since the companies needed to put up more cash collateral to back their losing positions. Unfortunately, many of them lost a lot of money in the process and were now out of business. Economists believed that this led to the beginning of the global financial crisis. The oil price dropped down sharply during that time as well as the gold price; however, the movement of the U.S. dollar index appeared inversely during the oil shock. After the shock in 2007–08, both gold and oil prices rallied gradually but the U.S. dollar index got weakened. However, the same situation, sharp fall of oil prices in 2007–08, seems to happen again in 2014–2015, so this should deserve special attention.

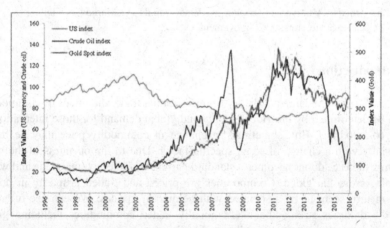

Fig. 1. Weighted value of US dollar against major currencies (US dollar index), Brent spot FOB US$/BBL (Crude oil index), and Gold spot index, M4/1996 – M4/2016

Additionally, Joy [10] is the one who noted the negative relationship between gold price and the U.S. currency. His work provides an explanation of the negative relationship between these two variables; a weaker dollar makes gold cheaper, increases demand for gold, which in turn drives up the gold price. Therefore, this kind of cause-and-effect events gives gold and the dollar their negative relationship.

Even though the above picture shows the co-movements between gold and oil prices which seem to inversely relate to the U.S. dollar, no clear relationship is apparent here. We still cannot see how much crude oil depends on the currency, and the same for gold prices. Therefore, the main objective of this paper is to find the dependence structures between gold, oil prices, and the US currency. We then need an appropriate method to model the dependence structures among these three variables; therefore, the modeling dependence will be discussed through the following section.

2   Literature: An Overview of Modeling Dependence

There are many methods used to measure the dependence between variables of interest whereas the classical methods such as the Pearson's correlation coefficient and Spearmans rho are no longer sufficient since they have the limitations about linear correlation and a strong assumption of normal distribution. In fact, the world's economy has faced recession or crisis many times over the past century. This event leads to a different degree of dependence between economic variables and the variables may not be normally distributed. In addition, the financial and economic data are considered to be far from the normal distribution thus using the parametric approach which imposes the normal distribution assumption becomes inflexible [8]. Another efficient way to estimate the dependence between variables is called Copula. The strength of copula approach allows us to model dependency between variables which have different marginal distribution.

However, such researchers as Romano [14] and Kang [11] suggest that the financial data is asymmetric and has heavy tail; therefore, the dependency should not be estimated by multivariate Gaussian copula. To get away from this problem, many studies consider the conventional multivariate t copula such as the works of Kang [11], and Autchariyapanitkul et al. [2]. Nevertheless, those studies seem not to provide the accurate result because they still get into the strong restriction of the same dependence structure between pair of variables. Therefore, another type of copula named Vine copula was introduced to deal with this problem. The vine copula is a more flexible class of dependence model since it allows us to select different bivariate copulas (blocks). Vine copula has been used widely in various researches, especially in finance. After Aas et al. [1] introduced the vine copula in 2009; they still developed an extension of vine copula and then brought in the canonical vine (C-vine) and the drawable vine (D-vine) copulas, which allow various dependence structures between different pair copulas, which in turn give a good starting point for modeling the dependence structure of high-dimensional data.

More recently, some copula studies, such as the works of Stoeber and Czado [16], and Pastpipatkul et al. [13], show that a combination of multivariate copula and the hidden Markov structure can capture regimes switches presented in the dependence structure of the data. This method accounts for the real economic phenomena which changes over time. Technically, the variances do not remain constant over time, which in turn lead to different degree of dependency across different economic behaviors (regimes) such as economic boom and recession.

Why do we need to consider several types of copula? The reason is that copulas have been applied and developed in various ways in which we never know which type (and also family) of copula will converge to the true dependence structure of the data. It is necessary to take into account all possible types of copula and decide on which copulas make sense and have natural interpretation for the data. Making a decision on copula is important because if we select the wrong or inappropriate copula for the data, it will give biased result [5]. Therefore, we consider various types of the copula that are (i) multivariate copula, (ii) vine copula, and (iii) Markov switching copula (MS) to measure the dependence structure of our variables under consideration: gold, crude oil, and the U.S. currency.

Until this stage, we may say that this paper will make two primary contributions. First, we take advantages from various copulas to model the dependence structure of economic variables, which in turn makes it particularly attractive for relatively high dimensional applications. Second, the dependency between gold prices, oil prices, and the U.S. currency will be measured by the best-fit copula functions.

For an overview of this paper, the next section will describe the basic copulas including the multivariate copula, CD-vine copula, and the Markov switching copula. Then, the data used in this paper will be described through Sect. 4, and in turn followed by the analysis of the results: measuring dependency between the investigated variables through Sect. 5. The final section will give the conclusions.

3 Methodology

The basic idea of copulas is to join any given marginal distributions of random variables, then form multivariate distribution function. Prior to explaining how we connect these two marginal distributions using copula function, we had better know how to get the marginal distribution. Therefore, this section begins with constructing marginal distribution by fitting standard residuals data with univariate GARCH model, and then continues to describe our key tool of this work 'copulas' in the following parts.

3.1 GARCH Models for Univariate Distributions

To construct the marginal distribution of each random variable, we employ a univariate ARMA(p,q)-GARCH(m,n) specification, where p is the order of the autoregressive part, q is the order of the moving average part, m is the order of the GARCH term h^2, and n is the order of the ARCH term ε^2. We use this model to construct the marginal because the data used are financial and economic data, which are known to be heteroscedastic and often autocorrelated, thus proper filtration is required. Following the notations above, the model ARMA(p,q)-GARCH(m,n) can take the form as:

$$y_t = \phi_0 + \sum_{i=1}^{p} \phi_i y_{t-i} + \sum_{j=1}^{q} \theta_j \varepsilon_{t-j} + \varepsilon_t \tag{1}$$

$$\varepsilon_t = h\eta_t \tag{2}$$

$$h_t^2 = \alpha_0 + \sum_{i=1}^{m} \alpha_i \varepsilon_{t-i}^2 + \sum_{j=1}^{n} \beta_j h_{t-j}^2, \tag{3}$$

where Eqs. (1) and (3) are the conditional mean and variance equation, respectively. The term ε_t represents the residual term which consists of the standard variance h_t and the standardized residual η_t, which is assumed to have normal distribution, Student-t distribution, skewed-t distribution, skewed normal distribution, generalized exponential distribution, and skewed generalized exponential distribution. The best-fit ARMA(p,q)-GARCH(m,n) will give the standardized residuals, which in turn will be transformed into a uniform distribution in (0,1).

3.2 Basic Concepts of Copula

The term copula is described as the linkage or the multivariate distribution function for which the marginal distribution is uniform. Sklar's theorem states that the joint distribution $F(x_1,\ldots,x_n)$ can be written in terms of univariate marginal distribution $F_1(x_1),\ldots,F_n(x_n)$, where $x_1,\ldots,x_n$ is the random variable vector. Thus, the joint distribution is given by

$$F(x_1,\ldots,x_n) = C(F_1(x_1),\ldots,F_n(x_n)) = C(u_1,\ldots,u_n). \tag{4}$$

We can see the term C in the above equation which represents copula distribution function of n-dimensional random variable with uniform marginal [0, 1]. Copula here works as a dependence structure between variables. When C is absolutely continuous, the joint density is given by

$$c(u_1,..,u_n) = \frac{\partial^n C(F_1(x_1),\ldots,F_n(x_n))}{\partial u_1\ldots\partial u_n}, \tag{5}$$

otherwise C could be singular or else. Note that c represents the multivariate copula density function and $u_i = F_i(x_i)$

3.3 CD-vine Copulas

Prior to the general form of C-vine and D-vine copulas, we had better start with the Pair copula construction (PPCs) which is introduced by Aas et al. [1]. Let's consider the case of n random variables with marginal functions $F_1,..,F_n$ and corresponding density c. By recursive conditioning, the n dimension joint density is given by

$$f(x_1,\ldots,x_d) = f(x_1)f(x_2|x_1)f(x_3|x_1,x_2)\ldots f(x_d|x_1,x_2,\ldots,x_{d-1}). \tag{6}$$

Following the example of three-dimensional random variables case,

$$f(x_2|x_1) = c_{12}(F_1(x_1), F_2(x_2))f_2(x_2)$$

$$f(x_3|x_1, x_2) = c_{23|1}(F_{2|1}(x_2|x_1), F_{3|1}(x_3|x_1)) \cdot c_{13}F_1(x_1), F_3(x_3) \cdot f_3(x_3),$$

so we get

$$f(x_1, x_2, x_3) = c_{23|1}(F_{2|1}(x_2, x_1), F_{3,1}(x_3|x_1)) \, c_{12}(F_1(x_1), F_2(x_2))c_{13}(F_1(x_1), F_3(x_3))f_1(x_1)f_2(x_2)f_3(x_3),$$

where c_{12}, c_{13}, and $c_{23|1}$ are the pair copulas which can be chosen separately and do not depend on the other pairs. The function $F_i(\cdot)$ is the cumulative distribution function (cdf) of x_i, and the function $c_{ij}(\cdot)$ is the copula density of (x_i, x_j).

Bedford and Cooke [3] especially concerned that in the joint density function as shown by Eq. (6) exists many such iterative PCCs. So, they introduce a graphical model called C-vine tree and D-vine tree to deal with this concern. The structure of C-vine tree shows a dependency between each variable and the first root nod, which in turn be modeled by bivariate copula. Following this brief explanation, the joint probability density function of n-dimension for the C-vine is given by

$$f(x_1, \ldots x_n) = \prod_{k=1}^{n} f_k(x_k) \prod_{j=1}^{n-1} \prod_{i=1}^{n-j} c_{i,i+j|i+1,\ldots,i+j-1}\left(F(x_i|x_{i+1}, \ldots x_{i+j-1}), F(x_{i+j}|x_{i+1}, \ldots, x_{i+j-1})|\theta_{j,j+i||1:(j-1)}\right), \quad (7)$$

where θ is a pair copula parameter. On the other hand, the D-vine tree is constructed by choosing a specific order of the variables. It shows a connection, for example, of first and second variables, second and third variables, third and fourth variables, and so on. Thus, the joint probability density function of n-dimension for D-vine is simply given by

$$f(x_1, \ldots x_n) = \prod_{k=1}^{n} f_k(x_k) \prod_{j=1}^{n-1} \prod_{i=1}^{n-j} c_{j,j+i|1,\ldots,j-1}\left(F(x_j|x_1, \ldots x_{j-1}), F(x_{j+i}|x_1, \ldots, x_{j-1})|\theta_{i,i+j|(i+1):(j+i-1)}\right). \quad (8)$$

Some important notation of both C-vine and D-vine methods are described as follows. The term $f_k(x_k)$ as shown in Eqs. (7) and (8) denotes the marginal density of x_k, where $k = 1, \ldots, d$. The terms $c_{i,i+j|i+1,\ldots,i+j-1}$ as shown in Eq. (10) and $c_{j,j+i|1,\ldots,j-1}$ in Eq. (11) represent the bivariate copula densities of each pair-copula in C-vine and D-vine, respectively.

3.4 Multivariate Copulas

There are two important classes of copulas namely Elliptical copula and Archimedean copula where each of them have their own family members. In this study, we consider the symmetric Gaussian and Student-t copulas of the Elliptical class and the asymmetric dependence Clayton, Gumbel, Joe, and Frank of the Archimedean class.

3.4.1   Elliptical Copulas
The general form of multivariate Elliptical copula is given by

$$C(u_1, \ldots, u_n \| \rho) = \Phi(\Phi^{-1}(u_1), \ldots, \Phi^{-1}(u_n); \rho), \tag{9}$$

where the function $\Phi(\cdot)$ is either the multivariate normal distribution function for Normal copula or multivariate Student t distribution with v degrees for Student-t copula. The term ρ is the correlation matrix $n \times n$ of the dependence parameters with interval $[-1, 1]$. The function Φ^{-1} is the inverse cumulative distribution function of the univariate standard normal and student-t for Normal and Student-t Copulas, respectively [9].

3.4.2 Archimedean Copulas

The multivariate Archimedean copula takes the form

$$C(u_1, \ldots, u_n | \theta) = \Phi(\sum_{j=1}^{n} \Phi^{-1}(u_j); \theta), \tag{10}$$

where the term $\Phi(\cdot)$ is the Laplace transform of univariate family of distribution which can be Clayton, Gumbel, Joe, and Frank. The term Φ^{-1} is an inverse cumulative distribution function of the univariate standard distribution. In the multivariate case, the exchangeable dependence is selected as a type of the symmetric positive definite matrix characterizing the elliptical copula and its range becomes narrower as the dimension increases. We elaborate the analysis in a case of $n > 2$, thus the dependence parameter θ is restricted to be $[0, +\infty)$ for Clayton, $[1, +\infty)$ for Gumbel and Joe, and $(0, \infty)$ for Frank copulas (for more details, see Nelson [12]).

3.5 Markov Switching Copulas

In any economy, it is true that the trend or economic behavior changes over time. We can simply separate economic phenomenon into two stages namely expansion (growth) and recession (contraction). The Markov switching (MS) model was introduced by Hamilton in 1989 [7] to capture this nonlinear behavior and the complex dynamic patterns of the economy. As we are dealing with the real economic data, we should take into account this nonlinear behavior as well. In addition, technically, we believe that the degree of the dependence parameter should not be constant over time due to the economic phenomenon. Therefore, this study applies the basic idea of the Markov switching into the multivariate copula and vine copula by extending one regime copula to two-regime Markov switching copula covering the two important stages of economy which are expansion and recession.

Briefly, the main idea of the Markov switching is that the dependence parameter $\Theta = \{\theta, \rho\}$ is governed by an unobserved variable (state/regime) denoted by S_t. The state variable here is assumed to have two regimes (k = 2) i.e. the high dependence regime (expansion) and the low dependence regime (recession). Thus, the joint distribution of n-dimensions $(x_1, \ldots, x_n)$ conditional on the state variable S_t is given by

$$(x_1, \ldots, x_n | \Theta^{S_t}; S_t = i) \sim C_t^{S_t}(u_1, \ldots, u_n; |\Theta^{S_t}), \qquad i = 1, 2, \tag{11}$$

where $i = 1, 2$ denotes the high dependence regime and the low dependence regime, respectively. Any parameter that depends on the state variable will be denoted by $(\cdot)^{S_t}$. The dependence parameter Θ^{S_t} follows one common Markov chain that determines the regime. Due to the Markov property, we assume that S_t is governed by first order Markov process such that it can be completely characterized by its transition matrix P which is given by

$$P_{ij} = \Pr(S_{t+1} = j | S_t = i) \quad \text{and} \quad \sum_{ij=1}^{k=2} p_{ij} = 1 \quad i, j = 1, 2. \tag{12}$$

The term p_{ij} means the probability of switching from regime i to regime j. Suppose we have two regimes, the transition probabilities matrix P is given by

$$P = \begin{bmatrix} p_{11} & p_{12} \\ p_{21} & p_{22} \end{bmatrix}. \tag{13}$$

In this study, the complete likelihood function of the two-regime MS multivariate copula and vine copula can be specified as [6]

$$l(\Theta^{S_t} | u_1, \ldots, u_n) = \sum_{S_t=1}^{2} c(u_1, \ldots, u_n | \Theta^{S_t}) \Pr(S_t | \Theta^{S_t}), \tag{14}$$

where the term $c(\cdot)$ is the density function of the multivariate copula and vine copula. The term $\Pr(S_t | \Theta^{S_t})$ represents filtered probabilities at $S_t = \{1, 2\}$ obtained from the Hamilton's filter [7] which is shown below

$$\Pr(S_t = i | \Theta_t) = \frac{c_t(u_1, \ldots, u_n | S_t = i, \Theta_{t-1}^{S_t=i}) \Pr[S_t = i | \Theta_{t-1}^{S_t=i}]}{\sum_{i=1}^{2} c_t(u_1, \ldots, u_n | S_t = i, \Theta_{t-1}^{S_t=i}) \Pr[S_t = i | \Theta_{t-1}^{S_t=i}]} \quad ; \quad i = 1, 2$$

By using this filtering process, we are able to get the filtered probabilities of each state given all information set up at time t.

4 Data

The data are monthly log returns of gold spot index (Gold), Brent crude oil index (Oil), and the U.S. dollar index (US index) spanning from M4/1996 to M4/2016, totaling 240 observations. We collected all the data from Thomson Reuters DataStream, from Financial Investment Center (FIC), Faculty of Economics, Chiang Mai University. The original series of gold, oil, and the U.S. currency are plotted through Fig. 1 (see Introduction) where we can see that oil and gold prices tend to move together while the U.S. currency seems to relate negatively to those prices.

Table 1. Summary of descriptive statistics

	US index	Oil	Gold
Mean	0.00012	0.00301	0.00479
Median	0.00188	0.01934	0.00405
Maximum	0.06467	0.26262	0.16526
Minimum	−0.04790	−0.37134	−0.20949
Std. Dev.	0.01685	0.10832	0.04839
Skewness	−0.03540	−0.66672	−0.10778
Kurtosis	3.54604	3.67991	4.90688
Jarque-Bera	3.03197	22.40340	36.82663
Probability	0.21959	0.00001	0.00000
ADF-test	0.00000	0.00000	0.00000

Source: Calculation.

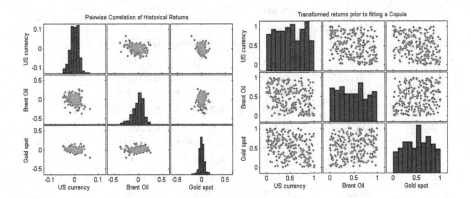

Fig. 2. Pairwise correlation of returns and transformed returns

Table 1 displays the descriptive statistics where the considered variables have a kurtosis above 3 and their asymmetry coefficients (Skewness) are all negative. In addition, Fig. 2 as shown below also presents the pairwise correlation of the data. It is found that the distribution of the data is skew to the right; meaning that the marginal distributions of this data tend to have a heavy tail to the left. Furthermore, Table 1 also shows that the null hypothesis of Jarque-Bera test for normality is rejected at 99 % confidence level, and the null hypothesis of Augmented Dickey-Fuller (ADF) test for stationarity is also rejected at 95 % confidence level, meaning that the data are stationary and non-normal distributions for all series.

5   Analysis of the Results

As this paper aims to measure the dependence structure between gold, oil, and the U.S. currency, we employ a powerful tool called copula with various types. The main idea of this section is to find one which is the best fit to the data where the candidates consist of the multivariate copulas, vine copulas, and the Markov switching copulas. However, prior to selecting the right copula, we have to go through the first subsection constructing marginal distribution.

5.1   Marginal Distribution by ARMA-GARCH Model

We need to consider the marginal distribution for adjusting the return distribution because the return series have volatility clustering. Therefore, we employ the ARMA

Table 2. Results of marginal distribution by ARMA-GARCH(1,1)

	US index	Oil	Gold
Mean equation			
ω_i	0.0012***	0.0013***	0.0016***
AR(1)	−0.3238***	−0.0063***	−0.5134***
AR(2)	0.1105***	0.4489***	0.0016***
AR(3)	−0.6389***	−0.0797***	−0.4276***
AR(4)	−0.3197***	0.2456***	0.1871***
AR(5)	−0.2807***	0.6777***	0.8203***
AR(6)	−0.4428***	−0.1386***	0.0089***
AR(7)	0.1388***	−0.2493***	
MA(1)	0.8800***	−0.0437***	0.7053***
MA(2)	0.0012***	−0.1011***	0.0056***
MA(3)	0.6361***	0.4643***	0.6336***
MA(4)	0.8957***	−0.6383***	−0.0026***
MA(5)	0.5763***	−0.7395***	−0.9976***
MA(6)	0.4719***	0.2701***	
Variance equation			
β_i	0.0001***	0.0014***	0.00003***
α	0.0061***	0.2845***	0.0352***
β	0.6810***	0.5668***	0.9445***
df		1.0000***	
γ	1.0320***	0.7414	1.0910
KS test (prob.)	0.9010	0.9030	0.8930
$Q^2(10)$ (prob.)	0.5510	0.6810	0.9550
AIC	−5.6371	−1.8981	−3.5606

Source: Calculation.
Note: "*," "**," and "***" denote rejections of the null hypothesis at the 10 %, 5 %, and 1 % significance levels, respectively.

(p,q)-GARCH(1,1) to obtain the standardized residual of each variable. As mentioned in Sect. 3.1, this study considers normal, student-t, skewed normal, skewed-t, generalized exponential, and skewed generalized exponential distributions with GARCH (1, 1).We then select the most appropriate ARMA(p,q)-GARCH(1,1) specification using a classical way of selecting model, the Akaike Information Criterion (AIC) in which the minimum value of AIC is preferred.

Table 2 shows the results of ARMA(p,q)-GARCH(1,1) model that best describes the data. We found that the U.S. dollar and gold indexes are well captured by skewed normal distribution with ARMA(7,6)-GARCH(1,1) and skewed normal distribution with ARMA(6,5)-GARCH(1,1), respectively. The oil index is well described by the skewed student-t distribution with the ARMA(7,6)-GARCH(1,1). Additionally, the Kolmogorov–Smirnov (KS) test is used as the uniform test for the transformed marginal distribution functions of standardized residuals. The result shows that none of the KS test accepts the null hypothesis, meaning that all marginal distributions are uniform in the interval [0,1]. The Ljung–Box test ($Q^2(10)$) is also employed to test autocorrelations of standardized residuals of the data. We found that there is no rejection of the null hypothesis, meaning no autocorrelation exists. Additionally, the correlations of the transformed returns or marginal distributions with uniform [0,1] are illustrated in Fig. 2.

5.2   Model Selection

As we consider various types of copula model i.e. the multivariate copulas, vine copulas and the Markov switching (MS) copulas, with two main classes of copula family namely Elliptical and Archimedean copulas, then this section is about selecting

Table 3. Model selection

Model type	LL	AIC	BIC
Multivariate normal copula	31.70	−57.40	−46.96
Multivariate student-t copula	32.15	−56.31	−42.38
Multivariate clayton copula	−21.99	44.19	44.45
Multivariate joe copula	−9.79	21.58	24.19
Multivariate frank copula	−27.63	57.26	59.86
Multivariate gumbel copula	−17.90	37.80	40.41
MS multivariate normal copula	35.59	−55.18	−27.33
MS multivariate student-t copula	35.49	−50.98	−16.17
MS multivariate clayton copula	−270.09	548.19	562.12
MS multivariate joe copula	12.32	−1.59	12.35
MS multivariate frank copula	−23.79	55.59	69.51
MS multivariate gumbel copula	−86.24	180.49	194.41
Mix CD-vine copula	**35.17**	**−64.34**	**−53.90**
MS mix CD-vine copula	−1135.62	2287.25	2315.09

Source: Calculation.

the best copula model among all candidates. We employ the Akaike Information Criterion (AIC) and also the Bayesian Information Criterion (BIC) to strengthen the results where the model having minimum value of AIC and BIC is preferred.

Table 3 presents the values of AIC, BIC, and log-likelihood (LL) of each copula type consisting of multivariate copula, vine copula and MS copula. Importantly, the case of vine copula here is denoted by *Mixed CD-vine copula*, which can be either C-vine or D-vine. The word 'Mixed' in this paper means each pair in the copula construction can be joined by different copula family. According to these results, we found that the minimum values of AIC and BIC are −64.34 and −53.90 (bold numbers), respectively, while the value of log-likelihood (LL) is equal to 35.17. This means the Mixed CD-vine copula is the best for this data.

Consider Table 4 below, it provides the sum of empirical Kendall's tau and shows that the US index has the strongest dependency in terms of the absolute value of empirical pairwise Kendall's tau. Therefore, in the case of C-vine, we determine the US index as the first root node and order the rest as follows: the US index (order 1), oil (order 2), and gold (order 3). Similarly, in D-vine, we found that the following order oil, the US index, and gold, has the biggest value of the sum of Kendall's tau. Both C-vine and D-vine here have the same structure of pair copulas; therefore, it is reasonable to calculate either C-vine or D-vine structure for this analysis.

Although we obtained from Table 3 that the Mixed CD-vine copula is the preferred model, we also found that C-vine and D-vine here give the same structure for this data. Therefore, we will select only the C-vine copula for inference on our data.

Table 4. Sum of empirical Kendall's tau

	US index	Oil	Gold
US index	1	−0.2220	−0.2146
Oil	−0.2220	1	0.1028
Gold	−0.2146	0.1028	1
Sum of absolute value	1.4366	1.3248	1.3147

Source: Calculation.

5.3   Estimates of the C-Vine Copula

This section presents the benchmark results for the dependency between gold price, oil price, and the U.S. currency. We constructed the dependency by the C-vine copula which is selected methodically from the previous section. We computed the dependence parameters using R environment with a package named CD-vine [4] which provides functionality of elliptical (Gaussian and Student-t) and Archimedean (Clayton, Gumbel, Frank, Joe, BB1, BB6, BB7 and BB8) copulas to cover a large range of possible dependence structures. The results are illustrated in Table 5

As for the dependence result of C-vine copula given in Table 5, it is found that Normal and Rotated Gumbel (270 °) are the best pair copula families due to the lowest values of AIC and BIC. The co-movement of every pair is statistically significant, with the second pair US-Gold having the largest dependence (Kendall's tau =−0.2284),

Table 5. Estimates of the C-vine copulas

Copula family	Parameter	Lower and upper tail dependence	Kendall's tau value	AIC	BIC
$C_{1,2}$ normal	−0.3390***	0	-0.2202	--26.9903	−23.5097
$C_{1,3}$ rotated gumbel (270 °)	−1.2960***	0	-0.2284	−37.7229	−34.2422
$C_{2,3|1}$ Normal	0.0843***	0	0.0537	0.3687	3.8494
Log-likelihood	35.1722	Sum of AIC (BIC)		−64.3445 (−53.9025)	
Other pair copulas					
$C_{1,3|2}$ rotated gumbel (270 °)	−1.2501***	0	--0.2001	−29.1387	−25.6581
$C_{1,2|3}$ rotated gumbel (270 °)	−1.1866***	0	--0.1574	18.4393	−14.9586
$C_{2,3}$ normal	0.2070***	0	0.1351	−8.1192	−4.6385

Note: "***" denote a rejection of the null hypothesis at the 1 % significance levels, and '1' denotes the US index, '2' denotes oil index, and '3' denotes gold index.

followed by the first pair US-Oil (Kendall's tau = −0.2202) and the third pair Oil-Gold conditional on US index (Kendall's tau = 0.0537).

The results are consistent with what we found in the real economy. The value of U. S. dollar typically has the inverse relationship with commodities over time including crude oil and gold; when the value of dollar increases against a basket of other major currencies, the prices of commodities typically decrease, and conversely. And we found the significant evidence supporting this theoretical idea that is the negative values of Kendall's tau. The crude oil, in fact, is a fairly significant direct and indirect cost input to the production of gold. The gold prices do not necessary tick higher for every tick lower in the crude oil price, but we found a positively conditional relationship among them. It does make sense if we consider both oil and gold just the traded commodities which follow the value of the dollar, and then the prices of crude oil should move along with the gold prices.

Additionally, we measure the dependence structures of other pair copulas, i.e. US-Gold conditional on oil and US–Oil conditional on gold, since we want to examine whether or not the oil prices have influence on the dependence structure between the U. S. currency and gold prices, and the same for another case. The bottom half of Table 5 provides the results of two conditional pair copulas $C_{1,3|2}$ and $C_{1,2|3}$, and unconditional pair copula $C_{2,3}$. We found some interesting point from this extension that is the Kendall's tau of $C_{1,3|2}$ and $C_{1,2|3}$, which equal to −0.2001 and −0.1574, respectively, decreased when the conditional variables are taken into account. For example, the Kendall's tau of $C_{1,2}$ equal to −0.2202 while $C_{1,2|3}$ equal to −0.1574, meaning that a relationship between U.S. dollar and oil prices itself is much stronger than that between U.S. dollar and oil prices conditional on gold prices.

Apart from that, we also found the Kendall's tau of $C_{2,3}$ is greater than that of $C_{2,3|1}$. This means a relationship between gold and oil prices itself is much stronger than the same relationship conditional on the value of dollar. This result implies that there is some complicated correlation between gold and oil which is deeper than being just a commodity in the same basket and moves in lockstep. The possible explanation is that the prices of crude oil partly account for inflation. Increase in the prices of crude oil results in an increase, for example, in gasoline prices which in turn are more costly to transport goods and their prices then move higher. This kind of event is so-called 'the cost push inflation' which happens when we experience rising prices due to higher costs of production and raw materials. The second part of the causal link is that gold tends to appreciate with rising inflation. Therefore, the increase in crude oil prices can eventually transfer into the increase in gold prices and cause their prices to move together.

6   Conclusion

This paper aims to examine the dependence structure between gold prices, oil prices, and the U.S. currency, as well as tries to find the best way of measuring the dependencies among these three variables. We consider the multivariate copulas, vine copulas, and the Markov switching copulas as the candidate models, but we empirically found that the Mixed CD-vine copula is the best fit for our data set.

According to the estimated results, we found evidence that gold and oil prices are negatively related to the value of dollar but the relationship between gold and oil itself is strongly positive, based on the values of Kendell's tau. These results are consistent with their correlations in the real economy. Additionally, as we tried to measure the dependence structures of other pair copulas, we found that some complicated relationships between these three variables arise when they are given condition by another variable. Investors and policy makers should take into account deeply this complicated correlations especially in gold and oil prices and closely monitor these variables.

References

1. Aas, K., Czado, C., Frigessi, A., Bakken, H.: Pair-copula constructions of multiple dependence. Insur. Math. Econ. **44**(2), 182–198 (2009)
2. Autchariyapanitkul, K., Chanaim, S., Sriboonchitta, S.: Portfolio optimization of stock returns in high-dimensions: a copula-based approach. Thai J. Math., pp. 11–23 (2014)
3. Bedford, T., Cooke, R.M.: Vines: a new graphical model for dependent random variables. Ann. Stat. **30**, 1031–1068 (2002)
4. Brechmann, E.C., Schepsmeier, U.: Modeling dependence with C-and D-vine copulas: the R-package CDVine. J. Stat. Softw. **52**(3), 1–27 (2013)
5. Durrleman, V., Nikeghbali, A., Roncalli, T.: Which copula is the right one (2000)
6. Hofert, M., Mächler, M., McNeil, A.J.: Likelihood inference for archimedean copulas in high dimensions under known margins. J. Multivar. Anal. **110**, 133–150 (2012)
7. Hamilton, J.D.: A New approach to the economic analysis of nonstationary time series and the business cycle. Econometrica **57**(2), 357–384 (1989)

8. Halulu, S.: Quantifying the risk of portfolios containing stocks and commodities (Doctoral dissertation, Bogaziçi University) (2012)
9. Joe, H., Hu, T.: Multivariate distributions from mixtures of max-infinitely divisible distribtions. J. Multivar. Anal. **57**(2), 240–265 (1996)
10. Joy, M.: Gold and the US dollar: Hedge or haven? Financ. Res. Lett. **8**(3), 120–131 (2011)
11. Kang, L.: Modeling the dependence structure between bonds and stocks: a multidimensional copula approach. Indiana University Bloomington (2007)
12. Nelson, R.B.: An Introduction to Copulas, vol. 139. Springer, Heidelberg (2013)
13. Pastpipatkul, P., Yamaka, W., Sriboonchitta, S.: Analyzing financial risk and co-movement of gold market, and Indonesian, Philippine, and Thailand stock markets: dynamic copula with markov-switching. In: Huynh, V.-N., Kreinovich, V., Sriboonchitta, S. (eds.). SCI, vol. 622, pp. 565–586. Springer, Heidelberg (2016). doi:10.1007/978-3-319-27284-9_37
14. Romano, C.: Calibrating and simulating copula functions: an application to the Italian stock market. Risk Management Function, Capitalia, Viale U. Tupini, p. 180 (2002)
15. Shafiee, S., Topal, E.: An overview of global gold market and gold priceforecasting. Resour. Policy **35**, 178–189 (2010)
16. Stoeber, J., Czado, C.: Detecting regime switches in the dependence structure of high dimensional financial data. arXiv preprint arXiv:1202.2009 (2012)

Modeling and Forecasting Interdependence of the ASEAN-5 Stock Markets and the US, Japan and China

Krit Lattayaporn, Jianxu Liu[✉], Jirakom Sirisrisakulchai,
and Songsak Sriboonchitta

Faculty of Economics, Chiang Mai University, Chiang Mai, Thailand
liujianxu1984@163.com

Abstract. A benefit of portfolio diversification has been designated as evidence showing a low level of stock market interdependence. Therefore, this study aims at examining the interdependence of ASEAN-5 stock markets and the US, Japan, and China in order to increase portfolio diversification benefit among those countries. We proposed the time-varying copula-based VAR model to measure the interdependence and the transmission of stock price movement. Also, a forecasting Kendall's tau method was proposed to check robustness of the copula-based model. The main findings of this study revealed that the dynamic Kendall's tau between the US and Indonesia, the US and Malaysia displayed tiny values. It indicates existence of opportunities to diversify an international portfolio. Moreover, the dynamic dependences also indicate that the interdependence between ASEAN-5 and China have been remaining limited. The results of IRFs showed that US had the strongest impact to ASEAN-5 while Indonesia and Malaysia had the lowest response to US, Japan, and China. In addition, the robustness check indicates that our prediction is precise.

Keywords: Time-varying copulas · Stock market interdependence · ASEAN-5 · Diversification benefit

1 Introduction

The benefit of international portfolio diversification relates to the interdependence of the equity market. According to Kearney and Lucey [5], the stronger integration of international stock markets, the lower the diversification benefits. Besides, Stulz [12] has defined a stock market integration as a perfect correlation of returns. Thus, investors may obtain an additional benefit of portfolio diversification when the co-movement of stock market returns are diminutive. There are several studies that have investigated the stock market interdependence of the ASEAN-5 to the US, Japan, and China. Following Shabri Abd et al. [9], they used co-integration approach with Generalized Method of Moments to test

© Springer International Publishing AG 2016
V.-N. Huynh et al. (Eds.): IUKM 2016, LNAI 9978, pp. 508–519, 2016.
DOI: 10.1007/978-3-319-49046-5_43

the benefit of diversifying international portfolio over the ASEAN-5 to the US and Japan. As a result, Indonesia was independent from both of US and Japan; Malaysia was more integrated with Japan rather than the US; Thailand was independent of the US but was related to Japan; The Philippines was more related to the US than Japan. Meanwhile, Singapore had a strong relationship with the US and Japan. Whereas Lim [6], which employed co-integration tests, revealed that the US market had the significant influence on all ASEAN- 5 markets. As for the study from ASEAN-5 to China, it is clear that the co-movement has been increasing from 1994 and remained in the limit by using a recursive co-integration method (Chien et al., [2]). As a result, the interdependence between ASEAN-5 and the US, Japan, and China are of considerable interest to international financial market researchers, policy makers, and investors. Since the shortcoming of co-integration approach is linear relationship between returns of stock markets, this study employed a Kendall's tau as an interdependence measurement which is probably more superior than a linear dependence from the previous studies. Therefore, the objectives of this study are to measure and forecast the interdependence of the ASEAN-5 stock markets to the US, Japan, and China; to investigate transmission of stock price movement from the US, Japan, and China; and to demonstrate an existing of the portfolio diversification benefit.

In order to achieve those objectives, the vector autoregressive model (VAR) has been considered. Because an atheoritical property allows us to treat every variable as an endogenous variable, majority of studies employ VAR model to investigate a financial market. Moreover, the impulse respond function (IRF) can be estimated with VAR model to measure response of dependence variables. Following Beirne and Gieck [1], they employed a global VAR model with a global IRF to assess the interaction between financial markets. However, there is evidence clarify that the stock market correlation is not homogeneous (Madaleno and Pinho, [7]). To overcome this problem, we employed the Gaussian kernel density function to apply time-varying copula-based VAR model. The model was used to measure and forecast the dynamic non-linear dependence between the ASEAN-5 stock markets and the US, Japan, and China. Meanwhile, IRFs were used to measure the transmission of stock price movement of ASEAN-5 and the US, Japan, and China.

The contribution of this study is threefold: First, the copula-based VAR model with a kernel density function was employed to investigate the stock market interdependence of the ASEAN-5 stock markets to the US, Japan, and China. Second, the student's t copula family was used to describe a dynamic non-linear dependence. We employed the rolling window method to forecast the dynamic non-linear dependence, and used the copula ratio to verify the predicted performance. Finally, the implications of this study demonstrate the opportunity of diversifying international portfolio between the ASEAN-5 stock markets and the US, Japan, and China.

The reminder of this study organizes as following. In Sect. 2, we present a time-varying copula-based VAR model in details. Section 3 presents a data descriptive and an empirical result. Section 4 presents a conclusion.

2   Methodology

This study utilizes the time-varying copula-based VAR model with copula ratio to achieve research purposes. Moreover, our model also displays several main advantages. First, we can construct the joint distribution with adequate flexibility. Second, the estimated dependence parameter is invariant under strictly monotonic transforms. The specification of VAR model, copulas, and copula ratio are introduced in the following sub-section.

2.1   VAR Model

VAR model is a simultaneous model that has developed from the univariate autoregressive model to multivariate time series model. VAR is a useful tool to describe a dynamic impact on the economy and the financial data. In addition, VAR model is an atheoretic model in which each equation is estimated separately by ordinary least squared. Therefore, there is no distinction between endogenous and exogenous variables as the distinction may affect to the estimation result. VAR model with degree k or VAR(k) is shown below:

$$y_{i,t} = \psi_{i,0} + \sum_{j=1}^{k} \psi_{1,ij} y_{1,t-j} + ... + \sum_{j=1}^{k} \psi_{m,ij} y_{m,t-j} + \epsilon_{i,t}, \tag{1}$$

where $y_{i,t}$ is a stationary random variable at time t while $i \in (1, ..., m)$ represents each random variable. $\psi_{i,0}$ is defined as a constant term. $\psi_{1,ij}$ is a coefficient of variable 1 that regresses on variable i with lag j. $\epsilon_{i,t}$ is a non-autocorrelated residual with zero mean that is generated from $y_{i,t}$. Even though, the VAR model parameters are difficult to be interpreted as a lack of theoretical support, the IRF is proposed to overwhelm this limitation.

2.2   Copulas

Copulas are mathematical tools that are commonly used in finance to help identify economic capital adequacy, market risk, credit risk, and operational risk. The advantage of using the copula is their joint distribution can be decomposed into marginal distributions and dependence structure. The term of the copula is defined as a Sklar's theorem. It states that all multivariate joint distribution can be expressed as a copula function. Let $y = (y_1, y_2)$ be a joint distribution of random variable function H, then the copula can be expressed as follows:

$$H(y_1, y_2) = C(F_1(y_1), F_2(y_2)), \tag{2}$$

where C is a copula function. F_1 and F_2 are marginal distribution functions that can be replaced by Gaussian kernel cumulative distribution function. It is given as follows:

$$\hat{F}_i(\epsilon_{i,t}) = \frac{1}{n \cdot h} \sum_{t=1}^{n} K \left(\frac{\epsilon - \epsilon_{i,t}}{h} \right), \tag{3}$$

where K is the standard Gaussian cumulative distribution function. h is a smoothing parameter that equals to $1.06\hat{\sigma}n^{-1/5}$. $\hat{\sigma}$ is a standard deviation and n is a number of observation. We employ kernel density estimation to transform a residual from VAR model (ϵ) into uniform [0,1], which generates a smoother marginal distribution.

Different copulas have different characteristics, such as upper tail dependence, lower dependence, positive and negative dependences, etc. Also, the dependence parameters are different for each copula family. For example, Gaussian copula has one parameter, ρ, that reflects linear correlation between variables. While student's t copula not only has the parameter ρ, but also has parameter degree of freedom, both of which represent linear correlation and tail dependence. Therefore, we used Gaussian, student's t, Clayton, Gumbel, Frank, Joe, BB1, BB6, BB7, BB8 and survival copula family in our study. Moreover, Kendall's tau, rank correlation, is usually used to describe non-linear dependence structure in copulas. The advantage of Kendall's tau is appropriate to our study (Embrechts et al., [3]). The Kendall's tau can be defined in terms of copulas as shows below:

$$\tau_{1,2} = 4 \int_0^1 \int_0^1 C(u_1, u_2) dC(u_1, u_2) - 1, \tag{4}$$

where C is a copula, and $u_i = \hat{F}_i(\epsilon_i)$ is from a Gaussian kernel cumulative distribution function.

The dynamic dependence may be more suitable than the static dependence. Following Wu et al. [13], we assume that the dependence parameters depend on the past dependence and historical data $(u_{1,t-1} - 0.5)(u_{2,t-1} - 0.5)$. The dependence is higher than the last period when both of $u_{1,t-1}$ and $u_{2,t-1}$ are varied from 0.5. In this study, the time-varying student's t copula is defined as:

$$\rho_t^* = \alpha + \beta\rho_{t-1}^* + \gamma(u_{1,t-1} - 0.5)(u_{2,t-1} - 0.5), \tag{5}$$

where ρ_t^* is a logistic transformation that is used to confine the dependence parameter in $[-1,1]$. It can be rewritten as: $\rho_t^* = -\ln\left(\frac{1-\rho_t}{\rho_t+1}\right)$ where ρ_t is a dependence parameter of student's t copula. β is confined to $(-1,1)$, represented the level of persistence to dependence structure. γ shows a significant level of the latest information. The other dependence parameter of student's t copula is the degrees of freedom (DoF). However, it remains invariant in a dynamic form. Also, the dynamic Kendall's tau can be expressed as: $\tau_t = \frac{2}{\pi}\arcsin(\rho_t)$.

In copula estimation, we used maximum likelihood method to estimate the dependence parameters of copulas, and select the best one in terms of Schwartz Information Criterion (SIC). Besides, the time-varying copula family is also adopted from the static copula family.

2.3   Robustness Check

Following Sirikanchanarak et al. [10], copula ratio (CR) is used to evaluate a forecasting ability of superior time-varying copulas. The CR is defined as

$$CR = \frac{C_p(u_{1,t+1}, u_{2,t+1}; \hat{\Theta}_{t+1})}{C_e(u_{1,t+1}, u_{2,t+1})}, \tag{6}$$

where C_p is a superior time-varying copula and C_e is an empirical copula. $\hat{\Theta}_{t+1}$ is the estimated parameters of superior time-varying copula at $t+1$. The closer CR to one, the higher prediction performance. This study simulates 10,000 values of superior time-varying copula distribution that can be expressed as:

$$CR_q = \frac{C_p(u_{1,q}, u_{2,q}; \hat{\Theta}_{t+1})}{C_e(u_{1,t+1}, u_{2,t+1})}, \tag{7}$$

where $q = 1, 2, \dots, 10,000$, and $u_{1,q}$ and $u_{2,q}$ are generated by copula simulation. If the copula ratio is in a boundary of confident 95 % interval then our prediction is effective.

3   Data and Empirical Results

3.1   The Data

We use the data set that composes of Shanghai Stock Exchange Composite Index (CH), Jakarta Stock Exchange Composite index (ID), Tokyo Price Index (JP), the Financial Times - Kuala Lumpur Bursa Malaysia Composite Index (ML), Philippine Stock Exchange index (PH), the Financial Times - Straits Time Index (SG), Stock Exchange of Thailand index (TH), and Standard and Poor's 500 index (US). All indices cover from January 1, 2000 to October 31, 2015, and collected from Bloomberg database. Also, the asset returns are calculated by using the difference between logarithmic closing prices for each index. Moreover, we adopt Occam's razor method to overcome the missing observation problem due to different stock market holidays. Following Shabri Abd et al. [9], the missing observations are replaced by using the previous daily closed prices. Next, we divide the data set into two periods. The in-sample extracts for the period from January 1, 2000 to October 31, 2013. The out-of-sample has a span from November 1, 2013 to October 31, 2015. Table 1 demonstrates the descriptive statistics of our data set. The average log-return is positive for all the series except for JP. The standard deviation as an unconditional volatility measurement shows that there is a relative similarity of volatility across national stock markets, excluding ML. For all series except PH, the data distribution are relatively similar to each other regarding a negative skewness coefficients. It implies that the left tail of probability density function is longer than the right tail. Finally, the Jarque-Bera test statistic strongly rejects null hypothesis of normality at the significant level of 1 % suggesting that every series do not follow a normal distribution. It persists to kurtosis values that the kurtosis of normal distribution does not exceed 3.

Table 1. Descriptive statistics of daily log-return

	CH	ID	JP	ML	PH	SG	TH	US
Mean	0.0002	0.0003	0.0000	0.0001	0.0002	0.0000	0.0002	0.0001
Maximum	0.0940	0.0762	0.1286	0.0450	0.1618	0.0728	0.1058	0.1096
Minimum	−0.0926	−0.1095	−0.1001	−0.0998	−0.1309	−0.0863	−0.1606	−0.0947
Std. Dev	0.0134	0.0116	0.0116	0.0070	0.0109	0.0095	0.0115	0.0105
Skewness	−0.3203	−0.7741	−0.4376	−0.9780	0.3700	−0.4517	−0.8552	−0.2169
Kurtosis	11.0888	13.6749	13.1609	19.2016	27.2054	12.5736	18.2193	15.9895
Jarque-Bera	15861.9	28030.6	25057.8	64160.6	141284.9	22277.6	56507.2	40694.6

3.2   The Results of VAR Model

We turn to construct the VAR model with lag length selection. Ozcicek and McMillin [8] suggested that the SIC is an appropriate criteria of VAR asymmetric lag model. The results of SIC displayed that VAR(1) is the most suitable. In our study, the IRF was then applied with the VAR model to investigate the transmission of stock price movement. Figure 1 illustrates an accumulated impulse respond function. The responses of all ASEAN-5 to China, Japan, and the US have occurred with one period lag. The US has the greatest impact to all ASEAN-5 where the level of response is ranked between 0.0018 to 0.0032. This result advocates Sok-Gee et al. [11] and Lim [6] that US has the stronger impact than Japan. The shock from China is smaller than the US but they have a positive effect unlike Japan. Indonesia is the only country that has a negative response to Japan. Besides, we can observe that Malaysia has the lowest impact to China, Japan, and the US. While Indonesia responds to Japan and the US as the second smallest impact.

3.3   Copulas Estimation Results

The residuals from VAR(1) are transformed into uniform [0,1] by using Gaussian kernel density estimations. The maximum likelihood is employed to estimate the copula. Then, SIC is proposed to select the best copula family. For static copulas, the student's t copula family had the best performance among all pairs. Therefore, we employed the student's t copula family to estimate time-varying models. In terms of the value of SIC, the time-varying student's t copula had a better performance than the static student's t copula. Table 2 displays the estimation results of time-varying student's t copula. β represents to the impact of the last period dependence parameter while γ explains the degree of the latest historical information. The last period dependence parameter had a positive impact to dynamic dependence parameter (CH,ID; CH,ML; CH,PH; CH,SG; CH,TH; JP,ID; JP,PH; JP,SG; US,ID; US,SG and US,TH). (JP,ML) is an only pair that β had a negative value showing an opposed impact from the latest period dependence parameter. Furthermore, the value of all significant β is closed to 1. It indicates that the historical dependence structure plays an important role in time-varying dependence parameters. Besides, there are only

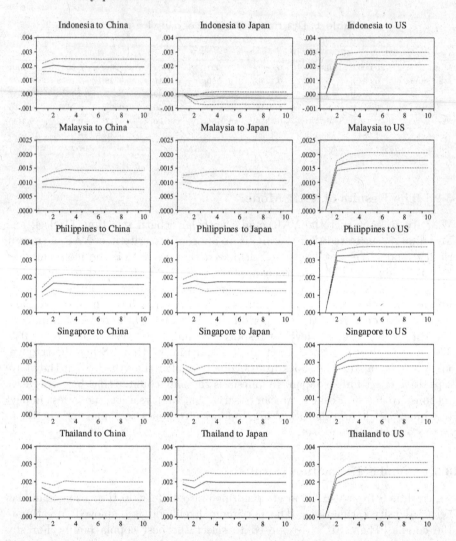

Fig. 1. Impulse response of ASEAN-5 stock market return and the return of US, Japan and China

six pairs (CH,ID; CH,TH; JP,ML; JP,SG; US,ML and US,SG) in which γ was statistically significant at 90 % confident level. Especially, the γ in (US,ML) was the largest implying a short-run information response is greater than others. Finally, the interdependence of (JP,TH) and (US,PH) can be well-described by a static dependence since the time-varying copula parameters (β and γ) were insignificant.

Figure 2 illustrates an empirical result of a dynamic Kendall's tau, including the out-of-sample forecasting. As for the holistic view, the values of Kendall's tau were mostly positive for all pairs, which imply concordances between stock

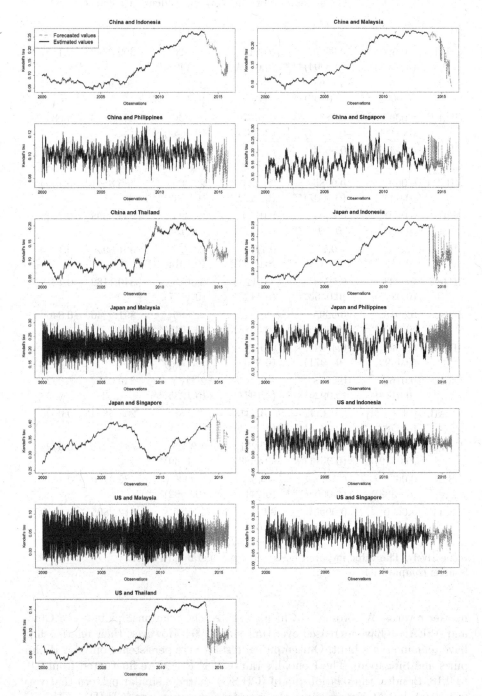

Fig. 2. Plot of estimated and forecasted Kendall's tau

Table 2. The estimation of time-varying student's t copulas

Pairs	α	β	γ	DoF	LL	SIC
CH,ID	−0.0001 (0.0001)	0.9995 (0.0004)***	0.0311 (0.0112)**	2.5820 (0.1404)***	309.3742	−0.1157
CH,ML	0.0001 (0.0002)	0.9995 (0.0007)***	0.0186 (0.0124)	2.7200 (0.1453)***	300.5022	−0.1122
CH,PH	0.0360 (0.0381)	0.8867 (0.1163)***	0.1089 (0.1633)	2.7596 (0.1539)***	231.4219	−0.0849
CH,SG	0.0148 (0.0125)	0.9634 (0.0278)***	0.2457 (0.1426)	2.5905 (0.1350)***	352.2935	−0.1327
CH,TH	0.0001 (0.0002)	0.9985 (0.0007)***	0.0434 (0.0172)*	2.5380 (0.1303)***	285.8811	−0.1064
JP,ID	0.0001 (0.0003)	0.9995 (0.0006)***	0.0139 (0.0121)	2.3394 (0.1135)***	533.5398	−0.2045
JP,ML	1.2162 (0.1122)***	−0.7527 (0.1301)***	0.7401 (0.2743)**	2.5222 (0.1214)***	461.6986	−0.1760
JP,PH	0.0233 (0.0215)	0.9634 (0.0353)***	−0.1269 (0.1120)	2.8015 (0.1577)***	340.1182	−0.1279
JP,SG	0.0018 (0.0006)**	0.9995 (0.0005)***	−0.0329 (0.0088)***	2.1885 (0.1016)***	891.9546	−0.3464
JP,TH	0.3568 (0.4699)	0.4621 (0.7071)	0.0791 (0.2863)	2.5554 (0.1284)***	442.6838	−0.1685
US,ID	0.0165 (0.0118)	0.8678 (0.0926)***	−0.3173 (0.1977)	2.4185 (0.1235)***	249.6357	−0.0921
US,ML	0.1564 (0.0582)**	−0.1342 (0.2915)	−0.9318 (0.2931)**	2.2886 (0.1108)***	268.5854	−0.0996
US,PH	0.0624 (0.0549)	−0.5155 (0.3511)	−0.2512 (0.2683)	3.2026 (0.1739)***	141.8318	−0.0494
US,SG	0.0676 (0.0383)	0.8508 (0.0908)***	−0.5567 (0.2415)*	2.0000 (0.0962)***	482.9410	−0.1844
US,TH	0.0000 (0.0006)	0.9995 (0.0040)***	0.0166 (0.0471)	2.3289 (0.1153)***	312.2805	−0.1169

Note: The superscripts *, ** and *** represent a significance level of 10%, 5%, 1% respectively. The value in parentheses represent a standard error.
Source: computation

market returns. According to Chien et al. [2], the relationships between China and ASEAN-5 have increased over time since 1994. However, their relationships have remained in a limit. Our empirical results were persistence only for Philippines and Singapore. The Kendall's tau of (CH,PH) were fluctuated from 0.07 to 0.13. Besides, the relationship of (CH,SG) shared a similar pattern that vary from 0.09 to 0.30. Nevertheless, the interdependence results (CH,ID), (CH,ML), and (CH,TH) are different with Chien et al. [2]. The dependence parameter of

(CH,ID) and (CH,ML) increased gradually since 2000 and reached its peak in 2013. On the other hand, (CH,TH) rose sharply in 2008 and then started to decrease slowly. Hence, the return of Chinese stock markets was more integrated with Indonesia, Malaysia, and Thailand than Singapore and Philippines.

The co-movement between ASEAN-5 and the US, and between ASEAN-5 and Japan were essential as Kearney [4] stated that U.S. and Japanese stock markets had more impact on Global stock market than European stock markets. The empirical results showed that (US,ID) and (US,ML) had an obvious portfolio diversification benefit. The dependence parameters of (US,ID) and (US,ML) displayed the fluctuated values, which were close to zero. It was persistent to Shabri Abd et al. [9] that US had an insignificant effect on both of Indonesia and Malaysia. However, the dependences of (JP,ID), (JP,SG), and (US,TH) gradually increased from the initial point, which indicates a decreasing benefit in portfolio diversification. For example, the dependence of (JP,ID) gradually rose from 2003 to 2013 with a peak of 0.29, corresponding to an increasing in their bilateral trade. It also rejected Shabri Abd et al. [9] that Indonesia is independence from Japan. We can conclude that (US,ID) and (US,ML) were the best pairs to increase portfolio diversification benefit during 2000 to 2013. Meanwhile, (CH,ID), (CH,ML), (CH,TH), (JP,ID), (JP,SG), and (US,TH) demonstrated a diminishing of portfolio diversification benefit.

Table 3 displays a forecasted validation of Kendall's tau by using the copula ratio. We can observe that all pairs satisfy the robustness check. The values of copula ratio of each observation were mainly located between the upper and the lower boundaries of 95 % interval. Moreover, the number of violated observa-

Table 3. The result of copula ratio

Pair	Number of violation	Violation percentage	Copula ratio	Empirical copula	95 % interval
CH,ID	18	2.47 %	1.013	0.312	[0.167, 22.104]
CH,ML	27	3.70 %	1.021	0.313	[0.269, 33.041]
CH,PH	23	3.15 %	0.994	0.313	[0.146, 19.570]
CH,SG	19	2.60 %	0.965	0.324	[0.208, 24.759]
CH,TH	25	3.42 %	0.981	0.319	[0.180, 22.534]
JP,ID	15	2.05 %	0.970	0.331	[0.100, 10.954]
JP,ML	16	2.19 %	0.978	0.329	[0.150, 16.254]
JP,PH	11	1.51 %	0.985	0.326	[0.124, 14.718]
JP,SG	9	1.23 %	1.016	0.347	[0.111, 10.402]
US,ID	15	2.05 %	0.981	0.277	[0.089, 12.647]
US,ML	25	3.42 %	0.986	0.286	[0.075, 10.459]
US,SG	12	1.64 %	0.984	0.301	[0.053, 6.7558]
US,TH	15	2.05 %	0.990	0.297	[0.061, 7.9323]

tions were less than 5 % of the total forecast (730 observations). It is clear that the prediction of all 13 pairs were valid. As we can see in Fig. 2, the forecasted dependence of (CH,ID), (CH,ML), (CH,TH), and (US,TH) showed an existence of portfolio diversification benefit. The forecasted Kendall's tau of (CH,ID) began to descend after the peak, as well as (CH,ML). The forecasted dependence of (CH,TH) sunk slightly, then vary from 0.09 to 0.16. The forecasted dependence of (US,TH) decreased dramatically, and began to vary between 0.05 and 0.15. Increasingly, the forecasted dependence of (US,ID) and (US,ML) still demonstrated a similar distribution to the estimated dependences. Finally, we can see that (US,ID) and (US,ML) were still the best pairs to increase portfolio diversification benefit during 2013 to 2015.

4 Conclusion

The interdependence among national stock markets are measured by using VAR model with IRFs and copula functions. Since financial correlation is not homogeneous and linear correlation that may not assess perfectly, the time-varying copula with Kendall's tau is employed to overcome the limitation.

The result reveals that the US had the largest impact to ASEAN-5. This result advocates Sok-Gee et al. [11] and Lim [6] that US had a stronger influencing impact to ASEAN-5 stock market than Japan and China. Malaysia had the lowest responses to the US, Japan, and China while Indonesia responded to the US and Japan as the second smallest portion. The IRFs of (JP,ML), (US,ML), (JP,ID), and (US,ID) were consistent with the result of Shabri Abd et al. [9]. It implies the US and Japan had an insignificant effect on Malaysia and Indonesia. The results of IRFs were also persistent with the dynamic Kendall's tau, indicates a low level of stock market concordances between the US and Indonesia, and between the US and Malaysia. They showed that dependence parameters of (US,ID) and (US,ML) were close to zero. It implies there exist an opportunity of portfolio diversification even though the level of Kendall's tau is tiny and volatile. In addition, the prediction result reveals that there is an obvious drop in the forecasted parameters of (CH,ID), (CH,ML), (CH,TH), and (US,TH). It implies that there are alternative ways to increase the benefit of portfolio diversification. Also, the dependence structures between ASEAN-5 and China are consistent with Chien et al. [2], which means the co-movement between them remain limited.

Finally, the extent of dependence structures between ASEAN-5 and the US, Japan, and China may carry some important implications for the macro stabilization policies and the financial policies of multinational corporations. The policies, which related to stock market imbalances, will depend crucially on the extent of stock market interdependence. Besides, the findings of this study also provided the information to exploit potential benefits of international diversification.

References

1. Beirne, J., Gieck, J.: Interdependence and contagion in global asset markets. Rev. Int. Econ. **22**(4), 639–659 (2014)
2. Chien, M.S., Lee, C.C., Hu, T.C., Hu, H.T.: Dynamic asian stock market convergence: evidence from dynamic cointegration analysis among china and asean-5. Econ. Model. **51**, 84–98 (2015)
3. Embrechts, P., McNeil, A., Straumann, D.: Correlation and dependence in risk management: properties and pitfalls. In: Dempster, M.A.H. (ed.) Risk Management: Value at Risk and Beyond, pp. 176–223. Cambridge University Press, Cambridge (2002)
4. Kearney, C.: The determination and international transmission of stock market volatility. Glob. Finance J. **11**(1–2), 31–52 (2000)
5. Kearney, C., Lucey, B.M.: International equity market integration: Theory, evidence and implications. Int. Rev. Financ. Anal. **13**(5), 571–583 (2004)
6. Lim, L.K.: Convergence and interdependence between asean-5 stock markets. Math. Comput. Simul. **79**(9), 2957–2966 (2009)
7. Madaleno, M., Pinho, C.: International stock market indices comovements: a new look. Int. J. Finan. Econ. **17**(1), 89–102 (2011)
8. Ozcicek, O., McMillin, W.D.: Lag length selection in vector autoregressive models: symmetric and asymmetric lags. Appl. Econ. **31**(4), 517–524 (1999)
9. Shabri Abd. Majid, M., Kameel Mydin Meera, A., Azmi Omar, M.: Interdependence of asean-5 stock markets from the us and japan. Glob. Econ. Rev. **37**(2), 201–225 (2008)
10. Sirikanchanarak, D., Liu, J., Sriboonchitta, S., Xie, J.: Analysis of transmission and co-movement of rice export prices between Thailand and Vietnam. In: Huynh, V.-N., Kreinovich, V., Sriboonchitta, S. (eds.) Causal Inference in Econometrics. SCI, vol. 622, pp. 333–346. Springer, Heidelberg (2016). doi:10.1007/978-3-319-27284-9_21
11. Sok-Gee, C., Karim, M.A., Karim, M.: Volatility spillovers of the major stock markets in asean-5 with the us and japanese stock markets. Int. Res. J. Finan. Econ. **44**, 161–172 (2010)
12. Stulz, R.: A model of international asset pricing. J. Finan. Econ. **9**(4), 383–406 (1981)
13. Wu, C.C., Chung, H., Chang, Y.H.: The economic value of co-movement between oil price and exchange rate using copula-based garch models. Energy Econ. **34**(1), 270–282 (2012)

Statistical Methods

Need for Most Accurate Discrete Approximations Explains Effectiveness of Statistical Methods Based on Heavy-Tailed Distributions

Songsak Sriboonchitta[1], Vladik Kreinovich[2(✉)], Olga Kosheleva[2], and Hung T. Nguyen[1,3]

[1] Faculty of Economics, Chiang Mai University, Chiang Mai, Thailand
songsakecon@gmail.com
[2] University of Texas at El Paso, 500 W. University, El Paso, TX 79968, USA
{vladik,olgak}@utep.edu
[3] Department of Mathematical Sciences, New Mexico State University, Las Cruces, NM 88003, USA
hunguyen@nmsu.edu

Abstract. In many practical situations, it is effective to use statistical methods based on Gaussian distributions, and, more generally, distribution for which tails are *light* – in the sense that as the value increases, the corresponding probability density tends to 0 very fast. There are many theoretical explanations for this effectiveness. On the other hand, in many other cases, it is effective to use statistical methods based on *heavy-tailed* distributions, in which the probability density is asymptotically described, e.g., by a power law. In contrast to the light-tailed distributions, there is no convincing theoretical explanation for the effectiveness of the heavy-tail-based statistical methods. In this paper, we provide such a theoretical explanation. This explanation is based on the fact that in many applications, we approximate a continuous distribution by a discrete one. From this viewpoint, it is desirable, among all possible distributions which are consistent with our knowledge, to select a distribution for which such an approximation is the most accurate. It turns out that under reasonable conditions, this requirement (of allowing the most accurate discrete approximation) indeed leads to the statistical methods based on the power-law heavy-tailed distributions.

1 Formulation of the Problem

Many effective statistical methods are based on light-tailed probability distributions. In many practical situations, we encounter Gaussian distributions. Their ubiquity can be explained by the fact that in many of these situations, the corresponding random quantity is the result of joint effect of many different factors, and according to the Central Limit Theorem, the distribution of the sum of many independent small random variables is close to Gaussian; see, e.g., [24].

© Springer International Publishing AG 2016
V.-N. Huynh et al. (Eds.): IUKM 2016, LNAI 9978, pp. 523–531, 2016.
DOI: 10.1007/978-3-319-49046-5_44

This argument explains why many actual distributions are close to Gaussian. For example, an empirical study of measuring instruments shows that for about 60 % of them, the probability distribution of the measurement error is close to Gaussian; see, e.g., [19,20]. In such situations – when we have enough data about the probability distribution to know that it is close to Gaussian – it is reasonable to use statistical methods based on Gaussian distributions.

Interestingly, however, statistical methods based on Gaussian distributions have been effectively used in much more general situations, when we only have a *partial* information about the probability distribution – e.g., when we only know the first two moments of the distribution; see, e.g., [21]. The effectiveness of Gaussian methods in such situations comes from the use of the Maximum Entropy approach [9]: if we have only partial information about the probability distribution – i.e., if several different probability density functions $\rho(x)$ are consistent with our knowledge – then it is reasonable to select the distributions with the largest uncertainty, i.e., with the largest value of the entropy $S = -\int \rho(x) \cdot \ln(\rho(x)) \, dx$. Often, the only information that we have about the probability distribution is its first two moments E and M_2, i.e., we only know that $\int \rho(x) \, dx = 1$, $\int x \cdot \rho(x) \, dx = E$, and $\int x^2 \cdot \rho(x) \, dx = M_2$. If we apply the Lagrange multiplier method to maximize entropy under these three constraints, we thus reduce the original constraint optimization problem to the unconstrained problem of maximizing the following functional

$$-\int \rho(x) \cdot \ln(\rho(x)) \, dx \; + \lambda_0 \cdot \left(\int \rho(x) \, dx - 1 \right) + \lambda_1 \cdot \left(\int x \cdot \rho(x) \, dx - E \right)$$
$$+ \lambda_2 \cdot \left(\int x^2 \cdot \rho(x) \, dx - M_2 \right).$$

Differentiating this expression with respect to each unknown $\rho(x)$ and equating the resulting derivative to 0, we conclude that

$$- \ln(\rho(x)) - 1 + \lambda_0 + \lambda_1 \cdot x + \lambda_2 \cdot x^2 = 0,$$

hence $\rho(x) = \exp((\lambda_0 - 1) + \lambda_1 \cdot x + \lambda_2 \cdot x^2)$. One can check that for the constraints to be satisfied, this function has to have the usual Gaussian form

$$\rho(x) = \frac{1}{\sqrt{2\pi} \cdot \sigma} \cdot \exp \left(-\frac{(x-a)^2}{2\sigma^2} \right),$$

where $a = E$ and $\sigma = \sqrt{M_2 - E^2}$.

For the Gaussian distribution, the probability density has *light tails* in the sense that as x increases, the probability density tends to 0 fast: so fast that for every k, we have a finite moment $\int x^k \cdot \rho(x) \, dx$.

Similarly, many other effective statistical methods are based on light-tailed distributions.

Some effective statistical methods are based on heavy-tailed distributions. On the other hand, in many empirical situations, it turns out to be efficient to use statistical methods based on *heavy-tailed* distributions, i.e., distributions for which the probability density $\rho(x)$ tends to 0 much slower – so that some moments become infinite; see, e.g., [3–6, 8, 13–17, 23, 25, 26]. A typical case is when asymptotically, the probability density function has a power law distribution $\rho(x) = C \cdot x^{-\alpha}$ for some $\alpha > 0$.

Natural question. If usual arguments lead to the effectiveness of statistical methods based on light-tailed distributions, how can we explain the effectiveness of statistical methods based on the heavy-tailed distributions?

What we do in this paper. In this paper, we provide a possible answer to the above question: namely, we show that the need for most accurate discrete approximations naturally leads to statistical methods based on heavy-tailed distributions.

2   Need for Most Accurate Discrete Approximations: The Main Idea

Need for discrete approximations: an example. Many real-life situations are well-described by *random walk* models; such models are especially ubiquitous in econometrics; see, e.g., [12] and references therein.

In the general random walk, each component $x(t+1)$ of the state at the next moment of time $t+1$ is obtained from the state $x(t)$ at the previous moment of time t by adding a random step $r(t)$: $x(t+1) = x(t) + r(t)$. In many cases, the random steps $r(t)$ and $r(t')$ corresponding to different moments of time $t \neq t'$ are independent.

In some applications, the empirical data is well-described by the simplest type of random walk, in which, for some constant $r_0 > 0$, the random step is equal either to r_0 or to $-r_0$, with probability 0.5 of each of these two values. In other applications, this simple model is not sufficient, so we need to use a more complex model, in which the step takes three, four, or more values with different probabilities. In the limit, when the number n of values tends to infinity, we get a continuous description, in which a step $r(t)$ is distributed according to some probability density function $\rho(x)$.

Resulting need for the most accurate discrete approximations. The more accurately we can approximate the actual continuous distribution with an n-point one, the more accurate are the discrete-approximation models (e.g., the corresponding random walk models).

Since random walk models are widely used, it makes sense to use the existence of the most accurate discrete approximations as an alternative criterion for selecting a probability distribution in situations when we only have partial information about the probabilities – i.e., in which several different probability distributions are consistent with our knowledge.

This is our main idea. This is our main idea, an idea that, as we show in this paper, will lead to an explanation for power-law distributions. To come up with this explanation, we first need to formulate our idea in precise terms.

3   Towards Formalizing the Main Idea

Approximating a continuous distribution by discrete ones: analysis of the problem. When we approximate a continuous probability distribution by a discrete one, we thus:

- approximate the actual random variable r which can take, in principle, any value from the real line,
- by a discrete variable that can only take a finite number of values

$$r_1, \ldots, r_n.$$

The approximating value r_i is, in general, different from the actual value r, so there is the approximation inaccuracy $\delta \overset{\text{def}}{=} |r_i - r|$. To describe continuous distributions which allow the most accurate discrete approximations, we need to formalize what it means, for a given probability distribution and for a given n, to select the most accurate discrete approximation.

Rational decision making: reminder. It is known that, in general, decision of a rational decision maker can be described as maximizing the expected value of an appropriate objective function $u(x)$ called *utility* [7,11,18,22]. This function is determined modulo a general linear transformation $u \to a \cdot u + b$ for some $a > 0$ and b.

Maximizing utility is equivalent to minimizing *disutility* $U \overset{\text{def}}{=} -u$. Disutility is also defined modulo a general linear transformation $U \to a \cdot U + b$.

Let us apply the general ideas of rational decision making to our situation. To apply the general ideas of rational decision making to our case, we need to describe the disutility $U(\delta)$ caused by inaccuracy δ.

To describe this disutility, we can take into account that in most case, the numerical value r of the corresponding quantity depends on the choice of a measuring unit.

- For a geometric random walk, the value r represents a distance, whose numerical value depends on our choice of a distance unit – e.g., meters or feet.
- For a financial random walk, e.g., for the financial random walk describing the stock market index, the value r represents the price, and its numerical value depends on the monetary unit – e.g., dollars or Euros.

If we choose a new measuring unit which is k times smaller than the original one, then the numerical value of the corresponding quantity increases by a factor of k: $r \to k \cdot r$. For example, if we use centimeters instead of meters, then in centimeters, the distance of $r = 2$ m takes the value $k \cdot r = 100 \cdot 2$ cm.

The choice of a measuring unit is usually rather arbitrary. It is therefore reasonable to require that the disutility function $U(\delta)$ not depend on the choice of a measuring unit. Of course, we cannot simply require that the disutility function does not change at all, i.e., that $U(k \cdot \delta) = U(\delta)$ for all k and δ, since that would imply that $U(\delta) \equiv$ const. However, we can take into account the fact that the disutility function is defined modulo some linear transformation. If we fix $U(0) = 0$, this still leaves us with a transformation $U \to a \cdot U$. We can therefore require that when we re-scale the unit for measuring the original quantity, the disutility function remains the same modulo an appropriate linear transformation, i.e., that for every $k > 0$, there exist a value $a(k)$ for which, for every $\delta \geq 0$, we have $U(k \cdot \delta) = a(k) \cdot U(\delta)$.

Small changes in accuracy should lead to small changes in utility. In mathematical terms, this means that the disutility function $U(\delta)$ must be continuous. It is known that for a continuous function, the above functional equation implies that $U(\delta) = C \cdot \delta^\beta$ for some values $C > 0$ and β; see, e.g., [1]. The larger the inaccuracy δ, the larger the disutility. So, we must have $\beta > 0$.

Towards formalizing the problem. Let us use this disutility function to describe the expected disutility of approximating the original continuous variable with a discrete one. Since the disutility increase with inaccuracy, once the points $r_1 < \ldots < r_n$ are selected, we should assign, to each value r, the point r_i for which the inaccuracy is the smallest possible, i.e., the value which is the closest to r. Thus, for every i, the point r_i is assigned to all the values r from the interval $\left[\dfrac{r_{i-1} + r_i}{2}, \dfrac{r_i + r_{i+1}}{2} \right]$.

For large n, we can describe the selection of the points $r_1 < r_2 < \ldots < r_n$ by describing the frequency $\rho_0(r)$, i.e., the number of points per unit length. The overall number of points is n, so we have $\int \rho_0(r)\, dr = n$. In this case, the length of each intervals $[r_i, r_{i+1}]$ is approximately equal to $\dfrac{1}{\rho_0(r)}$, and similarly, the length of the interval $I_i \stackrel{\text{def}}{=} \left[\dfrac{r_{i-1} + r_i}{2}, \dfrac{r_i + r_{i+1}}{2} \right]$, which is composed of the halves of $[r_{i-1}, r_i]$ and of $[r_i, r_{i+1}]$, is also approximately equal to $\dfrac{1}{\rho_0(r)}$. On this interval, the average value of inaccuracy is proportional to the interval width and thus, the average value of the disutility is proportional to the β-th power of this width, i.e., to $U_i \approx \dfrac{1}{(\rho_0(r_i))^\beta}$.

The overall expected value E_U of the disutility – the one that we need to minimize – is equal to $E_U = \sum\limits_{i=1}^{n} p_i \cdot U_i$, where p_i is the probability that the original random variable r occurs in the interval I_i. Here, $p_i = \int_{I_i} \rho(r)\, dr$, where $\rho(r)$ is the probability density of the original random variable. Thus,

$$E_U = \sum_{i=1}^{n} \frac{1}{(\rho_0(r_i))^\beta} \cdot \int_{I_i} \rho(r)\, dr.$$

By moving the term $\dfrac{1}{(\rho_0(r_i))^\beta}$ inside the intervals, we get

$$E_U = \sum_{i=1}^{n} \int_{I_i} \frac{1}{(\rho_0(r_i))^\beta} \cdot \rho(r)\, dr.$$

For large n and narrow intervals, we have $r_i \approx r$ and thus,

$$E_U \approx \sum_{i=1}^{n} \int_{I_i} \frac{1}{(\rho_0(r))^\beta} \cdot \rho(r) \, dr.$$

The intervals I_i cover the whole real line. Thus, the sum of integrals of the same function over all intervals I_i is simply the integral over the whole real line:

$$E_U \approx \int \frac{1}{(\rho_0(r))^\beta} \cdot \rho(r) \, dr.$$

So, we arrive at the following precise reformulation of the best discrete optimization problem:

– we know the probability density $\rho(r)$;
– we want to find the density $\rho_0(r)$ of the distribution of the discrete points as the one that minimizes the integral $\int \frac{1}{(\rho_0(r))^\beta} \cdot \rho(r) \, dr$ under the constraint $\int \rho_0(r) \, dr = n$.

After that, we will need to select, in each class of probability distributions, a distribution for which this best-case expected disutility has the smallest possible value.

4   Solving the Resulting Optimization Problem

Solving the resulting optimization problem: first step. Let us first find the optimal function $\rho_0(r)$ corresponding to a given probability density function $\rho(r)$.

To find this function $\rho_0(r)$, we must solve the above constraint optimization problem. For this problem, the Lagrange multiplier method leads to minimizing the following objective function:

$$\int \frac{1}{(\rho_0(r))^\beta} \cdot \rho(r) \, dr + \lambda \cdot \left(\int \rho_0(r) \, dr - n \right).$$

Differentiating this expression with respect to each unknown $\rho_0(r)$ and equating the derivative to 0, we conclude that

$$-\beta \cdot \frac{\rho(r)}{(\rho_0(r))^{\beta+1}} + \lambda = 0,$$

hence $(\rho_0(r))^{\beta+1} = \text{const} \cdot \rho(r)$, with $\text{const} = \dfrac{\lambda}{\beta}$. Thus,

$$\rho_0(r) = c \cdot (\rho(r))^{1/(\beta+1)},$$

for some constant c. This constant can be determined from the condition that $\int \rho_0(r)\, dr = n$. Substituting the above expression into this condition, we conclude that

$$\rho_0(r) = n \cdot \frac{(\rho(r))^{1/(\beta+1)}}{\int (\rho(s))^{1/(\beta+1)}\, ds}.$$

Solving the resulting optimization problem: second step. Now that we know which point density $\rho_0(r)$ is optimal for the given probability density function $\rho(r)$, we need to find the probability density function $\rho(r)$ for which the corresponding best-case disutility function attains the smallest possible value.

To find such $\rho(r)$, let us first use our result of solving the first-step optimization problem to come up with an explicit (and thus, easier-to-minimize) expression for the best-case average disutility.

Substituting the above expression for $\rho_0(r)$ into the formula for E_U, we get

$$E_U = \int \frac{\rho(r)}{(\rho_0(r))^{1/(\beta+1)}}\, dr = \frac{\int (\rho(r))^{1/(\beta+1)}\, dr}{\left(\int (\rho(r))^{1/(\beta+1)}\, dr\right)^{\beta}} = \frac{1}{\left(\int (\rho(r))^{1/(\beta+1)}\, dr\right)^{\beta-1}}.$$

Thus, depending on whether $\beta < 1$ or $\beta > 1$, minimizing the best-case expected disutility is equivalent to either minimizing or maximizing the integral

$$J \stackrel{\text{def}}{=} \int (\rho(r))^{1/(\beta+1)}\, dr.$$

Comment. It is worth mentioning that the resulting objective function for selecting a probability distribution is one of the scale-invariant objective functions described in [10]. This is not surprising since we based our derivation on the ideas of invariance with respect to selecting a measuring unit – which, in mathematical terms, is exactly scale invariance.

Solving the resulting optimization problem: final step. Now, in a class of probability density functions which are consistent with our knowledge, we need to find the ones for which the above expression J is optimal. In particular, if our knowledge consists of a moment-related constraint $\int |r|^k \cdot \rho(r)\, dr = M_k$, then we need to optimize the expression J under this constraint and the additional constraint $\int \rho(r)\, dr = 1$. For this constraint optimization problem, the Lagrange multiplier method leads to the need for optimizing the following expression:

$$\int (\rho(r))^{1/(\beta+1)}\, dr + \lambda_0 \cdot \left(\int \rho(r)\, dr - 1\right) + \lambda_k \cdot \left(\int |r|^k \cdot \rho(r)\, dr - M_k\right).$$

Differentiating this expression with respect to each unknown $\rho(r)$ and equating the derivative to 0, we conclude that

$$\frac{1}{\beta+1} \cdot (\rho(r))^{-\beta/(\beta+1)} = -\lambda_0 + \lambda_k \cdot |r|^k,$$

hence

$$\rho(r) = \frac{1}{(C_0 + C_1 \cdot |r|^k)^{1+1/\beta}},$$

for some constants C_0 and C_1.

Asymptotically, when $|r|$ increases, we have

$$\rho(r) \sim |r|^{-\alpha}$$

for $\alpha = k \cdot \left(1 + \dfrac{1}{\beta}\right)$. Thus, *we indeed have an explanation for the effectiveness of statistical methods based on heavy-tailed distributions.*

Comment. In this paper, we concentrated on the use of heavy-tail-based statistical methods in econometric (and similar) applications. However, such methods are also effectively used in many other application areas, e.g., in the analysis of graphs and networks; see, e.g., [2] and references therein. It would be interesting to analyze to what extent similar ideas can explain the effectiveness of these methods in network studies and in other application areas.

Acknowledgments. We acknowledge the partial support of the Center of Excellence in Econometrics, Faculty of Economics, Chiang Mai University, Thailand. This work was also supported in part by the National Science Foundation grants HRD-0734825 and HRD-1242122 (Cyber-ShARE Center of Excellence) and DUE-0926721, and by an award "UTEP and Prudential Actuarial Science Academy and Pipeline Initiative" from Prudential Foundation.

The authors are thankful to the anonymous referees for valuable suggestions.

References

1. Aczél, J.: Lectures on Functional Equations and Their Applications. Dover, New York (2006)
2. Barabasi, A.-L.: Network Science. Cambridge University Press, Cambridge (2016)
3. Beirlant, J., Goegevuer, Y., Teugels, J., Segers, J.: Statistics of Extremes: Theory and Applications. Wiley, Chichester (2004)
4. Chakrabarti, B.K., Chakraborti, A., Chatterjee, A.: Econophysics and Sociophysics: Trends and Perspectives. Wiley-VCH, Berlin (2006)
5. Chatterjee, A., Yarlagadda, S., Chakrabarti, B.K.: Econophysics of Wealth Distributions. Springer-Verlag Italia, Milan (2005)
6. Farmer, J.D., Lux, T. (eds.) Applications of statistical physics in economics and finance, a special issue of the Journal of Economic Dynamics and Control, vol. 32(1), pp. 1–320 (2008)
7. Fishburn, P.C.: Utility Theory for Decision Making. Wiley, New York (1969)
8. Gomez, C.P., Shmoys, D.B.: Approximations and randomization to boost CSP techniques. Ann. Oper. Res. **130**, 117–141 (2004)
9. Jaynes, E.T., Bretthorst, G.L.: Probability Theory: The Logic of Science. Cambridge University Press, Cambridge (2003)
10. Kreinovich, V., Kosheleva, O., Nguyen, H.T., Sriboonchitta, S.: Why some families of probability distributions are practically efficient: a symmetry-based explanation. In: Huynh, V.N., Kreinovich, V., Sriboonchitta, S. (eds.) Causal Inference in Econometrics, pp. 133–152. Springer Verlag, Cham (2016)
11. Luce, R.D., Raiffa, R.: Games and Decisions: Introduction and Critical Survey. Dover, New York (1989)

12. Malkiel, B.G.: A Random Walk Down Wall Street: The Time-Tested Strategy for Successful Investing. W. W. Norton & Company, New York (2007)
13. Mandelbrot, B.: The variation of certain speculative prices. J. Bus. **36**, 394–419 (1963)
14. Mandelbrot, B.: The Fractal Geometry of Nature. Freeman, San Francisco (1983)
15. Mandelbrot, B., Hudson, R.L.: The (Mis)behavior of Markets: A Fractal View of Financial Turbulence. Basic Books, New York (2006)
16. Markovich, N. (ed.): Nonparametric Analysis of Univariate Heavy-Tailed Data: Research and Practice. Wiley, Chichester (2007)
17. McCauley, J.: Dynamics of Markets, Econophysics and Finance. Cambridge University Press, Cambridge (2004)
18. Nguyen, H.T., Kosheleva, O., Kreinovich, V.: Decision making beyond Arrow's 'impossibility theorem', with the analysis of effects of collusion and mutual attraction. Int. J. Intell. Syst. **24**(1), 27–47 (2009)
19. Novitskii, P.V., Zograph, I.A.: Estimating the Measurement Errors. Energoatomizdat, Leningrad (1991). (in Russian)
20. Orlov, A.I.: How often are the observations normal? Ind. Lab. **57**(7), 770–772 (1991)
21. Rabinovich, S.G.: Measurement Errors and Uncertainty: Theory and Practice. Springer Verlag, Berlin (2005)
22. Raiffa, H.: Decision Analysis. Addison-Wesley, Reading (1970)
23. Resnick, S.I.: Heavy-Tail Phenomena: Probabilistic and Statistical Modeling. Springer-Varlag, New York (2007)
24. Sheskin, D.J.: Handbook of Parametric and Nonparametric Statistical Procedures. Chapman and Hall/CRC, Boca Raton, Florida (2011)
25. Stoyanov, S.V., Racheva-Iotova, B., Rachev, S.T., Fabozzi, F.J.: Stochastic models for risk estimation in volatile markets: a survey. Ann. Oper. Res. **176**, 293–309 (2010)
26. Vasiliki, P., Stanley, H.E.: Stock return distributions: tests of scaling and universality from three distinct stock markets. Phy. Rev. E Stat. Nonlinear Soft Matter Phy. **77**(3), Pt. 2, Publ. 037101 (2008)

A New Method for Hypothesis Testing Using Inferential Models with an Application to the Changepoint Problem

Son Phuc Nguyen[1]([✉]), Uyen Hoang Pham[1], Thien Dinh Nguyen[1], and Hoa Thanh Le[2]

[1] University of Economics and Law,
Khu pho 3, Linh Xuan, Thu Duc, Ho Chi Minh City, Vietnam
sonnp@uel.edu.vn
[2] Ho Chi Minh City University of Science,
227 Nguyen Van Cu st., Ho Chi Minh City, Vietnam

Abstract. Hypothesis testing, which has been studied since the time of Fisher, Neyman and Pearson, is a fundamentally important task in statistics. As of now, the classical p-value has been extensively used for more than half a century. However, a number of serious drawbacks have been documented over the years. With the flourish of data science recently, there is a growing demand for better approaches. In this paper, we propose a novel method for hypothesis testing based on the inferential models by Martin and Liu. Our approach not only avoids all major weaknesses of the classical p-value but also provides considerable flexibility in perform testing. Besides, in this regard, the inferential model has some advantages over the popular Bayesian framework.

As for application, the hazard rate estimation in the changepoint problem is investigated with the Down Jones index data. In particular, explicit computations are performed, and followed by a set of graphs at the changepoints.

Keywords: Inferential models · Changepoints · Hypothesis testing · Random sets · Discrete time series

1 Introduction

1.1 p-value Misuses, Misinterpretations and Some Alternatives

Since the American Statistical Association's statement on p-value was issued in March 2016, see [1], there have been a lot of discussions on various aspects of the misuses and misinterpretations of p-value in statistical data analysis, see for example [2]. The main criticism is that p-values are not the probabilities that null hypotheses are true. Therefore, a p-value by itself should not be used to perform any statistical testing.

Although a number of alternatives for p-value have also been suggested, many researchers in statistics and data analysis believe there is no quick fix for

© Springer International Publishing AG 2016
V.-N. Huynh et al. (Eds.): IUKM 2016, LNAI 9978, pp. 532–541, 2016.
DOI: 10.1007/978-3-319-49046-5_45

the problem of p values. The following excerpt by Andrew Gelman is extracted from [3]

> In summary, I agree with most of the ASAs statement on p-values but I feel that the problems are deeper, and that the solution is not to reform p-values or to replace them with some other statistical summary or threshold, but rather to move toward a greater acceptance of uncertainty and embracing of variation.

Most proposals to replace or supplement p-values are based on the Bayesian approach. In particular, Bayes factor plays a central role in these methods, see [6]. Bayes factor allows us to consider both the null and the alternative hypotheses at the same time. More precisely, let D be the observed data. The Bayes factor K is defined as

$$K = \frac{P(D \mid H_0)}{P(D \mid H_1)} \tag{1}$$

So, the Bayes factor determines which hypothesis the observed data D support better. However, the Bayes factor depends on the prior densities on the parameter space. As pointed out a lot of times in the literature, in order for the Bayes factor to be useful, **proper** prior densities are required. To ensure properness of the densities, the full prior, likelihood and posterior have to be in detail, and complete sensitivity analysis has to be performed. In reality, those tasks are usually very hard to carry out, and extremely computationally demanding.

In this paper, we propose a new alternative approach for hypothesis testing based on the inferential models, see [7], that does not require specifying proper densities of parameters, hence, to some extend, our approach offers a more objective and less computationally expensive method for statistical hypothesis testing.

There are two key ideas. The first one is to make more use of the sampling model to create auxiliary variables which do not depend on the desired parameters. The second idea is that, instead of real-valued variables, use set-valued variables by setting up a random set, and then, compute two functions, namely the belief and plausibility functions to express the upper and lower bounds of how likely an assertion (hypothesis) is.

1.2   Introduction to Inferential Model

In a classical inference problem in statistics, one has an unknown parameter (or many unknown parameters) of interest, and one tries to utilize experience, in the form of some observed data, to obtain certain knowledge about the desired parameter. In the classical approach, one relies on repeated sampling from the set of all possible datasets to form probabilistic assessments (for instance, confidence interval) about the parameter. These probabilistic measurements do not depend on the observed data, thus, their credibility is questionable. The Bayesian framework, on the other hand, is a systematic approach to acquire measures of uncertainty which are data dependent; therefore, provides greater flexibility in interpreting what the observed data imply about the parameter. However, in

Bayesian method, one has to provide a prior belief about the unknown parameter in the form of a prior probability distribution on the parameter space. This actually imposes relative importance on the parameter space (even in the case of the so-called uninformative prior). For this reason, there is a quest for probabilistic inference without any prior specific. Early effort in this direction includes Fisher's fiducial inference and the Dempster-Shafer theory. Inheriting these seminal work, Liu and Martin (see [7,12]) have developed the inferential models, a novel framework for prior-free statistical inference.

In brief, inferential models (IM) start with a sampling model a which generates data x dependent on certain unknown parameter θ. Then, the IM associates data x, parameter θ, some auxiliary variable u and a valid predictive random set S via the sampling model a in order to produce prior-free, postdata probabilistic measures of uncertainty about the unknown parameter θ. Below are the three steps of an IM. See [7] for the details

A-step. Associate the unknown parameter θ with each possible (x, u) pair to obtain a collection of sets $\Theta_x(u)$ of candidate parameter values.

P-step. Let $u^\star$ be the unobserved value of the auxiliary variable u. Predict $u^\star$ with a valid predictive random set S.

C-step. Combine $X = x$, $\Theta_x(u)$, and S to obtain a random set $\Theta_x(S) = \cup_{u \in S}\Theta_x(u)$. Then, for any assertion $\mathsf{A} \subseteq \Theta$, compute the probability that the random set $\Theta_x(S)$ is a subset of A as a measure of the available evidence in x supporting A.

1.3 Introduction to the Changepoint Problem

A data sequence is assumed to be generated by a process depending on parameters. Changepoints are abrupt variations in the generative parameters of the data sequence of interest. In reality, changepoints often represent important events in the underlying process. Therefore, they are very useful in modeling and prediction of time series in application areas such as finance, biometrics, and robotics. Recently, there have been considerable interest in detecting changepoints in time series with a number of different approaches including frequentist and bayesian methods. However, there is still a great demand for further research since changepoints are difficult to identify in noisy data streams.

In this work, we propose a novel method for changepoint prediction based on the inference models. In our case, the hazard rate h is assumed to be an unknown constant, and it can be decided, at any time period t, whether a changepoint happens or not. Moreover, an explicit parameterized random set is built, detailed computations are carried out, and criteria for hypothesis testing of the hazard rate h are provided.

1.4 Summary of Our Contributions

In this paper, we propose an alternative method for statistical hypothesis testing utilizing the inferential models, a framework for statistical inferences invented by Martin and Liu, see [7,12].

For application, we apply the inferential models to introduce a new method to do statistical inference for the hazard rate h in the changepoint problem.

2   Main Results

2.1   The Model Setup

In this work, we assume that the hazard rate h is an unknown constant, and that we know the locations of the changepoints up to any time period t. The purport of all of the assumptions is that we know for certain the changepoint count a_t up to time t which allows us to model (discrete) time series using a Bernoulli process as follows

Let y_t be a Bernoulli random variable such that

$$
\begin{aligned}
y_t = 1 \quad &\text{if a changepoint happen at time } t \\
y_t = 0 \quad &\text{if a changepoint doesn't happen at time } t
\end{aligned}
\tag{2}
$$

Note that (y_t) can be considered as an independent Bernoulli process (in discrete time) with a rate equal to the hazard rate h. Therefore, the hazard rate h can be inferred as the rate of the process given the binary observations. In other words, the inferred prediction of the hazard rate for time t can be written as $\tilde{h}_t = P(y_t = 1 \mid y_{1:t-1})$ where, by definition, $y_{1:t-1} = \{y_1, \ldots, y_{t-1}\}$.

For the inferential models, we sample u_i from the uniform distribution on the interval $[0, 1]$, and use the model

$$
y_i = I_{[0,h]}(u_i), \quad i = 1, \ldots, t
\tag{3}
$$

Putting altogether, we have an the auxiliary variable u which is a vector

$$
\begin{aligned}
u &= (u_1, \ldots, u_t) \\
u &\sim \text{Uniform}([0, 1]^t)
\end{aligned}
\tag{4}
$$

The selected random set is as follows

$$
S(u) = [A_1(u), B_1(u)] \times \cdots \times [A_t(u), B_t(u)]
\tag{5}
$$

where

$$
\begin{aligned}
A_i(u) &= u_i - \omega u_i \\
B_i(u) &= u_i + \omega(1 - u_i)
\end{aligned}
\tag{6}
$$

where ω is a prespecified parameter in $[0, 1]$

Using the model (3), the focal element of u with respect to the random set S will be

$$
M_t(u, S) = \{u_{(a_t)} - \omega u_{(a_t)}, \ u_{(a_{t+1})} + \omega(1 - u_{(a_{t+1})})\}
\tag{7}
$$

where $u_{(n)}$ denotes the nth order statistic of u. By a standard result, $u_{(n)}$ is a beta random variable with distribution $B(n, t + 1 - n)$.

2.2 Hypothesis Testing

In this section, we will go over the two most popular tests, namely the one-tailed and two-tailed hypothesis testing similar to those in classical statistics. Note that tests can be extended easily to the general case $h \in A$ where A is a subset of the parameter space $[0, 1]$.

Define $F_{B(\alpha, \beta)}$ to the beta distribution function with the shape parameters $\alpha, \beta \in \mathbb{R}$ and $\alpha, \beta > 0$.

<u>Case 1</u>: Consider the one-tailed hypothesis $A = \{h \leq h_0\}$, with the random set as in (5), the belief function for the assertion A has the following closed-form formula:

$$
\begin{aligned}
\mathrm{Bel}_t(A) &= \mu\Big(\{u \in [0, 1]^t \mid M_{a_t}(u, S) \subseteq A\}\Big) \\
&= \mu\Big(\{u \in [0, 1]^t \mid u_{(a_t+1)} + \omega(1 - u_{(a_t+1)}) \leq h_0\}\Big) \\
&= F_{B(a_t+1, t-a_t)}\left(\frac{h_0 - \omega}{1 - \omega}\right)
\end{aligned}
\tag{8}
$$

Similarly, the plausibility function is as follows

$$
\begin{aligned}
\mathrm{Plau}_t(A) &= 1 - \mathrm{Bel}_t(A^c) \\
&= F_{B(a_t, t+1-a_t)}\left(\frac{h_0}{1 - \omega}\right)
\end{aligned}
\tag{9}
$$

<u>Case 2</u>: In the same way, the hypothesis $A = \{h \geq h_0\}$ has the belief function:

$$
\begin{aligned}
\mathrm{Bel}_t(A) &= \mu\Big(\{u \in [0, 1]^t \mid M_{a_t}(u, S) \subseteq A\}\Big) \\
&= \mu\Big(\{u \in [0, 1]^t \mid u_{(a_t)} - \omega u_{(a_t)} \geq h_0\}\Big) \\
&= 1 - F_{B(a_t, t+1-a_t)}\left(\frac{h_0}{1 - \omega}\right)
\end{aligned}
\tag{10}
$$

Similarly, the plausibility function:

$$
\begin{aligned}
\mathrm{Plau}_t(A) &= 1 - \mathrm{Bel}_t(A^c) \\
&= 1 - F_{B(a_t+1, t-a_t)}\left(\frac{h_0 - \omega}{1 - \omega}\right)
\end{aligned}
\tag{11}
$$

<u>Case 3</u>: Consider the two-tailed hypothesis $A = \{h = h_0\}$, this is the point hypothesis, and the calculation leads to $\mathrm{Bel}_t(A) = 0$. However, the plausibility function is still nontrivial

$$
\begin{aligned}
\mathrm{Plau}_t(A) &= 1 - \mathrm{Bel}_t(A^c) \\
&= 1 - \Big[\mu\big(\{u \mid u_{(a_t+1)} + \omega(1 - u_{(a_t+1)}) \leq h_0\}\big) + \mu\big(\{u \mid u_{(a_t)} - \omega u_{(a_t)} \geq h_0\}\big)\Big] \\
&= 1 - \Big[F_{B(a_t+1, t-a_t)}\left(\frac{h_0 - \omega}{1 - \omega}\right) + 1 - F_{B(a_t, t+1-a_t)}\left(\frac{h_0}{1 - \omega}\right)\Big] \\
&= F_{B(a_t, t+1-a_t)}\left(\frac{h_0}{1 - \omega}\right) - F_{B(a_t+1, t-a_t)}\left(\frac{h_0 - \omega}{1 - \omega}\right)
\end{aligned}
\tag{12}
$$

The next question to be considered is how to make a decision on whether to reject or to not reject the hypothesis. In this paper, we suggest the following procedure:

Let α be a significance level, $\alpha \in [0, 1]$.

Procedure A:

- If $\text{Plau}_t(\mathbf{A}) \leq \alpha$ then reject the hypothesis **A**. Note that $\text{Bel}_t(\mathbf{A}) \leq \text{Plau}_t(\mathbf{A})$, thus, $\text{Bel}_t(\mathbf{A}) \leq \alpha$. The conclusion here is that we rule out the hypothesis **A**, and believe in the alternative hypothesis $\mathbf{A}^c$.
- In the case $\text{Plau}_t(\mathbf{A}) > \alpha$ and $\text{Bel}_t(\mathbf{A}) < \alpha$, the interpretation is that we cannot be sure whether **A** is believable (i.e. an appropriate hypothesis) since $\text{Bel}_t(\mathbf{A})$ is small; however, **A** is still plausible (i.e. we also cannot rule **A** out) since $\text{Plau}_t(\mathbf{A})$ is large. Here, if we must choose either the hypothesis **A** or the alternative hypothesis $\mathbf{A}^c$ then compare the two belief functions $\text{Bel}_t(\mathbf{A})$ and $\text{Bel}_t(\mathbf{A}^c)$, the hypothesis with larger belief function is better supported by the observed data (t, a_t).

Remark 1. 1. The indecisive case comes from the fact that any dataset X can be partitioned into three parts
   (a) *Part one:* All $x \in X$ that support the hypothesis **A**
   (b) *Part two:* All $x \in X$ that support the alternative hypothesis $\mathbf{A}^c$.
   (c) *Part three:* All $x \in X$ that support neither **A** nor $\mathbf{A}^c$. The data in this group is sort of *neutral* data that show no preference for **A** or $\mathbf{A}^c$
2. The belief function $\text{Bel}_t(\mathbf{A})$ is designed to measure *part one*, the set of data that support the assertion **A**, while the plausibility function $\text{Plau}_t(\mathbf{A})$ measures both *part one* and *part three* altogether, meaning the set of data that are not against the assertion **A**
3. Dempster (see [11], p. 375), while comparing between p-values and plausibilities, has pointed out a similar result as in the second point in this remark.

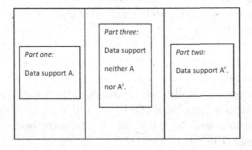

3   An Application

In this section, we provide an application in making prediction on the hazard rate h in the changepoint problem. Recall that the hazard rate h captures how often a changepoint happens in a time series.

The data in this study are the Dow Jones Industrial Average (DJIA) collected from Bloomberg. The historical data from May/2016 to June/2016 (2047 observations). Considering the mean of the series, the number of changepoints are determined by approximate (BinSeg) methods. A changepoint is marked as the first observation of the new segment. There are 5 change points of DJIA Index from May, 2016 to June, 2016 (2047 observations). Changepoints are at $t = 89, 312, 608, 1179, 1453$.

The figure below shows the data and the positions of the changepoints (Fig. 1).

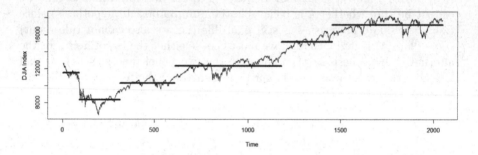

Fig. 1. DJIA Index, May, 2016 to June, 2016

As we can see from the figure, changepoints seldom happen, therefore the hazard rate h in this set of data is close to zero. That leads us to choose a family of asymmetric random set S depending on a parameter ω as in formula (5). On one hand, for the lower endpoints of the predictive intervals, different values of the parameter ω yield almost indistinguishable values for the plausibility and the belief functions. This is somewhat justified since we do not have much room for prediction when h is so close to zero. On the other hand, the upper endpoints show a totally different scenario; higher values of ω clearly predict the rate h to be farther from zero (Fig. 2).

In what follows, h_0 varies in $[0, 1]$; computations for the plausibility and the belief functions (as functions of h_0) of the lower-tail hypothesis $\mathtt{A} = \{h \leq h_0\}$ and the two-tailed hypothesis $\mathtt{A} = \{h = h_0\}$ are carried out for all five changepoints in the dataset. There are totally five figures for the five changepoints. On the left of each figure are the graphs of the belief and the plausibility functions for the lower tail hypothesis and on the right are the corresponding graphs for the two-tailed hypothesis (Fig. 3).

For the random set S, $\omega = 0$, 0.2, and 0.5 are selected. Each value produces a pair of graphs for the belief and the plausibility functions. Note that, when $\omega = 0$, the random set S is a singleton random set, and the prediction is exactly that of the famous Dempster-Shafer theory, see [11]. In addition, $\omega = 0$ is the most conservative prediction, staying close to the observed ratio $\frac{a_t}{t}$ while $\omega = 0.5$ is the most liberal prediction, allowing values considerably higher than the observed ratio $\frac{a_t}{t}$ (Fig. 4).

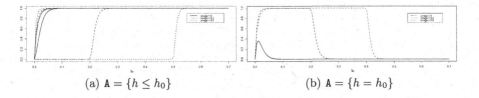

(a) $\mathtt{A} = \{h \leq h_0\}$ (b) $\mathtt{A} = \{h = h_0\}$

Fig. 2. Changepoint 1 at $t = 89$

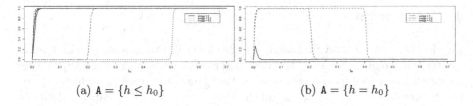

(a) $\mathtt{A} = \{h \leq h_0\}$ (b) $\mathtt{A} = \{h = h_0\}$

Fig. 3. Changepoint 2 at $t = 312$

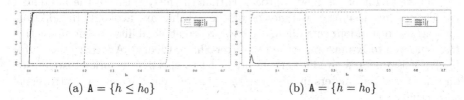

(a) $\mathtt{A} = \{h \leq h_0\}$ (b) $\mathtt{A} = \{h = h_0\}$

Fig. 4. Changepoint 3 at $t = 608$

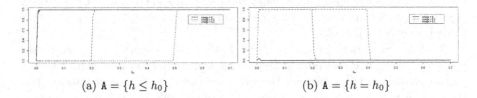

(a) $\mathtt{A} = \{h \leq h_0\}$ (b) $\mathtt{A} = \{h = h_0\}$

Fig. 5. Changepoint 4 at $t = 1179$

Our final remark is that graphs of the two functions can be drawn at any time period t, whether t is a changepoint or non-changepoint. Each time the estimated rate $\tilde{h}_t$ is updated. Thus, this framework is suitable for studying streaming time series where we would like to have updated predictions after each discrete time step (Figs. 5 and 6).

(a) $A = \{h \leq h_0\}$ (b) $A = \{h = h_0\}$

Fig. 6. Changepoint 5 at $t = 1453$

4   Conclusion

To address the need of an alternative for the classical p-value approach in hypothesis testing, we propose using the inferential models with two functions to express our prediction of a hypothesis A; namely, the plausibility functions which capture the data evidence not against A, and the belief functions which capture data evidence supporting A. Moreover, we adopt **Procedure A** on page 6 as the criteria for rejecting hypotheses.

One highlight of the new approach is that the mechanism of the inferential models ensures that the weaknesses of the p-value are avoided. In addition, making use of random sets allows us to have great flexibility in our prediction (as compared to, for instance, the Dempster-Shafer theory). Another remarkable highlight is that our predictions do not need to specify any prior distribution on the desired parameters in advance, thus, circumvent a bayesian controversial subjectivity.

We also present an application in estimation of the hazard rate in the changepoint problem. In particular, we select a parametric random set appropriate for the low-hazard-rate Down Jones index data in hand. Then, a set of plausibility and belief graphs are drawn. If desired, conclusions can be drawn based on the **Procedure A** on page 6.

Another theoretical important point is that the random set chosen in the application is a *valid* random set, (see [7] for more details on valid random sets), so any prediction is reasonable. However, we have not yet taken into account the *efficiency* of random set in this research. In the collection of all random sets, stochastic dominance serves as a partial ordering, so, optimal random sets are well defined; in fact, they can be approximated using stochastic approximation algorithms. This will be a topic for our further research. Our goal is to find a random set which produces exact coverage probability for any assertion.

As for applications, we would like to explore the changepoint problem where the hazard rate h is not a constant, but rather a random variable. This direction involves a lot of uncertainties, hence, requires a method to combine all of them to produce meaningful estimations.

Acknowledgements. We would like to express our deep gratitude to professor Hung T. Nguyen of New Mexico State University/Chiang Mai university for bringing the inferential models to our attention, for his encouragements, and for numerous helpful discussions.

References

1. Wasserstein, R.L., Lazar, N.A.: The ASA's statement on p-values: context, process, and purpose. Am. Stat. **70**, 129–133 (2016)
2. Goodman, S.: A dirty dozen: twelve P-value misconceptions. Semin. Hematol. **45**, 135–140 (2008). Elsvier
3. Gelman, A.: P values and statistical practice. Epidemiology **24**, 69–72 (2012)
4. Martin, R., Liu, C.: A note on p-values interpreted as plausibilities. Stat. Sin. **24**, 1703–1716 (2014)
5. Martin, R.: Plausibility functions and exact frequentist inference. J. Am. Stat. Assoc. **110**(512), 1552–1561 (2015)
6. Wetzels, R., Grasman, R., Wagenmakers, E.J.: A default Bayesian hypothesis test for ANOVA designs. Am. Stat. **66**(2), 104–111 (2012)
7. Martin, R., Liu, C.: Inferential Models: Reasoning with Uncertainty, vol. 145. CRC Press, New York (2015)
8. Martin, R., Zhang, J., Liu, C.: Dempster-Shafer theory and statistical inference with weak beliefs. Statistical Science, 72–87 (2010)
9. Wilson, R.C., Nassar, M.R., Gold, J.I.: Bayesian online learning of the hazard rate in change-point problems. Neural Comput. **22**(9), 2452–2476 (2010)
10. Adams, R.P., MacKay, D.J.C.: Bayesian online changepoint detection. arXiv preprint arXiv:0710.3742 (2007)
11. Dempster, A.P.: The Dempster-Shafer calculus for statisticians. Int. J. Approx. Reason. **48**(2), 365–377 (2008)
12. Martin, R., Liu, C.: Inferential models: a framework for prior-free posterior probabilistic inference. J. Am. Stat. Assoc. **108**(501), 301–313 (2013)
13. Martin, R.: Random sets and exact confidence regions. Sankhya A **76**(2), 288–304 (2014)

Confidence Intervals for the Ratio of Coefficients of Variation in the Two-Parameter Exponential Distributions

Patarawan Sangnawakij, Sa-Aat Niwitpong[✉], and Suparat Niwitpong

Applied Statistics, King Mongkut's University of Technology North Bangkok,
Bangkok 10800, Thailand
patarawan.s@gmail.com, sa-aat.n@sci.kmutnb.ac.th

Abstract. Ratio of coefficients of variation is one of statistical measurements used in many fields of applied research. However, the problem in statistical inference of the ratio of coefficients of variation has been little studied. In this paper, two new confidence intervals for this measure in the two-parameter exponential distributions are introduced based on the method of variance of estimates recovery (MOVER) and the generalized confidence interval (GCI). We use Monte carlo simulation to conduct the performance of the estimators. The results indicate that the coverage probabilities of the confidence interval based on the MOVER and the GCI maintain the nominal coverage level. The GCI has shorter expected length than the MOVER in most cases. In addition, real-world data are analyzed to illustrate the findings of the paper.

Keywords: Coefficient of variation · Generalized confidence interval · Method of variance of estimates recovery · Two-parameter exponential distribution · Simulation

1 Introduction

Estimation of the ratio of coefficients of variation is frequently used in applied research. In economics, the estimated ratio of coefficients of variation of staple food balance sheet, e.g. balance sheet of sugar and corn, or rice and soybean, is reported in order to indicate food shortage and price disparity in Indonesia [1]. In medical genetics, the estimated ratio of coefficients of variation on the expression level of proteins and that of mRNA are used to investigate the additional contribution of translation of gene age in humans [2]. Furthermore, it is applied in many applications, see e.g. [3,4].

Although the ratio of coefficients of variation is often used as a point estimator, its interval estimator has also been studied in the recent past. For example, the construction of confidence intervals for the ratio of coefficients of variation is concerned in the normal distribution and other continuous distributions, including the delta-lognormal and the gamma distributions [5–9]. However, in many

© Springer International Publishing AG 2016
V.-N. Huynh et al. (Eds.): IUKM 2016, LNAI 9978, pp. 542–551, 2016.
DOI: 10.1007/978-3-319-49046-5_46

areas of research, the exponential distribution plays a role similar to the normal distribution, particularly two-parameter case [10].

The distribution of interest mentioned before is the two-parameter exponential distribution. This probability model is often used to represent the time to failure in several studies, such as lifetime, survival, and reliability analysis [11,12]. Moreover, it is usually applied in practical statistical models of environmental pollution [13–15]. The probability density function of the two-parameter exponential distribution, denoted as $Exp(\lambda, \theta)$, is given by

$$f_X(x; \lambda, \theta) = \frac{1}{\lambda} \exp\{-(x - \theta)/\lambda\}, \quad x > \theta, \lambda > 0, \theta \in R, \tag{1}$$

where λ and θ are the scale and location parameters, respectively [16]. Recently, Sangnawakij and Niwitpong [17] proposed new confidence intervals for the single and the difference of coefficients of variation in the two-parameter exponential distributions. However, the problem of the ratio of coefficients of variation for this distribution has not yet been studied for constructing the confidence interval.

Therefore, the purpose of this paper is to introduce new confidence intervals for the ratio of coefficients of variation in the two-parameter exponential distributions. The methods for building these confidence intervals are the method of variance of estimates recovery and the generalized confidence interval. The details of these ideas are described in Sects. 3 and 4, respectively. We evaluate the performance of the proposed confidence intervals in terms of coverage probability and expected length through simulations, and then show the numerical results in Sect. 5. In Sect. 6, the computation on a real-world example is illustrated. Finally, concluding remarks are provided in Sect. 7.

2   Notations

Let $X = (X_1, X_2, \ldots, X_n)$ and $Y = (Y_1, Y_2, \ldots, Y_m)$ be two random samples of size n and m drawn from $Exp(\lambda, \theta)$ and $Exp(\lambda_1, \theta_1)$, respectively. The mean and variance of X are $E(X) = \lambda + \theta$ and $Var(X) = \lambda^2$, respectively. Moreover, we have the mean and variance of Y, $E(Y) = \lambda_1 + \theta_1$ and $Var(Y) = \lambda_1^2$. Since the coefficient of variation is defined by the ratio of the standard deviation of the distribution to the mean, the ratio of coefficients of variation in the two-parameter exponential distributions can be written as

$$\xi = \frac{\tau}{\tau_1} = \frac{\lambda/(\lambda + \theta)}{\lambda_1/(\lambda_1 + \theta_1)}, \tag{2}$$

where τ and τ_1 are the coefficients of variation of X and Y, respectively.

To estimate the point estimator of ξ, maximum likelihood estimation is applied. For variable X, the maximum likelihood estimators for λ and θ are $\hat{\lambda} = \bar{X} - X_{(1)}$ and $\hat{\theta} = X_{(1)} = \min(X_1, X_2, \ldots, X_n)$, respectively. Here, $\bar{X} = \sum_{i=1}^{n} X_i/n$ is the estimated mean of X. Similarly, we also obtain the estimators for λ_1 and θ_1 from variable Y. Therefore, the estimated ratio of coefficients of variation for ξ is given as

$$\hat{\xi} = \frac{\hat{\tau}}{\hat{\tau}_1} = \frac{(\bar{X} - X_{(1)})/\bar{X}}{(\bar{Y} - Y_{(1)})/\bar{Y}}, \tag{3}$$

where $\hat{\tau}$ and $\hat{\tau}_1$ are the estimated coefficients of variation for τ and τ_1, respectively.

3   Confidence Interval Based on the Method of Variance of Estimates Recovery

In this section, we first briefly review the concept of the method of variance of estimates recovery (MOVER) for establishing the confidence interval. The idea of this method is to find separate confidence intervals for two single parameters, recover the variance estimates, and then form the confidence interval for the function of parameters of interest, for example $\varepsilon + \varepsilon_1$ and $\varepsilon/\varepsilon_1$ [6,19].

This approach is based on the central limit theorem. In this paper, we focus on the confidence interval for the parameter of ratio function. Following Donner and Zou [6], the confidence interval for $\varepsilon/\varepsilon_1$ is given by

$$[L_r', U_r'] = \left[\frac{\hat{\varepsilon}\hat{\varepsilon}_1 - \sqrt{(\hat{\varepsilon}\hat{\varepsilon}_1)^2 - l'u_1'(2\hat{\varepsilon} - l')(2\hat{\varepsilon}_1 - u_1')}}{u_1'(2\hat{\varepsilon}_1 - u_1')}, \right.$$
$$\left. \frac{\hat{\varepsilon}\hat{\varepsilon}_1 + \sqrt{(\hat{\varepsilon}\hat{\varepsilon}_1)^2 - u'l_1'(2\hat{\varepsilon} - u')(2\hat{\varepsilon}_1 - l_1')}}{l_1'(2\hat{\varepsilon}_1 - l_1')} \right], \tag{4}$$

where $\hat{\varepsilon}$ and $\hat{\varepsilon}_1$ are the point estimators, and $[l', u']$ and $[l_1', u_1']$ are the confidence intervals for ε and ε_1, respectively. Since the parameter function that we are interested is the ratio of two coefficients of variation, the general form in Eq. (4) is applied to find the confidence interval for ξ.

Now, we set $\hat{\varepsilon} = \hat{\tau}$, $\hat{\varepsilon}_1 = \hat{\tau}_1$, $l' = L_{mov}$, $u' = U_{mov}$, $l_1' = L_{mov1}$, and $u_1' = U_{mov1}$ in Eq. (4), where $[L_{mov}, U_{mov}]$ and $[L_{mov1}, U_{mov1}]$ are the confidence intervals based on the MOVER for τ and τ_1, respectively, introduced by Sangnawakij and Niwitpong [17].

Therefore, the $(1 - \gamma)100\%$ confidence interval based on the MOVER for $\xi = \tau/\tau_1$, $CI_{r1} = [L_{rmov}, U_{rmov}]$, is given by

$$CI_{r1} = \left[\frac{\hat{\tau}\hat{\tau}_1 - \sqrt{(\hat{\tau}\hat{\tau}_1)^2 - L_{mov}U_{mov1}(2\hat{\tau} - L_{mov})(2\hat{\tau}_1 - U_{mov1})}}{U_{mov1}(2\hat{\tau}_1 - U_{mov1})}, \right.$$
$$\left. \frac{\hat{\tau}\hat{\tau}_1 + \sqrt{(\hat{\tau}\hat{\tau}_1)^2 - U_{mov}L_{mov1}(2\hat{\tau} - U_{mov})(2\hat{\tau}_1 - L_{mov1})}}{L_{mov1}(2\hat{\tau}_1 - L_{mov1})} \right]. \tag{5}$$

4   Generalized Confidence Interval

The generalized confidence interval is a method for building confidence intervals for complex parametric functions in many common applications [20]. This approach was introduced by Weerahandi [18] based on the concept that the confidence interval is obtained by inverting the distribution of a generalized pivotal quantity. The generalized pivotal quantity attributed to Weerahandi is described as follows.

Let X be a random sample with probability density function $f_X(x; \Delta)$, the observed value x, and vector of unknown parameters $\Delta = (\lambda, \theta)$. $R(X; x, \Delta)$ is a generalized pivotal quantity for parameter of interest, if it satisfies the following two properties:

(i) When x is fixed, the function $R(X; x, \Delta)$ has a probability distribution free from unknown parameters.
(ii) The observed value of $R(X; x, \Delta)$, denoted as $r(x; x, \Delta)$, does not depend on nuisance parameters.

Here, we consider the pivotal quantities in the two-parameter exponential distributions. From $X \sim \mathrm{Exp}(\lambda, \theta)$ and $Y \sim \mathrm{Exp}(\lambda_1, \theta_1)$, it is straightforward to see that

$$W_1 = \frac{2n(\hat{\theta} - \theta)}{\lambda}, \quad W_2 = \frac{2n\hat{\lambda}}{\lambda}, \quad V_1 = \frac{2m(\hat{\theta}_1 - \theta_1)}{\lambda_1}, \quad \text{and} \quad V_2 = \frac{2m\hat{\lambda}_1}{\lambda_1},$$

where W_1 and W_2 follow chi-square distributions with 2 and $2n - 2$ degrees of freedom, respectively, and V_1 and V_2 have chi-square distributions with 2 and $2m - 2$ degrees of freedom, respectively. It is easy to verify that the distributions of these pivots do not depend on unknown parameters.

Using the above information, the equivalent of the ratio of coefficients of variation ξ is derived as

$$\begin{aligned}
\xi &= \frac{\lambda}{\lambda + \theta} \Big/ \frac{\lambda_1}{\lambda_1 + \theta_1} \\
&\equiv \left[1 + \left(\frac{\hat{\theta} W_2}{2n\hat{\lambda}} - \frac{W_1}{2n} \right) \right]^{-1} \Big/ \left[1 + \left(\frac{\hat{\theta}_1 V_2}{2m\hat{\lambda}_1} - \frac{V_1}{2m} \right) \right]^{-1} \\
&= \left[1 + \frac{1}{2m} \left(\frac{\hat{\theta}_1 V_2}{\hat{\lambda}_1} - V_1 \right) \right] \Big/ \left[1 + \frac{1}{2n} \left(\frac{\hat{\theta} W_2}{\hat{\lambda}} - W_1 \right) \right].
\end{aligned} \tag{6}$$

Hence we obtain the generalized pivotal quantity for ξ:

$$R^{**}(X, Y; x, y, \xi) = \left[1 + \frac{1}{2m} \left(\frac{y_{(1)} V_2}{\bar{y} - y_{(1)}} - V_1 \right) \right] \Big/ \left[1 + \frac{1}{2n} \left(\frac{x_{(1)} W_2}{\bar{x} - x_{(1)}} - W_1 \right) \right], \tag{7}$$

where $x_{(1)}$, $y_{(1)}$, $\bar{x}$, and $\bar{y}$ are observed values of $X_{(1)}$, $Y_{(1)}$, $\bar{X}$, and $\bar{Y}$, respectively. The generalized pivotal quantity in Eq. (7) has the distribution free from unknown parameters and its observed value $r^{**}(x, y; x, y, \xi) = \xi$ does not depend on on nuisance parameters. Thus, it can be used to establish the confidence interval. Note that Eq. (7) is similar to the generalized pivotal quantity presented by Sangnawakij and Niwitpong [17], who studied the generalized confidence interval for the difference of coefficients of variation in two two-parameter exponential distributions.

Therefore, the $(1 - \gamma)100\%$ generalized confidence interval for $\xi = \tau/\tau_1$, $CI_{r2} = [L_{rgci}, U_{rgci}]$ is given by

$$CI_{r2} = [R^{**}(\gamma/2), R^{**}(1 - \gamma/2)], \tag{8}$$

where $R^{**}(\gamma/2)$ is the $(\gamma/2)100$th percentile of $R^{**}(X, Y; x, y, \xi)$. This confidence interval can be computed using the software package.

The performance of the proposed confidence intervals presented in Eqs. (5) and (8) is investigated by the coverage probability and expected length via Monte Carlo simulation in the next section.

5   Simulation Results

To assess statistical accuracy of the interval estimator, we evaluate the performance of the proposed confidence intervals in terms of coverage probability and expected length using the R statistical program [21]. The estimated coverage probability and expected length are given by

$$\text{C.P.} = \frac{c(L \leq \xi \leq U)}{M} \quad \text{and} \quad \text{E.L.} = \frac{\sum_{h=1}^{M}(U_h - L_h)}{M},$$

respectively, where M is the number of all simulation runs and $c(L \leq \tau \leq U)$ is the number of simulation runs for τ that lies within the confidence interval. Generally, we choose a confidence interval with a coverage probability greater than or close to the nominal coverage level and a short length interval.

The simulation study is conducted in various situations. Attributes of a Monte Carlo technique are specified as follows. The data are generated from two two-parameter exponential distributions, denoted as $\text{Exp}(\lambda, \theta)$ and $\text{Exp}(\lambda_1, \theta_1)$, where λ and λ_1 are fixed at 1.00. θ and θ_1 are adjusted to get the required coefficients of variation, where $(\tau, \tau_1) = (-0.05, -0,05), (-0.05, -0.10), (-0.05, -0.30), (-0.20, 0.20), (-0.20, 0.30), (-0.20, 0.50), (0.05, 0.05), (0.05, 0.10), (0.05, 0.30), (0.10, 0.10), (0.10, 0.20),$ and $(0.10, 0.40)$. The nominal coverage level of $1 - \gamma = 0.95$, the number of simulation runs $M = 10,000$, and 5,000 pivotal quantities for the generalized pivotal computations are used. The coverage probabilities and expected lengths of the 95% confidence intervals for the ratio of coefficients of variation are shown in Table 1.

The results indicate that the coverage probabilities of the confidence interval based on the MOVER, CI_{r1}, are much greater than the nominal coverage level at 0.95 in several cases. However, the coverage probabilities of CI_{r1} are slightly smaller than 0.95 when $n = 500$ and $(\tau, \tau_1) = (-0.05, -0.05), (-0.05, -0.10),$

Table 1. The coverage probability (C.P.) and the expected length (E.L.) of the 95 % confidence intervals for the ratio of coefficients of variation in the two-parameter exponential distributions

n	m	τ	τ_1	C.P.		E.L.	
				CI_{r1}	CI_{r2}	CI_{r1}	CI_{r2}
30	30	−0.05	−0.05	0.9500	0.9480	1.2535	1.2050
		−0.05	−0.10	0.9620	0.9551	0.6246	0.6079
		−0.05	−0.30	0.9653	0.9500	0.2350	0.2223
		−0.20	0.20	0.9998	0.9495	2.5050	1.2067
		−0.20	0.30	0.9999	0.9487	1.6977	0.7743
		−0.20	0.50	0.9888	0.9514	1.0947	0.4371
		0.05	0.05	0.9728	0.9524	1.2236	1.0507
		0.05	0.10	0.9768	0.9479	0.6170	0.5154
		0.05	0.30	0.9900	0.9500	0.2096	0.1542
		0.10	0.10	0.9812	0.9501	1.2307	0.9921
		0.10	0.20	0.9861	0.9490	0.6216	0.4700
		0.10	0.40	0.9964	0.9482	0.3220	0.2112
50	50	−0.05	−0.05	0.9530	0.9480	0.8803	0.8774
		−0.05	−0.10	0.9521	0.9552	0.4455	0.4515
		−0.05	−0.30	0.9540	0.9580	0.1574	0.1638
		−0.20	0.20	0.9943	0.9510	1.2529	0.8785
		−0.20	0.30	0.9961	0.9511	0.8447	0.5638
		−0.20	0.50	0.9985	0.9489	0.5257	0.3156
		0.05	0.05	0.9682	0.9473	0.8789	0.7866
		0.05	0.10	0.9702	0.9513	0.4383	0.3814
		0.05	0.30	0.9867	0.9490	0.1498	0.1150
		0.10	0.10	0.9790	0.9502	0.8806	0.7410
		0.10	0.20	0.9845	0.9498	0.4442	0.3511
		0.10	0.40	0.9963	0.9493	0.2293	0.1575
100	100	−0.05	−0.05	0.9590	0.9517	0.5847	0.5967
		−0.05	−0.10	0.9520	0.9522	0.2949	0.3063
		−0.05	−0.30	0.9470	0.9514	0.2691	0.3128
		−0.20	0.20	0.9773	0.9484	0.6939	0.5897
		−0.20	0.30	0.9836	0.9495	0.4674	0.3787
		−0.20	0.50	0.9925	0.9570	0.2909	0.2129
		0.05	0.05	0.9677	0.9481	0.5866	0.5397
		0.05	0.10	0.9717	0.9480	0.2940	0.2631
		0.05	0.30	0.9842	0.9531	0.0999	0.0789
		0.10	0.10	0.9740	0.9491	0.5871	0.5089
		0.10	0.20	0.9835	0.9510	0.2968	0.2417
		0.10	0.40	0.9943	0.9508	0.1529	0.1085

(continued)

Table 1. (*continued*)

n	m	τ	τ_1	C.P.		E.L.	
				CI_{r1}	CI_{r2}	CI_{r1}	CI_{r2}
200	200	−0.05	−0.05	0.9540	0.9506	0.4063	0.4791
		−0.05	−0.10	0.9530	0.9510	0.2804	0.3322
		−0.05	−0.30	0.9468	0.9501	0.1783	0.2166
		−0.20	0.20	0.9664	0.9500	0.4411	0.4069
		−0.20	0.30	0.9752	0.9507	0.2975	0.2615
		−0.20	0.50	0.9857	0.9507	0.1850	0.1470
		0.05	0.05	0.9643	0.9541	0.4030	0.3757
		0.05	0.10	0.9679	0.9508	0.2020	0.1832
		0.05	0.30	0.9838	0.9483	0.0686	0.0550
		0.10	0.10	0.9712	0.9488	0.4044	0.3555
		0.10	0.20	0.9810	0.9506	0.2036	0.1682
		0.10	0.40	0.9932	0.9516	0.1049	0.0755
500	500	−0.05	−0.05	0.9414	0.9505	0.2511	0.2609
		−0.05	−0.10	0.9310	0.9508	0.1259	0.1336
		−0.05	−0.30	0.9148	0.9506	0.0429	0.0489
		−0.20	0.20	0.9573	0.9510	0.2628	0.2539
		−0.20	0.30	0.9682	0.9494	0.1770	0.1628
		−0.20	0.50	0.9797	0.9484	0.1102	0.0916
		0.05	0.05	0.9589	0.9512	0.2509	0.2357
		0.05	0.10	0.9652	0.9501	0.1257	0.1149
		0.05	0.30	0.9848	0.9495	0.0427	0.0345
		0.10	0.10	0.9695	0.9500	0.2515	0.2231
		0.10	0.20	0.9799	0.9488	0.1267	0.1055
		0.10	0.40	0.9924	0.9489	0.0653	0.0475

and $(-0.05, -0.30)$. Meanwhile, the generalized confidence interval, CI_{r2}, has coverage probabilities close to 0.95 in all cases of the study. In terms of expected length, it was found that two proposed confidence intervals have short length intervals, where the expected lengths of CI_{r2} are slightly narrower than those of CI_{r1} in general.

6   Real Data Example

The data on the time (weeks) to relapse of the patients with acute leukemia treated by a drug 6-mercaptopurine and a placebo [22] are used to illustrate the proposed estimators. We compute the basic statistics, and present them in Table 2. The confidence interval based on the MOVER, CI_{r1}, is evaluated using the formula in Eq. (5) with information from the paper of Sangnawakij and Niwitpong [17]. Then, we use Eq. (8) to compute the generalized confidence interval, CI_{r2}.

Table 2. The results of the time to relapse of children from a drug 6-mercaptopurine and a placebo

Drug 6-mercaptopurine	Placebo
$n = 21$	$m = 21$
$\bar{x} = 17.10$	$\bar{y} = 8.67$
$\hat{\lambda} = 11.10$	$\hat{\lambda}_1 = 7.67$
$\hat{\theta} = 6.00$	$\hat{\theta}_1 = 1.00$
$\hat{\tau} = 0.65$	$\hat{\tau}_1 = 0.88$

It can be concluded that the 95 % confidence interval based on the MOVER is $CI_{r1} = [-0.31, 1.80]$ with the length of interval of 2.11. The generalized confidence interval is $CI_{r2} = [0.59, 0.88]$ with the interval length of 0.29. That means CI_{r2} more accurately estimate the ratio of coefficients of variation for these data. This is because its interval length is shorter than that of CI_{r1}. Clearly, these results support the simulation study presented in the previous section.

7   Conclusions

The ratio of coefficients of variation is a statistical tool applied in many practical applications in order to compare the diversity of the data from two groups on the same study. Extending the paper of Sangnawakij and Niwitpong [17], we focus on the ratio of coefficients of variation of the exponential distributions in the case of two parameters. The confidence intervals for the ratio of coefficients of variation in this paper are proposed based on two methods, the method of variance of estimates recovery (MOVER) and the generalized confidence interval (GCI). The former estimator is estimated using the formulas, while the latter confidence interval is computed based on the software package.

The performance of the proposed confidence intervals is assessed in terms of coverage probability and expected length through simulations. It can be concluded that the coverage probabilities of the proposed estimators are satisfactory. The confidence interval based on the MOVER has coverage probability greater than the nominal coverage level in many situations. The GCI hits the target level in all cases of the study, no matter what the coefficients of variation are positive values or not. The expected lengths of these two proposed confidence intervals decrease if sample size increases, where the GCI has slightly shorter expected lengths than the MOVER. This result is also confirmed by the computation on a real-data example. Therefore, we recommend the GCI and the confidence interval based on the MOVER as confidence intervals for the ratio of coefficients of variation in two two-parameter exponential distributions.

Acknowledgments. The first author gratefully acknowledges the financial support from Faculty of Applied Sciences, King Mongkut's University of Technology North Bangkok. We are also grateful to the referees for the valuable suggestions, which lead to improve the quality of this paper.

References

1. Jati, K.: Staple food balance sheet, coefficient of variation, and price disparity in indonesia. J. Adv. Manage. Sci. **2**, 65–71 (2014)
2. Popadin, K.Y., Gutierrez, A.M., Lappalainen, T., Buil, A., Steinberg, J., Nikolaev, S.I., Lukowski, S.W., Bazykin, G.A., Seplyarskiy, V.B., Ioannidis, P., Zdobnov, E.M., Dermitzakis, E.T., Antonarakis, S.E.: Gene age predicts the strength of purifying selection acting on gene expression variation in humans. Am. J. Hum. Genet. **95**, 660–674 (2014)
3. DiGregorio, D.A., Nusser, Z., Silver, R.A.: Spillover of glutamate onto synaptic AMPA receptors enhances fast transmission at a cerebellar synapse. Neuron **35**, 521–533 (2002)
4. Shiina, K., Tomiyama, H., Takata, Y., Yoshida, M., Kato, K., Nishihata, Y., Matsumoto, C., Odaira, M., Saruhara, H., Hashimura, Y., Usui, Y., Yamashina, A.: Overlap syndrome: additive effects of COPD on the cardiovascular damages in patients with OSA. Respir. Med. **106**, 1335–1341 (2012)
5. Verrill, S., Johnson, R.A.: Confidence bounds and hypothesis tests for normal distribution coefficients of variation. Commun. Stat.-Theory Methodol. **36**, 2187–2206 (2007)
6. Donner, A., Zou, G.Y.: Closed-form confidence intervals for functions of the normal mean and standard deviation. Stat. Meth. Med. Res. **21**, 347–359 (2010)
7. Buntao, N., Niwitpong, S.: Confidence intervals for the ratio of coefficients of variation of delta-lognormal distribution. Appl. Math. Sci. **7**, 3811–3818 (2013)
8. Niwitpong, S., Wongkhao, A.: Confidence intervals for the difference and the ratio of coefficients of variation of normal distribution with a known ratio of variances. Int. J. Math. Trends Technol. **29**, 13–20 (2016)
9. Sangnawakij, P., Niwitpong, S.-A., Niwitpong, S.: Confidence intervals for the ratio of coefficients of variation of the gamma distributions. In: Huynh, V.-N., Inuiguchi, M., Denoeux, T. (eds.) IUKM 2015. LNCS (LNAI), vol. 9376, pp. 193–203. Springer, Heidelberg (2015). doi:10.1007/978-3-319-25135-6_19
10. Salem, S.A., El-Glad, F.A.: Bayesian prediction for the range based on a two-parameter exponential distribution with a random sample size. Microelectron. Reliab. **33**, 623–632 (1993)
11. Hahn, G.J., Meeker, W.Q.: Statistical Interval: A Guide for Practitioners. Wiley, Hoboken (1991)
12. Thiagarajah, K., Paul, S.R.: Interval estimation for the scale parameter of the two-parameter exponential distribution based on time-censored data. J. Stat. Plann. Infer. **59**, 279–289 (1997)
13. Dijk, A.I.J.M., Meesters, A.G.C.A., Schellekens, J., Bruijnzeel, L.A.: A two-parameter exponential rainfall depth-intensity distribution applied to runoff and erosion modelling. J. Hydrol. **300**, 155–171 (2005)
14. Berger, A., Melice, J.L., Demuth, C.: Statistical distributions of daily and high atmospheric SO_2 concentrations. Atmos. Environ. **16**, 2863–2877 (1982)
15. Lu, H., Fang, G.: Estimating the emission source reduction of PM_{10} in central Taiwan. Chemosphere **54**, 805–814 (2004)

16. Lawless, J.F.: Statistical Models and Methods for Lifetime Data. Wiley, New York (2003)
17. Sangnawakij, P., Niwitpong, S.: Confidence intervals for coefficients of variation in two-parameter exponential distributions. (Accepted for publication in Communications in Statistics: Simulation and Computation)
18. Weerahandi, S.: Generalized confidence intervals. J. Am. Stat. Assoc. **88**, 899–905 (1993)
19. Zou, G.Y., Huang, W., Zhang, X.: A note on confidence interval estimation for a linear function of binomial proportions. Comput. Stat. Data Anal. **53**, 1080–1085 (2009)
20. Roya, A., Bose, A.: Coverage of generalized confidence intervals. J. Multivar. Anal. **100**, 1384–1397 (2009)
21. Notes on R: A programming environment for data analysis and graphics, http://cran.r-project.org/
22. Freireich, T.R., Gehan, E., Frei, E., Schroeder, L.R., Wolman, I.J., Anbari, R., Burgert, E.O., Mills, S.D., Pinkel, D., Selawry, O.S., Moon, J.H., Gendel, B.R., Spurr, C.L., Storrs, R., Haurani, F., Hoogstraten, B., Lee, S.: The Effect of 6-mercaptopurine on the duration of steroid induced remissions in acute leukemia: a model for evaluation of other potentially useful therapy. Blood **21**, 699–716 (1963)

Simultaneous Fiducial Generalized Confidence Intervals for All Differences of Coefficients of Variation of Log-Normal Distributions

Warisa Thangjai[✉], Sa-Aat Niwitpong, and Suparat Niwitpong

Faculty of Applied Science, Department of Applied Statistics, King Mongkut's University of Technology North Bangkok, Bangkok 10800, Thailand
wthangjai@yahoo.com, {sa-aat.n,suparat.n}@sci.kmutnb.ac.th

Abstract. This paper proposes a simultaneous fiducial generalized confidence intervals approach to construct simultaneous confidence intervals for all pairwise comparisons of coefficients of variation of log-normal distributions. A Monte Carlo simulation was conducted to evaluate the performance, coverage probability and average length, of the simultaneous confidence intervals. Simulation results show that the simulated coverage probabilities are above the nominal confidence level for all sample sizes. A real data example is provided to illustrate our approach.

Keywords: Coefficient of variation · Log-normal distribution · Simultaneous fiducial generalized confidence intervals

1 Introduction

A coefficient of variation is a useful statistical measure to compare several variables expressed in different units in practical applications. The coefficient of variation is defined as the ratio of the standard deviation to the mean. The coefficient of variation is widely used in various areas of science, medicine, economics, finance and life insurance. Even though the estimated coefficient of variation is a useful measure, perhaps the greatest use is for point estimation to construct a confidence interval for the population quantity. The confidence interval provides information respecting the population value of the quantity much more than the point estimate. Confidence intervals of coefficients of variation of normal distributions have been investigated by several researchers, for example, Vangel [1], Tian [2], and Wong and Wu [3].

The log-normal distribution is widely used to describe the distribution of positively right-skewed data; in particular, it is used to model data relevant to occupational hygiene and to model biological data; see Joulious and Debarnot [4] and Shen et al. [5]. Comparisons of several log-normal distributions have applications in biological or pharmaceutical statistics, for example, in bioequivalence studies when comparing several drug formulations; see Hanning et al. [6] and Schaarschmidt [7].

© Springer International Publishing AG 2016
V.-N. Huynh et al. (Eds.): IUKM 2016, LNAI 9978, pp. 552–561, 2016.
DOI: 10.1007/978-3-319-49046-5_47

In the literature, Hanning et al. [6] proposed simultaneous fiducial generalized confidence intervals for ratios of means of log-normal distributions. Sadooghi-Alvandi and Malekzadeh [8] constructed simultaneous confidence intervals for ratios of means of several log-normal distributions based on parametric bootstrap approach.

The goal of this paper is to provide a novel approach for the simultaneous confidence intervals estimation of coefficients of variation of log-normal distributions based on independent samples. Weerahandi [9] introduced the concepts of generalized pivotal quantity (GPQ) and generalized confidence intervals (GCIs). Hannig et al. [10] identified a subclass of generalized pivotal quantities (GPQs) and called them fiducial generalized pivotal quantity (FGPQ). The resulting confidence intervals were fiducial generalized confidence intervals (FGCIs). Many researchers have successfully used the concept of FGPQ to obtain FGCIs or simultaneous fiducial generalized confidence intervals (SFGCIs); for example, see Hanning et al. [6], Abdel-Karim [11], Chang and Huang [12], and Kharrati-Kopaei et al. [13]. But, from our knowledge, no SFGCIs approach exists for the simultaneous confidence intervals of coefficients of variation on several independent log-normal samples. Therefore, SFGCIs is proposed as a novel approach for simultaneous confidence intervals for all differences of coefficients of variation of log-normal distributions.

This paper is organized as follows. In Sect. 2, the theory and computational procedures to construct the simultaneous confidence intervals are described. Simulation results are presented to evaluate the coverage probabilities and average lengths of the SFGCIs approach in Sect. 3. In Sect. 4, the proposed approach is illustrated using an example. Finally, concluding remarks are given in Sect. 5.

2   Simultaneous Fiducial Generalized Confidence Intervals

Recall that a random variable Y is said to be log-normal if $X = \log Y$ is normally distributed, say, $N\left(\mu, \sigma^2\right)$. When $\mu \neq 0$, $\frac{\sigma}{\mu}$ is called the coefficient of variation of X. Since

$$\sigma_Y^2 = \left(\exp\left(\sigma^2\right) - 1\right)\exp\left(2\mu + \sigma^2\right)$$

and

$$\mu_Y = \exp\left(\mu + \left(\frac{\sigma^2}{2}\right)\right),$$

the coefficient of variation of Y is

$$\frac{\sigma_Y}{\mu_Y} = \frac{\sqrt{\left(\exp\left(\sigma^2\right) - 1\right)\exp\left(2\mu + \sigma^2\right)}}{\exp\left(\mu + \left(\frac{\sigma^2}{2}\right)\right)}$$

$$= \frac{\exp\left(\mu + \frac{\sigma^2}{2}\right)\sqrt{\left(\exp\left(\sigma^2\right) - 1\right)}}{\exp\left(\mu + \left(\frac{\sigma^2}{2}\right)\right)} = \sqrt{\exp\left(\sigma^2\right) - 1}.$$

Let Y_i, $i = 1, 2, \ldots, k$ be random samples from log-normal distributions, i.e., $X_i = \log Y_i \sim N\left(\mu_i, \sigma_i^2\right)$. Let

$$\theta_i = \sqrt{\exp\left(\sigma_i^2\right) - 1}$$

be the coefficient of variation of Y_i. We are interested in constructing simultaneous confidence intervals for the differences $\theta_i - \theta_l$, $i \neq l$, $i, l = 1, 2, \ldots, k$.

Let Y_{ij}, $i = 1, 2, \ldots, k$; $j = 1, 2, \ldots, n_i$ be random samples from the Y_i.

We follow Hannig et al. [10], and Chang and Huang [12]: let X^* denote a normal random variable independent of X having the same distribution as X and let
$$R(X, X^*, \mu, \sigma^2) = (R_1(X, X^*, \mu, \sigma^2), R_2(X, X^*, \mu, \sigma^2), \ldots, R_k(X, X^*, \mu, \sigma^2)),$$
where $R_i(X, X^*, \mu, \sigma^2)$'s are real functions of X, X^*, μ and σ^2, for $i = 1, 2, \ldots, k$. For real function $g_1, g_2, \ldots, g_l$, the random quantities $g_1(R(X, X^*, \mu, \sigma^2)), \ldots, g_l(R(X, X^*, \mu, \sigma^2))$ are said to be SFGPQs for $g_1(\sigma^2), g_2(\sigma^2), \ldots, g_l(\sigma^2)$ if the following two conditions are satisfied:

(i) The conditional distribution of $R(X, X^*, \mu, \sigma^2)$, given $X = x$, does not depend on μ and σ^2.
(ii) $R(x, x, \mu, \sigma^2) = \sigma^2$, for every x.

Let $\bar{X}_i$ and S_i^2 denote the sample mean and variance for log-transformed data $X_{ij} = \log Y_{ij}$ ($j = 1, 2, \ldots, n_i$) for the i-th sample and let $\bar{x}_i$ and s_i^2 denote the observed sample mean and variance respectively. It is well known that

$$\bar{X}_i \sim N\left(\mu_i, \frac{\sigma_i^2}{n_i}\right),$$

and

$$\frac{(n_i - 1)S_i^2}{\sigma_i^2} = V_i \sim \chi^2_{n_i - 1}; i = 1, 2, \ldots, k$$

where V_i is chi-square distribution with degree of freedom $n_i - 1$. We can rewrite

$$S_i^2 = \frac{V_i \sigma_i^2}{(n_i - 1)}.$$

Let $\bar{X}_i^*$ and S_i^{2*} denote random variable independent of $\bar{X}_i$ and S_i^2 having the same distribution as $\bar{X}_i$ and S_i^2, respectively,

$$\bar{X}_i^* \sim N\left(\mu_i, \frac{\sigma_i^2}{n_i}\right),$$

and

$$\frac{(n_i - 1)S_i^{2*}}{\sigma_i^2} = V_i \sim \chi^2_{n_i - 1}; i = 1, 2, \ldots, k.$$

Then

$$S_i^{2*} = \frac{V_i \sigma_i^2}{(n_i - 1)}.$$

According to Hanning et al. [6], the generalized pivotal quantities (GPQs)

$$R_{\mu_i} = \bar{X}_i - \frac{S_i}{S_i^*}\left(\bar{X}_i^* - \mu_i\right); i = 1, 2, \ldots, k$$

and

$$R_{\sigma_i^2} = \frac{S_i^2}{S_i^{2*}}\sigma_i^2; i = 1, 2, \ldots, k.$$

The problem of constructing simultaneous confidence intervals for $\theta_{il} = \theta_i - \theta_l$ for all $i \neq l$ is

$$\theta_{il} = \theta_i - \theta_l = \sqrt{\exp(\sigma_i^2) - 1} - \sqrt{\exp(\sigma_l^2) - 1}.$$

The simultaneous fiducial generalized pivotal quantities (SFGPQs) for $\theta_i - \theta_l$; $i \neq l$, $i, l = 1, 2, \ldots, k$ is given by

$$R_{\theta_{il}} = R_{\theta_i}\left(X, X^*, \mu, \sigma^2\right) - R_{\theta_l}\left(X, X^*, \mu, \sigma^2\right)$$

$$= \sqrt{\exp\left(R_{\sigma_i^2}\right) - 1} - \sqrt{\exp\left(R_{\sigma_l^2}\right) - 1}$$

where $R_{\sigma_i^2} = \frac{S_i^2}{S_i^{2*}}\sigma_i^2 = \frac{(n_i-1)s_i^2}{V_i}$ and $R_{\sigma_l^2} = \frac{S_l^2}{S_l^{2*}}\sigma_l^2 = \frac{(n_l-1)s_l^2}{V_l}$.
Let

$$T = \max_{i \neq l}\left|\frac{\sqrt{\exp(S_i^2) - 1} - \sqrt{\exp(S_l^2) - 1} - (R_{\theta_i} - R_{\theta_l})}{\sqrt{V_{il}}}\right| \tag{1}$$

where

$$V_{il} = \frac{S_i^4 \exp(2S_i^2)}{2(n_i - 1)(\exp(S_i^2) - 1)} + \frac{S_l^4 \exp(2S_l^2)}{2(n_l - 1)(\exp(S_l^2) - 1)}, \tag{2}$$

is a consistent estimator of the variance of $\sqrt{\exp(S_i^2) - 1} - \sqrt{\exp(S_l^2) - 1}$; see Theorem 1.

Then $100(1 - \alpha)\%$ two-sided SFGCIs for pairwise differences θ_{il}, $i \neq l$ of coefficients of variation of more than two independent log-normal distributions are

$$(L_{il}, U_{il}) = \sqrt{\exp(S_i^2) - 1} - \sqrt{\exp(S_l^2) - 1} \pm d_{1-\alpha}\sqrt{V_{il}} \tag{3}$$

where $d_{1-\alpha}$ is the $(1 - \alpha)$-th quantile of the conditional distribution of T given $X = x$. The value of $d_{1-\alpha}$ is easily obtained by Monte Carlo methods.

Theorem 1. *Let* $X = (X_1, X_2, \ldots, X_n)$ *and* $X = \log Y \sim N\left(\mu, \sigma^2\right)$ *where* μ, σ^2 *are respectively population mean and population variance of* X. *Then the estimator of* θ *is* $\hat{\theta} = \sqrt{\exp(S^2) - 1}$, *the variance of* $\hat{\theta}$ *is*

$$var\left(\hat{\theta}\right) = \frac{\sigma^4 \exp(2\sigma^2)}{2(n - 1)(\exp(\sigma^2) - 1)}.$$

Proof. The function of coefficient of variation is defined as

$$g(\sigma^2) = \sqrt{\exp(\sigma^2) - 1}.$$

Then

$$g'\left(\sigma^2\right) = \frac{\exp\left(\sigma^2\right)}{2\sqrt{\exp\left(\sigma^2\right) - 1}}.$$

Since

$$var(s^2) = \frac{2\sigma^4}{(n-1)}.$$

Using delta method for the variance of $g\left(s^2\right) = \sqrt{\exp\left(s^2\right) - 1}$ we get

$$var\left(g\left(s^2\right)\right) \approx \left(g'\left(\sigma^2\right)\right)^2 var(s^2) = \frac{\sigma^4 \exp\left(2\sigma^2\right)}{2\left(n-1\right)\left(\exp\left(\sigma^2\right) - 1\right)},$$

Casella and Berger [14].

Theorem 2. *Let $X_{ij} = \log Y_{ij} \sim N\left(\mu_i, \sigma_i^2\right)$, $i = 1, 2, \ldots, k$, $j = 1, 2, \ldots, n_i$, and let $\theta_i = \sqrt{\exp\left(\sigma_i^2\right) - 1}$. Assume that the ratio $\frac{n_i}{N} \to r_i \in (0,1)$ as $N \to \infty$ for each i, where $N = n_1 + n_2 + \ldots + n_k$. Then*

$$P\{\theta_i - \theta_l \in \left(\hat{\theta}_i - \hat{\theta}_l \pm d_{1-\alpha}\sqrt{V_{il}}\right), \forall i \neq l\} \to 1 - \alpha.$$

Proof. As in Hanning et al. [6,10], we have

$$P\{\theta_i - \theta_l \in \left(\hat{\theta}_i - \hat{\theta}_l \pm d_{1-\alpha}\sqrt{V_{il}}\right), \forall i \neq l\} = P\{\max_{i \neq l}\left|\frac{\hat{\theta}_i - \hat{\theta}_l - (\theta_i - \theta_l)}{\sqrt{V_{il}}}\right| \leq d_{1-\alpha}\}$$

$$= P\{Q_n \leq d_{1-\alpha}\}.$$

Then $N\left(\hat{\theta} - \theta\right) \xrightarrow{D} Z$, where $\xrightarrow{D}$ denotes convergence in distribution and $Z = (Z_1, \ldots, Z_k)$ are independent with Z_i distributed as a normal distribution with mean zero and variance $\frac{\sigma_i^2}{r_i}$.

Accoring to Ferguson [15], by Slutsky's theorem, it follows that $Q_n \xrightarrow{D} Q$, where

$$Q = \max_{i \neq l}\left|\frac{Z_i - Z_l}{\sqrt{\left(\frac{\sigma_i^2}{r_i}\right) + \left(\frac{\sigma_l^2}{r_l}\right)}}\right|.$$

By Skorohod's theorem ; see Billingsley [16], the random variables Y_n and Y on a common probability space are distributed as Q_n and Q, respectively, and $Y_n \to Y$. Assume that $Q_n \to Q$. Similarly, $T\left(X, X^*, \mu, \sigma^2\right) \xrightarrow{D} Q^*$, where

$$Q^* = \max_{i \neq l}\left|\frac{Z_i^* - Z_l^*}{\sqrt{\left(\frac{\sigma_i^2}{r_i}\right) + \left(\frac{\sigma_l^2}{r_l}\right)}}\right|,$$

where Z_i^*'s are random variable independent of Z_i's having the same distribution as Z_i's. Assume that $T\left(X, X^*, \mu, \sigma^2\right) \rightarrow Q^*$. The limiting distribution of $T\left(X, X^*, \mu, \sigma^2\right)$ is continuous and $d_{1-\alpha}(X) \rightarrow q_{1-\alpha}$, where $q_{1-\alpha}$ is the $(1-\alpha)$-th quantile of the distribution of Q^*. Therefore,

$$P(Q_n \leq d_{1-\alpha}) \rightarrow P(Q \leq q_{1-\alpha}) = P(Q^* \leq q_{1-\alpha}) = 1 - \alpha, as N \rightarrow \infty.$$

Then the $100\left(1-\alpha\right)\%$ two-sided simultaneous confidence intervals have

$$P\{\theta_i - \theta_l \in \left(\hat{\theta}_i - \hat{\theta}_l \pm d_{1-\alpha}\sqrt{V_{il}}\right), \forall i \neq l\} \rightarrow 1 - \alpha,$$

implies that

$$P\left(L_{il} \leq \theta_{il} \leq U_{il}, \forall i \neq l\right) \rightarrow 1 - \alpha.$$

The value of $d_{1-\alpha}$ for two-sided SFGCIs can be estimated using Monte Carlo procedure as follows:

Algorithm 1.
input : x_{ij}, $i = 1, 2, \ldots, k$; $j = 1, 2, \ldots, n_i$
output: $d_{1-\alpha}$
begin
  for $g = 1$ **to** m **do**
    Generate V_i from chi-square distribution with degree of freedom $n_i - 1$;
    Compute R_{θ_i}, $i = 1, 2, \ldots, k$;
    Compute T in (1);
  end
  Obtain an array of T;
  Rank the array of T from small to large;
  Compute the $(1 - \alpha)$-th quantile is $d_{1-\alpha}$;
end

3   Simulation Studies

A simulation study was performed to evaluate the coverage probabilities and average lengths of SFGCIs. Simultaneous confidence interval is satisfactory when the values of coverage probability are at least or close to the nominal confidence level $(1 - \alpha)$ and also has the short average length.

In this simulation study, the number of samples used are $k = 3$ with the sample sizes $n_1 = n_2 = n_3 = n = 10, 20, 30, 40$ and 50, the population mean of normal data within each sample $\mu_1 = \mu_2 = \mu_3 = \mu = 1$, and the population standard deviation $\sigma_1 = \sigma_2 = \sigma_3 = \sigma = 0.05, 0.10, 0.15, 0.20, 0.29, 0.39$ and 0.47. The number of simulation runs is $M = 5000$ and thus 2500 R_{θ_i}'s are obtained for each of 5000 simulations.

The coverage probability of SFGCIs can be estimated using procedure as follows:

Algorithm 2.
input : $M, m, k, n_i, \mu_i, \sigma_i$
output: The coverage probability
begin
    Compute $\theta_i = \sqrt{\exp(\sigma_i^2) - 1}$;
    Compute $\theta_i - \theta_l$, $i \neq l$, $i, l = 1, 2, \ldots, k$;
    for $h = 1$ to M do
        Generate x_{ij} from normal distribution with mean μ_i and variance σ_i^2;
        Compute $\bar{x}_i$ and s_i^2, $i = 1, 2, \ldots, k$;
        Use Algorithm 1 to find $d_{1-\alpha}$;
        Construct (L_{il}, U_{il}) in (3);
        if $L_{il} \leq \theta_i - \theta_l \leq U_{il}$ then
            Probability=1;
        else
            Probability=0;
        end
    end
    Obtain an array of probabilities;
    Compute average coverage probability;
end

Table 1 presents the coverage probabilities and average lengths for $k = 3$. The coverage probabilities of SFGCIs approach are increased as well as the average lengths, when the value of σ increases. For large sample sizes, the SFGCIs are the closest to the nominal confidence level of 0.95.

4   An Empirical Verification

Simultaneous confidence intervals for coefficients of variation are useful in many situations. In finance, a test of the equality of the coefficients of variation for two stocks can help to determine if the two stocks possess the same risk or not. In a diet study, when the intent is to compare the variability in the ratio of total/HDL cholesterol with the variability in vessel diameter change, a comparison of standard deviation makes no sense because cholesterol and vessel diameter are measured in different scales. A sensible comparison can use the coefficient of variation since the coefficient of variation measures the relative spread of data and therefore adjusts the scale. A confidence interval for difference $\theta = \tau_r - \tau_v$ is useful in this situation, where τ_r denotes the coefficient of variation in the ratio of total/HDL cholesterol and τ_v denotes the coefficient of variation in vessel diameter change. For multiple sample cases, we are interested in obtaining simultaneous confidence intervals for all pairwise differences of coefficients of variation.

A real data example was previously considered by Schaarscmidt [7] and Sadooghi-Alvandi and Malekzadeh [8]. The data, originally given by Hand et al. [17], contained 57 observations of nitrogen bound bovine serum albumin in

Table 1. The coverage probabilities (CP) and average lengths (AL) of approximately 95 % of two-side simultaneous fiducial generalized confidence intervals for coefficients of variation of the log-normal (based on 5000 simulations): 3 sample cases.

n	σ	SFGCI			
		CP	AL		
			$\theta_2 - \theta_1$	$\theta_3 - \theta_1$	$\theta_3 - \theta_2$
10	0.05	0.9999	0.1049	0.1048	0.1050
	0.10	0.9999	0.2118	0.2114	0.2117
	0.15	1.0000	0.3249	0.3258	0.3235
	0.20	1.0000	0.4480	0.4492	0.4474
	0.29	1.0000	0.6990	0.6983	0.7003
	0.39	1.0000	1.0647	1.0627	1.0668
	0.47	1.0000	1.4771	1.4759	1.4745
20	0.05	0.9951	0.0614	0.0615	0.0615
	0.10	0.9954	0.1240	0.1237	0.1237
	0.15	0.9965	0.1887	0.1885	0.1890
	0.20	0.9961	0.2575	0.2571	0.2561
	0.29	0.9978	0.3896	0.3918	0.3916
	0.39	0.9997	0.5697	0.5714	0.5714
	0.47	0.9995	0.7422	0.7447	0.7435
30	0.05	0.9907	0.0476	0.0476	0.0475
	0.10	0.9911	0.0960	0.0958	0.0957
	0.15	0.9913	0.1455	0.1458	0.1458
	0.20	0.9924	0.1974	0.1976	0.1975
	0.29	0.9941	0.2978	0.2983	0.2981
	0.39	0.9961	0.4310	0.4297	0.4305
	0.47	0.9960	0.5547	0.5555	0.5526
40	0.05	0.9894	0.0401	0.0401	0.0401
	0.10	0.9881	0.0807	0.0808	0.0808
	0.15	0.9883	0.1226	0.1224	0.1226
	0.20	0.9897	0.1657	0.1662	0.1657
	0.29	0.9889	0.2506	0.2507	0.2505
	0.39	0.9920	0.3593	0.3589	0.3589
	0.47	0.9939	0.4603	0.4607	0.4609
50	0.05	0.9851	0.0353	0.0353	0.0353
	0.10	0.9891	0.0711	0.0710	0.0711
	0.15	0.9875	0.1079	0.1078	0.1078
	0.20	0.9872	0.1460	0.1459	0.1459
	0.29	0.9895	0.2197	0.2198	0.2202
	0.39	0.9907	0.3143	0.3149	0.3145
	0.47	0.9923	0.4026	0.4021	0.4022

3 groups of mice: normal mice (group 1), alloxan-induced diabetic mice (group 2), and alloxon-induced diabetic mice treated with insulin (group 3) with $n_1 = 20$, $n_2 = 18$, and $n_3 = 19$. The original data was found to be correct; Schaarschmidt [7] showed that the data are consistent with the assumption of log-normality. The sample mean (sample variance) of the data were 4.859 (0.927), 4.867 (0.850), and 4.397 (0.696) for group 1, group 2, and group 3, respectively. The coefficients of variation were 1.236, 1.157, and 1.003 for group 1, group 2, and group 3, respectively. Using the SFGCIs approach of log-normal distributions, the SFGCIs for difference of the coefficients of variation $\theta_2 - \theta_1$ was $(-1.677, 1.520)$ with the length of interval 3.197. The SFGCIs for difference of the coefficients of variation $\theta_3 - \theta_1$ was $(-1.889, 1.423)$ with the length of interval 3.312. The SFGCIs for difference of the coefficients of variation $\theta_3 - \theta_2$ was $(-1.445, 1.136)$ with the length of interval 2.580. The results support the simulation results in the previous section.

5   Discussions and Conclusions

For the first time, this paper aims at constructing simultaneous confidence intervals for all differences of coefficients of variation of log-normal distributions based on SFGCIs approach. From the evaluation results, the SFGCIs approach was found to provide more coverage probabilities than nominal confidence level at 0.95 for all sample sizes. The constructed confidence intervals had correct asymptotic coverage probabilities matching the paper by Hanning et al. [6]. Therefore, the SFGCIs can be successfully used to construct simultaneous confidence intervals for all differences of coefficients of variation of log-normal distributions.

Results in this investigation were similar to the results in research papers by Hanning et al. [6], Abdel-Karim [11], Chang and Huang [12], and Kharrati-Kopaei et al. [13]. Further research will be conducted to try and find other approaches for comparison.

Acknowledgments. The first author gratefully acknowledges the financial support from the following agencies: Science Achievement Scholarship of Thailand; Faculty of Applied Science, King Mongkut's University of Technology North Bangkok; Graduate College, King Mongkut's University of Technology North Bangkok.

References

1. Vangel, M.G.: Confidence intervals for a normal coefficient of variation. J. Am. Stat. Assoc. **50**, 21–26 (1996)
2. Tian, L.: Inferences on the common coefficient of variation. Stat. Med. **24**, 2213–2220 (2005)
3. Wong, A.C.M., Wu, J.: Small sample asymptotic inference for the coefficient of variation: normal and nonnormal models. J. Stat. Plan. Infer. **104**, 73–82 (2002)
4. Joulious, S.A., Debarnot, C.A.M.: Why are pharmacokinetics data summarize by arithmetic means? J. Biopharm. Stat. **10**, 55–71 (2000)

5. Shen, H., Brown, L., Hui, Z.: Efficient estimation of log-normal means with application to pharmacokinetics data. Stat. Med. **25**, 3023–3038 (2006)
6. Hannig, J., Lidong, E., Abdel-Karim, A., Iyer, H.: Simultaneous fiducial generalized confidence intervals for ratios of means of lognormal distributions. Austrian J. Stat. **35**, 261–269 (2006)
7. Schaarschmidt, F.: Simultaneous confidence intervals for multiple comparisons among expected values of log-normal variables. Comput. Stat. Data Anal. **58**, 265–275 (2013)
8. Sadooghi-Alvandi, S.M., Malekzadeh, A.: Simultaneous confidence intervals for ratios of means of several lognormal distributions: a parametric bootstrap approach. Comput. Stat. Data Anal. **69**, 133–140 (2014)
9. Weerahandi, S.: Generalized confidence intervals. J. Am. Stat. Assoc. **88**, 899–905 (1993)
10. Hannig, J., Iyer, H., Patterson, P.: Fiducial generalized confidence intervals. J. Am. Stat. Assoc. **101**, 254–269 (2006)
11. Abdel-Karim, A.: Applications of generalized inference. Unpublished Doctoral dissertation, Colorado State University, Fort Collins, Colorado (2005)
12. Chang, Y.P., Huang, W.T.: Simultaneous fiducial generalized confidence intervals for all pairwise comparisons of means. Int. J. Inf. Manage. Sci. **20**, 459–467 (2009)
13. Kharrati-Kopaei, M., Malekzadeh, A., Sadooghi-Alvandi, M.: Simultaneous fiducial generalized confidence intervals for the successive differences of exponential location parameters under heteroscedasticity. Stat. Prob. Lett. **83**, 1547–1552 (2013)
14. Casella, G., Berger, R.L.: Statistical Inference, 2nd edn., p. 242. Duxbury Press, California (2002)
15. Ferguson, T.S.: A Course in Large Sample Theory, 6 edn., p. 6. Chapman and Hall/CRC, London (1996)
16. Billingsley, P.: Probability and Measure, 3 edn., p. 333. Wiley, New York (1995)
17. Hand, D.J., Daly, F., McConway, K., Lunn, D., Ostrowski, E.: A Handbook of Small Data Sets, 1st edn. Chapman and Hall/CRC, London (1994)

Confidence Intervals for Common Variance of Normal Distributions

Narudee Smithpreecha[✉], Sa-Aat Niwitpong, and Suparat Niwitpong

Applied Statistics, King Mongkut's University of Technology North Bangkok,
Bangkok 10800, Thailand
Nsmithpreecha@gmail.com, {sa-aat.n,suparat.n}@sci.kmutnb.ac.th

Abstract. This paper presents a construction of confidence intervals for the common variance of normal distributions based on generalized confidence intervals, and then compares the results with a large sample approach. A Monte Carlo simulation was used to evaluate the coverage probability and average length of confidence intervals. Simulation studies showed that the generalized confidence interval approach provided much better confidence interval estimates than the large sample approach. Two real data examples are exhibited to illustrate our approaches.

Keywords: Common variance · Generalized confidence interval · Large sample approach · Normal distribution

1 Introduction

The construction of confidence intervals for a normal variance are well known and simple to apply and attracted a great deal of attention from researchers. An investigation of the history and development of constructing confidence intervals for a normal variance was given in Cohen [1], and he constructed confidence intervals for the variance that had the same length as the usual minimum length interval but greater coverage probability. Analogously, Shorrock [2,3] presented an improved interval based on Stein's technique and a smooth version of Cohen's interval using Brewster and Zidek's technique. Stein-type improvements of confidence intervals for the normal variance with unknown mean were also obtained by Nagata [4]. Casella [5] constructed a class of intervals each of which improved both coverage probability and size over the usual interval. Lastly, Kubokawa [6] presented a unified approach to the variance estimation problem. There are many researchers that are also interested in the estimation of variance; see e.g., Shorrock and Zidek [7]. Sarkar [8] constructed the shortest confidence interval and Iliopoulos and Kourouklis [9] presented a stein-type interval for generalized variances.

The motivation of this paper comes from an analysis of variance (ANOVA), which are used to compare several means. Under the assumption of analysis of variance are normality, homogeneity of variance, and independence of errors. If the quantitative data of the sample n observations from k populations come

V.-N. Huynh et al. (Eds.): IUKM 2016, LNAI 9978, pp. 562–573, 2016.
DOI: 10.1007/978-3-319-49046-5_48

from a different time or space and in experimental situations have repeated many times. In this case if the variances are homogeneous, what is the best way for construction the confidence interval estimation of common variance to obtain a single estimation? Therefore, interval estimation procedures regarding common variance of normal distributions are interesting.

The practical and theoretical of developing procedures for interval estimation of common variance based on several independent normal samples are important. Thus, the goal of this paper was to provide two approaches for the confidence interval estimation of common variance derived from several independent samples from normal distributions. The generalized confidence interval and large sample confidence interval concept will be used for the end evaluation. The approach is based on the concepts of generalized confidence intervals. The notions of generalized confidence intervals are proposed by Weerahandi [10]. The generalized confidence interval approach has been successfully used to construct the confidence interval for many common parameters and since then these ideas have been applied to solve many statistical problems, for example, Tian [11], Tian and Wu [12], Krishnamoorthy [13], and Ye et al. [14]. However, the generalized confidence interval approach used to construct these confidence interval estimations for the common variance are also interesting. To our knowledge, there are no previous works on inferences on common variance referring to normal distributions with a generalized confidence interval approach compared with the large sample approach.

The remainder of the paper is organized as follows. Section 2 introduces the basic properties of normal distribution. Section 3 presents the generalized variable approach developed and describes computational procedures. Section 4 presents simulation results to evaluate the performances of generalized confidence interval approach and the large sample approach on coverage probabilities and average lengths. Section 5 illustrates the proposed approaches with real examples. Finally, conclusions are given in Sect. 6.

2 Properties of Normal Distribution

If the random variable X follows the normal distribution, that is $X \sim N\left(\mu, \sigma^2\right)$. The probability density function of X is given by

$$f\left(x\right) = \frac{1}{\sigma\sqrt{2\pi}} exp\left(-\frac{\left(x-\mu\right)^2}{2\sigma^2}\right), -\infty < x < \infty.$$

The maximum likelihood estimators (MLE) of μ and σ^2 are $\widehat{\mu}$ and $\widehat{\sigma}^2$ respectively,

where $$\widehat{\mu} = \overline{X} = \frac{1}{n}\sum_{i=1}^{n} X_i, \quad \widehat{\sigma}^2 = \frac{1}{n}\sum_{i=1}^{n}\left(X_i - \overline{X}\right)^2.$$

The estimator $\widehat{\sigma}^2$ is the sample variance of the sample $(X_1, X_2, ..., X_n)$. In practice, another estimator is often used instead of the $\widehat{\sigma}^2$. This other estimator is denoted S^2, and is also called the sample variance. The estimator S^2 differs from $\widehat{\sigma}^2$ by having $(n-1)$ instead of n in the denominator

Thus
$$S^2 = \frac{n}{n-1}\widehat{\sigma}^2 = \frac{1}{n-1}\sum_{i=1}^{n}(X_i - \overline{X})^2.$$

The estimator S^2 is an unbiased estimator of the underlying parameter σ^2, whereas $\widehat{\sigma}^2$ is biased.

Theorem 1. *Suppose $X \sim N(\mu, \sigma^2)$, where μ, σ^2 are respectively population mean and population variance of X. Then the estimator of σ^2 is $\widehat{\sigma}^2 = S^2$, the variance of $\widehat{\sigma}^2$ is*

$$var(\widehat{\sigma}^2) = \frac{2\sigma^4}{n-1}.$$

Proof. Let $X_1, X_2, ..., X_n$ be an independent and identically distributed random variables with mean μ and variance σ^2, then $\overline{X}$ and S^2 are unbiased estimators of μ and σ^2 :

where
$$\overline{X} = \frac{1}{n}\sum_{i=1}^{n}x_i, \quad S^2 = \frac{1}{n-1}\sum_{i=1}^{n}(x_i - \overline{X})^2.$$

Also, by the LehmannScheff theorem the estimator S^2 is uniformly minimum variance unbiased (UMVU). In finite samples both S^2 and $\widehat{\sigma}^2$ have scaled chi-squared distribution with $(n-1)$ degrees of freedom

so,
$$S^2 \sim \frac{\sigma^2}{n-1}\cdot \chi^2_{n-1}, \quad \widehat{\sigma}^2 \sim \frac{\sigma^2}{n}\cdot \chi^2_{n-1},$$

then sampling distribution of $\dfrac{(n-1)S^2}{\sigma^2}$ is chi-square with $n-1$ degrees of freedom. For the chi-square distribution, it turns out that the mean and variance are $E(\chi^2_{n-1}) = n-1, var(\chi^2_{n-1}) = 2(n-1)$.

We can use this to get the mean and variance of S^2

$$E(S^2) = E\left(\frac{\sigma^2\chi^2_{n-1}}{(n-1)}\right) = \frac{\sigma^2}{n-1}(n-1) = \sigma^2,$$

$$var(S^2) = var\left(\frac{\sigma^2\chi^2_{n-1}}{(n-1)}\right) = \frac{\sigma^4}{(n-1)^2}2(n-1) = \frac{2\sigma^4}{n-1}.$$

Hence,
$$var(\widehat{\sigma}^2) = var(S^2) = \frac{2\sigma^4}{n-1}.$$

3   The Confidence Interval Approaches of the Common Variance

3.1   The Generalized Confidence Interval Approach

The generalized confidence intervals (GCI) are based on the simulation of a known generalized pivotal quantity (GPQ). Weerahandi [10] introduced the concept of a generalized pivotal quantity for a parameter θ as follows:

Suppose that $X_{ij} \sim N\left(\mu_i, \sigma_i^2\right)$, for $i = 1, 2, ..., k, j = 1, 2, ..., n_i$ are a random samples from a distribution which depends on a vector of parameters $\theta = \left(\theta, \underset{\sim}{\nu}\right)$ where θ is the parameter of interest and $\underset{\sim}{\nu}$ is a vector of nuisance parameters. A generalized pivot $R\left(\underset{\sim}{X}, \underset{\sim}{x}, \theta, \underset{\sim}{\nu}\right)$ for interval estimation, where $\underset{\sim}{x}$ is an observed value of $\underset{\sim}{X}$, as a random variable having the following two properties:

1. $R\left(\underset{\sim}{X}, \underset{\sim}{x}, \theta, \underset{\sim}{\nu}\right)$ has a distribution free of the vector of nuisance parameters $\underset{\sim}{\nu}$.
2. The observed value of $R\left(\underset{\sim}{X}, \underset{\sim}{x}, \theta, \underset{\sim}{\nu}\right)$ is θ.

Let R_α be the 100α-th percentile of R. Then R_α becomes the $100\left(1 - \alpha\right)\%$ lower bound for θ and $\left(R_{\alpha/2}, R_{1-\alpha/2}\right)$ becomes a $100\left(1 - \alpha\right)\%$ two-side generalized confidence interval for θ.

Generalized Variable Approach. Consider k independent normal populations with a common variance θ. Let $X_{i1}, X_{i2}, ..., X_{in_i}$ be a random sample from the i-th normal population as follows:

$$X_{ij} \sim N\left(\mu_i, \sigma_i^2\right), for \; i = 1, 2, ..., k, j = 1, 2, ..., n_i.$$

Thus
$$\theta = \sigma_i^2.$$

Let S_i^2 denote the sample variance for data X_{ij} for the i-th sample and let s_i^2 denote the observed sample variance respectively. From

$$\frac{(n_i - 1) S_i^2}{\sigma_i^2} = V_i \sim \chi_{n_i-1}^2,$$

so,
$$\sigma_i^2 = \frac{(n_i - 1) S_i^2}{V_i} \quad where \; V_i \sim \chi_{n_i-1}^2.$$

where V_i is χ^2 variates with degrees of freedom and n_i-1, we have the generalized pivot

$$R_{\sigma_i^2} = \frac{(n_i - 1) s_i^2}{V_i} \sim \frac{(n_i - 1) s_i^2}{\chi_{n_i-1}^2}. \tag{1}$$

The generalized pivotal quantity for estimating θ based on the i-th sample is

$$R_\theta^{(i)} = R_{\sigma_i^2}. \tag{2}$$

From the i-th sample, the maximum likelihood estimator of θ is

$$\widehat{\theta}^{(i)} = \widehat{\sigma}_i^2, \quad where \ \widehat{\sigma}_i^2 = S_i^2. \tag{3}$$

The large sample variance for $\widehat{\theta}^{(i)}$ is

$$var\left(\widehat{\theta}^{(i)}\right) = var\left(\widehat{\sigma}_i^2\right) = var\left(S_i^2\right) = \frac{2\sigma_i^4}{n_i - 1}, \quad see \ Theorem \ 1. \tag{4}$$

The generalized pivotal quantity, we propose for the common variance θ is a weighted average of the generalized pivot $R_\theta^{(i)}$ based on k individual samples as; see, Ye et al. [14],

$$R_\theta = \frac{\sum\limits_{i=1}^{k} R_w R_\theta^{(i)}}{\sum\limits_{i=1}^{k} R_{w_i}}, \tag{5}$$

where

$$R_{w_i} = \frac{1}{R_{var(\widehat{\theta}^{(i)})}}, \tag{6}$$

$$R_{\mathrm{var}(\hat{\theta}^{(i)})} = \frac{2\left(R_{\sigma_i^2}\right)^2}{n_i - 1}. \tag{7}$$

That is, $R_{\mathrm{var}(\hat{\theta}^{(i)})}$ is $\mathrm{var}(\hat{\theta}^{(i)})$ with σ_i^2 replaced by $R_{\sigma_i^2}$.

Computing Algorithms. For a given data set X_{ij} for $i = 1, 2, \ldots, k$, $j = 1, 2, \ldots, n_i$, the generalized confidence intervals for θ can be computed by the following steps.

1. Compute $\bar{x}_i$ and s_i^2 for $i = 1, 2, \ldots, k$.
2. Generate $V_i \sim \chi_{n_i-1}^2$ and then calculate $R_{\sigma_i^2}$ from (1) for $i = 1, 2, \ldots, k$.
3. Calculate $R_\theta^{(i)}$ from (2) for $i = 1, 2, \ldots, k$.
4. Repeat steps 2, calculate R_{w_i} from (6) and (8) for $i = 1, 2, \ldots, k$.
5. Compute R_θ following (5).
6. Repeat step 2–5 a total m times and obtain an array of R_θ's.
7. Rank this array of R_θ's from small to large.

The 100α-th percentile of R_θ's, $R_\theta(\alpha)$, is an estimate of the lower bound of the one - sided $100(1-\alpha)\%$ confidence interval and $(R_\theta(\alpha/2), R_\theta(1 - \alpha/2))$ is a two - sided $100(1 - \alpha)\%$ confidence interval.

3.2 The Large Sample Approach

The large sample estimate of normal variance is a pooled estimate of the common normal variance defined as

$$
\widehat{\theta} = \frac{\sum_{i=1}^{k} \dfrac{\widehat{\theta}^{(i)}}{var\left(\widehat{\theta}^{(i)}\right)}}{\sum_{i=1}^{k} \dfrac{1}{var\left(\widehat{\theta}^{(i)}\right)}},
\tag{8}
$$

where $\widehat{\theta}^{(i)}$ is defined in (3) and $var\left(\widehat{\theta}^{(i)}\right)$ is an estimate of $var\left(\widehat{\theta}^{(i)}\right)$ in (4) with σ_i^2 replaced by s_i^2, respectively.

Hence, the large sample solution for confidence interval estimation is

$$
\left(\widehat{\theta} - z_{1-\alpha/2} \sqrt{\frac{1}{\sum_{i=1}^{k} \dfrac{1}{var\left(\widehat{\theta}^{(i)}\right)}}}, \widehat{\theta} + z_{1-\alpha/2} \sqrt{\frac{1}{\sum_{i=1}^{k} \dfrac{1}{var\left(\widehat{\theta}^{(i)}\right)}}} \right).
\tag{9}
$$

Computing Algorithms. For a given data set X_{ij} for $i = 1, 2, ..., k$, $j = 1, 2, ..., n_i$, the generalized confidence intervals for θ can be computed by the following steps.

1. Compute $\overline{x}_i$ and s_i^2 for $i = 1, 2, ..., k$.
2. Calculate $var\left(\widehat{\theta}^{(i)}\right)$ from (4) for $i = 1, 2, ..., k$.
3. Compute $\widehat{\theta}$ following (8).
4. Calculate confidence interval estimation from (9) for $i = 1, 2, ..., k$.

4 Simulation Studies

A simulation study was performed to estimate the coverage probabilities and average lengths of the common variance of the normal distributions for various combinations of the number of samples $k = 2$ and $k = 6$, the sample sizes $n_1 = ... = n_k = n$, the values used for sample sizes were 10, 30, 50,100 and 200 the population mean of normal data within each sample 1, and the population standard deviation $\sigma = 0.10, 0.20, 0.30, 0.40, 0.50, 0.60, 0.70, 0.80, 0.90, 1.00$ and 2.00. In this simulation study, we compared two methods, comprising of our proposed procedure generalized confidence interval approach and the large sample approach. For each parameter setting, 5000 random samples were generated, 2500 R_θ's were obtained for each of the random samples.

Tables 1 and 2 present the coverage probabilities and average lengths for 2 and 6 sample cases respectively. In 2 and 6 sample cases, the generalized

Table 1. Empirical coverage probabilities (CP) and length of approximate 95 % two side confidence bounds for common variance of normal distributions (based on 5000 simulations): 2 sample cases.

n	σ	GCI approach		Large sample approach	
		CP	Length	CP	Length
10	0.10	0.9156	0.0147	0.7602	0.0113
	0.20	0.9282	0.0588	0.7628	0.0453
	0.30	0.9198	0.1321	0.7486	0.1015
	0.40	0.9242	0.2368	0.7572	0.1820
	0.50	0.9306	0.3687	0.7588	0.2842
	0.60	0.9248	0.5330	0.7512	0.4103
	0.70	0.9248	0.7176	0.7526	0.5519
	0.80	0.9242	0.9473	0.7562	0.7311
	0.90	0.9230	1.1958	0.7524	0.9210
	1.00	0.9212	1.4668	0.7468	1.1298
	2.00	0.9268	5.8861	0.7530	4.5365
30	0.10	0.9440	0.0077	0.8762	0.0069
	0.20	0.9336	0.0311	0.8658	0.0277
	0.30	0.9312	0.0699	0.8532	0.0623
	0.40	0.9370	0.1241	0.8672	0.1106
	0.50	0.9328	0.1941	0.8632	0.1726
	0.60	0.9436	0.2806	0.8696	0.2497
	0.70	0.9364	0.3793	0.8626	0.3379
	0.80	0.9402	0.5007	0.8618	0.4453
	0.90	0.9430	0.6295	0.8712	0.5615
	1.00	0.9382	0.7800	0.8654	0.6949
	2.00	0.9360	3.1166	0.8630	2.7751
50	0.10	0.9418	0.0058	0.8936	0.0054
	0.20	0.9434	0.0235	0.9018	0.0217
	0.30	0.9438	0.0529	0.8988	0.0488
	0.40	0.9444	0.0941	0.9014	0.0870
	0.50	0.9360	0.1474	0.8946	0.1361
	0.60	0.9428	0.2127	0.9014	0.1963
	0.70	0.9438	0.2893	0.9020	0.2672
	0.80	0.9398	0.3768	0.8950	0.3473
	0.90	0.9494	0.4778	0.9006	0.4414
	1.00	0.9426	0.5897	0.9000	0.5441
	2.00	0.9464	2.3638	0.9000	2.1832

(continued)

Table 1. (*continued*)

n	σ	GCI approach		Large sample approach	
		CP	Lenght	CP	Lenght
100	0.10	0.9416	0.0040	0.9224	0.0038
	0.20	0.9454	0.0162	0.9190	0.0155
	0.30	0.9436	0.0366	0.9230	0.0349
	0.40	0.9402	0.0650	0.9178	0.0620
	0.50	0.9456	0.1016	0.9242	0.0969
	0.60	0.9482	0.1468	0.9234	0.1399
	0.70	0.9508	0.1996	0.9272	0.1903
	0.80	0.9448	0.2604	0.9186	0.2484
	0.90	0.9424	0.3298	0.9212	0.3149
	1.00	0.9444	0.4062	0.9240	0.3876
	2.00	0.9504	1.6329	0.9286	1.5557
200	0.10	0.9448	0.0028	0.9326	0.0027
	0.20	0.9512	0.0113	0.9400	0.0110
	0.30	0.9472	0.0255	0.9366	0.0248
	0.40	0.9454	0.0452	0.9332	0.0440
	0.50	0.9504	0.0707	0.9394	0.0689
	0.60	0.9506	0.1020	0.9426	0.0994
	0.70	0.9500	0.1389	0.9376	0.1353
	0.80	0.9480	0.1811	0.9362	0.1764
	0.90	0.9496	0.2293	0.9386	0.2233
	1.00	0.9502	0.2836	0.9352	0.2761
	2.00	0.9480	1.1331	0.9346	1.1039

confidence interval approach and the large sample approach provide the under-estimates coverage probabilities for most of the scenarios, especially when the sample size is small. Additionally, the coverage probabilities of the generalized confidence interval approach are better than the large sample approach for all sample sizes, especially when the sample size is small. In overall, the generalized confidence interval approach and the large sample approach have the coverage probabilities close to the nominal level when the sample size increases. In this case, there is no need to see the average lengths from two intervals since the large sample approach provide the coverage probability below the generalized confidence interval approach for almost cases. Finally, it was discovered that the generalized confidence interval approach provided much better results over the large sample approach in terms of coverage probabilities.

Table 2. Empirical coverage probabilities (CP) and length of approximate 95 % two side confidence bounds for common variance of normal distributions (based on 5000 simulations): 6 sample cases.

n	σ	GCI approach		Large sample approach	
		CP	Lenght	CP	Lenght
10	0.10	0.6156	0.0073	0.4614	0.0056
	0.20	0.6234	0.0296	0.4560	0.0227
	0.30	0.6196	0.0666	0.4672	0.0513
	0.40	0.6082	0.1178	0.4514	0.0903
	0.50	0.6086	0.1844	0.4528	0.1418
	0.60	0.6074	0.2653	0.4574	0.2044
	0.70	0.6082	0.3622	0.4488	0.2783
	0.80	0.6180	0.4746	0.4554	0.3648
	0.90	0.6170	0.6001	0.4580	0.4625
	1.00	0.6034	0.7393	0.4482	0.5680
	2.00	0.6098	2.9526	0.4588	2.2705
30	0.10	0.7840	0.0045	0.7238	0.0038
	0.20	0.7880	0.0180	0.7318	0.0154
	0.30	0.7818	0.0405	0.7230	0.0346
	0.40	0.7794	0.0722	0.7274	0.0616
	0.50	0.7824	0.1127	0.7190	0.0960
	0.60	0.7752	0.1624	0.7156	0.1385
	0.70	0.7850	0.2213	0.7210	0.1887
	0.80	0.7756	0.2888	0.7194	0.2468
	0.90	0.7848	0.3654	0.7254	0.3120
	1.00	0.7846	0.4509	0.7334	0.3855
	2.00	0.7730	1.8012	0.7196	1.5387
50	0.10	0.8344	0.0034	0.8080	0.0030
	0.20	0.8314	0.0137	0.8016	0.0122
	0.30	0.8344	0.0309	0.8074	0.0276
	0.40	0.8374	0.0550	0.8124	0.0491
	0.50	0.8274	0.0857	0.7984	0.0766
	0.60	0.8326	0.1236	0.8062	0.1106
	0.70	0.8366	0.1685	0.8004	0.1505
	0.80	0.8380	0.2199	0.8124	0.1966
	0.90	0.8408	0.2787	0.8110	0.2491
	1.00	0.8356	0.3435	0.8026	0.3070
	2.00	0.8446	1.3751	0.8146	1.2294

(*continued*)

Table 2. (*continued*)

n	σ	GCI approach		Large sample approach	
		CP	Lenght	CP	Lenght
100	0.10	0.8894	0.0023	0.8806	0.0022
	0.20	0.8888	0.0094	0.8764	0.0088
	0.30	0.8930	0.0213	0.8846	0.0199
	0.40	0.8878	0.0377	0.8784	0.0354
	0.50	0.8784	0.0590	0.8712	0.0554
	0.60	0.8918	0.0851	0.8808	0.0798
	0.70	0.8876	0.1159	0.8798	0.1087
	0.80	0.8948	0.1514	0.8840	0.1420
	0.90	0.8902	0.1914	0.8814	0.1795
	1.00	0.8878	0.2366	0.8786	0.2218
	2.00	0.8844	0.9456	0.8778	0.8871
200	0.10	0.9088	0.0016	0.9072	0.0015
	0.20	0.9158	0.0065	0.9146	0.0063
	0.30	0.9188	0.0147	0.9184	0.0142
	0.40	0.9146	0.0262	0.9124	0.0253
	0.50	0.9132	0.0409	0.9120	0.0396
	0.60	0.9120	0.0590	0.9106	0.0570
	0.70	0.9220	0.0804	0.9210	0.0776
	0.80	0.9154	0.1049	0.9128	0.1014
	0.90	0.9172	0.1328	0.9134	0.1283
	1.00	0.9148	0.1638	0.9140	0.1583
	2.00	0.9110	0.6562	0.9066	0.6337

5 An Empirical Application

In this section, two real data examples are exhibited to illustrate the generalized confidence interval approach and the large sample approach. The first data set compares two different procedures for the shear strength for steel plate girders. Data for nine girders for two of these procedures, Karlsruhe method and Lehigh method [15]. Using the data from Table 3, for test the hypothesis of equal mean treatment effects. Under the assumption are normality, homogeneity of variance, and independence of errors. The Shapiro - Wilk normality test indicate that the two sets of data come from normal populations and the variances were homogeneous by Levene's test. The sample variances of the normal data were 0.0213 and 0.0024 for Karlsruhe method and Lehigh method respectively. Using the generalized confidence interval approach, the generalized confidence interval for the overall variance was (0.0012, 0.0104) with the length of interval 0.0092. In

Table 3. The 18 observations of the shear strength for steel plate girders.

Karlsruhe method			Lehigh method		
1.186	1.151	1.322	1.061	0.992	1.063
1.339	1.200	1.402	1.062	1.065	1.178
1.365	1.537	1.559	1.037	1.086	1.052

Table 4. The 50 observations of blood Sugar Levels (mg/100 g) for 10 Animals from Each of five Breeds (A–E).

A		B		C		D		E	
124	120	111	129	117	148	104	119	142	149
116	110	101	122	142	141	128	106	139	150
101	127	130	103	121	122	130	107	133	149
118	106	108	122	123	139	103	107	120	120
118	130	127	127	121	125	121	115	127	116

comparison, the confidence interval by the large sample approach was (0.0003, 0.0050) with the length of interval 0.0047.

The second example was blood sugar levels (mg/100g) measured from ten animals of five different breeds [16]. The results are presented in Table 4, for test the hypothesis of equality of means for the five breeds. The data on the five set were tested from normal populations by Shapiro - Wilk normality test and the variances were homogeneous by Levene's test. The sample variances of the normal data were 84.0000, 124.6667, 126.5444, 101.1111 and 173.1667 for breeds A, B, C, D and E respectively. Using the generalized confidence interval approach, the generalized confidence interval for the overall variance was (56.9113, 156.9829) with the length of interval 100.0716. In comparison, the confidence interval by the large sample approach was (62.6062, 155.0373) with the length of interval 92.43106.

6   Discussion and Conclusions

This paper has presented a simple approach to construct confidence intervals for the common variance of normal distributions. The proposed confidence intervals were constructed by two approaches, the generalized confidence interval and large sample approaches. The generalized confidence interval approach provided coverage probability close to nominal level 0.95 and is better than the large sample approach for all sample sizes. The average lengths increased when the value of σ increased for both approaches. The results indicated that the confidence interval for the common variance of normal distributions based on the generalized confidence interval approach is better than confidence interval based on the large sample approach. In conclusion, the generalized confidence interval can be successfully used to estimate the common variance of normal distributions. This conclusion supports the research papers of Tian [11], Tian and Wu [12], Krishnamoorthy [13] and Ye et al. [14].

Acknowledgments. The first author gratefully acknowledges the financial support from Faculty of Applied Science and Graduate College of King Mongkuts University of Technology North Bangkok of Thailand.

References

1. Cohen, A.: Improved confidence intervals for the variance of a normal distribution. J. Am. Stat. Assoc. **67**, 382–387 (1972)
2. Shorrock, G.: A minimax generalized bayes confidence interval for a normal variance. Ph.D. dissertation, Dept. Statistics, Rutgers Univ (1982)
3. Shorrock, G.: Improved confidence intervals for a normal variance. Ann. Stat. **18**, 972–980 (1990)
4. Nagata, Y.: Improvements of interval estimations for the variance and the ratio of two variances. J. Jpn. Stat. Soc. **19**, 151–161 (1989)
5. Casella, G., Goutis, C.: Improved invariant confidence intervals for a normal variance. Ann. Statist. **19**, 2015–2031 (1991)
6. Kubokawa, T.: A unified approach to improving equivariant estimators. Ann. Stat. **22**, 290–299 (1994)
7. Shorrock, R.W., Zidek, J.V.: An improved estimator of the generalized variance. Ann Stat. **4**(3), 629–638 (1976)
8. Sarkar, S.K.: On improving the shortest length confidence interval for the generalized variance. J. Multivar. Anal. **31**, 136–147 (1989)
9. Iliopoulos, G., Kourouklis, S.: On improved interval estimation for the generalized variance. J. Stat. Plan Infer. **66**, 305–320 (1998)
10. Weerahandi, S.: Generalized confidence intervals. J. Am. Stat. Assoc. **88**, 899–905 (1993)
11. Tian, L.: Inferences on the common coefficient of variation. Stat. Med. **24**, 2213–2220 (2005)
12. Tian, L., Wu, J.: Inferences on the common mean of several log-normal populations: the generalized variable approach. Biometrical J. **49**, 944–951 (2007)
13. Krishnamoorthy, K., Lu, Y.: Inference on the common means of several normal populations based on the generalized variable method. Biometrics **59**, 237–247 (2003)
14. Ye, R.D., et al.: Inferences on the common mean of several inverse gaussian populations. Comput. Stat. Data Anal. **54**, 906–915 (2010)
15. Montgomery, C.D.: Design and Analysis of Experiments, p. 57. Wiley, New York (2001)
16. Rencher, A.C., Schaalje, B.G.: Linear Models in Statistics, p. 373. Wiley, New Jersey (2008)

Confidence Intervals for Common Mean of Normal Distributions with Known Coefficient of Variation

Sukritta Sodanin[✉], Sa-Aat Niwitpong, and Suparat Niwitpong

Applied Statistics, King Mongkut's University of Technology North Bangkok,
Bangkok 10800, Thailand
sodanin@gmail.com, {sa-aat.n,suparat.n}@sci.kmutnb.ac.th

Abstract. This paper presents confidence intervals for common mean of normal distributions with known coefficient of variation based on three Adjusted MOVER approaches ($M1$, $M2$ and $M3$) compared with large sample approach. The coverage probability and expected length of confidence intervals were then evaluated by the Monte Carlo simulation. The results showed that M3 provided much better performance than the others in terms of coverage probability. Two data sets are given to illustrate the confidence interval approaches.

Keywords: Confidence interval · Common mean · Normal distribution · Coefficient of variation · Method of variance estimates recovery

1 Introduction

The confidence interval for common mean has been used by several researchers. For example, confidence intervals for the common mean of several inverse Gaussian populations based on a combined confidence distribution were constructed by Liu et al. [1]. Tian and Wu [2] proposed a new approach for the confidence interval estimation and hypothesis testing of the common mean of several log-normal populations using the concept of generalized variable. The estimation of mean from a normal distribution with known coefficient of variation was used in the field of medical, biological and chemical experiments; see e.g., Searls [3], Brazauskas and Ghorai [4].

Under several situations of experiment, there are common practices to collect data in different settings. As a result, the problem involving estimation of the common parameter mean in normal samples with known coefficient of variation is interesting. Inference procedures about the common normal means with known coefficient of variation are of practical and theoretical importance. Thus, the goal of this paper is to develop a novel approach for the confidence interval estimation for the common normal mean with known coefficient of variation derived from several independent samples.

Construction of confidence interval for parameter of distributions has been successful when using the method of variance estimates recovery (MOVER).

© Springer International Publishing AG 2016
V.-N. Huynh et al. (Eds.): IUKM 2016, LNAI 9978, pp. 574–585, 2016.
DOI: 10.1007/978-3-319-49046-5_49

Several researchers have used the MOVER approach to construct confidence intervals; see e.g., Donner and Zou [5], Li et al. [6], Suwan and Niwitpong [7], Wongkhao [8] and Niwitpong [9]. We will apply the MOVER approach to construct confidence interval for common mean on several normal samples with known coefficients of variation. This approach is called the Adjusted MOVER. It is the first time the Adjusted MOVER approach has been used for the confidence interval for the common mean on several normal samples with known coefficients of variation. These confidence intervals are CI_s, CI_m and CI_t ($M1$, $M2$ and $M3$) compared with the large sample approach.

This paper is divided into the following sections; the methods of common mean for constructing confidence intervals are described in Sect. 2. The simulation results are presented to evaluate the performance of the proposed in Sect. 3. The illustrations to support the simulation results are explained in Sect. 4. Finally, there are discussion and conclusion in Sect. 5.

2   The Common Mean of Normal Distribution with Known Coefficient of Variation

2.1   The Large Sample Approach

Consider k independent normal populations with a common mean θ.

Let $X_{i1}, X_{i2}, ..., X_{in}$ be a random sample from the i-th normal population as follows:

$$X_{ij} \sim N(\mu_i, \sigma_i^2)$$

Thus we have $\theta = \mu_i$, where $i = 1, 2, \dots, k$.

Let $\bar{x}_i$ and s_i^2 denote the sample mean and variance for data X_{ij} ($j = 1, 2, \dots, n_i$) for the i-th sample and let $\bar{x}_i$ and s_i^2 denote the observed sample mean and variance. From the i-th sample, the maximum likelihood estimator of θ is

$$\hat{\theta}^{(i)} = \hat{\mu}_i, \text{ where } \hat{\mu}_i = \bar{x}_i.$$

According to Searls [3], the effective estimator of mean with known coefficient of variation ($\tau = \sigma/\mu$) can be defined as

$$\bar{x}^* = (n + \tau^2)^{-1} \sum_{j=1}^{n} X_j = \frac{n\bar{x}}{n + \tau^2}, \quad \sigma^2 = \frac{ns^2}{(n + \tau^2)^2}$$

Thus, we have

$$\hat{\theta}^{(i)} = \tilde{x}_i = \frac{n_i \bar{x}_i}{n_i + \tau_i^2} \tag{1}$$

and

$$var(\hat{\theta}^{(i)}) = var(\tilde{x}_i) = \frac{n_i \sigma_i^2}{(n_i + \tau_i^2)^2}. \tag{2}$$

The large sample estimate of normal mean is a pooled estimate of the common mean from k populations which is proposed by Tian and Wu (2007). It is defined as

$$\hat{\theta}_i = \sum_{i=1}^{k} \frac{\hat{\theta}^{(i)}}{var(\hat{\theta}^{(i)})} / \sum_{i=1}^{k} \frac{1}{var(\hat{\theta}^{(i)})} \tag{3}$$

where $\theta^{(i)}$ and $var(\theta^{(i)})$ are defined in (1), (2) with μ_i and σ_i^2 replaced by $\bar{x}_i$ and s_i^2, respectively. Hence, the large sample solution for confidence interval estimation is

$$\left(\hat{\theta}_i - z_{1-\alpha/2}\sqrt{1/\sum_{i=1}^{k} 1/var(\hat{\theta}^{(i)})}, \ \hat{\theta}_i + z_{1-\alpha/2}\sqrt{1/\sum_{i=1}^{k} 1/var(\hat{\theta}^{(i)})}\right) \tag{4}$$

2.2 The Adjusted Method of Variance Estimates Recovery Approach (The Adjusted MOVER)

The method of variance estimates recovery approach (the MOVER) is the alternative method to construction confidence intervals for functions of parameters. The MOVER was presented by Donner and Zou [5]. This method obtains the variance estimator of each parameter component separately for the lower and upper confidence limits. The method is based on the central limit theorem (CLT). Consider the construction of the $100(1-\alpha)\%$ two-sided confidence interval (L,U) for parameter $\theta_1 + \theta_2$, under the assumption of independence between the point estimates $\hat{\theta}_1$ and $\hat{\theta}_2$, the lower limit L and the upper limit U are given by

$$(L,U) = (\hat{\theta}_1 + \hat{\theta}_2) \mp z_{\alpha/2}\sqrt{var(\hat{\theta}_1) + (\hat{\theta}_2)}, \tag{5}$$

where $var(\hat{\theta}_1)$, $var(\hat{\theta}_2)$ are unknown variances of $\hat{\theta}_1$, $\hat{\theta}_2$, respectively. Let (l_i, u_i) be the $100(1 - \alpha)\%$ confidence interval for θ_i, where $i = 1, 2$. It seemed that the value of $l_1 + l_2$ is close to L and the value of $u_1 + u_2$ is close to U. By the central limit theorem, we have

$$(l_i, u_i) = \hat{\theta}_i \mp z_{\alpha/2}\sqrt{\widehat{var}\left(\hat{\theta}_i\right)}. \tag{6}$$

Thus, the estimated variance at $\hat{\theta}_i = l_i$ is equal to

$$\widehat{var}\left(\hat{\theta}_i\right) = \frac{(\hat{\theta}_i - l_i)^2}{z_{\alpha/2}^2}, \tag{7}$$

and the estimated variance at $\hat{\theta}_i = u_i$ is equal to

$$\widehat{var}\left(\hat{\theta}_i\right) = \frac{(u_i - \hat{\theta}_i)^2}{z_{\alpha/2}^2}. \tag{8}$$

Suppose k independent normal populations with a common mean θ. Consider the construction of the $100(1 - \alpha)\%$ two-sided confidence interval (L,U) for parameter $\theta_1 + \ldots + \theta_k$.

We now extend the results of Donner and Zou [5] to find the common mean confidence interval for $\hat{\theta} = \sum_{i=1}^{k} W_i \hat{\theta}^{(i)} / \sum_{i=1}^{k} W_i$ which is the best unbiased estimator for θ. Thus, $(L, U) = (\hat{\theta}_1 + \ldots + \hat{\theta}_k) \mp z_{\alpha/2} \sqrt{var\left(\hat{\theta}_1\right) + \ldots + var\left(\hat{\theta}_k\right)}$ where $z_{\alpha/2}$ is the upper $\alpha/2$ quintiles of the standard normal distribution. Suppose the $100(1 - \alpha)\%$ two-sided confidence interval for $\hat{\theta}_i$ is given by (l_i, u_i), $i = 1, \ldots, k$. The lower limit L is in the neighborhood of $l_1 + \ldots + l_k$.

From (6)–(8), we have $(l_i, u_i) = \hat{\theta}_i \mp z_{\alpha/2} \sqrt{\widehat{var}\left(\hat{\theta}_i\right)}$, $i = 1, 2, \ldots, k$. The estimated variance at $\hat{\theta}_i = l_i$ is equal to

$$\widehat{var}\left(\hat{\theta}_i\right) = \frac{(\hat{\theta}_i - l_i)^2}{z_{\alpha/2}^2}, \quad i = 1, 2, \ldots, k,$$

and the estimated variance at $\hat{\theta}_i = u_i$ is equal to

$$\widehat{var}\left(\hat{\theta}_i\right) = \frac{(u_i - \hat{\theta}_i)^2}{z_{\alpha/2}^2}, \quad i = 1, 2, \ldots, k.$$

We use the method of the large sample for normal mean with pooled estimate of the common mean from k populations. Therefore, the Adjusted MOVER approach applied to use with the concepts of a large sample is,

$$\hat{\theta} = \sum_{i=1}^{k} \frac{\hat{\theta}^{(i)}}{var(\hat{\theta}^{(i)})} / \sum_{i=1}^{k} \frac{1}{var(\hat{\theta}^{(i)})}$$

where

$$var(\hat{\theta}^{(i)}) = \frac{1}{2} \left(\frac{(\hat{\theta}^{(i)} - l_i)^2}{z_{\alpha/2}^2} + \frac{(u_i - \hat{\theta}^{(i)})^2}{z_{\alpha/2}^2} \right), \quad i = 1, 2, \ldots, k. \tag{9}$$

Therefore, the Adjusted MOVER limits the confidence intervals are as follows,

$$L_i = \hat{\theta} - \dot{z}_{\alpha/2} \sqrt{1 / \sum_{i=1}^{k} 1 / var(\hat{\theta}^{(i)})}$$

$$L_i = \hat{\theta} - z_{\alpha/2} \sqrt{1 / \sum_{i=1}^{k} 1 / \frac{(\hat{\theta}^{(i)} - l_i)^2}{z_{\alpha/2}^2}} \tag{10}$$

and

$$U_i = \hat{\theta} + z_{\alpha/2} \sqrt{1 / \sum_{i=1}^{k} 1 / var(\hat{\theta}^{(i)})}$$

$$U_i = \hat{\theta} + z_{\alpha/2} \sqrt{1 / \sum_{i=1}^{k} 1 / \frac{(u_i - \hat{\theta}^{(i)})^2}{z_{\alpha/2}^2}}. \tag{11}$$

Consequently, L_i, U_i are defined in Eqs. (10), (11). We have three confidence intervals for normal mean with known coefficient of variation described by Niwitpong [9]. There are CI_s, CI_m, CI_t which are presented in Eqs. (12)–(14). According to Niwitpong [9], the confidence intervals are

$$CI_s = (l_{1i}, u_{1i}) = \left(\bar{X}_i^* - z_{\alpha/2} \frac{\sqrt{n_i} S_i}{(n_i + \tau_i^2)}, \ \bar{X}_i^* + z_{\alpha/2} \frac{\sqrt{n_i} S_i}{(n_i + \tau_i^2)} \right) \qquad (12)$$

where $\bar{X}_i^* = (n_i + \tau_i^2)^{-1} \sum_{i=1}^{k} X_{ij}$,

$$CI_m = (l_{2i}, u_{2i}) = \left(\hat{\theta}_i - z_{\alpha/2} \sqrt{\frac{S_i^2}{n_i(1 + 2\tau_i^2)}}, \ \hat{\theta}_i + z_{\alpha/2} \sqrt{\frac{S_i^2}{n_i(1 + 2\tau_i^2)}} \right) \qquad (13)$$

where $\hat{\theta}_i = \sqrt{4\tau_i^2 S_i^2 + (1 + 4\tau_i^2)\bar{X}_i^2} - \bar{X}_i / 2\tau_i^2$ and

$$CI_t = (l_{3i}, u_{3i}) = \left(\bar{X}_i^* - c \frac{\sqrt{n_i} S_i}{(n_i + \tau_i^2)}, \ \bar{X}_i^* + c \frac{\sqrt{n_i} S_i}{(n + \tau_i^2)} \right) \qquad (14)$$

where $\bar{X}_i^* = (n_i + \tau_i^2)^{-1} \sum_{i=1}^{k} X_{ij}$, $c = t_{\alpha/2, n_i - 1}$.

Substituting (12)–(14) into (10)-(11). Therefore, we have three confidence intervals for common mean of normal distribution with known coefficient of variation using the Adjusted MOVER. Accordingly, we get these three confidence intervals (L_{1i}, U_{1i}), (L_{2i}, U_{2i}) and (L_{3i}, U_{3i}) and we defined as M1, M2 and M3 as follows: For CI_s: substituting (12) in (10)-(11), then we get (L_{1i}, U_{1i}).

We set

$$l_{1i} = \bar{X}_i^* - z_{\alpha/2} \frac{\sqrt{n_i} S_i}{(n_i + \tau_i^2)} \quad and \quad u_{1i} = \bar{X}_i^* + z_{\alpha/2} \frac{\sqrt{n_i} S_i}{(n_i + \tau_i^2)},$$

thus we have

$$L_{1i} = \hat{\theta} - z_{\alpha/2} \sqrt{ 1 / \sum_{i=1}^{k} 1 / \frac{(\hat{\theta}^{(i)} - l_{1i})^2}{z_{\alpha/2}^2} }, \qquad (15)$$

$$U_{1i} = \hat{\theta} + z_{\alpha/2} \sqrt{ 1 / \sum_{i=1}^{k} 1 / \frac{(u_{1i} - \hat{\theta}^{(i)})^2}{z_{\alpha/2}^2} }. \qquad (16)$$

Similarly, For CI_m: substituting (13) in (10)–(11), then we get (L_{2i}, U_{2i}). We set

$$l_{2i} = \hat{\theta}_i - z_{\alpha/2} \sqrt{\frac{S_i^2}{n_i(1 + 2\tau_i^2)}} \quad and \quad u_{2i} = \hat{\theta}_i + z_{\alpha/2} \sqrt{\frac{S_i^2}{n_i(1 + 2\tau_i^2)}},$$

thus we have

$$L_{2i} = \hat{\theta} - z_{\alpha/2}\sqrt{1/\sum_{i=1}^{k} 1/\frac{(\hat{\theta}^{(i)} - l_{2i})^2}{z_{\alpha/2}^2}}, \tag{17}$$

$$U_{2i} = \hat{\theta} + z_{\alpha/2}\sqrt{1/\sum_{i=1}^{k} 1/\frac{(u_{2i} - \hat{\theta}^{(i)})^2}{z_{\alpha/2}^2}}. \tag{18}$$

Similarly, For CI_t: substituting (14) in (10)-(11), then we get (L_{3i}, U_{3i}). we set

$$l_{3i} = \bar{X}_i^* - c\frac{\sqrt{n_i}S}{(n_i + \tau_i^2)} \quad and \quad u_{3i} = \bar{X}_i^* + c\frac{\sqrt{n_i}S}{(n + \tau_i^2)},$$

thus we have

$$L_{3i} = \hat{\theta} - z_{\alpha/2}\sqrt{1/\sum_{i=1}^{k} 1/\frac{(\hat{\theta}^{(i)} - l_{3i})^2}{z_{\alpha/2}^2}}, \tag{19}$$

$$U_{3i} = \hat{\theta} + z_{\alpha/2}\sqrt{1/\sum_{i=1}^{k} 1/\frac{(u_{3i} - \hat{\theta}^{(i)})^2}{z_{\alpha/2}^2}}. \tag{20}$$

In this study, we constructed the estimation of confidence intervals based on a large sample (LS) and the Adjusted MOVER method (M1, M2 and M3) to find the coverage probabilities (CP) and expected lengths (EL).

2.3 Computing Algorithms

For a given data set X_{ij} for $i = 1, 2, \ldots, k$, the confidence intervals for common mean based on a large sample approach and the Adjusted MOVER approach for θ can be computed by the following steps:

Common Mean Based on Large Sample Approach

1. Compute $\bar{x}_i$, s_i and τ_i for $i = 1, 2, \ldots, k$.
2. Compute $\theta^{(i)}$, $\hat{\theta}$ and $var(\theta^{(i)})$ for $i = 1, 2, \ldots, k$.
3. Construct confidence interval from Eq. (4).

Common Mean Based on the Adjusted MOVER Approach

1. Compute $\bar{x}_i$, s_i and τ_i for $i = 1, 2, \ldots, k$.
2. Compute $\theta^{(i)}$, $\hat{\theta}$ and $var(\theta^{(i)})$ for $i = 1, 2, \ldots, k$.
3. Construct confidence interval from Eqs. (10), (11), calculate (l_{1i}, u_{1i}) based on CI_s (M1), then compute (L_{1i}, U_{1i}).
4. Similarly, calculate (l_{2i}, u_{2i}) based on CI_m (M2), then compute (L_{2i}, U_{2i}).
5. Similarly, calculate (l_{3i}, u_{3i}) based on CI_t (M3), then compute (L_{3i}, U_{3i}).

3   Simulation Studies

In this section, Monte Carlo simulation study is performed to evaluate the coverage probabilities (CP) and expected lengths of all confidence intervals. Confidence intervals for common mean based on large sample approach (LS) compared with the Adjusted MOVER method of CI_s, CI_m, CI_t (M1, M2, M3), respectively. In this simulation study, the number of samples $k = 2$ and $k = 6$, sample sizes, $n_1 = \ldots = n_k = n_i$, $n_i = 10, 30, 50, 100$, the population mean of normal data within each sample $\mu_1 = \ldots = \mu_k = 1$, and the population standard deviation $\sigma = 0.01, 0.03, 0.05, 0.07, 0.09, 0.10, 0.30, 0.50, 0.70, 0.90$ and 1.00, $\tau = 0.01, 0.03, 0.05, 0.07, 0.09, 0.10, 0.30, 0.50, 0.70, 0.90$ and 1.00. For each parameter setting with a number of simulation runs, M $= 5{,}000$ and a coefficient of confidence interval at $1 - \alpha = 0.95$.

Tables 1 and 2 presents the coverage probabilities and expected lengths for 2 and 6 sample cases, respectively. The results in many situations show that the coverage probabilities of the confidence interval based on M3 are close to the nominal confidence level at 0.95. Coverage probabilities of the M3 approach are much closer to the nominal confidence level than the coverage probabilities of the LS, M1 and M2 approaches. Results show that the coverage probabilities of the M3 approach perform well. Therefore, the confidence intervals based on the M3 approach have an exact confidence interval. All of the approaches (the Adjusted MOVER method: M1, M2 and the large sample approach) have coverage probabilities close to the nominal confidence level of 0.95 when the sample size increases. The coverage probabilities tend to decrease when τ increases, but the expected lengths increase as the value of τ increases.

Regarding the coverage probabilities of confidence interval, based on a large sample (LS) and the Adjusted MOVER method (M1, M2, M3), performance is good in terms of moderated or larger sizes, $n \geq 50$. These coverage probabilities are not different with an equally large sample size up to $n \geq 100$. In the overall image, the coverage probabilities tend to rise when the sample size increases. The Adjusted MOVER method of M3 provides much better confidence interval estimates than other approaches. Therefore, the coverage probabilities of the Adjusted MOVER method M3 are more appropriate than the other approaches. Note that these simulation results are similar to those of Niwitpong [9], who studied the CI_s, CI_m, CI_t confidence intervals of normal mean with known coefficient of variation.

4   An Empirical Application

In this section, two data examples are exhibited to illustrate the confidence interval approaches. The first data set contained drying times in hours from two different brands of latex paint from Walpole et al. [10]. Using the data from Table 3, the sample mean (sample variance) of the normal data were 3.8200 (0.6074) and 4.9400 (0.5683) for paint A and paint B. Suppose coefficients of variation of these two data sets were 0.2040 and 0.1526, respectively. Using the

Table 1. Empirical coverage probabilities (CP) and expected lengths (EL) of approximately 95 % of two-side confidence bounds for mean of the normal distribution: 2 sample cases. (based on 5000 simulations)

n	τ	Large sample		M1 (CI_s)		M2 (CI_m)		M3 (CI_t)	
		CP	EL	CP	EL	CP	EL	CP	EL
10	0.01	0.8928	0.0082	0.8928	0.0082	0.8926	0.0082	0.9324	0.0094
	0.03	0.8986	0.0244	0.8990	0.0244	0.9002	0.0244	0.9388	0.0282
	0.05	0.8980	0.0409	0.8984	0.0409	0.8972	0.0407	0.9348	0.0472
	0.07	0.8968	0.0572	0.8974	0.0572	0.8968	0.0568	0.9340	0.0660
	0.09	0.9028	0.0735	0.9030	0.0735	0.9010	0.0728	0.9404	0.0849
	0.10	0.8972	0.0817	0.8954	0.0817	0.8958	0.0807	0.9330	0.0943
	0.30	0.8990	0.2436	0.9014	0.2435	0.8824	0.2189	0.9392	0.2811
	0.50	0.8958	0.3986	0.8984	0.3988	0.8572	0.3063	0.9352	0.4604
	0.70	0.8806	0.5428	0.8920	0.5457	0.8158	0.3535	0.9324	0.6300
	0.90	0.8586	0.6659	0.8858	0.6791	0.7900	0.3850	0.9246	0.7842
	1.00	0.8284	0.7095	0.8810	0.7381	0.7790	0.4035	0.9234	0.8526
30	0.01	0.9372	0.0049	0.9372	0.0049	0.9372	0.0049	0.9458	0.0052
	0.03	0.9340	0.0148	0.9336	0.0148	0.9332	0.0148	0.9462	0.0155
	0.05	0.9394	0.0248	0.9386	0.0248	0.9374	0.0247	0.9500	0.0259
	0.07	0.9330	0.0346	0.9324	0.0346	0.9324	0.0344	0.9438	0.0361
	0.09	0.9370	0.0446	0.9370	0.0446	0.9364	0.0441	0.9472	0.0465
	0.10	0.9304	0.0494	0.9306	0.0494	0.9294	0.0488	0.9420	0.0516
	0.30	0.9420	0.1483	0.9424	0.1482	0.9248	0.1323	0.9510	0.1547
	0.50	0.9346	0.2450	0.9350	0.2450	0.8956	0.1838	0.9460	0.2557
	0.70	0.9262	0.3404	0.9272	0.3406	0.8578	0.2112	0.9394	0.3554
	0.90	0.9250	0.4320	0.9266	0.4326	0.8140	0.2278	0.9386	0.4515
	1.00	0.9248	0.4779	0.9318	0.4793	0.8198	0.2374	0.9462	0.5002
50	0.01	0.9470	0.0039	0.9472	0.0039	0.9478	0.0039	0.9536	0.0040
	0.03	0.9386	0.0116	0.9386	0.0116	0.9380	0.0116	0.9422	0.0119
	0.05	0.9396	0.0193	0.9396	0.0193	0.9384	0.0193	0.9468	0.0198
	0.07	0.9392	0.0271	0.9392	0.0271	0.9386	0.0269	0.9442	0.0277
	0.09	0.9510	0.0348	0.9508	0.0348	0.9490	0.0344	0.9558	0.0357
	0.10	0.9394	0.0386	0.9402	0.0386	0.9402	0.0381	0.9456	0.0396
	0.30	0.9458	0.1160	0.9444	0.1160	0.9306	0.1034	0.9506	0.1189
	0.50	0.9410	0.1925	0.9398	0.1925	0.8966	0.1440	0.9462	0.1974
	0.70	0.9452	0.2687	0.9438	0.2687	0.8564	0.1650	0.9506	0.2755
	0.90	0.9376	0.3423	0.9378	0.3425	0.8278	0.1782	0.9442	0.3511
	1.00	0.9372	0.3785	0.9400	0.3789	0.8210	0.1839	0.9470	0.3885
100	0.01	0.9454	0.0028	0.9456	0.0028	0.9454	0.0028	0.9492	0.0028
	0.03	0.9446	0.0083	0.9442	0.0083	0.9446	0.0082	0.9472	0.0084
	0.05	0.9508	0.0138	0.9502	0.0138	0.9500	0.0137	0.9546	0.0139
	0.07	0.9460	0.0193	0.9462	0.0193	0.9454	0.0192	0.9488	0.0195
	0.09	0.9436	0.0248	0.9438	0.0248	0.9408	0.0245	0.9464	0.0251
	0.10	0.9504	0.0275	0.9504	0.0275	0.9496	0.0271	0.9516	0.0279
	0.30	0.9472	0.0826	0.9482	0.0826	0.9388	0.0735	0.9504	0.0836
	0.50	0.9438	0.1374	0.9436	0.1374	0.9078	0.1023	0.9466	0.1391
	0.70	0.9450	0.1919	0.9464	0.1919	0.8626	0.1161	0.9496	0.1942
	0.90	0.9434	0.2460	0.9426	0.2460	0.8344	0.1266	0.9450	0.2491
	1.00	0.9416	0.2721	0.9446	0.2722	0.8190	0.1301	0.9470	0.2756

Table 2. Empirical coverage probabilities (CP) and expected lengths (EL) of approximately 95 % of two-side confidence bounds for mean of the normal distribution: 6 sample cases. (based on 5000 simulations)

n	τ	Large sample		M1 (CI_s)		M2 (CI_m)		M3 (CI_t)	
		CP	EL	CP	EL	CP	EL	CP	EL
10	0.01	0.8842	0.0046	0.8842	0.0046	0.8840	0.0046	0.9268	0.0053
	0.03	0.8768	0.0137	0.8774	0.0137	0.8756	0.0136	0.9206	0.0158
	0.05	0.8826	0.0228	0.8818	0.0228	0.8820	0.0227	0.9256	0.0264
	0.07	0.7582	0.0319	0.8792	0.0319	0.8788	0.0316	0.9238	0.0368
	0.09	0.8832	0.0411	0.8818	0.0410	0.8768	0.0405	0.9248	0.0474
	0.10	0.8800	0.0456	0.8776	0.0456	0.8726	0.0448	0.9190	0.0526
	0.30	0.8844	0.1357	0.8826	0.1355	0.8346	0.1160	0.9270	0.1563
	0.50	0.7520	0.2225	0.8758	0.2218	0.7306	0.1470	0.9196	0.2561
	0.70	0.6706	0.3034	0.8552	0.3029	0.6114	0.1496	0.9100	0.3499
	0.90	0.7948	0.3701	0.8474	0.3768	0.5370	0.1480	0.8980	0.4356
	1.00	0.7530	0.3957	0.8502	0.4129	0.5126	0.1512	0.8992	0.4775
30	0.01	0.9384	0.0028	0.9384	0.0028	0.9388	0.0028	0.9450	0.0030
	0.03	0.9342	0.0085	0.9344	0.0085	0.9346	0.0085	0.9444	0.0089
	0.05	0.9362	0.0141	0.9364	0.0141	0.9344	0.0141	0.9448	0.0148
	0.07	0.7582	0.0198	0.9304	0.0198	0.9292	0.0197	0.9400	0.0207
	0.09	0.9324	0.0255	0.9322	0.0255	0.9310	0.0252	0.9442	0.0266
	0.10	0.9346	0.0283	0.9352	0.0283	0.9320	0.0278	0.9460	0.0295
	0.30	0.9406	0.0847	0.9388	0.0847	0.9064	0.0729	0.9476	0.0884
	0.50	0.7520	0.1406	0.9380	0.1405	0.8160	0.0918	0.9480	0.1466
	0.70	0.6706	0.1952	0.9284	0.1952	0.6808	0.0917	0.9416	0.2037
	0.90	0.9206	0.2480	0.9268	0.2481	0.5740	0.0883	0.9336	0.2589
	1.00	0.9154	0.2737	0.9228	0.2742	0.5666	0.0894	0.9360	0.2861
50	0.01	0.9424	0.0022	0.9424	0.0022	0.9424	0.0022	0.9480	0.0023
	0.03	0.9372	0.0067	0.9368	0.0067	0.9362	0.0066	0.9436	0.0068
	0.05	0.9430	0.0111	0.9434	0.0111	0.9422	0.0111	0.9480	0.0114
	0.07	0.7582	0.0155	0.9438	0.0155	0.9426	0.0154	0.9498	0.0159
	0.09	0.9458	0.0200	0.9458	0.0200	0.9440	0.0197	0.9504	0.0205
	0.10	0.9402	0.0222	0.9400	0.0222	0.9362	0.0219	0.9496	0.0228
	0.30	0.9388	0.0666	0.9380	0.0666	0.9106	0.0574	0.9448	0.0682
	0.50	0.7520	0.1105	0.9394	0.1105	0.8216	0.0724	0.9458	0.1133
	0.70	0.6706	0.1540	0.9314	0.1540	0.6796	0.0716	0.9376	0.1579
	0.90	0.9312	0.1968	0.9330	0.1968	0.5966	0.0699	0.9388	0.2018
	1.00	0.9294	0.2177	0.9312	0.2178	0.5584	0.0690	0.9370	0.2233
100	0.01	0.9406	0.0016	0.9406	0.0016	0.9408	0.0016	0.9442	0.0016
	0.03	0.9448	0.0048	0.9448	0.0048	0.9452	0.0048	0.9480	0.0048
	0.05	0.9436	0.0079	0.9440	0.0079	0.9426	0.0079	0.9470	0.0080
	0.07	0.7582	0.0111	0.9494	0.0111	0.9490	0.0110	0.9518	0.0112
	0.09	0.9456	0.0143	0.9460	0.0143	0.9438	0.0141	0.9478	0.0144
	0.10	0.9420	0.0159	0.9420	0.0159	0.9408	0.0156	0.9456	0.0161
	0.30	0.9474	0.0475	0.9466	0.0475	0.9190	0.0409	0.9500	0.0481
	0.50	0.7520	0.0792	0.9446	0.0792	0.8410	0.0519	0.9476	0.0802
	0.70	0.6706	0.1106	0.9418	0.1106	0.7098	0.0514	0.9446	0.1119
	0.90	0.9392	0.1415	0.9390	0.1416	0.6008	0.0494	0.9434	0.1433
	1.00	0.9440	0.1569	0.9406	0.1569	0.5678	0.0490	0.9450	0.1589

confidence interval of the large sample approach, for the upper and lower bound was (4.1243, 4.6716) with the length of interval at 0.5472. In comparison, the confidence interval for M1 was (4.1180,4.6653) with the length of interval 0.5472, the confidence interval for M2 was (4.1232, 4.6548) with the length of interval 0.5316 and the confidence interval for M3 was (4.0922, 4.6911) with the length of interval 0.5988.

The second data set contained the observations on the yield of a chemical process using five batches of raw material selected randomly. (The data on the batch variance component) from Walpole et al. [10]. Using the data from Table 4, the sample mean (sample variance) of the normal data were 7.9429 (1.4895), 8.5286 (1.7557), 11.5286 (9.2590), 8.3429 (3.9295) and 10.7571 (3.9062) for five batches of raw material 1, 2, 3, 4, and 5, respectively. Suppose coefficients of variation of these five data sets were 0.1537, 0.1554, 0.2639, 0.2376 and 0.1837, respectively. Using the confidence interval of the large sample approach, for the upper and lower bound was (8.2343, 9.3159) with the length of interval 1.0815. In comparison, the confidence interval for M1 was (8.2157, 9.2971) with the length of interval 1.0815, the confidence interval for M2 was (8.2616, 9.3146) with the length of interval 1.0530, the confidence interval for M3 was (8.0814, 9.4316) with the length of interval 1.3502. Therefore, the results from above examples support our simulation results.

Table 3. The 30 observations of drying times from two different brands of latex paint.

Latex paint (drying times in hours)									
Paint A					Paint B				
3.5	2.7	3.9	4.2	3.6	4.7	3.9	4.5	5.5	4.0
2.7	3.3	5.2	4.2	2.9	5.3	4.3	6.0	5.2	3.7
4.4	5.2	4.0	4.1	3.4	5.5	6.2	5.1	5.4	4.8

Table 4. The 35 observations of a chemical process using five batches of raw material

Batch of raw material (variance component)				
1	2	3	4	5
9.7	10.4	15.9	8.6	9.7
5.6	9.6	14.4	11.1	12.8
8.4	7.3	8.3	10.7	8.7
7.9	6.8	12.8	7.6	13.4
8.2	8.8	7.9	6.4	8.3
7.7	9.2	11.6	5.9	11.7
8.1	7.6	9.8	8.1	10.7

5 Discussion and Conclusions

The study is carried out to investigate the performance of the confidence intervals for common mean of normal distribution with known coefficient of variation based on the large sample approach (LS) and the Adjusted MOVER method: M1, M2 and M3. The coverage probabilities and expected lengths perform well when sample sizes are sufficiently large. The coverage probabilities are close to the nominal confidence level at 0.95 when sample size increases. The coverage probabilities tend to decrease when τ increases, but the expected lengths increase as the value of τ increases.

The coverage probabilities of M3 are much closer to the nominal confidence level at 0.95 than other approaches. The coverage probabilities of M3 are satisfactorily stable. Therefore, the confidence intervals based on M3 have an exact confidence interval. The coverage probabilities of M3 provides much better results than LS, M1 and M2, in terms of the coverage probabilities which are similar to the findings in research papers of Suwan and Niwitpong [7] and Niwitpong [9]. In conclusion, the coverage probabilities of the Adjusted MOVER method M3 are more appropriate than the other approaches.

In this study, the coverage probabilities of larger sample size are better than the coverage probabilities of smaller sample sizes. Although, these methods are not suitable for small sample sizes ($n \leq 30$). Therefore, this problem should be solved by other methods until the desired results are achieved.

Acknowledgments. The authors would like to thank the Faculty of Applied Sciences and Graduate College, King Mongkut's University of Technology North Bangkok for partial financial support.

References

1. Liu, X., et al.: Combining inferences on the common mean of several inverse Gaussian distributions based on confidence distribution. Stat. Prob. Lett. **105**, 136–142 (2015)
2. Tian, L., Wu, J.: Inferences on the common mean of several log-normal populations: the generalized variable approach. Biom. J. **49**, 944–951 (2007)
3. Searles, D.T.: The utilization of a known coefficient of variation in the estimation procedure. J. Am. Stat. Assoc. **59**, 1225–1226 (1964)
4. Brazauskas, V., Ghorai, J.: Estimating the common parameter of normal models with known coefficients of variation: a sensitivity study of asymptotically efficient estimators. J. Stat. Com. **77**, 663–681 (2007)
5. Donner, A., Zou, G.: Closed-form confidence intervals for functions of the normal mean and standard deviation. Stat. Methods Med. Res. **21**, 1–13 (2010)
6. Li, H.Q., et al.: Confidence intervals for ratio of two poisson rates using the method of variance estimates recovery. Comp. Stat. **29**, 869–889 (2014)
7. Suwan, S., Niwitpong, S.: Interval estimation for a linear function of variances of nonnormal distributions that utilize the Kurtosis. Appl. Math. Sci. **7**, 4909–4918 (2013)

8. Wongkhao, A.: Confidence intervals for parameters of normal distribution. Ph.D. Thesis. KMUTNB, Bangkok (2014)
9. Niwitpong, S.: Confidence intervals for the normal mean with known coefficient of variation. Far. East. J. Math. Sci. **97**, 711–727 (2015)
10. Walpole, R.E., et al.: Probability and Statistics for Engineers and Scientists, 9th edn., pp. 295–549. Prentice Hall, Pearson Education Inc., Boston (2012)

Pair Trading Rule with Switching Regression GARCH Model

Kongliang Zhu[✉], Woraphon Yamaka, and Songsak Sriboonchitta

Faculty of Economics, Chiang Mai University, Chiang Mai, Thailand
258zkl@gmail.com, woraphon.econ@gmail.com, songsakecon@gmail.com

Abstract. Pairs trading strategy is a famous strategy and commonly taken by many investors. There are various approaches to define the pairs trading signal which is the important part of the strategy. This study aims to propose an alternative approach, Markov Switching Regression GARCH model, to specify the trading signal for stock pair taking into account the structural change in the pair return. We applied our proposed model to the Stock Exchange of Thailand and the result shows our pairs trading strategy is relatively more effective for financial investment management compared with the single mean return from individual stock method.

Keywords: Pairs trading · Markov switching · GARCH · SET50 Index

1 Introduction

Today, pairs trading continues to remain an important quantitative method of speculation strategy since its invention at Morgan Stanley in 1987. Pairs trading is a trading strategy which is work by identifying two stocks whose prices have high correlation. The key advantage of this strategy is that it can be used to gain profit under different market conditions, including periods when the equity market goes up, down, or oscillating between a relatively narrow range, along with low or high volatilities [16]. When the price relation is broken, short the winner and buy the loser. If the past is a good mirror of the future, the prices of two stocks will converge to a mean and the arbitrageur will profit. Like the statistical arbitrage strategy, pairs trading is a market-neutral strategy that matches a long position and a short position of a two stocks that are correlated.

There exists a wide range of different researches on pairs trading such as a distance method, co-integration approach and stochastic spread method. These are three main methods applied in pairs trading strategy. Firstly, the distance method involves calculating the sum of squared deviations between two normalized stock prices as the criteria to select pairs and form trading opportunities. It was first used in the study by Gatev et al. [14] who found average annualized excess returns over 10 % based on the daily data from 1962 to 2002 in the US market. Later, Perlin [13] extended the analysis to investigate the profitability and risk of the pairs trading strategy for Brazilian stock market. Do and Faff [5]

© Springer International Publishing AG 2016
V.-N. Huynh et al. (Eds.): IUKM 2016, LNAI 9978, pp. 586–598, 2016.
DOI: 10.1007/978-3-319-49046-5_50

extended the original analysis of Gatev et al. [14] to June 2008 and found to be profitable for a long period of time, albeit at a declining rate. Secondly, Vidyamurthy [19] suggested a co-integration approach and described how to apply this method to pairs trading. If two stocks are co-integrated with each other, they should theoretically have a narrow spread in long-term equilibrium; and investors can attempt changing their portfolio to take a profit when co-integrated assets depart from their equilibrium. Miao [16] developed high frequency and dynamic pairs trading system using the two-stage correlation and co-integration approach. Chiu and Wong [4] derived the optimal trading strategy in a closed-form solution by investigating time-consistent mean-variance portfolio strategies for co-integrated assets in a continuous-time economy. Thirdly, Elliott et al. [7] proposed the stochastic spread method which applies a Kalman filter to estimating a parametric model of the spread, in which the spread is assumed to follow the Vasicek model. Do et al. [5] extended the stochastic spread method into the stochastic residual spread method to overcome the defects of the former method. Although these methods are likely to have appropriate result, there is a question on the linear model assumption. Many studies mentioned that the financial data might have a non-linear behaviour and that they are often found to switch between different regimes. Thus, the conventional linear method for pair spread data might fail to identify potential arbitrage opportunities [6] and might cause simple pairs trading signals to be wrong [3]. In the most recent literature, we found some studies proposed to use non-linear models such as threshold model and Markov Switching model. Many previous studies have suggested that both Markov-switching model and threshold model provide a better performance to the stock returns when compare with the linear models. Bock and Mestel [3] develop a useful trading rules for pairs trading to solve the problem related to phases of imbalance when the deviation of the price stock spread may temporarily or persistently endure. They also mentioned that the regime-switching rule for pairs trading generate a positive returns and hence it can be employed as an alternative model to traditional pairs trading rules. Yang et al. [20] combined the Markov regime-switching and Vasicek models with a mean-reverting strategy, and compare the model with conventional methods using 12 months of S&P 500 index daily price data. They found that his proposed method provide the best performance in a simple portfolio and that the shorter the trading period and the higher the performance is obtained. In the case of threshold model, Chen et al. [6] proposed a three-regime threshold GARCH (generalized autoregressive conditional heteroskedasticity) model to capture asymmetries in the average return, volatility level, mean reversion in the pair spread and also proposed to use the threshold value to determine the pairs trading strategy, say used as trading entry and exit signals. They found that the model can detect the regime change in the stock price and also provide a good trading signal, leading to reap adequate profits from the Dow Jones 30 stocks. Based on these previous studies, we also expect our data to have a non linear behavior; therefore, the non-linear model should be used for specifying the trading signal. In this study, we aim to extend the Threshold GARCH of Chen et al. [6] into Markov Switching regression GARCH since the threshold models have some limitations as discussd in Kuan [11]. Kuan [11] noted that first the non-linear optimization algorithms are difficult to find the global or the optimal solution in the parameter space. Second, the threshold models are proposed to describe certain nonlinear patterns of data and hence may not be so flexible. Moreover, the threshold model allows the parameter to change the regime on only occasion and exogenous changes. However, Kuan [11] suggested that the Markov switching model is more suitable for explaining correlated data that exhibit different behavior in unusual economic condition. Thus, instead of using Threshold GARCH model, this study proposed a Markov Switching regression GARCH as an alternative tool for determining pairs trading signals. Based on our best knowledge, Markov Switching regression GARCH has not been applied to explain pairs trading strategies before and thus one of our contributions will be an alternative model for pairs trading strategy.

The remainder of this study proceeds as follows. Section 2 introduces the Markov switching regression GARCH with different error distributions. Maximum likelihood estimation is also briefly discussed in Sect. 3. Section 4 explains pairs trading strategy and identifies pairs trading signals. The preliminary empirical results are provided in Sect. 5. Conclusions are presented in Sect. 6.

2   Markov Switching Regression GARCH

Over long period, the financial series exhibit different behavior; signifying depression, recession, bull market, and bear market and resulting in a regime change. [8] proposed the Markov switching model to capture the behavior change in the data where the regime probabilities are obtained by the proposed Hamilton-filter ([8,9]). Furthermore, Bollerslev [2] who introduced GARCH (Generalized Autoregressive Conditionally Heteroskedasticity) model noted that the time series data present variable volatility over time, thus tending to show GARCH effects in the model. Therefore, the Markov switching model has been extended to GARCH in many studies, such as those by Haas et al. [10] and Marcucci [12]. Bauwens et al. [1] in order to gain more ability to capture some stylized facts of financial time series namely volatility of the data. The general form of the Markov switching regression GARCH(m, q) model can be written as

$$y_t = \varphi_{0,s(t)} + \sum_{i=1}^{k} \varphi_{1,s(t)} X_t + \varepsilon_{t,s(t)} \tag{1}$$

$$\varepsilon_{t,s(t)} = h_{t,s(t)} v_{t,s(t)}$$

$$h^2_{t,s(t)} = \omega_{s(t)} + \sum_{i=1}^{m} \alpha_{i,s(t)} \varepsilon^2_{t-i,s(t)} + \sum_{i=1}^{q} \beta_{j,s(t)} h^2_{t-j,s(t)} \tag{2}$$

where Eqs. 1 and 2 are the mean and variance equations, respectively, and they are allowed to switch across regime. y_t is a dependent variable and X_t is a matrix of independent variables. $h^2_{t,s(t)}$ is the state dependent conditional variance and state dependent $\omega_{s(t)} \geq 0$, $\alpha_{i,s(t)} \geq 0$, and $\beta_{j,s(t)} \geq 0$ to ensure the positive conditional variance, $h^2_{t,s(t)}$. In this variance equation, the state dependent unconditional variance can be computed by $\omega_{s(t)}/(1 - \alpha_{i,s(t)} - \beta_{j,s(t)})$. In addition, some distributions, such as normal, student-t, generalized error distribution (GED), skewed GED, skewed normal, and skewed student-t distributions are adopted for innovation $v_{j,s(t)}$.

The feature of the Markov switching model is the estimate parameters in both mean and variance equations can switch across different regimes or are state dependent according to the first order Markov process. This means that all parameters are governed by a state variable $s(t)$ which is assumed to evolve according to $s(t-1)$ with transition probability, p_{ij}, thus

$$p(s(t) = j|s(t-1) = i) = p_{ij}, \qquad \sum_{j=1}^{h} p_{ij} = 1, \qquad for \quad i = 1, ..., h \quad (3)$$

Usually these probabilities can be formed as transition matrix (Q)

$$
\begin{aligned}
p(s_t = 1|s_{t-1} = 1) &= p_{11} \\
p(s_t = 1|s_{t-1} = 2) &= p_{12} \\
p(s_t = 2|s_{t-1} = 1) &= p_{21} = \\
&\vdots \\
p(s_t = i|s_{t-1} = j) &= p_{ij}
\end{aligned}
\begin{bmatrix}
p_{ij} & p_{11} & \cdots & p_{1j} \\
\vdots & p_{22} & \cdots & \vdots \\
\vdots & \vdots & \ddots & \vdots \\
p_{i1} & \cdots & \cdots & p_{ij}
\end{bmatrix}
\quad (4)
$$

3 Maximum Likelihood Estimator for Markov Switching Regression GARCH

In this study, the Markov switching regression GARCH(1, 1) is considered since it is able to reproduce the volatility dynamics of financial data and most commonly employed in many studies. To estimate the parameter set in this model, the maximum likelihood method is used and the general form of the likelihood can be defined as

$$L(\theta_{s(t)}|y, X) = f(\theta_{s(t)}|y, X)Pr(s(t) = j) \quad (5)$$

where $f(\theta_{s(t)}|y, X)$ is the density function, $\theta_{s(t)}$ is state dependent parameter set of the model and $Pr(s(t))$ is the filtered probabilities in each regime. Note that the study adopts 6 different distributions for innovation $v_{j,s(t)}$, namely normal, student-t, generalized error distribution (GED), skewed GED, skewed normal, and skewed student-t distributions. Thus the density function $f(\theta_{s(t)}|y, X)$ in Eq. 5 can be written differently according to the distribution of the $v_{j,s(t)}$.

To estimate the filtered probability, $Pr(s(t) = j)$, the Hamiltons filter as proposed in Hamilton (1989) is employed where the formula of the filter can be written as

$$Pr(s(t) = j|\theta_t) = \frac{f(y, X_t|s(t) = j, \theta_{t-1})Pr(s(t) = j|\theta_{t-1})}{\sum_{j=1}^{h} f(y, X_t|s(t) = j, \theta_{t-1})Pr(s(t) = j|\theta_{t-1})} \quad (6)$$

where $f(y, X_t|s(t) = j, \theta_{t-1})$ is the density function of each regime (see, Perlin 2004).

4　Pairs Trading

To select the pair stock, Chen et al. [6] proposed to select the pair stock using the lowest value of the Minimum Squared Distance method (MSD) which is given as follows.

$$MSD = \sum_{t=1}^{n} (P_t^1 - P_t^2)^2 \qquad (7)$$

where P_t^1 and P_t^2 are the normalized stock price

$$P_t^i = (P_t^i - \bar{P}_t^i)/sd_i$$

where, sd_i is the standard deviation of stock i. The selected pairs are then used to calculate the spread return, rS_t, using the Markov switching regression GARCH$(1, 1)$ which can be written as.

$$stock_t^1 = \varphi_{0,s(t)} + \varphi_{1,s(t)} stock_t^2 + \varepsilon_{t,s(t)} \qquad (8)$$

$$\varepsilon_{t,s(t)} = h_{t,s(t)} v_{t,s(t)}$$

$$h^2{}_{t,s(t)} = \omega_{s(t)} + \sum_{i=1}^{m} \alpha_{i,s(t)} \varepsilon^2{}_{t-i,s(t)} + \sum_{i=1}^{q} \beta_{j,s(t)} h^2{}_{t-j,s(t)} \qquad (9)$$

The in-sample return stock will be used to compute a simple hedge ratio which can be defined by the coefficient of $stock_t^2$, $\varphi_{1,s(t)}$, and then we will apply this hedge ratio to compute the spread return. The spread return of the stock pair is constructed by

$$rS_t = stock_t^1 - \sum_{s(t)=j}^{2} \left[\varphi_{0,s(t)=j} + \varphi_{1,s(t)=j} stock_t^2\right]' \bullet [Pr(s(t) = j|\theta_t) \times Q] \quad (10)$$

where $Pr(s(t) = j|\theta_t) \times Q$ is the multiplying of filtered probability and transition matrix.

　　To define the trading rule, the obtained $(rS_1, ..., rS_T)$ from Eq. 10 is used to compute the mean (u) and standard deviation (sd) in order to get the threshold value where the upper and lower threshold values can be defined as $Uthres = u + sd$ and $Uthres = u - sd$, respectively. Note that when the pair spread return exceeds our upper threshold $(Uthres)$, we sell $stock_t^1$ and buy$stock_t^2$. Once the spread drops below our lower threshold $(Lthres)$, we buy $stock_t^1$ and sell $stock_t^2$.

　　Finally, the average return of pairs trading can be computed by

$$r_1 = \frac{1}{D}\left[-ln\frac{P_{sell}^1}{P_{buy}^1} + ln\frac{P_{sell}^2}{P_{buy}^2}\right], \quad r_2 = \frac{1}{D}\left[ln\frac{P_{sell}^1}{P_{buy}^1} - ln\frac{P_{sell}^2}{P_{buy}^2}\right]$$

where D is number of holding days.

5 Estimate Results

5.1 Data Description

The daily close prices of 30 stocks in the Stock Exchange of Thailand (SET) SET50 Index are used as an illustration. The data are obtained from Thomson Reuter data stream, Faculty of Economics, Chiang Mai University over a 12-year time periods, from January 1, 2004 to February 17, 2016, totally 3165 observations. The in-sample period is from December 18, 2015 to January 29, 2016. Before the estimation or our model, we transform all the daily data to be log-return and the Augmented Dickey Fuller test (ADF) is employed for stationary test and we found that all log-returns are stationary at the level.

Notice how we defined in-sample range. We will use the in-sample data to compute a simple hedge ratio and then we will apply this hedge ratio to find a spread return. In this study, we select 30 companies comprising Advanced Info Service(ADVANC), Banpu(BANPU), Bangkok Bank(BBL), Bangchak Petroleum(BCP), Bangkok Dusit Med.Svs(BDMS), Bumrungrad Hospital(BH), Central Plaza Hotel(CENTEL), CH KarnChangCH(CK), Charoen Pokphand Foods(CPF), Central Pattana(CPN), Delta Electronics(DELTA), Electricity Generating(EGCO), Intuch Holdings(INTUCH), IRPC(IRPC), Italian-Thai Developement(ITD), Jasmine International(JAS), Kasikorn Bank(KBANK), Krung Thai Bank(KTB), Minor InternationalMINT), PTT Exploration & PRDN(PTTEP), Robinson Department store(ROBINS), Siam Comercial Bank (SCB), Siam Cement(SCC), Siam City Cement(SCCC), Tipco Asphalt(TASCO), Thanachart Capital(TCAP), TMB BANK (TMB), TPI Polene (TPIPL), True Corporation(TRUE), THAI Union Frozen PRDS(TU). And the five best candidate stock pairs are selected for further investigation using the lowest MSD between two normalized stock prices.

Prior to illustrating the pairs trading strategy, we calculate the MSD for all possible pair stocks. The MSD is conducted here to select the first five stock pairs that provide the lowest MSD. We found the five pais trading candidates as presented in Table 1.

Table 1. Pair selection

Pair	Stock 1	Stock 2	MSD
1	SCB	KBANK	84.8114
2	CPN	CENTEL	128.0145
3	INTUCH	ADVANCE	164.37
4	CENTEL	BDSM	197.3506
5	CPN	BDSM	205.1679

We then fit a Markov switching regression model with GARCH effect to these five selected pair returns. Once the model is fitted, the upper and lower threshold values, which are calculated from the standard deviation of spread return of the stock pair, are used as trading entry and exit signals. In this study, we follow a line of literatures in the pairs trading strategy by specifying that if spread return is above or below the upper or lower threshold value, we then either short or long one stock and either long or short the other stock. Once the position is open and the spread falls back to the standard deviation line, the position is closed.

5.2   Model Selection

As we mentioned before, the study conducted six different error distributions, thus we compared these six distributions, namely Normal, Skew-normal, Student-t, Skew-T, GED and Skew-GED, in both two- and three- regime model. To select the best fit distribution for our models, the Akaike Information Criterion (AIC) and Bayesian Information Criterion (BIC) are employed to compare the performance of our proposed models. Table 2 provides evidence that student-t is the best fit distribution for all pairs. However, the heterogeneous results are obtained for regime selection. We found that CPN-CENTEL, CENTEL-BDSM AND BDSM-CPN pairs prefer 2-regime Markov switching regression GARCH$(1,1)$ while 3-regime Markov switching regression GARCH$(1,1)$ provides the best fit to SCB-KBANK and INTUCH-ADVANCE pairs.

5.3   Estimation of MS-reg-GARCH Model

Table 3 shows the estimated results of two and three regimes MS-reg-GARCH$(1,1)$ when the error term has student-t distribution for five stock pairs. The model provides two equations namely, mean equation and variance equation for two and three regimes. Consider the mean equation, we interpret $\theta_{1,s(t)=i}$ as the hedge ratio and the result shows that the hedge ratio of these pairs changes when the regime changes. This confirms our expectation that there exist a regime change and non-linear structure in the stock pair returns. The results in Table 3 also show that SCB-KBANK and INTUCH-ADVANCE follow a three-regime model while the two- regime model fits the other pairs. When we compared the value of $\theta_{1,s(t)=i}$, we observed that the value of hedge ratio of all pairs decreases when the pair moves to the higher regime, except for SCB-KBANK pair. This indicates that those pairs tend to have a weaker movement when the pair returns shift to the higher regime. The SCB-KBANK pair on the contrary will have a stronger co-movement as it shifts to the higher regime (from regime 1 to regime 3). Then, let we consider the variance equation in order to interpret the meaning of each regime. It is important to identify which of these regimes presents a high volatility and which regime presents a low volatility. To answer this question, we consider the persistence of volatility shocks for each regime. Generally, the volatility persistence can be measured by the sum $\alpha_{s(t)=i} + \beta_{s(t)=i}$ and the higher value of $\alpha_{s(t)=i} + \beta_{s(t)=i}$ corresponds to the higher unconditional variance of the process. According to Table 3, we obtain a different regime interpretation and the result from these variance equations can be interpreted in two cases. In the first case, the value of $\alpha_{s(t)=i} + \beta_{s(t)=i}$ in each regime decreases when the regime is

Table 2. Model selection

1 Regime

AIC/BIC	SCB-KBANK	CPN-CENTEL	INTUCH-ADVANCE	CENTEL-BDSM	BDSM-CPN
Normal	−21115.38	−20549.8	−20011.5	−21016.88	−19812.83
	−21085.03	−20549.8	−19975.07	−20986.53	−19776.41
student-t	−21113.38	−20566.98	−21952.04	−21504.72	−20566.9
	−21076.95	−20530.55	−21909.54	−21468.29	−20530.47
skew-T	−21351.44	−20735.94	−21756.88	−21425.48	−20735.44
	−21308.94	−20693.44	−21714.38	−21395.13	−20692.94
Skew-normal	−20450.44	−15819.16	−16276.87	−15057.39	−14737.1
	−20413.97	−15782.73	−16234.38	−15020.96	−14694.6
GED	−19761.44	−19670.74	−20392.5	−20025.82	−19669
	−19725.02	−19640.39	−20356.07	−19995.47	−19638.64
skew GED	−19798.56	−20079.5	−18278.24	−20373	−20079.82
	−19756.07	−20037	−18235.74	−20330.5	−20037.32

2 Regime

AIC/BIC	SCB-KBANK	CPN-CENTEL	INTUCH-ADVANCE	CENTEL-BDSM	BDSM-CPN
Normal	−21350.9	−20810.66	−21872.36	−21493.02	−20810.66
	−21278.05	−20737.81	−21799.51	−21420.17	−20737.81
student-t	−21340.32	**−21828.4**	−22746.72	**−22868.84**	**−21228.36**
	−21261.4	**−21743.41**	−22661.73	**−22783.85**	**−21143.37**
skew-T	−21359.24	−20631.48	−22522.7	−21535.06	−20370.98
	−21262.11	−20534.35	−22437.71	−21437.93	−20273.85
skew-normal	−19096.79	−20631.86	−21583.52	−21487.52	−20631.84
	−19017.86	−20546.87	−21498.53	−21390.39	−20546.85
GED	−21396.7	−20798.12	−21831.74	−21437.2	−20798.12
	−21317.78	−20713.13	−21746.75	−21352.21	−20713.13
skew GED	−15029.48	−20627.86	−21583.52	−21487.52	−20627.86
	−14944.49	−20530.73	−21498.53	−21390.39	−20530.73

3 Regime

AIC/BIC	SCB-KBANK	CPN-CENTEL	INTUCH-ADVANCE	CENTEL-BDSM	BDSM-CPN
Normal	−22313.72	−20601.18	−21599.46	−20981.16	−20592.94
	−22186.23	−20473.69	−21471.97	−20853.67	−20465.45
student-t	**−23951.52**	−21024.4	**−23283.14**	−21748.22	−21003.84
	−23805.82	−20878.7	**−23137.44**	−21602.52	−20858.14
skew-T	−21138.62	−21720.68	−23279.78	−21813.8	−21094.72
	−20974.71	−21556.77	−23115.87	−21649.89	−20930.81
skew-normal	−22294.4	−20558.28	−21502.06	−20993.22	−20603.96
	−22148.7	−20412.58	−21356.36	−20847.52	−20458.26
GED	−22244.98	−20529.62	−21533.18	−20192.46	−20572.98
	−22099.28	−20383.92	−21387.48	−20046.76	−20427.28
skew GED	−20599.96	−19932.18	−21616.8	−20192.46	−20133.58
	−20436.05	−19768.26	−21452.89	−20046.76	−19969.67

higher. We found that CPN-CENTEL, INTUCH-ADVANCE, CENTEL-BDSM, and CPN-BDSM stock pair returns are in this case. In the second case, the value of $\alpha_{s(t)=i} + \beta_{s(t)=i}$ in each regime increases when the regime is higher and there is only one pair, namely SCB-KBANK that corresponds to this second case. Thus, we can interpret the first regime of CPN-CENTEL, INTUCH-ADVANCE, CENTEL-BDSM, and CPN-BDSM stock pair returns as the highest persistence of volatility shock regime while for the second or third regime, we interpret as

Table 3. Estimation result of MS-reg-GARCH for the five pair returns

Parameter	SCB-KBANK	CPN-CENTEL	INTUCH-ADVANCE	CENTEL-BDSM	BDSM-CPN
$\theta_{0,s(t)=1}$	0.0002	0.0001	0.0001	0.0001	0.0002
$\theta_{1,s(t)=1}$	0.4392***	0.1614***	0.9227***	0.3434***	0.2532***
$\omega_{s(t)=1}$	0.0001*	0.0001**	0.0001*	0.0001**	0.0001**
$\alpha_{s(t)=1}$	0.0001	0.1006***	0.2125	0.1007	0.1008
$\beta_{s(t)=1}$	0.5436***	0.8001	0.6192***	0.8142	0.7451
$v_{s(t)=1}$	2.1000***	3.0993	5.4264***	3.0963***	3.0995***
$\theta_{0,s(t)=2}$	0.0001	0.0001	0.0003	−0.0001	0.0001
$\theta_{1,s(t)=2}$	0.0701*	0.0806***	0.5246*	0.1716	0.1265
$\omega_{s(t)=2}$	0.0001*	0.0001	0.0001*	0.0001	0.0001
$\alpha_{s(t)=2}$	0.0001	0.005	0.2584	0.0507	0.0054
$\beta_{s(t)=2}$	0.5436***	0.4004***	0.3491***	0.4004***	0.4005***
$v_{s(t)=2}$	2.1000***	2.1142***	2.8354***	2.1372***	2.1431***
$\theta_{0,s(t)=3}$	0.0002		0.0001		
$\theta_{1,s(t)=3}$	0.6058***		0.3883***		
$\omega_{s(t)=3}$	0.0005***		0.0001***		
$\alpha_{s(t)=3}$	0.0917***		0.3184***		
$\beta_{s(t)=3}$	0.7896*		0.3026*		
$v_{s(t)=3}$	2.841***		2.1613***		
p_{11}	0.8108***	0.9000***	0.6242***	0.9003***	0.9001***
p_{22}	0.8106***	0.9	0.8096***	0.8998***	0.9
p_{33}	0.9464***		0.9712***		

moderate or low volatility regime. On the contrary, in the case of SCB-KBANK pair, we can interpret the regime in the opposite direction to those four other pairs.

In a nutshell, our empirical analysis provides evidence of: (1) positive hedge ratio for all pairs in every regime; (2) the hedge ratio likely to be high in the high volatility regime and vice versa. Thus, we can say that our stock pairs exhibit a stronger movement when the market exhibits a high volatility. This evidence seems to be in line with those found in previous works undertaken for example by Tofoli et al. [18] and Karimalis and Nimokis [15], and Pastpipatkul et al. [17]. These studies reported an interesting result about the high co-movement between financial assets in the market downturn regime. This evidence is very important for investors because putting an investment in different period seems to face with a different market situation.

Moreover, the Table also provides the result of the transition matrix and shows that the regimes in all pair are persistent because the probability of staying in their own regime is larger than 80 %, while the probability of switching between these regimes is less than 20 %, except for p_{11} of INTUCH-ADVANCE pair. This indicates that only an extreme event can switch the pair returns to change between regimes.

5.4 Pairs Trading Strategy

In this section, we illustrate a trading signal of our 5 pair returns in Figs. 1, 2, 3, 4 and 5 and also provide a summary result of the returns from pairs trading

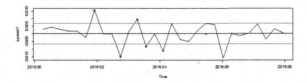

Fig. 1. SCB-KBANK pair spread return from December 18, 2015 to January 29, 2016 (Color figure online)

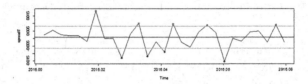

Fig. 2. CPN-CENTEL pair spread return from December 18, 2015 to January 29, 2016 (Color figure online)

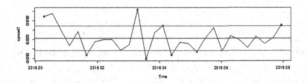

Fig. 3. INTUCH-ADVANCE pair spread return from December 18, 2015 to January 29, 2016 (Color figure online)

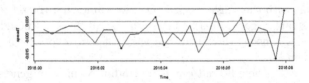

Fig. 4. CENTEL-BDMS pair spread return from December 18, 2015 to January 29, 2016 (Color figure online)

strategy. According to the following Figs. 1, 2, 3, 4 and 5, the results show the five pair returns spreads during the period of December 18, 2015 to January 29, 2016 and covering 30 trading days. Two blue lines in each figure are interpreted as threshold values which are used as trading entry and exit signals. Once the spread exceeds our upper threshold, we sell $stock_t^1$ and buy $stock_t^2$. Once the spread drops below our lower threshold, we buy $stock_t^1$ and sell $stock_t^2$.

From Figs. 1, 2, 3, 4 and 5, we summarize a trading strategy and the return of each stock pair during our in-sample period. Consider the number of trading of each pair which can be counted by counting the number of time the value of its spread exceeds either upper or lower threshold line. From Table 4, we found that

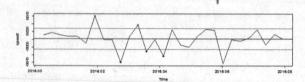

Fig. 5. CPN-BDMS pair spread return from December 18, 2015 to January 29, 2016 (Color figure online)

Table 4. Stock returns in five pairs and pair returns from December 18, 2015 to January 29, 2016

Pair	Stock1	Mean return	Stock2	Mean return	No. of trading	Pair return
1	SCB	3.32%	KBANK	3.60%	7	9.96%
2	CPN	−2.62%	CENTEL	−4.96%	9	5.50%
3	INTUCH	2.44%	ADVANCE	4.35%	8	16.95%
4	CENTEL	−4.96%	BDMS	−2.40%	9	−2.61%
5	CPN	−2.62%	BDMS	−2.40%	6	1.27%

there are 7.9 round trips trading on average from December 18, 2015 to January 29, 2016. In addition, we found that our trading signals contribute a positive return to all stock pairs, except for CENTEL-BDMS pair which presents a negative return during our in-sample period. Lets consider the individual pair return. We found that INTUCH and ADVANCE provides the highest pair return, followed by SCB-KBANK, CPN-CENTEL, and CPN-BDSM. When we compared our pair returns with the single stock return, we found that the returns from our pairs trading strategy generate a higher return in all cases. Although CENTEL-BDSM generates a negative return, the value of loss is lower when compared with single mean loss.

Finally, we can conclude that our pairs trading signal can generate a higher return when compared with the single mean return of individual stock. Thus, the obtained trading signal which was computed under the Markov switching approach works well in our application study.

6   Conclusions and Future Research

In this paper, Markov Switching regression GARCH is proposed to explain pairs trading strategy. The correlation measurement was first applied to all possible stock pair combinations. The study employed a Minimum Squared Distance method (MSD) to measure the distance between the two normalized stock price series and the first five with lowest MSD are proposed for our application. Then, the 5 selected pairs are then used to calculate the spread return, using the Markov switching regression GARCH. We found that CPN-CENTEL, CENTEL-BDSM and BDSM-CPN are preferred for 2-regime Markov switching regression

GARCH(1,1) while 3-regime Markov switching regression GARCH(1,1) provides the best fit to SCB-KBANK and INTUCH-ADVANCE pairs. The spreads are used for computing the mean and standard deviation in order to get the threshold value where the upper and lower threshold values can be defined as $Uthres = u+sd$ and $Uthres = u-sd$, respectively. Following the trading rule, we find that there are 7.9 round trips trading on average in the 30 trading days from the period December 18, 2015 to January 29, 2016. The average 5 pairs profit is 6.20 % where INTUCH and ADVANCE pair performs the highest return. In future research, Copula approach with regime switching can be applied to the pairs trading strategy and it would be useful for capturing the marginal distributions as well as the dependency structure between the stock returns. With a better understanding of the joint distribution of the two stocks, practitioners could gain preferential entry positions and have more trading opportunities.

References

1. Bauwens, L., Preminger, A., Rombouts, J.V.: Theory and inference for a Markov switching GARCH model. Econometrics J. **13**(2), 218–244 (2010)
2. Bollerslev, T.: Generalized autoregressive conditional heteroskedasticity. J. Econometrics **31**(3), 307–327 (1986)
3. Bock, M., Mestel, R.: A regime-switching relative value arbitrage rule. In: Fleischmann, B., Borgwardt, K.-H., Klein, R., Tuma, A. (eds.) Operations Research Proceedings 2008, pp. 9–14. Springer, Heidelberg (2009)
4. Chiu, M.C., Wong, H.Y.: Dynamic cointegrated pairs trading: meanvariance time-consistent strategies. J. Comput. Appl. Math. **290**, 516–534 (2015)
5. Do, B., Faff, R., Hamza, K.: A new approach to modeling and estimation for pairs trading. In: Proceedings of 2006 Financial Management Association European Conference, pp. 87–99 (2006)
6. Chen, C.W.S., Chen, M., Chen, S.-Y.: Pairs trading via three-regime threshold autoregressive GARCH models. In: Huynh, V.-N., Kreinovich, V., Sriboonchitta, S. (eds.) Modeling Dependence in Econometrics. AISC, vol. 251, pp. 127–140. Springer, Heidelberg (2014). doi:10.1007/978-3-319-03395-2_8
7. Elliott, R.J., Van Der Hoek, J., Malcolm, W.P.: Pairs trading. Quant. Financ. **5**(3), 271–276 (2005)
8. Hamilton, J.D.: A new approach to the economic analysis of nonstationary time series and the business cycle. Econometrica **57**, 357–384 (1989)
9. Hamilton, J.D.: Time Series Analysis. Princeton University Press, Princeton (1994)
10. Haas, M., Mittnik, S., Paolella, M.S.: A new approach to Markov-switching GARCH models. J. Financ. Econometrics **2**(4), 493–530 (2004)
11. Kuan, C.M.: Lecture on the markov switching model. Inst. Econ. Acad. Sinica, 1–30 (2002)
12. Marcucci, J.: Forecasting stock market volatility with regime-switching GARCH models. Stud. Nonlinear Dyn. Econometrics **9**(4), 1–55 (2005)
13. Perlin, M.: MS-Regress-the MATLAB package for Markov regime switching models. SSRN 1714016 (2015)
14. Gatev, E., Goetzmann, W.N., Rouwenhorst, K.G.: Pairs trading: performance of a relative-value arbitrage rule. Rev. Financ. Stud. **19**(3), 797–827 (2006)

15. Karimalis, E.N., Nomikos, N.: Measuring: systemic risk in the European banking sector: a copula CoVaR approach. Working paper, Cass City College, London (2014)
16. Miao, G.J.: High frequency and dynamic pairs trading based on statistical arbitrage using a two-stage correlation and cointegration approach. Int. J. Econ. Financ. **6**(3), 96 (2014)
17. Pastpipatkul, P., Yamaka, W., Sriboonchitta, S.: Analyzing financial risk and co-movement of Gold market, and Indonesian, Philippine, and Thailand stock markets: dynamic copula with Markov-switching. In: Huynh, V.-N., Kreinovich, V., Sriboonchitta, S. (eds.) Causal Inference in Econometrics. SCI, vol. 622, pp. 565–586. Springer, Heidelberg (2016). doi:10.1007/978-3-319-27284-9_37
18. Tfoli, P.V., Ziegelmann, F.A., Silva Filho, O.C.: A comparison study of copula models for European financial index returns. In: 34 Meeting of the Brazilian Econometric Society (2012)
19. Vidyamurthy, G.: Pairs: Trading: Quantitative Methods and Analysis, vol. 217. Wiley, New York (2004)
20. Yang, J.W., Tsai, S.Y., Shyu, S.D., Chang, C.C.: Pairs trading: the performance of a stochastic spread model with regime switching-evidence from the S&P 500. Int. Rev. Econ. Financ. **43**, 139–150 (2016)

Econometric Applications

An Empirical Confirmation of the Superior Performance of MIDAS over ARIMAX

Tanaporn Tungtrakul[(✉)], Natthaphat Kingnetr, and Songsak Sriboonchitta

Faculty of Economics, Chiang Mai University, Chiang Mai, Thailand
tanapornecon@gmail.com

Abstract. A new model (MIDAS) for forecasting economic data has been proposed recently to take into account of more relevant information than the ARIMAX model. However, in order to convince econometricians to use MIDAS instead of ARIMAX, at least some evidence from empirical studies is needed, to indicate that the forecast performance of MIDAS is superior than that of ARIMAX. Our present work precisely aims at filling this need. In doing so, we also investigate a practical problem: Forecasting export growth in Thailand. Our empirical findings confirm that the MIDAS regression model significantly improves the forecast accuracy when compared with the ARIMAX model.

Keywords: Forecast performance · ARMAX · MIDAS · Export · Thailand

1 Introduction

The Autoregressive Integrated Moving Average (ARIMA) model is one of the well-known approaches to forecast time series since the model was first introduced by Box and Jenkins in 1976 [4]. However, Claveria et al. [5] pointed out that ARIMA model cannot capture certain structure breaks in macroeconomic series. They found that including additional variables into the model could provide better predictive accuracy of the macroeconomic variables for European union member countries in comparison with the ARIMA model without the leading indicator. Therefore, the Autoregressive Integrated Moving Average with explanatory variable (ARIMAX) model should be employed. The ARIMAX model has been extensively used to forecast various economic situations such as the number of tourists [1,15], traffic counts and flows [8,20], and Thailand's export [13]. Their findings suggest that the ARIMAX model could outperform ARIMA model. In addition, the ARIMAX model reduces to the ARMAX model, when variables of interest are stationary without being differenced.

Nevertheless, the ARMAX model requires variables of interest to be measured with the same frequency. This could be troublesome when one works on macroeconomic series because several variables are available at different frequencies. For instance, gross domestic product (GDP) is measured quarterly, while statistics on the unemployed is collected monthly. Although one may transform

© Springer International Publishing AG 2016
V.-N. Huynh et al. (Eds.): IUKM 2016, LNAI 9978, pp. 601–611, 2016.
DOI: 10.1007/978-3-319-49046-5_51

the high frequency series to match the lower one, there may be some information loss due to aggregation of the series to reduce frequency.

To overcome this problem, Ghysels et al. [9] proposed the new forecasting approach called the Mixed Data Sampling (MIDAS) regression model. The MIDAS model allows us to incorporate high frequency data to forecast the lower frequency series. In addition, the MIDAS approach is very useful for combining a number of indicators in a single model rather than considering them individually for forecasting. Furthermore, the predictive ability of the indicators in comparison with an autoregression is stronger, and the lags of explanatory variables are specified in a more parsimonious way [7]. Due to its attractive properties, the MIDAS model has been applied in financial economics [11] and macroeconomics [6,7,14]. Hence, it is of practical interests to investigate the predictive performance of ARIMAX and MIDAS, and in this paper, we do that by considering the forecasting of export growth based on some macroeconomic variables.

Export is an important factor that can drive a country's economy. There are several theories regarding the determinants of export. The important factor in the equation of export is real exchange rate. According to the international trade theory, currency depreciation may affect export through the devaluation of the real exchange rate, which improves competitiveness [3]. Gylfason and Risager [12] also found that depreciation of currency improves current accounts in developing countries. Baharumshah [3] found that the real exchange rate has an impact on exports. Inflation and Gross Domestic Product (GDP) are also important factors with regard to the growth of export. The impact of inflation is that if a country's inflation rate increases more relative to its trade-partner countries, the export value is expected to decrease. Prachowny [17] found that the relationship between aggregate export price and inflation is negative. As for the case of GDP, Srivastava and Green [18] found that the GDP of the exporting country is a powerful explanatory variable in bilateral trade in 45 exporting nations and 82 importing nations. Yu and Zietlow [21] examined the determinants of bilateral trade in Asia-Pacific countries. Their results show that market size, as approximated by GDP is a significant indicator of bilateral trade.

The objective of this paper is to compare the forecast performance between the ARMAX and the MIDAS regression model on the example of the growth rate of export in Thailand. The important factors of export growth such as the inflation rate, GDP, and exchange rate are employed to be leading indicators for the export growth. The main results of this study will be useful for government in managing the macroeconomic factors via controlling exchange rate, targeting inflation and GDP growth to be favourable for export growth. It is also important for supporting appropriate decisions of producers, investors, and traders.

The organization of this paper is as the follows. Section 2 describes the scope of the data used in this study. Section 3 provides the methodology of this study. Section 4 discusses the empirical results. Finally, the conclusion of this study is drawn in Sect. 5.

2   Data

The data of this study consist of Thailand's quarterly export and GDP, as well as monthly exchange rate and inflation rate, obtained from the Bank of Thailand. Three trade partner countries are considered in this study, China, Japan, and US. The data during the period of 2002Q1 to 2015Q1 are used for model selection and estimation, whereas the data from 2015Q2 to 2016Q1 are used for forecast evaluation. Here, we consider GDP and the inflation rate as the quarterly indicators and the exchange rate as the monthly indicator for exports. All variables, except for inflation rate, are transformed into growth rate using the following formula:

$$g_{i,t} = \frac{x_{i,t} - x_{i,t-1}}{x_{i,t-1}} \times 100 \tag{1}$$

where $g_{i,t}$ is the growth rate of variable i at time t; $x_{i,t}$ and $x_{i,t-1}$ denote the value of variable i in the current period and the previous period respectively.

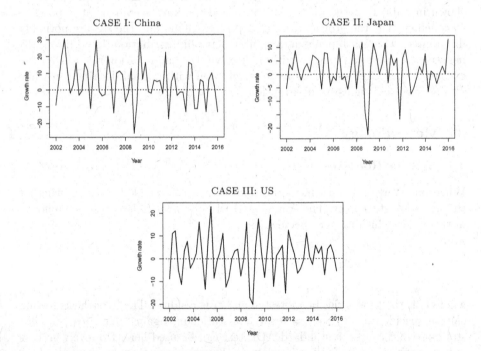

Fig. 1. Plots of Thailands growth in export to different countries

Figure 1 shows Thailand's growth in export with regard to each country over time. It can be seen that the values of export growth with respect to all the three countries fluctuate overtime with significant drop in 2009 and 2012 because of the financial crisis in the US.

Table 1. Descriptive statistics of series

| | Export | | | QGDP | Inflation rate | Exchange rate | | |
	China	Japan	US			CHY	JPY	USD
Mean	3.753	1.303	1.077	1.704	2.408	0.021	−0.031	−0.118
Median	4.821	1.743	1.997	1.890	2.474	−0.060	−0.186	−0.171
Maximum	30.767	12.821	22.864	10.637	7.549	4.446	7.384	3.517
Minimum	−25.907	−22.505	−20.065	−8.329	−2.824	−3.041	−7.564	−3.549
Std. dev	12.150	6.905	9.301	2.643	2.121	1.348	2.299	1.377
Skewness	0.113	−0.967	−0.010	−0.745	−0.024	0.341	0.227	0.155
Kurtosis	2.846	4.644	2.757	8.006	3.372	3.241	4.163	3.091
Period	2002Q1-2016Q1					2002M01-2016M03		

We also provide the descriptive statistics for each variable in this study, as shown in Table 1. Based on the skewness and the kurtosis values, it is possible to conclude that the growth rates of export to China and the US are normally distributed, whereas Japan's data exhibits left tail distribution. In addition, the negative skewness can be seen with quarterly GDP and inflation rate. As for the exchange rate, CHY and USD fit relatively well with normal distribution while JPY experiences right-tail distribution.

3   Methodology

3.1   A Unit Root Test

When one is working on time series data, it is recommended to check whether the series are stationary. The Augmented Dickey-Fuller (ADF) test is employed in this study which can be specified as

$$\Delta y_t = \alpha_0 + \alpha_1 y_{t-1} + \sum_{i=1}^{p} \alpha_{2i} \Delta y_{t-i} + \varepsilon_t \tag{2}$$

where y_t is the time series being tested and ε_t is residual. The hypothesis testing can be specified as $H_0 : \alpha_1 = 0$ for non-stationary against $H_1 : |\alpha_1| < 1$ for stationary. Of course, a standard ADF test can be used here. However, in view of recent criticisms of p-values involving in hypothesis testing, we employed here an alternative testing method based on the use of Akaike information criterion (AIC) that was introduced by Anderson et al. [2]. The idea is to convert the null hypothesis testing problem of the ADF test into model selection problem. Then, one obtains the Akaike weight based on AIC and calculates the probability for each hypothesis as

$$Prob(H_i|data) = \frac{\exp\left(-\frac{1}{2}\Delta_i\right)}{\exp\left(-\frac{1}{2}\Delta_0\right) + \exp\left(-\frac{1}{2}\Delta_a\right)} \tag{3}$$

where $i = 0$ for H_0 and $i = a$ for H_a; $\Delta_i = AIC_i - \min AIC$. The one with higher probability will be chosen.

3.2 Autoregressive Moving Average with Explanatory Variables (ARMAX)

As pointed out by Nguyen et al. [16], the linear model can handle the forecasting series of interest surprisingly well. If we are interested in forecasting a certain variable that is affected by other variables so-called leading indicators. Then, the ARMAX model could provide reasonable prediction. The ARMAX model allows us to forecast the series of interest using information from its lags and leading indicators. The ARMAX model is generally written as

$$y_t = \alpha_0 + \sum_{i=1}^{p} \alpha_i y_{t-i} + \sum_{j=1}^{q} \lambda_j \varepsilon_{t-j} + \gamma z_t + \varepsilon_t \tag{4}$$

where y_t is a dependent variable; α_0 is a constant term; α_i is ith autoregressive parameter; λ_j is jth moving average parameter; ε_t is the error term; and z_t is a leading indicator with a parameter γ.

3.3 MIDAS Model

Ghysels et al. [9] proposed the Mixed Data Sampling (MIDAS) approach to deal with the various frequencies in multivariate models. Particularly, a MIDAS regression tries to deal with a low-frequency variable by using higher frequency explanatory variables as a parsimonious distributed lag. Moreover, it also does not use any aggregation procedure and allow long lags in distributed lag function with only small number of parameters that have to be estimated [7]. The basic MIDAS model is given by

$$y_t = \alpha + \beta B(\theta) x_t^{(m)} + \varepsilon_t \tag{5}$$

where $x_t^{(m)}$ is an exogenous variable measured at higher frequency than y_t. $B(\theta)$ controls the polynomial weight and this term allows us to smooth past values of $x_t^{(m)}$ by using the polynomial $B(\theta)$ that is

$$B(\theta) = \sum_{k=1}^{K} b_k(\theta) L^{(k-1)/m} \tag{6}$$

where K is the number of data points on which the regression is based. L is the lag operator such that $L^{(k-1)/m} x_t^{(m)} = x_{t-(k-1)/m}^{(m)}$, and $b_k(\theta)$ is the weight function that can be in various forms. In this study, we choose the normalized exponential Almon lag polynomial weight proposed by Ghysels et al. [10] which can be specified as

$$b_k(\theta) \equiv b_k(\theta_1, \theta_2) = \frac{\exp(\theta_1 k + \theta_2 k^2)}{\sum\limits_{k=1}^{K} \exp(\theta_1 k + \theta_2 k^2)} \tag{7}$$

where the parameter θ is also part of the estimation problem and can be influenced by the last K values of $x_t^{(m)}$. This weight specification is attractive due to its weight restriction that weights are non-negative and it has been widely used in MIDAS literature.

In addition, the autoregressive models are often applied in macroeconomic, such as forecasting in export growth. Stock and Watson [19] suggested that combination of autoregressive dynamic in models and incorporate leading indicators as explanatory variables is better than using individual indicator model for forecasting. Hence, the MIDAS model for forecasting series of interest (y_t), augmented with pth-order autoregressive component and exogenous variables (z_t) sampled at same frequency as y_t, can be specified as

$$y_t = \alpha_0 + \sum_{i=1}^{P} \alpha_p y_{t-p} + \sum_{r=1}^{R} \gamma_r z_{r,t} + \beta B(\theta) x_t^{(m)} + \varepsilon_t \tag{8}$$

3.4 Model Specifications for Estimation and Forecast

In order to forecast the export growth rate with respect to each trade-partner country using the ARMAX(p, q) model, the equation for the task is

$$EXP_{f,t} = c + \sum_{i=1}^{p} \alpha_i EXP_{f,t-i} + \sum_{j=1}^{q} \beta_j \varepsilon_{t-j} + \gamma_1 GDP_t + \gamma_2 INF_t + \gamma_3 EXC_{f,t} + \varepsilon_t \tag{9}$$

where f is trade-partner country, c denotes the constant term, $EXP_{f,t}$ is export growth to trade-partner country f, GDP_t is Thailand's GDP growth rate, INF_t denotes Thailand's inflation rate, and $EXC_{f,t}$ is the growth of currency exchange with respective trade partner country f. The optimal amount of AR term (p) and MA term (q) are selected based on Akaike Information Criteria (AIC).

In case of MIDAS model, the export growth, GDP growth, and inflation rate are collected quarterly, whereas the growth of exchange rate is measured monthly. This means the ratio between high and low frequency variable is $m = 3$. Therefore, the MIDAS model for this study can be specified as

$$EXP_{f,t} = c + \sum_{i=1}^{p} \alpha_i EXP_{f,t-i} + \gamma_1 GDP_t + \gamma_2 INF_t + \beta B(\theta) EXC_{f,t}^{(3)} + \varepsilon_t \tag{10}$$

where $B(\theta) = \sum_{k=1}^{K} b_k(\theta) L^{(k-1)/3}$ and $b_k(\theta)$ is the normalized exponential Almon weight function. We allow the maximum possible data point of monthly indicators up to $K = 24$. This means the model can incorporate up to the last two years of monthly data for predicting the export growth. We employ the Akaike information criterion to choose the appropriate data points. In addition, we will denote the MIDAS model with the selected p lags of dependent variable and K data points of a high frequency indicator as the MIDAS(p, K) model.

4   Empirical Results

In this section, we show the results of ADF unit root tests, both original and alternative approaches, followed by the discussion on the forecasting performance comparison between the ARMAX model and the MIDAS regression model.

Table 2. ADF unit root tests

Series	ADF test (original)				ADF test (AIC approach)	
	Lag	ADF statistic	Prob.	Inference	$Prob(H_1)$	Inference
QGDP	0	−8.405	0	stationary	0.599	stationary
Inflation	1	−4.658	0	stationary	0.539	stationary
Export						
China	1	−7.826	0	stationary	0.592	stationary
Japan	1	−6.997	0	stationary	0.577	stationary
US	1	−10.611	0	stationary	0.636	stationary
Exchange rate						
CHY	0	−8.529	0	stationary	0.543	stationary
JPY	0	−9.654	0	stationary	0.553	stationary
USD	0	−8.652	0	stationary	0.544	stationary

Note: If $Prob(H_1)$ is greater than 0.5, the stationary model is preferred.

The results of ADF unit root tests from both conventional approach and alternative one using AIC are reported in Table 2. It can be seen that they reach the same conclusion that series are stationary. Therefore, we conclude that all series are stationary and appropriate for further analysis.

Since we are interested in comparing forecasting performance between ARMAX model and MIDAS regression model, the first step is to see how they perform graphically. 4 periods from 2015Q2 to 2016Q1 are forecasted using the data during the period of 2002Q1 to 2015Q1 for model specification and estimation. The model specifications for the ARMAX and the MIDAS models are based on the Akaike Information Criteria (AIC). The maximum lag of export growth is 4, and the maximum data point for high frequency series is 24. Figure 2 provides

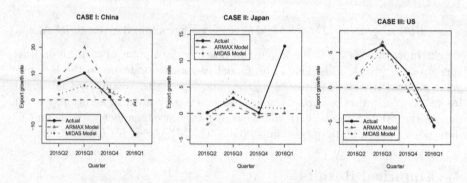

Fig. 2. Forecast and actual export growth 2015Q2 to 2016Q1

Table 3. Out-of-sample forecast of selected ARMAX models

Period	China		Japan		US	
	Actual	Forecast	Actual	Forecast	Actual	Forecast
2015Q2	6.326	8.426	0.151	−2.105	4.176	1.435
2015Q3	10.181	19.994	2.878	1.597	6.009	6.494
2015Q4	1.214	3.161	0.134	−0.667	1.997	−0.564
2016Q1	−13.074	−1.978	12.821	0.115	−5.427	−4.593
Selected model	ARMAX (4,3)		ARMAX(2,0)		ARMAX(2,2)	

Note: 1. ARMAX (p, q) indicate the p AR terms and q MA terms.
2. The model selection is based on AIC.

a forecasting comparison plot between actual values, the ARMAX model, and the MIDAS model.

Tables 3 and 4 show results in more detail for ARMAX and MIDAS models respectively. According to these results, we can roughly see that in case of China and Japan the plot of ARMAX and MIDAS models do not seem to follow the actual export growth, whereas both models could predict the export growth to US relatively well.

Next, we evaluate the forecasting performance of each model using root mean square error (RMSE). The results are shown in Table 5. Starting with the ARMAX model, we can see that the performance of the ARMAX model become worse as the forecast horizon increases for most cases. China and Japan exhibit wider forecasting error in 2016Q1 than in the rest of periods, while it is opposite for the US. It seems that the ARMAX model could explain the trend of Thailand's export to US relatively well. The sudden changes in export growth to China and Japan in 2016Q1 may be influenced by other factors which the ARMAX model in this study could not capture. Hence, this cause the RMSE in both cases to become largest when considering 4 forecasting periods.

In the case of MIDAS models, it can be seen that the forecasting error is in the same manner as in the ARMAX model. The RMSE may become higher as

Table 4. Out-of-sample forecast of selected MIDAS models

Period	China		Japan		US	
	Actual	Forecast	Actual	Forecast	Actual	Forecast
2015Q2	6.326	2.128	0.151	−1.032	4.176	1.264
2015Q3	10.181	5.456	2.878	4.068	6.009	5.321
2015Q4	1.214	3.747	0.134	1.149	1.997	1.089
2016Q1	−13.074	−0.975	12.821	1.011	−5.427	−5.646
MIDAS specification	MIDAS (2,22)		MIDAS (1,24)		MIDAS (4,24)	

Note: 1. MIDAS (p, K) indicate the p lags of dependent variable and K data points of high frequency variable.

2. The model selection is based on AIC.

Table 5. Out-of-sample forecast evaluation

Forecast horizon	RMSE		RMSE ratio (r)
	ARMAX	MIDAS	
Panel I: China			
1	**2.100**	4.198	1.999
2	7.096	**4.469**	0.630
3	5.902	**3.931**	0.666
4	7.543	**6.942**	0.920
Panel II: Japan			
1	2.257	**1.184**	0.525
2	1.835	**1.187**	0.647
3	1.568	**1.132**	0.722
4	6.497	**5.986**	0.921
Panel III: US			
1	**2.741**	2.912	1.062
2	**1.968**	2.116	1.075
3	2.184	**1.805**	0.827
4	1.937	**1.567**	0.809

Note:

1. Forecast horizon indicates the number of out-of-sample forecasting periods.

2. $r < 1$ means the MIDAS model performing better than the ARMAX model and vice versa.

3. The lower RMSE values are in **bold**.

the forecasting period increases for China and Japan, while it is lower for the case of US. This may imply that the selected leading indicators in this study may work quite well for predicting the export growth to the US.

Lastly, we calculate the ratios of RMSEs of export growth rate between the ARMAX model and the MIDAS model for each country to see how much they differ in forecasting performance. The RMSE ratio (r) can be specified as

$$r = \frac{RMSE_{MIDAS}}{RMSE_{ARMAX}}$$

When the ratio r is lower than one, it indicates that the selected MIDAS model could outperformed the selected ARMAX model. The results presented in Table 5 show that the MIDAS regression model provides better forecast accuracy as most of the RMSE ratios are less than 1. The ARMAX model was able to do better for short period forecasting in the case of China and the US, but the MIDAS model was better when forecasting further away. In the case of Japan, the MIDAS model improves the forecasting performance in every period. Although the improvement seems to be moderate, the MIDAS regression model can still improve the overall forecasting accuracy. Therefore, it is possible to conclude that the MIDAS model can provide us with greater forecasting precision than the ARMAX model.

5   Conclusion

In this paper, we carried out a comparison of the forecasting performance between the ARMAX model and the MIDAS model for quarterly growth rate of exports from Thailand to three countries, namely China, Japan, and the US. Exchange rate, inflation rate and GDP were used as the leading indicators for the prediction. We employed MIDAS model which allow us to exploit the monthly information of exchange rate to be used directly to forecast without transforming it into quarterly frequency. This could not be done in the ARMAX model. Our results showed that the MIDAS model could outperform the ARMAX model for most cases based on the RMSE and the RMSE ratios. However, due to limitation of data this study focused on four-period forecasting and only a single type of the MIDAS model. The ability of other types of the MIDAS model to forecast longer horizons for the case of Thailand remains to be seen.

Acknowledgement. We are grateful for financial support from "Puay Ungpakoyn Centre of Excellence in Econometrics", Faculty of Economics, Chiangmai University.

References

1. Akal, M.: Forecasting Turkey's tourism revenues by ARMAX model. Tour. Manag. **25**(5), 565–580 (2004)
2. Anderson, D.R., Burnham, K.P., Thompson, W.L.: Null hypothesis testing: problems, prevalence, and an alternative. J. Wildl. Manag. **64**(4), 912–923 (2000)
3. Baharumshah, A.Z.: The effect of exchange rate on bilateral trade balance: new evidence from Malaysia and Thailand. Asian Econ. J. **15**(3), 291–312 (2001)
4. Box, G.E.P., Jenkins, G.M., Reinsel, G.C., Ljung, G.M.: Time Series Analysis: Forecasting and Control. Wiley, New York (2015)
5. Claveria, O., Pons, E., Ramos, R.: Business and consumer expectations and macroeconomic forecasts. Int. J. Forecast. **23**(1), 47–69 (2007)

6. Clements, M.P., Galvao, A.B.: Macroeconomic forecasting with mixed-frequency data. J. Bus. Econ. Stat. **26**(4), 546–554 (2008)
7. Clements, M.P., Galvao, A.B.: Forecasting US output growth using leading indicators: an appraisal using MIDAS models. J. Appl. Econom. **24**(7), 1187–1206 (2009)
8. Cools, M., Moons, E., Wets, G.: Investigating the variability in daily traffic counts through use of ARIMAX and SARIMAX models: assessing the effect of holidays on two site locations. Transportation Research Record: J. Transp. Res. Board (2136), 57–66 (2009)
9. Ghysels, E., Santa-Clara, P., Valkanov, R.: The MIDAS touch: mixed data sampling regression models. In: CIRANO (2004)
10. Ghysels, E., Sinko, A., Valkanov, R.: MIDAS regressions: further results and new directions. Econom. Rev. **26**(1), 53–90 (2007)
11. Ghysels, E., Valkanov, R.I., Serrano, A.R.: Multi-period forecasts of volatility: direct, iterated, and mixed-data approaches. In: EFA 2009 Bergen Meetings Paper (2009)
12. Gylfason, T., Risager, O.: Does devaluation improve the current account? Eur. Econ. Rev. **25**(1), 37–64 (1984)
13. Kongcharoen, C., Kruangpradit, T.: Autoregressive integrated moving average with explanatory variable (ARIMAX) model for Thailand export. In: 33rd International Symposium on Forecasting, South Korea, pp. 1–8 (2013)
14. Kuzin, V., Marcellino, M., Schumacher, C.: MIDAS vs. mixed-frequency VAR: nowcasting GDP in the euro area. Int. J. Forecast. **27**(2), 529–542 (2011)
15. Lim, C., Min, J.C.H., McAleer, M.: Modelling income effects on long and short haul international travel from Japan. Tour. Manag. **29**(6), 1099–1109 (2008)
16. Nguyen, H.T., Kreinovich, V., Kosheleva, O., Sriboonchitta, S.: Why ARMAX-GARCH linear models successfully describe complex nonlinear phenomena: a possible explanation. In: Huynh, V.-N., Inuiguchi, M., Demoeux, T. (eds.) IUKM 2015. LNCS, vol. 9376, pp. 138–150. Springer, Heidelberg (2015). doi:10.1007/978-3-319-25135-6_14
17. Prachowny, M.F.J.: The relation between inflation and export prices: an aggregate study. A Working Paper 27. Queen's University (1970)
18. Srivastava, R.K., Green, R.T.: Determinants of bilateral trade flows. J. Bus. **59**(4), 623–640 (1986)
19. Stock, J.H., Watson, M.W.: Has the business cycle changed and why? In: NBER Macroeconomics Annual 2002, vol. 17, pp. 159–230. MIT press, Cambridge (2003)
20. Williams, B.: Multivariate vehicular traffic flow prediction: evaluation of ARIMAX modeling. Transportation Research Record: J. Transp. Res. Board (1776), 194–200 (2001)
21. Yu, C.-M.J., Zietlow, D.S.: The determinants of bilateral trade among Asia-Pacific countries. ASEAN Econ. Bull. **11**(3), 298–305 (1995)

Modelling Co-movement and Portfolio Optimization of Gold and Global Major Currencies

Methas Rattanasorn, Jianxu Liu[✉], Jirakom Sirisrisakulchai,
and Songsak Sriboonchitta

Faculty of Economics, Chiang Mai University, Chiang Mai, Thailand
liujianxu1984@163.com

Abstract. This study aims to investigate the co-movement of gold and global major currencies. We propose several time-varying copula-based GARCH models to measure the co-movement of gold and exchange rate returns and construct the optimized portfolio under copula-based models with modern portfolio theory. The empirical results show that all of the exchange rates are positively correlated with gold, except for USD. With the exception of AUD and GBP, all other exchange rates exhibit a significantly time-varying dependence which is beneficial for portfolio diversification opportunities. We construct optimization problems based on modern portfolio theory and mean-variance portfolio. Our results suggest that the maximum Sharpe ratio portfolio outperformed both an equally weighted portfolio and a conventional Markowitz portfolio model. Finally, USD and gold are the best portfolio and their cumulative return of investment is about 20 % over five years.

Keywords: Co-movement · GARCH · Gold · Exchange rates · Portfolio optimization · Time-varying copulas

1 Introduction

Gold is commonly seen as a safe asset because its value is not easily wielded. It also has common characteristics with money which it has been used as a medium of exchange in the past. Gold has its own value and may not be directly affected by interest rate decisions, monetary policy implementation or other actions of central banks. Unlike gold, the value of currencies are actually determined by interest rates. On one hand, a currency is more appreciated when an interest rate is increased. On the other hand, a reduction of interest rate would normally depreciate a currency. Changes in currency value certainly affect inter-market investor's portfolios both directly and indirectly. Santis and Gerard [5] have found an evidence and suggested that currency risk is significant and representing a large portion of the total risk faced by inter-market investors. Hence, the dependence structures of gold and currency value needs to be concerned. Sjaastad and Scacciavillani [17] have provided theoretical and empirical evidences

© Springer International Publishing AG 2016
V.-N. Huynh et al. (Eds.): IUKM 2016, LNAI 9978, pp. 612–623, 2016.
DOI: 10.1007/978-3-319-49046-5_52

on the relationship between gold and exchange rates. This is considered as the foundation of studies on co-movement of gold and exchange rates.

Gold is not only regarded as a standard medium of exchange but also has been used as a hedge against inflation, as well as preserved of value during high uncertainty situations such as war and crisis. Considering the effectiveness of gold as a hedge and haven against uncertainties, it has been applied in the financial risk management fields in terms of diversifying asset.

Majorities of recent studies considered using the GARCH-type model originated by Bollerslev [3] to investigate co-movements of financial data. The GARCH model normally assumed the conditional distribution to follow a normal or student-t distribution. But, as we know that in financial markets, returns of financial assets are found to be leptokurtic, asymmetric, volatility clustering and non-linear dependence (Ang and Chen [1]; Erb et al. [6]; Login and Solnik [11]). Therefore, the assumption of normally distributed and linearly correlated returns would mislead us in co-movement modelling between gold and exchange rates returns. In order to overcome these deficiencies, we employ the copula functions into our model.

Markowitz [12] had introduced the modern portfolio theory called "mean-variance portfolio" and is now accepted as the standard portfolio selection model. However, under the assumption that returns undergoing the normal distribution. It brings us to the conclusion that normal mean-variance portfolio objectives may not be efficient enough. Modern researchers came up with a more developed portfolio constraint such as Khrokhmal et al. [9], proposed a model optimizing the portfolio returns with Conditional Value-at-Risk constraint.

Therefore, combining these shortcomings and imperfections, we came up with the intention to model co-movement of global major currencies with gold based on the copula-GARCH model under various non-normal conditional distributions namely the skewed student-t and the skewed generalized error distribution. Moreover, we apply a static and time-varying copula functions in co-movement modelling. Furthermore, we also optimize the portfolio problems based on modern portfolio theory with different risk measurement aspects.

The remainder of this research paper is organized as follows: in the next section we outline our methodology. In Sects. 3 and 4 we present data and empirical estimation results. Finally, we conclude the study in Sect. 5.

2   Methodology

2.1   GARCH Model Specification

We adopt the GARCH model proposed by Bollerslev [3] to measure volatilities of gold and exchange rate returns as the GARCH model allows for a longer memory and more flexible lag structures. Basically, the GARCH (1,1) model can be expressed as

$$r_t = \mu + \sigma_t z_t, \tag{1}$$

$$\varepsilon_t = \sigma_t z_t, \tag{2}$$

$$\sigma_t^2 = \omega + \alpha \varepsilon_{t-1}^2 + \beta \sigma_{t-1}^2, \tag{3}$$

where r_t is the return r at time t, σ_t^2 denotes the conditional variance of r_t. z_t is the standardized residuals and can be assume as any distributions.

Bollerslev [3] suggests the use of conditional student-t distribution in order to appropriately characterize the non-normality characteristics of the financial data. In this study, we consider using Skewed student-t distribution (SSTD) proposed by Fernandez and Steel [7] and Skewed generalized error distribution (SGED) proposed by Theodossiou [18] as recent researches such as Jondeau and Rockinger [8] pointed out that these skewed distribution better described the financial data than the normal and student-t distribution (see Liu et al. [10]).

2.2 Copula Function

A copula function is a popular statistical tool used for asymmetric dependences modelling in finance. It is used to connect margins into a multivariate distribution function. Since Pearson's linear correlation coefficient is inappropriate to model financial data, copula function flexibly allows the dependence structures and marginal behaviors of random variables to be captured independently. Thus, We employ a copula-based GARCH model to investigate the co-movement of global major currencies with gold. Various methods of constructing copula functions are referred to the work of Nelsen [13]. A copula function is based on Sklar's theorem, which states that a copula function C is existed if

$$F_{XY}(x, y) = C(F_X(x), F_Y(y)), \tag{4}$$

where $F_{XY}(x, y)$ denotes the joint distribution of X and Y. Copula function is not affected by the monotonic transformation as it is related to the quantile function of marginal distributions rather than the primitive variables. Contrastingly to copula function, using linear correlation coefficient to model dependence structures would not be sufficient when the joint distribution of two variables are not close to the elliptical distributions. Following Patton [14], the conditional copula function can be expressed as

$$F_{XY|W}(x, y \,|\, w) = C(F_{X|W}(x| w), C(F_{Y|W}(y| w) |W), \tag{5}$$

where W denotes the conditioning variables. $F_{X|W}(x| w)$ and $F_{Y|W}(y| w)$ is the conditional distribution of $X|W = w$ and $Y|W = w$ respectively.

Another essential properties of copula functions are the tail dependences. It measures the asymptotic likelihood of two variables whether they are likely to move up or down at a particular time. In order to be more efficient to capture all of the dependence structures of gold and exchange rate returns, we employ various families of copula function namely: Gaussian, Student-t, Clayton, Gumbel, Frank, Joe, BB1, BB6, BB7 and BB8 copula function. Each one of them has its unique characteristic of dependence structure. For instance, the Gaussian copula is symmetric, and it cannot capture tail behavior of the data; student-t copula surprisingly can capture more of the case of heavy tail dependence. Moreover,

Clayton copula accounts for a stronger left tail than the right one while Gumbel copula is said to be the opposite of Clayton because it accounts for stronger right tail but weak left tail dependence. The rotation scenarios of each copula family are also account in this study, where the details of copula rotation we recommend readers to see Cech [4]. According to his work, the "survival" copula is simply 180° rotation of a corresponding copula function.

It is complicated to identify causes of a change in copula function parameters (Patton [14]). As mentioned earlier that the financial data are found to be non-linear, asymmetric and time-varying. We employ a time-varying copula function to our study as it will be more reasonable to capture dynamic dependences of gold and exchange rate returns. The time-varying copula function flexibly allows the dependence parameter(s) to change over time. This allow us to investigate the dynamic dependence structures. In this study we follow the concept of Patton [14], Bartram et al. [2], Wu et al. [19] and Liu et al. [10] in time-varying copula function and consider the use of former proposed types of time-varying copula functions. The example of corresponding time-varying copula are as follows: The time-varying Gaussian copula function are expressed as

$$\rho_t = \tilde{\Lambda}\{\omega + \beta_1\rho_{t-1} + ... + \beta_p\rho_{t-p} + \alpha(u_{t-1} - 0.5)(v_{t-1} - 0.5)\}, \qquad (6)$$

where ϕ^{-1} is the inverse of the cumulative distribution function of the univariate standard normal distribution. u and v is the marginal distributions. ρ_t is a time-varying dependence parameter and $\tilde{\Lambda}$ is a logistic transformation defined as $\tilde{\Lambda} = \frac{(1-e^{-x})}{(1+e^{-x})}$.

The main purpose of using logistic transformation is to restrict the correlation coefficient ρ to lie between the range $(-1, 1)$.

The time-varying Gumbel can be expressed as

$$\tau_t = \Lambda\{\omega + \beta_1\tau_{t-1} + ... + \beta_p\tau_{t-p} + \alpha(u_{t-1} - 0.5)(v_{t-1} - 0.5)\}, \qquad (7)$$

where $\Lambda = (1 + e^{-x})^{-1}$. This is to restrict the dependence parameter τ to lie in between $(-1, 1)$.

As BBX copula consists of two dependence parameters, we assume both parameters in each copula function to be time variant. Thus, The time-varying form of BB1, BB7 and BB8 take the form of

$$\theta_t = H\{\omega + \beta_1\theta_{t-1} + ... + \beta_p\theta_{t-p} + \alpha(u_{t-1} - 0.5)(v_{t-1} - 0.5)\}, \qquad (8)$$

$$\delta_t = H\{\omega + \beta_1\delta_{t-1} + ... + \beta_p\delta_{t-p} + \alpha(u_{t-1} - 0.5)(v_{t-1} - 0.5)\}, \qquad (9)$$

where H is an appropriate logistic transformation function. The BBX copula takes the similar form with Gumbel copula. Therefore, H in time-varying BBX is the same with Λ in time-varying gumbel. When we focus on the formula of two parameters for time-varying BBX copula, the only difference is that the H for BB8 copula in the formula of parameter θ is equal to $\tilde{\Lambda}$. For the evaluation and the selection of a preferable copula family, we select the best one through the lowest value of Bayesian Information Criterion (BIC).

2.3   Portfolio Optimization Problems

Quantification of risk and dependence structures would be pointless for investment if it is unable to determine which asset and what portion of that asset should be included in a portfolio to maximize returns and minimize the risk of investment. Our portfolio optimization problems are based on the "Mean-Variance portfolio" proposed by Markowitz [12]. We forecast the assets weight for a portfolio consisting of gold and exchange rate at the one-period ahead $(t + 1)$ by applying a copula-based GARCH model under preferred conditional distribution where the formulation can be expressed as

$$r_{i,t+1} = \hat{\mu}_{i,t+1} + \varepsilon_{i,t+1}, \tag{10}$$

$$\varepsilon_{i,t+1} = \hat{\sigma}_{i,t+1} \times z_{i,t+1}, \tag{11}$$

$$\sigma^2_{i,t+1} = \omega_i + \hat{\alpha}_i \hat{\varepsilon}^2_{i,t} + \hat{\beta}_i \hat{\sigma}^2_{i,t}, \tag{12}$$

where $r_{i,t+1}$ is the log-return of asset i at time $t + 1$. $\sigma^2_{i,t+1}$ is the conditional variance. $\varepsilon_{i,t+1}$ is the standardized residual which assumed to be conditionally distributed as SGED or SSTD. Therefore, a return of a portfolio consisted of i assets is given by

$$R_{i,t+1} = w_1 r_{1,t+1} + w_2 r_{2,t+1} + ... + w_n r_{n,t+n}, \tag{13}$$

where w_1, w_2 denotes the weight of asset 1, asset 2 until assets n of a portfolio.

The conventional Markowitz maximize return mean-variance portfolio can be expressed as

$$\max R^T w, \tag{14}$$

subjected to $\sum_{i=1}^{n} w_i = 1$, and $w^T \Sigma w \leq \sigma^2$, where w_i denote the vector of portfolio weight of asset i and restricted to be positive (Long position only). $R^T w$ is the expected return of a portfolio. Σ is defined as the variance-covariance matrix. So, the term $w^T \Sigma w$ represents the variance of a portfolio return. Therefore, the variance of an efficient portfolio must not exceed the variance of an individual asset.

As the risk of investment is referred to the standard deviation and variance of a portfolio return, hence, the minimize risk mean-variance portfolio can be expressed as

$$\min w^T \Sigma w, \tag{15}$$

subjected to $\sum_{i=1}^{n} w_i = 1$, and $R^T w \geq \bar{R}$, where w_i denote the vector of portfolio weight of asset i and restricted to be positive (Long position only). $w^T \Sigma w$ represents the variance of a portfolio return. $R^T w$ is the expected optimized portfolio return which an expected return of an efficient portfolio must not be less than the mean return $\bar{R}$.

As mentioned in the Sect. 1, an ordinary risk may not be appropriate in the sense of non-normality characteristics of returns of financial data, therefore, the minimize expected shortfall portfolio can be expressed as

$$\min F_\alpha(w, \beta) = \alpha + \frac{1}{q(1 - \beta)} \sum_{k=1}^{q} [-w^T R - \alpha]^+, \tag{16}$$

subjected to $\sum_{i=1}^{n} w_i = 1$, where q is the number of samples generated by Monte Carlo simulation, α denote the Value-at-risk and β is the significance level (normally set at 0.95 or 0.99).

The Sharpe ratio introduced by Sharpe [16]. It measures a reward-to-variability ratio. The maximize Sharpe ratio portfolio can be expressed as

$$\max \frac{R^T w - r_f}{w^T \Sigma w}, \tag{17}$$

subjected to $\sum_{i=1}^{n} w_i = 1$, where the term $\frac{R^T w - r_f}{w^T \Sigma w}$ is defined as the Sharpe ratio of a portfolio and r_f is the return of risk-free asset.

The selection and evaluation of portfolio performances are according to their cumulative return generated to a portfolio. The portfolio with relatively highest cumulative return is considered as the best performing portfolio.

3   Data and Descriptive Statistics

We empirically examined the co-movement of gold prices and exchange rate returns using the weekly data obtained from Bloomberg. We also forecast and construct portfolio problems in order to specify the best optimization technique. The gold prices are measured in terms of USD per ounce and the exchange rates data used in this study are priced against the USD. The exchange rates data are as follows: the Euro (EUR), the Australian Dollar (AUD), the British pound (GBP), the Swiss franc (CHF), the Norwegian krone (NOK), the Japanese yen (JPY). Additionally, we consider the use of Broad Trade Weighted Exchange Index (TWEXB) following the work of Reboredo [15] to investigate the relationship between gold and the US dollar. The data was collected weekly from 31 December 1999 to 30 October 2015 and separated into two period namely in-the-sample period from 31 December 1999 to 31 December 2010 and the out-of-sample period from 7 January 2010 to 30 October 2015. According to Reboredo [15], the use of weekly data benefited our work as the daily or high frequency data are usually affected by white noises and drifts. This made the process of modelling dependence structures and marginal distributions to be more complicated. In order to ensure that the data is stationary, we covert the data into natural logarithmic returns.

Descriptive statistics for the logarithmic return data for gold prices and exchange rates are reported in Table 1. The excess kurtosis presented for all

Table 1. Descriptive statistics for gold and exchange rate returns

	EUR	AUD	GBP	CHF	NOK	JPY	USD	GOLD
Mean	0.0004	0.0007	−0.0001	0.0009	0.0005	0.0004	−0.0003	0.0027
Median	0.0011	0.0025	0.0005	0.0011	0.0018	0.0002	−0.0006	0.0048
Maximum	0.0499	0.0702	0.0519	0.0656	0.0673	0.0758	0.0412	0.1320
Minimum	−0.0605	−0.1852	−0.0835	−0.0387	−0.0642	−0.0459	−0.0293	−0.0868
Std. dev.	0.0147	0.0197	0.0140	0.0148	0.0169	0.0146	0.0070	0.0253
Skewness	−0.2995	−1.8527	−0.6282	0.1653	−0.3861	0.3286	0.6913	−0.1842
Kurtosis	4.0684	18.1443	6.5313	3.3771	3.7572	4.2224	7.2690	5.0390
Jarque-Bera	35.9454	5823.7850	336.5776	6.0257	28.0237	46.1492	482.4200	102.8567
Corr. gold	0.4474	0.3622	0.3293	0.4587	0.4777	0.2181	−0.4521	1.0000

data series ranging from 3.4 to 18.1. This is consistent with the previous empirical studies about financial returns characteristics. Moreover, the Jarque-Bera statistics are large and significant at 99% level for most of the data series confirming the non-normality characteristic of the unconditional distribution. The linear correlation coefficient of exchange rates with gold show a positive value except for the USD, Since gold is priced in USD per ounce.

4   Empirical Results

4.1   Results of Marginal Models

Table 2 shows the results of estimated BIC of two different conditional distributions of a GARCH (1,1) model, SGED and SSTD. The preferred conditional distributions are selected based on their BIC, where the lower BIC indicated the better performances. Most of the exchange rates data are fitted by the SGED except for JPY and USD. For gold, SSTD is found to be better fitted.

Table 2. Estimated BIC for gold and exchange rate return

	EUR	AUD	GBP	CAD	NOK	JPY	USD	GOLD
SGED	−5.6352	−5.3195	−5.8300	−5.5474	−5.3469	−5.6092	−7.2240	−4.6048
SSTD	−5.6307	−5.3095	−5.8267	−5.5378	−5.3452	−5.6125	−7.2292	−4.6113

As after we got the marginal distributions with the most preferable conditional distribution, various families of static copula functions are employed in order to investigate the dependence structures and tail behaviors of exchange rates and gold. We then use a preferable copula family and apply time-varying approaches to investigate further dynamic dependences that changes over time. There are 40 families of copula function involved in the study where Table 3 represents the best fitted copula family of each corresponding asset pairs and the estimated time-varying copula parameters. The result shows that all of the exchange rates data are best fitted by the survival BB8 copula except for JPY

Table 3. Estimated parameters of time-varying copula

Asset pairs	Family	α	β	γ	δ	LL	BIC
EUR, Gold	sBB8	0.1491	0.7364[a]	1.9677[a]	0.7421[a]	89.2732	−159.4833
		(0.1218)	(0.1237)	(0.7671)	(0.1167)		
AUD, Gold	sBB8	0.1183	0.4572	1.9614	0.9029[a]	90.0803	−161.0974
		(0.1784)	(0.4086)	(1.0181)	(0.0525)		
GBP, Gold	sBB8	0.2746	−0.1009	1.5132	0.8216[a]	52.4062	−85.7494
		(0.4509)	(0.4016)	(1.2270)	(0.1137)		
CHF, Gold	sBB8	0.1315	0.8132[a]	1.0765	0.7133[a]	86.6863	−154.3095
		(0.1105)	(0.1036)	(0.5709)	(0.1369)		
NOK, Gold	sBB8	0.0848	0.75425[a]	0.5537	0.8616[a]	80.9783	−142.8935
		(0.1487)	(0.3205)	(0.8356)	(0.0722)		
JPY, Gold	sGumbel	−0.2867	0.8487[a]	2.4677[a]		30.6713	−42.2796
		(0.1266)	(0.0703)	(1.0188)			
USD, Gold	rBB8	0.2304	0.6458[a]	0.9204	−0.8173[a]	101.7223	−184.3815
		(0.2328)	(0.2863)	(0.7850)	(0.1024)		

Note: [a]represents a rejection of null hypothesis at 5 % significant level, Standard errors are reported in brackets. The "s" stands for survival and "r" for rotated 270°.
Source: computation

which the survival Gumbel copula is found to be more preferable. It implies that only JPY returns exhibit a stronger right-tail dependence. Additionally, a stronger right-tail dependence indicates that during the market upturn, the JPY is more likely to give positive returns. Moreover, we can see that the autoregressive parameter β is significant for most of the data except for GBP and AUD indicating that the rest of the exchange rates has time-varying dependences except for AUD and GBP. Figure 1 shows estimated values of static and time-varying of Kendall's tau coefficient. We can see that the values of Kendall's tau are always positive, except for USD. In addition, the time-varying Kendall's tau of JPY is always fluctuates lower than a static Kendall's tau, it suggests that the actual dependence of JPY and gold may be lower when the time is in consideration.

4.2 Portfolio Optimization Results

This section we forecast assets weight for a portfolio consisting of gold and exchange rate at the one-period ahead (t + 1) by applying a copula-based GARCH model under preferred conditional distribution reported in Table 2.

The Optimization results of portfolios are presented in Table 4, where it is separated into two parts. Panel A of Table 4 represents cumulative returns of portfolios with different objective functions. We can see that under the maximization of Sharpe ratio objective, most of the portfolio yields the highest cumulative returns except for GBP and NOK where the conventional Markowitz portfolio gave the highest returns. There are Three profitable portfolios namely:

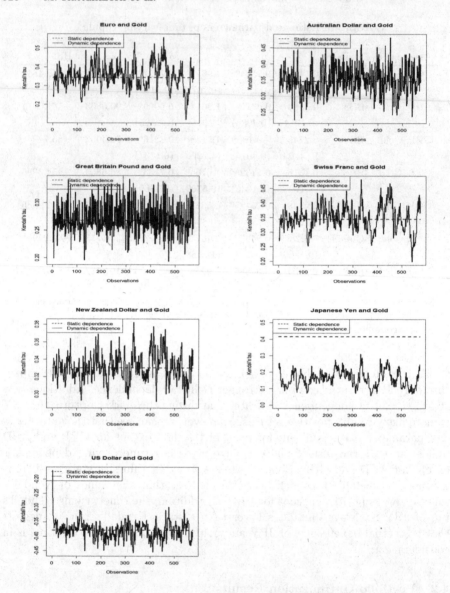

Fig. 1. Estimated values of Kendall's tau coefficient

CHF and USD of maximum Sharpe ratio objective and USD of a conventional Markowitz model with cumulative returns of 3.5, 20 and 19 % respectively. According to the reported cumulative returns, the maximum Sharpe ratio portfolio outperformed other types of optimization problems except for GBP and NOK which the maximize return portfolio is better. In addition, our proposed portfolios outperformed both equally weighted asset and conventional Markowitz portfolios. Figure 2 illustrates the cumulative returns of USD portfolio under

Table 4. Portfolio optimization results

Panel A: Cumulative returns	EUR	AUD	GBP	CHF	NOK	JPY	USD
Maximum return	0.7988	0.7391	0.9668	0.9777	0.7173	0.7091	1.1917
Minimum risk, ES	0.7890	0.7244	0.8683	0.9857	0.7170	0.7146	0.9697
Maximum Sharpe ratio	0.8927	0.7643	0.8760	1.0352	0.6694	0.7232	1.2020
Panel B: Statistical summary							
Minimum	0.0000	0.0000	0.7764	0.0000	0.4410	0.0000	0.0000
1st. Quantile	0.0000	0.0000	0.9364	0.0000	0.7596	0.0000	0.0000
Median	0.7142	0.5310	0.9958	0.4977	0.8592	0.6636	0.7028
Mean	0.5400	0.4953	0.9614	0.4951	0.8232	0.5214	0.5243
3rd. Quantile	1.0000	1.0000	1.0000	1.0000	0.9046	1.0000	1.0000
Maximum	1.0000	1.0000	1.0000	1.0000	1.0000	1.0000	1.0000

Note: This table reports cumulative returns of all portfolios (Panel A) and a statistical summary of exchange rate's weight of the best performing portfolio (Panel B). The estimated weights data are available upon request.
Source: computation

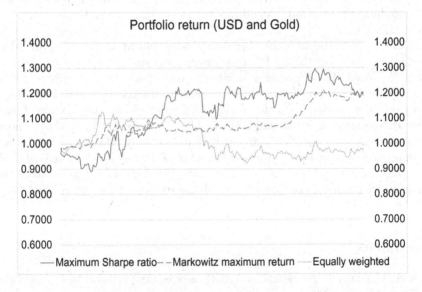

Fig. 2. USD Portfolio cumulative returns

three different objective functions. Even though we faced a financial crisis, we can see that the cumulative returns of maximum Sharpe ratio and a conventional Markowitz portfolio gradually increases over time.

Panel B of Table 4 represents a statistical summary of exchange rate's weight of the best performing portfolios. We can see that the overall mean of the weight data are close to 0.5 except for GBP and NOK implying that the maximum Sharpe ratio portfolio allocates assets more efficiently than equally weighted portfolio.

5   Conclusion

For decades, researchers have been trying to figure out the role of gold as a hedge instrument against uncertainties. However, majority of the studies have tested properties of gold as a hedge and haven against USD movements. This research, we came up with the intention to model co-movement of gold and global major currencies using a copula-based GARCH model. The model is known to be capable to capture non-normal characteristics of financial asset returns. Our results pointed out that most of the exchange rates are positively correlated with gold except for USD which is reasonably right and consistent with most of the past studies. On one hand, we statically investigate co-movements by copula functions, the results suggest that all of the exchange rates are best fitted by the survival BB8 copula except for USD and JPY which the 270° rotated BB8 and survival Gumbel are more preferable in terms of BIC respectively. On the other hand, by applying time-varying copula functions, our results found that most of the exchange rates exhibit significantly time-varying dependences, except for AUD and GBP.

More importantly, we construct portfolio optimization problems based on the modern portfolio theory by using several constraints and objective functions. We found that the maximize Sharpe ratio portfolio outperformed both an equally weighted portfolio and a conventional Markowitz model. In addition, There are three profitable portfolios namely: CHF, Gold and USD, Gold of maximum Sharpe ratio objective and USD, Gold of a conventional Markowitz model with cumulative returns of 3.5, 20 and 19 % respectively. The reason why others are unprofitable may due to the depreciation of currency values against USD and a crash of gold price in the past few years. Finally, this research restricts for a long position only. Therefore, we would like to suggest further researchers to relax the restriction.

References

1. Ang, A., Chen, J.: Asymmetric correlations of equity portfolios. J. Financ. Econ. **63**(3), 443–494 (2002)
2. Bartram, S.M., Taylor, S.J., Wang, Y.H.: The Euro and European financial market dependence. J. Bank. Financ. **31**(5), 1461–1481 (2007)
3. Bollerslev, T.: Generalized autoregressive conditional heteroskedasticity. J. Econom. **31**(3), 307–327 (1986)
4. Cech, C.: Copula-based top-down approaches in financial risk aggregation. Available at SSRN 953888 (2006)
5. De Santis, G., Gerard, B.: How big is the premium for currency risk? J. Financ. Econ. **49**(3), 375–412 (1998)

6. Erb, C.B., Harvey, C.R., Viskanta, T.E.: Forecasting international equity correlations. Financ. Anal. J. **50**(6), 32–45 (1994)
7. Fernández, C., Steel, M.F.: On Bayesian modeling of fat tails and skewness. J. Am. Stat. Assoc. **93**(441), 359–371 (1998)
8. Jondeau, E., Rockinger, M.: The Copula-GARCH model of conditional dependencies: an international stock market application. J. Int. Money Financ. **25**(5), 827–853 (2006)
9. Krokhmal, P., Palmquist, J., Uryasev, S.: Portfolio optimization with conditional value-at-risk objective and constraints. J. Risk **4**, 43–68 (2002)
10. Liu, J., Sriboonchitta, S., Nguyen, H.T., Kreinovich, V.: Studying volatility and dependency of Chinese outbound tourism demand in Singapore, Malaysia, and Thailand: a vine Copula approach. In: Huynh, V.-N., Kreinovich, V., Sriboonchitta, S. (eds.) Modeling Dependence in Econometrics. AISC, vol. 251, pp. 295–311. Springer, Heidelberg (2014). doi:10.1007/978-3-319-03395-2_17
11. Longin, F., Solnik, B.: Is the correlation in international equity returns constant: 1960–1990? J. Int. Money Financ. **14**(1), 3–26 (1995)
12. Markowitz, H.: Portfolio selection. J. Financ. **7**(1), 77–91 (1952)
13. Nelsen, R.B.: An Introduction to Copulas. Lecture Notes in Statistics, vol. 139. Springer, New York (1999)
14. Patton, A.J.: Modelling asymmetric exchange rate dependence. Int. Econ. Rev. **47**(2), 527–556 (2006)
15. Reboredo, J.C.: Is gold a safe haven or a hedge for the US dollar? Implications for risk management. J. Bank. Financ. **37**(8), 2665–2676 (2013)
16. Sharpe, W.F.: Mutual fund performance. J. Bus. **39**(1), 119–138 (1966)
17. Sjaastad, L.A., Scacciavillani, F.: The price of gold and the exchange rate. J. Int. Money Financ. **15**(6), 879–897 (1996)
18. Theodossiou, P.: Skewed generalized error distribution of financial assets and option pricing. Multinatl. Financ. J. **19**(4), 223–266 (2015)
19. Wu, C.C., Chung, H., Chang, Y.H.: The economic value of co-movement between oil price and exchange rate using Copula-based GARCH models. Energy Econ. **34**(1), 270–282 (2012)

Does Asian Credit Default Swap Index Improve Portfolio Performance?

Chatchai Khiewngamdee[(⊠)], Woraphon Yamaka, and Songsak Sriboonchitta

Faculty of Economics, Chiang Mai University, Chiang Mai, Thailand
getliecon@gmail.com, woraphon.econ@gmail.com

Abstract. This study aims to find whether adding Asian Credit Default Swap (CDS) index will improve the portfolio performance. We introduce a new approach namely Markov switching copula approach to estimate the dependence between asset returns and evaluated the risks of the portfolio by using Value at Risk and Expected Shortfall. The empirical results show that the risk level in high dependence regime is less than in low dependence regime. We also find that including Asian CDS index in portfolio clearly increases the portfolio return.

Keywords: Credit Default Swap · Markov switching · Copula · Value at Risk · Expected Shortfall

1 Introduction

Since the Global Financial Crisis in 2008, policymakers around the world have tried to stimulate their economy by introducing an ultralow interest rate policy, or even a negative interest rate, led by such major economies as US, Europe and Japan. Consequently, investors who invest in bonds earn very low returns and investing in stocks or commodities may lead to a higher risk because of stock market turmoil and tumbling commodity prices. Credit Default Swap (CDS) contract, surprisingly, yields a higher return and provides more efficient hedge than other financial instruments such as bonds and equities, particularly in Asian Credit Default Swap (CDS) market excluding Japan.

The outstanding performance in Asian CDS market has drawn attention of many investors and hedge funds around the globe. This causes the CDS market to be even more liquid and grow rapidly. Nowadays, there is no doubt that many investors including investment banks and hedge funds enjoy trading CDS and partly put their portfolio into CDS contracts. Hence, CDS has proved to be a useful tool for portfolio management and speculation. It is also likely to remain an important instrument in the financial market (Pinsent [1]). However, there are only few studies focusing on portfolio containing CDS in Asian market.

In portfolio optimization, many studies showed that applying copula and regime switching concept to econometric models can improve the accuracy of the model estimation. For instance, Bai and Sun [2], Halulu [3], Boubaker and

© Springer International Publishing AG 2016
V.-N. Huynh et al. (Eds.): IUKM 2016, LNAI 9978, pp. 624–636, 2016.
DOI: 10.1007/978-3-319-49046-5_53

Sghaier [4] and Kakouris and Rustem [5], to optimize portfolio allocations, investigated using the copula-based models which were found to outperform the traditional Markowitz approach. Patton [6] reasoned that copula-based models can increase the flexibility and reduce the computational complexity of the estimating models. Additionally, Tu [7] and Ma et al. [8] applied regime switching in asset pricing model and asset allocation. The results show that the regime switching model is more realistic than a single regime model as in reality market faces more than one economic scenario.

Furthermore, for portfolio measurement, Kaura [9] suggested that optimal portfolios can be computed by the portfolio risk and hence Value at Risk (VaR) and Expected Shortfall (ES) will be used. In order to compute the VaR and ES of a portfolio, it is important to specify the assumption on the distribution of asset prices change (Cotter and Longin [10]) and estimate the dependency or correlation between the assets. Thus, the copula function is used to estimate the dependence of our variables. However, the conventional copula model does not take into account the behavior change in the economic or financial market. Therefore, to overcome this problem, we extend the Markov Switching model of Hamilton [11] to the copula model in order to make it more flexible in measuring the dependency of the variables in different market conditions.

The objective of this paper is to find out the optimal weights of a diversified portfolio consisting of stocks, commodities, and bond. We also aim to examine whether adding Asian CDS index can improve our portfolio performance. In doing so, we first model the co-movement of and the dependence among the Asian CDS spreads, Hang Seng Futures index, Brent crude oil futures price, gold futures price and US bond return using the Markov Switching copula (MS-copula) approach which was introduced by Rodriguez [12] and Okimoto [13]. The model takes the advantage of the Markov Switching model and allows the dependence parameter to vary across the economic states or regimes, according to Markov process. Then, we estimate the volatility of the portfolios based on the concepts of VaR and ES. Finally, we extend the obtained expected return and ES to find the optimal weights of these portfolios.

The remainder of this study is constructed as follows. In Sect. 2, we describe the methodologies used in this study. Section 3 presents the data used and provides the empirical results and the final section gives conclusions.

2 Methodology

2.1 Basic Concepts of Copula

According to Sklars theorem [14], the joint distribution function of n-dimension variables can be decomposed into copula function C. Formally, for any joint distribution $H(x_1, ..., x_n)$ with marginal distribution function $F_1(x_1), ..., F_n(x_n)$, there exists a unique Copula C such that

$$H(x_1, ..., x_n) = C(F_1(x_1), ..., F_n(x_n)). \tag{1}$$

However, if $F_1(x_1), ..., F_n(x_n)$ are continuous marginal distribution function and C is a copula, H in Eq. (1) is a joint distribution function with marginal $F_1(x_1), ..., F_n(x_n)$. If H is known, we obtain

$$C(u_1, ..., u_n) = H(F_1^{-1}(u_1), ..., F_n^{-1}(u_n)), \tag{2}$$

where $C(u_1, ..., u_n)$ is a multivariate copula function whose marginal distribution $u_1, ..., u_n = F_1(x_1), ..., F_n(x_n)$ are uniform in the $[0.1]$ interval and $F_1^{-1}(u_1), ..., F_n^{-1}(u_n)$ are the quantile functions of the marginal distributions.

In this study, the maximum likelihood estimation is conducted to obtain the copula parameters thus, the likelihood function of the joint density of copula can be obtained by differentiating Eq. (1) with respect to $x_1, ... x_n$, such that

$$f(u_1, ..., u_n) = c(F_1(u_1), ..., F_n(u_n)) \cdot \prod_{i=1}^{n} f(u_i), \tag{3}$$

where $C(F_1(u_1), ..., F_1(u_n))$ is the multivariate copula density function and $f(u_i)$ is the density function of each marginal distribution.

The above approach allows us to decompose a multivariate distribution function into marginal distributions of each random variable and link each marginal distribution using the copula function. In this study, we use various copula families such as the Gaussian copula, Student-t copula, Frank copula, Joe copula, Gumbel copula and Clayton copula to link the proper marginal distributions which can be obtained by the ARMA-GARCH model. Then the proper copula function is chosen to describe the dependence structure between n-dimensional variables. There are two important classes of copula, namely Elliptical copulas and Archimedean copulas. The symmetric Gaussian and t-copulas are the families of the copula in Elliptical copulas which are simply the copulas of elliptical contoured distribution. The density functions of these copulas are provided in Patton [15], Hofert et al. [16] and Wang [17].

2.2 Markov Switching Model

Since the financial time series exhibit the different degree of dependence over time. The studies by Stöber and Czedo [18], Pastpipatkul et al. [19] and Chollete et al. [20] found that there is different degree of dependence between different market conditions namely, market upturn and market downturn. Thus, the Markov Switching copula is considered to deal with the stylized facts observed in the dependence structure of multivariate financial return series and to investigate whether there exists different dependence behavior along sample periods.

To measure the dependence structure in different market conditions, the Markov Switching model (MS) of Hamilton [11] is applied in the copula model. The MS model allows dependence parameters to switch between k regimes. In essence, the model assumes that the hidden underlying process or state of economy can influence the financial time series (Stöber and Czedo [18]). Let $\psi = \{\theta \text{ or } R\}$ be the matrix of dependence of the copula parameter, which is governed by an unobserved state variable (S_t). Thus, the different possible dependence can then be obtained in different regime. In this paper, lets consider $S_t = 1, 2$ where $S_t = 1$ is a high dependence regime and $S_t = 2$ is a the low dependence regime. Thus, the joint distribution of n-dimensional $(x_1, ..., x_n)$ conditional on S_t can be defined as

$$(x_1, ..., x_n | \psi_{S_t}; S_t = i) \sim C_t^{S_t}(u_1, ..., u_n | \psi_{S_t}), \qquad i = 1, 2, \qquad (4)$$

where $i = 1, 2$ denotes high dependence and low dependence regime, respectively. The dependence parameter ψ_{S_t} is a state dependence parameter. In this study, we assume that S_t is governed by first order Markov chain such that the probability in this state depends on the previous state's

$$P_{ij} = Pr(S_{t+1} = j | S_t = i) \quad and \quad \sum_{ij=1}^{k=2} P_{ij} = 1, \qquad i, j = 1, 2, \qquad (5)$$

where p_{ij} is the probability of switching from regime i to regime j. In general, the complete log likelihood function of the two-regime (k=2) MS multivariate copula can be written as

$$L(\psi_{S_t} | u_1, ..., u_n) = \sum_{k=1}^{2} \left[\sum_{i=1}^{T} \sum_{j=1}^{n} \ln (f_{ji}(u_i)) \right.$$
$$\left. + \sum_{i=1}^{T} \ln c(F_1, (x_{i,1}), ..., F_n, (x_{i,n})) Pr(S_t | \psi_{S_t}) \right], \qquad (6)$$

where $f_{ji}(u_i)$ is the density of each marginal distribution, $c(.)$ is a the multivariate copula density and $Pr(S_t = j | \psi_{S_t})$ is the filtered probabilities which is used in an important process of filtering the estimated coefficients into two different regimes. In computation of ML estimator, it is difficult to estimate the large unknown parameter thus the two-stage maximum likelihood (ML) approach called inference function margins (IFM) is employed to estimate our model (Patton [15]). Thus, in the first stage, we estimate and select the parameters of the marginal distributions that provide the best fit for the individual variables using ARMA-GARCH model and then plug the marginal distributions into MS copula model for estimated parameters. Therefore, the log likelihood function for our model can be rewritten from Eq. (6) as

$$L(\psi_{S_t} | u_1, ..., u_n) = \sum_{k=1}^{2} \left[\sum_{i=1}^{T} \ln c(F_1, (x_{i,1}), ..., F_n, (x_{i,n})) Pr(S_t | \psi_{S_t}) \right]. \qquad (7)$$

As we mentioned before that $Pr(S_t = j | \psi_{S_t})$ are filtered probabilities which are important in the process to separate the dependence parameters into two regimes, hence we employ the Hamiltons filter as proposed in Hamilton [11] and the algorithm can be explained as in the following.

1. Given an initial value of transition probabilities matrix Q which are the probabilities of switching between regimes.

$$Q = \begin{bmatrix} 0.8 & 0.2 \\ 0.2 & 0.8 \end{bmatrix}$$

2. Updating the probability of each state by the following formula, for $l = 1, 2$

$$Pr(S_t = l | \psi_{S_{t=l}}) = \frac{c_t(u_1, ..., u_n | S_t = l, \psi_{s_{t-1}=l}) Pr[S_t = l | \psi_{s_{t-1}=l}]}{\sum_{k=1}^{2} c_t(u_1, ..., u_n | S_t = k, \psi_{s_{t-1}=k}) Pr[S_t = k | \psi_{s_{t-1}=k}]} \quad (8)$$

3. Iterating step 1 and 2 for $t = 1, ..., T$. We can obtain the filter probabilities.

2.3 Value at Risk with Copula

Value at Risk (VaR) and conditional Value at Risk or Expected Shortfall (ES) have been widely used to measure risk since the 1990s. The VaR of portfolio can be written as

$$VaR_\alpha = inf\{l \in R : P(L > l) \leq 1 - \alpha\}, \quad (9)$$

where, α is a confidence level with a value $[0,1]$ which presents the probability of Loss L exceeding l but not larger than $1 - \alpha$. The alternative method ES, which is the extension of the VaR approach, can remedy two problems of VaR (Halulu [3]). Firstly VaR measures only percentiles of profit-loss distribution which is difficult to control for non-normal distribution. Secondly, VaR is not sub-additive. ES can be written as

$$ES_\alpha = E(L|L > VaR_\alpha). \quad (10)$$

To find the optimal portfolios, Rockafellar and Uryasev [21] introduced the portfolio optimization by calculating VaR and extending VaR to optimized ES. The approach focused on the minimizing of ES to obtain the optimal weights of a large number of instruments. In other words, we can write the problem as in the following.

The objective function is to

$$Minimize \quad ES_\alpha = E(L|L > VaR_\alpha) \quad (11)$$

Subject to $\sum_{i=1}^{n} w_i \cdot r_i = R_p, \sum_{i=1}^{n} w_i = 1, 0 \leq w_i \leq 1, i = 1, 2, .., n$

where R_p is an expected return of the portfolios, w_i is a vector of weight portfolio, and r_i is the return of each instrument.

3 Dataset and Empirical Results

As we mentioned that the IFM method is conducted here thus we need to estimate the marginal distribution of each variable, through univariate ARMA(p, q)-GARCH(m, n) which can be described as

$$y_t = a_0 + \sum_{i=1}^{p} b_i y_{t-i} + \sum_{j=1}^{q} \phi_j \varepsilon_{t-j} + \varepsilon_t, \quad (12)$$

$$\varepsilon_t = h\eta_t,$$

$$h_t^2 = c + \sum_{i=1}^{m} \alpha_i \alpha_{t-i}^2 + \sum_{j=1}^{n} \beta_j h_{t-j}^2, \tag{13}$$

where Eqs. 12 and 13 are the conditional mean and variance equation, respectively. ε_t is the residual term which consists of the standard variance, h_t, and the standardized residual, η_t. The best-fit ARMA(p, q)-GARCH(m, n) will give the standardized residuals and there are transformed into a uniform distribution in $(0, 1)$ using the cumulative distribution function before plugging in the MS copula model.

In this study, we use the daily log returns of iTraxx Asia ex-Japan Credit Default Swap (CDS), Hang Seng futures index (HS), Gold futures price (Gold), 10-year US government bond (Bond), and Brent crude oil futures price (Brent) from Thomson Reuter Database. We collect over the period from April 1st, 2011 to March 31st, 2016, totaling 1186 observations.

Table 1. Summary statistics

	CDS	**HS**	**Brent**	**Gold**	**Bond**
Mean	0.00025	−0.000117	−0.000934	−0.000137	4.58E-05
Median	−0.001899	0.000246	−0.000384	−0.000163	0.000418
Maximum	0.219331	0.056909	0.104162	0.07494	0.065757
Minimum	−0.123978	−0.076758	−0.124928	−0.098206	−0.10967
Std. Dev.	0.031172	0.012877	0.020481	0.011834	0.006618
Skewness	1.336147	−0.248941	−0.126257	−0.655119	−3.570068
Kurtosis	10.30411	6.019899	7.759199	10.41276	77.71005
Jarque-Bera	2994.316***	463.6999***	1124.33***	2804.956***	278812.4***
ADF-test	−34.5710***	−34.0610***	−38.4293***	−35.6057***	−36.1286***

*, **, and *** denote 10 %, 5 %, and 1 % significance levels, respectively.

Table 1 summarizes the statistics for our variable returns. We can see that all the variables have a kurtosis above 6, indicating that extreme futures returns are frequent. The negative values of the skewness suggest the probability of large decrease in futures returns except for CDS. Furthermore, the normality of these returns is strongly rejected by the JarqueBera test with 99 % significance. Therefore, all returns show clear traits of non-normality as confirmed by the BeraJarque test, skewness and kurtosis. Consequently, we assume that the skewed Student-t distribution, skewed-normal distribution, and the skewed generalized exponential distribution (GED) are chosen for our marginal model. Finally, according to the Augmented Dickey Fuller (ADF) test, all returns are stationary.

Table 2. Results of ARMA-GARCH(1,1)

	CDS	HS	Brent	Gold	Bond
Mean equation					
ϕ_0	0.0017	−0.0006***	−0.0003	−0.0003	0.00001***
AR(1)	−0.9558***	0.0135***	0.8612***	−0.7737***	0.8689***
AR(2)	0.0558***	−0.4898***	−0.4323**	−0.0625***	
AR(3)	−0.2429***	−0.0677***	−0.0684*		·
AR(4)	−0.5693***	−0.7554***	0.0739**		
AR(5)	0.0212*	0.4483***			
AR(6)		−0.0785***			
MA(1)	1.0000***	0.0238***	−0.9521***	0.6999***	−0.9885***
MA(2)	−0.0471**	0.4446***	0.5081**		
MA(3)	0.1980***	0.0456***			
MA(4)	0.5597***	0.7270***			
MA(5)		−0.5026***			
Variance equation					
α_0	0.00005***	0.00002	0.00001	0.00001**	0.00005
α_1	0.2005***	0.0509**	0.0623***	0.0301***	0.1800**
β_1	0.7565***	0.9373***	0.9379***	0.9533***	0.7329***
df	1.1170***	5.7100***	4.9100***	1.0510***	
γ	1.2450***	0.8983***	0.9166***	0.9817***	
KS test (prob.)	0.6119	0.511	0.8134	0.8547	0.9924
LM-test (prob.)	0.8544	0.7276	0.9823	0.6838	0.9986
AIC	−4.4297	−6.0726	−5.3738	−6.2421	−7.3493

*, **, and *** denote 10 %, 5 %, and 1 % significance levels, respectively.

Table 2 shows that the ARMA(5, 4)-GARCH(1, 1) and ARMA(2,1)-GARCH (1, 1) with skewed GED well capture CDS and Gold, while ARMA(6, 5)-GARCH(1, 1) and ARMA(4, 2)-GARCH(1, 1) with skewed Student-t provide the best marginal fit to the HS and Brent, respectively. For Bond, we found that it is well captured by an ARMA(1, 1)-GARCH(1, 1) with normal distribution. Prior to estimate our model, the goodness of fit is evaluated by two different tests. Firstly, the uniform test, KolmogorovSmirnov (KS) test, is employed for the transformed standardize residuals. And the result show that we can accept the null hypothesis of uniform distribution in all margins. Secondly, the ARCH-LM test is employed for the autocorrelation test on standardized residuals and the results confirm that there is no autocorrelation in the residuals of the marginal models.

In order to find whether CDS can improve portfolio performance, we conduct test on two portfolios namely, CDS portfolio and non-CDS portfolio and then compare their risks and returns. Firstly, we choose the best fit copula function

Table 3. Results of model comparison

	CDS portfolio			Non-CDS portfolio		
	LL	AIC	BIC	LL	AIC	BIC
Gaussian copula	313.243	−582.486	−470.725	104.556	−181.112	−109.992
Student-t copula	**328.046**	**−608.089**	**−486.168**	**112.453**	**−192.905**	**−111.625**
Clayton copula	−56.625	121.251	141.571	−1175.256	2358.512	2378.832
Joe copula	−5.447	18.882	39.202	−9.515	27.029	47.349
Frank copula	−1.751	358.127	378.448	−57.03	122.061	142.381
Gumbel copula	−87.75232	183.505	203.825	−21.488	50.976	71.296

Table 4. Empirical Markov switching Student-t copula parameters

	CDS portfolio		Non-CDS portfolio	
	High dependence regime	Low dependence regime	High dependence regime	Low dependence regime
$C_{HS,Brent}$	−0.35 (−0.05)	0.09 (−0.03)	0.18 (−0.03)	0.14 (−0.07)
$C_{HS,Gold}$	0.18 (−0.11)	0.18 (−0.03)	0.03 (−0.03)	−0.02 (−0.07)
$C_{HS,Bond}$	0.45 (−0.05)	0.06 (−0.03)	0.18 (−0.03)	0.15 (−0.07)
$C_{Brent,Gold}$	0.22 (−0.01)	−0.06 (−0.03)	−0.12 (−0.04)	−0.09 (−0.09)
$C_{Brent,Bond}$	−0.41 (−0.05)	−0.15 (0.00)	−0.32 (−0.03)	0.23 (−0.08)
$C_{Gold,Bond}$	−0.18 (−0.07)	0.18 (−0.04)	0.08 (−0.04)	0.35 (−0.07)
$C_{CDS,HS}$	−0.47 (−0.04)	−0.49 (−0.02)		
$C_{CDS,Brent}$	−0.38 (−0.04)	−0.14 (−0.03)		
$C_{CDS,Gold}$	0.57 (−0.08)	0.07 (−0.04)		
$C_{CDS,Bond}$	−0.23 (−0.06)	−0.04 (−0.04)		
Average	**-0.06**	**-0.04**	**0.01**	**0.12**
df	25.42 (−5.21)	15.24 (−4.32)	23.53 (−5.41)	11.73 (−4.09)
Probability	Regime1	Regime2	Regime1	Regime2
Regime1	0.9872	0.0041	0.9983	0.0048
Regime2	0.0128	0.9959	0.0017	0.9952
Duration	78.12	243.9	588.23	208.33

Note: the numbers in parentheses are standard deviation.

to estimate the dependence structure between variables returns in these two portfolios. The results, as shown in Table 3, show that Student-t copula is the best copula function for both CDS and non-CDS portfolios according to the lowest AIC and BIC values.

In portfolio diversification strategy, combining multiple assets with low positive dependence and/or high negative dependence would reduce the risk of the portfolio. In other words, the lower value of dependence parameters, the lower risk in portfolio. In this section, we estimate the matrix of dependence parameters of two regimes using Student-t copula for CDS portfolio and non-CDS portfolio. The results in Table 4 presents that, for CDS portfolio, the dependence parameters of $C_{CDS,HS}$, $C_{CDS,Brent}$, $C_{CDS,Bond}$, $C_{HS,Brent}$, $C_{Brent,Bond}$ and $C_{Gold,Bond}$ have a negative sign in high dependence regime while $C_{HS,Gold}$, $C_{HS,Bond}$, $C_{Brent,Gold}$ and $C_{CDS,Gold}$ show a positive sign. However, the size of dependence parameters of $C_{CDS,HS}$, $C_{HS,Gold}$ and $C_{Gold,Bond}$ in high dependence regime is smaller than that in low dependence regime. Moreover, the sign of dependence parameters of $C_{HS,Brent}$, $C_{Brent,Gold}$ and $C_{Gold,Bond}$ changes between regimes.

The dependence parameters for non-CDS portfolio show that, in high dependence regime, $C_{HS,Brent}$, $C_{HS,Gold}$, $C_{HS,Bond}$ and $C_{Gold,Bond}$ have a positive sign while $C_{Brent,Gold}$ and $C_{Brent,Bond}$ show negative sign. For $C_{Gold,Bond}$, however, the size of dependence parameter is larger in low dependence regime. In addition, $C_{HS,Gold}$ and $C_{Brent,Bond}$ dependence sign switches between regimes.

In summary, the average values of dependence parameters in high dependence regime are lower than in low dependence regime both for CDS and non-CDS portfolios. This suggests that the risk of the portfolios should be lower in high dependence regime. In addition, both regimes are persistent as the probabilities of shifting between regimes are very low (Fig. 1).

Table 5 reports the estimation of VaR and ES with 1%, 5% and 10% loss probability on equally weighted CDS and non-CDS portfolio in two regimes. The results show that both the average losses based on VaR and ES in low dependence regime are larger than those in high dependence regime as we expected in the previous section, since the average values of dependence parameters in low dependence regime are greater than those in high dependence regime. Apart from this, most of the average loss values in CDS portfolio are lower than its counterpart; only VaR at 5% and 10% and ES at 10% present higher average loss values.

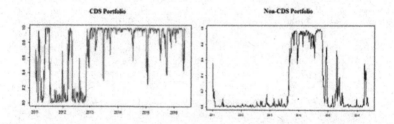

Fig. 1. Filtered probabilities of being low dependence regime for CDS and non-CDS portfolio. The figure shows that CDS portfolio return mostly took place in low dependence regime, while non-CDS portfolio return was in low dependence regime only during mid-2013 to 2015 period.

Table 5. Value at Risk and Expected Shortfall estimation

Portfolio		High dependence regime			Low dependence regime		
		1%	5%	10%	1%	5%	10%
CDS	VaR	−1.64%	−1.05%	−0.80%	−1.82%	−1.19%	−0.91%
	ES	−2.08%	−1.42%	−1.17%	−2.24%	−1.59%	−1.31%
Non-CDS	VaR	−1.67%	−1.02%	−0.73%	−2.24%	−1.30%	−0.98%
	ES	−2.18%	−1.43%	−1.14%	−2.89%	−1.88%	−1.50%

Table 6. Optimal weights for CDS portfolio

Portfolio	CDS	HS	Brent	Gold	Bond	Return	Risk
High dependence regime							
1	0.107	0.287	0.000	0.075	0.531	0.007 %	1.926 %
2	0.294	0.189	0.000	0.000	0.517	0.021 %	3.282 %
3	0.681	0.000	0.000	0.000	0.319	0.040 %	6.682 %
4	0.840	0.000	0.000	0.000	0.160	0.047 %	7.950 %
5	1.000	0.000	0.000	0.000	0.000	0.054 %	9.257 %
Low dependence regime							
1	0.218	0.142	0.000	0.283	0.357	0.003 %	2.382 %
2	0.265	0.007	0.000	0.363	0.365	0.007 %	3.056 %
3	0.509	0.000	0.000	0.482	0.010	0.013 %	4.960 %
4	0.836	0.000	0.000	0.164	0.000	0.020 %	7.603 %
5	1.000	0.000	0.000	0.000	0.000	0.024 %	9.042 %

Table 7. Optimal weights in non-CDS portfolio

Portfolio	HS	Brent	Gold	Bond	Return	Risk
High dependence regime						
1	0.177	0.000	0.269	0.553	-0.001 %	1.773 %
2	0.147	0.000	0.199	0.654	0.001 %	1.930 %
3	0.097	0.000	0.133	0.771	0.002 %	2.154 %
4	0.081	0.000	0.059	0.860	0.003 %	2.429 %
5	0.000	0.000	0.000	1.000	0.005 %	2.743 %
Low dependence regime						
1	0.173	0.000	0.080	0.747	-0.001 %	2.238 %
2	0.166	0.000	0.042	0.792	0.001 %	2.332 %
3	0.163	0.000	0.002	0.835	0.003 %	2.435 %
4	0.084	0.000	0.000	0.916	0.005 %	2.582 %
5	0.000	0.000	0.000	1.000	0.007 %	2.773 %

Tables 6 and 7 present the optimal weight for each asset which will help minimize the risk of the portfolio for any desired rate of return in CDS and non-CDS portfolio, respectively. The estimated results show that, at the same rate of return, the levels of risk in low dependence regime are larger than in high dependence regime for both CDS and non-CDS portfolios. For instance, at 0.007 % of CDS portfolio return, the risk levels in high and low dependence regime are 1.926 % and 3.056 %, respectively. Furthermore, CDS portfolio obviously outperforms non-CDS portfolio as its levels of risk are lower at any fixed rate of return compared to non-CDS portfolio. These results are consistent with

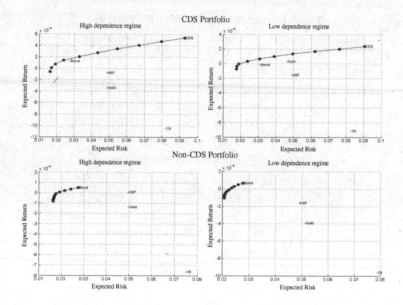

Fig. 2. Efficient frontiers, which presents the set of optimal portfolios, of CDS and non-CDS portfolio in high and low dependence regime. CDS portfolio generates much greater returns and higher risks compared to non-CDS portfolio.

the VaR and ES results of equally weighted portfolios in previous section. The results also show that including CDS in portfolio can increase the range of investment possibilities and thus investors can achieve higher returns but also risks. For instance, while non-CDS portfolio yields maximum return and risk only at 0.007 % and 2.773 %, respectively, the CDS portfolio can provide up to 0.054 % rate of return and 9.257 % of risk (Fig. 2).

4   Conclusion

In this paper, we aim to prove whether including Asian Credit Default Swap (CDS) index can improve the portfolio performance. We construct two portfolios, with and without CDS index, namely CDS and non-CDS portfolio. Other assets in our portfolios consist of stocks, commodities and bond. We employ Markov switching copula approach to estimate the dependence parameters between asset returns. Then, we calculate the risk of our portfolios by applying Value at Risk (VaR) and Expected Shortfall (ES) concepts. We finally obtain expected return and risk to find the optimal weights of the portfolios.

We find that Student-t copula is the best fit copula function for both CDS and non-CDS portfolios. Additionally, the VaR and ES values in high dependence regime are less than in low dependence regime on equally weighted portfolios. The reason is that the average values of dependence parameters in high dependence regime are less than those in low dependence regime. The results also report the lower average loss values in CDS portfolio compared to non-CDS portfolio. Moreover, CDS portfolio clearly outperforms non-CDS portfolio as its levels of risk are lower at any desired rate of return compared to non-CDS portfolio. Finally, we find that including Asian CDS index can increase both our portfolio return and risk since it extends the range of investment possibilities.

Acknowledgement. We are grateful for financial support from Graduate School, Chiang Mai University.

References

1. Pinsent, W.: Credit default swaps: an introduction. Investopedia (2015)
2. Bai, M., Sun, L.: Application of copula and copula-CVaR in the multivariate port-folio optimization. In: Chen, B., Paterson, M., Zhang, G. (eds.) ESCAPE 2007. LNCS, vol. 4614, pp. 231–242. Springer, Heidelberg (2007). doi:10.1007/978-3-540-74450-4
3. Halulu, S.: Quantifying the risk of portfolios containing stocks and commodities (Doctoral dissertation, Bogazii University) (2012)
4. Boubaker, H., Sghaier, N.: Portfolio optimization in the presence of dependent financial returns with long memory: a copula based approach. J. Bank. Finan. **37**(2), 361–377 (2013)
5. Kakouris, I., Rustem, B.: Robust portfolio optimization with copulas. Eur. J. Oper. Res. **235**(1), 28–37 (2014)
6. Patton, A.J.: A review of copula models for economic time series. J. Multivar. Anal. **110**, 4–18 (2012)
7. Tu, J.: Is regime switching in stock returns important in portfolio decisions? Manag. Sci. **56**(7), 1198–1215 (2010)
8. Ma, Y., MacLean, L., Xu, K., Zhao, Y.: A portfolio optimization model with regime-switching risk factors for sector exchange traded funds. Pac. J. Optim. **7**(2), 281–296 (2011)
9. Kaura, V.: Portfolio optimisation using value at risk (Doctoral dissertation, Masters thesis, Imperial College London, 2005.[cited at p. 14]) (2005)
10. Cotter, J., Longin, F.M.: Implied correlation from VaR. SSRN 996080 (2007)
11. Hamilton, J.D.: A new approach to the economic analysis of nonstationary time series and the business cycle. Econometrica J. Econometric Soc. **57**(2), 357–384 (1989)
12. Rodriguez, J.C.: Measuring financial contagion: a copula approach. J. Empirical Finan. **14**(3), 401–423 (2007)
13. Okimoto, T.: New evidence of asymmetric dependence structures in international equity markets. J. Financ. Quant. Anal. **43**(3), 787–816 (2008)
14. Sklar, M.: Fonctions de rpartition n dimensions et leurs marges. Universit Paris **8**, 229–231 (1959)
15. Patton, A.J.: Modeling asymmetric exchange rate dependence. Int. Econ. Rev. **47**(2), 527–556 (2006)
16. Hofert, M., Machler, M., McNeil, A.J.: Likelihood inference for archimedean copula in high dimensions under known margins. J. Multivar. Anal. **110**, 133–150 (2012)
17. Wang, Y.: Numerical approximations and goodness-of-fit of copulas (2012)
18. Stöber, J., Czado, C.: Detecting regime switches in the dependence structure of high dimensional financial data. arXiv preprint http://arxiv.org/abs/1202.2009 (2012)

19. Pastpipatkul, P., Yamaka, W., Sriboonchitta, S.: Analyzing financial risk and co-movement of gold market, and indonesian, philippine, and thailand stock markets: dynamic copula with markov-switching. In: Huynh, V.-N., Kreinovich, V., Sriboonchitta, S. (eds.) Causal Inference in Econometrics. SCI, vol. 622, pp. 565–586. Springer, Heidelberg (2016). doi:10.1007/978-3-319-27284-9_37
20. Chollete, L., Heinen, A., Valdesogo, A.: Modeling international financial returns with a multivariate regime-switching copula. J. Financ. Econometrics **7**(4), 437–480 (2009)
21. Rockafellar, R.T., Uryasev, S.: Conditional value-at-risk for general loss distributions. J. Bank. Finan. **26**(7), 1443–1471 (2002)

A Copula-Based Stochastic Frontier Model and Efficiency Analysis: Evidence from Stock Exchange of Thailand

Phachongchit Tibprasorn[✉], Somsak Chanaim, and Songsak Sriboonchitta

Faculty of Economics, Chiang Mai University, Chiang Mai, Thailand
phachongchit_t@cmu.ac.th

Abstract. This study applies the concept of stochastic frontier model (SFM) to estimate stock efficiencies of the top 50 companies with the highest market capitalization in the Stock Exchange of Thailand (SET50). We decompose the actual return deviation from its expected return into a stochastic noise and inefficiency term, and use copula approach to join these two error components. Four copulas are considered, and the most appropriate copula is selected using the lowest AIC. The empirical results show that the majority of average return efficiencies (60 % of stocks) lie between 0.9571 and 0.9999, suggesting that most of the stocks are quite efficient. However, the overall average of all return efficiencies is found to be 0.7313, indicating that stock price does not reflect all relevant information in the market.

Keywords: Stochastic frontier · Copula · Return efficiency · Capital asset pricing model · Financial econometrics

1 Introduction

In stock market, efficient market hypothesis (EMH) developed by Fama [1] states that stock is always traded at its fair value. It is not possible for investors to either purchase stock at overly low price or sell overvalued. Thus, stock return should be impossible to outperform the market return. And most related studies conclude that nobody can beat the market consistently on a risk-adjusted basis because efficient market reacts to new information instantaneously, and that the deduction in discount rate is already reflected in its price, Fama et al. [2, 3]. Nevertheless, whether stock market is efficient remains puzzled after some phenomena such as the stock market bubbles occurred in the past decade and some investors or funds look like being able to beat the market consistently.

In the real world, investment in stock is prone to risks arising from financial policy, financial crisis, exchange rates, political disturbance and natural disaster which may increase the chance of price variations. Future returns are not nevertheless guaranteed by past performance. But, it is hard for the investors to make an accurate decision about what's stock to buy or the price to pay if they

© Springer International Publishing AG 2016
V.-N. Huynh et al. (Eds.): IUKM 2016, LNAI 9978, pp. 637–648, 2016.
DOI: 10.1007/978-3-319-49046-5_54

do not receive enough information about risk and performance. This will lead to market inefficiency.

In order to test whether stock market is efficient, Fama suggests that it needs an asset pricing theory which explains how stock price is determined. The first and famous model which is used to price the risky securities is the Capital Asset Pricing Model (CAPM) introduced by Sharpe [4], Treynor [5], and Lintner [6]. CAPM displays the relationship between the expected excess returns on individual assets and their systematic risk, β. This relationship claims that the expected excess returns on any stock is directly proportional to its β. According to Jensen [7], however, his alpha coefficient (α) shows the difference between the actual excess return and the expected excess return. This value should be equal to zero even negative if the investors cannot access all information at the same time. But, there are many studies showing the inadequacy of α value. As shown in the study of Miller and Scholes [8], they found that α depends in a systematic way on its β. The stock with low β tends to have positive α, and in contrary to the high β. In addition, Jensen et al. [9] suggested that if we select the stocks by considering their high β values, this would increase a positive bias in beta of portfolio because the stocks which are entered the first tend to have positive measurement errors in their approximated beta ($\widehat{\beta}$) and lead to a negative bias in portfolio's approximated alpha.

Apart from the use of β, we also apply the concept of stochastic frontier model (SFM) suggested by Aigner et al. [10] to estimate the efficient frontier a stock return can reach, and to capture the deviation of the actual return from its frontier which is defined as the return inefficiency. This value indicates the level of efficiency that stock price can reflect information existing in the market, which can be used to determine the stock performance of individual investor, that is the main goal of this paper. And we can use the results to select stocks into a portfolio.

Anneart et al. [11] applied the SFM to determine whether the asset prices are efficient by decomposing the deviation of mutual fund return from its expected return into a stochastic noise and an efficiency error term. If the efficiency error term equals to 100 percent, it means that the fund is not underperforming. Hasan et al. [12] used the Cobb-Douglas stochastic frontier to estimate the technical efficiency of each company group in Bangladesh Stock Market. And Luo et al. [13] used the stochastic frontier model to estimate the pricing efficiency of initial public offering in the ChiNext market. All of their studies follow the strong assumption that the stochastic noise and the technical inefficiency are independent. Now, we can relax this strong assumption by using copula to find the joint distribution of them (see, Tibprasorn et al. [14], Amsler et al. [15], Burns [16], Smith [17] and Wiboonpongse et al. [18], etc.).

For the setting of the error components, we assume the stochastic noise (v) is normally distributed, and the inefficiency measure (u) is assigned in 2 cases including half-normal and exponential distributions. A fair degree of the dependence between v and u are measured using four copulas, and we select the most appropriate copula by considering the lowest Akaike Information Criterion

(AIC). For estimating all parameters, we use a likelihood function and approximate integral by using the maximum simulated likelihood (MSL) technique (see, Burns [16], Wiboonpongse et al. [18] and Greene [19,20], etc.).

We study the return efficiencies of top 50 companies with the highest market capitalization in the Stock Exchange of Thailand (SET50). The empirical results show that our extended models can be used strategically for selecting stocks into the portfolio compared with the classical SFM. We can justify that the stocks with their $\beta > 1$ do not indicate the high level of average return efficiency.

The remainder of this paper is arranged as follows. Section 2 talks about the knowledge of copula approach and stochastic frontier model. Section 3 presents the implementation to stock market. Section 4 shows the empirical results. And the last section concludes the study.

2   Copula and Stochastic Frontier Model

2.1   Copula

Copula is a famous approach, which is used to describe the dependence structure between random variables especially in the case of different marginal distributions. This approach is derived from the Sklar's Theorem which suggested that joint marginal distribution of each random variable can be linked by a copula function.

Let X and Y be random variables with continuous distribution functions F and G, $U_1 = F(X)$ and $U_2 = G(Y)$ are uniformly distributed on the interval $[0,1]$. Thus, the joint distribution function H with marginal distribution function $F(X)$ and $G(Y)$ is denoted as, for $x, y \in \mathbb{R}$

$$\begin{aligned} H(x,y) &= p(X \leqslant x, Y \leqslant y) \\ &= p(F(X) \leqslant F(x), G(Y) \leqslant G(y)) \\ &= p(U_1 \leqslant F(x), U_2 \leqslant G(y)) \\ &= C(F(X), G(Y)). \end{aligned} \tag{1}$$

In the above equation, the C is a copula function of two marginals $F(x)$, and $G(y)$. In addition, the strength of dependence between random variables is controlled by either parametric or non-parametric copula function. However, different copulas capture different kinds of dependencies.

With the concept of copula we provide Gaussian, Frank, Clayton and Independence copulas to construct the dependence structure and measure a fair degree of dependence. The definition of copulas used in this study is shown in Table 1.

Table 1. Definition of copulas.

Name of copula	$C_\theta(u_1, u_2)$	Parameter θ
Gaussian	$\Phi_\Sigma(\phi^{-1}(u_1), \phi^{-1}(u_2))$	$\theta \in (-1, 1)$
Frank	$-\frac{1}{\theta}log(1 + \frac{(e^{-\theta u_1}-1)(e^{-\theta u_2}-1)}{e^{-\theta}-1})$	$\theta \in R \setminus \{0\}$
Clayton	$(u_1{}^{-\theta} + u_2{}^{-\theta} - 1)^{-1/\theta}$	$\theta \in [-1, \infty) \setminus \{0\}$
Independence	$u_1 u_2$	

where $\phi(u_1)$ and $\phi(u_2)$ are the cdf of univariate Gaussian with mean 0 and variance 1, and Φ_Σ is the cdf of a standard bivariate Gaussian with mean 0 and covariance matrix $\Sigma = \begin{bmatrix} 1 & \theta \\ \theta & 1 \end{bmatrix}$.

2.2 Copula-Based Stochastic Frontier Model

The stochastic frontier model (SFM) suggested by Aigner et al. [10] is given by

$$Y = f(X; \beta) + (V - U), \tag{2}$$

where Y is the output variable and X is the input variable; β is associated with a vector of unknown parameters to be estimated; the stochastic noise V is the symmetric normal distribution error; and the technical inefficiency U is the non-negative error (one-sided error). The error components, V and U, are assumed to be independent. Following the above equation, the technical efficiency is given by

$$TE = \exp(-U). \tag{3}$$

However, there may exist the dependency between the error components. Thus, by Sklar's theorem, the joint cdf of U and V is

$$H(u, v) = Pr(U \leqslant u, V \leqslant v) \tag{4a}$$

$$= C_\theta(F_U(u), F_V(v)), \tag{4b}$$

where $C_\theta(\cdot, \cdot)$ is denoted as the bivariate copula with unknown parameter θ. Following Smith [17], transforming (U, V) to (U, ξ) as the probability density function (pdf) of (U, ξ) then we get

$$h(u, \varepsilon) = f_U(u)f_V(u + \varepsilon)c_\theta(F_U(u), F_V(u + \varepsilon)), \tag{5}$$

where $f_U(u)$ and $f_V(v)$ are the marginal density of $H(u, v)$, the composite error $\xi = \varepsilon(-\infty < \varepsilon < \infty)$, $v = u + \varepsilon$ and $c_\theta(\cdot, \cdot)$ is the copula density of $C_\theta(\cdot, \cdot)$. Thus, the pdf of ε is obtained by

$$h_\theta(\varepsilon) = \int_0^\infty h(u, \varepsilon)du \tag{6a}$$

$$= \int_0^\infty f_U(u)f_V(u + \varepsilon)c_\theta(F_U(u), F_V(u + \varepsilon))du. \tag{6b}$$

Smith [17] argued that it is very difficult to find the closed-form expression for finding the pdf of ε because there are very few densities of ε for estimating the maximum likelihood. Thus, we employ the maximum simulated likelihood (MSL) technique mentioned in Wiboonpongse [18] to approximate integral of $h_\theta(\varepsilon)$. The pdf of ε is obtained by

$$h_\theta(\varepsilon_i) \approx \frac{1}{J} \sum_{j=1}^{J} \frac{h(u_j, \varepsilon_i)}{f(u_j)}, \tag{7}$$

where $f(u_j)$ is the pdf of u_j and $j = 1, ..., J$ is a sequence of random drawn from the distribution of u_j.

Therefore, the likelihood function for copula-based stochastic frontier model is represented by

$$L(\beta, \sigma_v, \sigma_u, \theta) = \prod_{i=1}^{N} h_\theta(y_i - x_i'\beta) = \prod_{i=1}^{N} h_\theta(\varepsilon_i), \tag{8}$$

where σ_v and σ_u are the scale parameters of marginal distribution of V and U, θ is the parameter of copula and $i = 1, ..., N$ is the number of observations.

Taking a natural logarithm, the log-likelihood function for copula-based stochastic frontier model becomes

$$\ln L(\beta, \sigma_v, \sigma_u, \theta) = \sum_{i=1}^{N} \ln h_\theta(\varepsilon_i) \approx \sum_{i=1}^{N} \ln \frac{1}{J} \sum_{j=1}^{J} \frac{h(u_j, \varepsilon_i)}{f(u_j)}. \tag{9}$$

Following Battese and Coelli [21], the technical efficiency of each copula (TE_θ) can be computed by

$$TE_\theta = E[\exp(-U)|\varepsilon] \tag{10a}$$

$$= \frac{1}{h_\theta(\varepsilon)} \int_0^\infty \exp(-u)h(u, \varepsilon)du \tag{10b}$$

$$= \frac{\int_0^\infty \exp(-u)f_U(u)f_V(u+\varepsilon)c_\theta(F_U(u), F_V(u+\varepsilon))du}{\int_0^\infty f_U(u)f_V(u+\varepsilon)c_\theta(F_U(u), F_V(u+\varepsilon))du}. \tag{10c}$$

and using Monte Carlo integration. Thus, we obtain

$$TE_\theta = \frac{\sum_{i=1}^{N} \exp(-u_i)f_U(u_i)f_V(u_i+\varepsilon_i)c_\theta(F_U(u_i), F_V(u_i+\varepsilon_i))}{\sum_{i=1}^{N} f_U(u_i)f_V(u_i+\varepsilon_i)c_\theta(F_U(u_i), F_V(u_i+\varepsilon_i))}. \tag{11}$$

Note that u_i follows the cumulative distribution of U. Since $U \geqslant 0$, the distribution of U must be positive. Therefore, it is convenient to assume U to be either half-normal or exponential distribution. In the case of half-normal distribution, parameter θ is restricted to $u_i \in [0, \infty)$. The density function is given by

$$f_U(u; \sigma_u) = \frac{2}{\sqrt{2\pi\sigma_u^2}} \exp\{-\frac{u^2}{2\sigma_u^2}\}. \tag{12}$$

For the case of exponential distribution, parameter θ is restricted to $u_i \in [0, \infty)$ and scale parameter $\lambda \in (0, \infty)$. The density function follows

$$f_U(u; \lambda) = \lambda e^{-\lambda u}. \tag{13}$$

And v_i follows the distribution of V assumed to be distributed as $N(0, \sigma_v^2)$. The density can be written as following

$$f_V(v; \sigma_v) = \frac{1}{\sqrt{2\pi\sigma_v^2}} \exp\{-\frac{v^2}{2\sigma_v^2}\}. \tag{14}$$

3   Copula-Based Stochastic Frontier Implementation

To compare the return efficiencies across individual stocks, we start with the Jensens alpha (α_s) by Jensen [7] as a benchmark for presenting the relationship between the expected excess return (risk premium) and their systematic risk (market risk). This relationship claims that the expected excess return on any stock is directly proportional to its systematic risk, β. For any stock s the β is calculated by

$$\alpha_s = R_s - E(R_s) = R_s - \beta_s E(R_m), \tag{15}$$

where R_s is the actual excess return of stock s, $E(R_s)$ is the expected excess return on stock s, $E(R_m)$ is its systematic risk and is the expected excess return on market.

In accordance with the efficient market hypothesis (EMH), α_s should be equal to zero even negative if the investors cannot access all information at the same time. But, many previous studies showed the inadequacy of this relation. Especially in the study of Miller and Scholes, they found that α_s depends in a systematic way on its β_s. The stock with high β_s tends to has negative α_s, and in contrary to the low β_s. In addition, Jensen et al. [9] suggested that if we select the stocks by considering their high β_s, this would increase a positive bias in beta of portfolio because the stocks which are entered the first tend to have positive measurement errors in their approximated beta ($\widehat{\beta}_s$) and lead to a negative bias in portfolio's approximated alpha.

Apart from the use of β_s, our methodology is constructed to capture the deviation of the actual excess return from the efficient frontier a stock can reach. This method can be used strategically for selecting stocks in the portfolio. Following the stochastic frontier model in Eq. (2), the empirical version of SFM with specification of time series regression CAPM can be represented as

$$E(R_{s,t}) = \beta_{0,s} + \beta_{1,s} E(R_{m,t}) + V_{s,t} - U_{s,t}, \tag{16}$$

where $E(R_{s,t})$ is the expected excess return on stock s for week t; $\beta_{0,s}$ is the intercept term of stock s; $E(R_{m,t})$ is the expected return on market for week .t; $V_{s,t}$ is stochastic noise capturing measurement error of stock s for week t and $U_{s,t}$ is non-negative error capturing the inefficiency of stock s for week t.

The value of inefficiency is defined in terms of return inefficiency which shows the failure of actual excess return to achieve the efficient excess return due to some out of control events such as asymmetric information, and it indicates the level of efficiency that stock price can reflect all information in market.

Four copulas, Gaussian, Frank, Clayton and Independence, are proposed to capture a fairly good degree of the dependence between $V_{s,t}$ and $U_{s,t}$. We use the likelihood function to obtain all parameters, and approximate integral by using the maximum simulated likelihood (MSL) technique. Therefore, the return efficiency of stock s at week t for each copula $(RE_\theta)_{s,t}$ is given by

$$(RE_\theta)_{s,t} = E[\exp(-U_{s,t})|\varepsilon], \tag{17}$$

where $s = 1, \dots 50$, s is the sequence of individual stocks. The average return efficiency of individual stocks $(\overline{RE}_s)$ is as follows:

$$\overline{RE}_s = \frac{1}{T} \sum_{t=1}^{T} (RE_\theta)_{s,t}, \tag{18}$$

where T is the number of weeks of each stock. And the overall average of all return efficiencies $(\overline{RE})$ is given by

$$\overline{RE} = \frac{1}{50} \sum_{s=1}^{50} \overline{RE}_s. \tag{19}$$

4   Empirical Results

We use several real data and Akaike Information Criterion (AIC) to present that our copula-based stochastic frontier with the specification of time series CAPM provides better fit than the classical stochastic frontier model. The data contains the top 50 companies with the highest market capitalization in the Stock Exchange of Thailand (SET50), and we provide the SET50 index as a benchmark. The excess stock return and excess market return are computed using the weekly prices collected by Thomson Reuters. Thailand Government 10-Year Bond Yield collected by Datastream is used as a risk free rate. All weekly data are taken from March 2009 to March 2016, except for the stocks with fewer returns.

By applying the model in Sect. 3, the average return efficiency of individual stock $(\overline{RE}_s)$ in SET50 is estimated, and the frequency histogram of $\overline{RE}_s$ is presented in Figs. 1 and 2, respectively. According to empirical results shown in Figs. 1 and 2, the values of $\overline{RE}_s$ are between 0.1431 and 0.9999. And the majority of average return efficiencies (60 % of stocks) are between 0.9571 and 0.9999 (see Fig. 1), suggesting that most of the stocks under this study are quite efficient. However, the overall average of all return efficiencies is found to be 0.7313 (see Fig. 2), indicating that stock price does not reflect all relevant information in the market.

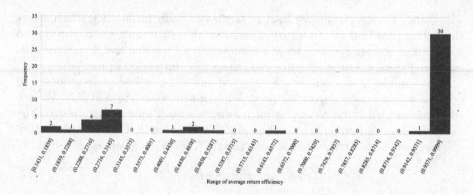

Fig. 1. The frequency histogram of individual stocks' average return efficiency.

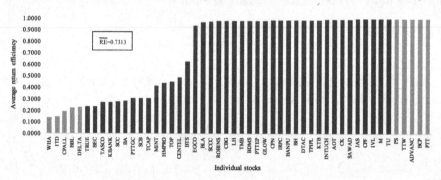

Fig. 2. The bar chart of individual stocks' average return efficiency.

According to $\overline{RE}_s$ values as shown in Fig. 2, the top five highest values are PTT Public Company Limited (PTT), Bangchak Petroleum Public Company Limited (BCP), Advanced Info Service Public Company Limited (ADVANC), TTW Public Company Limited (TTW) and Pruksa Real Estate Public Company Limited (PS). And the top five lowest values are WHA Corporation Public Company Limited (WHA), Italian-Thai Development Public Company Limited (ITD), CP All Public Company Limited (CPALL), Bangkok Bank Public Company Limited (BBL) and Delta Electronics (Thailand) Public Company Limited (DELTA). A detailed result analysis is presented in Tables 2 and 3.

According to AIC computed from the choice models as mentioned in Sect. 3, the best model of top five highest $\overline{RE}_s$ value is based on Frank copula in which the return inefficiency ($U_{s,t}$) distributes as half-normal for the cases of PTT, BCP, TTW and PS. Their $\overline{RE}_s$ values are 0.9999, 0.9999, 0.9989 and 0.9987, respectively. For the case of ADVANC, the best model is Clayton copula in which $U_{s,t}$ follows half-normal distribution, and its $\overline{RE}_s$ value is 0.9990 (see Table 2).

For the top five lowest $\overline{RE}_s$ values, the best model is based on Clayton copula in which $U_{s,t}$ distributes as exponential for the cases of WHA, ITD, CPALL and

Table 2. Empirical results of top five individual stocks with the highest average return efficiency.

Parameter	PTT	BCP	ADVANC	TTW	PS
β_o	−1.3878 (0.2094)	−1.4341 (0.2237)	−0.0287 (0.8759)	−0.4261 (0.1533)	−0.3976 (0.4077)
β_1	1.1380 (0.0532)	0.6915 (0.0552)	0.7021 (11.6786)	0.4150 (0.0426)	1.1183 (0.0868)
σ_v	3.4395 (0.2407)	4.4037 (0.3010)	3.4792 (17.3602)	3.0226 (0.1549)	5.1123 (0.5363)
σ_u	0.0001 (0.0000)	0.0001 (0.0000)	0.0012 (0.0078)	0.0013 (0.0001)	0.0016 (0.0002)
λ	−	−	−	−	−
θ	7.5524 (1.0948)	7.6740 (1.0912)	0.0000 (0.0802)	8.3318 (1.1363)	7.2558 (8.6208)
$\overline{RE}_s$	0.9999	0.9999	0.9990	0.9989	0.9987
Distribution of u	Half normal	Half normal	Half normal	Half normal	Half normal
Name of Copula	Frank	Frank	Clayton	Frank	Frank
Log L	−623.5008	−710.3963	−859.2281	−837.9721	−943.5520
AIC	1257.0017	1430.7925	1728.4561	1685.9443	1897.1041
Data description					
$\overline{R}_s$	0.1616 %	0.3780 %	0.2233 %	0.1965 %	0.5166 %
S.D.	3.8964 %	3.7089 %	3.9405 %	2.8152 %	5.5582 %
Max	12.1100 %	14.2700 %	12.8500 %	10.6800 %	19.3700 %
Min	−13.5800 %	−16.7100 %	−21.6000 %	−13.0400 %	−19.2600 %
$\overline{R}_m$	0.2974 %	0.2974 %	0.2974 %	0.2974 %	0.2974 %
$\overline{R}_f$	0.0675 %	0.0675 %	0.0675 %	0.0675 %	0.0675 %
No. Obs	369	369	369	369	369

BBL. Their $\overline{RE}_s$ values are 0.1431, 0.1514, 0.1940 and 0.2291, respectively. For the case of DELTA, the best model is based on Clayton copula in which $U_{s,t}$ follows half-normal distribution, and its $\overline{RE}_s$ value is 0.2333 (see Table 3).

Clearly, these empirical results reflect that our extended models can be used strategically for selecting the appropriate stocks in the market. We can justify that the individual stocks with their $\beta_1 > 1$, presenting the possibility of higher rate of return but greater risk, do not imply the high $\overline{RE}_s$ such as WHA and ITD. Therefore, we should select stocks into the portfolio by considering both β_1 and $\overline{RE}_s$.

Table 3. Empirical results of top five individual stocks with the lowest average return efficiency.

Parameter	WHA	ITD	CPALL	BBL	DELTA
β_o	3.6388 (3.0835)	2.8808 (8.9088)	3.2762 (1.4473)	2.9667 (0.2347)	1.9629 (0.1980)
β_1	1.2899 (3.2103)	1.4165 (5.5116)	0.6655 (0.0331)	0.9595 (0.4033)	0.5858 (0.0307)
σ_v	7.6185 (4.6409)	6.9783 (6.4404)	4.8616 (0.6662)	3.7281 (0.5372)	5.0496 (0.0803)
σ_u	– –	– –	– –	– –	2.6650 (0.1546)
λ	3.4754 (3.0417)	3.4780 (16.7444)	3.1280 (0.6099)	3.1967 (1.0488)	
θ	3.5879 (3.8381)	5.6615 (1.2936)	4.3411 (0.2399)	5.3884 (0.3057)	6.1734 (0.5817)
$\overline{RE}_s$	0.1431	0.1514	0.194	0.2291	0.2333
Distribution of u	Normal/ Exponential	Normal/ Exponential	Normal/ Exponential	Normal/ Exponential	Normal/ Half normal
Name of Copula	Clayton	Clayton	Clayton	Clayton	Clayton
Log L	−555.5795	−1099.6896	−952.9363	−805.2133	−974.7938
AIC	1121.1589	2209.3793	1915.8727	1620.4265	1959.5876
Data description					
$\overline{R}_s$	0.6799 %	0.3186 %	0.5407 %	0.2344 %	0.5765 %
S.D.	7.0853 %	6.5001 %	3.8487 %	3.4041 %	4.1929 %
Max	28.5200 %	23.7300 %	15.6200 %	14.7600 %	15.9600 %
Min	−20.3200 %	−26.1600 %	−12.6900 %	−10.3900 %	−15.0500 %
$\overline{R}_m$	0.0160 %	0.2974 %	0.2974 %	0.2974 %	0.2974 %
$\overline{R}_f$	0.0634 %	0.0675 %	0.0675 %	0.0675 %	0.0675 %
No. Obs	176	369	369	369	369

5   Conclusions

Many previous studies mostly focus on computing excess stock returns and consider only the systematic risk (β). However, our study provides a new method to estimate the frontier of stock return and efficiency of stock by applying stochastic frontier model (SFM) with the specification of time series CAPM. Our study is laid within the following scopes. Firstly, the strong assumption of classical SFM is relaxed by using copula approach to present the dependence structure of error components, $V_{s,t}$ and $U_{s,t}$. In this study, four copulas are employed to show the degree of the dependence, such as Gaussian copula, Frank copula, Clayton copula and Independence copula. The marginal distribution of $V_{s,t}$ is assigned to be normal, and $U_{s,t}$ is designed to be either half-normal or exponential for more suitable distribution. Secondly, the pdf of composite error is obtained by using the maximum simulated likelihood technique. Thirdly, the Akaike Information Criterion (AIC) is used to select the appropriate copula family for joining the two error components.

Finally, we apply our model to analyze the stock returns of the top 50 companies with the highest market capitalization in SET50. The empirical results show that our copula-based stochastic frontier model provides better fit than the classical stochastic frontier model considering by the measure of AIC. The values of average return efficiency are predominantly between 0.9571 and 0.9999, indicating that most of the individual stock returns are quite efficient which means that their returns are not under or overestimated in terms of return efficiency. However, the overall average of all return efficiencies is 0.7313. Thus, we can conclude that the stock price does not reflect all relevant information in the market. The top five highest average return efficiencies are PTT, BCP, ADVANC, TTW and PS, respectively. While, WHA, ITD, CPAll, BBL and DELTA are the top five lowest average return efficiencies. Moreover, we can justify that the stock with its β_1 greater than one does not indicate its high level of average return efficiency. Clearly, our extended models can be used strategically for selecting stocks in the market. To compute the excess stock returns, any other asset pricing models can be applied so the other models could be extended in the future study.

Acknowledgement. we are grateful for financial support from Rajamangala University of Technology Lanna and we would like to thank Mr. Woraphon Yamaka, Phd. student in Economics program and research assistant of Center of Excellence in Econometrics, Faculty of Economics, Chiang Mai University, for his comments and suggestions.

References

1. Fama, E.F.: The behavior of stock-market prices. J. Bus. **38**(1), 34–105 (1965). http://www.jstor.org/stable/2350752
2. Fama, E.F., Fisher, L., Jensen, M., Roll, R.: The adjustment of stock prices to new information. Int. Econ. Rev. **10**(1), 1–21 (1969). doi:10.2139/ssrn.321524. http://www.jstor.org/stable/2525569
3. Fama, E.F., French, K.R.: Size, value, and momentum in international stock returns. J. Financ. Econ. **105**, 457–472 (2012)
4. Sharpe, W.F.: Capital asset prices: a theory of market equilibrium under conditions of risk. J. Financ. **19**(3), 425–442 (1964)
5. Treynor, J. L.: Jack Treynor's Toward a Theory of Market Value of Risky Assets (1962). Available at SSRN: http://ssrn.com/abstract=628187 or http://dx.doi.org/10.2139/ssrn.628187
6. Lintner, J.: The valuation of risk assets and the selection of risky investments in stock portfolios and capital budgets. Rev. Econ. Stat. **47**(1), 13–37 (1965). doi:10.2307/1926735. http://www.jstor.org/stable/1924119

7. Jensen, M.C.: The performance of mutual funds in the period 1945–1964. J. Financ. **23**(2), 389–416 (1968)
8. Miller, M.H., Scholes, M.: Rates of return in relation to risk: a reexamination of some recent findings. In: Studies in the Theory of Capital Markets, vol. 23 (1972)
9. Jensen, M.C.: Studies in the Theory of Capital Market. Praeger Publishers Inc., New York (1972)
10. Aigner, D., Lovell, K., Schmidt, P.: Formulation and estimation of stochastic frontier function models. J. Econometrics **6**(1), 21–37 (1977)
11. Annaert, J., Broeck, J.V.D., Vennet, R.V.: Determinants of mutual fund underperformance: a Bayesian stochastic frontier approach. Eur. J. Oper. Res. **151**(3), 617–632 (2003)
12. Hasan, M.Z., Kamil, A.A., Mustafa, A., Baten, M.A.: Stochastic frontier model approach for measuring stock market efficiency with different distributions. Plos ONE **7**(5), e37047 (2012)
13. Luo, C., Ouyang, Z.: Estimating IPO pricing efficiency by Bayesian stochastic frontier analysis: the CliNext market case. Econ. Model. **40**, 152–157 (2014)
14. Tibprasorn, P., Autchariyapanitkul, K., Chaniam, S., Sriboonchitta, S.: A copula-based stochastic frontier model for financial pricing. In: Huynh, V.-N., Inuiguchi, M., Denoeux, T. (eds.) IUKM 2015. LNCS (LNAI), vol. 9376, pp. 151–162. Springer, Heidelberg (2015). doi:10.1007/978-3-319-25135-6_15
15. Amsler, C., Prokhorov, A., Schmidt, P.: Using copulas to model time dependence in stochastic frontier models. Econometrics Rev. **33**(5–6), 497–522 (2014)
16. Burns, R.: The Simulated Maximum Likelihood Estimation of Stochastic Frontier Models with Correlated Error Components. The University of Sydney (2004)
17. Smith, M.D.: Stochastic Frontier Models with Correlated Error Components. The University of Sydney (2004)
18. Wiboonpongse, A., Liu, J., Sriboonchitta, S., Denoeux, T.: Modeling dependence between error components of the stochastic frontier model using copula: application to inter crop coffee production in Northern Thailand frontier. Int. J. Approximate Reasoning **65**, 33–34 (2015)
19. Greene, W.H.: Simulated likelihood estimation of the normal-gamma stochastic frontier function. J. Prod. Anal. **19**(2), 179–190 (2010)
20. Greene, W.H.: A stochastic frontier model with correction for sample selection. J. Prod. Anal. **34**(1), 15–24 (2010)
21. Battese, G.E., Coelli, T.J.: Prediction of firm-level technical efficiencies with a generalized frontier production function and panel data. J. Econometrics **38**(3), 387–399 (1988)

Economic Growth and Income Inequality: Evidence from Thailand

Paravee Maneejuk[(✉)], Pathairat Pastpipatkul,
and Songsak Sriboonchitta

Faculty of Economics, Chiang Mai University, Chiang Mai, Thailand
mparavee@gmail.com, ppthairat@hotmail.com

Abstract. Evidence from this study shows a significant relationship between growth and income inequality in which inequality creates a negative impact on the growth, but the impact of economic growth on income inequality is non-linear. This result captures the Kuznets Hypothesis, finding the level of growth that can reduce unequal distribution of income in Thailand. Finally, this study also discovers a certain level of inflation that can help reduce income inequality effectively.

Keywords: Economic development · Gini coefficients · Kuznets Hypothesis · Simultaneous equations · Kink regression · Copula

1 Introduction

The relationship between income disparity and economic growth has been a critical issue in macroeconomics over several decades. Countries need economic growth to improve standard of living, reduce poverty, and also to survive in the modern economy. But in the meantime a rise in growth can also cause the inequality which may lead to a long-standing problem which in turn can become destructive to the growth.

While we seem to consider the inequality as a consequence of economic growth, some economists tend to argue that inequality may affect the growth at the same time and perhaps it is necessary to propel the growth. 'Two sides of the same coin' is what Ostry and Berg, IMF staff, used to define the effect of inequality on economic growth through their paper in 2011 [13]. They reckoned that the impact of unequal distribution of income is split into two sides. Inequality can influence the growth negatively and become a problem for society. For example, it can obstruct the human capital accumulation especially in education. That is because individuals having higher income can invest more in education than poor people can and have lots of choices in occupation. Recently, IMF has released one more discussion paper about inequality and growth entitled 'Causes and Consequences of Income Inequality: A Global Perspective' [5]. One part of this paper refers to a negative effect of inequality on growth and reckons that inequality can also leads to political and economic instability, resulting in corruption, resource misallocation, and nepotism. In contrast, inequality can also be good for economy since it provides incentives for people to invest and save for their future livelihood and also for entrepreneurs to run their businesses successfully [10].

© Springer International Publishing AG 2016
V.-N. Huynh et al. (Eds.): IUKM 2016, LNAI 9978, pp. 649–663, 2016.
DOI: 10.1007/978-3-319-49046-5_55

Inequality does not instantly propel economy but without inequality economy is like socialism where the growth of economy hardly exists.

Trade-off between increasing growth and reducing inequality

While growth is needed, inequality still matters. Some economists suggest that an improvement in income distribution and economic growth can be achieved at the same time [2]. Simon Kuznets, an American economist who received the Nobel Memorial Prize in Economic Sciences in 1971, hypothesizes that as a country develops, the economic inequality would first increase and then later decrease [9]. This hypothesis is graphed as an inverted-U shape called 'Kuznets Curve' where the horizontal x-axis is income per capita and the vertical y-axis is economic inequality. This inverted-U shape curve implies that income inequality following the curve is gradually increased resulting from economic evolution. That is, at the initial level of low economic growth, income inequality is also low. Then, as countries seek to increase the rate of economic growth, industrial sector often plays a key role to achieve the goal. This event leads to a big gap between rich and poor, and hence income inequality increases. According to the Kuznets Hypothesis, the level of inequality keeps rising as long as the economic growth increases. However, according to the same hypothesis, inequality will be decreased after the growth reaches some certain level, a threshold, in which the economy is developed efficiently resulting in an increase in average income, an improvement in social welfare of society, higher educated populations, and so on.

Motivated by this reasoning, this paper aims to investigate whether the Kuznets Hypothesis holds in the case of some developing country like Thailand where Thai government has been trying to promote the economic growth and especially with a concern about inequality at the same time. This is crucially important because, if we could prove that the hypothesis holds and the result could illustrate the threshold accurately, it means we can find the proper rate of economic growth that will not hurt income distribution in the long term. Our findings should be useful for the Thai government and policy makers and can contribute to longer-run benefits for society.

2   Econometric Modeling Framework

Investigating the inverted U-shaped curve or the so-called Kuznets curve is a profoundly complicated problem and difficult to do in the real world because economic growth and income inequality themselves actually depend on many factors. Therefore, an appropriate tool for investigating this phenomenon with special concern in the real mechanism of growth and income inequality becomes so important. As in the literature, numerous studies failed to find the Kuznets curve, especially the empirical studies on time series (see, e.g., Gallup [6]). Therefore, this paper presents a novel tool of investigating the Kuznets curve that is the simultaneous kink equation (SKE) model. We employ the idea of Kink regression with unknown threshold as introduced in Hansen [7] for identifying the unequal effect of growth on the distribution of income.

2.1   Introduction to the SKE Model

This study deals with a simultaneous relationship between income inequality and economic growth and speculates some covariate variables can produce effects that would be considered both negative and positive. Therefore, Kink regression model with unknown threshold is applied to the simultaneous equation model in order to capture our considerations. *What is the kink regression?* Briefly, it is simply a regression discontinuity model. The regression function itself is continuous but the slope is discontinuous at a threshold point, thereby a kink [7]. With that in mind, we have the simultaneous kink equation (SKE) model as an econometric model for this work. The SKE model can take the form as:

$$y_{1,t} = \alpha_1 + \sum_{i=1}^{G} \beta_{1i}^-(Y_{-i,t} - r_{1i})_- + \sum_{i=1}^{G} \beta_{1i}^+(Y_{-i,t} - r_{1i})_+ + \sum_{j=1}^{k} \phi_{1j}^-(X_{i,t} - r_{1j})_- + \sum_{j=1}^{k} \phi_{1j}^+(X_{i,t} - r_{1j})_+ + \sigma_1 \varepsilon_{1,t}$$

$$\vdots \tag{1}$$

$$y_{m,t} = \alpha_m + \sum_{i=1}^{G} \beta_{mi}^-(Y_{-i,t} - r_{mi})_- + \sum_{i=1}^{G} \beta_{mi}^+(Y_{-i,t} - r_{mi})_+ + \sum_{j=1}^{k} \phi_{mj}^-(X_{i,t} - r_{mj})_- + \sum_{j=1}^{k} \phi_{mj}^+(X_{i,t} - r_{mj})_- + \sigma_m \varepsilon_{m,t}$$

The model above contains m-simultaneous kink equations where $(y_{1,t}, \ldots, y_{m,t})$ and $(\varepsilon_{1,t}, \ldots, \varepsilon_{m,t})$ are $T \times 1$ vector of dependent variables and error terms, respectively. $X_{i,t}$ is a $T \times k$ matrix of k exogenous regressors, and $Y_{-i,t}$ is a $T \times G$ matrix of G endogenous regressors on the right-hand side of the ith equation. Note that exogenous and endogenous regressors can simply be viewed as independent variables of the model. The exogenous regressors refer to $X_{i,t}$. The endogenous regressors refer to the dependent variable of other equations $Y_{-i,t}$ that also work as an independent variable for this i^{th} equation. Therefore, this structure is called the simultaneous equation model. The term $(\sigma_1, \ldots, \sigma_m)$ are the scale parameters of margin $1, \ldots, m$ and $(\varepsilon_{1,t}, \ldots, \varepsilon_{m,t})$ are the margin errors that follow some distributions. Following Hansen [7], we generate negative and positive functions of real number a as $(a)_- = \min(a, 0)$ and $(a)_+ = \max(a, 0)$, respectively, in order to separate the regressor variables into two regimes. For i^{th} equation, the coefficients with respect to the right hand side variables are equal to $(\beta_{1i}^-, \ldots \beta_{mi}^-)$ for the values of $Y_{-i,t}$ less than certain values $(r_{1i}, \ldots, r_{mi})$, and $(\phi_{1j}^-, \ldots, \phi_{mj}^-)$ for the values of $X_{i,t}$ less than $(r_{1j}, \ldots, r_{mj})$. Conversely, the coefficients will be $(\beta_{1i}^+, \ldots, \beta_{mi}^+)$ and $(\phi_{1j}^+, \ldots, \phi_{mj}^+)$ if the values of $Y_{-i,t}$ and $X_{i,t}$ are larger than $(r_{1i}, \ldots, r_{mi})$ and $(r_{1j}, \ldots, r_{mj})$, respectively, and yet the regression function is continuous in all variables [7].

2.2   Modelling Dependence with Copulas

In our SKE model, it is assumed to have a correlation between the errors across m - simultaneous kink equations; therefore a joint distribution or dependency is somehow needed. We are thinking about the term 'Copulas' which are the best way to give us a joint cumulative distribution function. Therefore, we employ the copulas to measure the nonlinear dependence structures with different marginal distributions as in the suggestions of Wichitaksorn et al. [18] and Pastpipatkul et al. [14]. The estimation of our proposed model 'the Copula based SKE' will be discussed later.

What is Copulas? It is a parametrically specified joint distribution which is generated from any given marginal distribution [17]. The main advantage of using copulas is to separate the marginal behavior and dependence structure of variables from their joint distribution function [11]. This paper applies both bivariate Elliptical Copulas and Archimedean Copulas to model the dependence structure in the SKE model. Here, our model has two equations, the economic growth and the inequality equations, $m = 2$, in which u_1 and u_2 are the marginal distributions of ε_1 and ε_2, respectively. Let $u_1 = F(\varepsilon_1)$ and $u_2 = G(\varepsilon_2)$. According to Sklar's Theorem (see Nelson [12]), the marginals can be inserted into any copula function C taken a form as $C(u_1, u_2) = C(F(\varepsilon_1), G(\varepsilon_2))$, where u_1 and u_2 are uniformly distributed and bounded on [0,1].

Elliptical Copulas. Two common elliptical copulas considered here are the Gaussian and Student's t. By considering the bivariate Gaussian Copula which is parameterized by the linear correlation coefficient ρ, the function form can be defined as

$$C^N(u_1, u_2 \mid \rho) = \Phi_\rho(\Phi_\rho^{-1}(u_1), \Phi_\rho^{-1}(u_2)) \tag{2}$$

where Φ_ρ is the bivariate Gaussian distribution function and Φ_ρ^{-1} is the inverse univariate Gaussian distribution function. Next, the bivariate Student's t distribution is expressed in terms of

$$C^T(u_1, u_2 \mid \kappa, v_c) = T_{\kappa, v}(T_{\kappa, v_c}^{-1}(u_1), T_{\kappa, v_c}^{-1}(u_2)) \tag{3}$$

where $T_{\kappa, v}$ is the bivariate Student's t distribution function and $T_{\kappa, v}^{-1}$ is the inverse univariate Student's t distribution function. Parameters v and κ are degree of freedom and dependence parameter, respectively.

Archimedean Copulas. Following Cherubini et al. [3], Archimedean copula can be defined as

$$C(u_1, u_2) = \Phi^{-1}(\Phi(u_1) + \Phi(u_2)) \tag{4}$$

where Φ is a strictly generator with Φ^{-1} completely monotonic on $[0, \infty)$. Let's consider the bivarite Archimedean class consisting of Clayton, Gumbel, Frank, and Joe copulas.

1. Clayton Copula. The Clayton copula is usually referred to in the bivariate case. It has an ability to capture lower tail dependence. The closed from of this copula is given by

$$C_\theta^{Cl}(u_1, u_2 : \theta) = (u_1^{-\theta} + u_2^{-\theta} - 1)^{-1/\theta} \tag{5}$$

where θ is a degree of dependence, $0 < \theta < \infty$. If $\theta \to \infty$ the Clayton Copula will converge to the monotonicity Copula with positive dependence, but if $\theta = 0$ then the marginal distributions become independence.

2. Gumbel Copula. The Gumbel copula is employed to model asymmetric dependence of marginals. It is used to capture an upper tail dependence and weak lower tail dependence. The form of bivariate Gumbel copula is given by

$$C_\theta^G(u_1, u_2 : \theta) = \exp(-[(-\log u_1)^\theta + (-\log u_2)^\theta]^{1/\theta}) \tag{6}$$

where the copula parameter θ is restricted on the interval $[1,\infty)$.

3. Frank Copula. The Copula function of Frank Copula can be defined by

$$C_\theta^F(u_1, u_2, \theta) = -\theta^{-1} \log\left\{1 + \frac{(e^{-\theta u_1} - 1)(e^{-\theta u_2} - 1)}{e^{-\theta} - 1}\right\} \tag{7}$$

where the copula parameter θ is restricted on the interval $(-\infty,\infty)$.

4. Joe Copula. The Joe copula is defined by

$$C_\theta^J(u_1, u_2, \theta) = 1 - ((1 - u_1)^\theta + (1 - u_2)^\theta)^{1/\theta} \tag{8}$$

where the copula parameter θ is restricted on the interval $[1,\infty)$.

2.3 Estimation Technique

As we construct two-equation SKE model to represent growth and inequality, the bivariate copula is considered as a joint for these two equations. Prior to the model estimation, we need to check the stationary of the data so that we use the Augmented Dickey-Fuller unit roots test. The estimation technique we employ here is the maximum likelihood estimation where the log likelihood function of our model can be defined by

$$\begin{aligned}
&\log L(y_1, Y_{-1}, y_2, Y_{-2}, X_1, X_2, u_1, u_2 | \Theta_1, \Theta_1, \theta) \\
&= \Sigma_{i=1}^T [(\log l(y_1, Y_{-1}, X_1 | \Theta_1) + \log(y_2, Y_{-2}, X_2 | \Theta_2) + \log c(u_1, u_2 | \theta)]
\end{aligned} \tag{9}$$

The terms $\Theta_1 = \left\{\alpha_1, \beta_{1i}^-, \beta_{1i}^+, r_{1i}, r_{1i}, \phi_{1j}^-, \phi_{1j}^+\right\}$, $\Theta_2 = \left\{\alpha_2, \beta_{2i}^-, \beta_{2i}^+, r_{2i}, r_{2i}, \phi_{2j}^-, \phi_{2j}^+\right\}$ and θ are the estimated parameters of the growth equation, the income inequality equation, and copula parameter, respectively. In this study, we consider four different marginal distributions namely Normal, Student's t, skewed Normal, and skewed Student's t distributions. Copula families, i.e. Gaussian, Student's t, Clayton, Gumbel, Frank, and Joe, are employed to join these two different marginals or to create a joint distribution function of a bivariate random variable with the univariate marginal distribution.

3 Simulation Study

We conduct a simulation study to explore the performance and accuracy of our proposed *Copula based SKE* model. The Monte Carlo simulation is employed to simulate dependence parameter for the Copulas. For the Elliptical copulas, we set the true value for the correlation coefficients ρ (Gaussian) and κ (Student's t) equal to 0.5 and the additional degree of freedom parameter v_c equal to 4. Then, for the Archimedean copulas, the dependence parameter of Frank and Joe copulas are set as $\theta = 2$ and dependence parameter of Clayton and Gumbel copulas are set as $\theta = 3$. Then, the obtained uniform data are transformed to $\varepsilon_{1,t}$ and $\varepsilon_{2,t}$ using quantile function of marginal distribution choices in which σ_1 and σ_2 are equal to 1, the additional degree of freedom parameter of quantile function of the Student's t distribution v is equal to 6 and the skewed parameter for skewed distribution is set as $\gamma = 1$. In this section, we are not going to examine all subsets of the cases in this simulation study; instead we are especially considering three comprehensive cases shown as follows:

Case 1 :   Copula based SKE with Normal and Student's t margins
Case 2 :   based SKE with Skewed Student's t and Student's t margins
Case 3 :   Copula based SKE with Normal and Skewed Normal margins.

Thus, the simulation model takes the following form:

$$y_{1,t} = \alpha_1 + \phi_{11}^-(X_{1,t} - r_{11})_- + \phi_{11}^+(X_{1,t} - r_{11})_+ + \sigma_1\varepsilon_{1,t}$$
$$y_{2,t} = \alpha_2 + \beta_{21}^-(y_{1,t} - r_{21})_- + \beta_{21}^+(y_{1,t} - r_{21})_+ + \sigma_2\varepsilon_{2,t}. \tag{10}$$

Prior to generate a dependent variable $y_{1,t}$, we randomly simulate an independent variable $X_{1,t}$ from $N(15,10)$ and set the kink point r_{11} equal to 15, the true value for α_1 equal to 1. Moreover, the values for coefficients ϕ_{11}^- and ϕ_{11}^+ are set to be -0.5 and 2, respectively. Similarly, to simulate $y_{2,t}$ in the second equation, first, we have to generate the value of its independent variable, $y_{1,t}$. However, as we are dealing with the simultaneous equations model, the dependent variable of the first equation also plays a role as an independent variable for the second equation at which $y_{1,t}$ has been simulated already from the first equation. In addition, we set the value for the kink point r_{21} equal to 20, the true value for α_2 equal to 1, and for coefficients ϕ_{21}^- and ϕ_{21}^+ equal to 0.1 and -2, respectively.

To save the space, we decide to show the simulation result only for the first case that is the model with Normal and Student's t margins. Table 1 shows the results of the Monte Carlo simulation investigating the maximum likelihood estimation of the Copula based SKE model. We found that our model can perform very well through simulation study. The mean parameters are very close to their true values with low and acceptable standard errors. For example, the mean value of the coefficient α_1 is 0.90 with standard error equals 0.13 while the true value is 1. This result demonstrates the accuracy of estimation and the same for the other cases. Overall, the Monte Carlo simulation suggests that our model 'Copula based SKE' is reasonably accurate.

Table 1. Normal and student's t margins

Copula		Gaussian		Student-t		Joe	
Parameter	True	Estimates	S.E.	Estimates	S.E.	Estimates	S.E.
α_1	1	0.90	0.13	0.94	0.12	1.01	0.11
ϕ_{11}^+	2	2.00	0.01	1.99	0.01	1.99	0.01
ϕ_{11}^-	-0.5	-0.48	0.02	-0.50	0.02	-0.52	0.01
σ_1	1	0.99	0.50	0.86	0.04	0.97	0.05
α_2	1	0.97	0.14	0.78	0.14	1.07	0.13
β_{21}^+	-2	-1.98	0.01	-.99	0.01	-1.97	0.01
β_{21}^-	-0.1	-0.10	0.01	-0.08	0.01	-0.09	0.01
σ_2	1	1.08	0.10	0.86	0.05	0.97	0.07
v_2	6	4.54	1.25	10.95	6.67	5.06	1.30
r_{11}	20	19.92	0.09	20.01	0.08	20.02	0.07
r_{21}	15	15.11	0.09	15.21	0.10	15.07	0.08
ρ	0.5	0.54	0.05	–	–	–	–
κ	0.5	–	–	0.35	0.06	–	–
θ	2	–	–	–	–	2.05	0.18
Copula		Clayton		Gumbel		Frank	
Parameter	True	Estimates	S.E.	Estimates	S.E.	Estimates	S.E.
α_1	1	1.03	0.08	1.05	0.10	1.06	0.12
ϕ_{11}^+	2	2.00	0.01	2.01	0.01	2.00	0.01
ϕ_{11}^-	-0.5	-0.50	0.01	-0.49	0.01	-0.51	0.01
σ_1	1	0.97	0.04	1.12	0.06	0.91	0.05
α_2	1	0.90	0.10	1.17	0.09	1.11	0.12
β_{21}^+	-2	-2.01	0.01	-2.00	0.01	-2.00	0.01
β_{21}^-	-0.1	-0.08	0.01	-0.11	0.01	-0.12	0.01
σ_2	1	0.96	0.06	1.67	0.49	1.14	0.11
v_2	6	5.92	1.49	5.57	0.45	4.47	1.44
r_{11}	20	20.01	0.04	19.94	0.05	20.05	0.08
r_{21}	15	15.05	0.06	14.94	0.04	14.91	0.03
$\theta(\theta_{Frank})$	3(2)	3.02	0.35	2.61	0.32	1.97	0.47

Source: Calculation.

4   Model Specification and Variables

To test a relationship between income inequality and economic growth, we use the following two equations which are modeled simultaneously. The first equation represents the economic growth and the second equation represents the income inequality which is measured by the best known index Gini coefficient[1]. A key point of this model

---

[1] The Gini coefficient is a number on a scale measured from 0 to 1, where 0 represents perfect equality and 1 represents total inequality [15].

is that the growth equation contains Gini coefficient and the GDP growth is also entered into the Gini equation to find a direct link between these two factors.

At this stage, we construct this model and decide which economic variables should be included by combining a lot of information from expert advices and literatures. The first part of the growth equation includes change in nominal GDP lagged by one period and *structural characteristic* (Str_t), where $\ln(Str_t) = \ln(S_t) - \ln(n_t + 0.75)$. The variables capturing structural characteristics were introduced by Sørensen and Whitta-Jacobsen in 2010 to include S_t, the country's average gross investment rate and n_t, the growth rate of its average labor force [16]. In brief, they suggest this term is basically needed because growth has its own long run path which is defined by the basic structural characteristics of the country. The next variable is *human capital accumulation* (HC_t) in terms of education and training, health status, and experience which is found to have greatly positive impact on economic growth and induce sustainable growth. Also, *institution* should be considered as a cause of the long-run growth [1]. The institution in a society is commonly related to such social mechanisms as democracy, religion, property rights, and rule of law. It determines the incentives and creates constraints in the economy. The last included variable in the growth equation is change in *government debt* $(\Delta Debt)$. We consider this term because the literature shows that debt also matters for economic growth.

$$\begin{aligned} \Delta GDP_t = {} & \alpha_1 + \beta_{11}\Delta GDP_{t-1} + \beta_{12}\ln(Str_t) + \beta_{13}\ln(HC_t) + \beta_{14}\ln(Inst_t) \\ & + \beta_{15}Gini_t + \beta_{16}\Delta Debt_t + u_1 \end{aligned} \quad (11)$$

$$\begin{aligned} Gini_t = {} & \alpha_2 + \beta_{21}\Delta GDP_t + \beta_{22}\ln(GDP_t^{pc}) + \beta_{23}\ln(PS_t) + \beta_{24}\ln(Inf_t) \\ & + \beta_{25}\Delta Edu_t + \beta_{26}\Delta Unemp_t + u_2 \end{aligned} \quad (12)$$

Next, we will discuss about potential sources influencing income inequality. It is not easy to choose which factors should be included in the Gini equation because inequality is somewhat complicated and unclear to define. However, following the study of Kaasa [8], we are able to classify the factors affecting inequality into four groups as follows. The first group is economic growth and general development of a country where we select the growth in GDP (ΔGDP_t) and *GDP per capita* (GDP_t^{pc}) to represent this group and to describe a particular country's development level. The second group refers to the political factors which are defined as privatization and the public sector [8]. Privatization is believed to cause wealth concentration which finally leads to income inequality. Therefore, the bigger *share of private sector* may imply the higher income inequality, thereby the variable PS_t. The third group is macroeconomic factors. We especially consider *inflation rate* (Inf_t) and *the number of unemployed people* $(Unemp_t)$ to represent this group. The last group is demographic factors. We focus especially on some part of demographic development that is education because education is greatly important for reducing uneven income [5], thereby a variable *expenditure on education* (Edu_t).

5 Exploring the Link Between Growth and Inequality Using Thailand's Data

First of all, we'd better explain about data used in this study. We have a quarterly data set of all considered variables, spanning from 1993:Q1 to 2015:Q4. The economic growth data comprises GDP, a structural characteristics, human capital index, institution index, and government debt. In addition, the income inequality data consists of the Gini coefficient, GDP per capita, private sector share of GDP, inflation rate, government expenditure on education, and rate of unemployed labor. We collected all the data from Thomson Reuters DataStream, from Financial Investment Center (FIC), Faculty of Economics, Chiang Mai University, except the Gini coefficients which were collected from the National Statistical Office of Thailand. Additionally, we apply the Augmented Dickey-Fuller (ADF) unit roots test to the subsets of data to examine whether or not the data series contains a unit root, before we estimate the model. We do not show the result here but we found that all variables passed the test at level with probability equal to zero, meaning all of them are stationary.

5.1 Testing for a Kink (Threshold) Effect

This section is conducted to explore whether or not the kink (or threshold) exists with respect to our model. We employ the *Likelihood ratio test (LR-test)* to test the kink effect following a recommended algorithm of Hansen [7]. The kink effect on a relationship between response variable, y, and its covariate, x_i, is examined as a single equation. This algorithm is kept using for each pair of the covariate and response variable. We assume the null hypothesis is linear regression and the alternative is kink regression. The results are shown in Table 2.

Table 2. Results of likelihood ratio test

	Covariate					
Growth equation ΔGDP_t	ΔGDP_{t-1}	$\ln(Str_t)$	$\ln(HC_t)$	$\ln(Inst_t)$	$Gini_t$	$\Delta Debt_t$
	0.067	0.009	1.538	0.765	0.008	0.016
Inequality equation $Gini_t$	ΔGDP_t	$\ln(GDP_t^{pc})$	$\ln(PS_t)$	$\ln(Inf_t)$	ΔEdu_t	$\Delta Unemp_t$
	20.96***	1.054	0.814	27.91***	0.302	1.098

Source: Calculation.
Note: "***" denote that the null hypothesis of no regime shift is rejected at the 1 % significance level.

Assuming the null hypothesis is true, Table 2 shows that only the covariates ΔGDP_t and $\ln(Inf_t)$ reject the null hypothesis of linearity at the 1 % level, and hence their outcomes are said to be in favor of the kink regression. The result implies that the kink test is conclusive regarding the question of whether or not there is a regression kink effect on the economic growth and income inequality due to these covariates.

Therefore, the SKE model with specified kink effect can take the form as in the equations that follow. This model is set similarly to Eqs. (11) and (12) but the slope with respect to the GDP growth and inflation are discontinuous since they have a kink at $\Delta GDP = r_1$ and $\ln(Inf_t) = r_2$ where we treat the parameters r_1 and r_2 as the kink points which need to be estimated.

$$
\begin{aligned}
\Delta GDP_t = {} & \alpha_1 + \beta_{11}\Delta GDP_{t-1} + \beta_{12}\ln(Str_t) + \beta_{13}\ln(HC_t) + \beta_{14}\ln(Inst_t) \\
& + \beta_{15}Gini_t + \beta_{16}\Delta Debt_t + \sigma_1\varepsilon_{1t}
\end{aligned}
\tag{13}
$$

$$
\begin{aligned}
Gini_t = {} & \alpha_2 + \phi_{21}^-(\Delta GDP_t - r_1)_- + \phi_{21}^+(\Delta GDP_t - r_1)_+ + \beta_{22}\ln(GDP_t^{pc}) \\
& + \beta_{23}\ln(PS_t) + \phi_{24}^-(\ln(Inf_t) - r_2)_- + \phi_{24}^+(\ln(Inf_t) - r_2)_+ + \beta_{25}\Delta Edu_t \\
& + \beta_{26}\Delta Unemp_t + \sigma_2\varepsilon_{2t}
\end{aligned}
\tag{14}
$$

5.2   Selecting Copulas for SKE Model

This part is about selecting a Copula that is best-fit for the data. Given a set of Copula families, we then select the Copula using the Akaike Information Criterion (AIC) and Bayesian Information Criterion (BIC). Again, this paper is interested in some well-known families of Copula, i.e. Gaussian, Student's t, Clayton, Gumbel, Frank, and Joe, as described in Sect. 5.2, and it assumes four different distributions for the marginals, namely Normal, Student's t, skewed Normal, and skewed Student's t distributions. Therefore, we have got sixteen cases for a pair of marginal distributions of the growth and inequality equations, respectively.

According to the results shown in Table 3, the minimum values of AIC and BIC are -796.8 and -741.3 (bold numbers), respectively, from which Gaussian Copula is chosen to be a linkage between normal margin of the growth equation and Student's t margin of the inequality equation.

5.3   Estimates of Copula Based SKE Model

To explore the relationship between growth and income inequality in Thailand, Eqs. (13) and (14) are estimated by MLE and the results are presented in Table 4. In general, the model can perform well across the data sets and it can capture a nonlinear effect of some variables. Most of the parameters are rightly signed and statistically significant at the conventional levels. The estimation of the growth equation, shown at the beginning of Table 4, reveals that only some variables suggested by the empirical growth literatures, are significant in the case of Thailand. It is found that Thailand's economic growth depends significantly on the structural characteristics and the Gini coefficient, but the impact of other variables on the economic growth are not significantly found.

Table 3. AIC and BIC criteria for model choice

Marginal distribution	Class of Copula					
	Gaussian	Student's t	Clayton	Gumbel	Frank	Joe
[1] Normal/Normal	−785.9	−779.6	−700.4	−234.2	−787.0	−750.1
	−732.9	−724.1	−647.4	−181.2	−734.1	−697.2
[2] Normal/Student's t	**−796.8**	−787.4	−682.6	−759.6	−781.5	−592.9
	−741.3	−729.4	−627.1	−704.1	−726.0	−537.4
[3] Normal/Skewed Normal	−518.9	−637.7	−248.7	−333.9	−572.9	−424.2
	−463.4	−579.7	−193.2	−278.5	−522.5	−368.7
[4] Normal/Skewed Student's t	3,030.8	1,978.9	2,374.5	2,927.5	3,203.3	3,269.5
	3,088.8	2,039.5	2,432.5	2,985.5	3,261.3	3,327.5
[5] Skewed Normal/Normal	−196.9	−474.6	−358.3	−153.8	−442.8	−320.6
	−141.4	−416.6	−302.8	−98.3	−387.3	−265.2
[6] Skewed Normal/Student's t	−336.3	−691.9	−480.7	−439.4	−535.3	−493.7
	−278.3	−631.4	−422.7	−381.4	−477.3	−435.7
[7] Skewed Normal/Skewed Normal	−541.6	−724.5	−347.6	2,610.4	−481.4	2,627.2
	−483.6	−664.0	−289.6	2,668.4	−423.4	2,685.2
[8] Skewed Normal/Skewed Student's t	−496.2	1,631.6	1,352.1	−26.0	−594.7	−495.5
	−435.7	1,694.7	1,412.6	34.5	−534.2	−435.0
[9] Student's t/Normal	−608.8	−358.1	−504.0	−753.0	−281.5	−424.2
	−555.8	−305.2	−451.1	−700.1	−226.0	−368.7
[10] Student's t/Student's t	−699.1	−776.7	−659.2	−634.8	−488.8	−443.3
	−641.1	−716.2	−601.2	−576.8	−430.8	−385.3
[11] Student's t/Skewed Normal	−565.8	−772.9	−724.1	1,304.7	−337.5	130.6
	−507.8	−712.4	−666.1	1,362.7	−279.5	188.6
[12] Student's t/Skewed Student's t	−762.5	−753.1	−517.9	−718.6	−750.6	−587.1
	−702.0	−690.0	−457.4	−658.1	−690.1	−526.6
[13] Skewed Student's t/Normal	−252.0	−436.4	−707.4	−176.4	−257.1	−238.1
	−194.0	−375.9	−649.4	−118.4	−199.1	−180.1
[14] Skewed Student's t/Student's t	−525.1	−717.0	−578.5	−667.6	−637.5	−347.4
	−464.6	−654.0	−518.0	−607.0	−576.9	−286.9
[15] Skewed Student's t/Skewed Norma	5,164.7	4,998.3	6,316.7	1,354.5	4,665.2	3,734.9
	5,225.2	5,061.3	6,377.2	1,415.0	4,725.7	3,795.4
[16] Skewed Student's t/Skewed Student's t	6,714.1	4,969.0	4,428.8	6,054.3	4,103.3	−384.5
	6,777.1	5,034.5	4,491.8	6,117.3	4,166.3	−321.5

The structural characteristic is an essential economic variable and needed for the economy to preserve an existence of convergence. Economists believe that the growth rates have to converge to some steady state equilibrium growth path, however, the convergence will occur due to some important condition that is a negative sign of a coefficient of the characteristics variable in order to let the conditional convergence occur [16]. Our result provides numerical evidence consistent with this condition, the coefficient is −0.0004 meaning that Thailand's GDP growth converges significantly to a country-specific long run growth path with speed 0.0004. Importantly, we also found a negative relationship between growth and the Gini coefficient. The coefficient is equal to −0.1147, meaning that an increase in the income inequality − measured by Gini coefficient- leads to GDP growth decline in Thailand. In contrast, if we could reduce the Gini coefficient just by 1 %, Thailand's economic growth could be boosted over 0.11 %.

As also reported in Table 4, income inequality in Thailand depends significantly on GDP growth, inflation, and unemployment. We found that GDP growth has both negative and positive impacts on income inequality, as well as the impacts of inflation. This happens due to the kink effect and we will discuss completely about this effect later. Unemployment is found to be positively correlated with income inequality. We found that the coefficient of unemployment is 0.0106, meaning that an additional 1 % of the growth rate of unemployed worker leads to 0.01 % increase in Gini coefficient. This estimate result confirms the previous works and also the empirical literatures as reported in Sect. 2. However, we failed to find significance for other important variables across this data set, such as GDP per capita, share of private sector, and expenditure on education.

A specific capability of our method allows us to preserve the nonlinear impacts of GDP growth and inflation on income inequality. The impacts are split into two groups based on kink points. In Fig. 1 (a) we display the relationship between Gini coefficient and GDP growth through a regression line, on which the vertical axis is GDP growth and the horizontal axis is income inequality measured by the Gini coefficient. We can see that the regression shows a small positive slope for low GDP growth with a significant kink around 0.0074 (0.74 %), displayed as the blue dot, switching to a negative slope for GDP growth beyond that kink value. This means GDP growth displays a positive impact on the Gini coefficient when the growth is below 0.74 % per quarter. The coefficient of this positive part is 0.0501, meaning that an additional 1 % of GDP growth leads to 0.05 % increase in Gini coefficient. On the other hand, the same Fig. 1 (a) also shows a negative impact on the Gini coefficient when the growth is over 0.74 % per quarter. The coefficient of GDP growth in the negative trend is 0.0206, meaning that an increase in GDP growth by 1 % can cause a reduction in Gini coefficient by 0.02 %.

In Fig. 1 (b) we illustrate the relationship between Gini coefficient and inflation through another regression line. We can see that the regression shows a steeply positive slope for low inflation with a significant kink around 3.2419 (3.24 %), displayed as the blue dot, switching to a smaller positive slope for the inflation beyond that kink value. The result observes the positive relationship between inflation and income inequality, in which the higher inflation the more income inequality. However, our work can do more than that; we find a salient result that the impacts of inflation on inequality are split into two levels based on the kink point. In the first stage, at any level of inflation below 3.24 %, an additional 1 % of inflation leads to 0.013 % increase in Gini coefficient. But in the second stage, the effect is much less than the first stage. We found that an additional 1 % of inflation just leads to 0.005 % (0.0047) increase in Gini coefficient. This result is consistent with the finding of Crowe [4], CEP researcher and

Table 4. Estimation results of Copula based SKE model

Variables	Estimated value	S.E. ($\times 10^{-2}$)	Confident interval	
			2.5 %	97.5 %
Growth				
Intercept	0.0524	(2.15)***	0.0094	0.0955
GDP growth lagged by one period	0.0512	(10.11)	−0.1511	0.2536
Structural Characteristic	**−0.0004**	(0.01)***	−0.0008	−0.0001
Human Capital Accumulation	0.0015	(0.44)	−0.0073	0.0104
Institution	0.0009	(0.20)	−0.0031	0.0049
Gini Coefficient	**−0.1147**	(1.37)***	−0.1422	−0.0872
Government debt	0.0007	(0.28)	−0.0051	0.0064
Sigma (σ_1)	0.0210	(0.15)***	0.0179	0.0240
Income Inequality				
Intercept	0.4059	(0.41)***	0.3977	0.4142
GDP growth (positive part)	**0.0501**	(2.21)***	−0.2412	0.3414
GDP growth (negative part)	**−0.0206**	(0.51)***	−0.1605	0.1193
GDP per capita	0.0001	(0.05)	−0.0008	0.0011
Share of Private Sector	−0.0114	(27.69)	−0.5653	0.5424
Inflation (positive part)	**0.0131**	(0.53)***	−0.1295	0.1557
Inflation (negative part)	**0.0047**	(0.13)***	−0.0659	0.0753
Expenditure on Education	−0.0519	(4.94)	−0.1508	0.0468
Unemployment	**0.0106**	(0.63)*	−0.0019	0.0233
Kink point in GDP growth (r_1)	0.0074	(2.81)***	−0.0486	0.0635
Kink point in inflation (r_2)	3.2419	(25.12)***	2.7394	3.7445
Sigma (σ_2)	0.0697	(8.34)	1.5408	0.2367
Degree of freedom (v_2)	2.1068	(27.07)	1.5408	2.6729
Joint Estimate				
Dependency (ρ)	0.2409	(27.07)*	−0.3005	0.7824

Source: Calculation.

Note: "*," "**," and "***" denote rejections of the null hypothesis at the 10 %, 5 %, and 1 % significance levels, respectively.

IMF's officer who first mentions about positive relationship between inflation and inequality that should be separated into two levels.

But, from our opinions, we think it is true that the greater inflation may cause the higher income inequality; but what we suggest here is to use the proper level of inflation to reduce income inequality in the case of Thailand. Our experiment found that in the first stage, under the kink value 3.24 %, the elasticity of the Gini coefficient with respect to the inflation is low; therefore a small decrease in the inflation rate results in a large decrease in the Gini coefficient. In contrast, the elasticity is much higher in the second stage, meaning that even the central bank provides a big decrease in the inflation rate; we can observe just a minor decrease in the Gini coefficient. Hence, it is reasonable to preserve the inflation rate below that kink value.

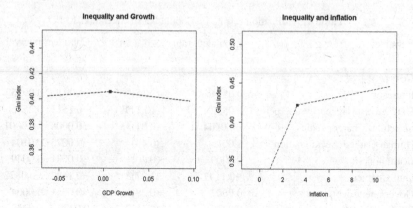

Fig. 1. Plot of Gini coefficient and its covariates estimated by Copula based Kink regression (Color figure online)

6 Conclusion

This study attempts to explore the relationship between income inequality and economic growth with special focus on Thailand. We propose Copula based simultaneous kink equations (SKE) model to investigate this relationship. The estimate results show that Thailand's economic growth depends significantly on the structural characteristics and the Gini coefficient. For income inequality, it is found that the Gini coefficient depends significantly on GDP growth, inflation, and unemployment, in which the first two variables create the kink effects on the Gini coefficient.

Let us explain about our considerable issue, income inequality and growth. We found that the Gini coefficient creates a negative impact on the growth; on the other hand, the growth can create both negative and positive impacts on the Gini coefficient due to the kink effect. This result is corresponding to the Kuznets Hypothesis finding the level of growth that could reduce income inequality. From this data set, we experimentally found that the growth beyond the kink value, 0.74 % per quarter, can reduce income inequality in Thailand. But more than that, we also discover an alternative way to reduce income inequality that is the inflation. Our study found that at any level of inflation under the kink value 3.24 %, a small decrease in the inflation rate results in a large decrease in the Gini coefficient. Therefore, it is reasonable to preserve the inflation rate below that kink value.

References

1. Acemoglu, D., Johnson, S., Robinson, J.A.: Institutions as a fundamental cause of long-run growth. In: Handbook of economic growth, vol. 1, pp. 385–472 (2005)
2. Boonyamanond, S.: Can equality and growth be simultaneously achieved? Chulalongkorn J. Econ. **19**(2), 135–160 (2007)

3. Cherubini, U., Luciano, E., Vecchiato, W.: Copula Methods in Finance. Wiley, Chichester (2004)
4. Crowe, C.W.: Inflation, Inequality, and Social Conflict (No. 6-158). International Monetary Fund (2006)
5. Dabla-Norris, M.E., Kochhar, M.K., Suphaphiphat, M.N., Ricka, M.F., Tsounta, E.: Causes and consequences of income inequality: a global perspective. International Monetary Fund (2015)
6. Gallup, J.L.: Is there a Kuznets curve. Portland State University (2012)
7. Hansen, B.E.: Regression kink with an unknown threshold. J. Bus. Econ. Stat. (just-accepted) (2015). doi:10.1080/07350015.2015.1073595
8. Kaasa, A.: Factors influencing income inequality in transition economies. University of Tartu Economics and Business Administration Working Paper Series (18) (2003)
9. Kuznets, S.: Economic growth and income inequality. Am. Econ. Rev. **45**(1), 1–28 (1955)
10. Lazear, E., Rosen, S.: nRank order tournaments as optimal salary schemeso. J. Polit. Econ. **89**(5), 841 (1981)
11. Mahfoud, M., Michael, M.: Bivariate Archimedean copulas: an application to two stock market indices. BMI Paper (2012)
12. Nelson, R.B.: An Introduction to Copulas, vol. 139. Springer Science & Business Media, New York (2013)
13. Ostry, J.D., Berg, A.: Inequality and unsustainable growth: two sides of the same coin? (No. 11/08). International Monetary Fund (2011)
14. Pastpipatkul, P., Maneejuk, P., Wiboonpongse, A., Sriboonchitta, S.: Seemingly unrelated regression based copula: an application on thai rice market. In: Huynh, V.-N., Kreinovich, V., Sriboonchitta, S. (eds.) Causal Inference in Econometrics, pp. 437–450. Springer International Publishing, Switzerland (2016)
15. Ray, D.: Development Economics. Princeton University Press, Princeton (1998)
16. Sørensen, P.B., Whitta-Jacobsen, H.J.: Introducing Advanced Macroeconomics: Growth and Business Cycles. McGraw-Hill higher education, New York (2010)
17. Trivedi, P.K., Zimmer, D.M.: Copula Modeling: An Introduction for Practitioners. Now Publishers Inc., Hanover (2007)
18. Wichitaksorn, N., Choy, S.T.B., Gerlach, R.: Estimation of bivariate copula-based seemingly unrelated Tobit models. Discipline of Business Analytics, University of Sydney Business School. NSW (2006)

Thailand's Export and ASEAN Economic Integration: A Gravity Model with State Space Approach

Pathairat Pastpipatkul[✉], Petchaluck Boonyakunakorn, and Songsak Sriboonchitta

Faculty of Economics, Chiang Mai University, Chiang Mai, Thailand
ppthairat@hotmail.com, petchaluck_boon@cmu.ac.th

Abstract. The purpose of this paper is to study bilateral behavior of trading partners of Thailand in AEC and capture advantageous characteristics of two-sided trades between them. Gravity model is commonly used for investigating bilateral trade. However, log linear transformation of gravity model could cause bias resulting in heteroscedasticity. Furthermore, one of the assumptions of OLS or panel estimators has to be homoscedasticity which leads to the inconsistency of parameter estimator. In the gravity model, coefficients are also assumed to be constant over time meanwhile it tends to be volatility in Thai export earnings. Hence constant coefficient of gravity model might not be adequate in explaining the behavior of Thailand's export. As an alternative approach, state space approach with gravity model tends to be more appropriate. It does not only need no homogeneous assumption, but also determines the changes in the coefficients over time. Therefore the estimation from state space approach capture more characteristics of trade in each country and also provides more accurate parameter estimators.

Keywords: Gravity model · State space model · Bilateral trade · AEC

1 Introduction

The Thai economy significantly depends on exports which accounts for more than two-thirds of Thailand's GDP. One of its most export markets is ASEAN (Association of Southeast Asian Nations). At present Thai exports to ASEAN countries account for approximately 25 % of total exports. Sustainable cooperation with other countries in order to expand the potential of Thai exports is needed. The most important objective of ASEAN is to promote economic and political cooperation among its 10 members, Brunei, Cambodia, Indonesia, Laos, Malaysia, Myanmar, Philippines, Singapore, Thailand and Vietnam. In addition, there is an establishment of the ASEAN Economic Community (AEC) in 2015. The principle purpose of the AEC is to transform ASEAN into a competitive region, a single market with free movement of goods, investment, service, skilled labor, and capital [10]. For Thailand, AEC can bring many benefits since it can enhance exports. Moreover, it also provides Thai investors with competitive advantages in import of raw materials, components, and other production inputs by reducing and eliminating trade barriers [12].

© Springer International Publishing AG 2016
V.-N. Huynh et al. (Eds.): IUKM 2016, LNAI 9978, pp. 664–674, 2016.
DOI: 10.1007/978-3-319-49046-5_56

Thailand's export mainly depends on global economic conditions. Its exports to all AEC members have fluctuated over the last 25 years. During this period there are two major global crises which produced enormous impact on the export. The first is the Asian Crisis during 1997–1998 when Thailand lost value in exports and struggled in serious debt crises. Another cause was the slowdown in global demand and the collapse of global trade in 2008 which brought about a fall in Thai exports. Then there is sustainable resilience in Thailand's exports. Nevertheless, there seems to be changing point around in 2011 or 2012. After that Thailand's exports have fluctuated again with a downward trend until now. The following Fig. 1 shows the dramatic fall in the volume of Thai exports to the AEC members during the mentioned periods with the fluctuating trend.

THAILAND'S EXPORTS

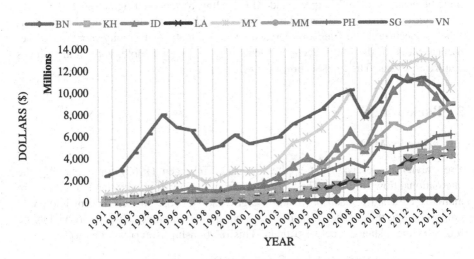

Fig. 1. Thailand's exports to AEC countries during the period 1991–2015

To maintain the competitive abilities in international trade, a study of the characteristics of the economic factors in international trade between the members becomes crucial and calls for more attention to be investigated with econometric approaches. Gravity model has been becoming increasingly popular and has also been applied worldwide for investigating the relationships in international trade. Therefore in this paper we apply gravity model to study the relationship between Thailand exports and other AEC member. However, gravity model needs log- linear process for transformation and the process could produce biased estimations resulting in heteroscedasticity [14]. In OLS or panel estimators, it is assumed that variance of the error term is constant. When there is presence of heteroscedasticity, it leads to biased parameter estimator. In addition, there are differences in individual unobserved factors, which account for individual country heterogeneity. Thus it will cause heteroscedasticity [16]. Serlenga and Shin [13] has also found that ignoring the unobserved country heterogeneity can lead to inconsistent estimates of bilateral trade relationships.

Furthermore, conventional approach in the gravity model assumes parameters to be constant over time. While there is volatility in export earnings including covariates namely GDP, population. Hence a constant coefficient of gravity model may not be a good explanation of the behavior of Thailand's exports. For all the aforementioned reasons, state space approach with gravity model might be more appropriate than other approaches because it is able to determine the changes in the coefficients over time and has no homogeneity assumption. It also has an advantage over classical regression in terms of better fit to the data because it considers the time dependence between the observations of a time series. Consequently, residuals are much closer to independent random values than in classical linear regression [3]. This helps capture better the characteristics of trade in each country and also provides more accurate parameter estimators. Hence we employ the state space approach with gravity model.

This paper addresses capturing the description of the relationship in international trade between Thailand's exports and AEC trading partners during period 1991–2015. It provides more realistic interpretation in trade. The paper is divided into 5 sections as follows. Section 2 provides an overview of an estimation of gravity model as a theoretical basis for the study. Section 3 demonstrates the methodology. Section 4 shows empirical results. The final section is conclusion.

2   Background

2.1   Gravity Model

The gravity model was first applied by Jan Tinbergen (1962) for explaining the international flows of trade. The idea comes from the law of gravity that two bodies are mutually attracted to each other and the gravitational force between them is directly proportional to the product of their masses m_1 and m_2 and inversely proportional to the square of the distance r between them. This relationship is expressed as follows:

$$F = G\frac{m_1 m_2}{r^2} \tag{1}$$

With the law of gravity equation, it is adapted for investigating bilateral trade. According to Eq. 1, the concept of bilateral trade can be described as attraction of trade flows or exports from country i to country j in place of gravitational force (F). The value of trade between two countries is proportional to the product of the two countries of their economic size and inversely proportional to their distance. Anderson and Van Wincoop [2] proposed the basic equation of international trade between countries i and j.

$$Trade_{ij} = G\frac{GDP_i GDP_j}{DISTANCE_{ij}}, \tag{2}$$

where $Trade_{ij}$ is trade flow between country i and j, GDP_i and GDP_j are gross domestic products which refer to economic sizes of country. $DISTANCE_{ij}$ is the distance between the trading countries and G is an empirical physical constant gravitational effects.

A log-linear form is used in the gravity model in order to estimate the parameters, and the corresponding estimable equation, is as follows:

$$\ln Trade_{ij} = \ln G + \beta_1 \ln GDP_i + \beta_2 \ln GDP_j + \beta_3 \ln DISTANCE_{ij}, \quad (3)$$

where $\ln G$ is the intercept, β_1, β_2 and β_3 are coefficients (elasticities) to be estimated. These coefficients can determine which these factors influence the volume of bilateral trade. Moreover, mass in Eq. 1 can be associated with both GDP and population (POP). The population is interpreted in terms of market size. The gravity trade equation becomes:

$$\ln Trade_{ij} = \ln G + \beta_1 \ln GDP_i + \beta_2 \ln GDP_j + \beta_3 POP_i + \beta_4 POP_j + \\ \beta_5 \ln DISTANCE_{ij} \quad (4)$$

As regards correlation, the expected coefficient of bilateral trade is predicted to be a positive with national income, population and negative with distance. Then many researchers employ gravity model to international trade flows to explore the many issues in regional and location economics. Furthermore, gravity model has added more explanatory variables such as border, culture, etc. in order to better capture the factors that affect bilateral trade.

3 Methodology

3.1 State Space Model and Applications

State Space model (SSM) estimates time varying coefficients through the Kalman filter. It was developed for engineering field in order to track or control objects which are moving over time. Then state space model has been adopted in numerous fields including economics in order to deal with time-varying parameter model. According to economic structure, several macroeconomic variables have structural change, nonlinearities which need time-varying parameter model. It will extend a conventional approach to capture time varying structure of such linear model as Markov regime switching model and stochastic volatility models. Especially in finance fields, Kalman filter has been widely employed. Adam and Gyamfi [1] used it as a method to study the time-varying of African stock market integration with the global stock market. It is also used in the analysis of Galati et al. [6] in order to capture the behavior of financial cycles varying over time.

3.2 Model for Longitudinal Data

In this paper we apply log-linear gravity model with time-varying coefficients in order to estimate the varying relationship of bilateral trade. Meanwhile, gravity model is based on longitudinal data so we employ gravity model with longitudinal data.

Now suppose $Y_{i,t}$ denote a $N \times 1$ multivariate time series which is evolving over time and which can be characterized in terms of unobservable state vector with dimension $m \times 1$. In order to investigate the varying relationship across time, we need a joint model for the variate process. After that we apply Kalman filter to set up a time-varying parameter model, where the variance of $V_{i,t}$ is depending on time. The Kalman filter is a case of the general state space model consisting of two equations. The first equation is called the measurement or observation equation which describes the vector of observation $Y_{i,t}$ through signal $\theta_t^{(i)}$.

$$Y_{1,t} = \theta_t^{(i)} X + V_{i,t}, \quad V_{i,t} \sim N(0, V_i) \tag{5}$$

where $i = 1, \ldots, n$, $Y_{i,t}$ is dependent variable which is measurement of $\theta_t^{(i)}$, $V_{i,t}$ is a vector of disturbances, X is exogenous regressor, $\theta_t^{(i)}$ is unobservable state vector which is varying coefficients in economic aspect. Equation (4) is called the transition or state equation. It defines the evolution of the state space over time by using the first order Markov structure. It describes the transition of state variables from time t to time $t + 1$, where is G unknown. The variances might be different but with the same time-invariant matrices G and X; that is

$$\theta_t^{(i)} = G\theta_{t-1}^{(i)} + W_t^{(i)}, \quad W_t^{(i)} \sim N(0, W_i) \tag{6}$$

In order to explore the varying correlation between variable Y and X, the dynamic linear model with longitudinal data can be expressed as below

$$Y_{1,t} = \alpha_{i,t} + \beta_{i,t} X_t + V_{i,t}, \quad V_{i,t} \sim N(0, \sigma_i^2) \tag{7}$$

$$\alpha_{i,t} = G_1 \alpha_{i,t-1} + W_{1t,i}, \quad W_{1t,i} \sim N\left(0, \sigma_{w_{1,i}}^2\right) \tag{8}$$

$$\beta_{i,t} = G_2 \beta_{i,t-1} + W_{2t,i}, \quad W_{2t,i} \sim N\left(0, \sigma_{w_{2,i}}^2\right) \tag{9}$$

The observed variable $Y_{i,t}$ merely depend on both its parameter $\eta_i = \left(\sigma_i^2, \sigma_{w_{1,i}}^2, \sigma_{w_{2,i}}^2\right)$ and its state vector $\theta_t^{(i)} = (\alpha_{i,t}, \beta_{i,t})$. Supposing that the variable $Y_{i,t}$ is conditionally independent given the state processes $\left(\theta_t^{(i)}, \ldots, \theta_t^{(m)}\right)$ and the parameters $\eta_1, \ldots, \eta_m$, a dependence among variable $Y_{1,t}$ can be estimated by the joint probability of state processes and of the parameters [11].

In order to estimate the coefficients of the model, a recursive function of maximum likelihood algorithm is applied. The first step is to use the Kalman filter to obtain filtered estimates of the coefficients. The fact is that it is a minimum mean-squared error estimator aiming to improve the accuracy of the present estimation. The second is to obtain smooth coefficient estimates by using the information from the entire measurement [4].

3.3   Kalman Filter and Gravity Model

In conventional gravity model approach, it is assumed that coefficient parameters are constant whereas the relationship between Thailand's exports and various macroeconomic variables namely GDP evolves over time. Therefore in order to detect relationships between them, we apply dynamic linear regression model by Kalman filter approach which allows for time varying regression coefficients. This is an extension of a gravity model in which coefficients of explanatory variables vary over time. The observation equation can be written as

$$\ln EXPORT_{ij,t} = \beta_{0,t} + \beta_{i1,t} \ln GDP_{i,t} + \beta_{j1,t} \ln GDP_{j,t} + \beta_{i2,t} \ln POP_{i,t} +$$
$$\beta_{j2,t} \ln POP_{j,t} + \beta_{ij1,t} \ln D_{ij,t} + V_{ij,t}. \tag{10}$$

For $i = 1, \ldots, N$ where i refers to Thailand, $j = 1, \ldots, N$ where j refers to partner countries, $ij = 1, \ldots, N$ where ij refers to between two countries, $t = 1, \ldots, T$, $V_{ij,t}$ is the vector of normal distribution with zero mean, and constant variance, where $EXPORT_{ij,t}$ corresponds to the total value of Thai exports goods and services to countries in AEC at year t. $GDP_{i,t}$ and $GDP_{j,t}$ stand for total value of final goods and services being produced within country for specific period of time (a year). $POP_{i,t}$ and $POP_{j,t}$ correspond to population in Thailand and partner countries at time t respectively. $D_{ij,t}$ is the measured distance between Thailand and partner countries.

3.4   Data

Our dataset consists of a real data panel from the period 1991–2014 for 10 observed countries, which are Brunei Darussalam (BN), Cambodia (KH), Indonesia (ID), Laos (LA), Malaysia (MY), Myanmar (MM), the Philippines (PH), Thailand (TH), Singapore (SG), and Vietnam (VN). This data panel is strongly balanced, with no missing data. Together, the dataset yields 240 complete observations. Data regarding the exports of Thailand are collected from the foreign trade statistics of Thailand. Data on GDP and population are from the World Bank database. The distance data are from the CEPII GeoDist dyadic dataset [8].

4   The Results

The coefficients generated by Kalman filter are reported in Figs. 2, 3, 4, 5 and 6 which describe the varying coefficients of each explanatory variable.

Figures 2 and 3 show that coefficients of intercepts and Thailand's GDP respectively are varying through time, furthermore all trading partners have their own coefficients. On average, the highest coefficient is Vietnam, whereas the lowest beta coefficient is Brunei Darussalam, which can be attributed to the volume of goods imported from Thailand.

Thailand's exports to all trading partners are clearly consistent with Thailand's GDP in general showing that the movement of its GDP and the movements of exports

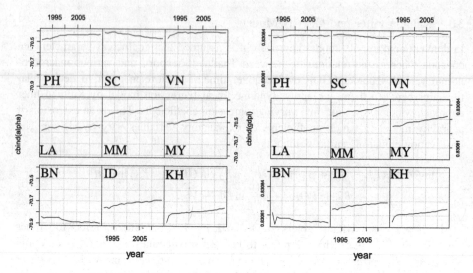

Fig. 2. Coefficients of intercept

Fig. 3. Coefficients of $GDP_{i,t}$

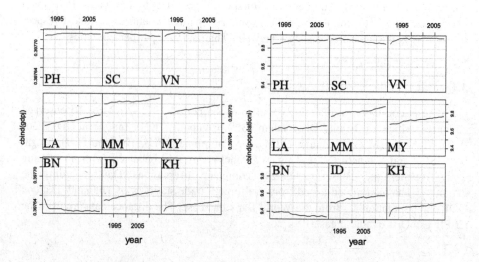

Fig. 4. Coefficients of $GDP_{j,t}$

Fig. 5. Coefficients of $POP_{i,t}$

are positively related. The result of the significant upward trends in coefficient movements for Laos, Myanmar, Malaysia, Indonesia and Cambodia explains that the relationships between Thailand's GDP to Thailand's exports to the countries increase over time. In contrast, Brunei and Singapore with their downward trends have an opposite result.

From the Fig. 4, the results show that the correlations between GDPs of the trading partners and Thailand's exports are moving in the same direction. These show that higher GDPs of the trading partners are consistent with import growth from Thailand.

However, Brunei and Singapore produce opposite results because of their negative trend in their coefficients. The latter results describe that Brunei and Singapore import less from Thailand when their GDPs are higher. According to Markusen [8], this observation indicates that capital rich countries tend to trade more with other capital rich countries. This is because high income consumers are likely to consume larger-budget-shares intensive goods. As Brunei and Singapore are the top richest countries in Southeast Asia, they tend to import more from other capital rich countries instead of Thailand.

Figure 5 shows that the coefficient of Thai population is positive. This means that an increase in the population increases exports to all trading partners as more population could enhance the production supply. Figure 6 shows that the coefficients of populations of trading partners have a negative sign, and this result is in line with the finding of Ekanayake et al. [5]. Importer population has a negative effect on exports, indicating that more populated countries import less than small countries. There are significant downward trends in the Philippines, Malaysia, and Cambodia, indicating that when the populations of the Philippines, Malaysia, and Cambodia are increasing over the time, they tend to import from Thailand less and less. This might be because when countries have more population, they are likely to have a high level of self-sufficiency since a large population promotes division of labor, and this means that there is an economy of scale in production and they are able to produce a greater variety of goods.

Figure 7 reveals that the coefficient for distance variable as a constant is also varying over times. One of the main reasons is that distance variable is normally denoted by a proxy which is transportation cost which can fluctuate in sensitivity with the change in fuel price. As a result, the effect of distance is varying according to transportation cost.

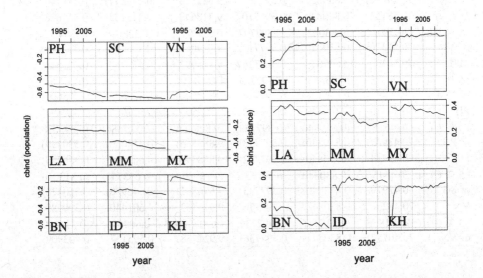

Fig. 6. Coefficients of $POP_{j,t}$ **Fig. 7** Coefficients of $D_{ij,t}$

After applying Kalman filter approach, we have time-varying coefficient for each variable. However we do not know the mean of true parameter and its distribution. Therefore we apply bootstrap sampling in order to estimate the mean coefficient. This coefficient provides more robust estimate of the true parameter [9]. And then using bootstrapping to create a 95 % confidence interval for the mean coefficient.

In Table 1 we use six decimal places in order to identify the different coefficients in each country. The results show that all coefficients fall into the level of the 95 % confident interval. This indicates that all explanatory variables significantly influence Thailand's export. In addition, all the coefficients show that only figures of trading partners' population related to Thailand's exports are negative compared to the rest which are positive.

Table 1. The mean coefficient and 95 % confidence interval of each explanatory variable.

Country	Constant	$\ln GDP_{i,t}$	$\ln GDP_{j,t}$	$\ln POP_{i,t}$	$\ln POP_{j,t}$	$\ln D_{ij,t}$
BN	−70.87754	0.830807	0.39763	9.368458	−0.087464	0.077177
95 % CI of BN	(−70.88487, −70.86989)	(0.830806, 0.830807)	(0.397628, 0.397632)	(9.360345, 9.376712)	(−0.088967, −0.085941)	(0.054385, 0.100184)
KH	−70.79459	0.830807	0.39763	9.368551	−0.087470	0.294299
95 % CI of KH	(−70.80567, −70.78514)	(0.830806, 0.830807)	(0.397628, 0.397632)	(9.360647, 9.376796)	(−0.088966, −0.085889)	(0.268529, 0.310602)
ID	−70.79459	0.830817	0.397661	9.550875	−0.206411	0.294299
95 % CI of ID	(−70.80567, −70.78514)	(0.830817, 0.830818)	(0.397659, 0.397664)	(9.540105, 9.561229)	(−0.214843, −0.198234)	(0.268529, 0.310603)
LA	−70.65382	0.830822	0.397684	9.627582	-0.255593	0.361602
95 % CI of LA	(−70.65841, −70.64942)	(0.830822, 0.830822)	(0.397689, 0.397687)	(9.622051, 9.632858)	(−0.261058, −0.249811)	(0.352252, 0.371746)
MY	−70.58121	0.830829	0.397710	9.718649	−0.313133	0.363678
95 % CI of MY	(−70.58952, −70.57326)	(0.830828, 0.830829)	(0.397707, 0.397712)	(9.707978, 9.729060)	(−0.328919, −0.298042)	(0.354102, 0.373167)
MM	−70.50482	0.83083	0.397726	9.815993	−0.437760	0.292648
95 % CI of MM	(−70.51453, −70.49499)	(0.830835, 0.830838)	(0.397724, 0.397727)	(9.804017, 9.828140)	(−0.451485, −0.423578)	(0.280464, 0.305609)
PH	−70.44519	0.830840	0.397732	9.882893	−0.571016	0.312894
95 % CI of PH	(−70.45157, −70.43942)	(0.830840, 0.830840)	(0.397732, 0.397733)	(9.876466, 9.888385)	(−0.587884, −0.554680)	(0.293638, 0.330439)
SG	−70.43714	0.830840	0.397732	9.888398	−0.652291	0.336775
95 % CI of SG	(−70.44456, −70.42979)	(0.830840, 0.830840)	(0.397731, 0.397733)	(9.880307, 9.896361)	(−0.657551, −0.646988)	(0.313046, 0.360428)
VN	−70.41236	0.830842	0.397736	9.915563	−0.601440	0.390985
95 % CI of VN	(−70.41856, −70.40752)	(0.830841, 0.830842)	(0.397735, 0.397736)	(9.908699, 9.920738)	(−0.609695, −0.595137)	(0.374009, 0.403909)

Moreover, Thailand's population is the first most important variable in exports to all trading partners as the coefficients of Thailand's population are in the range of 9.368458 to 9.915563. Thailand's GDP is the second most important variable with the coefficient values falling in the range of 0.830807 to 0.830842.

Meanwhile the coefficient of distance should be negative because transportation cost is used as a proxy of distance. It tends to increase with distance; nevertheless the empirical figures show positive sign with the lowest coefficient value compared to the other variables. It implies that distance has the least influence on exports.One of the reasons is that the efficiency of transportation has been improving due to high technology which leads to lower transportation cost. Wu [15] also found that the progress of technology can reduce the transportation cost.

5   Conclusion

In this paper, we investigate bilateral behavior of trading partners of Thailand in AEC and capture advantageous characteristics of two-sided trades between them by gravity model. Meanwhile in conventional gravity model, parameters are assumed to be constant over time. Therefore, gravity model could produce inconsistent estimations when there is presence of heteroscedasticity. Furthermore there is volatility in exports, a constant coefficient of gravity model may not be a good explanation of the behavior of Thailand's exports. Therefore, state space approach might be more appropriate than other approaches because it is able to determine the change in the coefficients over time and has no homogeneity assumption.

The results show that all coefficients are varying through time. In brief, that all explanatory variables significantly influence Thailand's exports. In addition, all the coefficients show that only figures of trading partners' population related to Thailand's exports are negative compared to the rest which are positive. Thailand's population is the first most significant variable in exports to all trading partners, and Thailand's GDP is the second most important variable. Meanwhile the coefficients of distance are positive, but it has the least influence on exports. One of the reasons is that the efficiency of transportation has been improving due to high technology which leads to lower transportation cost. Further research in this area may include more variables such as tariff, exchange rate which might also affect Thai exports in order to better understand the relationship in international trade between Thailand's exports and AEC trading partners.

References

1. Adam, A.M., Gyamfi, E.N.: Time-varying world integration of the African stock markets, a Kalman filter approach (2015)
2. Anderson, J.E., Van Wincoop, E.: Gravity with gravitas: a solution to the border puzzle. Am. Econ. Rev. **93**(1), 170–192 (2003)

3. Commandeur, J.J., Koopman, S.J.: An Introduction to State Space Time Series Analysis. OUP, Oxford (2007)
4. Durbin, J., Koopman, S.J.: Time Series Analysis by State Space Methods, vol. 38. Oxford University Press, Oxford (2012)
5. Ekanayake, E.M., Mukherjee, A., Veeramacheneni, B.: Trade blocks and the gravity model: a study of economic integration among Asian developing countries. J. Econ. Integr. **25**, 627–643 (2010)
6. Galati, G., Hindrayanto, I., Koopman, S.J., Vlekke, M.: Measuring financial cycles with a model-based filter. Empirical evidence for the United States and the euro area (2016)
7. Markusen, J.R.: Explaining the volume of trade: an eclectic approach. Am. Econ. Rev. **76**, 1002–1011 (1986)
8. Mayer, T., Zignago, S.: Notes on CEPII's distances measures. The GeoDist database (2011)
9. Miller, D.P.: Bootstrap 101. Obtain robust confidence intervals for any statistic. In: SUGI, vol. 29, pp. 193–29 (2004)
10. Pangestu, M., Ing, L.Y.: ASEAN. Regional integration and reforms. Asian economic papers. In: Working Papers. DP-2015-69 (2015)
11. Petris, G., Petrone, S., Campagnoli, P.: Dynamic Linear Models, pp. 31–84. Springer, New York (2012)
12. Saggi, K., Yildiz, H.M.: Bilateral trade agreements and the feasibility of multilateral free trade. Rev. Int. Econ. **19**(2), 356–373 (2011)
13. Serlenga, L., Shin, Y.: Gravity models of this intra-EU trade application of the Haushan-Taylor estimation in heterogenous panels with common time-specific factors. Edinburgh School of Economics. Edinburgh School of Economics Discussion Paper. 88 (2004)
14. Silva, J.S., Tenreyro, S.: The log of gravity. Rev. Econ. Stat. **88**(4), 641–658 (2006)
15. Wu, H.: Re-visiting the distance coefficient in gravity model (2015). arXiv preprint arXiv: 1503.05283
16. Zietz, J.: Heteroskedasticity and neglected parameter heterogeneity. Oxf. Bull. Econ. Stat. **63** (2), 263–273 (2001)

Volatility Hedging Model for Precious Metal Futures Returns

Roengchai Tansuchat[(⊠)], Paravee Maneejuk,
and Songsak Sriboonchitta

Faculty of Economics, Chiang Mai University, Chiang Mai, Thailand
roengchiatan@gmail.com, mparavee@gmail.com

Abstract. This study attempts to evaluate appropriately the optimal hedge ratio and hedging effectiveness between spot and futures returns of precious metals with a special concern in gold, silver, and platinum. We employ the Markov switching dynamic copula model to measure the dependence structure between spot and futures returns of these precious metals and then evaluate the hedging strategies. Evidence from this study can bring about the contribution to the discussion on this area.

Keywords: Markov switching copula · Time-varying dependence · Hedging strategy · Hedge ratio · Hedging effectiveness

1 Introduction

Precious metal has an important role as investment and a store of value, and has economic value higher than the typical metal. Currently, precious metals are traded not only in the spot market (cash market or physical market) where assets are sold for cash and delivered immediately as a raw material for jewelry and specific industries but also in financial market as a diversification asset against stocks, because they are considered as a safe-haven asset under financial and economic instability especially during global financial crisis. The price of precious metal in spot market is called spot price. However, precious metals still have price and other risks due to fluctuations in the US dollar, inflation rate, international political situations, changes in supply and demand, major crises in the regional and global economy, etc.

Futures contract is an agreement between two parties or entities where the buyer and the seller agree to trade asset, such as tangible asset or financial asset, for delivery in the future. Futures contracts are one of the most common derivatives that investors can reduce or "hedge" the risk of adverse price movements and market volatility by taking an offsetting position, or make a profit from price speculation. This market has remarkable feature that it has a low capital investment, and is not related to other spot market. Therefore, the investor usually uses futures contract to hedge price risk of the underlying precious metal in spot market.

Hedging is a financial tool to reduce risk or minimize the negative impact of the unfavorable price changes in underlying asset by offsetting probability of loss from fluctuations in the prices of commodities. The hedge ratio is one of the most widely used measures to make hedging strategies, and derived based on the minimum variance of spot portfolio. In order to hedge this spot price risk, the minimum variance hedge ratio (or optimal hedge ratio) is the ratio of futures position relative to the spot position that minimizes the risk or variance of the position (see Brooks [3] and Brooks et al. [4]). If the change of ratio in spot price and futures price are the same, it represents a hedge perfectly. However, in fact, both of these prices have a different behavior in changing, which causes the risk, but there is a high correlation between them. In order to deal with the potential risk, the hedging strategy is considered. It develops the basis of the spot and futures price relation derived from econometric model estimation. In the literature, the bivariate GARCH model, Extreme value theory, and Copula model have been popularly applied to measure the risk, volatility spillover, dependency, and correlation in capital asset markets.

On precious metal market, most of the previous studies modeled volatility dynamics, risk measures, volatility transmission, and hedge ratio. For example, Hammoudeh et al. [6], Kumar [8], Arouri et al. [1] and Basher and Sadorsky [2] studied hedge ratio between precious metals such as gold and other financial assets or exchange rate by using bivariate and multivariate GARCH. McCown and Shaw [11] examined the investment potential and risk-hedging characteristics of platinum, palladium, and rhodium by analyzing returns on their spot prices and comparing them with gold, crude oil, and stocks by applying multifactor model regressions. However, there is no previous research about hedge ratio evaluation for precious metal futures index. The objective of this research is to evaluate the optimal hedge ratio and hedging effectiveness between spot and futures returns of precious metals with a special concern in gold, silver, and platinum.

2   Methodology

To be a contribution to the discussion on hedge ratio and hedging effectiveness of precious metals, this paper employs the dynamic copula model in conjunction with the idea of structure change namely the Markov switching approach to model the dependence between spot and futures returns and then evaluate the optimal hedge ratio and hedging effectiveness sequentially. We consider the dynamic copula as introduced by Patton [14] due to the use of time variation in the dependency that allows us to gain more efficiency from the estimate result. Additionally, we also apply the Markov switching to the dynamic copula model as suggested by Tofoli et al. [15] and Pastpipatkul et al. [13] because we desire to capture the real market phenomena, say a bull market and a bear market, or the asymmetric dependence structure of the considered variables. Because of this reason, the dependence structure will be determined by a hidden Markov chain with two states.

To estimate a dynamic copula with Markov Switching, we adopt three estimation steps. In the first step, we estimate the marginal distributions using the ARMA-GARCH model. In the second step, we model residuals using generalized Pareto distribution (GPD) in order to estimate the better extreme tails of distribution. In the third step, the Markov switching dynamic copula is used for the analysis of the dependence structure between spot and futures returns of precious metals. Finally, we will use the obtained conditional variance from the first step and dynamic dependence coefficient from the third step to create a time-varying hedge ratio and hedging effectiveness.

2.1 GARCH-EVT-Copula Model

In this study, we adopt the univariate standard ARMA(p,q)-GARCH(1,1) model for, the returns to obtain the marginal distributions which is defined as follows:

$$y_t = \alpha + \sum_{i=1}^{p} \beta_i y_{t-i} + \sum_{i=1}^{q} \varphi_i \varepsilon_{t-i} + \varepsilon_t \tag{1}$$

$$\varepsilon_t = h_t z_t \tag{2}$$

$$h_t^2 = \omega + \gamma \varepsilon_{t-1}^2 + \delta h_{t-1}^2. \tag{3}$$

The terms mean (α), autoregressive coefficients (β_i), and moving average coefficients (φ_i) are the estimated parameters of the mean equation (Eq. 1). The terms mean (ω), ARCH term (γ), and GARCH term (δ) are the unknown parameters of conditional variance equation (Eq. 3). The terms y_t and h_t^2 denote the spot or futures returns and conditional variance, respectively. Note that, the conditions $\omega > 0$, $\alpha_1 > 0$, $\beta_1 > 0$ and $\alpha_1 + \beta_1 < 1$ are restricted to ensure the positive and stationary nature of h_t^2. ε_t is the white noise process at time t and z_t is iid standardized residual with a unit variance. The study assumes that ε_t is student-t distribution with zero mean, σ^2 variance, and degree of freedom v, i.e., $\varepsilon_t \sim t(0, \sigma^2, v)$. This paper applied the Extreme value theory (EVT) by using the Generalized Pareto distribution (GPD) to estimate for the upper and lower tails, and the Gaussian kernel estimate for estimating the remaining part, as defined below:

$$F(z_t) = \begin{cases} \frac{N_L}{N}\left(1 + \frac{\xi_L(m_L - z_t)}{v_L}\right)^{-1/\xi_L}, & z_t \leq m_L \\ 1 - \frac{N_U}{N}\left(1 + \frac{\xi_R(m_R - z_t)}{v_R}\right)^{-1/\xi_R}, & z_t > m_L \end{cases}$$

where ξ_L and ξ_R are the shape parameters for lower and upper thresholds, m_L and m_R, v_L and v_R are the scale parameters. The choice of the optimal threshold is important; this study chose the threshold level at the 10th percentile of the samples.

2.2 Conditional Copula Model

Patton [14] extended the Sklar theorem to construct a joint conditional distribution function using copula approach. In considering bivariate case, let H be the joint conditional distribution of $(x_1, x_2)|W$ with marginal $F_1(x_1|W)$ and $F_2(x_2|W)$, where W is the conditioning variable with support Ω, then the joint conditional copula C is

$$H(x_1, x_2 | W) = C(F_1(x_1 | W), F_2(x_2 | W)). \tag{4}$$

If F_1 and F_2 are the continuous function, then C is unique. In contrast, if $F_1(x_1 | W)$ and $F_2(x_2 | W)$ are the conditional distribution, C will be a conditional copula. Thus, we can compute the conditional density function by twice differentiating (Eq. 4) with respect to x_1 and x_2.

$$f_1(x_1 | W) \cdot f_2(x_2 | W) \cdot c(u_1, u_2 | W) \tag{5}$$

where $u_1 = F_1(x_1 | \omega)$ and $u_2 = F_2(x_2 | \omega)$, and these marginal distributions are uniform in the [0,1].

2.3   Regime Switching Dynamic Copula

Recently, many empirical studies have found that financial returns tend to exhibit different patterns of dependence over time. Thus, following Tofoli et al. [15], the time-varying copula of Patton [14] was extended to the Markov Switching of Hamilton [5] and proposed a Markov-switching copula with time-varying dependence to model dependence parameter. The study allows the copula dependence parameter to vary across the economic states or regimes, say the upturn market (regime 1) and downturn market (regime 2). Thus, dependence parameter (θ_t) is assumed to be governed by an unobserved variable (S_t). Thus, when the regime switching is taken into account, and θ_t conditioned on past observations, Tofoli et al. [15] proposed to employ ARMA(1,10) process to construct the time-varying dependence.

$$\theta_{S_t,t} = \Lambda(a_{S_t} + b\theta_{S_t,t-1} + \varpi \Gamma_t) \tag{6}$$

where $\Lambda(\square)$ is the logistic transformation which is used to keep the value of the copula dependence variable in the its interval, a_{S_t} and b_{S_t} are the intercept term, b_{S_t} is the autoregressive coefficient; and Γ_t is the forcing variable that forces $\theta_{S_t,t}$ to be on the copula bound, as defined in the following:

$$\Gamma_t = \begin{cases} \frac{1}{10} \sum_{j=1}^{10} F_1^{-1}(u_{1,t-j}) F_2^{-1}(u_{2,t-j}) & elliptical \\ \frac{1}{10} \sum_{j=1}^{10} |u_{1,t-j} - u_{2,t-j}| & Archimedean \end{cases} .$$

For simplicity, the unobservable regime (s_t) is governed by the first order Markov chain, meaning that the probability of this time t is governed by $t-1$, hence, we can write the following transition probabilities (p):

$$p_{ij} = \Pr(S_t = i | S_{t-1} = j) \mathfrak{M} \text{ and } \sum_{j=1}^{2} p_{ij} = 1 \ , \ i,j = 1,2 \tag{7}$$

where p_{ij} is the probability of switching from regime i to regime j, and these transition probabilities can be formed in a transition matrix Q, as follows:

$$Q = \begin{bmatrix} P_{11} & P_{12} = 1 - P_{11} \\ P_{21} = 1 - P_{22} & P_{22} \end{bmatrix}. \tag{8}$$

Recently, there are many functional forms of copula being proposed to join the marginal distributions. In this study, five conditional copula families comprising Gaussian copula, Student-t copula, Gumbel copula, Clayton copula, and Symmetrized Joe–Clayton (SJC) copula are compared and the best fit copula family is selected to measure the dependence of spot and futures returns. They are as follows.

1. Gaussian copula

$$C_N(u_1, u_2 | \theta_{S_t}^N) = \int_{-\infty}^{\Phi^{-1}(u_1)} \int_{-\infty}^{\Phi^{-1}(u_2)} \frac{1}{2\pi \sqrt{1 - \left(\theta_{S_t}^N\right)^2}} \exp\left\{ -\frac{x_1^2 - 2\theta_{S_t}^N x_1 x_2 + x_2^2}{2(1 - \left(\theta_{S_t}^N\right)^2)} \right\} dx_1 dx_2$$

$$\tag{9}$$

The Gaussian copula is a linear correlation with symmetric dependence because the upper and the lower dependences are equal and lie in interval $[-1, 1]$. $\Phi^{-1}(u)$ is an inverse cumulative normal distribution. Its dynamic dependence $(\theta_{S_t,t}^N)$ equation can be written as

$$\theta_{S_t,t}^N = \Lambda(a_{S_t} + b\theta_{S_t,t-1}^N + \varpi \cdot \frac{1}{10}\sum_{j=1}^{10} \Phi^{-1}(u_{1,t-j})\Phi^{-1}(u_{2,t-j})). \tag{10}$$

2. Student-t copula

The Student-t copula has a linear correlation coefficient $[-1,1]$ and has symmetrical tail dependence. In this bivariate case, it has the following form:

$$C_T(u_1, u_2 | \theta_{S_t}^T, v) = \int_{-\infty}^{t_v^{-1}(u_1)} \int_{-\infty}^{t_v^{-1}(u_2)} \frac{1}{2\pi \sqrt{1 - \left(\theta_{S_t}^T\right)^2}} \left\{ 1 + \frac{x_1^2 - 2\theta_{S_t}^T x_1 x_2 + x_2^2}{v(1 - \left(\theta_{S_t}^T\right)^2)} \right\}^{-\frac{v+2}{2}} drds,$$

$$\tag{11}$$

where $\theta_{S_t}^T$ is the linear dependence parameter, and v is the degree of freedom. $t_v^{-1}(u)$ is the cumulative student-t distribution. Its dynamic equation can be written as

$$\theta_{S_t,t}^T = \Lambda(a_{S_t} + b\theta_{S_t,t-1}^T + \varpi \cdot \frac{1}{10}\sum_{j=1}^{10} T_v^{-1}(u_{1,t-j})T_v^{-1}(u_{2,t-j})). \tag{12}$$

3. Gumbel copula

Gumbel copula is an asymmetrical tail dependence copula and has only upper tail dependence $[1, \infty]$. Thus, its functional form can be written as:

$$C_G(u_1, u_2 | \theta_{S_t}^G) = \exp(-((-\log u_1)^{\theta_{S_t}^G} + (-\log u_2)^{\theta_{S_t}^G})^{1/\theta_{S_t}^G}) \tag{13}$$

and the dynamic equation is given as follows:

$$\theta_{S_t,t}^G = \Lambda(a_{S_t} + b\theta_{S_t,t-1}^G + \varpi \cdot \frac{1}{10}\sum_{j=1}^{10}|u_{1,t-j} - u_{2,t-j}|). \tag{14}$$

4. Clayton copula
Clayton copula also has asymmetrical tail dependence. However, it shows only upper tail dependence. Thus, its functional form can be written as:

$$C_C(u_1, u_2 | \theta_{S_t}^C) = (u_1^{-\theta_{S_t}^C} + u_2^{-\theta_{S_t}^C} - 1)^{-1/\theta_{S_t}^C} \tag{15}$$

and the dynamic equation is given as follows:

$$\theta_{S_t,t}^C = \Lambda(a_{S_t} + b\theta_{S_t,t-1}^C + \varpi \cdot \frac{1}{10}\sum_{j=1}^{10}|u_{1,t-j} - u_{2,t-j}|). \tag{16}$$

5. Symmetrized Joe–Clayton copula
The symmetrized Joe–Clayton (SJC) copula has both upper and lower tail dependence where the lower tail dependence is different from the upper tail dependence and the function can be written in the following form:

$$C_{SJC}(u_1, u_2 | \lambda_{S_t,U}, \lambda_{S_t,L}) = \frac{1}{2} \cdot (C_{JC}(u_1, u_2 | \lambda_{S_t,U}, \lambda_{S_t,L}) + C_{JC}(1 - (1 - u_1, 1 - u_2 | \lambda_{S_t,U}, \lambda_{S_t,L}) + u_1 + u_2 - 1), \tag{17}$$

where C_{JC} is the Joe–Clayton copula, known as the BB7 copula, given by

$$C_{JC}(u_1, u_2 | \lambda_{S_t,U}, \lambda_{S_t,L}) = 1 - \{[1 - (1 - u_1)^\kappa]^{-\gamma} + [1 - (1 - u_2)^\kappa]^{-\gamma} - 1\}^{-1/\gamma})^{-1/\kappa}, \tag{18}$$

with $\kappa = 1/\log_2(2 - \tau^U)$, $\gamma = -1/\log_2(\tau^U)$, and $\tau^U, \tau^L \in (0,1)$. $\lambda_{S_t,U}$ and $\lambda_{S_t,L}$ are the upper and lower tail dependences which are not dependent on each other. Its dynamic functional form can be written as:

$$\lambda_{S_t,L,t} = \Lambda(a_{S_t,L} + b_L\lambda_{S_t,L,t-1} + \varpi_L \cdot \frac{1}{10}\sum_{j=1}^{10}|u_{1,t-j} - u_{2,t-j}|) \tag{19}$$

$$\lambda_{S_t,U,t} = \Lambda(a_{S_t,U} + b_U\lambda_{S_t,U,t-1} + \varpi_U \cdot \frac{1}{10}\sum_{j=1}^{10}|u_{1,t-j} - u_{2,t-j}|). \tag{20}$$

2.4   Time-Varying Hedge Ratio and Hedging Effectiveness

In this sub section, the time-varying hedge ratio and hedge effectiveness as proposed in Lai et al. [9] and Park and Switzer [12], respectively. The time-varying hedge ratio is computed by

$$HR_{S_t} = \theta_{S_t} \frac{h_{s,t} h_{f,t}}{h_{f,t}^2},$$ (21)

where $h_{s,t}$ and $h_{f,t}$ are the conditional variance of spot and futures returns, respectively. θ_{S_t} is the conditional dependence parameter obtained from the Markov-switching copula with time-varying dependence model. For the time-varying hedge effectiveness which is the percentage of risk reduction can be computed by

$$HE_{S_t} = 1 - \frac{h_{R_{h,S_t}}^2}{h_{R_s}^2}$$ (22)

where $h_{R_{h,S_t}}$ is the conditional variance of the return of hedge portfolio R_{h,S_t}, which is defined as $R_{h,S_t} = R_s - HR_{S_t} \times R_f$, where R_f and R_s are returns of spot and futures, respectively. If HE_{S_t} is close to 1, hedge effectiveness is higher; on the other hand the portfolio risk is reduced. Finally, we can construct the expected time-varying hedge ratio and hedge effectiveness by two following equations.

$$E(HR_t) = \sum_{j=1}^{2} [HR_{S_t=j}]' \cdot [\Pr(S_t = j | \Psi_t) \times Q]$$ (23)

$$E(HE_t) = \sum_{j=1}^{2} [HES_{t=j}]' \cdot [\Pr(S_t = j | \Psi_t) \times Q]$$

where $\Pr(S_t = j | \Psi_t) \times Q$ is the multiplying of filtered probability and transition matrix, where Ψ_t is all information available at time t.

3   Data

The daily logarithmic returns of spot and futures, R_s and R_f, are collected from Thomson Reuters DataStream, from Financial Investment Center (FIC), Faculty of Economics, Chiang Mai University. We work with the futures and spot prices of gold, platinum, and silver in the New York Mercantile Exchange (NYMEX). The study period was from the period of April 1, 2009 to June 17, 2016, covering 1,883 observations. Table 1 gives summary statistics of all data series. Moreover, the table also provide the Jarque Bera probability(JB-prob) and Augmented Dickey Fuller probability (ADF-prob) to test the normality and stationary of all returns. The results show that all returns are not normally distributed, but there are stationary at 1 % level.

Table 1. Descriptive Statistics

	GOLDs	SILs	PLATs	GOLDf	SILf	PLATf
Mean	0.00018	0.0001	−8.06E-05	0.0001	0.0001	−7.89E-05
Median	0.0000	0.0006	0.0000	0.0000	9.85E-05	0.0000
Maximum	0.0479	0.0632	0.0437	0.0461	0.0770	0.0452
Minimum	−0.0959	−0.1384	−0.0657	−0.0983	−0.1951	−0.0709
Std. Dev.	0.0107	0.0193	0.0123	0.0109	0.0204	0.0129
Skewness	−0.5812	−0.8554	−0.2991	−0.8425	−0.9244	−0.2966
Kurtosis	8.8134	8.1222	4.6186	9.3384	9.5691	4.4065
JB-prob	0.0000	0.0000	0.0000	0.0000	0.0000	0.0000
ADF-prob	0.0000	0.0000	0.0000	0.0000	0.0000	0.0000

Source: Calculation.

Note: The subscripts "s and f" represent returns of spot and futures returns, respectively.

3.1 Marginal Distributions

This study uses ARMA-GARCH to model the volatilities of the precious metal returns while the employing Generalized Pareto Distribution (GPD) to model the tails of the distribution. Table 2 shows the estimated parameters for mean and variance equations of ARMA-GARCH with student-t distribution and the tails from GPD of all data series. The appropriate lag for ARMA(p,q)-GARCH(1,1) model is chosen by the minimum values of Akaike information criterion (AIC) and Bayesian information criterion (BIC). The results as shown in Table 2 indicate that the estimated equations of both gold spot and silver spot are ARMA(1,1)-GARCH(1,1), gold futures and platinum futures are ARMA(2,2)-GARCH(1,1), whereas the model ARMA(3,3)-GARCH(1,1) is the best one for platinum spot and silver futures.

In the variance equation, it is observed that the ARCH parameter α and the GARCH parameter β for all the series are statistically significant, meaning that all the returns experience the ARCH and GARCH effects. Finally, from the EVT equation, we can see that the upper tail of all the series exhibited the negative value of shape parameter ξ_R, therefore, the upper tail for all returns are with short-tailedness. On the other hand, a strong tailedness is obtained in lower tail of the returns since there exists a positive value of ξ_L, except for the futures and spot of platinum.

Moreover, Table 2 also reports the p-value from the autocorrelation ARCH-LM test and the uniform Kolmogorov–Smirnov test (KS-test). We found that the *p*-value of the ARCH-LM test and KS-test for all returns show an insignificant result, thus we can confirm that our standardized residuals have no autocorrelation and the cumulative probability transformation of the standardized residuals are uniform distribution within interval [0,1].

Table 2. Estimate of ARMA(q,p)-GARCH (1,1) and EVT

Equation		GOLDs	SILs	PLATs	GOLDf	SILf	PLATf
Mean	μ	0.000 (000)	0.000 (0.000)	0.000 (0.000)	0.000 (0.000)	0.001 (0.000)	0.000 (0.000)
	AR(1)	−0.908 (0.152)	−0.898 (0.260)	−0.603 (0.026)	1.484 (0.015)	1.389 (0.019)	0.910 (0.048)
	AR(2)	–	–	−0.720 (0.019)	−0.670 (0.014)	−0.246 (0.094)	−0.924 (0.066)
	AR(3)	–	–	0.277 (0.019)	–	−0.280 (0.092)	–
	MA(1)	0.901 (0.157)	0.900 (0.258)	0.654 (0.013)	−1.523 (0.003)	−1.444 (0.002)	−0.889 (0.036)
	MA(2)	–	–	0.765 (0.000)	0.712 (0.003)	0.325 (0.084)	0.919 (0.069)
	MA(3)	–	–	−0.241 (0.004)	–	0.263 (0.083)	–
Variance	ω	0.000 (0.000)	0.000 (0.000)	0.000 (0.000)	0.000 (0.000)	0.000 (0.000)	0.000 (0.000)
	α	0.025 (0.004)	0.027 (0.003)	0.036 (0.022)	0.027 (0.006)	0.029 (0.007)	0.042 (0.008)
	β	0.962 (0.004)	0.972 (0.003)	0.954 (0.025)	0.962 (0.005)	0.970 (0.005)	0.945 (0.011)
	df	4.827 (0.635)	3.978 (0.337)	8.124 (1.887)	3.884 (0.582)	3.549 (0.290)	8.668 (1.554)
EVT	ξ_L	0.0143 (0.0248)	0.0407 (0.0340)	−0.1048 (0.0267)	0.0609 (0.0339)	0.0631 (0.0365)	-0.1434 (0.0273)
	υ_L	0.7238 (0.0301)	0.6919 (0.0331)	0.8587 (0.0365)	0.6621 (0.0313)	0.0669 (0.0333)	0.8988 (0.0386)
	ξ_R	−0.1311 (0.0260)	−0.1220 (0.0300)	−0.1969 (0.0202)	−0.1259 (0.0257)	−0.1172 (0.0309)	−0.1975 (0.0222)
	υ_R	0.8255 (0.0349)	0.7449 (0.0333)	0.8823 (0.0338)	0.7907 (0.0336)	0.7374 (0.0337)	0.8812 (0.0345)
LM-test(p-val)		0.8001	0.5124	0.5248	0.8841	0.9641	0.2124
KS-test(p-val)		0.2441	0.2581	0.9996	0.8041	0.5554	0.7485

Source: Calculation.
Note: The value in parenthesis is standard deviation.

3.2 Model Selection

Prior to the estimation, selecting the appropriate copula is our first concern. We select the best-fit copula family for the Markov switching copula model using the AIC and BIC criteria. The results are reported in Table 3. We found that Gumbel copula can appropriately describe the dependence structure between futures and spot of gold as well as silver, while the Clayton copula is an appropriate linkage for the pair of futures and spot of platinum.

Table 3. AIC and BIC for Each Pair Copula

AIC/BIC	GOLDf-GOlDs	SILf- SILs	PLATf- PLATs
Gaussian	1358.4	3521.0	3284.1
	1391.6	3554.3	3317.4
Student- t	1371.0	3522.4	3214.1
	1415.3	3566.7	3257.5
Clayton	1296.2	3029.7	**2819.6**
	1329.5	3063.0	**2852.9**
Gumbel	**987.9**	**2683.1**	3098.9
	1021.2	**2716.3**	3132.2
SJC	1379.7	3571.9	3172.3
	1435.1	3627.3	3227.7

Source: Calculation.

3.3   The Estimates of the Markov Switching Dynamic Copula

The estimated parameters from the Markov switching dynamic copula for the three different pairs of futures and spot returns are reported in Table 4. The intercept coefficient of regime 1 is denoted by $\alpha_{(S_t=1)}$ and regime 2 denoted by $\alpha_{(S_t=2)}$. Since the intercept term of regime 1 is smaller than regime 2, we interpret regime 1 as the low dependence regime and regime 2 as the high dependence regime. Consider the autoregressive coefficient b; there appears a negative sign for every pair indicating that those correlations are persistent over time. For the coefficient φ which represents the distance from the perfect correlation, the results also provide a negative sign for every pair. That means the greater distance from the perfect correlation can bring about a decrease of their dependences. Additionally, Table 4 also provides at its bottom the transition probabilities for all pairs. The probabilities p_{11} and p_{22} refer to the probability of staying in the same regime, for example p_{11} means the probability of remaining in

Table 4. Estimated Parameters from Markov-switching Dynamic Copula

	$C_{GOLDf,GOLDs}$	$C_{SILf,SILs}$	$C_{PLATf,PLATs}$
$\alpha_{(S_t=1)}$	1.4789	2.3049	2.8237
	(0.1574)	(0.1823)	(0.1466)
$\alpha_{(S_t=2)}$	3.2975	4.4824	5.1044
	(0.4028)	(0.2479)	(0.1834)
b	−0.2325	−0.1639	−0.1750
	(0.0723)	(0.0186)	(0.0065)
ϖ	−1.7270	−6.6606	−13.8511
	(0.5352)	(1.4005)	(1.2013)
p_{11}	0.6418	0.8407	0.8200
	(0.3301)	(0.2227)	(0.2106)
p_{22}	0.8805	0.7031	0.6350
	(0.3327)	(0.1992)	(0.2071)

Source: Calculation

regime 1. We observe that both regimes are rather persistent due to the high values of p_{11} and p_{22}.

Figure 1 shows the time-varying dependences between spot and futures returns of gold (a), silver (b), and platinum (c). The dependences of gold and silver are obtained from the Markov switching dynamic Gumbel copula while the dependence of platinum is the one based on dynamic Clayton copula. The dependence values are very different from one regime to the other where the first regime is the high dependence regime and the second one is the low dependence regime. The estimated time-varying dependence of gold seems fluctuate in small range over time whereas the dependence of silver is mildly vacillating. However, the dependence of platinum happens to fluctuate the most.

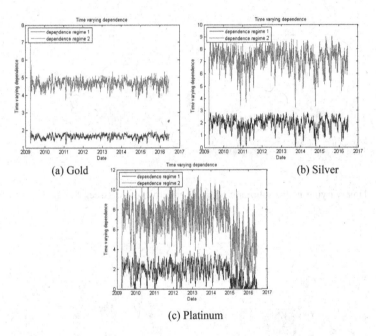

(a) Gold   (b) Silver

(c) Platinum

Fig. 1. Time-varying dependences between spot and futures returns

3.4   Hedge Ratio and Hedging Effectiveness of the Three Precious Metals

To achieve the goal of this study, we extend the estimated results and find the effectiveness and hedge ratio regarding the hedging strategy. We employ the time-varying hedge ratio and hedging effectiveness with regime switching to measure the percentage of risk reduction from pair hedging strategy with special concern in gold, silver, and platinum futures and spots returns. Note that the data used in this analysis is spanning from the years 2009 to 2016 (see Sect. 3).

Figures 2, 3 and 4 illustrate the optimal expected hedge ratio (HR) and hedge effectiveness (HE) of the three precious metals: gold, silver, and platinum, respectively. We found the similar patterns among these precious metals in which the values of HR of gold, silver and platinum increase sharply and then decrease during the periods from

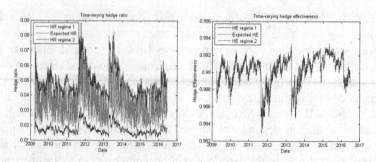

Fig. 2. Time-varying hedge ratio and hedging effectiveness between Gold spot and futures

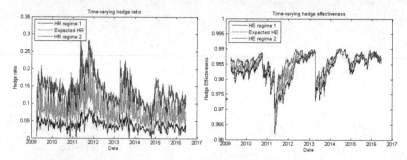

Fig. 3. Time-varying hedge ratio and hedging effectiveness between Silver spot and futures

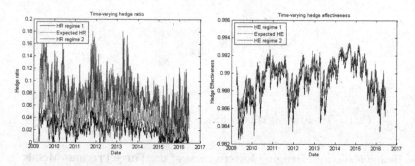

Fig. 4. Time-varying hedge ratio and hedging effectiveness between Platinum spot and futures

June 2011 to December 2012, and from March 2013 to October 2013. In the meantime, the values of HE happen to follow the opposite direction of HR. It drops down dramatically and reaches the lowest values around 0.983 (again 0.985) before rising up during those periods of time.

Surprisingly, these two periods of time coincide with the time of the US subprime mortgage crisis and the European debt crisis. It implies that the severe economic crisis brought about the greatly expected HR of these precious metals in which the expected HR of platinum fluctuated the most based on this series. Or otherwise, the crisis can have a negative impact on the expected HE as we can see obviously the drop down of the expected HE during these two crises. The possible interpretation is that when the investors face with the higher risk they will tend to increase their purchasing of futures contracts to reduce the risk. These results are consistent with the findings of Lien and Yang [10] and Hsu et al. [7] who studied the hedge ratio for non-precious metals. Therefore, we may conclude that the hedging strategy in terms of the HR and the HE are the same for precious and non-precious metals due to the evidences from this data series. Additionally, the values of expected HE are low during the economic crisis, implying that the portfolio risk increases during that time and this type of hedging strategy seems not to work well in this situation. But, when we consider the other period, we found the expected HR of these three precious metals are low and the expected value of HE becomes bigger. It indicates that the hedging strategy of these precious metals seems to work well in the market upturn.

4 Conclusion

The absence of researches regarding the hedging strategies for precious metals allows us to generate the study to fill up this area. The study attempts to evaluate the optimal hedge ratio and hedging effectiveness between spot and futures returns of precious metal namely gold, silver, and platinum. We employ the Markov switching dynamic copula model to measure the dependence structure between spot and futures returns of the precious metals and to evaluate the hedging strategies. The optimal hedge ratio (HR) and hedge effectiveness (HE) are measured to answer our research problems and to build the contribution to the discussion on this area. The benchmarking results based on this data series show that, when the precious metal markets face the severe economic crisis, the value of HR increases while the value of HE decreases. But, the estimated values of HR are found to be lower and HE higher when we are in the market upturn. Therefore, we can interpret our results that the hedging strategy for futures and spot of these three precious metals work well only in the market upturn. We thus suggest that the investors and the risk managers should employ hedge ratios to reduce the risk and to make profits only when the market is in the upturns.

Acknowledgement. The authors are grateful to the Graduate School, Chiang Mai University and Puay Ungphakorn Centre of Excellence in Econometrics, Faculty of Economics, Chiang Mai University for the financial support.

References

1. Arouri, M.E.H., Laiani, A., Nguyen, D.K.: World gold prices and stock returns in china: insights for hedging and diversification strategies. Energy Model. **44**, 273–282 (2015)
2. Basher, S.A., Sadorsky, P.: Hedging emerging market stock prices with Oil, Gold, VIX, and Bonds: a Comparison between DCC, ADCC, and GO-GARCH. Energy Econ. **54**, 235–247 (2016)

3. Brooks, C.: Introductory Econometrics for Finance. Cambridge University Press, Cambridge (2014)
4. Brooks, C., Henry, O.T., Persand, G.: The effect of asymmetries on optimal hedge ratios. J. Bus. **75**(2), 333–352 (2002)
5. Hamilton, J.D.: A new approach to the economic analysis of nonstationary time series and the business cycle. Econometrica J. Econometric Soc. **57**, 357–384 (1989)
6. Hammoudeh, S.M., Yuan, Y., McAleer, M., Thompson, M.A.: Precious metals-exchange rate volatility transmissions and hedging strategies. Int. Rev. Econ. Finan. **19**, 633–647 (2010)
7. Hsu, W.H., Liau, Y.S., Yang, J.J.: Evidence on hedge ratios changes around the subprime mortgage crisis. Global J. Bus. Res. **4**(3), 61–71 (2010)
8. Kumar, D.: Return and volatility transmission between gold and stock sectors: application of portfolio management and hedge effectiveness. IIMB Manage. Rev. **26**, 5–16 (2014)
9. Lai, Y., Chen, C.W.S., Gerlach, R.: Optimal dynamic hedging via copula-threshold-GARCH models. Math. Comput. Simul. **79**, 2609–2624 (2009)
10. Lien, D., Yang, L.: Hedging with chinese metal futures. Global Finan. J. **19**, 123–138 (2008)
11. McCown, J.R., Shaw, R.: Investment Potential and Risk Hedging Characteristics of Platinum Group Metals. Quarterly Review of Economics and Finance (2016)
12. Park, T.H., Switzer, L.N.: Bivariate GARCH estimation of the optimal hedge ratios for stock index futures: a note. J. Futures Markets, pp. 61–67 (1995)
13. Pastpipatkul, P., Yamaka, W., Sriboonchitta, S.: Analyzing financial risk and co-movement of gold market, and indonesian, philippine, and thailand stock markets: dynamic copula with markov-switching. In: Huynh, V.-N., Kreinovich, V., Sriboonchitta, S. (eds.) Causal Inference in Econometrics. SCI, vol. 622, pp. 565–586. Springer, Heidelberg (2016). doi:10.1007/978-3-319-27284-9_37
14. Patton, A.J.: Modelling asymmetric exchange rate dependence*. Int. Econ. Rev. **47**(2), 527–556 (2006)
15. Tofoli, P.V., Ziegelmann, F.A., Silva Filho, O.C.: A comparison Study of Copula Models for European Financial Index Return (2013)

What Firms Must Pay Bribes and How Much? An Empirical Study of Small and Medium Enterprises in Vietnam

Thi Thuong Vu[1] and Chon Van Le[2(✉)]

[1] University of Danang, Kontum Campus, 704 Phan Dinh Phung, Kontum, Vietnam
thuong.vt@vnp.edu.vn
[2] International University - VNU HCMC,
Quarter 6, Linh Trung, Thu Duc District, Ho Chi Minh City, Vietnam
lvchon@hcmiu.edu.vn

Abstract. This paper uses panel data from the Small and Medium Enterprise Survey in Vietnam from 2005 to 2013 to investigate the incidence and size of corruption in Vietnam. The Heckman's two-step model is employed to take into account censored nature of the data on bribes and sample selection bias. We find strong evidence that the propensity to bribe as well as bribe amounts are highly positively correlated with interaction level with government officials, firms' ability to pay, and regulatory-type burdens imposed on firms. In addition, firms without official business registration licenses are more likely to avoid paying the informal costs. These results are robust when lagged values of profit are used as instruments for profit.

Keywords: Bribery · Government bureaucrats · Small and medium enterprises

JEL Codes: C34 · D22 · G38

1 Introduction

Corruption occurs in both developed and developing countries in various degrees and has impacted almost all parts of society (Rohwer 2009). Amundsen (1999) described corruption as "a disease, a cancer that eats into the cultural, political and economic fabric of society, and destroys the functioning of vital organs". And the World Bank (2011b) considered corruption as "among the greatest obstacles to economic and social development".

In Vietnam, this problem has provoked growing alarm due to the government's failure to reduce corruption over the past years. The Corruption Perceptions Index (CPI) 2014, conducted by Transparency International (2014), ranked Vietnam 119 out of 175 countries globally and 18 out of 28 countries in the Asia Pacific region. It is worthy to note that Vietnam's CPI score was unchanged in

© Springer International Publishing AG 2016
V.-N. Huynh et al. (Eds.): IUKM 2016, LNAI 9978, pp. 689–700, 2016.
DOI: 10.1007/978-3-319-49046-5_58

three consecutive years from 2012 to 2014 whereas a positive change was recorded in neighboring countries. In addition, according to a report of Provincial Competitiveness Index (PCI) 2014 (VCCI 2015), Vietnam witnessed a significant decline in controlling informal costs that firms had to pay. Specifically, the proportion of firms that paid bribes jumped from 41 % in 2013 to 66 % in 2014. And this informal payment cost more than 10 % of revenues of one among every ten firms.

The study of corruption has attracted considerable attention in the last two decades, focusing on two segments: the determinants of corruption and the effect of corruption on growth. Most of these studies share three common features. First, they have primarily based on cross-country analyses. Second, they use perceptive data rather than quantitative data. Third, the interpretation of corruption is based on a function of macro-level factors such as countries' policies or institutional environment. However, the interpretation grounded on cross-country analyses and perceptive data has some drawbacks. On one hand, exploiting perception data on corruption can suffer from bias (Batra et al. 2003; Malomo 2013). On the other, macro-level determinants of corruption constrain the interpretation of variation in corruption within a country (Svensson 2003).

To avoid these problems, Svensson (2003) exploited quantitative data on corruption, derived from the 1998 Ugandan enterprise survey which was designed to collect a representative sample of private enterprises, operating in manufacturing and processing sectors. In similarity with Svensson (2003), using a panel data set from the Small and Medium Enterprise (SME) Surveys between 2005 and 2013, this paper investigates the corruption in Vietnam, particularly what firms must pay bribes and how much they pay. The Heckman's two-step model is applied to take into account censored nature of the data on bribes and sample selection bias. We find strong evidence that the propensity to bribe as well as the amount of bribes are highly positively correlated with interaction level with public officials, firms' ability to pay and regulatory-type burdens imposed on firms. In addition, firms without official business registration licenses are more likely to avoid paying the informal costs. These results are robust when lagged values of profit are used as instruments for profit.

The paper is structured as follows. Section 2 presents a literature review on corruption, incidence and levels of bribery. Section 3 outlines data and the econometric model. Section 4 discusses regression results. Conclusions follow in Sect. 5.

2   Incidence and Level of Bribery

Previous firm-level studies indicate that the incidence and level of bribery are affected by the interaction level with public officials and firm characteristics. In these studies, there is some probability that a firm faces a request for a bribe from corrupt officials who may take actions to either benefit or hurt the firm's business. A firm's propensity to pay a bribe depends on several factors which can be classified into three groups, namely, (i) control right of public officials over a firm, (ii) bargaining power and (iii) visibility.

(i) Control right hypothesis. Tanzi (1998) shows that the bribery may arise from the burden of *regulations*. The existence of these regulations along with the monopoly power of government officials in controlling these activities gives officials a good opportunity to extract bribes from those who need the authorization or permits. The author argues that the emergence of these regulations requires frequent and direct contact between government officials and citizens. To cope with these regulations, citizens have to spend a large amount of time which can be reduced considerably if they agree to make informal payments. Svensson (2003) also finds that enterprises usually have to pay the informal cost when they deal with public officials who have the power to affect their business. Without that informal payment, firms are likely to spend more time and more money on accountants and specialized service providers handling regulations.

To measure the burden of regulations, Svensson (2003) uses types of *taxes* that firms have to pay while Malomo (2013) uses the percentage of sales declared for tax purposes. Svensson (2003) proves a significant and positive relationship between two variables. Similarly, Malomo (2013) reports that companies that spend higher percentages of their sales on taxes are more likely to pay bribes.

According to Lecraw (1984) and Luo (2007), firms that serve mostly in domestic markets are likely to interact more intensively with local suppliers, customers, employees, and government officials. Repeated exposure to the local environment may increase firms' potential legal vulnerabilities related to various regulatory requirements. Moreover, Kobrin (1987) asserts that exporting firms are characterized with higher technical and managerial capabilities which are strengthened over time through learning and innovation that they can acquire in the global market. These firms' high levels of competitiveness help relax their dependence on domestic markets, hence their susceptibility to government corruption.

There is another argument about the negative relationship between *exports* and the likelihood of bribery. In developing countries and especially those with balance of payments problems, export activities are greatly appreciated because of their contributions to foreign exchange earnings and job generation (Grosse 1996; Vernon 1971; UNCTAD 2006). Public officials are thus more likely to reduce their bribe demands (Lee et al. 2010).

Pfeffer and Salancik (1978) show that the tendency to bribery is associated with the extent of firms' *dependence on the government*. Those companies whose revenues largely come from state contracts tend to be more vulnerable to rent-seeking practices (Hillman 2005). Using a dummy variable to indicate whether the government is a firm's customer, Rand and Tarp (2012) and Malomo (2013) find that firms whose business depends more on government contracting are more involved in corruption.

Firms' dependence on the government also takes the form of using publicly provided goods and services. According to Tanzi (1998), in most countries, the state sector supplies many goods, services, and resources, but sometimes at below-market prices. It probably leads to shortages which require rationing made by public officials. And they may take advantage of their authority in giving favored access to the limited supply to firms that are willing to pay bribes.

(ii) Bargaining hypothesis. Svensson (2003) builds a bargaining model to explain variation across firms' bribery. A firm's informal payment depends on that firm's "ability to pay", measured by its present and expected profitability and that firm's "refusal power", determined by its technology choice. Higher current and expected future profits weaken the firm's bargaining position, while a technology that yields low sunk costs and higher operation costs has an inverse impact. However, empirical results of this study show no evidence of those links between firms' ability to pay bribes and their power to avoid them and their incidence of corruption. Malomo (2013) and Rand and Tarp (2012) also test these relationships. Malomo (2013) observes that profits and sunk costs do not influence whether or not firms have to pay informal costs but Rand and Tarp (2012) find significant and positive relationships among them.

(iii) Visibility hypothesis. It is argued that a firm's propensity to bribe is affected by its visibility which is proxied by firm size and its formal or informal status. However, there is no consensus about the impact of firm size on bribery. Beck et al. (2005) demonstrate that smaller firms are less likely harassed by rent-seeking bureaucrats and other institutional problems than their larger counterparts. In the meanwhile, Hellman and Schankerman (2000) claim that smaller firms pay a higher proportion of their income as bribes and are subject to higher frequency of bribe requests. In addition, weak political resources of smaller firms do not allow them to shape regulations in their favor. As a consequence, they become more vulnerable to government extortion (Bennedsen et al. 2009; Harstad and Svensson 2011).

Regarding formal or informal status, Rand and Tarp (2012) suggest that smaller and informal firms seem to be less exposed to government regulations and corruption. Sharing similar views, Dabla-Norris and Koeda (2008) find that as the cost of complying with taxes, bribes, and burdensome regulations increases, more firms choose to operate informally. Nevertheless, Rand and Tarp (2012) also point out an opposite effect of informality on bribe incidence. In particular, if firms may seek the benefits of informality, they are willing to offer an informal payment to maintain or achieve the informal status. Therefore, the net effect of informality is an open question.

3   Data and Econometric Model

a.Data. This study uses a firm-level panel from the Small and Medium Enterprise (SME) Survey in Vietnam from 2005 to 2013. The survey is conducted every two years by the Central Institute for Economic Management, the Ministry of Planning and Investment, the Institute of Labour Science and Social Affairs (the Ministry of Labour, Invalids and Social Affairs of Vietnam), the Department of Economics at the Copenhagen University, and the Embassy of Denmark in Vietnam. The survey covers over 2,500 enterprises in ten provinces, i.e. Hanoi, Phu Tho, Ha Tay, Haiphong, Nghe An, Quang Nam, Khanh Hoa, Lam Dong, Ho Chi Minh City, and Long An, a large proportion of these firms being repeats from previous years. Firm-level variables are defined in Table 1. Since we would

Table 1. Firm-level variable definitions

Variable	Description
Bribe$_{it}$	A dummy equals 1 if firm i in year t pays bribes and 0 otherwise
ln(Bribe/Labor)$_{it}$	Log of reported informal cost per employee
Firm_size$_{it}$	Log of total employment of firm i in year t
ln(Sunk_cost)$_{it}$	Log of Capital/Labor ratio
ln(Profit/Labor)$_{it}$	Log of profit per employee
ln(Export/Labor)$_{it}$	Log of export revenue per employee
ln(Import/Labor)$_{it}$	Log of import value per employee
Government_aid$_{it}$	A dummy equals 1 if firm i in year t receives financial or technical assistance from government and 0 otherwise
Tax/Sales$_{it}$	Percentage of firm i's sales that is paid for tax purposes
Regulations_time$_{it}$	Percentage of firm i's managerial time spent dealing with government regulations each month
Informality$_{it}$	A dummy equals 1 if firm i in year t does not have an official business registration license and 0 otherwise
ln(GvntSale/Labor)$_{it}$	Log of revenue per employee that comes from government procurement contracts
ln(GvntSupplies/Labor)$_{it}$	Log of expenditure per employee on goods and services supplied by the government

like to investigate those firms that operate at least in three consecutive years, the sample ends up with 1,753 firms and 7,139 observations.

On average, a small and medium enterprise in Vietnam bribes VND 9 million. Bribes are used mainly to gain access to public services (27.8 %), to deal with tax collectors (26.6 %), and to win government contracts (11.7 %). However, 18.2 % of the firms declare receiving financial or technical assistance from the government.

Table 2 presents some descriptive statistics of key independent variables. For log variables, we add one to level variables before taking the natural logarithm of them so that their minimum value of 0 would produce a value of 0 for log variables. It is noteworthy that firms depend on the government much more as a source of input supplies for firms' operations than as a demander for firms' output. Moreover, firm managers spend on average 9 % of their working hours dealing with government regulations. In some exceptional cases, they have to devote their whole working time to handling regulations.

b. Econometric Model. Bribe amounts are observed only for firms that must pay bribes to government officials. These firms normally form a non-randomly selected sample from the SME surveys. The OLS regression of the bribe payment on various variables could produce biased estimates due to missing data on those firms that do not bribe. This problem, called sample selection bias (Heckman 1979), can be corrected by Heckman's two-step model. It comprises outcome equation and selection equation:

Table 2. Descriptive statistics of key variables

	Obs.	Mean	Std. dev	Min	Max
Firm_size or ln(Employment)	7139	1.930	1.127	0.000	7.848
ln(Sunk_cost) or ln(K/L)	7139	10.224	1.946	0.000	16.812
ln(Profit/Labor)	7139	9.559	1.103	0.000	15.783
ln(Export/Labor)	7139	0.547	2.896	0.000	19.836
ln(Import/Labor)	7139	0.424	2.578	0.000	20.279
Tax/Sales	7139	0.730	0.687	0.000	3.502
Regulations_time	7139	8.929	20.821	0.000	100.000
ln(GvntSale/Labor)	7139	0.788	3.325	0.000	19.586
ln(GvntSupplies/Labor)	7139	11.481	5.432	0.000	21.131

Source: Authors' calculation.

$$
\begin{aligned}
b_i &= \mathbf{w}_i'\boldsymbol{\gamma} + \varepsilon_i, \\
d_i &= 1 \quad \text{if} \quad E(d_i^*|b_i > 0) = \mathbf{x}_i'\boldsymbol{\beta} + u_i > a, \\
d_i &= 0 \quad \text{if} \quad E(d_i^*|b_i = 0) = \mathbf{x}_i'\boldsymbol{\beta} + u_i \leq a, \quad i = 1, ..., n,
\end{aligned}
\tag{1}
$$

where b_i is bribe amount paid by firm i (observed only for firms that must pay bribes), $\mathbf{w}_i$ are observed variables related to firm i's characteristics such as its ability to pay bribes and its bargaining power, a is the minimum criterion for a firm's informal payment (if total characteristics $\mathbf{x}_i$ of firm i are below this criterion, that firm does not have to make an informal payment), $d_i = 1$ if firm i bribes and 0 otherwise, $\boldsymbol{\gamma}$ and $\boldsymbol{\beta}$ are unknown parameter vectors, ε_i and u_i are two error terms that are assumed to have a bivariate normal distribution. The outcome equation in (1) represents the desire relationship between the bribe amount and its underlying factors in the population. The selection equation takes into account the non-representative nature of the non-random sample.

There are two approaches to estimating the model (1), that is, the Heckman maximum likelihood procedure and the Heckman two-step procedure, among which the second is more frequently used (Cameron and Trivedi 2009). Accordingly, the first step is to estimate the selection equation via probit over the whole sample to obtain estimates of $\boldsymbol{\beta}$. The model (1) can be stated with the bivariate normal distribution assumption as

$$
\begin{aligned}
b_i &= \mathbf{w}_i'\boldsymbol{\gamma} + \varepsilon_i, \\
d_i &= 1(\mathbf{x}_i'\boldsymbol{\beta} + u_i > a) \\
d_i &= 0(\mathbf{x}_i'\boldsymbol{\beta} + u_i \leq a) \\
\begin{bmatrix} \varepsilon_i \\ u_i \end{bmatrix} &\sim N\left[\mathbf{0}, \begin{pmatrix} \sigma_\varepsilon^2 & \sigma_{\varepsilon u} \\ \sigma_{u\varepsilon} & 1 \end{pmatrix}\right], \quad i = 1, ..., n,
\end{aligned}
\tag{2}
$$

where σ_u^2 is normalized to 1. Under the assumption of bivariate normal distribution for the two error terms which implies independence between the errors and the regressors, the model (2) can be rewritten as

$$b_i = \mathbf{w}_i'\boldsymbol{\gamma} + \sigma_{\varepsilon u}\lambda(\mathbf{x}_i'\boldsymbol{\beta}) + \xi_i,$$
$$d_i = 1(\mathbf{x}_i'\boldsymbol{\beta} + u_i > a) \qquad (3)$$
$$d_i = 0(\mathbf{x}_i'\boldsymbol{\beta} + u_i \leq a), \quad i = 1, ..., n,$$

where $\sigma_{\varepsilon u}$ is the covariance between ε and u, and $\lambda(\mathbf{x}_i'\boldsymbol{\beta})$ is the inverse Mills ratio which is implied by the bivariate normality of $(\varepsilon_i, u_i)'$. The inverse Mills ratio is defined as

$$\lambda(\mathbf{x}_i'\boldsymbol{\beta}) = \frac{\phi(\mathbf{x}_i'\boldsymbol{\beta})}{\Phi(\mathbf{x}_i'\boldsymbol{\beta})},$$

where $\phi(\mathbf{x}_i'\boldsymbol{\beta})$ and $\Phi(\mathbf{x}_i'\boldsymbol{\beta})$ are the probability density function and the cumulative distribution function, respectively, of the univariate standard normal distribution $N(0, 1)$.

In the second step, the outcome equation is estimated by OLS in which the vector $\mathbf{w}_i$ and the constructed value of the inverse Mills ratio are the explanatory variables:

$$b_i = \mathbf{w}_i'\boldsymbol{\gamma} + \sigma_{\varepsilon u}\lambda(\mathbf{x}_i'\boldsymbol{\beta}) + \xi_i.$$

According to Heckman (1979), for identification purpose, the vector $\mathbf{x}_i$ should include at least one variable that does not show up in the vector $\mathbf{w}_i$.

In this paper, the Heckman's two-step model is:

(i) The probit step:

$$\text{Probit}(d = 1|\mathbf{x}) = \Phi(\mathbf{x}_i'\boldsymbol{\beta}) + u$$
$$= \beta_0 + \beta_1\ln(\text{Employment}) + \beta_2\ln(K/L) + \beta_3\ln(\text{Profit/Labor})$$
$$+\beta_4\ln(\text{Import/Labor}) + \beta_5\ln(\text{Export/Labor}) + \beta_6\text{Regulations\_time}$$
$$+\beta_7\ln(\text{GvntSale/Labor}) + \beta_8\ln(\text{GvntSupplies/Labor}) + \beta_9\text{Informality}$$
$$+\beta_{10}\text{Government\_aid} + \beta_{11}\text{Tax/Sales} + \sum_{j=1}^{13}\alpha_j\text{sector\_dummy}_j + \sigma_{\varepsilon u}\lambda(\mathbf{x}_i'\boldsymbol{\beta}) + u,$$

(ii) The OLS step:

$$\ln(\text{Bribe/Labor})|\mathbf{w} = \mathbf{w}_i'\boldsymbol{\gamma} + \sigma_{\varepsilon u}\lambda(\mathbf{x}_i'\boldsymbol{\beta}) + \xi_i$$
$$= \gamma_0 + \gamma_1\ln(K/L) + \gamma_2\ln(\text{Profit/Labor}) + \gamma_3\ln(\text{Import/Labor})$$
$$+\gamma_4\ln(\text{Export/Labor}) + \gamma_5\text{Regulations\_time} + \gamma_6\ln(\text{GvntSale/Labor})$$
$$+\gamma_7\ln(\text{GvntSupplies/Labor}) + \gamma_8\text{Informality} + \gamma_9\text{Government\_aid}$$
$$+\gamma_{10}\text{Tax/Sales} + \sum_{j=1}^{13}\delta_j\text{sector\_dummy}_j + \xi_i.$$

4   Empirical Results

The main results of the Heckman's two-step model are reported in Table 3. The estimated coefficient of the inverse Mills ratio, $\hat{\lambda}$, is statistically significant at 1 % level in the Heckman two-step procedure. It indicates that the estimation of the outcome equation will be biased without taking into account the fact that bribing firms do not form a randomly selected sample from the SME survey. Therefore, the Heckman's two-step model is required to get rid of the sample selection bias. Another concern is the existence of heteroscedasticity in panel regression which may render inferences from the estimated model unreliable. The maximum likelihood estimation (MLE) that clusters at the firm level is employed (columns (2) and (4)). However, the two procedures produce similar results. We shall interpret the results under the Heckman two-step procedure.

A firm's visibility proxied by its number of employees and informality status is significant at 1% level and has the expected signs. Larger firms will face higher probabilities of paying bribes while smaller ones, especially those not having business registration licenses, are more likely to avoid corruption. Moreover, informal firms pay 55% less bribes than their formal counterparts, other things being equal. It is consistent with the hypothesis by Rand and Tarp (2012) that the Vietnamese firms with an informal standing are less prone to corruption.

As firm managers have to spend more time dealing with various government regulations, their firms tend to pay bribes and pay bigger amounts. In addition, when the burden of regulations which is manifested in diverse types of taxes imposed on firms become heavier, the incidence and size of bribes both increase.

Firms' engagement in international trade have different impacts. Table 3 implies that exporters are less likely to pay bribes than firms which serve mostly in domestic markets. While domestic oriented firms are exposed more intensively to numerous regulatory requirements, exporting firms have higher technical and managerial capabilities, and hence stronger bargaining power against the government. In contrast, imports do not have a significant influence on firms' bribery.

Furthermore, greater dependence on the government as a client or a supplier is associated with firms' more acute vulnerability to rent-seeking practices. When a firm's sales to or purchases from the government per employee increase by 1%, it is 2% more inclined to be involved in corruption and its bribe payment is 0.03% and 0.06% bigger. An interesting point is that although government aid does not affect the incidence of bribery, it affects considerably bribe amounts. Firms that receive financial or technical assistance from the government pay 28% more than those who do not.

Regarding the bargaining hypothesis, it is found empirically that greater "ability to pay" (or higher profit per employee) impairs a firm's bargaining position, and stronger "refusal power" (or lower capital-labor ratio) does the opposite. As capital per employee which represents sunk cost associated with chosen technology declines by 1%, the probability that firms have to pay bribes falls by 1%. On the contrary, when firms make 1% more profit per employee, they are 13% more likely to pay bribes and their amount is 0.31% larger. Additionally, estimated coefficients of sector dummies suggest that the propensity to bribe varies among industries in Vietnam.

One issue is that how the above results would change if there is a feedback from bribery to profits? Politicians and bureaucrats may compete for rent by selling government favor such as subsidies, discretionary tax relief, and other forms of regulations which become primary determinants of firm profitability and

Table 3. Heckman's Two-Step Model for Bribery

Variables		Incidence of Bribery		Bribe Amount	
		Heckman two-step	Heckman vce (cluster)	Heckman two-step	Heckman vce (cluster)
		(1)	(2)	(3)	(4)
Constant		−2.433***	−2.606***	0.077	0.628
		(−11.97)	(−10.88)	(0.18)	(1.23)
ln(Employment)		0.351***	0.384***		
		(16.61)	(19.70)		
ln(K/L)		0.012**	0.011*	0.014	0.012
		(2.03)	(1.65)	(1.33)	(1.05)
ln(Profit/Labor)		0.134***	0.144***	0.310***	0.295***
		(7.68)	(6.16)	(9.99)	(5.56)
ln(Import/Labor)		−0.006	−0.010	−0.010	−0.012
		(−0.72)	(−0.92)	(−0.68)	(−0.74)
ln(Export/Labor)		−0.018***	−0.020***	0.001	−0.002
		(−2.72)	(−2.69)	(0.13)	(−0.20)
Regulations_time		0.003***	0.003***	0.004**	0.002
		(3.20)	(3.75)	(1.84)	(1.37)
ln(GvntSale/Labor)		0.018***	0.020***	0.029***	0.023**
		(3.50)	(3.77)	(3.02)	(2.18)
ln(GvntSupplies/Labor)		0.015***	0.012**	0.056***	0.052***
		(3.11)	(2.41)	(6.28)	(5.82)
Informality		−0.567***	−0.542***	−0.798***	−0.591***
		(−12.26)	(−11.85)	(−5.70)	(−5.77)
Government_aid		−0.024	−0.015	0.246***	0.216**
		(−0.52)	(−0.32)	(2.78)	(2.45)
Tax/Sales		0.026***	0.027***	0.069***	0.062***
		(3.57)	(3.56)	(5.03)	(4.32)
Mills ratio					
	lambda	1.576***			
		(9.60)			
rho = 0					
	chi2(1)		213.250		
	Prob>chi2		0.000		
Number of observations		7,139	7,139		
Censored observations		5,236	5,236		
Uncensored observations		1,903	1,903		
Wald chi2(23)		270.170	278.240		
Prob>chi2		0.000	0.000		

Notes: z-statistics are in parentheses.
Sector dummies are not reported.
*, **, *** denote 10 %, 5 %, and 1 % level of significance, respectively.
Source: Authors' calculation.

Table 4. Heckman's two-Step model for bribery, using instrument variables

Variables		Incidence of bribery		Bribe amount	
		Heckman two-step, IV	Heckman vce, IV	Heckman two-step, IV	Heckman vce, IV
		(1)	(2)	(3)	(4)
Constant		−5.068***	−5.362***	−4.674***	−4.480***
		(−9.76)	(−8.27)	(−6.05)	(−3.50)
ln(Employment)		0.388***	0.408***		
		(16.96)	(19.02)		
ln(K/L)		0.009	0.009	0.006	0.006
		(1.62)	(1.28)	(0.66)	(0.51)
ln(Profit/Labor)		0.422***	0.448***	0.885***	0.882***
		(7.68)	(6.33)	(10.82)	(6.05)
ln(Import/Labor)		−0.013	−0.016	−0.021	−0.022
		(−1.45)	(−1.48)	(−1.64)	(−1.35)
ln(Export/Labor)		−0.016**	−0.016**	0.010	0.008
		(−2.39)	(−2.18)	(0.96)	(0.66)
Regulations_time		0.003***	0.003***	0.003	0.002
		(3.13)	(3.71)	(1.56)	(1.33)
ln(GvntSale/Labor)		0.023***	0.025***	0.035***	0.032***
		(4.35)	(4.61)	(3.93)	(3.09)
ln(GvntSupplies/Labor)		0.001	−0.002	0.025***	0.023**
		(0.21)	(−0.28)	(2.70)	(2.12)
Informality		−0.549***	−0.536***	−0.618***	−0.547***
		(−11.81)	(−11.59)	(−4.70)	(−5.20)
Government_aid		0.018	0.032	0.339***	0.320***
		(0.38)	(0.68)	(4.03)	(3.63)
Tax/Sales		0.025***	0.025***	0.060***	0.058***
		(3.36)	(3.40)	(4.77)	(4.17)
Mills ratio					
	lambda	1.255***			
		(8.34)			
rho = 0					
	chi2(1)		137.69		
	Prob>chi2		0.000		
Number of observations		7,139	7,139		
Censored observations		5,236	5,236		
Uncensored observations		1,903	1,903		
Wald chi2(23)		310.02	280.57		
Prob>chi2		0.000	0.000		

Notes: z-statistics are in parentheses.
Sector dummies are not reported.
*, **, *** denote 10 %, 5 %, and 1 % level of significance, respectively.
Source: Authors' calculation.

then rent-seeking would become widespread (Mbaku 1992). However, according to causal empiricism, the regulatory process is dominated by large firms that have political power rather than small firms. Most firms in this sample are small. Therefore, it is difficult to prove the feedback from corruption to profits.

In addition, Svensson (2003) argues that it is questionable when treating profits as exogenous. As a robustness test, he uses two sets of instrument variables for profits. Similarly, this paper uses lagged values of profit per employee as an instrument variable for profit per employee. Table 4 summarizes main results of the Heckman two-step regression using instrument variable technique.

The results are highly consistent with those in Table 3. In regard to the incidence of bribery, seven out of nine variables in Table 3 continue to be significant at 1 % level and have the same signs, except capital-labor ratio and government supplies per employee which are no longer significant. Most estimated coefficients of statistically significant variables are larger than those in Table 3. Referring to factors underlying bribe amounts, all significant variables in Table 3 remain good regressors in explaining the variation in bribe amounts.

5 Conclusions

This paper uses a panel data set from the Small and Medium Enterprise Survey in Vietnam from 2005 to 2013 to investigate the incidence and size of corruption in Vietnam. The Heckman's two-step model is employed to take into account censored nature of the data on bribes and sample selection bias. We find strong evidence that the propensity to bribe as well as bribe amounts are highly correlated with firm characteristics and regulation structure.

In particular, smaller firms and those not having business registration licenses are more likely to avoid corruption and to pay less bribes than their formal counterparts. As the burden of regulations proxied by managers' time spent on coping with regulations and percentage of sales spent on taxes becomes heavier, firms tend to agree to bribing. In addition, more frequent and intensive interaction with corrupt government officials, measured by sales to and purchases from the government per employee and government assistance, is associated with more rent-seeking practices. Those firms that earn higher profits and higher capital-labor ratios have weaker bargaining positions against government bureaucrats. However, exporting firms' bargaining power is much stronger. To disentangle the two-way causal relationship between bribes and profits, this paper uses lagged values of profit as instruments. The above results are robust to different econometric specifications.

References

Amundsen, I.: Political corruption: an introduction to the issues. CMI Working Paper (1999)

Batra, G., Kaufmann, D., Stone, A.: Investment Climate Around the World: Voices of the Firms from the World Business Environment Survey. World Bank, Washington, DC (2003)

Beck, T., Demirguc-Kunt, A.S.L.I., Maksimovic, V.: Financial and legal constraints to growth: does firm size matter? J. Finan. **60**(1), 137–177 (2005)

Bennedsen, M., Feldmann, S.E., Lassen, D.D.: Strong Firms Lobby, Weak Firms Bribe: A Survey-Based Analysis of the Demand for Influence and Corruption (2009)

Cameron, A.C., Trivedi, P.K.: Microeconometrics with Stata. Stata Press, Texas (2009)

Dabla-Norris, E., Koeda, J.: Informality and bank credit: evidence from firm-level data. IMF Working Papers WP/08/94 (2008)

Grosse, R.: The bargaining relationship between foreign MNEs and host governments in Latin America. Int. Trade J. **10**(4), 467–499 (1996)

Harstad, B., Svensson, J.: Bribes, lobbying, and development. Am. Polit. Sci. Rev. **105**(01), 46–63 (2011)

Heckman, J.J.: Sample selection bias as a specification error. Econometrica **47**(1), 153–161 (1979)

Hellman, J., Schankerman, M.: Intervention, corruption and capture: the nexus between enterprises and the state. Econ. Transit. **8**(3), 545–576 (2000)

Hillman, A.J.: Politicians on the board of directors: do connections affect the bottom line? J. Manage. **31**(3), 464–481 (2005)

Kobrin, S.J.: Testing the bargaining hypothesis in the manufacturing sector in developing countries. Int. Organ. **41**(4), 609–638 (1987)

Lecraw, D.J.: Bargaining power, ownership, and profitability of transnational corporations in developing countries. J. Int. Bus. Stud. **15**(1), 27–43 (1984)

Lee, S.H., Oh, K., Eden, L.: Why do firms bribe? Manage. Int. Rev. **50**(6), 775–796 (2010)

Luo, Y.: Are joint venture partners more opportunistic in a more volatile environment? Strateg. Manage. J. **28**(1), 39–60 (2007)

Malomo, F.: Factors influencing the propensity to bribe and size of bribe payments: evidence from formal manufacturing firms in West Africa. In: Pacific Conference For Development Economics (2013)

Mbaku, J.M.: Bureaucratic corruption as rent-seeking behavior. Konjunkturpolitik **38**(4), 247–265 (1992)

Pfeffer, J., Salancik, G.: The External Control of Organizations, A Resource Dependence Perspective. Harper and Row, New York (1978)

Rand, J., Tarp, F.: Firm-level corruption in Vietnam. Econ. Dev. Cult. Change **60**(3), 571–595 (2012)

Rohwer, A.: Measuring corruption: a comparison between the transparency international's corruption perceptions index and the world bank's worldwide governance indicators. CESifo DICE Rep. **7**(3), 42–52 (2009)

Svensson, J.: Who must pay bribes and how much? Evidence from a cross section of firms. Q. J. Econ. **118**(1), 207–230 (2003)

Tanzi, V.: Corruption around the world: causes, consequences, scope, and cures. IMF Staff Pap. **45**(4), 559–594 (1998)

TI (Transparency International): Global Corruption Report (2014)

UNCTAD: World Investment Report 2006: Transnational Corporations and the Internationalization of R&D. New York and Geneva, United Nations (2006)

VCCI: The Provincial Competitiveness Index (PCI) 2014. Labour Publishing House, Hanoi (2015)

Vernon, R.: Sovereignty at Bay: The Multinational Spread of US Enterprises. Basic Books, New York (1971)

World Bank: Governance and Anticorruption. World Bank, Washington, DC (2011b)

Analysis of Agricultural Production in Asia and Measurement of Technical Efficiency Using Copula-Based Stochastic Frontier Quantile Model

Varith Pipitpojanakarn, Paravee Maneejuk[✉], Woraphon Yamaka, and Songsak Sriboonchitta

Faculty of Economics, Chiang Mai University, Chiang Mai, Thailand
oakvarith@gmail.com, mparavee@gmail.com

Abstract. The purpose of this paper is to evaluate the efficiency of agricultural production in Asia and analyze the production function of Asian countries. Methodologically, we employ the stochastic frontier model with the concern about dependency between two-sided error term and one-sided inefficiency. Likewise, we try to improve the performance of the standard stochastic frontier model by applying quantile regression to the frontier production function. Therefore, this paper introduces the model called Copula-based stochastic frontier quantile model as an alternative tool for this issue. The accuracy of this model is proved through a simulation study before applying to the agricultural production data of Asia.

Keywords: Stochastic frontier · Frontier production function · Quantile regression · Copula · Technical efficiency

1 Introduction

It is true that food drives the world and access to adequate food is one of the primary concerns for most people on the globe. With that in mind, Asia is well-known as the best area for growing staple food such as vegetable, fruit, and wheat, in which many Asian countries are on the lists of top agricultural producers and top exporters. This makes agriculture one of the largest and most significant sectors for Asian economy.

As agricultural production sector is a major driving force for Asias economic growth, we have to take into account the productive efficiency of this sector. Motivated by this reasoning, we attempt to analyze the agricultural production function of Asian countries and then examine the technical efficiencies of this region. Why do we need to consider the technical efficiencies? There are a variety of reasons that technical efficiencies are critical. For instance, technical efficiencies can lead to a rising in agricultural productivity without increasing the resource base, meaning that we are able to produce more output from the same

© Springer International Publishing AG 2016
V.-N. Huynh et al. (Eds.): IUKM 2016, LNAI 9978, pp. 701–714, 2016.
DOI: 10.1007/978-3-319-49046-5_59

quantity of inputs. Likewise, the technical efficiency can bring about the producers competitiveness without increase in input factors (Bezat-Jarzebowska and Rembisz [1]). Most importantly, efficiency measurement is necessary not only for Asian countries, but also for other nations who aim to allocate effectively agricultural funds across heterogeneous farmers and maintain an adequate standard of living in rural communities (Kaditi and Nitsi [2]). This seems to make the technical efficiency important for nations, especially in the role of agricultural productivity growth.

The idea of technical efficiency was proposed by Farrell [3] through the use of frontier production function. His discovery spills over several extensions on estimation of the frontier production function as well as the measurement of technical efficiency. However, this paper takes into account a powerful model called stochastic frontier (SFM) which was proposed by Aigner et al. [4]. The main idea of this model is to find a linear relationship between output and input levels with two independent error terms representing inefficiency and the standard normal error, respectively. This model is widely used to assess technical efficiency of production units. The efficiency is simply measured by the parameters of the frontier production function; we calculate the distance between a country's actual level of production output and the maximum level of output given inputs, which is called the production frontier.

Apart from the use of the SFM, the study of Kaditi and Nitsi [2] pointed out that the SFM still makes a strong assumption for the functional form of the inefficiency distribution and is sensitive towards outliers, which in turn lead to a misspecification. Therefore, they employed a quantile regression as an alternative model to estimate the efficiency in agricultural sector. Various studies have followed this idea such as Duy [5], and Gregg and Rolfe [6]; they found that this approach is well-suited for efficiency estimations when they concern about heterogeneity in the different country-level data. Likewise, this approach also describes well the production of efficient producers or countries in different quantile level rather than on the average. With that in mind, this paper is trying to take the advantage of the quantile regression. But would rather not giving up using the SFM, we will put the use of quantile approach into the SFM and introduce the stochastic frontier quantile model as an alternative method for this issue. We believe that the quantile approach will provide new information to the SFM by estimating the whole percentile of production functions corresponding to different efficiency levels. (Bernini, Freo, and Gardini [7]).

Additionally, as the SFM contains two independent error terms i.e. the standard normal error and inefficiency, many studies concern about the validity of this independence assumption and try to explain it in many different ways. For example, Das [8] suggests that inefficiency at current time may depend on the noise at the previous time. Moreover, due to the misspecification of model, the standard normal error may contain some important variable, which in turn makes the inefficiency dependent. Therefore, this paper employs a well-known joint distribution function called copula to be a linkage between the two error terms as suggested by Smith [9] and Wiboonpongse et al. [10]. They empirically found that the independence assumption can be relaxed appropriately by the use of copulas which in turn allow us to explore the dependence structure of the error components.

Therefore, this paper suggests a new approach to the analysis of the frontier production function called Copula-based stochastic frontier quantile model which takes into account the impact of inputs on the production output and the efficiency scores of agriculture in different quantiles. In addition, this model also allows for the dependence between two error components through the ability of copula, which in turn makes this model more flexible and far from the grossly overestimates efficiency, as in the original SFM.

The outline of this paper is as follows. Section 2 we explain thoroughly about our proposed model and other necessary statistical properties related to our model including the basic idea of copula. Section 3 we discuss about the estimation technique and then some Monte Carlo experiments are reported Sect. 4. The empirical study of agricultural production in Asia is given in Sect. 5. Section 6 contains conclusion.

2   Methodology: An Introduction to the Stochastic Frontier Quantile Model

To construct the stochastic frontier quantile model (SFQM), three statistical approaches are considered: (i) the conventional stochastic frontier model (SFM), (ii) quantile regression with an Asymmetric Laplace distribution (ALD), and (iii) copula approach. These approaches will be discussed later.

As we mentioned previously, the SFM assumes the two error components, U and V, to have normal and positive distributions, respectively. These two errors represent noise and inefficiency of the SFM model. (Smith [9]) pointed out that U and V are dependent, so he suggested using copula to join these errors together to eliminate the weak independence assumption of the SFM. However, without considering heterogeneity in the country-level data, the SFM of (Smith [9]) may not be robust against outliers and cannot capture the extremes of distribution i.e. tail behavior of a probability distribution. Therefore, to overcome this problem, we abandon the normality assumption of the U in favor of the ALD; that is we extend the quantile regression to the SFM of (Smith [9]) and introduce a stochastic frontier quantile analysis (SFQM) model. Hence, this model becomes more flexible to the outlier and it can measure the relationship between output and input levels across efficiency quantiles. In addition, this model also provides the different slopes of parameters describing the production of Asian countries rather than average value.

2.1   Modelling the SFQM with Correlated Error Components

Consider a case of cross-section of countries; the stochastic frontier quantile model is given by the following equation where Y_i is the output level of the country i in which $i = 1, ..., I$, and X'_{ik} is a $I \times K$ matrix of K different input quantities of the $i - th$ country. The term β^ρ represents $(I \times K)$ matrix of estimated parameters of the input variables at quantile level denoted by ρ, such that $\rho \in [0, 1]$. The function $f(\cdot)$ is the imposed functional form of frontier such as the Cobb-Douglas production model. The term TE^ρ denotes technical efficiency across quantiles and E_i is the composed error term which will be discussed later.

$$Y_i = f(X'_{ik}\beta^\rho) \cdot TE^\rho$$
$$Y_i = X'_{ik}\beta^\rho + E_i, \quad i = 1, ..., I$$
$$E_i = U_i - V_i \tag{1}$$
$$U \sim ALD(0, \sigma_u^2, \rho)$$
$$V \sim ALD(0, \sigma_u^2, \rho)$$

The conditional quantile of given an input matrix is measured using the conditional quantile function denoted by

$$Q_{\overline{y}_i} = (\rho|X_{ik}) = \beta(\rho)X'_{ik} \tag{2}$$

In addition, the composed error term of the model is defined by $E_i = U_i - V_i$ where U_i is assigned as the Asymmetric Laplace distribution (ALD) with mean zero and variance σ_U^2 and V_i is a nonnegative random error that is truncated positive ALD with mean zero and variance σ_V^2 (see [11]). Therefore, when using ALD, we can get consistent estimation of the quantile function and obtain a different slope coefficients as well as the technical efficiency (TE) across different quantiles ρ. In the context of the SFQM, the technical efficiency or TE can be defined as the ratio of the observed output (Y_i) to the corresponding frontier output (Y_i^*) conditional on the levels of inputs used by the country at each quantile level. Thus, the technical efficiency across quantiles or TE^ρ is given by

$$TE^\rho = exp(X_{ik}\beta^\rho + V_i - U_i)/exp(X_{ik}\beta^\rho + V_i) \tag{3}$$

where U_i and V_i represent the noise and technical inefficiency, respectively. Most importantly, these two errors are assumed to be related in this case; therefore, the joint distribution of U_i and V_i then will be modelled by the copula approach, which in turn will be explained in the next section.

2.2   Copula Functions

By a theorem due to Sklar, copula is a powerful tool used for building multivariate distributions. Copula represents dependence structures among component variables. In this study, we consider the case of two variables that are U_i and V_i, with distribution functions F_1 and F_2, respectively. Suppose that both U_i and V_i are continuous, then the joint distribution function of a two-dimensional random vector and can be expressed as

$$H(u, v) = P(U \leq u, V \leq v) = P(F_1(U) \leq F_2(u), F_2(V) \leq F_2(v)) \tag{4}$$

Note that the marginal U_i and V_i are uniformly distributed on the interval $[0,1]$. The term $H(u,v)$ is a joint distribution of U_i and V_i evaluated at the point $(F_1(u), F_2(v)) \in [0,1]^2$. As such, it is of the form

$$H(u,v) = C(F_1(u), F_2(v)) \tag{5}$$

for some copula C, then the unique copula C is obtained as

$$C(u,v) = H(F_1^{-1}(u), F_2^{-1}(v)) \tag{6}$$

where F_i^{-1} is the quantile functions of marginal $i = 1,2$; and u_i, v_i are uniform $[0,1]$. In summary, a (bivariate) copula is a (restriction of) bivariate distribution with uni-form marginal on $[0,1]$. Any joint distribution function can be built up from the marginal distributions and copula. The copula of a joint distribution can be extracted from the joint distribution.

2.3 Copula-Based Stochastic Frontier Quantile Model

In recent years, the validity of independence assumption between the two error components of stochastic frontier model has been questioned, particularly in the context of inefficiency in a dynamic setup, Das [8]. Two statisticians Burns [12] and Smith [9] suggest relaxing this weak independence assumption by using the ability of copula joining two marginal distributions of U_i and V_i, and they prove that the copula can work well as a joint distribution for this case. Hence, this study decides to employ the copula to join our two error components. The joint density of U_i and V_i can be derived by the copula function, $C(u,v)$ so that

$$F(U_i, V_i) = C(F_1(u), F_2(v)) \tag{7}$$

This bivariate distribution function $F(U_i, V_i)$ can be obtained using the marginal distribution function $F_1(u)$ and $F_2(v)$ of u and v, and bivariate copula function $C(u,v)$. The corresponding bivariate copula density function can be obtained by differentiating Eq. (6) with respect to u and v as follows

$$\begin{aligned} f(u,v) &= \frac{\partial^2}{\partial u \partial v} C(F_1(u), F_2(v)) \\ &= f_1(u) f_2(v) C_\theta(F_1(u), F_2(v)) \end{aligned} \tag{8}$$

As in Eq. (7), the terms $F_1(u)$ and $F_2(v)$ denote the probability density function (pdf) of U_i and V_i, respectively. The term $C_\theta(F_1(u), F_2(v))$ is the density function of the copula. Since the inefficient V_i cannot be obtained directly through the SFQM, this study employs the simulated likelihood function, which is the intractable integrals appearing in the likelihood functions. It is actually expectation of a well behaved function of random V_i. Thus, we transform (U_i, V_i) to be (E_i, V_i) where $E_i = U_i - V_i$. Thus we can rewrite the Eq. (7) as

$$f(u,v) = f(v, v+e) = (f_1(v) f_2(v+e) c(F_1(v), F_2(v+e)) \tag{9}$$

According to Smith [9], the pdf of is given by

$$f(e) = \int_0^M f(v,e)du \tag{10}$$
$$= E_v(f(V_i + E_i)c_\theta(F_1(v), F_2(v+e)))$$

where

$$f(V_i + E_i) = \frac{p(1-p)}{\sigma_{(V_i - E_i)}} exp\left\{-\rho_p \frac{(V_i + E_i)}{\sigma_{(V_i - E_i)}}\right\} \tag{11}$$

where V_i is simulated from the positive truncated ALD with mean equal to zero and variance, σ_V^2. Consider the second density, the bivariate copula density for $V_i + E_i$ and V_i is contracted by either Elliptical copulas or Archimedean copula. In this study, we consider six copula functions namely, Normal copula, Student's t copula, Frank copula, Clayton copula, Gumbel copula, and Joe copula. The joint distribution or the copula function is uniform marginal thus the simulated $V_i + E_i$ and V_i are transformed by cumulative ALD and cumulative truncated ALD, respectively. One of the most important purposes of stochastic frontier analysis is to measure the technical efficiencies of countries at different quantile level (TE^ρ) based on the combination of input and the level of outputs. TE^ρ is the effectiveness of given set of inputs used to produce an output at each quantile level. It appears to be technically efficient if the countries produce a maximum output from the minimum of inputs. In fact, we cannot observe TE^ρ directly especially with different quantile levels, but by following the model of original TE given by Battese and Coelli [13], we are able to apply the quantile approach and copula to the original formula of TE. And hence, the formula of TE^ρ can be expressed as follows.

$$TE^\rho = E(exp(-V_i)|E_i = e)$$
$$= \frac{\sum_{i=1}^M exp(-V_i)f(V_i + E_i)c(F_i(V_i), F_2(V_i + E_i)|\theta)}{\sum_{i=1}^M f(V_i + E_i)c(F_i(V_i), F_2(V_i + E_i)|\theta)} \tag{12}$$

where $(U = V_i + E_i) \sim ALD(0, \sigma_U^2, \rho)$ and $V \sim ALD^+(0, \sigma_V^2)$.

3   Estimation of the Copula-based SFQM

In general, the estimation of the parameters of a copula model is done by inference function for margins method or IFM. However, V_i cannot be observed and estimated by the univariate likelihood function, say $f(V_i + E_i)$. Therefore, the estimation in our model has to be necessarily based on the full likelihood in Eq. (9). As we mentioned before, the two error components U_i and V_i are assumed to be related and the joint distribution of and can be constructed by employing the copula approach. Here, we employ six well-known families of copula consisting of Gaussian, Student's t, Frank, Joe, Gumbel, and Clayton copulas in which a brief summary of the property of each copula family is described in [10].

As discussed in the works of Smith [9] and Das [8], the model faces with multiple integrals in the likelihood. Thus they suggested employing a simulated likelihood estimation function to obtain the asymptotically unbiased simulators for the integrals. Therefore, the exact likelihood functions based on a sample $\{V_i^R\}_{r=1}^R$ of size R, where $V_i^R = (V_1^R, ..., V_N^R)$, is expressed by

$$L(\beta_k^\rho, \sigma_{(V+E)}, \sigma_V, \theta) = \sum_{i=1}^N \left(\frac{1}{R} \sum_{j=1}^R \log f(V_{ij} + E_{ij}) c(v_{ij}, (v_j + e_j) | \theta) \right) \quad (13)$$

Then, the log likelihood function as shown in Eq. (13) will be maximized using the BFGS algorithm, which in turn makes the likelihood consistent for every quantiles. (Snchez, Lachos, Labra [14]).

4   Monte Carlo Simulation Study

In this section, we employ a simple Monte Carlo simulation study to evaluate primarily the performance and accuracy of our proposed model, which takes the form as Eq. (1). In practice, we generate data from the ALD and half ALD distributions and employ six copula families as described in the previous section to model the dependence structure of the two error components of the SFQM. We start with simulation of uniform u and v by setting the true copula dependency θ equal to 0.5 for Gaussian, Student's t, and Clayton copulas and equal to 3 for the rest copulas i.e. Gumbel, Joe, and Frank. For the case of Student's t copula, we set the true value of the additional degree of freedom vf equal to 4. Then, the obtained u and v are transformed into $U_i \sim ALD(0, \sigma_U, \rho)$ and $V_i \sim ALD^+(0, \sigma_V, \rho)$ by the quantile function of ALD and half-ALD, where $\sigma_U = 1$ and $\sigma_V = 0.5$. The covariates X_1 and X_2 of the output Y_i are randomly simulated from $uni(0, 2)$. We assume the true parameter for the intercept term α^ρ to be 1.5 and the coefficients β^ρ to be 2 and -2 for all quantiles $\rho = (0.25, 0.5, 0.75)$, and then generate data set n = 100. For each data set, we can observe the performance and accuracy of our proposed model by comparing the true parameters with the estimated parameters.

Table 1 shows the results of the Monte Carlo simulation investigating the maximum likelihood estimation of the Copula-based SFQM. We found that our model can perform very well through simulation study. It is found that the estimated parameters are very close to their true values with acceptable standard errors shown as the values in the braces. For example, in the case that we use Gaussian copula as a joint distribution for U and V, the estimated values of the intercept term α are 1.1055, 1.7249, and 1.6081 for quantile 0.25, 0.5, and 0.75, respectively, while the true value is 1.5 for all quantiles. The estimated coefficients β_1 are equal to 1.6946, 1.8084, and 1.5332 for different quantiles while the true value is equal to 2. This result is acceptable and the same for other copulas, therefore, the Monte Carlo simulation suggests that our proposed Copula-based SFQM is reasonably accurate.

Table 1. Simulation result

Copula	Gaussian			Student's t			Gumbel		
Parameter/Quantile	0.25	0.5	0.75	0.25	0.5	0.75	0.25	0.5	0.75
α	1.1055	1.7249	1.6081	1.1055	1.4154	1.5066	1.4173	1.5797	1.6831
	(0.1626)	(0.1734)	(0.1552)	(0.1626)	(0.0011)	(0.3336)	(0.3839)	(0.0747)	(0.5178)
β_1	1.6946	1.8084	1.5332	1.6943	1.8182	1.6376	1.9093	1.9566	1.7714
	(0.182)	(0.2755)	(0.2020)	(0.1820)	(0.4298)	(0.4359)	(0.3499)	(0.1666)	(0.3784)
β_2	−2.4778	−2.8275	−2.2651	−2.4778	−2.3800	−2.2684	−2.1553	-2.1102	-2.1441
	(0.4029)	(0.4523)	(0.1183)	(0.4029)	(0.1089)	(0.4982)	(0.3899)	(0.2815)	(0.1794)
σ_u	0.9959	1.0506	0.9841	0.1006	0.9760	1.0563	1.1106	1.0839	1.2877
	(0.0111)	(0.0977)	(0.1936)	(0.0741)	(0.1008)	(0.0076)	(0.1126)	(0.0301)	(0.1804)
σ_v	0.4645	0.6863	0.6909	0.4959	0.4383	0.7240	0.4273	0.4354	0.5918
	(0.2645)	(0.0336)	(0.0635)	(0.0119)	(0.0058)	(0.1128)	(0.1130)	(0.0428)	(0.0699)
θ	0.3555	0.5359	0.6113	0.4555	0.5809	0.5551	3.1728	4.2498	2.5983
	(0.1395)	(0.0819)	(0.0207)	(0.0395)	(0.0224)	(0.0081)	(0.0500)	(0.4288)	(0.0276)
Copula	Frank			Joe			Clayton		
Parameter/Quantile	0.25	0.5	0.75	0.25	0.5	0.75	0.25	0.5	0.75
α	1.8021	1.4844	1.8172	1.5313	1.3962	1.2315	1.2596	1.9652	1.7974
	(0.5793)	(0.0909)	(0.3652)	(0.1669)	(0.1049)	(0.0415)	(0.3714)	(0.3417)	(0.4397)
β_1	1.9164	1.7469	1.5253	2.0347	1.8012	1.7219	1.3307	1.8310	1.5333
	(0.1597)	(0.5266)	(0.1515)	(0.1338)	(0.1050)	(0.0448)	(0.1929)	(0.2864)	(0.2941)
β_2	−2.2897	−2.3334	−2.2191	−1.9108	−2.2988	−2.2577	−2.0191	−2.4506	−2.3424
	(0.6428)	(0.5568)	(0.3973)	(0.1164)	(0.0812)	(0.0391)	(0.3636)	(0.1867)	(0.6668)
σ_u	0.8298	0.9317	1.0467	0.8409	1.3523	0.9704	1.0381	1.0368	1.0694
	(0.1768)	(0.1075)	(0.1461)	(0.0934)	(0.0602)	(0.0452)	(0.1044)	(0.1054)	(0.1648)
σ_v	0.5291	0.4025	0.8604	0.2975	0.6641	0.4323	0.5029	0.5044	0.4878
	(0.0799)	(0.0344)	(0.0031)	(0.0503)	(0.0851)	(0.0304)	(0.0977)	(0.0811)	(0.0148)
θ	3.100	2.3677	2.2796	2.6488	2.5491	4.0894	0.6544	0.6405	0.4887
	(0.2065)	(0.3190)	(0.2142)	(0.1945)	(0.0549)	(0.0430)	(0.0009)	(0.0076)	(0.0768)

Note: We assume the true value for intercept term α^{ρ} for all quantiles $\rho = (0.25, 0.5, 0.75)$ to be 1.5, the coefficients β^{ρ} to be 2 and −2, and $\sigma_u = 1$ and $\sigma_v = 0.5$. The true copula dependency θ is equal to 0.5 for Gaussian, Student's t, and Clayton copulas but it is equal to 3 for Gumbel, Joe, and Frank copulas.

5 Empirical Results: Agricultural Production Model for Asia

This part presents the benchmark result of this paper. We analyze the agricultural production function of Asian countries using our proposed model copula-based stochastic frontier quantile model which has been proved to be accurate through the simulation study.

5.1 Dataset

Prior to the estimated result, this part begins with the brief explanation of the data used in this paper. This paper considers three important input variables for estimating Asian production function, including labor, fertilizer, and agricultural area. Since we consider a cross-section of countries, we then collect the data in year 2013 when the data of every country are the most perfect and latest, from World Bank database and Thomson Reuters DataStream, from Financial

Investment Center (FIC), Faculty of Economics, Chiang Mai University, covering 44 countries in Asia.

Production output (Y) refers to the crop production index which shows agricultural production for each year relative to the base period 2004-2006. It includes all crops except fodder crops and is calculated from the underlying values in international dollars, normalized to the base period 2004-2006.

Labor (L) refers to the rural population. Due to some limited access to the data, we are unable to get the exact number of labor working in the agricultural sector. So, we decide to use the number of rural population as defined by national statistical offices to represent this variable since we believe that people living in rural areas have high possibility of working in agriculture.

Agricultural area (A) refers to the share of land area that is arable, under permanent crops, and under permanent pastures.

Fertilizer (F) refers to the fertilizer consumption measured by the quantity of plant nutrients used per unit of arable land. The variable covers nitrogenous, potash, and phosphate fertilizers (including ground rock phosphate), except traditional nutrients such as animal and plant manures.

5.2 Model Specification

To analyze the factors affecting agricultural output and measure the technical efficiency of Asian production, we considered the following production model. The model primarily takes the form of Cobb-Douglas production function where labor (L), fertilizer (F), and agricultural area (A) are inputs.

$$Y_i = \alpha L_i^{\beta_1^\rho} F_i^{\beta_2^\rho} A_i^{\beta_3^\rho} \tag{14}$$

Then, we transform Eq. (14) into a translog production frontier which takes the form as

$$\ln Y_i = \alpha + \beta_1^\rho \ln L_i + \beta_2^\rho \ln F_i + \beta_3^\rho \ln A_i + U_i - V_i. \tag{15}$$

In this study, we consider three quantile levels that are $\rho = (0.25, 0.5, 0.75)$ to represent three groups of agricultural countries in Asia as classified by the 2008 World Development Report of World Bank, namely (1) agriculture-based, (2) transforming, and (3) urbanized countries, respectively.

5.3 Model Selection

As a sequential estimation method, the copula has to be selected before estimation. Therefore, this part is also about selecting a copula that is best-fit for the data. We employ the Akaike Information Criterion (AIC) and the Bayesian Information Criterion (BIC) to pick the best copula among the copula families that we concern, i.e. Gaussian, Students t, Clayton, Gumbel, Frank, and Joe. The results are presented in Table 2.

Table 2 shows the values of AIC and BIC for each Copula-based stochastic frontier quantile model, where the minimum values are bold numbers. According

Table 2. AIC and BIC for each Copula-based SFQM

Copula	Quantile level		
	0.25	0.5	0.75
Gaussian	610.47	4.25	3.42
	635.48	30.82	28.43
Student's t	734.21	6.32	8.83
	762.79	36.69	37.41
Clayton	885.55	3.39	6.13
	907.56	29.96	31.14
Gumbel	2.567	5.01	6.92
	27.57	31.57	31.93
Frank	1.11	5.85	5.97
	26.06	32.43	30.98
Joe	5.92	3.17	0.68
	32.49	29.74	25.68

to both criteria, the best model for quantile level 0.25 is the one based on the Frank copula where the values of AIC and BIC are equal to 1.11 and 26.06, respectively. However, the best models for quantile levels 0.5 and 0.75 are the one based on the Joe copula since it has the minimum values of AIC and BIC as shown in the table.

5.4   Estimated Results of Copula-Based SFQM

This part presents the benchmark result of this paper, where we estimate the stochastic frontier model based on the copulas we chose in the previous section. The results are presented in Table 3. Technically, it is found that the dependence between error components exists since the estimated parameters of the copulas θ are significant for all quantiles at the 1 % level. Apart from that technical consideration, we found some interesting point that the estimated parameters are not so different across quantile levels. This means the impact of inputs, i.e. labor, fertilizer, and agricultural area, on the agricultural output are quite the same for all Asian countries. For example, an additional 1 % of labor leads to 0.013 % increase in agricultural output at the 0.25-quantile, 0.015 % and 0.018 % at the 0.5 and 0.75 quantiles, respectively.

Fertilizer is found to affect significantly only the first quantile which represents a group of agriculture-based countries. This seems to make sense because the other groups which are transforming ($\rho = 0.5$) and urbanized ($\rho = 0.75$) countries are able to access to get high-tech agricultural innovation such as solar power, hydroponics, and aeroponics. These smart technologies help farmers get more output and improve their crops, which in turn make fertilizer exert less influence on agricultural product.

The last input, agricultural area, seems to create the largest impact on the output compared with other factors, but the impacts are not different across quantiles. We found that an additional 1 % of area brings about 0.038 % increase in agricultural output at the 0.25 and 0.5 quantiles. This result corresponds with many agricultural reports in which Asia uses a very large area to grow crops; the quantity of output depends essentially on the lands used.

Table 3. Estimated parameters and standard errors for Copula-based SFQM.

Parameter	Quantile level		
	0.25	0.5	0.75
α	4.335***	4.557***	4.708***
	(0.018)	(0.048)	(0.094)
β_1^ρ	0.013***	0.015***	0.018*
	(0.006)	(0.008)	(0.011)
β_2^ρ	0.022*	0.026	0.027
	(0.011)	(0.026)	(0.032)
β_3^ρ	0.038***	0.038***	0.037
	(0.011)	(0.013)	(0.026)
σ_U	0.037***	0.086***	0.233***
	(0.008)	(0.018)	(0.025)
σ_V	0.058***	0.010	0.022
	(0.003)	(0.023)	(0.022)
θ	1.164***	3.144***	5.558***
	(0.077)	(0.058)	(0.336)

Note: *, **, and *** denote rejections of the null hypothesis at the 10 %, 5 %, and 1 % significance levels, respectively.

Additionally, Fig. 1 is constructed to illustrate the position of each country. We aim to find out which quantile level that a country fits most based on different inputs. Each of the quantile levels has meaning (See Sect. 5.2); that is the 0.25-quantile means agriculture-based country, the 0.5-quantile means transforming country, and the 0.75-quantile means urbanized country. To visualize the position, we plot the data of each input, i.e. area (top left), fertilizer (top right), and labor (bottom), against the level of output (crop production). Note that the data are log-transformed. The dot lines refer to the quantile lines for the 0.25, 0.5, and 0.75 quantiles, which are estimated from the copula-based SFQM. Finally, the result from the copula dependence shows the significant positive correlation exists between noise and inefficiency.

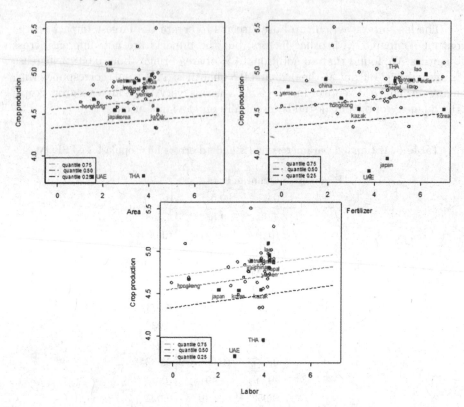

Fig. 1. Plot of production data fitting to quantile lines

5.5 Estimate of Technical Efficiency

This section presents the estimated result of the technical efficiency of Asian production. As we described in the end of Sect. 2.3, the technical efficiency is given by the ratio of observed output to maximum feasible output in which the value is equal to 1 means a country obtains the maximum feasible output. The value less than 1 refers to a shortfall of the observed output from maximum feasible output.

Figure 2 displays the values of technical efficiency at different quantiles, which are estimated by the copula-based SFQM. We found that the efficiencies are not the same for all quantile levels. The first quantile (0.25) representing the agriculture-based country has the lowest efficiency score in agricultural production where the range is 0.87 to 0.93 (average 0.90). The second and third quantiles representing the transforming and urbanized countries, respectively, have quite the same efficiency score with in the range around 0.91 to 0.99.

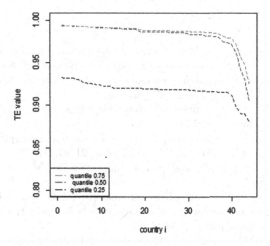

Fig. 2. Technical efficiency of Asian production at different quantiles

6   Conclusion

Since agricultural production sector has played a key role to the Asian economy, this paper attempts to analyze the agricultural production function of Asian countries and examine the technical efficiencies of this region. In methodology, we take advantage of the stochastic frontier model in terms of the dependency between the error term U and the non-negative inefficiency V. We employ the copulas to model this dependence structure, and then extend the quantile regression to the SFM with dependent error components to capture the tail behavior of a probability distribution. Therefore, we introduce the model named the Copula-based stochastic frontier quantile model as a frontier model for this issue.

To model the agricultural production in Asia, we consider labor, fertilizer, and agricultural area to be inputs and the crop production to be output. This methodology is applied to the agricultural data of 44 Asian countries. The results show that the impact of labor on production output is quite the same for all Asian countries, whereas fertilizer is found to have effect significantly only on the first quantile which represents a group of agriculture-based countries. Agricultural area creates the largest impact on output compared with other factors, but the effect sizes are not different across quantiles. For technical efficiency, we found that the 0.25-quantile has the lowest efficiency score in agricultural production within the range 0.87 to 0.93 while the other two quantiles have the efficiency score within the range 0.91 to 0.99. The overall results suggest that the agricultural sector in Asia is able to perform effectively in the long-run.

References

1. Bezat-Jarzbowska, A., Rembisz, W.: Efficiency-focused economic modeling of competitiveness in the agri-food sector. Procedia Soc. Behav. Sci. **81**, 359–365 (2013)
2. Kaditi, E.A., Nitsi, E.: Applying regression quantiles to farm efficiency estimation. In: 2010 Annual Meeting, pp. 25–27, July 2010
3. Farrell, M.J.: The measurement of productive efficiency. J. R. Stat. Soc. Ser. A (Gen.) **120**(3), 253–290 (1957)
4. Aigner, D.J., Lovell, C.A.K., Schmidt, P.: Formulation and estimation of stochastic frontier production function models. J. Econ. **6**, 21–37 (1977)
5. Duy, V.Q.: Access to credit and rice production efficiency of rural households in the Mekong Delta (2015)
6. Gregg, D., Rolfe, J.: The value of environment across efficiency quantiles: a conditional regression quantiles analysis of rangelands beef production in north Eastern Australia. Ecol. Econ. **128**, 44–54 (2016)
7. Bernini, C., Freo, M., Gardini, A.: Quantile estimation of frontier production function. Empirical Econ. **29**(2), 373–381 (2004)
8. Das, A.: Copula-based stochastic frontier model with autocorrelated inefficiency. Cent. Eur. J. Econ. Model. Econometrics **7**(2), 111–126 (2015)
9. Smith, M.D.: Stochastic frontier models with dependent error components. Econometrics J. **11**(1), 172–192 (2008)
10. Wiboonpongse, A., Liu, J., Sriboonchitta, S., Denoeux, T.: Modeling dependence between error components of the stochastic frontier model using copula: application to intercrop coffee production in Northern Thailand. Int. J. Approximate Reasoning **65**, 34–44 (2015)
11. Horrace, W.C., Parmeter, C.F.: A Laplace stochastic frontier model. Econometric Rev. 1–27 (2015)
12. Burns, R.C.J.: The simulated maximum likelihood estimation of stochastic frontier models with correlated error components, Unpublished Dissertation, Department of Econometrics and Business Statistics, The University of Sydney, Australia (2004)
13. Battese, G.E., Coelli, T.J.: A model for technical inefficiency effects in a stochastic frontier production function for panel data. Empirical Econ. **20**(2), 325–332 (1995)
14. Snchez, B.L., Lachos, H.V., Labra, V.F.: Likelihood based inference for quantile regression using the asymmetric Laplace distribution. J. Stat. Comput. Simul. **81**, 1565–1578 (2013)

Statistical and ANN Approaches in Credit Rating for Vietnamese Corporate: A Comparative Empirical Study

Hung Nguyen[1(✉)] and Tung Nguyen[2]

[1] HoaSen University, Ben Thanh Ward, Ho Chi Minh City, Vietnam
hung.nguyenba@hoasen.edu.vn
[2] Eximbank, North Office, Long Bien, Ha Noi, Vietnam
tung.nx@eximbank.com.vn

Abstract. Decisions to grant credit to customer is the most crucial part in credit business. In the recent years, advances in information technology have lessened the costs of acquiring, managing and analyzing data in an effort to build more efficient and powerful models for credit rating in credit risk management.

By using common methods as statistical and learning methods, credit scoring models are built to evaluate credit risk for real Vietnamese corporate data. We found that the logistic model give promising results compare with multivariate discrimination analysis, probit and neural network.

Keywords: Credit scoring · Credit risk · Risk assessment · Discriminant analysis · Neural network · Multi layer perceptron

1 Introduction

Commercial banks in Vietnam in turn have published their financial reports of the first 6 months in August 2014. Accordingly, the Non-Performing Loan (NPL) ratio is rising, with higher degree, especially for the big banks that try to keep this ratio below the safety level - 3 % to be inline with the announcement from the State bank of Vietnam (SBV). In Vietinbank, NPL has raised 150 % from 1 % of total debts to 2.53 % of total debts; Vietcombank to 3.09 %; MBBank to 3.1 %; SHB keep the pace of 4 %; PVCombank to 5.2 % and OceanBank to 4.84 %, according to Tri Thuc Tre.[1] Where as, in 2015, their activities in lending still growing in a high speed especially in case of VPBank, LienVietPostbank, BIDV, SHB (Figs. 1, 2 and 3).

Take another look at these bank's total bad debts in the first 9 months in 2015:

According to Tri Thuc Tre, the irrecoverable debt (debts in group 5) increases 24,1 % compare with the previous year, reach 24,498 billion Dong and consists

---

[1] Tri Thuc Tre: http://ttvn.vn/kinh-doanh.htm.

© Springer International Publishing AG 2016
V.-N. Huynh et al. (Eds.): IUKM 2016, LNAI 9978, pp. 715–726, 2016.
DOI: 10.1007/978-3-319-49046-5_60

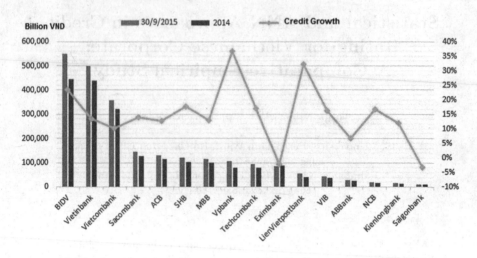

Fig. 1. Credit growth of vietnamese banks. Source: Tri Thuc Tre

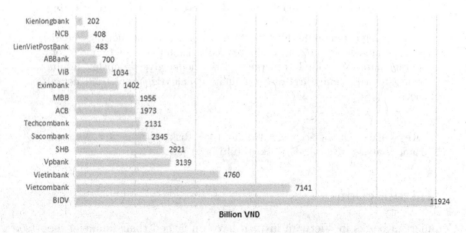

Fig. 2. Total bad debts of vietnamese banks. Source: Tri Thuc Tre

of over 50 % of total debts. Subprime debt (debts in group 3) increases 9.3 %
while doubtful debt (debts in group 4) decreases of 29.7 % compare with the end
of 2014

At the end of 3rd quarter of 2015, the bad debts ratios are reported lower
than the end of 2014:

The interesting thing is that 15 banks in the figure had reported NPL ratio
less than 3 % at the end of 3rd quarter 2015, while as the overall bad debts ratio
of banking sector in Vietnam is 2.93 %. This rise a question of how much bad
debts that the remaining 16 banks hold?

We can see that when banks in Vietnam is experiencing stagnation in bad
debt recovery phase, the scrutiny of loan classification to international-practice

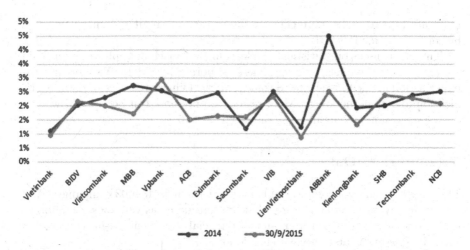

Fig. 3. NPL gap. Source: Tri Thuc Tre

loan classification is crucial for banks to ease the comparison and to enhance risk assessment.

With qualitative methods for classification models and credit assessment, especially using Expert System, the selection of most basic elements of credit rating system mainly depend on subjective viewpoints of experts instead of rely on the historical statistic data and econometrics analyst. This subjective internal credit rating results should not become the framework of the basic of building quantitative risk assessment metrics which help banks to compute more efficiently the expected loss and minimum capital requirement for risk provision. This lead to restrictions on portfolio risk management, credit evaluation, risk positioning, investment decisions, merge and acquisition, credit financing and partnership.

The quantitative approach in credit rating depends on statistic historical data of a bank's customer to evaluate their risk metrics including: Probability of Default (PD), Loss Given Default (LGD), Exposure at Default (EAD) and Expected Loss (EL), then, this objective credit rating model could help banks determine the operational status and develop prospect in the future to improve credit quality, contribute to the economic growth in a sustainable manner and prevent the financial crisis.

Accordingly, a knowledge tool is crucial in decision making regarding it's application for credit risk management. The efficiency and improvement of financial system play the most important role in economic development of countries. The better the banks act, the more economic development. In recent years, banks have given much efforts to solve these problems. Specifically, bad loans have been become difficult issues with banks in Vietnam. Credit management is an urgent task in Vietnam, where 10 largest banks have developed sophisticated systems in an attempt to make more efficient the process of credit risk management to be compliance for Basel II.[2]

. This study focus on statistical and learning approach in application for building credit rating models for Vietnamese corporate firm data. We organize this paper as follow: Sect. 2 devote to some empirical characteristics of data, Sect. 3 mention about methodology. Main results are presented in Sect. 4 and we conduct some comments about these models in Sect. 5.

2   Empirical Data

The data is given by a commercial bank in Vietnam with 1003 companies. With each company, we have 30 features that are considered as variables. All information's company is collected from annual financial statements of firms up to Dec 12, 2012. These variables are listed as follow:

– **Dependent variable**: Good and bad customers. A large historical loan sample on corporate portfolio is divided into two criteria, good loans and bad loans. In this paper, we aim to estimate probability of default or non-default customer. In this regard, non-default customer is a company that paybacks its loan and the interest at the due date while in contrast, default customer is a company that doesn't payback its loan at the due date. To distinguish between default and non-default customers in our models, we assign 0 for non-default customers and 1 to indicate default customers
– **Independent variables**: There are 30 variables that are defined as independent variables and listed in Table 1.

3   Methodology

3.1   Multivariate Discriminant Analysis

In credit assessment procedure, the objective of multivariate discriminant analysis (MDA) [4,5] is to differentiate default from non-default customers[3] as accurately as possible by a function of several independent creditworthiness factors (financial ratios, figures from annual financial statements). We classify a customer into one of several groups base on their individual characteristics. MDA is used primarily to classify and/or make prediction in problems where the dependent variable is binary or categorical variable, for example, default or non-default, male or female. Thus, the key step is to establish comprehensive groups classification

---

[2] IMF: 2016 Article IV Consultation—Press Release; Staff Report; and Statement by the Executive Director for Vietnam.
[3] In simplest case, we consider that a firm is either default or non-default. In the general case, we can divide firms into groups based on discriminant score.

Table 1. Financial variables

Total assets	Short-term assets	Money and equal things
Money and equal things	Short-term investments	Receivable
Inventories	Other Short-term assets	Long-term assets
Long-term Receivables	Fixed assets	Real estate
Long-term investments	Other long-term assets	Goodwill
Debts	Short-term debts	Long-term debts
Shareholders' equity	Minority holding	Revenue
Pure Revenue	Sales	Compound Profit
Financial Cost	Interest rate Cost	Business Profit
Other Profit	Profit	Basic Earning per Share
EBIT		

In MDA, a weighted linear combination of factors is formed in order to classify default or non-default customers as much discriminatory power as possible on the basis of the discriminant score D:

$$D = a_0 + a_1 K_1 + ... + a_n K_n. \tag{1}$$

where:
n: the number of financial ratios
K_i: the specific ratio value
a_i: ratio's coefficient within the rating functions.

With MDA models, the selection of ratios is the most important. Credit experts will use their experience in this works. We follow [2] to get use the discriminant function as follows:

$$D = a_0 + a_1 K_1 + a_2 K_2 + a_3 K_3 + a_4 K_4 + a_5 K_5. \tag{2}$$

where:
K_1: Working Capital/Total Assets
K_2: Retained Earning/Total Assets
K_3: Earning Before Interest and Taxes/Total Assets
K_4: Market Value Equity/Book value of total Debts
K_5: Sales/Total Assets

3.2 Regression Models

Regression models show the relationship of a binary variable with other independent variables. In practical credit assessment procedures, certain creditworthiness factors (independent variables) will help credit manager decide whether a customer could be classified as default or not (dependent binary variable)[4].

---

[4] We also use these model to classify customers to different groups.

Using regression models also enable us to calculate membership probabilities and thereby to determine default probabilities directly from the model function.

We present in this paper logistic and probit regression with the graphs of model functions and their mathematical formula shown in Fig. 4. Φ is the cumulative standard normal distribution function, and $\sum$ represents a linear combination of the financial factors:

$$\sum = b_0 + b_1 K_1 + \ldots + b_n K_n. \tag{3}$$

where:

n: the number of financial factors
K_i: the specific value of creditworthiness criteria i
b_i: factor's coefficient within the rating function (for $i = 1, \ldots, n$)

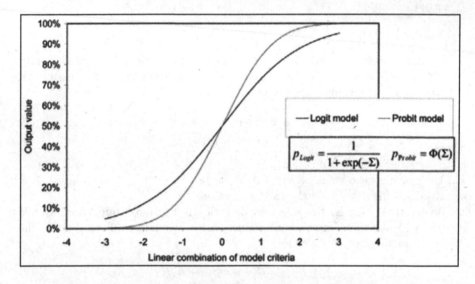

Fig. 4. Functional forms for logit and probit models

Logit Model: In binary classification using logistic model, the default probability p which is a given customer is classified as default or non-default is calculated as follow:

$$p_{Logit} = \frac{1}{1 + exp[-(\sum)]}. \tag{4}$$

Probit Model: In this models we calculate default probability by:

$$p_{Probit} = \Phi(\sum). \tag{5}$$

where the function Φ denotes cumulative standard normal distribution. We use maximum likelihood method to estimate the coefficients.

3.3   Artificial Neural Network

Neural networks provide a new alternative to MDA and logistic regression [1,3,5], especially in the credit assessment where the number of independent usually large and there exists complex non-linear relationships between dependent and independent variables.

The neural network architecture built for this research, is composed of an input layer, composed of 10 neurons, two hidden layers, composed respectively of 10 and 5 neurons, and one output layer, composed by score.

3.4   Discriminatory Power

The term "Discriminatory Power" refers to the essential ability of a ranking model to distinguish between default and non-default customers. In this study, non-default customers are divided into seven classes, from 1st class to 7th class, and default customers are divided into three next classes, from 8th class to 10th class. In this context, the categories non-default and default refer to whether a credit default occurs or does not occur over the forecasting horizon after the rating system has classified the case.

We present 6 essential measures [6] used in this study:

- Type I (α) and Type II (β) errors, where α error is case which is actually bad is not rejected and β error is case which is actually good is rejected. In banking industry, α error is seriously concerned and the costs associated with α and β errors are significantly difference.
- Receiver Operation Characteristic (ROC) Curve.
- Area under Curve (AUC).
- Gini Coefficient[5]   Gini Coefficient = 2AUC - 1
- Accuracy Ratio (AR).
- Sum-of-Square Error (SSE) Function:

$$E(t,y) = \frac{1}{2}\sum_{k=1}^{n}(t_k - y_k)^2. \tag{6}$$

4   Empirical Results

We construct a confusion matrix as follow:

Table 2. Confusion matrix

Predict	Actual	
	Default	Non-default
Bad	True Positive (TP)	False Positive (FP)
Good	False Negative (FN)	True Negative (TN)

---

[5] Gini, C. (1912).

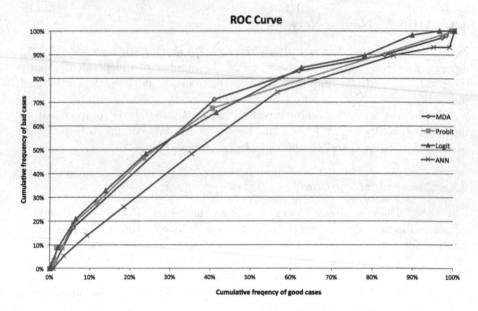

Fig. 5. ROC curves

where TP and TN are observations that correctly predicted as default and no-default respectively; FP is predicted default that actually non-default and a FN is a predicted non-default that actually defaults. The errors of the model are shown on the off diagonal of the matrix where FN represents α error and FP represents β error. Using this matrix, α error, β error and AR are computed as:

$$\alpha \text{ error} = \frac{\text{FN}}{\text{TP} + \text{FN} + \text{FP} + \text{TN}}; \tag{7}$$

$$\beta \text{ error} = \frac{\text{FP}}{\text{TP} + \text{FN} + \text{FP} + \text{TN}}; \tag{8}$$

$$\text{AR} = \frac{\text{TP} + \text{TN}}{\text{TP} + \text{FN} + \text{FP} + \text{TN}} \tag{9}$$

An alternative and better approach to examine the power of a classification model is ROC curve with its quantitative measure - AUC [7]. This is the probability that the model will rank a randomly chosen default customer higher than a randomly chosen non-default customer, i.e.:

$$AUC = P\left(X^+ > X^-\right).$$

where X^+ is the score for a positive instance (default customer) and X^- is the score for a negative instance (non-default customer) (Table 2).

ROC curve is independent with sample's distribution and show how the performance of an model vary with different cut-off thresholds [6]. Figure 5 presents the ROC curves of four models. One can see that the logit model shows a steeper curve along the lower range of scores on the left, hence performs better than the others since it closest to (0,1) and furtherest from true positive rate equal false positive rate in 8/10 class.

For the remaining discrimination power measurements, we have in Table 3 the statistics of four above models. With MDA model the estimated discriminant function is:

$$D = -2.39 - 0.89K_1 + 0.42K_2 + 0.43K_3 - 0.001K_4 - 0.14K_5.$$

In Table 3, for the case of MDA and ANN, the default rate per rating class for class 9 and 10 are zero and show that the α error is high, these two models fail to detect default firms for class 9 and 10. While probit model perform inconsistent with the ranking.

With logit model, despite the non-concave of ROC curve, one can see that default rate for each rating class increases steadily from class 1 to class 10. Hence, this rating system is clearly able to classify cases by default probability.

Measurement of discrimination power for our models are shown in Table 4. The boldfaces are the best results and the underline are the worsts one. Logit model perform best in 5/6 criteria here, it is also noted that, in banking practice,

Table 3. Models statistic

Class	Number of good cases				Number of bad cases				Default rate(%)			
	MDA	Logit	Probit	ANN	MDA	Logit	Probit	ANN	MDA	Logit	Probit	ANN
1	12	34	562	1	0	0	19	0	0.0	0.0	3.3	0.0
2	16	64	161	1	1	1	12	0	5.9	1.5	6.9	0.0
3	37	110	110	33	1	5	11	5	2.6	4.4	9.1	13.2
4	94	145	54	57	4	3	5	4	4.1	2.0	8.5	6.6
5	162	200	29	121	4	11	6	7	2.4	5.2	17.1	5.5
6	223	166	12	568	7	10	0	21	3.0	5.7	0.0	3.6
7	339	94	4	124	31	9	3	16	8.4	8.7	42.9	11.4
8	52	70	6	38	10	7	1	5	16.1	9.1	14.3	11.6
9	7	42	2	2	0	7	0	0	0.0	14.3	0.0	0.0
10	3	20	5	0	0	5	1	0	0.0	20.0	16.7	0.0

Table 4. Main results of all models

Model	AR	AUC	Gini Coefficient	α Error	SSE
MDA	89.03%	66.4%	0.304	5.16%	54.03
Logit	82.95%	**68.3%**	**0.342**	**4.58%**	**26.24**
Probit	**93.12%**	66.9%	0.312	5.67%	28.24
ANN	90.73%	56.2%	0.094	5.53%	27.29

since the loan portfolio almost have skew distribution toward the good customers, we focus on AUC and α error where the first one is independent with sample distribution and the latter show how a model performs in the downside that is detect bad loan applications.

For the case of ANN model, this poor performance might be of the imbalanced characteristic of input data affect in decreasing training effect on the default customers [9]; as a result, ANN tends to favor the majority class - good customer, this drawback is shown in Table 4 where ANN has high correctness accompany with smallest AUC.

5   Conclusion

Discriminant analysis is the first tool to be used in developing credit rating models [1–3]. Nonetheless, the implementation of MDA has been criticized because of its normal distribution assumptions on those financial ratios

Logistic regression, in this comparative study, has a number of strong points compare with MDA. It not only does not require normal distribution in input variables which enable logistic regression to undertake qualitative creditworthiness factors directly but also its result can be interpreted as the probability of group membership [8]. And, logistic regression model give more robust and more accurate results than those generated by discriminant analysis and ANN. One more note on that promising result of logit model is in its ease of modeling and link other crucial elements of Basel II[6] with PD. We suggest that Vietnamese bank with small portfolio of loans could find comfortable in establishing logit model for corporate's as well as retail's PD.

In credit scoring model, it does not determine important differences between the two models, regression model and artificial neural network if there are few extreme cases in the reference sample. This is not supported in this study, where results of regression model dominate artificial neural network's results.

In summary, the empirical results in this comparative study support the use of the logit model in classifying and inspection loan portfolio. The use of ANN in credit rating model for bank should be taken with carefully treatment in term of imbalanced dataset and loan assessment interpretation.

Further work will focus on both methodological and application issues. As to methodology, we are designing a new procedure for data processing and for parameter optimization of new qualitative creditworthiness characteristic of credit data. On the side of applications, we plan to assess the generalization capabilities of the networks and statistical methods by testing them on larger database and to investigate on the applicability of ensemble models. One more note on the dataset is that our dataset is very highly skewed toward the good case which is the natural properties of credit application document and the severe imbalanced characteristic of this dataset, of course, need deeper and carefully treatment to have stable credit scoring models in term of discrimination power.

---

[6] Basel Committee on Banking Supervision - Revisions to the Standardized Approach for credit risk.

Acknowledgment. We are immensely grateful to 3 anonymous reviewers for their comments on an earlier version of the manuscript, although any errors are our own and should not tarnish the reputations of these esteemed persons.

References

1. Rencher, A.C.: Methods of Multivariate Analysis, 2nd edn. Brigham Young University, Brigham (2002)
2. Altman, E.: Iltman : Financial ratios, discriminant analysis and the prediction of corporate bankruptcy. J. Finance, XXII I(4), 589–609 (1968)
3. Altman, E.I., Marco, G., Varetto, F.: Corporate distress diagnosis; Comparisons using linear discriminant analysis and neural networks (the Italian experience). J. Banking Finance **18**, 505–529 (1994)
4. Bierman, H., Hausman, W.H.: The credit granting decision. Manage. Sci. **16**, 519–532 (1970)
5. Buckley, J., James, I.: Linear regression with censored data. Biometrika **66**, 429–436 (1979)
6. Powers, D.M.W.: Evaluation: From precision, recall and F-measure to ROC, informedness, markedness correlation. J. Mach. Learn. Technol. **2**, 37–63 (2011)
7. Spackman, K.A.: Signal detection theory: Valuable tools for evaluating inductive learning. In: Proceedings of the Sixth International Workshop on Machine Learning, pp. 160–163. Morgan Kaufmann, San Mateo, CA (1989)
8. Marsland, Stephen: Learning, Machine: An Algorithmic Perspective. Massey University, Palmerston North, New Zealand (2009)
9. Ganganwar, V.: An overview of classification algorithms for imbalanced datasets. Int. J. Emerg. Technol. Adv. Eng. **4**, 42–47 (2012)

Further Reading

10. Lahsasna, A., Ainon, R.N., Wah, T.Y.: Credit scoring models using soft computing methods: a survey. Int. Arab J. Inf. Technol. **7**(2), 115–123 (2010)
11. Boyle, M., Crook, J.N., Hamilton, R., Thomas, L.C.: Methods for credit scoring applied to slow payers. In: Thomas, L.C., Crook, J.N., Edelman, D.B. (eds.) Credit Scoring and Credit Control, pp. 75–90. Oxford University Press, Oxford (1992)
12. Banasik, J., Crook, J.N., Thomas, L.C.: Does scoring a subpopulation make a difference? Int. Rev. Retail Distrib. Consum. Res. **6**, 180–195 (1996)
13. Banasik, J., Crook, J.N., Thomas, L.C.: Not if but when borrowers default. J. Oper. Res. Soc. **50**, 1185–1190 (1999)
14. Carter, C., Catlett, J.: Assessing credit card applications using machine learning. IEEE Expert **2**, 71–79 (1987)
15. Angelini, E., Tollo, G., Roli, A.: A neural network approach for credit risk evaluation. Q. Rev. Econ. Finance **48**(4), 733–755 (2008)
16. Hammer, P.L., Kogan, A., Lejeune, M.A.: A logical analysis of banks' financial strength ratings. Expert Syst. Appl. **39**(9), 7808–7821 (2012)
17. Ahn, H., Kim, K.-J.: Corporate credit rating using multi-class classification models with order information. World Acad. Sci. Eng. Technol. **5**, 161–177 (2011)
18. Majer, I.: Application scoring: Logit model approach and the divergence method compared, Working Paper No. 10, Warsaw School of Economics, Poland (2006)

19. De Andrés, J., Lorca, P., de Cos Juez, F.J.: Bankruptcy forecasting: a hybrid approach using fuzzy c-means clustering and multivariate adaptive regression splines (MARS). Expert Syst. Appl. **38**, 1866–1875 (2011)
20. Galindo, J., Tamayo, P.: Credit risk assessment using statistical and machine learning: Basic methodology and risk modeling applications. Comput. Econ. **15**(1–2), 107–143 (2000)
21. Narain, B.: Survival analysis and the credit granting decision. In: Thomas, L.C., Crook, J.N., Edelman, D.B. (eds.) Credit scoring and credit control, pp. 109–122. Oxford University Press, Oxford (1992)
22. Sheskin, D.: Handbook of Parametric and Nonparametric Statistical Procedures. CRC Press, Boca Raton (2004)
23. Pacelli, Vincenzo: Michele, A.: An artificial neural network approach for credit risk management. J. Intell. Learn. Syst. Appl. **3**, 103–112 (2011)
24. Huang, Z., Chen, H., Hsu, C.-J., Chen, W.-H., Wu, S.: Credit rating analysis with support vector machines and neural networks: A market comparative study. Decis. Support Syst. **37**, 543–558 (2004)

Author Index

Printed in the United States
By Bookmasters